AF845594

Coastal and Marine Hazards, Risks, and Disasters

Hazards and Disasters Series

Coastal and Marine Hazards, Risks, and Disasters

Series Editor

John F. Shroder
Emeritus Professor of Geography and Geology
Department of Geography and Geology
University of Nebraska at Omaha
Omaha, NE 68182

Volume Editors

Jean T. Ellis
Associate Professor
Department of Geography and
Marine Science Program
University of South Carolina
Columbia, SC

Douglas J. Sherman
Professor and Chair
Department of Geography
University of Alabama
Tuscaloosa, AL

ELSEVIER

AMSTERDAM • BOSTON • HEIDELBERG • LONDON • NEW YORK • OXFORD
PARIS • SAN DIEGO • SAN FRANCISCO • SINGAPORE • SYDNEY • TOKYO

Elsevier
Radarweg 29, PO Box 211, 1000 AE Amsterdam, Netherlands
The Boulevard, Langford Lane, Kidlington, Oxford OX5 1GB, UK
225 Wyman Street, Waltham, MA 02451, USA

Copyright © 2015 Elsevier Inc. All rights reserved.

No part of this publication may be reproduced or transmitted in any form or by any means, electronic or mechanical, including photocopying, recording, or any information storage and retrieval system, without permission in writing from the publisher. Details on how to seek permission, further information about the Publisher's permissions policies and our arrangements with organizations such as the Copyright Clearance Center and the Copyright Licensing Agency, can be found at our website: www.elsevier.com/permissions.

This book and the individual contributions contained in it are protected under copyright by the Publisher (other than as may be noted herein).

Notices
Knowledge and best practice in this field are constantly changing. As new research and experience broaden our understanding, changes in research methods, professional practices, or medical treatment may become necessary.

Practitioners and researchers must always rely on their own experience and knowledge in evaluating and using any information, methods, compounds, or experiments described herein. In using such information or methods they should be mindful of their own safety and the safety of others, including parties for whom they have a professional responsibility.

To the fullest extent of the law, neither the Publisher nor the authors, contributors, or editors, assume any liability for any injury and/or damage to persons or property as a matter of products liability, negligence or otherwise, or from any use or operation of any methods, products, instructions, or ideas contained in the material herein.

Library of Congress Cataloging-in-Publication Data
Application submitted

British Library Cataloguing in Publication Data
A catalogue record for this book is available from the British Library

ISBN: 978-0-12-396483-0

For information on all Elsevier publications
visit our web site at http://store.elsevier.com

ELSEVIER Book Aid International
Working together to grow libraries in developing countries

www.elsevier.com • www.bookaid.org

Contents

Contributors

Clarissa R. Anderson, Institute of Marine Sciences, University of California, Santa Cruz, CA, USA

Robert W. Brander, School of Biological, Earth and Environmental Sciences, UNSW Australia, Sydney, NSW, Australia

Caroline K. Cusack, Marine Institute, Oranmore, County Galway, Ireland

Robert J.N. Devoy, The Coastal and Marine Research Centre, Beaufort Research and the Department of Geography, University College Cork, Cork, Ireland

Hajo Eicken, Geophysical Institute, University of Alaska Fairbanks, Fairbanks, AK, USA

Jean T. Ellis, Department of Geography and Marine Science Program, University of South Carolina, Columbia, SC, USA

Eugene J. Farrell, Discipline of Geography, National University Ireland Galway, Galway, Ireland

Kazimierz K. Furmanczyk, University of Szczecin, Szczecin, Poland

Virginia D. Hansen, US Environmental Protection Agency, Gulf Breeze, FL, USA

Jim D. Hansom, School of Geographical and Earth Sciences, University of Glasgow, Glasgow, UK; Department of Geography, University of Canterbury, Christchurch, New Zealand

Kieran R. Hickey, Discipline of Geography, National University Ireland Galway, Galway, Ireland

Diane P. Horn, Birkbeck College, University of London, London, UK

Wensheng Jiang, Ocean University of China, Qingdao, China

Utku Kânoğlu, Department of Engineering Sciences, Middle East Technical University, Ankara, Turkey

Paul S. Kench, School of Environment, The University of Auckland, Auckland, New Zealand

Carmen Lacambra, Grupo Laera, Bogotá, Colombia

Andrew R. Mahoney, Geophysical Institute, University of Alaska Fairbanks, Fairbanks, AK, USA

Anna McIvor, Cambridge Coastal Research Unit, Department of Geography, University of Cambridge, Cambridge, UK

Iris Möller, Cambridge Coastal Research Unit, Department of Geography, University of Cambridge, Cambridge, UK

Stephanie K. Moore, Environmental and Fisheries Sciences, Northwest Fisheries Science Center, National Marine Fisheries Service, National Oceanic and Atmospheric Administration, Seattle, WA, USA

Robert J. Nicholls, Faculty of Engineering and the Environment, University of Southampton, Southampton, UK

Jonathan F. Nott, College of Science, Technology and Engineering, James Cook University, Cairns, Queensland, Australia

Susan D. Owen, School of Environment, The University of Auckland, Auckland, New Zealand

Jeremy Pile, Earth Observatory of Singapore, Nanyang Technological University, Singapore

Kelly Chinners Reiss, HT Odum Center for Wetlands, University of Florida, Gainesville, FL, USA

Tiffany M. Roberts Briggs, Department of Geosciences, Florida Atlantic University, Boca Raton, FL, USA

Anja M. Scheffers, Southern Cross GeoScience, Southern Cross University, Lismore, NSW, Australia

Douglas J. Sherman, Department of Geography, University of Alabama, Tuscaloosa, AL, USA

Joe Silke, Marine Institute, Oranmore, County Galway, Ireland

Mark Spalding, The Nature Conservancy, Department of Zoology, University of Cambridge, Cambridge, UK

Thomas Spencer, Cambridge Coastal Research Unit, Department of Geography, University of Cambridge, Cambridge, UK

Adam D. Switzer, Earth Observatory of Singapore, Nanyang Technological University, Singapore; Division of Earth Sciences, Nanyang Technological University, Singapore

Costas Synolakis, Viterbi School of Engineering, University of Southern California, Los Angeles, CA, USA; Technical University of Crete, Chania, Greece

Michelle C. Tomlinson, NOAA, National Centers for Coastal Ocean Science, East-West Highway, Silver Spring, MD, USA

Hans von Storch, Institute of Coastal Research, Helmholtz Zentrum Geesthacht, Geesthacht, Germany

Ping Wang, School of Geosciences, University of South Florida, Tampa, FL, USA

Editorial Foreword

General hazards, risks, and disasters: Hazards are processes that produce danger to human life and infrastructure. Risks are the potential or possibilities that something bad will happen because of the hazards. Disasters are that quite unpleasant result of the hazard occurrence that caused destruction of lives and infrastructure. Hazards, risks, and disasters have been coming under increasing strong scientific scrutiny in recent decades as a result of a combination of numerous unfortunate factors, many of which are quite out of control as a result of human actions. At the top of the list of exacerbating factors to any hazard, of course, is the tragic exponential population growth that is clearly not possible to maintain indefinitely on a finite Earth. As our planet is covered ever more with humans, any natural or human-caused (unnatural) hazardous process is increasingly likely to adversely impact life and construction systems. The volumes on hazards, risks, and disasters that we present here are thus an attempt to increase understandings about how to best deal with these problems, even while we all recognize the inherent difficulties of even slowing down the rates of such processes as other compounding situations spiral on out of control, such as exploding population growth and rampant environmental degradation.

Some natural hazardous processes such as volcanoes and earthquakes that emanate from deep within the Earth's interior are in no way affected by human actions, but a number of others are closely related to factors affected or controlled by humanity, even if however unwitting. Chief among these, of course, are climate-controlling factors, and no small measure of these can be exacerbated by the now obvious ongoing climate change at hand (Hay, 2013). Pervasive range fires and forest fires caused by human-enhanced or -induced droughts and fuel loadings, megaflooding into sprawling urban complexes on floodplains and coastal cities, biological threats from locust plagues, and other ecological disasters gone awry; all of these and many others are but a small part of the potentials for catastrophic risk that loom at many different scales, from the local to planet girdling.

In fact, the denial of possible planetwide catastrophic risk (Rees, 2013) as exaggerated jeremiads in media landscapes saturated with sensational science stories and end-of-the-world, Hollywood productions is perhaps quite understandable, even if simplistically shortsighted. The "end-of-days" tropes promoted by the shaggy-minded prophets of doom have been with us for centuries, mainly because of Biblical verse written in the early Iron Age during remarkably pacific times of only limited environmental change. Nowadays, however, the Armageddon enthusiasts appear to want the worst to validate their death desires and prove their holy books. Unfortunately we are all entering times when just a few individuals could actually trigger societal breakdown by error or terror, if Mother Nature does not do it for us first. Thus we enter contemporaneous times of considerable peril that present needs for close attention.

These volumes we address here about hazards, risks, and disasters are not exhaustive dissertations about all the dangerous possibilities faced by the ever-burgeoning human populations, but they do address the more common natural perils that people face, even while we leave aside (for now) the thinking about higher-level existential threats from such things as bio- or cybertechnologies, artificial intelligence gone awry, ecological collapse, or runaway climate catastrophes.

In contemplating existential risk (Rossbacher, 2013), we have lately come to realize that the new existentialist philosophy is no longer the old sense of disorientation or confusion at the apparently meaninglessness or hopelessly absurd worlds of the past, but instead an increasing realization that serious changes by humans appear to be afoot that even threaten all life on the planet (Kolbert, 2014; Newitz, 2013). In the geological times of the Late Cretaceous, an asteroid collision with Earth wiped out the dinosaurs and much other life; at the present time, in contrast, humanity itself appears to be the asteroid.

Misanthropic viewpoints aside, however, an increased understanding of all levels and types of the more common natural hazards would seem a useful endeavor to enhance knowledge accessibility, even while we attempt to figure out how to extract ourselves and other life from the perils produced by the strong climate change so obviously underway. Our intent in these volumes is to show the latest good thinking about the more common endogenetic and exogenetic processes and their roles as threats to everyday human existence. In this fashion, the chapter authors and volume editors have undertaken to show you overviews and more focused assessments of many of the chief obvious threats at hand that have been repeatedly shown on screen and print media in recent years. As this century develops, we may come to wish that these examples of hazards, risks, and disasters are not somehow eclipsed by truly existential threats of a more pervasive nature. The future always hangs in the balance of opposing forces; the ever-lurking, but mindless threats from an implacable nature, or the heedless bureaucracies countered only sometimes in small ways by the clumsy and often feeble attempts by individual humans to improve our little lots in life. Only through improved education and understanding will any of us have a chance against such strong odds; perhaps these volumes will add some small measure of assistance in this regard.

Coastal and marine hazards, risks, and disasters: The chapters presented in this volume show us that the hazards, risks, and disasters associated with marine environments include several chapters on tsunami, because that hazards and associated risks have generated a number of recent megadisasters that have played out for all to see in recent years with the extensive and eye-catching coastal inundations that killed hundreds of thousands of people in the Indian Ocean and western Pacific so much. Storm surges from the oceans, as well as sea-level rise from climate change are well featured, as are the so-called "rogue" or extreme waves, as well as various threats to or from reefs, mangrove swamps, marsh resources, rip currents/undertows, and sea ice. Perhaps the largest threat to humankind related to the sea, however, is extreme sea-level rise, which will inundate low-lying coastal areas worldwide in future decades as polar ice caps melt; this will displace millions of coastal dwellers with consequences that we can barely understand, although we all know that it will be most serious. Jean Ellis and Doug Sherman can be complimented for pulling together this mixture of important chapters on hazards, risks, and disasters connected with the marine environments.

John (Jack) Shroder
Editor-in-Chief
July 14, 2014

REFERENCES

Hay, W.W., 2013. Experimenting on a Small Planet: A Scholarly Entertainment. Springer-Verlag, Berlin, 983 p.

Kolbert, E., 2014. The Sixth Extinction: An Unnatural History. Henry Holt & Company, NY, 319 p.

Newitz, A., 2013. Scatter, Adapt, and Remember. Doubleday, NY, 305 p.

Rees, M., 2013. Denial of catastrophic risks. Science 339 (6124), 1123.

Rossbacher, L.A., October 2013. Contemplating existential risk. Earth, Geologic Column 58 (10), 64.

Chapter 1

Perspectives on Coastal and Marine Hazards and Disasters

Jean T. Ellis[1] **and Douglas J. Sherman**[2]
[1] *Department of Geography and Marine Science Program, University of South Carolina, Columbia, SC, USA,* [2] *Department of Geography, University of Alabama, Tuscaloosa, AL, USA*

ABSTRACT
This chapter introduces the *Sea and Ocean Hazards, Risks, and Disasters* volume in the *Hazards and Disasters* series. The chapters in this volume summarize numerous aspects of our knowledge of a suite of coastal and marine hazards and disasters. A common theme is the persistence and increase of the global threat to coastal populations driven by environmental change and continued population growth. Here, coastal hazards are organized according to two main themes: threats to human systems (mainly) from the geophysical hazards such as tsunami, storm surge, sea ice, sea level rise, shoreline change, rip currents, and threats to humans and loss of ecosystem services caused by biological/ecological systems such as mangroves, corals, marshes, and harmful algal blooms.

1.1 INTRODUCTION

Recent disasters remind us of the exposure and vulnerability of humans, their resources, and biological and geophysical systems in coastal and marine environments. Hurricane Sandy (in 2012), for example, devastated portions of New York and New Jersey. The Borough of Manhattan in New York had approximately 4.5 m of storm surge causing extensive flooding (NOAA-NWS, 2014) and a 150-m crane precariously dangled over West 57th Street that houses a major gas line, thus causing numerous evacuations (FDNY, 2014). The subway system was flooded, the New York Stock Exchange was closed for 2 days, and lower Manhattan was without power for days; 7–9 million were without power across the region (Spurlock, 2012; CDC, 2013). The economic impact of Sandy has been estimated at US$62 billion (Crutsinger, 2012) and 117 perished (CDC, 2013). Typhoon Haiyan (in 2013), known locally as Yolanda, was the strongest tropical cyclone ever recorded at landfall, with wind speeds of 280 km/h. The 600-km-wide storm swept across the Visayas

Coastal and Marine Hazards, Risks, and Disasters. http://dx.doi.org/10.1016/B978-0-12-396483-0.00001-7
Copyright © 2015 Elsevier Inc. All rights reserved.

(Figure 1.1), in the Philippines, killing 6,111; impacting over 16 million; forcing evacuation of 800,000; damaging over 1 million homes; and leaving over 5 million displaced (Chan et al., 2013; American Red Cross, 2014). In 2011, one of the most powerful earthquakes on record (magnitude 9.0) occurred near Japan. This, and the associated Tōhoku Tsunami, with wave heights of 40 m, killed almost 16,000 people; displaced an additional 340,000; left an estimated 4.4 million without electricity; and contributed to the partial meltdown and major nuclear disaster of the Fukushima power plant. Physical

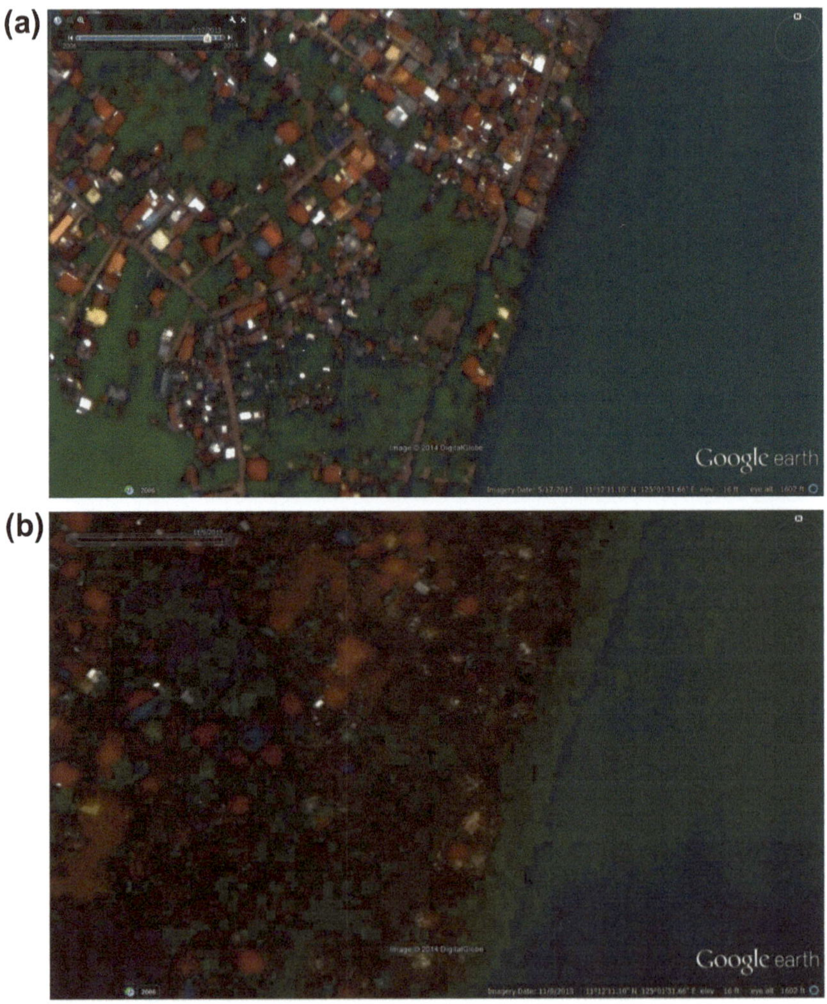

FIGURE 1.1 Impact of Typhoon Yolanda storm surge in the vicinity of Tacloban City, Philippines. (a) Google™ earth image in May, 2013, before Yolanda made landfall. (b) Google™ earth image in November, 2013, shortly after landfall.

damage from the earthquake and tsunami is estimated as exceeding US$300 billion (Nanto et al., 2011).

The magnitude of potential losses associated with coastal and marine hazards will continue to increase, despite the concerted efforts in many countries to reduce the risk. More people live in coastal environments than ever before, and predictions indicate that these populations will grow at increasingly disproportionate rates. The United Nations Environment Programme reports that more than one-third of the world's population lives in coastal areas, which is 4% of the Earth's land area (UNEP, 2006). Presently, in the United States, over 50% of the population resides within coastal-designated areas. From 1970 to 2010, populations in US coastal counties increased by 39% and in coastal watershed areas by 45% (NOAA-NOS, 2013). An additional 10% increase in coastal population is projected by 2020 (NOAA-NOS, 2013). Internationally, of the 30 megacities, or metropolitan areas with at least 10 million inhabitants, 16 are on the coast and an additional 7 are in a coastal region (or city center within 150 km). The top five megacities, in terms of their population, are in Asia and are coastal. Tokyo, with 34.8 million residents, is the largest megacity; one that has experienced rapid growth, especially since over half the city was destroyed during World War II bombings. The United Nations estimates that the fastest growing coastal city in terms of population from 2011 to 2015 will be Karachi, Pakistan, (currently the ninth largest megacity) with a 43% growth rate; note the 2000–2010 rate in this megacity exceeded 80%. The data most definitely reveal a "de-Westernizing" of the world's population (Kotkin and Cox, 2013).

In many coastal megacities the exposure of their populations to coastal and marine hazards is increasing simply as a consequence of providing basic human needs for survival, such as drinking water. Groundwater pumping in large coastal cities, such as Jakarta, Bangkok, and New Orleans, has already caused extensive and severe subsidence, increasing the frequency and magnitude of potential surge-induced inundation, for example. Subsidence in Jakarta has placed large, urbanized regions under threat, with almost half of the megacity below sea level. In a survey of environmental threats to megacities, Pelling and Blackburn (2013) cite reports that tidal flooding inundated up to 70% of the city and displaced more than 400,000 residents. Deltaic environments are at particular risk from a variety of processes that drive subsidence. Syvitski et al. (2009) estimated that 85% of the world's deltas experienced significant flooding since the turn of the millennium, and the rate and severity of inundation was predicted to increase. People and resources in these vulnerable environments will continue to be threatened and to experience loss from increased exposure to coastal flooding and inundation by storm surges and tsunamis, especially.

Predictions of the consequences of anticipated climate change also point to increased threat to coastal populations. These consequences include raised sea levels that will increase rates and severity of coastal flooding, degradation of coastal fresh water supplies, and increased coastal erosion and magnification of the impacts of other coastal and marine hazards (e.g., Nicholls et al., 2011).

It is not only the human systems that are under threat in coastal and marine environments. Pressures stemming from human development and from environmental change have also impacted the functions of many ecosystems, bringing some to potential tipping points. Mangrove systems are threatened by coastal development (Alongi, 2002), coral systems are threatened by ocean warming (Hughes et al., 2003; Lesser, 2007), and marsh systems are being affected by sea level rise (Kirwan et al., 2010). Changes to these systems are usually anticipated to lead to loss of ecosystem services, and thus cause indirect threat to human systems.

Decades of focused research, planning, and management efforts have been successful, in many instances, in reducing immediate threat to life and property. The risk associated with many coastal and marine hazards can be, and sometimes has been, reduced, especially along the coastlines of developed nations (e.g., Filatova et al., 2011). The absolute risks stemming from geophysical and biological systems, however, cannot be eliminated. Therefore the vulnerability of people moving to the coastal zone can only be expected to increase—even with increased attention to reducing risk. Measures of adaptation, such as protection, accommodation, and retreat can lessen the hazard impact and reduce the vulnerability (Nicholls, 2015), which is spatially and economically variable (Cutter et al., 2003).

It is this current and growing threat associated with coastal and marine hazards that underpins the motivation for this volume. The reviews and case studies presented represent broad-spectrum threats and disasters caused by geophysical and ecological changes and events. Some of the events are acute, such as tsunami or storm surge. Others are chronic, such as sea level rise or degradation of marsh or coral systems. All require our continued efforts to understand the nature of the related disasters, the sources of threat, and potential pathways toward the amelioration of vulnerability.

This volume is composed of 16 chapters summarizing much of our knowledge of coastal and marine hazards and disasters. We have organized the chapters in two main themes: threats to human systems (mainly) from the geophysical hazards such as tsunami, storm surge, sea ice, sea level rise, shoreline change, rip currents, and sea ice and threats to humans and loss of ecosystem services caused by biological/ecological systems such as mangroves, corals, marshes, and harmful algal blooms (HABs).

1.2 GEOPHYSICAL HAZARDS

Geophysical hazards originate from the operation of Earth systems. They include slow-onset, long-duration threats such as sea level rise; persistent threats such as rip currents and sea ice; and fast-onset, short-duration threats such as tsunami and storm surge. Most geophysical hazards are not amenable to reduction of the magnitude or frequency of the source events, although the impacts of the events can be mitigated through warning systems or protective adjustments to the hazards. In this section, we present reviews of the dynamics, characteristics, impacts, and mitigation strategies for many of the most common geophysical hazards. Many of these topics are also reviewed, with different foci, in the Coastal Geomorphology volume of the Treatise on Geomorphology, including the chapters on waves (Huntley, 2013) and barrier islands (McBride et al., 2013). Others are included in the Treatise on Coastal and Estuarine Science, with reviews of sea level change and coastal geomorphic response (Plater and Kirby, 2011).

1.2.1 Tsunamis

Some of the more (in)famous tsunamis, their origins, and their dynamics are discussed in multiple tsunami chapters: the 2004 Indian Ocean (Boxing Day) Tsunami, the 2011 Tōhoku Tsunami, and the Storegga Slides. Both Kânoğlu and Synolakis (in this volume) and Scheffers (in this volume) mention the recent explosion of tsunami research in the past 25 years. Specifically, Scheffers notes that prior to 2004, the year of the Indian Ocean (Boxing Day) Tsunami, only about 1,000 scientific papers had been published on the topic, but in the decade since then, more than 1,000 papers have been published.

Kânoğlu and Synolakis (in this volume) lead off the tsunami section with a chapter on their generation, characteristics, and dynamics, including a discussion on the recent advances in modeling. They posit that there was no critical mass of tsunami research prior to the 1990s and that 2004 Indian Ocean (Boxing Day) Tsunami was turning point regarding the science surrounding and the public awareness of tsunami. Five recent advances are discussed: (1) calculation of the directivity of tsunamis, (2) experiments with idealized seafloor deformations and the establishment of the solitary wave model, (3) analytical results and experiments with solitary wave run-up, (4) the demise of the solitary wave model and the realizations that the front wave in a tsunami train is dipole shaped, and (5) realization of the fact that tsunami generation by landslides resulted from moderate earthquakes. They also review the recent focus on tsunami mitigation and the development of inundation maps, especially for the US coast. Kânoğlu and Synolakis conclude by noting that it remains that most lives are saved from these high-impact and large-duration disasters with advanced understanding of ancestral knowledge, especially in remote and/or poorly developed regions.

Rather than broadly reviewing the gamut of research, Scheffers (in this volume) targets research subjects under debate regarding the signatures and occurrences of paleotsunamis: (1) tsunami generation from extraterrestrial impacts, (2) problems associated with determining the source of historic boulder movements and fine washover sediment deposition (strong storms or tsunami waves?), and (3) verification of flood myths, or tsunami reports, in historical documents. She concludes her chapter with a list of future research topics such as understanding paleotsunamis during high sea levels experienced during the interglacial periods and using archaeological techniques to ascertain tsunami histories and reasons for inland shifts in settlements (tsunami or sea level rise?). Throughout her chapter, Scheffers discusses the importance of discerning the impacts of sea level rise and tsunami for paleotsunami research; however, it is also critical to our understanding and the modeling of future impacts of sea level rise, which are topics of later chapters.

Farrell et al. (in this volume) review 10 tsunami case studies, grouped according to the tsunamigenic source: (1) earthquake (2004 Indian (Sumatra–Andaman) earthquake, 2011 Tōhoku earthquake, 1964 Alaska earthquake, 1960 Valdivia earthquake, 1946 Aleutian Islands earthquake, 1908 Messina–Reggio earthquake, 1755 Great Lisbon earthquake); (2) landslide (Storegga Slides 30,000 and 7200–7000 years before present, Papua New Guinea, 1998); and (3) volcano (Krakatoa 1883). The generation, size and extent, impact, relief effort, and aftermath of each tsunami are described. The case studies reveal that the size and extent of the tsunami are not necessarily correlated with the fatalities or damage extent.

1.2.2 Storm Surge

Storm surge is the increased sea surface elevation resulting from a tropical or extratropical cyclone moving toward the coast. Tropical storm intensity contributes to storm surge height, but it also related to forward speed, central pressure, and angle of approach (relative to the coast) of the storm and the local topography and bathymetry. Of all the geophysical risks to coastal regions, storm surges are the most severe (von Storch and Woth, 2008). The largest loss of life associated with tropical cyclones is caused by storm surge (Doocy et al., 2013).

Nott (in this volume) discusses the reconstruction of historic storm surges through the investigation of coral rubble, cheniers, sand beach ridges, shell ridges, gravel ridges, pumice ridges, sand splays, and washover deposits, with the oldest of the deposits being 7,000 years. Of these, only beach ridges are used to underpin modeling using Generalized Extreme Value theory and Bayesian analysis approach to statistically derive storm surge return periods (Nott and Jagger, 2013). Studies in the United States, Japan, and Australia reveal similar temporal patterns of quiescent and heightened tropical cyclone activity. Nott also summarizes the literature that discusses the relationship between El Niño and La Niña in the past 5,000 years to tropical cyclone activity in the southwest Pacific and Atlantic Oceans. Similar to Scheffers' chapter, understanding historical records aids our present-day efforts to future cast.

Horn (in this volume) suggests that we have a substantial understanding of the physical components of storm surges, including modeling, and the warning systems are continually improving, yet there continues to be substantial loss of life. The reason for increased loss of life relates to inadequate mitigation, acting to reduce storm surge severity and number, and adaptation, living with storm surge risk, efforts. The majority of her chapter focuses on adaptation measures that, in turn, may reduce the vulnerability to storm surge: (1) storm surge barriers, with discussions on Delta Works in the Netherlands; recent efforts in the United Kingdom and the Thames Estuary; the Mose system in Venice, Italy; the Hurricane and Storm Damage Risk Reduction System in New Orleans, USA; and proposed storm surge barriers to protect New York and New Jersey post Hurricane Sandy (2012); (2) storm surge warning; and (3) storm surge disaster risk reduction by decreasing exposure and vulnerability to storm surges, which are now increasing.

von Storch et al. (in this volume) present case studies on storm surge to demonstrate the variability between storm surges and tropical and extratropical storm intensity, erosion extent, and loss of life and property. For example, they discuss tropical-cyclone-generated storm surges at Qingdao, China. This region, which is important to China's economy, has experienced nine events from 1949 to 2003 with sea levels in excess of 5.1 m. The periodicity of storm surge activity, likely related to anthropogenic climate change, is demonstrated using Hamburg and the Elbe estuary. The final case study is the southern Baltic Sea coast, which has experienced several extratropical storm surges, with the largest being in 1913.

1.2.3 Sea Level Rise

The chapters, by Devoy (in this volume) and Nicholls (2015), center exclusively on sea level change (SLC), which is a well-studied topic that still requires substantial additional research. The most recent Intergovernmental Panel on Climate Change report (Church et al., 2013) reports that the current rate of sea level rise will continue or accelerate through the twenty-first century. Rising seas, from mass melting of glaciers, will result in coastal flooding, displacement of coastal residents, and economic loss.

Devoy provides a comprehensive discussion on the recent history of climate change and SLC, sea level measurements, and causes for change. He reviews the complexities associated with sea level trends and models using various time scales ranging from short-term (meteorological) to long-term (glacial eutasy) changes.

Sea level change becomes significant when its impacts increase the risk and vulnerabilities of coastal inhabitants, thus forcing adaptation measures, which is the focus of Nicholls' contribution. He first distinguishes global mean from relative sea level rise. The relationship between climate mitigation and SLR is detailed, followed by a comprehensive discussion on adaptation. Nicholls bases his discussion around a systems model (his Figure 1.1) that identifies the key interactions relating to SLR and the coastal zone.

1.2.4 Rip Currents

Rip currents are fast, offshore-flowing currents in the near-shore zone of many beaches that cause many drownings and near drownings. Brander (in this volume) first reviews the state of the science of rip current research and modeling. Next, he discusses rip currents as the significant risk that they pose to bathers; it is estimated that at least 80% and 89% of beach rescues in the United States and Australia, respectively, are related to rip currents. Brander argues that the lack of education about rip current hazards substantially contributes to the high number of rescues. He offers the importance of collaboration between researchers and beach safety organizations for advancing the science and improving education and awareness to the public.

1.2.5 Sea Ice

Eicken and Mahoney (in this volume) review the risks and disaster implications associated with direct and indirect effects of sea ice presence and human activity, using the Arctic and sub-Arctic as case studies. Hazards and associated risks of sea ice are discussed in three temporal categories: (1) long-term, or summer ice extent reduction; (2) near-term, or sea ice extent changes and coastal impacts (erosion and infrastructure); and (3) immediate, or sea ice hazards and human activities together, for example, shipping or resource development. They emphasize the importance of preventing the escalation of initial events that may lead to large-scale disasters that involve human activities, such as shipping or resource development. Further, the criticality of mapping the location and patterns of the changing sea ice extent and the ice age (older ice is thicker, therefore serves as a greater hazard) and deformation are all discussed. This chapter reveals that scant research has been conducted on spatially linking Artic sea ice, ecosystems, and human activities.

1.3 BIOLOGICAL/ECOLOGICAL HAZARDS

The degradation of the biological and/or ecosystem structure often increases the vulnerability of the coastal system to many of the geophysical hazards discussed above. Human interactions are largely responsible for damage to coral reefs, marshes, and mangroves, and the increasing prevalence of HABs. Accordingly, in this section we present reviews of mangroves, corals, marshes, and HABs. Many of these topics are also reviewed, with different foci, in the Coastal Geomorphology volume of the Treatise on Geomorphology, including the chapters on mangroves (Spencer and Möller, 2013) and coral systems (Kench, 2013). Others are included in the Treatise on Coastal and Estuarine Science, with a review of salt marsh restoration (Boorman and Garbutt, 2011).

1.3.1 Mangroves

McIvor et al. (in this volume) discuss the importance of considering natural defenses, such as mangroves where geographically permissible, as an alternative to hard engineering structures. It is estimated that maintenance of hard-engineered structures, which increase risk, average US$ 12–17 billion per annum (Hinkel et al., 2013). Similar to other contributions in this volume, they discuss how the 2004 Indian Ocean (Boxing Day) Tsunami increased interest in research, which in this case is regarding the efficacy of mangroves in reducing storm surge and protecting the disappearing mangrove forest. McIvor et al. provide a comprehensive review demonstrating the need for mangroves to reduce the risk of storm surge. They frame their argument according to the following: (1) direct and model-based evidence of storm surge reduction related to mangroves; (2) characteristics of mangroves and mangrove forests that contribute to storm surge reduction, such as the width of the mangrove forest, individual plant characteristics (diameter of aerial roots), local topography, and presence of channel and pools within the forest; (3) characteristics of the storm; and (4) reduction of surface winds due to mangroves.

1.3.2 Corals

Similar to the environmental conditions with mangroves, coral reefs demonstrate complex interactions between ecological and physical processes, with the former being better studied compared to the latter. After reviewing the building blocks and environmental limits of coral growth, Kench and Owen (in this volume) frame their chapter around the linkages between the biological and physical processes of the coral reef system, anthropogenic pressures, ecosystem services, and environmental controls. They also identify natural and anthropogenic stressors impacting reef system function and ecosystem services, with a recognition that reef health shall be identified according to Perry et al. (2008), which includes structural integrity and carbonate budget state in their assessment of reef health. Stressors presented are ocean temperatures, coral bleaching, ocean acidification, pollution, and changes in water quality. The impact of tsunami on coral reef systems and their vulnerability is also discussed. Coral reefs are a prime example demonstrating cumulative impacts, therefore integrative studies are important.

1.3.3 Marshes

Salt marshes, which cover approximately 40–80 million ha (Nellemann et al., 2009) provide substantial ecosystem services, as they protect the shoreline, support fisheries and wildlife habitat, improve water quality, and sequester carbon. However, as Hansen and Reiss (in this volume) discuss, marshes are under extreme threat from anthropogenic impacts, specifically agriculture and urban construction, from pollution and the impact of climate change and invasive species. An estimated 25% of historic salt marshes are lost (Secretariat of the Convention on Biological Diversity, 2010), with the loss approximating 38% in the United States (Gedan et al., 2011) and exceeding 50% in China (An et al., 2007). These threats need to be mitigated immediately using innovative approaches, including ecosystem-based services, to reduce the magnitude of future disasters.

1.3.4 Harmful Algal Blooms

Anderson et al. (in this volume) examine the devastation associated with HABs, indicating the imminent threat for seafood industries and human health. HABs are capable of killing fish, mammals, and birds and cause substantial human illness. They are composed of several taxa of phytoplankton, making them challenging to study because of the species diversity. In their chapter, Anderson et al. discuss pressures within the environment that promote HABs, strategies for preventing and/or controlling HABs, modeling, and case studies of coastal regions where monitoring, modeling, adaptive management, and/or governing is being implemented.

1.4 CONCLUSIONS

Natural hazards are a critical facet of human history, and will continue to be a challenge for future development. The chapters in this volume summarize numerous aspects of our knowledge of a suite of coastal and marine hazards and disasters. A common theme is the persistence and increase of the global threat to coastal populations driven by environmental change and continued population growth. It has been estimated that 50% of the world's population will reside within 100 km of the coast by 2050 (Small and Nicholls, 2003), and per annum, 46 million people are at risk from a storm surge (IPCC, 2001). Substantial investment in protecting population and resources in developed nations has dramatically reduced the loss of life, so that even extreme catastrophes seldom involve death tolls exceeding 10,000 lives. At the same time, however, the value of property losses has escalated disproportionately, with three of the five most costly natural disasters occurring in the past decade. In developing nations, the risk of catastrophic loss of life and property persists.

The coastal region is a complex system composed of stakeholders interested in scientific study and monitoring, policy, management, education, politics, recreation and tourism, coastal habitation, and natural systems. The variable uses often create conflict that limits the efficacy of coastal management and planning activities, thus potentially exacerbating the consequences of coastal and marine natural hazards. Although reducing the magnitude and frequency of the hazards is difficult to impossible, we attest that increased communication among stakeholders, additional research and constituent education, and improved coastal management will dramatically decrease the treat to humans.

REFERENCES

Alongi, D.M., 2002. Present state and future of the world's mangrove forest. Environ. Conserv. 39, 331–349.

American Red Cross, 2014. Red Cross Typhoon Haiyan Work in Philippines. http://www.redcross.org/news/article/Red-Cross-Sends-Support-to-Philippines-for-Typhoon-Response (accessed 15.06.14.).

An, S., Li, H., Guan, B., Zhou, C., Wang, Z., Deng, Z., Zhi, Y., Liu, Y., Xu, C., Fang, S., Jiang, J., Li, H., 2007. China's natural wetlands: past problems, current status, and future challenges. Ambio 36, 335–342.

Anderson, C.R., Moore, S.K., Tomlinson, M.C., Silke, J., and Cusack, C. Living with harmful algal blooms in a changing world: strategies for modelling and mitigating their effects in coastal marine ecosystems, in this volume.

Boorman, L.A., Garbutt, A., 2011. Restoration strategies for intertidal salt marshes. In: Wolanski, E., McLusky, D., Chicharo, L. (Eds.), Treatise on Estuarine and Coastal Science, Ecohydrology and Restoration, vol. 10. Academic Press, San Diego, CA, pp. 189–215.

Brander, R.W. Rip currents, in this volume.

CDC (Centers for Disease Control and Prevention), 2013. Deaths Associated with Hurricane Sandy – October-November 2012. Morbidity and Mortality Weekly Report (MMWR). http://www.cdc.gov/mmwr/preview/mmwrhtml/mm6220a1.htm (accessed 15.06.14).

Chan, E.Y.Y., Liu, S., Hung, K.C., 2013. Typhoon Haiyan and beyond. The Lancet 382 (9908), 1873. http://dx.doi.org/10.1016/S0140-6736(13)62415-0.

Church, J.A., Clark, P.U., Cazenave, A., Gregory, J.M., Jevrejeva, S., Levermann, A., Merrifield, M.A., Milne, G.A.R., Steven Nerem, R.S., Nunn, P.D., Payne, A.J., Pfeffer, W.T., Stammer, D., Unnikrishnan, A.S., 2013. Sea level change. In: IPCC Working Group I, Fifth Assessment Report.

Crutsinger, M., 2012. Sandy Economic Damage Worse than Expected at $62 Billion. Huffington Post. http://www.huffingtonpost.com/2012/11/29/sandy-economic-impact-damage_n_2214060.html (accessed 15 June 2014).

Cutter, S.L., Boruff, B.J., Shirley, W.L., 2003. Social vulnerability to environmental hazards. Soc. Sci. Q. 84 (2), 242–261.

Doocy, S., Dick, S., Daniels, A., Kirsch, T.D., April 16, 2013. The human impact of tropical cyclones: a historical review of events 1980–2009 and systematic literature review. PLOS Curr. Disasters. http://dx.doi.org/10.1371/currents.dis.f4deb457904936b07c09daa98ee8171a. Edition 1.

Devoy, R.J.N. Sea-level rise: Causes, Impacts and Scenarios for Change, in this volume.

Eicken, H., Mahoney, A.R. Sea Ice: Hazards, Risks and Implications for Disasters, in this volume.

Farrell, E.J., Ellis, J.T., and Hickey, K.R. Tsunami Case Studies, in this volume.

FDNY (Fire Department New York), 2014. Personal Communication. Battalion 6, Engine 16 and Tower Ladder 7.

Filatova, T., Mulder, J.P.M., van der Veen, A., 2011. Coastal risk management: how to motivate individual economic decisions to lower flood risk. Ocean Coastal Manage. 54, 164–172.

Gedan, K.B., Kirwan, M.L., Wolanski, E., Barbier, E.B., Silliman, B.R., 2011. The present and future role of coastal wetland vegetation in protecting shorelines: answering recent challenges to the paradigm. Clim. Change 106, 7–29.

Hansen, V.D. Reiss, K.C. Threats to Marsh Resources and Mitigation, in this volume.

Hughes, T.P., Baird, A.H., Bellwood, D.R., Card, M., Connolly, S.R., Folke, C., Grosberg, R., Hoegh-Guldberg, O., Jackson, J.B.C., Kleypas, J., Lough, J.M., Marshall, P., Nystrom, M., Palubmi, S.R., Pandolfi, S.R., Rosen, B., Roughgarden, J., 2003. Climate change, human impacts, and the resilience of coral reefs. Science 301, 929–933.

Huntley, D.A., 2013. Waves. In: Shroder, J., Sherman, D.J. (Eds.), Treatise on Geomorphology, Coastal Geomorphology, vol. 10. Academic Press, San Diego, CA, pp. 39–73.

Hinkel, J., Lincke, D., Vafeidis, A.T., Perrette, M., Nicholls, R.J., Tol, R.S.J., Marzeiong, B., Fettweish, X., Ionescu, C., Levermann, A., 2013. Coastal Flood Damage and Adaptation Costs under 21st Century Sea-level Rise. URL. www.pnas.org/cgi/doi/10.1073/pnas.1222469111.

Horn, D. Storm Surge Mitigation and Adaptation, in this volume.

IPCC (Intergovernmental Panel on Climate Change), 2001. Climate Change 2001: IPCC Third Assessment Report. Working Group II: Impacts, Adaptation and Vulnerability. Cambridge University Press, Cambridge.

Kench, P.S., 2013. Coral systems. In: Shroder, J., Sherman, D.J. (Eds.), Treatise on Geomorphology, Coastal Geomorphology, vol. 10. Academic Press, San Diego, CA, pp. 328–359.

Kench, P.S., Owen, S.D. Coral Reef Systems and Complexity of Hazards, in this volume.

Kirwan, M.L., Guntenspergen, G.R., D'Alpaos, A., Morris, J.T., Mudd, S.M., Temmerman, S., 2010. Limits on the adaptability of coastal marshes to rising sea level. Geophys. Res. Lett. 37 (23). http://dx.doi.org/10.1029/2010GL045489.

Kotkin, J., Cox, W., 2013. The world's fastest-growing megacities. New Geogr. http://www.newgeography.com/content/003630-the-worlds-fastest-growing-megacities (accessed 15.06.14.).

Kânoğlu, U. and Synolakis, C. Tsunami Dynamics, Forecasting, and Mitigation, in this volume.

Lesser, M.P., 2007. Coral reef bleaching and global climate change: can corals survive the next century? Proc. Natl. Acad. Sci. 104, 5259–5260.

McBride, R.A., Anderson, J.B., Buynevich, I.V., et al., 2013. Morphodynamics of barrier systems: a synthesis. In: Shroder, J., Sherman, D.J. (Eds.), Treatise on Geomorphology, Coastal and Submarine Geomorphology, vol. 10. Academic Press, San Diego, CA, pp. 166–244.

McIvor, A., Spencer, T., Spalding, M. Lacambra, C., Möller, I. Mangroves, Tropical Cyclones and Coastal Hazard Risk Reduction, in this volume.

Nanto, D.K., Cooper, W.H., Donnelly, J.M., 2011. Japan's 2011 Earthquake and Tsunami: Economic Effects and Implications for the United States. Congressional Research Service Report for Congress, Prepared for Members and Committees of Congress. http://fpc.state.gov/documents/organization/159785.pdf (accessed 15.06.14.).

Nellemann, C., Corcoran, E., Duarte, C., Valdés, L., De Young, C., Fonseca, L., Grimsditch, G. (Eds.), 2009. Blue Carbon. A Rapid Response Assessment. United Nations Environment Programme, GRID-Arendal.

Nicholls, R.J., 2015. Adapting to sea level rise, in this volume.

Nicholls, R.J., Marinova, N., Lowe, J.A., Brown, S., Vellinga, P., de Gusmão, D., Hinkel, J., Tol, R.S.J., 2011. Sea-level rise and its possible impacts given a 'beyond 4ºC world' in the twenty-first century. Philos. Trans. R. Soc. A 369, 161–181.

NOAA-NWS [National Oceanic and Atmospheric Administration – National Weather Service], 2014. Hurricane Sandy – October 29, 2012. http://www.weather.gov/okx/HurricaneSandy (accessed 15.06.14.).

NOAA-NOS [National Oceanic and Atmospheric Administration – National Ocean Service], 2013. National Coastal Population Report: Population Trends from 1970-2020. http://stateofthecoast.noaa.gov/features/coastal-population-report.pdf (accessed 15.06.14.).

Nott, J. Paleo-storm Surges and Inundations, in this volume.

Nott, J., Jagger, T., 2013. Deriving robust return periods for tropical cyclone inundations from sediments. Geophys. Res. Lett.. http://dx.doi.org/10.1029/2012GL054455.

Pelling, M., Blackburn, S., 2013. Case studies: governing social and environmental transformation in coastal megacities. In: Pelling, M., Blackburn, S. (Eds.), Megacities and the Coast. Routledge, New York, pp. 200–236.

Perry, C.T., Spencer, T., Kench, P.S., 2008. Carbonate budgets and reef production states: a geomorphic perspective on the ecological phase shift concept. Coral Reefs 27, 853–866.

Plater, A.J., Kirby, J.R., 2011. Sea-level change and coastal geomorphic response. In: Wolanski, E., McLusky, D., Flemming, B.W., Hansom, J.D. (Eds.), Treatise on Estuarine and Coastal Science, Estuarine and Coastal Geology and Geomorphology, vol. 3. Academic Press, San Diego, CA, pp. 39–72.

Scheffers, A.M. Paleo-tsunami research – current debate and controversies, in this volume.

Secretariat of the Convention on Biological Diversity, 2010. Global Biodiversity Outlook 3 (Montreal).

Small, C., Nicholls, R.J., 2003. A global analysis of human settlement in coastal zones. J. Coastal Res. 19 (3), 584–599.

Spencer, T., Möller, I., 2013. Mangrove systems. In: Shroder, J., Sherman, D.J. (Eds.), Treatise on Geomorphology, Coastal Geomorphology, vol. 10. Academic Press, San Diego, CA, pp. 360–391.

Spurlock, C., 2012. Hurricane Sandy New York City Power Outage Map: Thousand without Electricity in Metro Area. Huffington Post. http://www.huffingtonpost.com/2012/10/31/hurricane-sandy-new-york-city-power-outage-map_n_2050380.html (accessed 15.06.14.).

Von Storch, H., Jiang, W., Furmanczyk, K.K. Storm Surge Case Studies, in this volume.

Von Storch, H., Woth, K., 2008. Storm surges, perspectives and options. Sustainability Sci. 3, 33–44. http://dx.doi.org/10.1007/s11625-008-0044-2.

Syvitski, J.P.M., Kettner, A.J., Overeem, I., Hutton, E.W.H., Hannon, M.T., Brakenridge, G.R., Day, J., Vörösmarty, c., Saito, Y., Giosan, L., Nicholls, R.J., 2009. Sinking deltas due to human activities. Nat. Geosc. 2, 681–686.

(UNEP) United Nations Environmental Programme, 2006. Marine and Coastal Ecosystems and Human Well-being: A Synthesis Report Based on the Findings of the Millennium Ecosystem Assessment. United Nations Environment Programme, Nairobi.

Chapter 2

Tsunami Dynamics, Forecasting, and Mitigation

Utku Kânoğlu [1] **and Costas Synolakis** [2,3]

[1] *Department of Engineering Sciences, Middle East Technical University, Ankara, Turkey,* [2] *Viterbi School of Engineering, University of Southern California, Los Angeles, CA, USA,* [3] *Technical University of Crete, Chania, Greece*

ABSTRACT

Tsunamis had been earlier believed as extremely rare events, yet about one event per year has been reported in the past two decades, making them a more common extreme hazard. After the 2004 Indian Ocean tsunami, the need for substantial improvements in tsunami real-time and long-term forecasting capabilities, education, and development of tsunami-resilient communities became evident. Thereafter, there were substantial advances in tsunami science, i.e., significant advancements in warning methodologies, predisaster preparedness, and basic understanding of related phenomena. The 2011 Japan tsunami, broadcasted live to a stunned world audience, underscored the difficulties of implementing theoretical advances in applied hazard mitigation. Japan is possibly the most tsunami-ready nation on the Earth. Nonetheless, the size of the 2011 earthquake was largely unexpected and, in many instances, the floods penetrated several times the distances that had been anticipated in pre-event planning. Three years later, Japan is still recovering. A need exists for acquainting the broader scientific community on advances in prediction and mitigation in hopes that applied disaster preparedness improves.

2.1 INTRODUCTION

After the December 26, 2004, Indian Ocean (Boxing Day) tsunami, Huppert and Sparks (2006) wrote "It is likely that in the future, we will experience several disasters per year that kill more than 10,000 people." Their assessment was not far off, unfortunately, to wit the March 11, 2011, Japan tsunami (the Great East Japan Earthquake Disaster) that alone resulted in more than 20,000 casualties. Tsunamis and other coastal disasters have killed over 200,000 since Boxing Day 2004. Coastal communities are now extensively developed centers of substantial commercial activities that are also at risk.

Coastal and Marine Hazards, Risks, and Disasters. http://dx.doi.org/10.1016/B978-0-12-396483-0.00002-9
Copyright © 2015 Elsevier Inc. All rights reserved.

Tens of millions of people live in high-risk coastal communities around the world, and hundreds of thousands of tourists are at high-risk beaches at any given time. This appears yet another case of what Jackson (2006) has called *fatal attraction*; earthquakes occur in places where they would likely cause more casualties compared to earlier times because in some places the availability of water is linked to underlying faults and because many rural communities have grown much larger with mostly poor building standards.

Furthermore, geological hazards such as tsunamis are not only threats to the countries in whose territories they originate but also can cause wide-scale devastation across national boundaries, as dramatically shown during the 2004 Boxing Day tsunami which impacted at least 16 countries directly and tourists from many other countries (Synolakis and Kong, 2006). Among the casualties during the 2004 disaster were 428 Swedish people, out of a population of about 10 million.

The word tsunami made its grand debut in most world languages with the December 26, 2004 event. Yet the first historical inference of coastal inundation by tsunamis refers to the eruption of the Thera volcano in the Eastern Mediterranean (Marinatos, 1939), now believed to have occurred around 1620 BC. This precipitated the demise of the Minoans in Crete Island (Bruins et al., 2008). Referred to as tidal wave in English, the exact translation of tsunami from the Japanese is *harbor wave*. Probably, early observations of these unusual waves by eyewitnesses were in ports, as harbors were centers of commercial activity and the points of contact with the sea. In Japan, where historic records exist since the ninth century AD, these motions were often associated with tsunamis, hence the name. It is not uncommon that a relatively small tsunami entering a port or harbor can trigger substantial water level oscillations and reach substantial heights, and water motions can persist for hours, as most recently observed in many harbors along the Japanese coast, as well harbors in the Pacific coast of the United States after the 2011 Japan tsunami (Figure 2.1).

Tsunamis are generated by impulsive geophysical events of the seafloor and of the coastline, such as earthquakes and submarine and subaerial landslides. Volcanic eruptions and asteroid impacts are less common but more spectacular tsunami-generation mechanisms (Gisler, 2009; Morrison, 2006). Tsunamis are high-impact, long-duration disasters and often entail substantial human drama, as outlined in the Hollywood movie *The Impossible*. They are long waves with small steepness and evolve substantially, through spatial and temporal spreading from their source region, as suggested in the map of energy propagation of the Boxing Day tsunami (Titov et al., 2005a) (Figure 2.2); also see Figure 2.3 for the March 11, 2011, tsunami (Tang et al., 2012).

The determination of the terminal effects of tsunamis as they strike shorelines and coastal structures remains one of the quintessential problems in tsunami hazard mitigation. Since the Boxing Day tsunami, a new science of tsunami forecasting has emerged (Bernard and Robinson, 2009). The first extensively tested real-time forecasting methodology is now officially in use

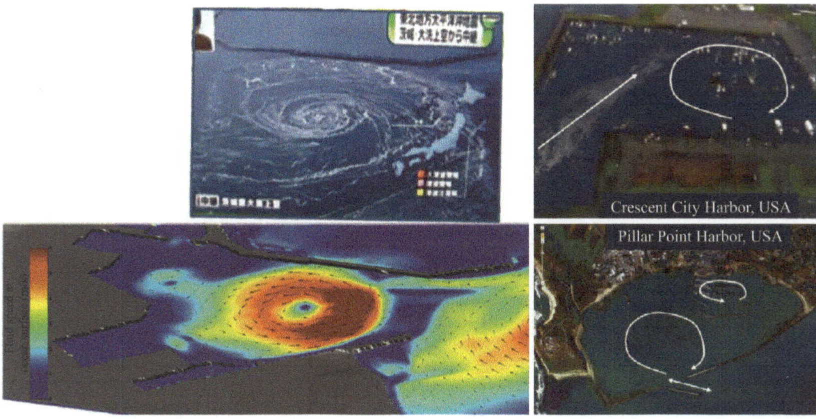

FIGURE 2.1 Significant tsunami currents were observed in many harbors during the March 11, 2011, tsunami. (Top left) View of whirlpool at Port of Oarai, Japan, taken from helicopter approximately at 17:54 (local time), i.e., 3 h 8 min after the earthquake; (bottom left) numerical results of Lynett et al. (2012) for the fluid speed of the tsunami in the Port of Oarai; after the 2011 Japan tsunami (top right) surge jetting in to the inner harbor of Crescent City, California; and (bottom right) Pillar Point Harbor, south of San Francisco, which experienced counterrotating eddies in the inner and outer basins. *After Lynett et al. (2012).*

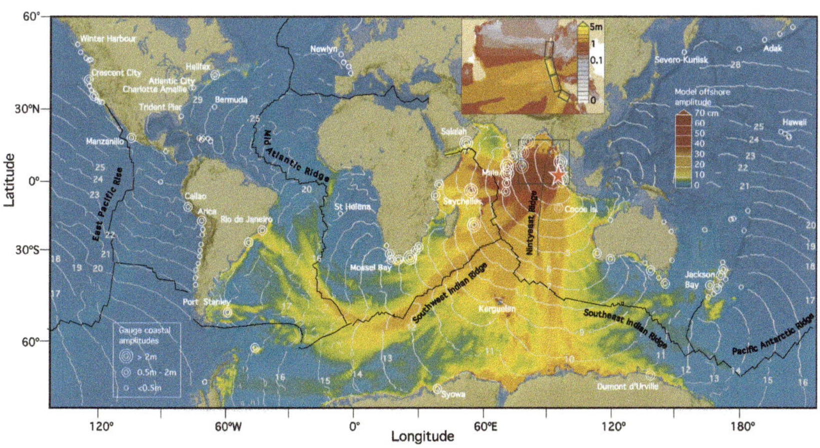

FIGURE 2.2 Global maximum tsunami heights of the Boxing Day tsunami computed from numerical model of Method of Splitting Tsunami (MOST) (Titov and González, 1997), after 44 h of propagation. Inset shows distribution of the slip among four subfaults (from south to north: 21 m, 13 m, 17 m, and 2 m) which provides best fit for satellite altimetry data and correlates well with seismic and geodetic data inversions, and the computed wave heights in the Bay of Bengal. Wave amplitudes, directionality, and global propagation patterns appear primarily determined by the orientation and intensity of the offshore seismic line source and subsequently by the trapping effect of midocean ridge topographic waveguides. Contours show computed tsunami travel times. Circles denote the selected tide gauge stations where amplitudes of tsunami are given in three range categories. *After Titov et al. (2005a).*

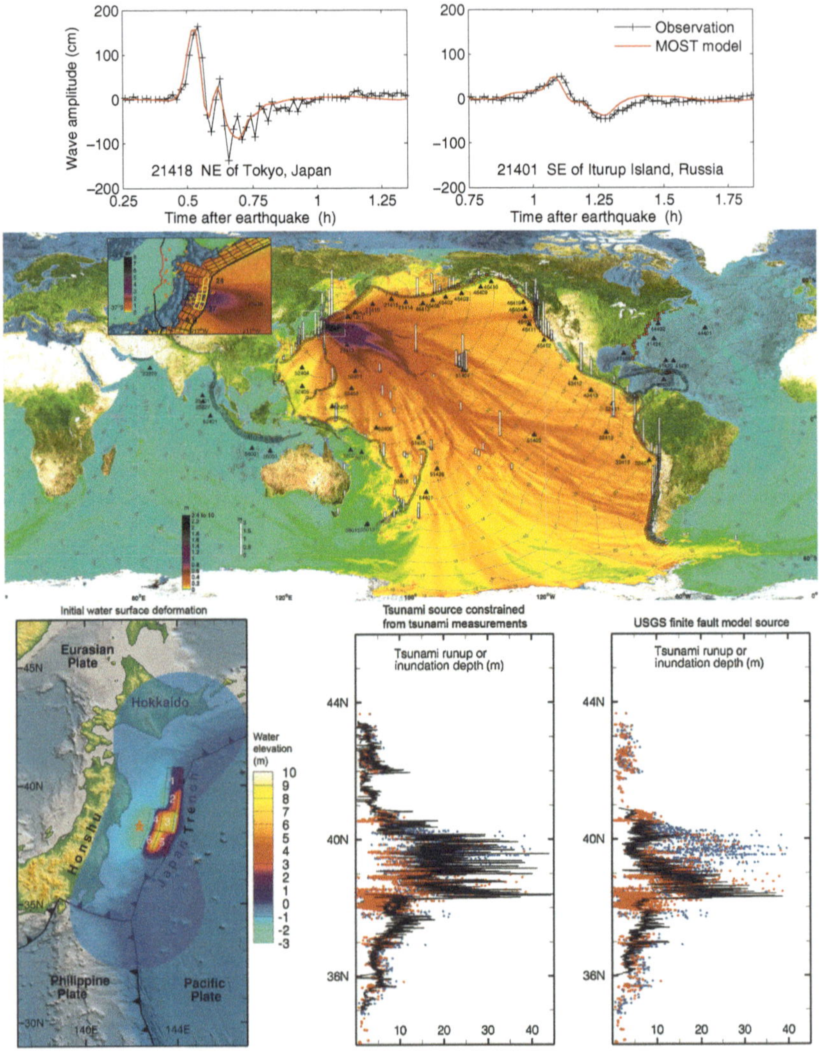

FIGURE 2.3 (Top) Deep-ocean Assessment and Reporting of Tsunamis (DART) measurements used by the United States National Oceanic and Atmospheric and Administration (NOAA) Center for Tsunami Research (NCTR) for the inversion of the March 11, 2011, Japan tsunami and the resulting waveforms at DARTs after the inversion. (Middle) Global maximum wave amplitudes for the source identified by the NCTR in real time. Inset shows the resultant unit sources. *After Tang et al. (2012).* (Bottom left) Initial tsunami source defined by the NCTR in real time for the event. Comparison of computed tsunami maximum wave amplitudes on land based on (bottom center) tsunami source constrained from DART measurements and (bottom right) the US Geological Survey (USGS) finite fault model source with measured tsunami heights and runup values (black lines, red dots, and blue dots denote computed MOST model maximum wave amplitudes (m), observed tsunami heights (m), and observed runup values (m), respectively). Although computed MOST model maximum wave amplitudes are given by Wei et al. (2013), the observed tsunami heights and runup values are given by Mori et al. (2011).

by tsunami warning centers (Vasily Titov, personal communication, June 8, 2013) and is rapidly becoming web-based (Burger et al., 2013). This technology is based on real-time data assimilation from measurements from tsunamographs: deep-ocean buoys also known as DARTs (Tang et al., 2012; Wei et al., 2008). Measurements from tsunamographs are used to better define the seafloor displacement that triggered the tsunami. Such hydrodynamic inversions are now as standard as seismic inversions. New markers have emerged that help to identify paleotsunamis and thus better infer the mechanisms of paleoearthquakes. High-end computational tools now allow for inundation predictions, even from submarine landslides. New analytical results help explain the scaling of vexing tsunami evolution problems. Most coastlines of highly developed nations along the Pacific now have inundation maps for pre-event planning (see section 2.4.1), as do at-high-risk cities in Indonesia, such as Padang. Several communities, including several in the United States, have earned the coveted *tsunami-ready* designation. With the exception of the Mediterranean, all the at-risk world oceans and seas are now covered with rapid warning procedures.

Tsunami studies are multi-disciplinary, at the cross-roads of geology, geophysics, geography, oceanography, coastal engineering, mathematics, and social science. Once initial conditions, offshore bathymetry, and onshore topography are known, the evolution of a tsunami can be calculated fairly accurately, at least until the time of its maximum inland penetration. Uncertainties arise in the specification of the initial condition, assessing the repeat interval of the phenomenon, and in developing mitigation measures, particularly given the variability of the human response. In addition, the long duration of tsunamis with multiple waves attacking target coastlines challenge rescue efforts, and might cause secondary effects such as fire, nuclear accidents (Satake, 2013), and debris flows (Lebreton and Borrero, 2013).

In early tsunami studies, none of the analytical or numerical results for predicting inundation had been validated with data from laboratory experiments, and no realization existed of the scales of tsunamis of geophysical interest. Up until the 1970s, no numerical computations existed, and tsunamis were studied as idealized periodic waves generated by uniform seafloor displacements. Most studies were analytical, and results were either in terms of complex formulas or asymptotics, yet they revealed little of the underlying physics and were of equally little practical interest.

The analytical results remained disjointed from the laboratory experiments, possibly because of the different ranges in wave steepness that analysis and experiments were referring to, the latter having been undertaken with far steeper waves than the former. No realization existed of what tsunamis resembled, beyond that they were long water waves. Even empirical relationships did not exist to relate the size of the wave offshore to the runup with inundation, and, what can one only still describe as confusion, remained as to the order of approximations used in the different analytical approaches.

Worse, the few historic photos of advancing tsunamis from the 1946 Aleutian and 1983 Japan Sea events were interpreted as suggestive of advancing turbulent bores, i.e., fluid structures notoriously difficult for computations. Most people who watched live footage from the 2011 Japan tsunami can visualize tsunamis better than the few tsunami experts in the early 1980s could.

Beginning in the 1980s, detailed modeling tools and high-resolution, small-scale laboratory experiments become available. Many of the existing results were synthesized into powerful analytical and numerical tools, and then validated with laboratory experiments, thus setting the stage for the advances in the now widely used tsunami forecast models (FMs). The 1990s showed rapid progress toward a more realistic understanding of the entire tsunami evolution process from generation to runup. Not only did large-scale laboratory data become available (Briggs et al., 1995; Liu et al., 1995) but also tsunamis started to be reported at a rate of almost one per year, debunking their extreme hazard status, at least for moderate events. Post-tsunami surveys started being carried out after every event, and observations helped to motivate advances.

With hindsight, no critical mass of scientists studying tsunamis existed until the early 1990s. The growth was precipitated through a series of landmark workshops which allowed for identification of the show stoppers and, later, for intermodel comparison (Synolakis and Okal, 2005). The first took place in Twin Harbors, Catalina Island, California, in 1990. This workshop brought together applied mathematicians and coastal engineers, and was geared more toward understanding the state-of-the-art of tsunami hydrodynamics with the emphasis on how to best validate numerical and analytical tools, then under development (Liu et al., 1991). The main conclusions were that the shallow water model was adequate for applications of geophysical interest and also that there was a pervasive need for laboratory data for long waves propagating in two directions to allow further progress in computational models. No realization existed of the importance of the initial condition, or of how an appropriate one could be derived from geological or geophysical considerations, and the solitary wave model remained the standard. All this was to change.

The deficiencies of threshold models[1] triggered rapid development of two-dimensional numerical inundation models, i.e., models that include shoreline motions in the tsunami evolution calculations. Comparisons between field data and model predictions are referred to as model verification and are a crucial part of any scientific modeling effort. Without comparison to real-world data no basis exists to accept the predictive capability of any model. To underline this process, another workshop was organized at Friday Harbor, Washington, in 1995 (Yeh et al., 1996). The focus was the validation and verification of

1. Refers to the models that interrupt the computation at some threshold offshore location, such as at 10 m water depth.

computational codes to predict runup and inundation. A conspicuous characteristic of almost all models presented was their gradual evolution from threshold one-dimensional propagation models, to one-dimensional inundation models, to threshold two-dimensional models, to two-dimensional inundation models. This evolution over a 10-year period allowed modelers to identify nuances and artifacts. The 1995 Friday Harbor workshop showed that given reasonable initial data, the predictions of runup heights were correct to the first order, and therefore attention shifted to defining realistic initial conditions for the computations.

In 1995, the Workshop on Seafloor Deformation Models took place in Santa Monica, California (Synolakis et al., 1997). There was then wide consensus that the seismic moment, the hypocentral location, and the dip and strike angles are reliably determinable in the short term for first-order initialization of hydrodynamic computations, and these estimates are sufficient for differentiating between small and large events. Eddie Bernard and Frank González presented NOAA's plans for the development of network of instruments to record the free-field signature of tsunamis in the deep ocean. These are tsunamographs, also now known as DARTs.

The quest for more realistic initial conditions was catapulted by the 1998 Papua New Guinea (PNG) tsunami, whose trigger, although controversial at first, was finally shown to have been a submarine landslide. Another workshop took place in 2000 at the University of Southern California, Los Angeles, California, to discuss submarine mass failures and their consequences. A companion North Atlantic Treaty Organization-sponsored Advanced Research Workshop on Underwater Ground Failures on Tsunami Generation, Modeling, Risk and Mitigation took place in Istanbul, Turkey, in 2001 with similar objectives but wider international representation (Yalciner et al., 2003).

In summary, at the dawn of the twenty-first century, the tsunami community had grown by a factor of two within a decade, and sufficient understanding existed of tsunami sources and computations to start developing inundation maps for emergency preparedness. The development of tsunami hydrodynamics and tectonic tsunamis is discussed by Synolakis and Bernard (2006) and by Shuto (2003) with a historical point of view. Synolakis and Kânoğlu (2009) discuss tsunami science from the perspective of development of validated and verified numerical models. Here, we will provide a brief summary of different visualizations of tsunamis, then proceed to discuss numerical codes.

2.2 SIGNIFICANT ADVANCES IN TSUNAMI SCIENCE BEFORE THE 2004 BOXING DAY TSUNAMI

As with all incompressible fluid motions, the evolution of tsunamis can be described by approximations of the Navier–Stokes equations. Even when these three coupled nonlinear differential equations are approximated,[2] exact solutions are only possible for a few idealized initial waveforms. These exact solutions are very useful in helping to validate numerical models and to understand the underlying phenomenology. Laboratory experiments that establish the ground truth are equally useful and of fundamental importance in benchmarking numerical procedures (Synolakis et al., 2008).

Before discussing the now-standard computations of tsunami impacts, we briefly describe what we feel were five seminal advances in the past three decades; the calculation of the directivity of tsunamis and experiments with idealized seafloor deformations resulting the establishment of the solitary wave model, the analytical results and experiments with solitary wave runup, the demise of the solitary wave model and the realizations that the front wave in a tsunami train is dipole shaped, development of large-scale laboratory and field benchmark tests, and the realization of tsunami generation by landslides resulted from moderate earthquakes.

1. The fact that impacts from tsunamis originating far-field were localized, and at times appeared capricious, was first explained by Ben-Menahem and Rosenman (1972). They used linear theory to calculate the two-dimensional radiation pattern from an underwater moving source and showed that tsunami energy radiates primarily in a direction normal to the rupturing fault. The concept of source directivity predated the identification of the 1700 Cascadia earthquake by Satake et al. (1996) and helped to explain far-field effect of the 1755 Lisbon (All Saints' Day) tsunami, as a combination of source directivity and focusing by irregular bathymetry (Woods and Okal, 1987; Satake, 1988). Okal et al. (2003) reported field observations of the 1946 Aleutian tsunami in the far-field, and concluded that a large slow earthquake and a landslide must have occurred concurrently to have caused the observed far-field distribution and near-field runup. Emile Okal used classic directivity arguments to show that the 2004 Boxing Day tsunami was caused by a long fault, in the early post-event days, when there were two competing scenarios for the fault motion. Directivity arguments are not helpful in estimating the coastal impact of tsunamis, beyond identifying qualitatively areas that may be at higher risk than others. To understand the shape of the waves on the free water surface generated by impulsive motions, Hammack (1973, 1972) used a novel generation method where one end of a laboratory channel had a short section that could be impulsively lifted or dropped. He measured precisely the evolution of the resulting waveforms over a constant depth. Hammack then related the initial wave to the wave motion at large distances, and

2. Peregrine (1966) provided a comprehensive exposition of the orders of approximation through a perturbation expansion.

predicted that certain initial conditions would generate a series of solitary waves at infinity.

It was thus clear, or so it seemed, that solitary waves were an appropriate model for tsunamis at least in the far-field, given that most seafloor disturbances would cause waves which eventually would become solitary. This was a great conceptual advance, and triggered a race between Caltech, Stanford, and Massachusetts Institute of Technology to find how solitary waves evolved over the continental shelf; see Goring (1978) and Madsen and Mei (1969). The stage was thus set for the calculation of the tsunami runup, the most relevant parameter for estimating impacts.

The runup (R in Figure 2.4) is the maximum elevation with respect to the initial shoreline that the wave reaches. The region between the maximum runup and the initial shoreline is the inundation. An inundation depth is the overland flow depth reached as the tsunami evolves over initially dry land, and of course it is time dependent. Measurements of flow depths after a tsunami usually identify the maximum flow depth reached at that location.

2. To calculate the evolution of a solitary wave and its runup, Synolakis (1987, 1986) solved the boundary value problem of the nonlinear shallow water wave (NSW) equations. He defined the latter as a long wave evolving over constant depth and then climbing up a sloping beach (Figure 2.4). His work resulted in the following scaling law—now known as *the runup law*,

$$R/d = 2.831\sqrt{\cot\beta}\,(H/d)^{5/4}.$$

The coefficient in this equation was calculated analytically. Synolakis proceeded with experiments at the Caltech Laboratory, then directed by Professor Fred Raichlen, to measure solitary wave runup. Synolakis (1987, 1986) compared the analytical predictions with laboratory measurements and found excellent agreement (Figure 2.5).

Synolakis (1987, 1986) introduced a solution to the seminal nonlinear transformation developed by Carrier and Greenspan (1958) which

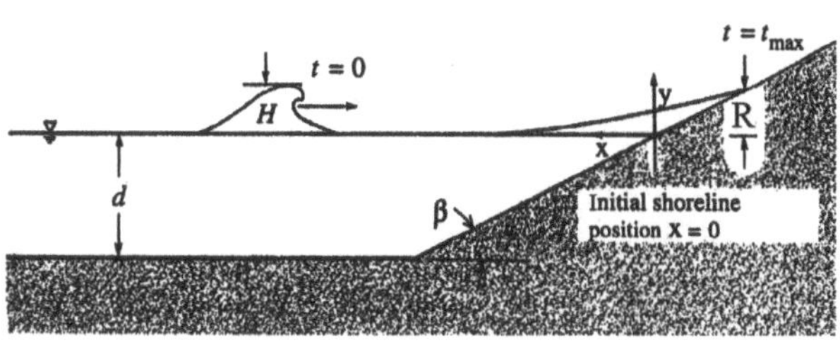

FIGURE 2.4 Definition sketch for the canonical problem, i.e., evolution of a long wave over a constant depth first, and then climbing a sloping beach.

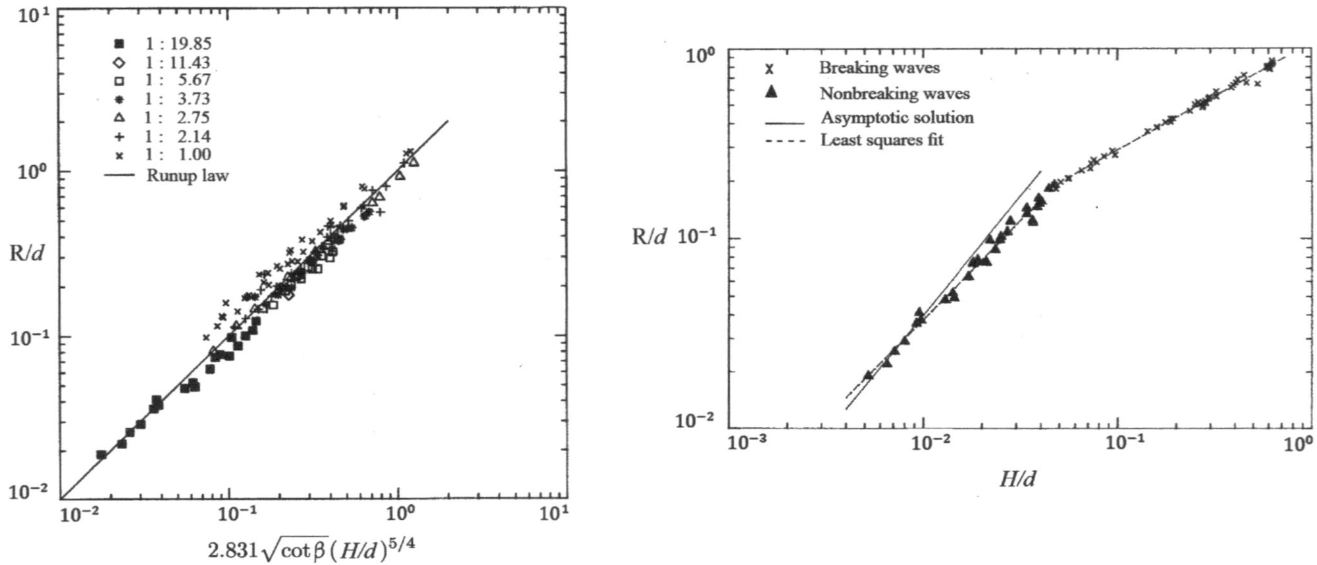

FIGURE 2.5 (Left) Laboratory data for maximum runup of nonbreaking waves climbing up different beach slopes: 1:19.85 (Synolakis, 1986); 1:11.43, 1:5.67, 1:3.73, 1:2.14, and 1:1.00 (Hall and Watts, 1953); and 1:2.75 (Pedersen and Gjevik, 1983). Solid line represents the runup law. (Right) Maximum runup of solitary waves over a 1:19.85 sloping beach in nonbreaking and breaking regimes (Synolakis, 1986).

transformed the NSW equations into a single linear equation solvable with standard methods. He proceeded to calculate the nonlinear evolution for complex spectral distributions, such as those of solitary waves, through the entire runup and rundown process for the canonical problem, and thus demonstrated what is now known as the runup invariance, i.e., for the same initial condition, linear and nonlinear theories predict mathematically identical results. Then laboratory data and his analytical solutions remain the standard for benchmarking numerical codes (Synolakis et al., 2008).

Despite the availability of analytical results and small-scale laboratory measurements, in the early 1990s, there were not even one-dimensional codes that would reliably predict runup. Although large-scale, two-dimensional experiments had been planned, and were in fact just starting at the Waterways Experiment Station in Vicksburg, Mississippi, tsunamis started being reported more frequently than before. These reports, often associated with descriptions of the resulting human drama, intensified the need for quality benchmark data, but also changed the focus from solitary to dipole-shaped N-wave, i.e., wave which has depression and elevation parts.

3. The September 1, 1992, Nicaragua tsunami launched a scientific decade of modern tsunami observations and surveys. More scientists participated in this survey than any other before 2004. Beginning in 1992, post-tsunami surveys have been carried out by the highly interdisciplinary International Tsunami Survey Teams (ITSTs), immediately following each event (Synolakis and Okal, 2005). The ITST scientists measured runup, overland flow depths, inundation distances, and on-occasion water currents, and they interviewed eyewitnesses to obtain information about the timing of the tsunami arrival and the community response. Three distinct features were observed during the Nicaragua tsunami:

   One, the parent earthquake was clearly a *tsunami earthquake*,[3] and the wave it triggered killed more than 160 people. Tsunami earthquakes had been studied by Newman and Okal (1998) who introduced a slowness parameter, $\Theta = \log_{10}(E^E/M_0)$, where E^E is the estimated energy carried by high-frequency seismic body waves and M_0 is seismic moment measured on long-period surface waves. This parameter can be determined in real time and defines how slow an earthquake is, i.e., whether it is unusually efficient in generating tsunamis.

   Two, the wave, despite having been generated very close to the shore, featured remarkable directivity in its impact. At El Transito, the most devastated site, runup values ranged up to 11 m; in the adjoining beach

3. A tsunami earthquake is a seism whose tsunami has far greater amplitude than expected from its conventional magnitude, and its characteristic slow motion is often reported as not felt by eyewitnesses (Okal, 1993; Kanamori, 1972).

Playa Hermosa, even the beach umbrellas had been left standing. Clearly, two-dimensional computations at a resolution of at least 50 m are likely necessary to resolve such differences. Not only were inundation models absent then but also such resolutions were not possible then with ordinary computers in reasonable times.

Three, the earthquake-triggered tsunami manifested itself first with a shoreline recession, which was the only precursor for the coastal residents. Sadly, they did not identify it as a tsunami, since the shaking from the $M_0 = 3.4 \times 10^{27}$ dyn cm parent earthquake was not felt by most coastal residents queried during the subsequent field survey.

This last observation led to a paradigm change, i.e., considering N-wave as an appropriate initial condition for tsunami modeling instead of solitary waves (Tadepalli and Synolakis, 1994). Tsunamis that cause the initial shoreline to retreat are now known as leading depression N-waves (LDNs). In less than 1 year after the 1992 Nicaragua tsunami, Tadepalli and Synolakis (1994) presented analytical results that showed that LDNs had higher impact than their mirror images, leading elevation N-waves (LENs). Tadepalli and Synolakis (1994) derived asymptotic results for different families of N-waves and showed that LDNs always runup higher than their companion LENs. LDNs had been believed hydrodynamically unstable; the crest was supposed to quickly overtake the trough. Controversy ensued despite additional reports of LDNs striking the south coast of Java (1994), Mindoro Island (1994), Shitokan Island (1994), and Manzanillo (1995).

In 1996, Tadepalli and Synolakis further refined the N-wave model and suggested an explanation to reconcile the reports of receding shorelines with prevailing inverse-scattering-based theory and the resulting solitary wave paradigm (Tadepalli and Synolakis, 1996). First, they showed that N-waves were one particular solution of the parent equations of motion when forcing the linear shallow water wave (LSW) equation with a step-function motion. They then showed N-waves of geophysical scales would have to evolve more than once around the perimeter of the Earth, to reach the idealized infinity of inverse scattering theory, i.e., trains of solitary waves. They suggested that because of the nature of the dipolar seafloor deformation in subduction zones, LDNs would strike the adjacent shoreline, while LENs would move toward the open ocean. The fact that the near-field manifestation of a tsunami is an LDN was finally settled in the 2004 Boxing Day tsunami (Satake, 2007), with hundreds of descriptions from eyewitnesses confirming it.

4. Before anyone had the chance to absorb the impacts of the Nicaragua tsunami, a 5.1×10^{27} dyn cm earthquake triggered a tsunami on December 12, 1992, in the eastern part of Flores Island, Indonesia. Runup heights ranged from 5 to 26 m along the coast. This event is mostly remembered by the catastrophe in the conical-shaped volcanic Babi Island located between the epicentral region and Flores Island, about 5 km directly north of Flores

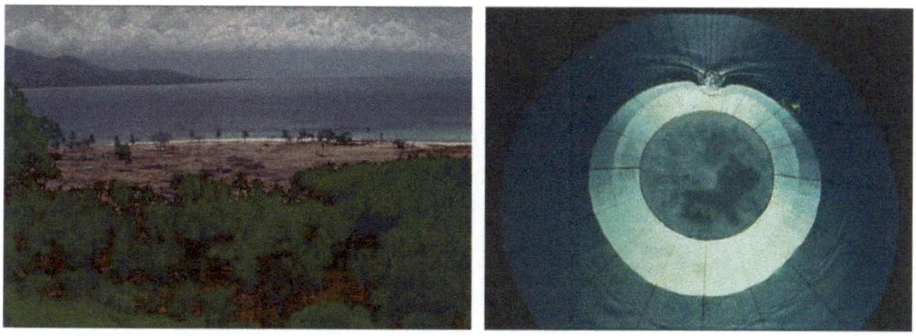

FIGURE 2.6 (Left) Catastrophe on the back side of Babi Island (Yeh et al., 1994). (Right) Top view of the laboratory manifestation of a solitary wave attacking a conical island to model the catastrophe in Babi Island in 1992 (Kânoğlu and Synolakis, 1998; Kânoğlu, 1998; Briggs et al., 1995; Liu et al., 1995). Top shows back of the island with respect to incoming wave in both insets.

coast. The tsunami attacked from the north of Babi, yet the impact was higher on the lee side of the island, destroying two fishing villages normally protected from swell waves (Yeh et al., 1994) (Figure 2.6). This raised some controversy as to whether the enhanced runup was due to some unidentified circular island effect or reflection of the tsunami off Flores.

By serendipity, as a result of the 1990 Catalina workshop, large-scale experiment had already been planned to measure the interaction of solitary waves with a conical island. A series of laboratory tests were undertaken for a laboratory model of a 7.2-m-base-diameter island with a slope angle of 14° (Kânoğlu and Synolakis, 1998; Kânoğlu, 1998; Briggs et al., 1995; Liu et al., 1995) (Figure 2.6). Solitary waves of different crest lengths were generated and their runup were measured. These experiments appear to be the first laboratory visualization of a tsunami catastrophe over realistic two-dimensional bathymetry and topography.

The experiments demonstrated that once the wave hits the circular island, the crest splits into two waves that propagate around the island having their crest perpendicular to the shoreline and collide behind it, in a spectacular demonstration of constructive interference (Yeh et al., 1994). The experimental data set has been used extensively for benchmarking of numerical codes (Titov and Synolakis, 1998), as described in Synolakis et al. (2008). Further, Kânoğlu and Synolakis (1998) were able to compute the runup around a conical island using entirely analytical tools, based on the Synolakis (1987) methodology, thus verifying the applicability of linear theory even in the two-dimensional problem.

On July 12, 1993, the Hokkaido-Nansei-Oki earthquake ($M_0 = 4.7 \times 10^{27}$ dyn cm) generated a tsunami that devastated the island of Okushiri, west of Hokkaido. Field measurements identified extreme runup value of 33 m at a river gully near Monai, and currents ranging to up 18 m/sec. The wave overflowed a 7-m-high protection wall and destroyed the town of Aonae. High-resolution seafloor bathymetry before and following the event allowed meaningful identification of the seafloor deformation. Titov and

Synolakis (1997) were able to successfully model the extreme runup. Here, it became clear that the details of the timing of seafloor deformation were not as important as some believed, at least compared to a fairly accurate description of the seafloor deformation. Numerical modeling of the event showed that a resolution of 5 m was necessary to predict the extreme runup. Modeling of this event has become another benchmark test for model validation.

5. The PNG tsunami on July17, 1998, resulted in over 2,100 fatalities, despite the relatively small size of the parent earthquake. It changed tsunami science on a scale analogous to what the 2004 megatsunami did. Exceptional runup heights, reaching 15 m at Arop, but concentrated on a 25-km stretch of coastline, were measured by Kawata et al. (1999). The excessive amplitude and concentration of the runup were incompatible with any modeling based on the excitation of the tsunami by a seismic dislocation and eyewitness reports generally indicated that the tsunami had arrived at least 10 min later than predicted by all acceptable models of propagation (Synolakis et al., 2002). In addition, the earthquake was not a tsunami earthquake. Finally, Okal (2003) identified the hydroacoustic record in Wake Island which demonstrated beyond doubt the characteristic signature of the landslide through comparisons with the acoustic signatures at the same locale from the earthquake and its main aftershock.

   Kawata et al. (1999) proposed that an underwater landslide generated the tsunami. Later, a number of hydrographic surveys (e.g., Sweet and Silver, 2003) identified a 4 km^3 slump contained in a bowl-shaped amphitheater, located 25 km away from the coast. Numerical modelings that used the landslide as the source of the tsunami successfully modeled the runup along the Sandaun coast (Synolakis et al., 2002; Heinrich et al., 2000). As a result, hazard posed by relatively moderate, approximately magnitude 7 level, earthquakes started being reexamined (Borrero et al., 2004; Bardet et al., 2003; Borrero et al., 2001) for their submarine landslide-generation potential.

   By the end of the 1990s, no analytical results existed for waves generated by submarine landslides, and empirical formulas produced vastly different predictions for the initial wave height, even for the same slide geometry. Conceptually, the physical problem is challenging for it involves a time-dependent boundary. Although a forcing term can be included in the shallow water wave equations with small additional computational cost, the timescale of the landslide motion remains undefined. If the slide is very close to the initial shoreline, the computations become more evolved, because now not only the shoreline but also the seafloor moves in the same region. To provide an analytical benchmark problem for numerical codes, Liu et al. (2003) developed an analytical solution for a subaerial landslide that starts moving from the shoreline. Again, this solution has become a benchmark for validating numerical codes (Synolakis et al., 2008).

By the 2004 Boxing Day tsunami, substantial advances had been achieved in the development of benchmarked numerical codes. California was the first to have started producing inundation maps for emergency planning (Eisner et al., 2001), and soon thereafter most Pacific states of the United States were following. Landslide tsunamis had started being considered, and increased emphasis was given to documenting and understanding tsunamis in the past 100 years.

2.3 DEVELOPMENTS SINCE THE 2004 BOXING DAY TSUNAMI

Summaries of the observations from the 2004 Boxing Day tsunami can be found in Synolakis and Bernard (2006), Synolakis and Kong (2006), and Geist et al. (2006). The death toll of at least 240,000 remains unfathomable even to specialists, and a few even compared its impact to the philosophy of science to that of the 1755 Great Lisbon earthquake, which triggered debates even in theology. Voltaire's *Candide* is an example of the fascination of eighteenth-century thinkers with this event. Here, we discuss developments in the past decade. We describe scaling laws that identify differences between landslides and tectonic tsunamis, focusing of N-waves and unusual amplification effects from resonance. In the last section, we discuss mitigation measures including inundation maps and real-time forecasts.

The 2004 Boxing Day tsunami prompted hundreds of photos, some found in cameras that were washed by the tsunami. In many of these photos people appear to pose in front of the advancing wavefront, which at the time of the picture taking was offshore. This was perplexing, why did these people not run away as they saw the tsunami advancing, and instead think they had the time to take pictures? One remarkable observation whose usefulness was not entirely recognized until these pictures were examined is the acceleration of the wavefront past the initial shoreline. As shown in Synolakis (1987), in the evolution of the front of the solitary wave, the wave first slows down as it climbs up the sloping beach, due to the reduction in depth; recall that the wave celerity $c \approx \sqrt{gh(x)}$, where $h(x)$ is the changing depth. Once the wave hits the initial shoreline, it slows down, but accelerates again, before decelerating to its ultimate on-land penetration point, the maximum wave runup. This appears to happen even far inland, for different reasons.

In a video taken near the Grand Mosque in Aceh, one can infer that the wavefront first moved at speeds less than 8 km/h, then accelerated to 35 km/h (Fritz et al., 2006). The same phenomenon is probably responsible for the mesmerization of victims during tsunami attacks, first noted in a series of photographs of the 1946 Aleutian tsunami approaching Hilo, Hawaii. The wavefront appears slow as it propagates initially past the shoreline, leading to a sense of false security; it appears as if humans can outrun it, and then the wavefront accelerates rapidly as the main disturbance arrives. A series of three photographs from the 1995 Manzanillo tsunami taken by a survivor which show three men being unable to outrun a very thin wavefront underscores this point (Borerro et al., 1997).

With the exception of recent tsunamis in the Solomon Islands, it is not common that two tsunamis triggered by comparable earthquakes strike the same region within a 5-year period. This happened in PNG in 1998 and 2002. Based on the measured differences between the 1998 and 2002 tsunamis, Okal and Synolakis (2004) fitted Gaussian distributions to the longshore runup measurements from real events and the computed runup values over idealized geometries using MOST to augment the historic data set. They defined two dimensionless parameters, one each for tectonic and landslide tsunamis. These invariants characterize the tsunami source, landslide versus dislocation, yet they are largely independent of the exact parameters describing the respective sources. Hence, they showed that these invariants can be used as discriminants to identify the nature of the tsunami source.

This analysis was useful, because coseismic landslides are not uncommon, yet it had been very difficult to differentiate from seismic recordings alone whether the main tsunami was triggered by the earthquake or the landslide, as the PNG slide demonstrated. This is even more so for studies of paleotsunamis. The geosciences have several scaling laws to allow for quantification of parameters that are otherwise grueling to infer.

Note that one fundamental observation linking the tsunami source with inundation, although never published as such, is George Plafker's. While surveying the aftermath of the 1964 Great Alaskan tsunami (Pararas-Carayannis, 1972), Plafker proposed that the maximum runup locally does not exceed twice the height of deformed seafloor offshore (Plafker, 1965; Plafker et al., 1969). This is now referred to as *the Plafker rule*, and has largely withstood the test of time, after being unrecognized for three decades (Synolakis and Okal, 2005). The Okal and Synolakis (2004) parameter is precisely what the empirical Plafker rule suggests. The 2004 tsunami did not contradict this rule, even at Longa near Banda Aceh.

It is instructive here to briefly discuss the history of landslide experiments, at least from the point of view of classical hydrodynamics. Wiegel (1955) performed landslide experiments using a wedge-shaped block sliding down a plane beach and, measuring the waves propagation offshore, concluded that only about 1 percent of the slide energy is converted to energy in the generated wave. However, his emphasis was on the shape of the wave moving offshore. In an effort to better understand the generation and runup of waves from submarine and subaerial slides, Raichlen and Synolakis (2003) conducted large-scale experiments at Oregon State University, Corvallis, Oregon (Figure 2.7). Following Wiegel (1955)'s experiments, they first used a vertical-face wedge block, and then semispherical and rectangular boxes of equal volume in the wave tank with the depth of 2.44 m. Two configurations of the

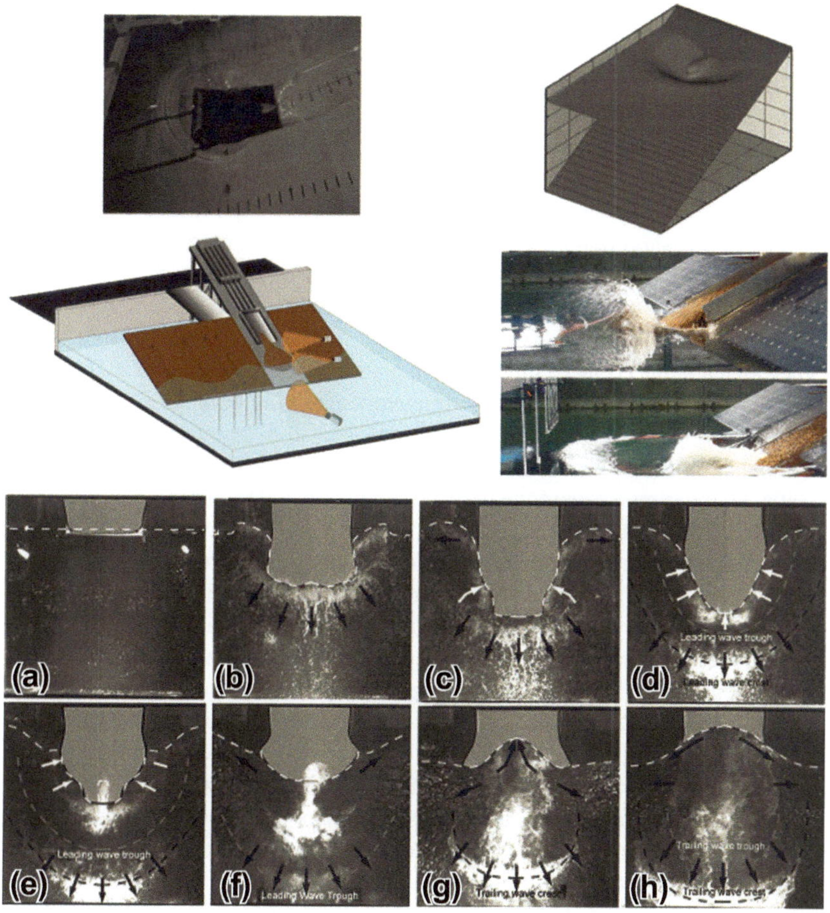

FIGURE 2.7 (Top left) A moving wedge as a model of submarine landslide (Raichlen and Synolakis, 2003). The initial wave from a submerged slide, immediately after motion started; note the leading depression N-wave. (Top right) Animation still from Liu et al. (2005)'s direct numerical simulation (DNS) modeling of the landslide experiments. (Middle left) The pneumatic landslide tsunami generator on the hillslope and (middle right top) granular landslide impact and (middle right bottom) the leading wave crest observed in experiments. *After Mohammed and Fritz (2012).* (Bottom) Tsunami wave generation by a three-dimensional deformable granular landslide: (a) initial impact of landslide front with water surface, (b) and (c) water displacement by the impact and surface depression, (d) surface restoration, (e) surface restoration and radial wave propagation, (f) surface collapse and generation of first trailing wave, (g) wave runup on the hillslope postcollapse, and (h) end of collapse of surface depression. *After Mohammed and Fritz (2012).*

wedge on the slope were used, one with the front face of the wedge vertical and another with the wedge turned end-for-end so that for this orientation the top and front faces were neither horizontal nor vertical. By varying the weight of the blocks, Raichlen and Synolakis (2003) were able to vary the initial acceleration, whereas the initial position varied from totally aerial to totally submerged. After its release, the sliding wedge generated a leading positive seaward propagating wave, but the water surface above the wedge is depressed and causes the shoreline to retreat first. The experiments revealed a depression forming over the wedge as the motion begins. For the aspect ratio of their boxes, they observed that wave generation became inefficient as the submergence of the blocks exceeded one block height. The results from these experiments defined one of the four benchmark problems in the 2004 National-Science-Foundation-supported Catalina workshop (Liu et al., 2008), and have since been used as another benchmark test for hydrodynamic model validation (Synolakis et al., 2008).

These experiments were the reality check and validation exercise of the full DNS solution of Liu et al. (2005) (Figure 2.7). However, hydrodynamic computations of the evolution of the free surface motions from landslide waves are very challenging. The seafloor is continuously deforming, and the impact of the solid slide with the water surface during subaerial generation may trigger local breaking. Breaking can also happen if the slide moves rapidly and if it is located very close to the surface. Further, the wedge geometry used in the experiments has sharp corners which trigger further instabilities, even in full DNS of the parent Navier–Stokes equations of motions. Hence, not only extreme near-shore or onshore landslide tsunami generation in the laboratory remains a vexing calculation but also simulations of prototype landslides are even more difficult, for the timing of the slide remains an important but poorly understood parameter, exquisitely dependent on several constitutive parameters of the slide material.

In terms of the work in deformable slides, Fritz (2002) investigated tsunamis generated by granular subaerial landslides in a two-dimensional, physical laboratory model, at the Swiss Federal Institute of Technology. The landslide material in the experiment matched the rockslide characteristics. Hermann Fritz and his collaborators determined that the relevant parameters for free surface wave generation were the granular slide mass, the slide impact velocity (v_s), the still water depth (h), and the slide thickness (Fritz et al., 2003a,b). Most of the impulse waves were located in the intermediate water depth regime, yet the propagation velocity of the leading wave crest closely followed the theoretical approximations for solitary waves, which are long waves. The slide Froude number ($F = v_s/\sqrt{gh}$, where g is the gravitational acceleration) was identified as the dominant parameter. The physical model results were then compared to the giant rockslide generated impulse wave which struck the shores of the Lituya Bay, Alaska, in 1958 (Miller, 1960). That slide produced runup in excess of 400 m along the steep fjord walls. A cross-section of Gilbert Inlet was rebuilt at small scale and the measured wave runup matched the trimline of forest destruction (Fritz et al., 2001). The Lituya Bay experiments highlighted the formation of an impact crater, as the slide plunged into the fjord, thus increasing the water displacement. The runup measurements revealed a short propagation distance and runup of a solitary wave exceeding the breaking limit by a factor of 1.5, likely due to the narrow Gilbert inlet and the steep headland. Recently, Mohammed and Fritz (2012) conducted three-dimensional experiments on tsunami generation by subaerial landslides in the tsunami wave basin of the Network for Earthquake Engineering Simulation at Oregon State University in Corvallis, Oregon, USA. They used a novel pneumatic landslide generator and showed a sequence of the wave generation by landslide impact (Figure 2.7). They characterized the landslide by its volume, front velocity, length, thickness, and width on the hillslope at impact. They determined that 1–15 percent of the landslide kinetic energy at impact is converted into the wave train energy. They also determined that the wave amplitudes, periods, and wavelengths are related to the landslide parameters at impact with the landslide Froude number being a dominant parameter. This experimental data set on tsunamis generated by three-dimensional, deformable, granular landslides could serve validation of numerical models.

One vexing question raised by the PNG landslide wave was whether the coastal evolution of landslide tsunamis could be effectively modeled with the NSW equations, or whether the Boussinesq equations are necessary—they are computationally far more complex. Landslide waves are steeper and shorter than tectonic tsunamis and disperse more rapidly. Lynett et al. (2003) compared results from Boussinesq and NSW models with the field measurements and concluded that while the near-shore evolution was predicted to be different, the runup predictions were not. In this regard, and given the considerable known uncertainties in characterizing the landslide motion, it appears that computations with Boussinesq-type models are an overkill when it comes to determining inundation, and likely misleading. Such computations are reminiscent of adding decimal digits in a computation where even the significant figures cannot be determined with certainty.

The field survey of the 2010 Mentawai tsunami revealed another counterintuitive observation for tsunami–island interaction (Hill et al., 2012). The runup behind small islands off the Mentawai in Indonesia was higher than in areas not sheltered by the islands. Stefanakis et al. (2013) used active learning methods and employed numerical solutions of the NSW equations for the bathymetry of a conical island off a shoreline with a plane beach and an incoming solitary waveform. They used seven parameters to describe the physical problem; even if they had only tried 10 values for each of the seven parameters, they would have required 100,000 simulations to determine which combinations produced higher runup along the continental coastlines sheltered by the island. With active learning, they only had to do 200 simulations to determine the global maximum. In none of their simulations did the island offer any additional protection. The *Economist* magazine in its review of the article commented on the dangers of insularity, and of how mainland residents should be informed of the additional hazard offshore islands posed in case of a tsunami attack. The situation may become even worse if resonant wave interactions occur (Stefanakis et al., 2011).

Stefanakis et al. (2011) numerically simulated the one-dimensional NSW equations to investigate the boundary value problem for plane and more complex bathymetries forcing with monochromatic waves, as well as virtual wave-gage recordings from real tsunami simulations. They observed resonant phenomena between the incident wavelength and the beach slope, which result in enhanced runup of nonleading waves. The resonance occurs due to incoming and reflected wave interactions, and the actual amplification ratio depends on the beach slope. They suggested that these phenomena can explain why it is not always the first wave that causes the highest runup, as well as why the tail of a single wave may produce leading-order runup values. However, when the bathymetry is more complex, it is not clear to what extent resonance is attributed to wave trapping and generation of harmonics. Resonant mechanisms are not limited to the plane beach paradigm (Ezersky et al., 2013) but can be observed in more complex bathymetries as well, thus suggesting that local runup amplification is not rare.

An interesting question that has arisen from field surveys is whether the enhanced runup observed in theoretical investigations for one-dimensional N-waves can explain real tsunami impacts with two-dimensional bathymetries. Recently, Kânoğlu et al. (2013) introduced a new analytical solution to study the propagation of a finite strip source over constant depth, using LSW theory. Their two-dimensional strip N-wave represents the waves generated by tectonic seafloor displacements. Kânoğlu et al. (2013)'s two-dimensional solution is not only exact but also general and allows the use of realistic initial waveforms. They showed the existence of focusing points for N-wave-type initial displacements, i.e., points where unexpectedly large wave heights may be observed (Figure 2.8). Focusing points exist using linear nondispersive, linear dispersive, nonlinear nondispersive, and weakly nonlinear weakly dispersive theories, all approximations of the parent Navier–Stokes equations. They discussed geophysical implications of their solution using the July17, 1998, PNG and the July 17, 2006, Java tsunamis (Fritz et al., 2007) as examples. Their results may also help to explain high runup values observed during the March 11, 2011, Japan tsunami, which are otherwise not consistent with existing scaling relationships. They concluded that N-waves generated by tectonic displacements feature focusing points, which may significantly amplify runup beyond what is often assumed from widely used scaling relationships.

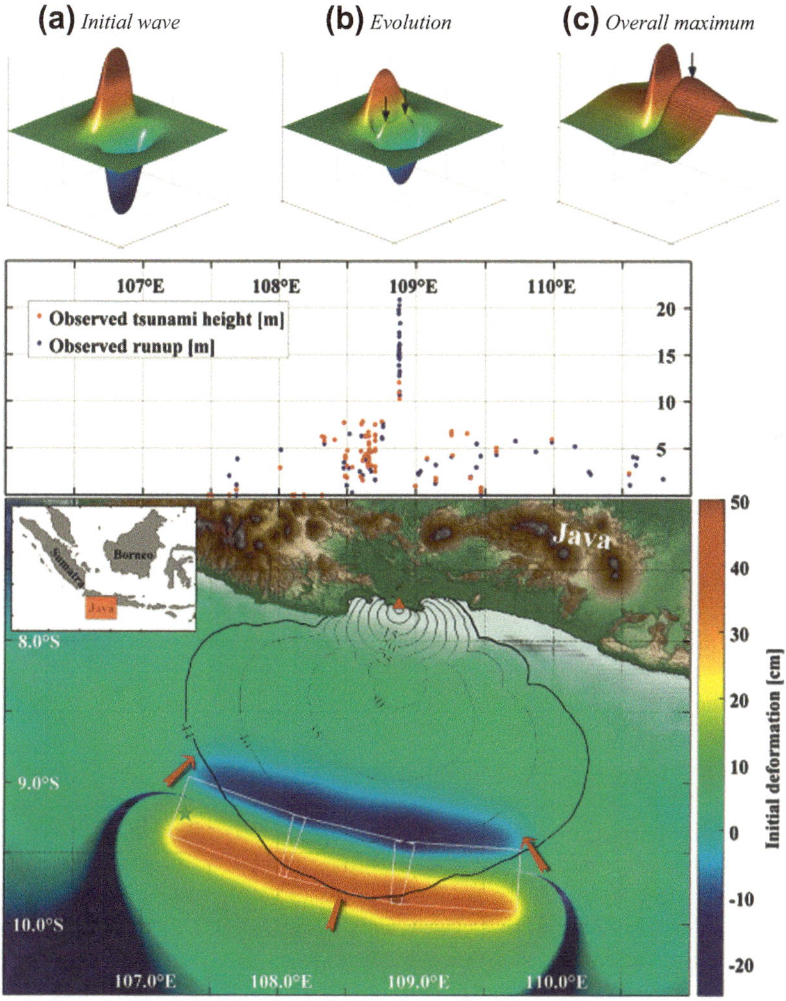

FIGURE 2.8 (Top) Definition sketch for focusing. Evolution of an N-wave source over a constant depth calculated using the MOST model: (a) initial wave, (b) evolution, and (c) maximum amplitude at each grid point for entire propagation. (Middle) The July 17, 2006, Java tsunami field runup and wave height measurements (Fritz et al., 2007) (red and blue dots denote observed tsunami heights (m) and observed runup values (m), respectively). (Bottom) Reverse tsunami travel time (RTTT) contours for Permisan (red filled triangle) where the runup exceeded 20 m, while runup values ranged from 5 to 7 m at the surrounding areas. Three of the NCTR's 100 by 50 km^2 tsunami source functions (white rectangles) are presented as a possible source mechanism considering focusing at Permisan. The red arrows indicate approximate locations of two positive waves from the sides of the depression and a positive wave from the center of the elevation, which arrive simultaneously to the focusing point (red filled triangle) considering the same RTTT (44 minutes) contour. *After Kânoğlu et al. (2013).*

2.4 MITIGATION MEASURES

It is evident that since 2004 substantial interest has occurred in developing tsunami mitigation plans for tsunami-prone regions worldwide (Synolakis and Bernard, 2006). Mitigation plans start with understanding the hazard and the development of flooding maps. The state-of-the-art numerical models used in inundation mapping have evolved through a painstaking process of careful validation and verification; see the validation and verification of the ones used for inundation mapping of the US coastlines in National Tsunami Hazard Mitigation Program (NTHMP) (2012). The substantial and nonstandard difficulty is with the prediction of tsunami inundation in what is referred to as the runup or the inundation computation. The tsunami wave evolution over dry land involves the interaction of three phases of matter (Liu et al., 1991). In finite-difference (FD)-type numerical solutions, this involves introducing additional grid points as the tsunami front evolves on the beach and runs up, the removal of these grid points as the wave runs down, and the repetition of this cycle as the next wave in the tsunami train approaches to the beach. Although used in the 1970s, finite element methods have been proved to be cumbersome and less flexible than FD solutions, for this problem. Although estimating tsunami evolution can be performed with most reasonable computational models, numerical solutions for tsunami inundation are notoriously unstable. Moreover, operational FMs used in real time require extensive additional verification with measurements from real tsunamis.

We briefly describe the inundation mapping effort, citing California as an example. Then, we summarize the forecasting methodology of the NCTR at the Pacific Marine Environmental Laboratory (PMEL), Seattle, Washington, i.e., how the inundation code MOST is utilized as the forecasting tool and how it is used by the two tsunami warning centers (TWCs) located in Hawaii and Alaska and operated by NOAA's National Weather Service (NWS). Finally, we discuss observations at harbors providing examples from recent tsunami events, because tsunami could disrupt commercial activities for a long time, causing huge economic losses for communities.

2.4.1 Inundation Maps for Emergency Preparedness

Inundation maps provide emergency managers in coastal communities the necessary tools to plan for and mitigate against tsunami disasters. Inundation maps are not only useful in assessing the population and facilities at risk, but are also helpful in planning for emergency response in the event of a tsunami warning. The preparation of inundation maps involves both the assessment of the near-field and far-field geologic hazards, their interpretation to derive appropriate initial conditions for numerical computations, and the calculation of the potential coastal inundation. For example, California has maps that demonstrate the maximum potential flooding both from local offshore sources and from tsunamis triggered by earthquakes around the Pacific Rim, i.e., what are referred to as far-field tsunamis. Very early inundation maps depicted flooding from past events that hit the region.

Houston and Garcia (1974a,b), in an attempt to develop flooding maps for the Federal Emergency Management Agency (FEMA), calculated the two-dimensional deep-ocean propagation of tsunamis from Chile and Alaska evolving toward the west coast of the United States. They argued that the only reliable data for defining source characteristics at that time were from the 1964 Alaskan and the 1960 Chilean earthquakes. They approximated the initial ground deformation by a hypothetical seafloor of ellipsoidal shape, about 960 km long with an aspect ratio of 1:5 and maximum vertical uplift of 8–10 m. They then divided the Aleutian trench into segments and calculated the wave evolution from each segment, and repeated the procedure for tsunamis from the Peru–Chile trench. They thus derived 100-year and 500-year tsunami runup heights. Borrero (2002) and Borrero et al. (2003) have argued that the 100-year hazard in California is dominated by distant events, hence the Houston and Garcia analysis is probably adequate. Given the realization of offshore landslide hazards after the PNG event, they argued that the 500-year hazard is dominated by local events, and therefore revised the 500-year Houston and Garcia estimates.

In 1996, the Tsunami Hazard Mitigation Federal/State Working Group submitted a report to the US Congress recommending the preparation of inundation maps for Alaska, California, Hawaii, Oregon, and Washington. The report led to mobilization of significant federal resources for tsunami hazards mitigation, and to the establishment of the US NTHMP program which provides resources to all five states for mitigating tsunami hazards. The NTHMP was the focus of a program review during the International Tsunami Symposium (ITS) held on August 5–7, 2001, in Seattle, Washington. This ITS marked the start of a comprehensive effort to produce inundation maps for the US coastlines. Here, as an example, we discuss the inundation maps for California.

The State of California has the most densely populated coastline in the Western United States. It has a very short historic record of tsunamis and earthquakes, extending to less than 150 years. In addition to vulnerability from tsunamis originating around the Pacific, faults occur just offshore, most of which have not been documented properly to allow for quantitative estimates of potential earthquakes. Evidence exists of historic landslides that likely produced large local tsunamis. As early as 1997, the California Coastal Region Administrator of the Governor's Office of Emergency Services (OES) through a series of workshops and publications informed local governments and emergency agencies of the plans to address tsunami hazards.

It was decided that the maps would include realistic worst-case scenarios to be identified in the mapping process. In 1998, the OES contracted to the Tsunami Research Center (TRC) of the University of Southern California,

Los Angeles, California, for the development of the first generation of inundation maps for the State. The TRC has since produced all generations of inundation maps for the state.

An interesting question was whether to provide emergency planners with estimates of inundation areas at different levels of risk. For example, one suggestion was to include a low- and a high-risk line on the inundation maps. Another question was whether to include separate lines for near-field and far-field events. After extensive discussions with emergency preparedness professionals in different jurisdictions across the state, it was felt that a single line representing a worst-case scenario was preferable. Lines identifying risk zones for near-field and far-field events might have proved to be cumbersome and confusing for the public. Inundation predictions for any given tsunami event are highly bathymetry and topography dependent and vary substantially along the coast. Since the location of historic sources is seldom accurately known, and future earthquakes are unlikely to be exact replicas of past events, a potential source was moved around within a range of uncertainty. Along California's flat coastlines, this relocation of the tsunami source resulted in relocation of the maximum runup along the coast.

When asked, emergency planners preferred to have a single value for every region identifying the maximum elevation that tsunami waves from the different local offshore sources would attain (Figure 2.9). This practice simplified the communication of the risk to the public and provided information that was easy to remember and implement in regional emergency preparedness. This way, a region could plan for tsunami evacuation areas above a certain minimum elevation across its jurisdiction. In the development of the maps, the highest inundation value among all different runs identified, and resulting map reflected this value. Borrero et al. (2003) found that most low-lying coastal areas could undergo high inundation, if the source was relocated appropriately within geological constraints. Thus the inundation maps for California do not represent the inundation from any particular event or characteristic earthquake, but the locus of maximum penetration distances from relocating worst-case scenarios.

This practice was vindicated in the 2011 Japan tsunami. The Fukushima Dai-ichi Nuclear Power Plant (NPP) was flooded by a tsunami estimated at the shoreline to have been 13 m. The tsunami took out the six emergency diesel generators that had been placed on a 6-m platform. Battery power lasted for only 8 h, and since connections with the electricity grid were lost, the worst NPP accident since Chernobyl materialized. Note that the plant suffered no structural damage from the earthquake directly; the accident resulted from the loss of power. Note also that another NPP located about 110 km north of Fukushima experienced no loss of power, because it had been fronted by a 14-m-high seawall, while Fukushima Dai-ichi only had a 3-m-high seawall.

Allegedly, this 3-m high seawall design height was calculated by considering the local inundation depth from the 1960 Chilean tsunami. The tsunami

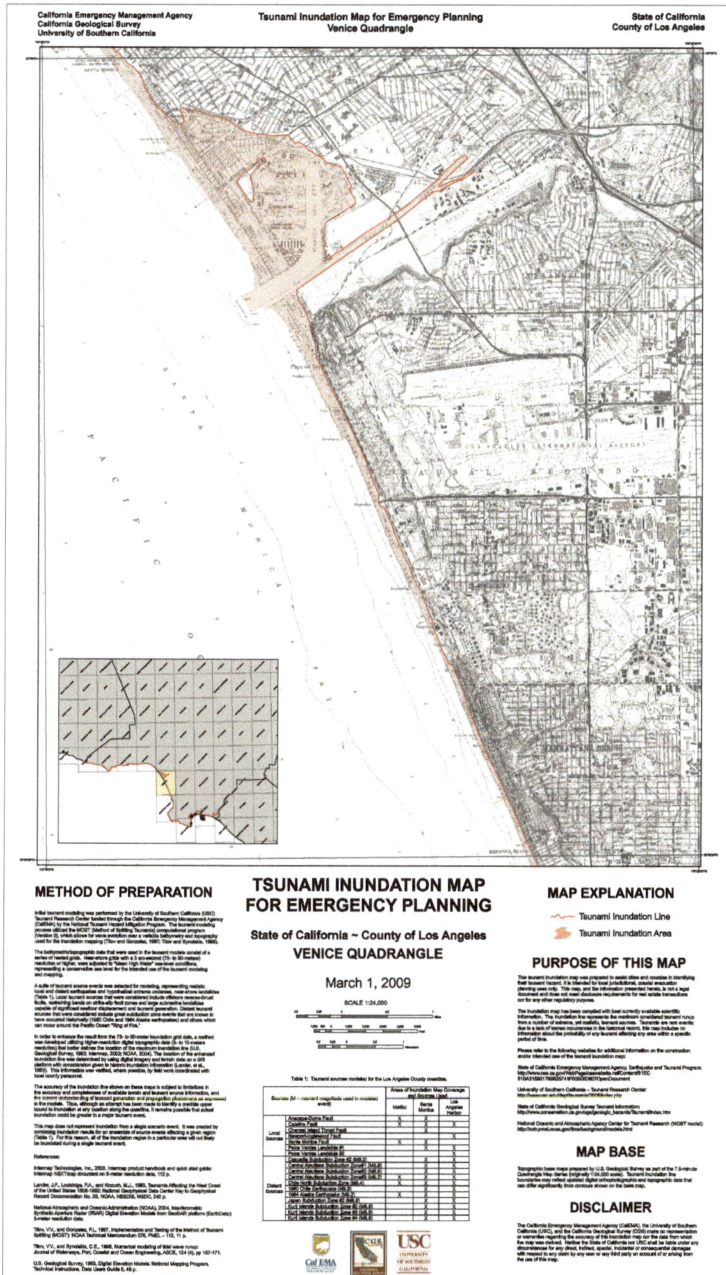

FIGURE 2.9 Tsunami inundation map for Venice Beach, California.

hazard planning of the owner of Fukushima Dai-ichi NPP, the Tokyo Electric Power Company (TEPCO) was based on estimates of inundation from an earthquake source in the vicinity of the plant and from a tsunami that originated from Chile. Northern Japan had experienced a 32-m runup from the 1896 Sanriku tsunami, yet the design inundation depth for the NPP was 6.1 m. The most recent assessment of the tsunami hazards for the Fukushima Dai-ichi NPP took place only a few months before the accident, in the aftermath of the 2010 Chilean event. Had TEPCO used the same analysis as California did for over a decade before 2011 (Eisner et al., 2001), the accident might have been avoided.

It is not difficult to underscore the guidelines for pre-tsunami emergency planning:

- The development of a tsunami plan requires a multidisciplinary approach and should involve local specialists such as emergency responders, planners, engineers, geologists, utilities, and community-based organizations.
- One of the most critical elements of a tsunami plan is the evacuation and traffic control plan. A far-field tsunami may allow several hours to evacuate. A near-field tsunami may require immediate self-evacuation through areas damaged by earthquake shaking. Each jurisdiction should analyze how much time an evacuation would require and build that into the decision-making procedure. Optimum evacuation routes should be identified.
- Inundation projections and resulting planning maps need to be informed with both far-field and near-field scenarios. The inundation maps should not only reflect the worst event on record but also strive to include all possible events from all sources. When human lives are at stake, there should be no hesitation to include extreme but realistic scenarios in disaster planning.
- Maps need to be produced by professionals with extensive experience who can interpret projections in the historical and geologic context of the region under consideration.

The availability of online numerical models that allow graphic projections for scenario events with the ease and learning curves of commercial computer games is a hazard on its own. So are newly minted numerical models that have not been benchmarked to allow the community to judge their veracity and their developers to identify shortcomings. As the TEPCO pre-event analysis for the Fukushima Dai-ichi painfully demonstrated, inundation projections, even if correct for one particular scenario, must to be put in context. For example, one needs to consider that tsunamis with heights at the shoreline of about 3 m occur, on average once a year, at least in the last 22 years, for which fairly reliable records exist and that a magnitude 8.5 earthquake just offshore may produce over 10 m inundation depths. Although the specific values depend on local bathymetry and topography, if a computer model suggests lower values, these need to be carefully evaluated before making their way into inundation maps or being used for the design of seawalls and other protective measures. Further, it is important to consider the capacity of an entire earthquake zone to trigger tsunamis. In a 1000-km-long subduction zone where an estimated 400-km segment ruptured a few centuries ago, but only smaller segments did since then, the design tsunami for all locations in the subduction zone should be based on the most severe event known.

As tsunami warnings become increasingly sophisticated and targeted, it is anticipated that within a decade real-time projections will exist of areas likely to be flooded, at least for tsunamis that originate from the far-field. These projections will guide evacuations, but planning needs to be based on evacuation maps prepared well in advance of the emergency. Evacuation maps should be updated at least once every decade as new information about geologic hazards arise, and to benefit from improved numerical models.

Probabilistic estimates can be useful in evaluating the relative impact of different hazards in a given time period, if undertaken with the same methodology. Even so, they are not useful when it comes to planning for fail-safe events.

2.4.2 Numerical Modeling for Real-Time Forecasting

The NCTR at the PMEL, Seattle, Washington, has been developing tsunami forecasting capabilities for the US coastlines and its territories (Tang et al., 2012, 2009; Wei et al., 2008) in support of the TWCs located in Hawaii and Alaska and operated by NOAA's NWS (Titov, 2009; Titov et al., 2005b). It further supports long-term forecast products such as inundation maps for hazard assessment and planning (González et al., 2005), tsunami hazard assessment (Uslu et al., 2013, 2010a,b), and probabilistic tsunami flooding maps (González et al., 2009; Tsunami Pilot Study Working Group, 2006). To achieve these forecasting capabilities, the NCTR integrates real-time deep-ocean observations of evolving tsunamis from several tsunamographs, a precomputed basinwide propagation database of wave amplitudes and flow velocities based on potential tsunamigenic unit sources, an inversion algorithm to refine the tsunami source based on deep-ocean observations using tsunamigenic unit source combination during an event, and, finally, high-resolution tsunami inundation FMs to provide accurate and timely forecasts for tsunami-prone coastal communities. After rigorous testing at the NCTR, the current version of the system has been distributed to the warning centers for operational tests. Recently, it has been accepted for operational use at the two TWCs, but is, nonetheless, under continuous testing and development at the NCTR.

A critical component of the forecast system is the real-time measurements from the network of tsunamographs, deployed at specific locations in the world's oceans (Figure 2.10) to provide rapid measurements of a passing

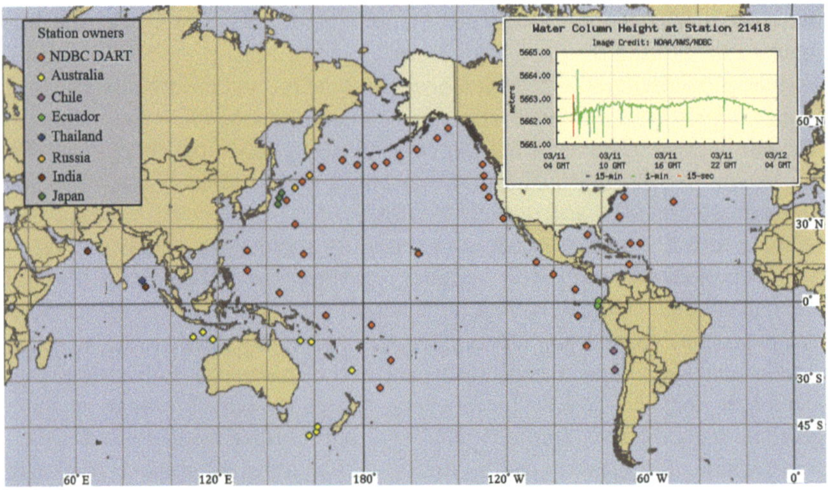

FIGURE 2.10 Network of tsunamographs (DARTs) deployed by the NOAA's National Data Buoy Center (NDBC) and several other countries. Inset shows the recording of the March 11, 2011, Japan tsunami at the tsunamograph 21418. Note that the measurement includes tide.

tsunami (Bernard et al., 2006).[4] The methodology behind tsunamograph site selection process is explained in Spillane et al. (2008). Before direct measurements of tsunamis in the open sea, the only available recordings were from tide gages, which are most often located in sheltered locations inside harbors or inlets. Tide-gage measurements reliably only provide arrival time and possibly the character of the first wave. Typically in a harbor, they record the harbor excitation, not necessarily the tsunami; harbors can be excited by wind waves, and sometimes amplify the incoming wave, dependent on its dominant frequency.

Both the short- and long-term forecasting methodologies of the NCTR use a precomputed propagation database of tsunami evolution from unit earthquakes, hence they are often called *unit sources*. These unit earthquakes have a magnitude $M_W = 7.5$ and are generated by 100-km-long, 50-km-wide fault planes with a slip value of 1 m. Unit sources are placed along the subduction zones of the Pacific, Atlantic, and the Indian Oceans, in several rows, depending on the width of the subduction zone (Gica et al., 2008). The ocean-bottom deformation is calculated with the Okada (1985) formulation for each unit source. The specific characteristics depend on the subduction zone geometry and known history of earthquakes. The NSW equations are used to calculate the propagation of tsunamis from each unit source for entire basin. They are then stored in a set of ocean-specific tsunami propagation databases.

4. http://www.ndbc.noaa.gov/dart/dart.shtml.

Currently, there are 1,725 precomputed unit source propagation model runs covering the world's oceans and they are maintained at the NCTR.[5] A similar concept is in use by the Bureau of Meteorology, Australia, see Greenslade et al. (2007), but, with a different approach. Rather than employing unit sources, the bureau uses specific earthquakes at specified locations. Comparison of predictions with the same inundation model MOST with boundary conditions from the two scenario databases is provided in Greenslade and Titov (2008). The same kind of database is in the process of development for the Mediterranean and the Aegean Sea (Kânoğlu et al., 2012).

Note that, usually, the seafloor movement is considered as an instantaneous vertical displacement, which can be directly translated into water surface as an initial condition. This is a good approximation for tectonic tsunamis, where the rupture velocity is substantially larger than the local tsunami speed over the deformation area (Arcas et al., 2013; Okal and Synolakis, 2003). It is of course less so for slowly moving landslides. The linearity of tsunami propagation in deep water allows scaling and/or combination of the precomputed propagation results of unit sources from the databases to create offshore scenario wave kinematics for the production of short- and long-term forecast products. These combinations/scaling provide the necessary initial/boundary conditions for the water elevation and flow velocities to initiate near-shore nonlinear inundation models close to shore, thus expediting forecasts.

For short-term forecasting products, as the tsunami propagates across the ocean and crosses tsunamograph observation sites, recorded sea levels are ingested into an inversion algorithm (Percival et al., 2011) to produce an improved estimate of the tsunami source for a given earthquake estimate, in real time—see Figure 2.3 for the March 11, 2011, Japan tsunami, for example. Then, given the offshore scenario based on this estimate, the NSW model MOST with three nested grids takes over to produce inundation predictions. The methodology is explained in Tang et al. (2009) in detail, and specific examples of real-time forecast results are summarized in Tang et al. (2012) and Wei et al. (2008) for the March 11, 2011, Japan and August 15, 2007, Peru events, respectively.

Numerical models are prominent in short- and long-term tsunami forecasting. The NCTR uses the MOST (Titov and González, 1997; Titov and Synolakis, 1998, 1997, 1995; Titov, 1997), which is a finite-difference model solving the NSW equations and is based on the method of characteristics and the fractional step method. It simulates transoceanic propagation and inundation over dry land. Also, MOST takes results from the propagation database as input and then, through a series of nested grids, it resolves the near-shore bathymetry and topography, estimates the water level, and predicts inundation at coastal sites of interest. MOST has gone through extensive validation and verification as outlined in Synolakis et al. (2008, 2007) using analytical

5. http://sift.pmel.noaa.gov.

solutions, experimental results, and field measurements and was successfully used for simulations of many historical tsunami events. Validation and verification of tsunami numerical models is essential, especially for those used in forecasting environment and design of critical structures such as NPPs (González et al., 2007).

At the NCTR, two types of models are developed for sites of interest, the high-resolution reference model (RM) and FM. RMs are used as the basis for the development of FMs, which provide an estimate of arrival times, wave heights, and flow velocities, while a tsunami is still propagating in the open ocean. FMs are optimized for their grid resolution and run-time, comparing their results with high spatial and temporal resolution RM results to make sure that results mimic the parameters of interest reasonably well. RMs and FMs utilize MOST and consist of three levels of telescoping grids with increasing resolution (Figure 2.11) to model the tsunami dynamics and inundation onto dry land. FMs are used to provide an estimate of wave arrival time, wave height, and inundation shortly after a tsunami event for tsunami-prone communities. They are optimized to run faster than real time, i.e., faster than a real tsunami would propagate from its origin to the target coastline of the FMs. FMs are designed to simulate 4 h of coastal tsunami dynamics in approximately 10 min. Consequently, they are designed to perform under very stringent time constraints, so they are stable and robust. FM results are verified with historical data, if available for the FM location. Tang et al. (2009) described the technical aspects of FM development, stability testing, and robustness. Currently, there are 75 FMs constructed for densely populated, at-risk coastal communities in the Pacific, Atlantic, and Caribbean for the US and its territorial coastlines. The FM developed for Port Angeles, Washington, is shown in Figure 2.11 with its prediction for the March 11, 2011, Japan and the November 28, 2012, Charlotte tsunamis (Kânoğlu, in review). Note that accurate forecasting of the tsunami impact on coastal communities largely relies on how well the bathymetry and topography are known and the use of validated and verified tsunami numerical modeling. The high spatial and temporal grid resolution necessary for modeling accuracy poses a challenge in run-time requirements for real-time forecasts.

After the 1964 Alaskan tsunami, the United Nations Educational, Scientific, and Cultural Organization (UNESCO)'s Intergovernmental Oceanographic Commission (IOC) formed its first Intergovernmental Coordination Group (ICG) for tsunami warnings and, in 1965, established a Tsunami Warning System in the Pacific region. Following the Indian Ocean tsunami of December 26, 2004, UNESCO/IOC established the ICG for the Indian Ocean Tsunami Warning and Mitigation System (IOTWS). The ICG/IOTWS recommended the establishment of a web-based community computational model to transfer modeling expertise and capability to, between, and within Indian Ocean countries. Following this recommendation, the NCTR developed the Community Modeling Interface for Tsunamis (ComMIT), which is a

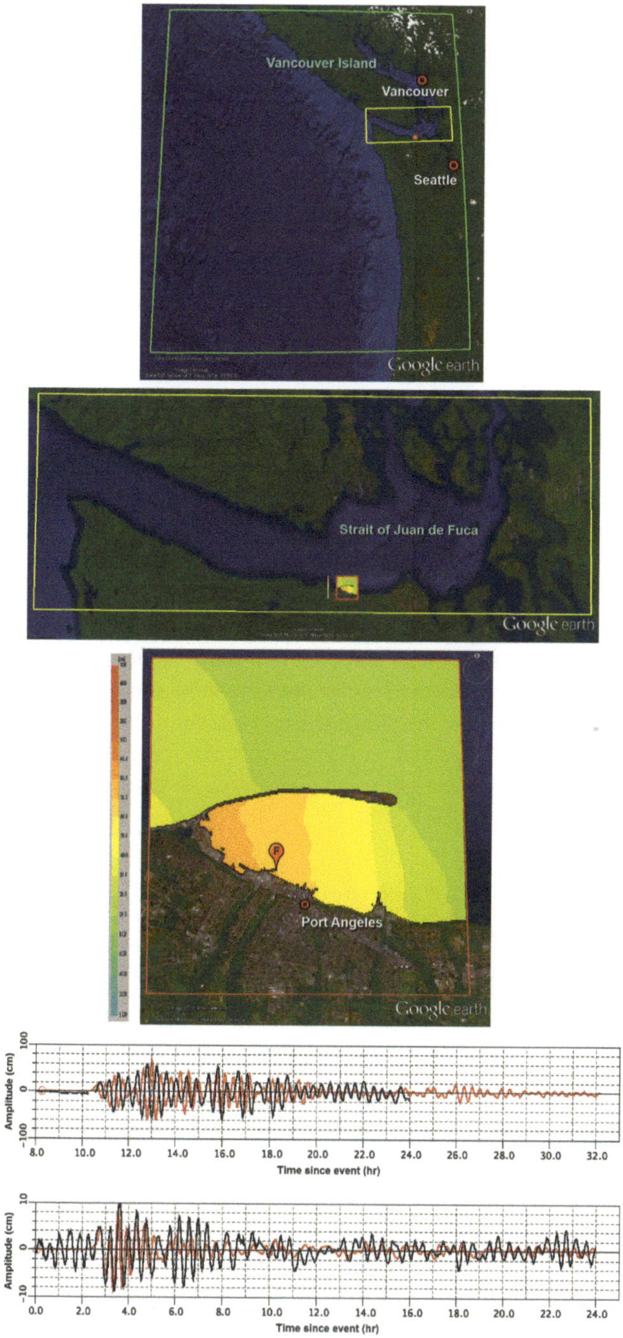

FIGURE 2.11 The NCTR's Port Angeles, Washington forecast model (Kânoğlu, in review). (Top three insets) Three nested grids in increasing resolution used in MOST modeling, including maximum wave amplitudes in the near-shore grid resulted from the 2011 Japan tsunami. (Bottom two insets) The comparisons of the tide-gage recordings (black) and the MOST model results (red) are presented for the March 11, 2011, Japan and the November 28, 2012, Charlotte tsunamis. Tide gage location (forecast point) is shown in the near-shore grid.

user-friendly interface to the MOST model. ComMIT is a tool that could be used for inundation mapping as well as real-time tsunami forecast applications and thus could be a critical tool for building tsunami-resilient communities in the region. Several training courses have been undertaken and many engineers are being trained in tsunami science, ComMIT, and how ComMIT could be applied in the development of tsunami-resilient communities, particularly within the Indian Ocean (Titov et al., 2011).

Synolakis and Bernard (2006) presented an overview of the state of knowledge of tsunami forecasting challenges, and they identified promising frontiers toward a global warning system. They wrote "Recent advances in the science of tsunami forecasting, communications technology and tsunami-resilient community development hold promise for a global tsunami mitigation system that would greatly reduce the impact of future tsunamis." Incidentally during the ITS in Novosibirsk, Russia (July 14–16, 2009), Satake et al. (2011b,a), a large earthquake ($M_W = 7.8$) occurred off the west coast of South Island, New Zealand, on July 15, 2009. Some personnel from the TWCs of Australia, Europe, New Zealand, and the United States, which issued tsunami information and warnings, were attending the meeting. Uslu et al. (2011) made the real-time assessment of tsunami remotely during the meeting using a similar tool to ComMIT. Again, during the Japan tsunami the NCTR modelers were in Tanzania to conduct another ComMIT training course. As soon as the tsunami reached the tsunamographs, they inverted the source and made that available to the ComMIT users around the world. Scientists at the NCTR, in New Zealand and in Turkey were able to use the inverted source and run several FMs to identify inundation at some locales along the Pacific coast.

2.4.3 Harbor and Basin Oscillations

Harbor oscillations occur when incident storm waves contain frequencies at directions of incidence which cause the harbor to amplify their amplitudes at select locations.[6] Harbor and basin oscillations can also be excited by incident tsunamis. Except for large ports, most harbors have natural frequencies that are longer than most storm waves and shorter than most tsunami waves. The possibility of resonance, however, needs to be carefully evaluated. Even if harbor oscillation do not result in casualties, the loss of use of harbor facilities even for a few days can result in substantial monetary losses, as one can extrapolate from the losses in Kobe 1995. The 1964 Alaskan tsunami caused an alleged US$1 million in damages in the Port of Long Beach. Notably, most of the damage was in the Cerritos Channel and was due to the tsunami-induced currents, not due to overtopping. The 1999 Izmit, Turkey,

6. For an extensive review of seiches and harbor resonance refer to Synolakis (2002).

earthquake caused a small tsunami and possibly harbor oscillations that caused an oil tanker to break its moorings, triggering a refinery fire that burned for days.

An interesting effect identified by Okal et al. (2006) in their survey of the 2004 Boxing Day tsunami in Salalah, Oman, is the tsunami-induced setup of substantial vortical motions in ports and harbors with long breakwaters or piers. These vortices are not necessarily associated with the classic concept of harbor resonance. They are a consequence of large current encountered at port entrances and quays. Small tsunami waves can induce disproportionally large rotational motions in far-field harbors. To wit, in the Port of Los Angeles, the 2010 Chilean tsunami featured less than 40-cm vertical excursions at surrounding coastlines, but, nonetheless, produced currents strong enough to render large fire department vessels difficult to control for periods of time. In another spectacular example, two nuclear submarines broke their mooring lines in Guam and wandered in the outer and inner basins out of control.

Although unusual amplification in a harbor hours after a leading tsunami wave arrived was the only understood consequence of resonance, other amplification phenomena are important. Recently, the 2011 Japan tsunami provided one of the most comprehensive observational and instrumental data set for harbor response to an incoming tsunami, see Lynett et al. (2012) for details. In the near-field, the Port of Oarai, Japan, provided one of the most vivid examples of rotational currents approximately 3 h after the earthquake (Figure 2.1). In far-field, it had severe effect in Crescent City, California. It is well known that Crescent City has a tendency to amplify far-field tsunami surges (Horrillo et al., 2008; Kowalik et al., 2008). The Crescent City Harbor experienced significant damage following the 1964 Great Alaska earthquake (Griffin, 1984) and following the November 15, 2006, Kuril Island tsunami (Dengler et al., 2008; Uslu et al., 2007). Nontrivial tsunami effects occurred in Crescent City caused by the September 29, 2009 Samoa tsunami (Wilson et al., 2011). Repairs from the Kuril tsunami were nearly complete when the harbor was severely damaged by the Japan tsunami. During the 2011 Japan tsunami, reconstructions of current velocities from security camera footage suggested up to 4 m/s maximal speed. However, experienced mariners who witnessed the event generally reported higher current velocity estimates (Dengler et al., 2011). In Humboldt Bay, roughly 100 km south of Crescent City, a current meter recorded a detailed time history of the event with a maximum current speed of 0.9 m/s. This was accurately simulated to first order using MOST, NOAA's tsunami FM (Dengler et al., 2011). At the south of San Francisco, Pillar Point Harbor, incoming tsunami induced a clockwise-rotating vortex in the outer harbor basin, and a counterclockwise rotation in the inner harbor as a result of the orientation of the harbor entrance and breakwaters with respect to incoming tsunami. The currents were estimated to be on the order of 2.6–5.1 m/s by local captains. Across the Pacific, several locations in New Zealand experienced strong and dangerous currents affecting recreational maritime activities. At the Port of Tauranga in the Bay of Plenty, three tide gages recorded water level fluctuations of the order of ± 0.5 m, whereas a current meter recorded maximum current speeds (tsunami + tide) exceeding 2.3 m/s, coinciding with the falling tide approximately 10 h after the tsunami's first arrival. After removal of the tidal component, the maximum tsunami current was on the order of 1 m/s, and it occurred ~3 h after the first wave arrival on the incoming tide. The strongest currents occurred during peak tsunami energy which persisted for ~8 h after first arrival. Although these currents were significant, operations were not disrupted at the Port of Tauranga. In the Galapagos Islands, strong currents were noted in all of the main harbors including Puerto Ayora. A tidal gage and a current meter recorded the event in Puerto Ayora Harbor. The maximum tsunami-induced currents were ~0.3 m/s associated with the first waves and currents persisted for more than 24 h after the arrival of the initial wave.

2.5 CONCLUSIONS

Despite substantial advances in tsunami science in 2000–2010, equally substantial challenges remain for improving tsunami hazard mitigation. Although at least one forecasting methodology had gone through extensive testing, and is now officially in use by the warning centers, standards must be established for tsunami science. Synolakis et al. (2008) discussed standards for tsunami modeling. Standards for forecasting methodologies need urgently to be established. In Europe, several warning centers occur, none of which has yet to issue an operational warning for a real event. If it happens, we fear that there will be substantial confusion with possibly contradictory or competing warnings from different national agencies.

In reviewing advances for this volume and elsewhere, it became clear that if the tsunami community appeared at first perplexed in the aftermath of the Boxing Day tsunami, it was not due to the failure of recognized hydrodynamic paradigms, much as certain geophysical paradigms and scaling laws failed, but continuing loss of life from new events, the worst possible surprise. What saves human lives in many remote areas of the world is ancestral knowledge and community preparedness, as demonstrated in the Solomon Island 2007 (Fritz and Kalligeris, 2008) and then again in 2010, and in Vanuatu 1998 (Caminade et al., 2000) tsunamis. All ITSTs should spend substantial time during their field work for public outreach and education as discussed in Fritz and Kalligeris (2008) as well as in Synolakis and Okal (2005).

Live broadcasting of the March 11, 2011, Japan event helped with the education of the public wherever ancestral knowledge does not exist. Synolakis and Bernard (2006) concluded with the statement "Before the next Sumatra-type tsunami strikes, we must resolve to create a world that can coexist with the tsunami hazard." Unfortunately, the 2011 Japan tsunami dramatically showed that we are not there yet. Never again should there be a repeat of the flawed TEPCO analysis for the safety of the Fukushima Dai-ichi NPP, which was primarily due to lack of familiarity with the context of numerical predictions and with how to appropriately identify scenario events that may impact the site. The accident was the result of *a cascade of stupid errors, almost impossible to ignore* as the New York Times puts it (The New York Times, 2011). We hope this would not be the case when next disaster strikes and the need occurs for educating the broader scientific community.

ACKNOWLEDGMENTS

This writing was partially supported by the project ASTARTE—Assessment, STrategy And Risk Reduction for Tsunamis in Europe, Grant 603839, 7th FP (ENV.2013.6.4-3 ENV.2013.6.4-3) to the Technical University of Crete and the Middle East Technical University (METU). We also acknowledge the partial support from the Scientific and Technological Research Council of Turkey (TÜBİTAK) project no. 109Y387 and the General Secretariat for Research and Technology, Greece (GSRT) project no. 10TUR/1-50-1 in the framework of the joint research program between Turkey and Greece. We also thank Naeimeh Sharghivand for her assistance with the figures. UK would like to thank to the NCTR for their support during the summer visits from 2004 to 2013.

REFERENCES

Arcas, D., Kânoğlu, U., Moore, C., Aydın, B., 2013. Effect of time-dependent rupture on tsunami generation. Abstract OS11D-1674 presented at 2013, Fall Meeting, AGU, San Francisco, CA, December 9–13.

Bardet, J., Synolakis, C., Davies, H., Imamura, F., Okal, E., 2003. Landslide tsunamis: recent findings and research directions. Pure Appl. Geophys. 160, 1793–1809.

Ben-Menahem, A., Rosenman, M., 1972. Amplitude patterns of tsunami waves from submarine earthquakes. J. Geophys. Res. 77, 3097–3128.

Bernard, E., Robinson, A. (Eds.), 2009. The Sea: Ideas and Observations on Progress in the Study of the Seas. Tsunamis, vol. 15. Harvard University Press, Cambridge, MA.

Bernard, E.N., Mofjeld, H.O., Titov, V., Synolakis, C.E., González, F.I., 2006. Tsunami: scientific frontiers mitigation, forecasting and policy implications. Philos. Trans. R. Soc. A: Math. Phys. Eng. Sci. 364, 1989–2006. Discussion Meeting on Extreme Natural Hazards, Royal Society, London, England, October 26–27, 2005.

Borerro, J., Ortiz, M., Titov, V., Synolakis, C., 1997. Field survey of Mexican tsunami produces new data, unusual photos. EOS, Trans. Am. Geophys. Union 78, 85–88.

Borrero, J., 2002. Analysis of Tsunami Hazards in Southern California. Ph.D. thesis University of Southern California, Los Angeles, California Los Angeles, California. Available from ProQuest, UMI 3094306.

Borrero, J., Dolan, J., Synolakis, C., 2001. Tsunamis within the eastern Santa Barbara channel. Geophys. Res. Lett. 28, 643–646.

Borrero, J., Legg, M., Synolakis, C., 2004. Tsunami sources in the Southern California bight. Geophys. Res. Lett. 31, L13211.

Borrero, J., Yalciner, A., Kânoğlu, U., Titov, V., McCarthy, D., Synolakis, C., 2003. Producing tsunami inundation maps: the California experience. In: Yalciner, A.C., Pelinovsky, E.N.,

Okal, E., Synolakis, C.E. (Eds.), Submarine Landslide and Tsunamis, NATO Science Series, IV. Earth and Environmental Sciences, vol. 21, pp. 315–326. Proceedings of the NATO Advanced Research Workshop on Underwater Ground Failures on Tsunami Generation, Modeling, Risk and Mitigation, Istanbul, Turkey, May 23–26, 2001.

Briggs, M., Synolakis, C., Harkins, G., Green, D., 1995. Laboratory experiments of tsunami runup on a circular island. Pure Appl. Geophys. 144, 569–593.

Bruins, H.J., MacGillivray, J.A., Synolakis, C.E., Benjamini, C., Keller, J., Kisch, H.J., Kleugel, A., van der Plicht, J., 2008. Geoarchaeological tsunami deposits at Palaikastro (Crete) and the Late Minoan IA eruption of Santorini. J. Archaeol. Sci. 35, 191–212.

Burger, E., Kamb, L., Nakamura, T., 2013. A web based tsunami-forecating tool: design approach and implementation. In: 29th Conference on Environmental Information Processing Technologies (Formerly IIPS), 93rd Annual Meeting of the AMS in Austin, TX.

Caminade, P., Charlie, D., Kânoğlu, U., Koshimura, S.-I., Matsutomi, H., Moore, A., Ruscher, C., Synolakis, C., Takahashi, T., 2000. Vanuatu earthquake and tsunami cause much damage, few casualties. EOS, Trans. Am. Geophys. Union 81, 641, 646–647.

Carrier, G., Greenspan, H., 1958. Water waves of finite amplitude on a sloping beach. J. Fluid Mech. 4, 97–109.

Dengler, L., Admire, A., Uslu, B., Crawford, G., Montoya, J., Wilson, R., 2011. Observed and modeled tsunami current velocities on California's North Coast. In: IUGG, Melbourne July 2011 (Abstract).

Dengler, L., Uslu, B., Barberopoulou, A., Borrero, J., Synolakis, C., 2008. The vulnerability of Crescent City, California, to tsunamis generated by earthquakes in the Kuril Islands region of the Northwestern Pacific. Seismol. Res. Lett. 79, 608–619.

Eisner, R., Borrero, J., Synolakis, C., 2001. Inundation maps for the state of California. In: Proceedings of the 2001 International Tsunami Symposium. NTHMP Review Session, Paper R-4.

Ezersky, A., Tiguercha, D., Pelinovsky, E., 2013. Resonance phenomena at the long wave run-up on the coast. Nat. Hazards Earth Syst. Sci. 13, 2745–2752.

Fritz, H., 2002. PIV applied to landslide generated impulse waves. In: Adrian, R.J., Durao, D.F.G., Heitor, M.V., Maeda, M., Tropea, C., Whitelaw, J.H. (Eds.), Laser Techniques for Fluid Mechanics. European Res Off; USA; USN; USAF; Luso-Amer Fdn Dev, pp. 305–320, 10th International Symposium on Laser Techniques for Fluid Mechanics, Calouste Gulbenkian FDN, Lisbon, Portugal, Jul 10–13, 2000.

Fritz, H., Hager, W., Minor, H., 2003a. Landslide generated impulse waves. 1. Instantaneous flow fields. Exp. Fluids 35, 505–519.

Fritz, H., Hager, W., Minor, H., 2003b. Landslide generated impulse waves. 2. Hydrodynamic impact craters. Exp. Fluids 35, 520–532.

Fritz, H., Hager, W., Minor, H.-E., 2001. Lituya Bay case: rockslide impact and wave run-up. Sci. Tsunami Hazards 19, 3–22.

Fritz, H.M., Borrero, J.C., Synolakis, C.E., Yoo, J., 2006. 2004 Indian ocean tsunami flow velocity measurements from survivor videos. Geophys. Res. Lett. 33, L24605.

Fritz, H.M., Kalligeris, N., 2008. Ancestral heritage saves tribes during 1 April 2007 Solomon Islands tsunami. Geophys. Res. Lett. 35, L01607.

Fritz, H.M., Kongko, W., Moore, A., McAdoo, B., Goff, J., Harbitz, C., Uslu, B., Kalligeris, N., Suteja, D., Kalsum, K., Titov, V., Gusman, A., Latief, H., Santoso, E., Sujoko, S., Djulkarnaen, D., Sunendar, H., Synolakis, C., 2007. Extreme runup from the 17 July 2006 Java tsunami. Geophys. Res. Lett. 34, L12602.

Geist, E., Titov, V., Synolakis, C., 2006. Tsunami: wave of change. Sci. Am. 294, 56–63.

Gica, E., Spillane, M., Titov, V.V., Chamberlin, C.D., Newman, J.C., 2008. Development of the Forecast Propagation Database for NOAA's Short-term Inundation Forecast for Tsunamis (SIFT). Technical Report OAR PMEL-139, NTIS: PB2008−109391. The National Oceanic and Atmospheric Administration.

Gisler, G., 2009. Chapter 6 − tsunami generation: other sources. In: Bernard, E.N., Robinson, A.R. (Eds.), The Sea: Ideas and Observations on Progress in the Study of the Seas, Tsunamis, vol. 15. Harvard University Press, Cambridge, MA, pp. 179−200.

González, F., Bernard, E., Dunbar, P., Geist, E., Jaffe, B., Kânoğlu, U., Locat, J., Mofjeld, H., Moore, A., Synolakis, C., Titov, V., Weiss, R., 2007. Scientific and Technical Issues in Tsunami Hazard Assessment of Nuclear Power Plant Sites. Technical Report OAR PMEL-136. The National Oceanic and Atmospheric Administration, Pacific Marine Environmental Laboratory, Seattle, WA.

González, F., Titov, V., Mofjeld, H., Venturato, A., Simmons, R., Hansen, R., Combellick, R., Eisner, R., Hoirup, D., Yanagi, B., Yong, S., Darienzo, M., Priest, G., Crawford, G., Walsh, T., 2005. Progress in NTHMP hazard assessment. Nat. Hazards 35, 89−110.

González, F.I., Geist, E.L., Jaffe, B., Kânoğlu, U., Mofjeld, H., Synolakis, C.E., Titov, V.V., Arcas, D., Bellomo, D., Carlton, D., Horning, T., Johnson, J., Newman, J., Parsons, T., Peters, R., Peterson, C., Priest, G., Venturato, A., Weber, J., Wong, F., Yalciner, A., 2009. Probabilistic tsunami hazard assessment at Seaside, Oregon, for near- and far-field seismic sources. J. Geophys. Res.: Oceans 114.

Goring, D., 1978. Tsunamis − the Propagation of Long Waves onto a Shelf. Technical Report KH-R-38. W. M. Keck Laboratory Hydraulics & Water Resources, California Institute Technology, Pasadena, California.

Greenslade, D.J.M., Simanjuntak, M., Chittleborough, J., Burbidge, D., 2007. A First Generation Realtime Tsunami Forecasting System for the Australian Region. Technical Report BMRC Research Report No: 126. Bureau of Meteorology, Australia.

Greenslade, D.J.M., Titov, V.V., 2008. A comparison study of two numerical tsunami forecasting systems. Pure Appl. Geophys. 165, 1991−2001.

Griffin, W., 1984. Crescent City's Dark Disaster: Crescent City, California March 27−28, 1964 Tsunami and 20 Years Later. Crescent City Printing Company, Crescent City, CA.

Hall, J., Watts, J., 1953. Report on Waves. Technical Report 33. Beach Erosion Board, USACE, p. 14.

Hammack, J., 1972. Tsunamis − a Model of Their Generation and Propagation. Technical Report KH-R-28. W. M. Keck Laboratory Hydraulics & Water Resources, California Institute Technology, Pasadena, California.

Hammack, J., 1973. Note on tsunamis − their generation and propagation in an ocean of uniform depth. J. Fluid Mech. 60, 769−799.

Heinrich, P., Piatanesi, A., Okal, E., Hebert, H., 2000. Near-field modeling of the July 17, 1998 tsunami in Papua New Guinea. Geophys. Res. Lett. 27, 3037−3040.

Hill, E.M., Borrero, J.C., Huang, Z., Qiu, Q., Banerjee, P., Natawidjaja, D.H., Elosegui, P., Fritz, H.M., Suwargadi, B.W., Pranantyo, I.R., Li, L., Macpherson, K.A., Skanavis, V., Synolakis, C.E., Sieh, K., 2012. The 2010 M-w 7.8 Mentawai earthquake: very shallow source of a rare tsunami earthquake determined from tsunami field survey and near-field GPS data. J. Geophys. Res.: Solid Earth 117, B06402.

Horrillo, J., Knight, W., Kowalik, Z., 2008. Kuril Islands tsunami of November 2006: 2. Impact at Crescent city by local enhancement. J. Geophys. Res.: Oceans 113, C01021.

Houston, J., Garcia, A., 1974a. Type 16 Flood Insurance Study: Tsunami Predictions for Pacific Coastal Communities. Technical Report WES Report H-74−3. US Army Corps of Engineers Waterways Experiment Station, Vicksburg, MS.

Houston, J., Garcia, A., 1974b. Type 19 Flood Insurance Study: Tsunami Predictions for Southern California. Technical Report HL-80–18. US Army Corps of Engineers Waterways Waterways Experiment Station, Vicksburg, MS.

Huppert, H.E., Sparks, R.S.J., 2006. Extreme natural hazards: population growth, globalization and environmental change. Philos. Trans. R. Soc. A: Math. Phys. Eng. Sci. 364, 1875–1888. Discussion Meeting on Extreme Natural Hazards, Royal Society, London, England, October 26–27, 2005.

Jackson, J., 2006. Fatal attraction: living with earthquakes, the growth of villages into megacities, and earthquake vulnerability in the modern world. Philos. Trans. R. Soc. A: Math. Phys. Eng. Sci. 364, 1911–1925. Discussion Meeting on Extreme Natural Hazards, Royal Society, London, England, October 26–27, 2005.

Kanamori, H., 1972. Mechanism of tsunami earthquakes. Phys. Earth Planet. Inter. 6, 346–359.

Kânoğlu, U., 1998. The runup of long waves around piecewise linear bathymetries. Ph.D. thesis University of Southern California Los Angeles, California. Available from ProQuest, UMI 9919061.

Kânoğlu, U. A Tsunami Forecast Model for Port Angeles, Washington. Technical Report NOAA OAR Special Report. Pacific Marine Environmental Laboratory Tsunami Forecast Series, in review.

Kânoğlu, U., Hoto, O., Kalligeris, N., Flouri, E., Aydın, B., Moore, C., Synolakis, C., 2012. Tsunami Propagation Database for the Mediterranean and Aegean Seas. Abstract NH33A-1666 presented at 2012, Fall Meeting, AGU, San Francisco, CA, December 3–7.

Kânoğlu, U., Synolakis, C., 1998. Long wave runup on piecewise linear topographies. J. Fluid Mech. 374, 1–28.

Kânoğlu, U., Titov, V.V., Aydın, B., Moore, C., Stefanakis, T.S., Zhou, H., Spillane, M., Synolakis, C.E., 2013. Focusing of long waves with finite crest over constant depth. Proc. R. Soc. A: Math. Phys. Eng. Sci. 469, 20130015.

Kawata, Y., Benson, B., Borrero, J., Borrero, J., Davies, H., de Lange, W., Imamura, F., Letz, H., Nott, J., Synolakis, C., 1999. Tsunami in Papua New Guinea was as intense as first thought. EOS, Trans. Am. Geophys. Union 80, 101–105.

Kowalik, Z., Horrillo, J., Knight, W., Logan, T., 2008. Kuril Islands tsunami of November 2006: 1. Impact at Crescent city by distant scattering. J. Geophys. Res.: Oceans 113, C01020.

Lebreton, L.C.M., Borrero, J.C., 2013. Modeling the transport and accumulation floating debris generated by the 11 March 2011 Tohoku tsunami. Mar. Pollut. Bull. 66, 53–58.

Liu, P., Lynett, P., Synolakis, C., 2003. Analytical solutions for forced long waves on a sloping beach. J. Fluid Mech. 478, 101–109.

Liu, P.L.-F., Cho, Y., Briggs, M., Kânoğlu, U., Synolakis, C., 1995. Runup of solitary waves on a circular island. J. Fluid Mech. 302, 259–285.

Liu, P.L.-F., Synolakis, C., Yeh, H., 1991. Report on the international workshop on long-wave run-up. J. Fluid Mech. 229, 675–688.

Liu, P.L.-F., Wu, T., Raichlen, F., Synolakis, C., Borrero, J., 2005. Runup and rundown generated by three-dimensional sliding masses. J. Fluid Mech. 536, 107–144.

Liu, P.L.-F., Yeh, H., Synolakis, C. (Eds.), 2008. Advanced Numerical Models for Simulating Tsunami Waves and Runup. Advances in Coastal and Ocean Engineering, vol. 10. World Scientific Publishing Co. Pte. Ltd, Singapore.

Lynett, P., Borrero, J., Liu, P., Synolakis, C., 2003. Field survey and numerical simulations: a review of the 1998 Papua New Guinea tsunami. Pure Appl. Geophys. 160, 2119–2146.

Lynett, P.J., Borrero, J.C., Weiss, R., Son, S., Greer, D., Renteria, W., 2012. Observations and modeling of tsunami-induced currents in ports and harbors. Earth Planet. Sci. Lett. 327, 68–74.

Madsen, O., Mei, C., 1969. Transformation of a solitary wave over an uneven bottom. J. Fluid Mech. 39, 781–791.

Marinatos, S., 1939. The volcanic destruction of Minoan Crete. Antiquity 13, 425–439.

Miller, D., 1960. Giant Waves in Lituya Bay, Alaska. Technical Report Professional Paper 354-C U.S. Geological Survey. Shorter contributions to general geology, 1959.

Mohammed, F., Fritz, H.M., 2012. Physical modeling of tsunamis generated by three-dimensional deformable granular landslides. J. Geophys. Res.: Oceans 117, C11015.

Mori, N., Takahashi, T., Yasuda, T., Yanagisawa, H., 2011. Survey of 2011 Tohoku earthquake tsunami inundation and run-up. Geophys. Res. Lett. 38, L00G14.

Morrison, D., 2006. Asteroid and comet impacts: the ultimate environmental catastrophe. Philos. Trans. R. Soc. A: Math. Phys. Eng. Sci. 364, 2041–2054. Discussion Meeting on Extreme Natural Hazards, Royal Society, London, England, October 26–27, 2005.

National Tsunami Hazard Mitigation Program (NTHMP), 2012 (Technical Report Boulder: U.S. Department of Commerce/NOAA/NTHMP. The National Oceanic and Atmospheric Administration Special Report). In: Proceedings and Results of the 2011 NTHMP Model Benchmarking Workshop.

Newman, A., Okal, E., 1998. Teleseismic estimates of radiated seismic energy: the E/M-0 discriminant for tsunami earthquakes. J. Geophys. Res.: Solid Earth 103, 26885–26898.

Okada, Y., 1985. Surface deformation due to shear and tensile faults in a half-space. Bull. Seismol. Soc. Am. 75, 1135–1154.

Okal, E., 1993. Seismology – predicting large tsunamis. Nature 361, 686–687.

Okal, E., 2003. T waves from the 1998 Papua New Guinea earthquake and its aftershocks: timing the tsunamigenic slump. Pure Appl. Geophys. 160, 1843–1863.

Okal, E., Plafker, G., Synolakis, C., Borrero, J., 2003. Near-field survey of the 1946 Aleutian tsunami on Unimak and Sanak Islands. Bull. Seismol. Soc. Am. 93, 1226–1234.

Okal, E., Synolakis, C., 2003. A theoretical comparison of tsunamis from dislocations and landslides. Pure Appl. Geophys. 160, 2177–2188.

Okal, E., Synolakis, C., 2004. Source discriminants for near-field tsunamis. Geophys. J. Int. 158, 899–912.

Okal, E.A., Fritz, H.M., Raad, P.E., Synolakis, C., Al-Shijbi, Y., Al-Saifi, M., 2006. Oman field survey after the December 2004 Indian ocean tsunami. Earthquake Spectra 22, S203–S218.

Pararas-Carayannis, G., 1972. A study of the source mechanism of the Alaska earthquake and tsunami of March 27, 1964. In: The Committee on the Alaska Earthquake of the Division of Earth Sciences (Ed.), The Great Alaska Earthquake of 1964: Seismology and Geodesy. National Academy of Sciences, Washington DC, pp. 249–258.

Pedersen, G., Gjevik, B., 1983. Run-up of solitary waves. J. Fluid Mech. 135, 283–299.

Percival, D.B., Denbo, D.W., Eble, M.C., Gica, E., Mofjeld, H.O., Spillane, M.C., Tang, L., Titov, V.V., 2011. Extraction of tsunami source coefficients via inversion of DART® buoy data. Nat. Hazards 58, 567–590.

Peregrine, D., 1966. Calculations of development of an undular bore. J. Fluid Mech. 25, 321–330.

Plafker, G., 1965. Tectonic deformation associated with 1964 Alaska earthquake – earthquake of 27 March 1964 resulted in observable crustal deformation of unprecedented areal extent. Science 148, 1675–1687.

Plafker, G., Kachadoorian, R., Eckel, E.B., Mayo, L.R., 1969. Effects of the Earthquake of March 27, 1964, on Various Communities. Technical Report Prof. Paper 542-G. US Geological Survey.

Raichlen, F., Synolakis, C., 2003. Runup from three dimensional sliding masses. In: Briggs, M., Koutitas, Ch (Eds.), Long Waves Symposium, pp. 247–256. Long Waves Symposium, August 25–27, 2003, Thessaloniki, Greece. In parallel with the XXX IAHR Congress.

Satake, K., 1988. Effects of bathymetry on tsunami propagation – application of ray tracing to tsunamis. Pure Appl. Geophys. 126, 27–36.

Satake, K., 2007. Tsunamis. In: Kanamori, H. (Ed.), Treatise on Geophysics, Earthquake Seismology, vol. 4. Elsevier, Oxford, pp. 483–511.

Satake, K., 2013. The 2011 Tohoku earthquake and tsunami hazard assessment. In: Probabilistic Safety Assessment and Management (PSAM) Topical Conference: In Light of the Fukushima Dai-ichi Accident, April 14–18, 2013. Tokyo, Japan (Unpublished presentation).

Satake, K., Rabinovich, A., Kânoğlu, U., Tinti, S., 2011a. Introduction to "Tsunamis in the World Ocean: past, present, and future. Volume I". Pure Appl. Geophys. 168, 963–968.

Satake, K., Rabinovich, A., Kânoğlu, U., Tinti, S., 2011b. Introduction to "Tsunamis in the World Ocean: past, present, and future. Volume II". Pure Appl. Geophys. 168, 1913–1917.

Satake, K., Shimazaki, K., Tsuji, Y., Ueda, K., 1996. Time and site of a giant earthquake in Cascadia inferred from Japanese tsunami records of January 1700. Nature 379, 246–249.

Shuto, N., 2003. Tsunamis of seismic origin: science, disasters and mitigation. In: Yalciner, A., Pelinovsky, E., Okal, E., Synolakis, C. (Eds.), Submarine Landslides and Tsunamis, NATO Science Series, IV. Earth and Environmental Sciences, vol. 21. Kluwer Academic Publishers, Dordrecht, The Netherlands, pp. 1–8. Proceedings of the NATO Advanced Research Workshop on Underwater Ground Failures on Tsunami Generation, Modeling, Risk and Mitigation, Istanbul, Turkey, May 23–26, 2001.

Spillane, M., Gica, E., Titov, V.V., Mofjeld, H.O., 2008. Tsunameter Network Design for the U.S. DART® *Arrays in the Pacific and Atlantic Oceans*. Technical Report OAR PMEL-143. The National Oceanic and Atmospheric Administration.

Stefanakis, T., Contal, E., Vayatis, N., Dias, F., Synolakis, C., 2013. Can Small Islands Protect Nearby Coasts from Tsunamis? an Active Experimental Design Approach, vol. 15. Abstract EGU2013-11314 presented at EGU General Assembly 2013, Vienna, Austria, April 7–12.

Stefanakis, T.S., Dias, F., Dutykh, D., 2011. Local run-up amplification by resonant wave interactions. Phys. Rev. Lett. 107, 124502.

Sweet, S., Silver, E., 2003. Tectonics and slumping in the source region of the 1998 Papua New Guinea tsunami from seismic reflection images. Pure Appl. Geophys. 160, 1945–1968. http://dx.doi.org/10.1007/s00024-003-2415-z.

Synolakis, C., 1986. The runup of long waves. Ph.D. thesis California Institute of Technology, Engineering and Applied Science Pasadena, California.

Synolakis, C., 1987. The runup of solitary waves. J. Fluid Mech. 185, 523–545.

Synolakis, C., 2002. Tsunami and seiche. In: Chen, W.-F., Scawthorn, C. (Eds.), Earthquake Engineering Handbook, New Directions in Civil Engineering. CRC Press, Oxford. http://dx.doi.org/10.1201/9781420042443.ch9.

Synolakis, C., Bardet, J., Borrero, J., Davies, H., Okal, E., Silver, E., Sweet, S., Tappin, D., 2002. The slump origin of the 1998 Papua New Guinea tsunami. Proc. R. Soc. A: Math. Phys. Eng. Sci. 458, 763–789. http://dx.doi.org/10.1098/rspa.2001.0915.

Synolakis, C., Kânoğlu, U., 2009. Chapter 8 – tsunami modelling: development of benchmarked models. In: Bernard, E.N., Robinson, A.R. (Eds.), The Sea: Ideas and Observations on Progress in the Study of the Seas, Tsunamis, vol. 15. Harvard University Press, Cambridge, MA, pp. 237–294.

Synolakis, C., Liu, P.L.-F., Carrier, G., Yeh, H., 1997. Tsunamigenic sea-floor deformations. Science 278, 598–600. http://dx.doi.org/10.1126/science.278.5338.598.

Synolakis, C.E., Bernard, E.N., 2006. Tsunami science before and beyond Boxing Day 2004. Philos. Trans. R. Soc. A: Math. Phys. Eng. Sci. 364, 2231–2265. http://dx.doi.org/10.1098/rsta.2006.1824. Discussion Meeting on Extreme Natural Hazards, Royal Society, London, England, October 26–27, 2005.

Synolakis, C.E., Bernard, E.N., Titov, V.V., Kânoğlu, U., González, F.I., 2007. Standards, Criteria, and Procedures for NOAA Evaluation of Tsunami Numerical Models. Technical Report OAR PMEL-135, NTIS: PB2007–109601. The National Oceanic and Atmospheric Administration.

Synolakis, C.E., Bernard, E.N., Titov, V.V., Kânoğlu, U., González, F.I., 2008. Validation and verification of tsunami numerical models. Pure Appl. Geophys. 165, 2197–2228.

Synolakis, C.E., Kong, L., 2006. Runup measurements of the December 2004 Indian Ocean tsunami. Earthquake Spectra 22, S67–S91. http://dx.doi.org/10.1193/1.2218371.

Synolakis, C.E., Okal, E.A., 2005. 1992-2002: perspective on a decade of post-tsunami surveys. In: Satake, K. (Ed.), Tsunamis: Case Studies and Recent Developments Advances in Natural and Technological Hazards Research, vol. 23, pp. 1–29. International Workshop on Tsunamis in the South Pacific, Wellington, New Zealand, September 25–26, 2003.

Tadepalli, S., Synolakis, C., 1994. The run-up of N-waves on sloping beaches. Proc. R. Soc. A: Math. Phys. Sci. 445, 99–112. http://dx.doi.org/10.1098/rspa.1994.0050.

Tadepalli, S., Synolakis, C., 1996. Model for the leading waves of tsunamis. Phys. Rev. Lett. 77, 2141–2144. http://dx.doi.org/10.1103/PhysRevLett.77.2141.

Tang, L., Titov, V.V., Bernard, E.N., Wei, Y., Chamberlin, C.D., Newman, J.C., Mofjeld, H.O., Arcas, D., Eble, M.C., Moore, C., Uslu, B., Pells, C., Spillane, M., Wright, L., Gica, E., 2012. Direct energy estimation of the 2011 Japan tsunami using deep-ocean pressure measurements. J. Geophys. Res.: Oceans 117, C08008. http://dx.doi.org/10.1029/2011JC007635.

Tang, L., Titov, V.V., Chamberlin, C.D., 2009. Development, testing, and applications of site-specific tsunami inundation models for real-time forecasting. J. Geophys. Res.: Oceans 114, C12025. http://dx.doi.org/10.1029/2009JC005476.

The New York Times, 2011. Japanese Rules for Nuclear Plants Relied on Old Science. By Norimitsu Onishi and James Glanz, March 26, 2011. Last accessed on February 21, 2014 at. http://www.nytimes.com/2011/03/27/world/asia/27nuke.html?pagewanted=all& r=1&.

Titov, V., 1997. Numerical modeling of long wave runup. Ph.D. thesis University of Southern California Los Angeles, California. Available from ProQuest, UMI 9733151.

Titov, V., 2009. Chapter 12 – tsunami forecasting. In: Bernard, E.N., Robinson, A.R. (Eds.), The Sea: Ideas and Observations on Progress in the Study of the Seas, Tsunamis, vol. 15. Harvard University Press, Cambridge, MA, pp. 371–400.

Titov, V., Rabinovich, A., Mofjeld, H., Thomson, R., González, F., 2005a. The global reach of the 26 December 2004 Sumatra tsunami. Science 309, 2045–2048. http://dx.doi.org/10.1126/science.1114576.

Titov, V., González, F., Bernard, E., Eble, M., Mofjeld, H., Newman, J., Venturato, A., 2005b. Real-time tsunami forecasting: challenges and solutions. Nat. Hazards 35, 41–58. http://dx.doi.org/10.1007/s11069-004-2403-3.

Titov, V., González, F., 1997. Implementation and Testing of the Method of Splitting Tsunami (MOST) Model. Technical Report ERL PMEL-112 NOAA/Pacific Marine Environmental Laboratory. p. 11.

Titov, V., Synolakis, C., 1995. Modeling of breaking and nonbreaking long-wave evolution and runup using VTCS-2. J. Waterw. Port Coastal Ocean Eng.—ASCE 121, 308—316. http://dx.doi.org/10.1061/(ASCE)0733-950X(1995)121:6(308).

Titov, V., Synolakis, C., 1997. Extreme inundation ows during the Hokkaido-Nansei-Oki tsunami. Geophys. Res. Lett. 24, 1315—1318. http://dx.doi.org/10.1029/97GL01128.

Titov, V., Synolakis, C., 1998. Numerical modeling of tidal wave runup. J. Waterw. Port Coastal Ocean Eng.—ASCE 124, 157—171. http://dx.doi.org/10.1061/(ASCE)0733-950X(1998)124:4(157).

Titov, V.V., Moore, C.W., Greenslade, D.J.M., Pattiaratchi, C., Badal, R., Synolakis, C.E., Kânoğlu, U., 2011. A new tool for inundation modeling: community modeling interface for tsunamis (ComMIT). Pure Appl. Geophys. 168, 2121—2131. http://dx.doi.org/10.1007/s00024-011-0292-4.

Tsunami Pilot Study Working Group, 2006. Seaside, Oregon Tsunami Pilot Study—Modernization of FEMA Flood Hazard Maps. Technical Report Contribution No. 2975 NOAA/OAR/PMEL, Seattle, WA.

Uslu, B., Borrero, J.C., Dengler, L.A., Synolakis, C.E., 2007. Tsunami inundation at Crescent city, California generated by earthquakes along the Cascadia subduction zone. Geophys. Res. Lett. 34, L20601. http://dx.doi.org/10.1029/2007GL030188.

Uslu, B., Eble, M., Arcas, D., Titov, V., 2013. Tsunami Hazard Assessment for the Commonwealth of the Northern Mariana Islands. Technical Report NOAA OAR Special Report. In: Tsunami Hazard Assessment Special Series, vol. 3.

Uslu, B., Eble, M., Chamberlin, C., 2010a. Tsunami Hazard Assessment for Guam. Technical Report NOAA OAR Special Report. In: Tsunami Hazard Assessment Special Series, vol. 1.

Uslu, B., Eble, M., Titov, V., Bernard, E., 2010b. Distant Tsunami Threats to the Ports of Los Angeles and Long Beach, California. Technical Report NOAA OAR Special Report. In: Tsunami Hazard Assessment Special Series, vol. 2.

Uslu, B., Power, W., Greenslade, D., Eble, M., Titov, V., 2011. The July 15, 2009 Fiordland, New Zealand tsunami: real-time assessment. Pure Appl. Geophys. 168, 1963—1972. http://dx.doi.org/10.1007/s00024-011-0281-7.

Wei, Y., Bernard, E.N., Tang, L., Weiss, R., Titov, V.V., Moore, C., Spillane, M., Hopkins, M., Kânoğlu, U., 2008. Real-time experimental forecast of the Peruvian tsunami of August 2007 for US coastlines. Geophys. Res. Lett. 35, L04609. http://dx.doi.org/10.1029/2007GL032250.

Wei, Y., Chamberlin, C., Titov, V.V., Tang, L., Bernard, E.N., 2013. Modeling of the 2011 Japan tsunami: lessons for near-field forecast. Pure Appl. Geophys. 170, 1309—1331. http://dx.doi.org/10.1007/s00024-012-0519-z.

Wiegel, R., 1955. Laboratory studies of gravity waves generated by the movement of a submerged body. Trans. Am. Geophys. Union 36, 759—774. http://dx.doi.org/10.1029/TR036i005p00759.

Wilson, R.I., Dengler, L.A., Goltz, J.D., Legg, M.R., Miller, K.M., Ritchie, A., Whitmore, P.M., 2011. Emergency response and field observation activities of geoscientists in California (USA) during the September 29, 2009, Samoa Tsunami. Earth Sci. Rev. 107, 193—200. http://dx.doi.org/10.1016/j.earscirev.2011.01.010.

Woods, M., Okal, E., 1987. Effect of variable bathymetry on the amplitude of teleseismic tsunamis — a ray-tracing experiment. Geophys. Res. Lett. 14, 765—768. http://dx.doi.org/10.1029/GL014i007p00765.

Yalciner, A., Pelinovsky, E., Okal, E., Synolakis, C. (Eds.), 2003. Submarine Landslides and Tsunamis. NATO Science Series, IV. Earth and Environmental Sciences, vol. 21. Kluwer Academic Publishers, Dordrecht, The Netherlands. Proceedings of the NATO Advanced Research Workshop on Underwater Ground Failures on Tsunami Generation, Modeling, Risk and Mitigation, Istanbul, Turkey, May 23–26, 2001.

Yeh, H., Liu, P.L.-F., Briggs, M., Synolakis, C., 1994. Propagation and amplification of tsunamis at coastal boundaries. Nature 372, 353–355. http://dx.doi.org/10.1038/372353a0.

Yeh, H., Liu, P.L.-F., Synolakis, C. (Eds.), 1996. Long-Wave Runup Models. World Scientific Publishing Co. Pte. Ltd, Singapore.

Chapter 3

Paleotsunami Research—Current Debate and Controversies

Anja M. Scheffers
Southern Cross GeoScience, Southern Cross University, Lismore, NSW, Australia

ABSTRACT
During 1989–2014, our knowledge and research focus on modern and paleotsunami events have increased exponentially. Before the catastrophic Indian Ocean Tsunami of 2004 AD, nearly 1000 scientific papers had discussed the topic, but since then, over another 1000 research articles have been published. The strong tsunami events of Chile, 2010, and particularly the Tohoku megatsunami in Japan of March 2011 will certainly advance tsunami research further. Hitherto, a large percentage of tsunami-related articles have highlighted the origin of tsunamis, their geophysics, tsunami wave progradation, tsunami risk evaluation, and mitigation. This contribution focuses on fields of debate that still exist, in particular the identification of paleotsunami sediments in geoarchives, the discussion of boulder movements induced either by strong storms or tsunami waves, the enigma of impact tsunamis during the Quaternary, and modeling versus field observations in paleotsunami science.

3.1 INTRODUCTION

One of the earliest research papers that attributed sedimentary evidence to paleotsunami events is the publication of Atwater (1987) and the detection of the so-called "Orphan Tsunami" of 1700 AD in Washington State, United States (with the source along the coast of Japan), and the article by Dawson et al. (1988) on the submarine Holocene Storegga slides, west of Norway. Since then several reviews of tsunami research have been published (e.g., Dawson and Shi, 2000; Synolakis and Okal, 2005; Dominey-Howes et al., 2006; Tappin, 2007; Dawson and Stewart, 2007; Shiki et al., 2008; Synolakis et al., 2008; Scheffers et al., 2009; Satake et al., 2011). The aim of this contribution is to highlight some aspects of palaeotsunami research that are contentious in the current debate. Catalogs on historical tsunamis have been

Coastal and Marine Hazards, Risks, and Disasters. http://dx.doi.org/10.1016/B978-0-12-396483-0.00003-0
Copyright © 2015 Elsevier Inc. All rights reserved.

published for many sections of the world's coastlines, for example, Altinok and Ersoy (2000) for Turkey; Baptista and Miranda (2009) for Portugal; Papadopoulos and Chalkis (1984), Soloviev (1990), Tinti and Maramai (1996), Soloviev et al. (2000), Papadopoulos and Fokaefs (2005), Ambraseys and Synolakis (2010) for the Mediterranean Sea; Lander et al. (2002) and O'Loughlin and Lamb (2003) for the Caribbean; Goff (2008), as well as Goff et al. (2010c) for New Zealand; Hamzah et al. (2000) for Indonesia; or the National Geophysical Data Center (NGDC) in Boulder, Colorado (2013), which aims to provide a global dataset for certain coastal areas and spans nearly 4,000 years with >2,000 tsunami events in different probability categories.

The NGDC dataset not only lists tsunamis based on historical reports but also those from "mythical" sources such as the collapse of Santorini volcano in the Aegean Sea (~1628 BC). Hitherto, this event attracted public attention, several publications, and also speculations, though substantial field evidence is lacking compared to other events, with the exception of deep sea tsunamites (*Augias turbidites*), which have been associated with the last Santorini caldera-forming event (Cita and Aloisi, 2000; Hieke and Werner, 2000; McKoy and Heiken, 2000; Dominey-Howes, 2004; Gutscher, 2005; Sironi and Rimoldi, 2005; Bruins et al., 2008; Goodman-Tchernow et al., 2009).

More problematic is the reliability of these ancient reports. The question that must be asked is how much of their content is simple fantasy or exaggeration, whether for political or religious reasons? The NGDC dataset classifies the reports that describe events before the year AD 1500 into different categories of reliability; nevertheless, an uncertainty remains. In some instances, these historical reports could be verified by field research and numerical dating, but so far, even events that have been classified as reliable have been elusive with regard to physical evidence as sedimentary signatures in the landscape.

On the other hand, paleotsunami research has unearthed and dated many tsunami events of often strong impacts during historical times, which are not listed in any of the catalogs. Whether this conundrum can be attributed to the rather young scientific discipline of tsunami research and the gap will close in the years to come, or the catalogs and conclusions of field event studies are false for a large proportion of cases, has to be elaborated by future and interdisciplinary research.

Well-studied examples of ancient tsunami events include that of Caesarea Maritima, Israel, in AD 115 (Reinhardt et al., 2006), and multiple tsunami impacts in the Lechaeon region north of Corinthos in Greece (Hadler et al., 2011). Often cited is the coseismic sudden uplift of western Crete in 365 AD with several written sources describing associated strong tsunamis in the eastern Mediterranean (Pirazzoli, 1986; Kelletat, 1991; Stiros, 2001; Stiros and Drakos, 2006; Scheffers, 2006; Shaw et al., 2008). In a European perspective, the Lisbon earthquake and tsunami of 1755 AD continue to receive attention because of the great number of fatalities, that is, >35,000 victims (Baptista et al., 1999; Martinez-Solares, 2001; Scheffers and Kelletat, 2005; Gracia et al., 2006; Lario et al., 2010; and many others). In contrast and quite surprisingly, the Messina earthquake and strong tsunami of 1908 AD (with >60,000 fatalities) is much less investigated.

Strong tsunami events outside of Europe have only attracted scientists in exceptional cases, such as the explosion of Krakatoa volcano in the Sunda Strait between the Indonesian Island of Sumatra and Java in 1883, or the 1960 Chile event where one of the strongest earthquakes since instrumental records began caused a destructive tsunami event (Eaton et al., 1961; Pfafker and Savage, 1970). A comparable seismic energy was released by the Andaman–Sumatra earthquake of December 26, 2004, the first megatsunami to be documented continuously with photographs, videos, and immediate postevent field research. The magnitude and the unpreparedness of the affected countries, with approximately 230,000 fatalities along thousands of kilometers of coastline in the northern Indian Ocean, triggered tsunami and paleotsunami research worldwide. This tsunami disaster forced political and scientific organizations worldwide to deal with tsunami as natural hazards and associated socioeconomic risks much more intensively than in the past.

Ten years have passed since that event, and hundreds of tsunami studies have been published, older tsunami events have been reinvestigated, new physical models for tsunami wave propagation have been developed, and warning systems have been discussed and established. Still, the Andaman–Sumatra earthquake event will have an impact on tsunami and paleotsunami research for many years to come (compare, e.g., Bishop et al., 2005; Lay et al., 2005; Lavigne et al., 2006; Tsuji et al., 2006; Richmond et al., 2006; Bahlburg and Weiss, 2007; Kelletat et al., 2007; Choowong et al., 2007; Goto et al., 2007; Sawai et al., 2009). The Tohoku Tsunami of March 2011 along the northeast coast of Japan had a similar earthquake energy released (9.3 on the Richter scale) to that in Andaman–Sumatra, but there were significantly fewer fatalities.

Seismic activity along plate boundaries (mostly subduction zones) accounts for >85 percent of the generation of tsunami events, including oceanwide megatsunamis. In several cases, it is difficult to discriminate a seismic origin from a volcanic induced tsunami event as active volcanoes are concentrated along the "Ring of Fire" and collision structures that are parallel to subduction zones. Volcanoes in the sea or near the coast may collapse as calderas (e.g., Santorini and Krakatoa), or breakdown because of steep slopes and a very rapid construction of the volcanic edifice. The latter has been reported from Mt Fogo on the Cape Verde Islands off West Africa (Figure 3.1) with a failure volume of >100 km^3 that occurred over 10,000 years ago, or collapses of Mt Etna on Sicily, Italy (Pareschi et al., 2006).

Advances in mapping sea floor topography have provided ample evidence for slides, slumps, or mass failures of (steep) slopes mobilized either by seismic shocks, oversteepening of the slope profile, or occasionally by gas hydrate

FIGURE 3.1 Scar of the collapsed Fogo volcano (8.5× 11.3 km wide) on the Cape Verde Archipelago off western Africa. The slide reaches nearly 15 km out into the sea. A new volcanic cone has grown within the slide area. *Image credits: ©Google earth 2014.*

eruptions. The largest of these submarine slides have been found (and partly dated) around the volcanic group of the Canary Islands, Spain (Carracedo et al., 1999; Masson et al., 2008) and the Hawaiian Islands in the central Pacific Ocean (e.g., Moore et al., 1989). They have been active during the Last Interglacial period of high sea level (~125,000 years ago) back to the early Quaternary, >1.2 Million years ago. One prominent Holocene slide that left deposits along many coastal sites around the northeastern Atlantic Ocean is the Storegga Slide event (in fact, probably three different slides) with the oldest around 7,200/8,200 BP and the youngest slide occurring about 1,500 years BP (Bondevik et al., 1997 a, b; Haflidason et al., 2004; Bryn et al., 2005; see also Figures 3.2 and 3.3). Examples from smaller submarine slides and tsunamis are described in Charalampakis et al. (2007) and Tinti et al. (2007).

Studies by the IFM-Geomar at Kiel University, Germany (cf. Kopp and Weinrebe, 2009) showed that seamounts on the subducting sea floor area may affect the overriding plate as instruments triggering faults and seismic activity that may initiate submarine slides with possible tsunami generation (Figure 3.4 as well as Brune et al., 2010).

Based on the knowledge of past submarine slides around volcanic islands, the question arose whether similar disastrous events may occur in the future. Computer simulations have been conducted to model a collapse of the western slope of La Palma island in the Canaries, Spain (Figure 3.5(a)–(c)), in particular the Cumbre Vieja region in the south, which developed an N–S-trending open fault in 1949. This site has the potential for many hundreds of cubic kilometers of land being set in motion, thereby triggering a megatsunami with high velocity over the deep water of the open Atlantic Ocean. The resulting tsunami would affect the east coast of the United States with wave heights surpassing 30 m and

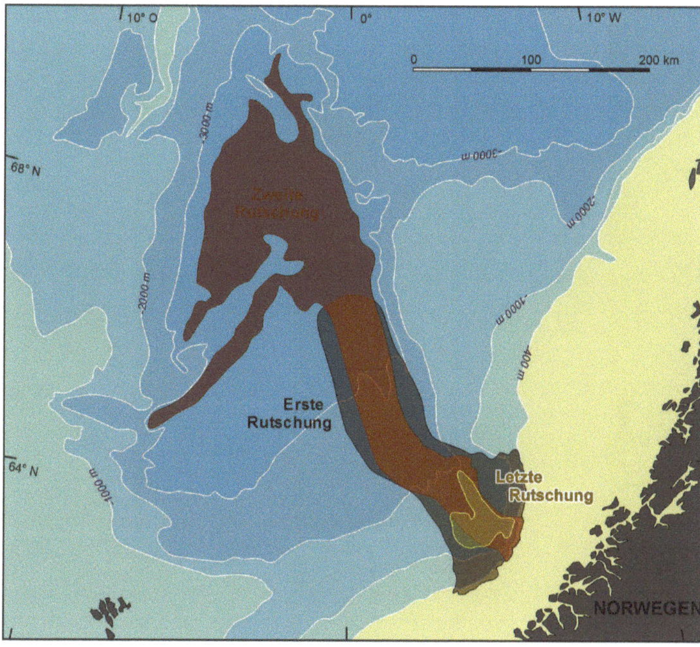

FIGURE 3.2 Three slides in the Storegga area west of southern Norway. They ran out for >300 km on a very flat sea floor. The two smaller slide sections have a volume of about 1,700 km^3, the largest of 3,880 km^3 (compare also Bondevik et al., 1997; Bondevik et al., 2003). The youngest slide is the smallest event, while the largest submarine slide was the second event.

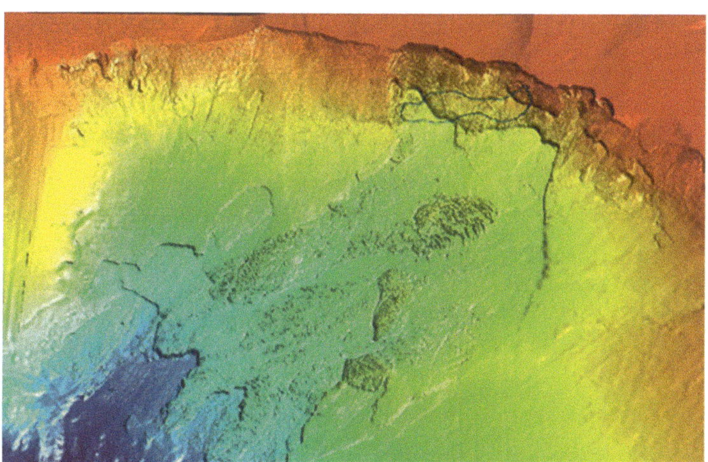

FIGURE 3.3 Upper section of the Storegga slide off western Norway. *Credit: British Geological Society http://www.bgs.ac.uk/research/marine/marineGeohazards.html.*

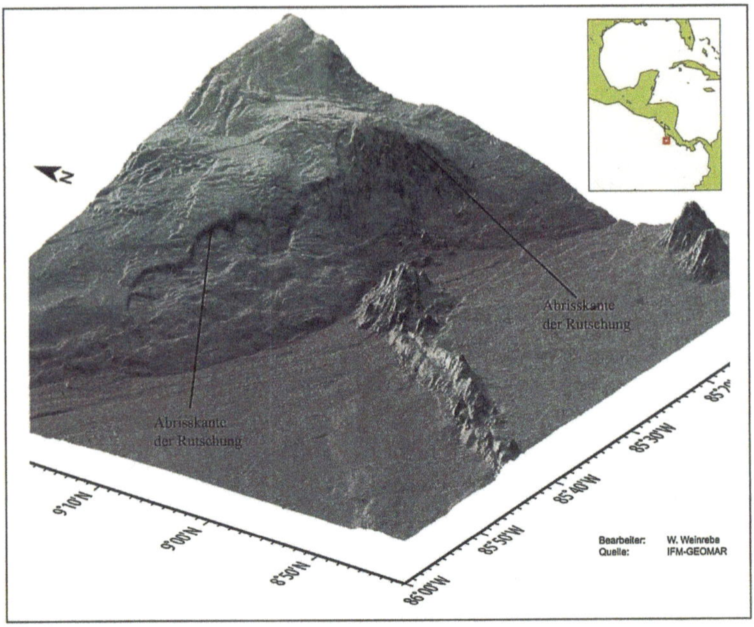

FIGURE 3.4 Submarine slides triggered by subduction of seamounts along the west coast of Costa Rica. *Credit: Kopp and Weinrebe (2009); based on data from IFM-Geomar, Kiel, Germany.*

also regions from the British Isles in the north to South America in the south (Ward and Day, 2001). Submarine slides in general are a constant risk to generate far-field strong tsunamis even outside tectonically active regions.

3.2 CURRENT RESEARCH SUBJECTS UNDER DEBATE

A review of the tsunami and paleotsunami literature reveals several topics that have attracted interest of researchers from different scientific disciplines with the following topics discussed most contentiously:

1. The possibility of extraterrestrial impacts and the forming of chevrons.
2. Identifying overwash and lagoon sediments of tsunamigenic origin and the control of older cores (commonly from geoarcheological studies) in respect to tsunami signatures.
3. The boulder movement problem (storms against tsunamis) including boulder density, mode of transport, and character of deposition.

3.2.1 Origin of Tsunamis from Extraterrestrial Impacts and the Chevron Debate

From the Near Earth Objects Project of the National Aeronautics and Space Administration (NASA) and the sophisticated devices for space observation,

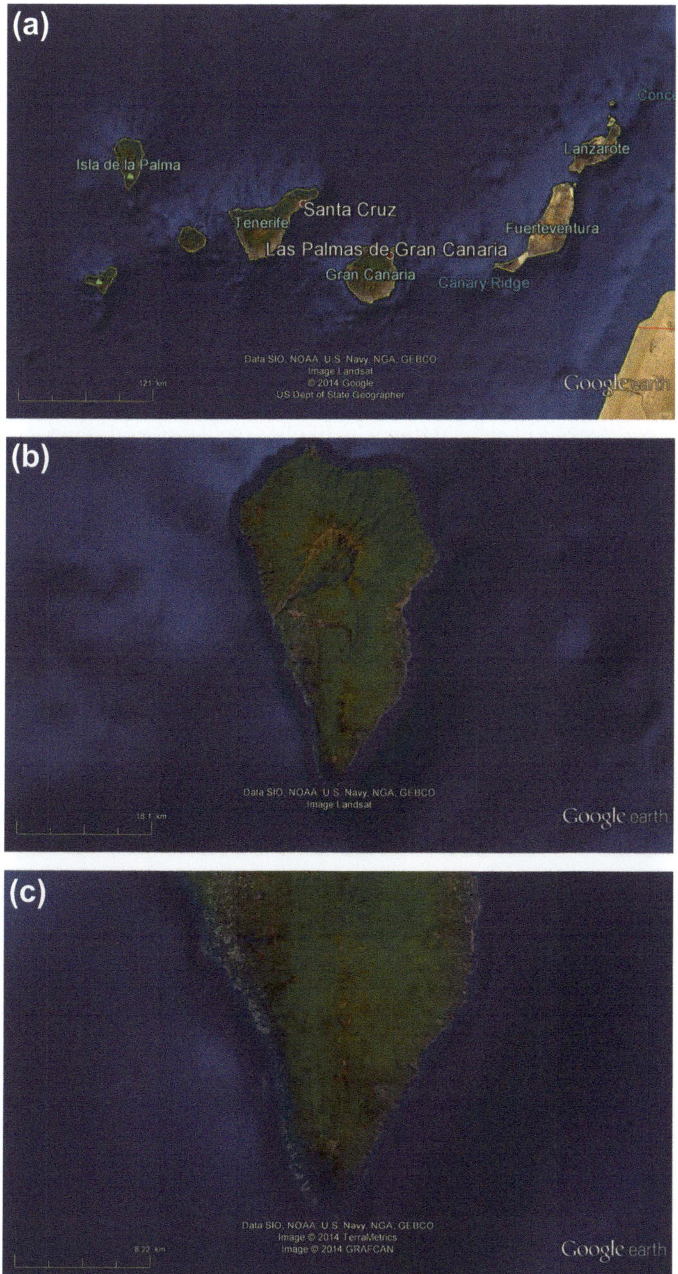

FIGURE 3.5 (a) The Canaries with La Palma Island in the northwest. Scene is 1,200 km wide *(Image credits: ©Google earth 2014.)*. (b) La Palma Island with the 15-km-long scar of the slide from Cumbre Nueva to the west *(Image credits: ©Google earth 2014.)*. (c) The 22-km-long southern section of La Palma island (Cumbre Vieja), up to 1880 m asl and 10 km wide, with a volume of >300 km^3. A possible failure of the western section along the crater line may trigger an oceanwide tsunami that could cross the Atlantic to the Caribbean and the east coast of the United States. *Image credits: ©Google earth 2014.*

we know that thousands of meteoroids and asteroids are moving in the asteroid belt between Mars and Jupiter, or may approach as comets from the Oort's Cloud or the Kuiper Belt. We are aware of nearly 200 impact craters on planet Earth, some of them billions of years old, though only about one dozen are land based from Holocene times. Therefore, it is highly likely that more than twice this number have collided into the oceans, and if they were large enough ($\geq$100 m in diameter), they would cause significant tsunamis affecting near and far coastlines. This potential has been calculated and modeled in tsunami research (Kristan-Tollmann and Tollmann, 1994; Hills and Mader, 1997; Powars, 2000; Abbott et al., 2006; Tester et al., 2007; Bryant et al., 2007; Bunch et al., 2008; Bryant, 2008; Goto, 2008; Bryant et al., 2010; Goff et al., 2010). Melted quartz and other shocked minerals, iridium peaks in fine sediments, or foraminifera attached to molten minerals, have been observed in coastal deposits likely associated with these (possible) events, and as a landform, chevron deposits have been described and cautiously attributed to cosmogenic tsunami events—although this hypothesis is controversial (Scheffers et al., 2008; Bourgeois and Weiss, 2009: but compare Figures 3.6–3.10).

3.2.2 Fine Washover Sediments Stored in Geobioarchives

The identification of tsunami deposits has long been a subject of scientific debate. A still unresolved problem is the type of fine tsunami sediments that

FIGURE 3.6 One to five km long sandy "chevrons" south of Shark Bay, Western Australia, starting at a near vertical cliff top 85–143 m asl. Tip of the small northern form is at 26° 55′ 18.89″ S and 113° 46′ 58.39″ E. At the base of the cliff and because of its plunging character, no sand deposit is available *(Credit: Google earth.)*. Dr Phillip Playford, geologist in Western Australia, just found megablocks up to 700 t nearby on Dirk Hartog island at 15 m above sea level and 250 m inland *(Source: Michelle Wheeler in "The West Australian", June 29, 2013.)*, and more sites with large boulders high on cliff tops can be found in that area.

FIGURE 3.7 Nearly 3 km long, young Pleistocene chevron (sand and coral debris, partly cemented) of the Exumas, Bahamas. *Image credits: ©Google earth 2014.*

FIGURE 3.8 Holocene chevron (coral rubble), a breach in the atoll rim of Rangiroa, Tuamotus, French Polynesia. *Image credits: ©Google earth 2014.*

can be accurately distinguished from fine sediments left by storm waves. In the early phase of paleotsunami research, the catalogs of signatures for tsunami deposits have been rather extensive, describing a large number of specifications for typical tsunami sedimentation of fine material. But, as a consequence of the rising number of contemporaneous inspections of the impacts of recent tsunamis, in particular that of the 2004 event, it became clear that tsunamis may leave sedimentation characteristics very similar or identical to those of storms during overwash (Figure 3.11). This may be the

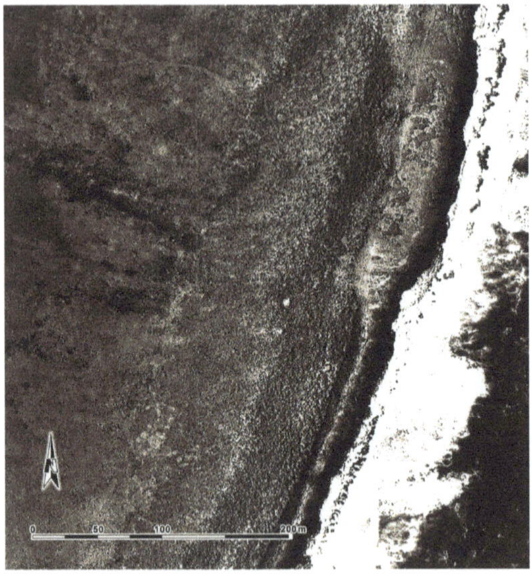

FIGURE 3.9 Oblique aerial photograph of two chevrons composed of coarse coral rubble on the east coast of Bonaire. The features reach about 300-m inland at +5 m asl and dated to approximately 5,000 years BP. *Credit: A. Scheffers.*

FIGURE 3.10 Washover deposits at the westernmost part of the Rhone delta (scene is 700 m wide). *Image credits: ©Google earth 2014.*

reason why during decades of corings in coastal sediments, pre-1990 AD tsunami layers have not been identified, simply because they did not differ substantially from other wave deposits. During 1994–2014, only three sedimentological features remained as diagnostic, and these are mentioned in most modern paleotsunami papers (besides the deposition of a generally

FIGURE 3.11 Washover deposit of shell and coral debris on a coastal dune with a silty "mud-cap," from tropical cyclone Laurence in 2009 (Wallal Downs, NW coast of Australia). *Photos: A. Scheffers.*

coarser grain size unit): (1) a sharp basal discordance/unconformity; (2) a mudcap; and (3) fining up sequences or a graded bedding that might not be finely developed (e.g., Engel et al., 2010). However, we found two of these "distinct" pieces of evidence of tsunami within storm washover deposits in Western Australia (Figure 3.11). Japanese colleagues, who have in the southern part of their country a high number of recent and historically confirmed tsunamis and cyclone landfalls, came to the conclusion that there is not a single signature that can be exclusively connected to a tsunami on one side, or a strong storm on the other side (see several contributions in Shiki et al. (2008)). Today, in combination with coarse onshore deposits and a crosscheck of numerical dating, paleotsunami deposits have quite often been identified (and sometimes connected to known historical events) in cores from lagoonal geobioarchives (Andrade, 1992; Nichol et al., 2003; Kelletat and Scheffers, 2004; Switzer et al., 2005; Vött et al., 2006; Hori et al., 2007; Switzer and Jones, 2008; Goff et al., 2009; Sawai et al., 2009; Vött et al., 2009; Fujino et al., 2010; Engel et al., 2010; Bahlburg and Spiske, 2011; Brill et al., 2011; Vött et al., 2011a,b; Scheffers et al., 2013, and many others).

As strong tsunamis tend to inundate wide flat landscapes with a rather strong flow, 2–6 m/s or even faster as detected from recent video tapes (e.g., in Thailand, 2004 or Japan, 2011), it is astonishing that ripple marks, as a typical surface pattern of flow, have been reported as developing almost exclusively from surfaces on Bonaire island (Figures 3.12 and 3.13). These ripple marks exhibit individual forms many meters wide and long and several decimeters high in multimodal sediments (sand to medium size

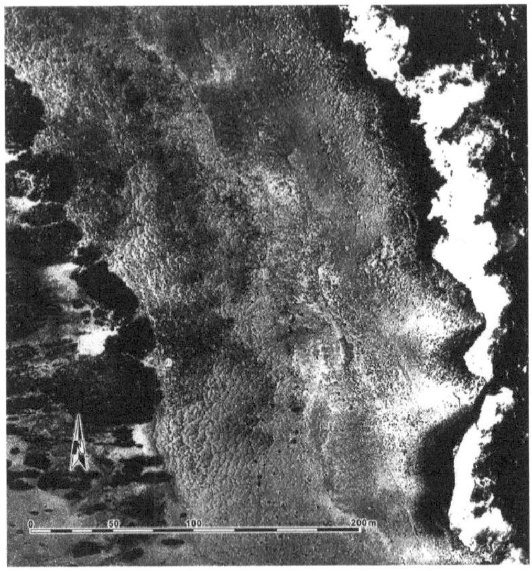

FIGURE 3.12 Ripple marks in onflow tsunami deposits, east coast of Bonaire, southern section of Washikemba. Track crossing is >200 m from the coastline (at right). The N–S extension of the ripple mark field is >1 km. Farther north in the Seru Grandi section, radiocarbon dating of coral fragments in a similar ripple field give possible dates for these features between about 1600 and 3200 BP (Scheffers, 2005).

FIGURE 3.13 Backwash in tsunami ripple marks, east coast of Bonaire, southern Washikemba section. The tsunami reached at least 500 m inland from the cliff. The backwash lobe is >350 m "long" to seaward and about 500 m wide at around +7 m asl. *From Scheffers (2005).*

coral debris), reaching a few hundred meters inland and nearly +10 m asl on the surfaces of a cliff and terrace. Even backwash lobes are clearly developed (Scheffers, 2005).

3.2.3 Boulder Movement: Observations, Calculations, Modeling and the Contrast to Storm-Wave Impacts

The discussion and debate on the origins and implications of coarse coastal sedimentary deposits, in particular large boulders, are also unresolved. The logical conclusion is that with the large flow depth of a strong tsunami, its high velocity passing the shoreline, its substantial inland inundation, and its long duration compared to storm-wave impacts, much larger boulders should be transported farther inland than by the strongest storm waves. However, this is not generally accepted (Weiss, 2012). For example, this can be seen in the paper of Bourrouilh-Le-Jan and Talandier (1985) on the Tuamotu boulders that have maximum weights of around 2000 metric tons. Also, there is the Tongan boulder that weighs, at a minimum, 1600 t (Frohlich et al., 2009), and that was found 9.4 m asl and at a distance of 130 m landward from the cliff top. The authors keep the conclusion open on the transport modus, tsunami, or cyclone. The argument against tsunamis is generally that no tsunamis have taken place in the respective area, but this is a weak argument that should merely be interpreted as no tsunami signatures have been found so far! In fact, during the megatsunamis of the last few decades, which are well investigated by task forces, the transport of coarse debris has rarely been described. Recent coastal scientific research has begun to address this void (compare, e.g. Nott, 2000, 2003a,b; Bryant and Nott, 2001; Bryant, 2008; Bryant et al., 1997; Whelan and Kelletat, 2002, 2003; Mhammdi et al., 2010; de Martini et al., 2010; Noormets et al., 2002, 2004; Scheffers, 2002, 2004, 2005, 2008; Scheffers and Kelletat, 2003a,b, 2005, 2006a,b; Scheffers and Scheffers, 2007; Scheffers et al., 2005, 2008, 2009a,b, 2010a,b; Scicchitano et al., 2007; Williams and Hall, 2004; Hall et al., 2006, 2010; Morton et al., 2008; Hansom and Hall, 2009; Etienne and Paris, 2010).

Fortunately, parallel to the descriptions resulting from field observations, models are being developed and tested on the transport energy of tsunami flow and extreme storm waves. The results of these models vary significantly. At least as important as theoretical tests and models are the real-time observations of the wave power associated with storm systems. Therefore, some of the recently published results concerning storm-wave transport are reviewed. Of particular interest should be observations and investigations at coastal sites that exhibit modern (or historical) strong tsunamis as well as strong cyclone landfalls. This is the case, for example, along the coastlines of southern Japan, where the best evidence for the comparison of the energy of both processes can be expected to be found (e.g., Goto and Imamura, 2007; Imamura et al., 2008; Goto et al., 2007, 2009, 2010).

First, we will refer to examples where strong storm waves have been observed and the transport of boulders onsite has been checked in close temporal proximity:

- Lorang (2000) observed a maximum boulder of 1.1 t moved by waves of 8 m at the coastline.
- Bartel and Kelletat (2003) visited the coastlines of Mallorca Island, Spain, after the strongest hurricane ever observed in the western Mediterranean in December 2001. They found the maximum fresh boulder transport caused by waves of 9 m immediately along the coast to be 2.5 t at +3.5 m asl with a transport distance (sliding) of 27 m (Figure 3.14). Another example of a

FIGURE 3.14 A 27-m long, slide track of a 2.5-t boulder at 3.5 m asl during a "Medicane" (hurricane cat. 2) in December 2001 on Mallorca island, Mediterranean Spain. *Photo: D. Kelletat.*

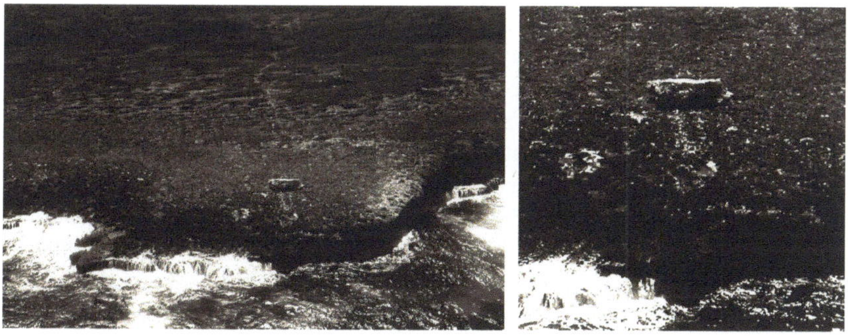

FIGURE 3.15 Sliding of a large old tsunami boulder for several meters in cliff top position (+4.5 m asl) on the east coast of Curacao by a storm before AD 2001. *Oblique aerial photograph 2001 by A. Scheffers.*

boulder slide on a rough surface has been documented from the east coast of Curacao (Figure 3.15; Scheffers, 2002).

- Saintilan and Rogers (2005) found that an 11.8-m deep water wave at the coast of NSW (Australia) had transported boulders weighing 9.7 t and 6 t a distance of 40 m onshore at an elevation of +2 m asl.
- Scheffers and Scheffers (2006) observed the impact of hurricane Ivan in 2004 along the cliff coasts of Bonaire (Netherlands Antilles, Caribbean), measuring hundreds of strong waves at with heights of at least 12 m, with an onshore bore flow. The largest fresh boulder broken from the cliff top weighed 6 t, and was found at +5 m asl after being transported inland for 30 m. Much larger boulders further inland were moved by the hurricane bore for a few meters, but were never transported uphill.
- Etienne and Paris (2010) investigated the boulder movement at the southern spit of the rocky Reykjanes peninsula south of Reykjavik, in a severely energetic storm-wave environment. They measured 47 transported boulders, and 10 had weights >4 t. They found the volumes of the two largest moved boulders to be 22 and 26 m^3, with maximum altitudes of deposition around 11–14 m and 17–18 m asl. The most extreme examples are as follows:
  - A boulder of 3 m^3 and 7.2 t at +6.5 m transported across 105 m.
  - A boulder of 26 m^3 and 70.3 t at +6 m that had been moved 65 m.
  - A boulder of 6 m^3 and 16.2 t at +8 m that moved 105 m.
- Suanez et al. (2009) studied the boulder dislocation by a “one in 30 years” storm (March 10, 2008) on Banneg Island of Brittany, France. Flow was detected at elevations up to +19 m, which is 5–10 m higher than the cliff tops. Boulders were dislocated at elevations that were +7.5 to +14.5 m, at a distance 14 m from the +9 m cliff top. The largest boulder in this position weighed 32 t, although the median boulder measurements were only

$0.19\ m^3$ and 0.6 t. The bore flow transported boulders of 0.3 to 1.4 t distances up to 50–90 m inland.

- Goto et al. (2009) observed boulder movements by strong typhoons on islands of southern Japan. As a consequence of a storm with open water wave heights of 11.9 m that occurred in association with spring tide, they found the average size of translocated boulders to be 3.7 t, but the largest was 127 t, occurring 29 m from the reef edge without a significant vertical dislocation. Boulders <47 t have been found up to 200 m from the reef edge. The boulders moved by typhoons are definitely much smaller, closer to the reef edge and in lower position than those from historical tsunamis. The measured change of bore flow velocity across the reef was from an initial 6.5 to 3.1 ms^{-1} across a distance of 260 m.
- Khan et al. (2010) observed and measured boulder movements from four recent hurricanes of category 4–5, which directly passed Jamaica. The maximum values for freshly emplaced boulders were found to be weights of 1.77–26.4 t at 0- to +2-m elevation, which had been moved distances up to 25 m. The maximum values were 4.48 t and found at a +0- to 25-m distance, and 26.4 t at a +0- to 6-m distance.

Existing boulders have been moved as larger fragments and for larger distances, for example, 0.08 to 79.74 t, +0 to +14 m up to 76 m distant from the shoreline, with maximum values observed:

79.74 t at +12- and 55-m distance moved 2 m.
50.4 t at +0- and 30-m distance moved 8.5 m.
5.39 t at +8- and 22-m distance was overturned at its place.
0.08 t at +14- and 76-m distance was overturned at its place.

The discussion on boulder movement onshore is difficult for several reasons:

- The first is the debate on boulder volume. Several authors just multiply the lengths of the three axes *a*, *b*, and *c*, which may, in any instance, yield volumes that are up to 200 percent greater than those obtained by multiplying mean axis length (e.g., Scheffers, 2002), or by determining the boulder volume by laser scanning or other modern methods (Engel and May, 2012). A test by Kelletat (2012, unpublished) of 16 boulders (mostly cuboid to platy forms) from Mallorca found that estimating boulder volume by simply multiplying the dimensions of *a*–*b*–*c* axes will give volumes that were between 80 percent and 215 percent too large, with a mean deviation of +154 percent!
- The second is the debate on bulk density, which gives (when multiplied by the volume) the mass that may be moved. Comparing measurements of bulk densities from reef rock and coral species therein (all from MIS

5e-terrace, i.e., Younger Pleistocene from the east coast of Bonaire), the following deviations and accordances have been found:

Spiske et al., 2008, Figure 3.7	Engel and May 2012, Table 3.1
Calcarenite: 2.7–2.75 g/cm^3	(Rudstone) matrix: 2.305 (1.95–2.59 g/cm^3)
Acropora palmata: 2.25–2.32 g/cm^3	Acropora palmata: 2.24 (2.05–2.37 g/cm^3)
	Acropora cervicornis: 2.31 g/cm^3
Montastrea sp.: 2.21 (1.90–2.38 g/cm)	Montastrea sp.: 1.74 (1.23–2.37 g/cm^3)
Diploria sp.: 1.70 (1.35–1.95 g/cm^3)	Diploria sp.: 1.92 (1.62–2.22 g/cm^3)

Kelletat (unpublished, 2012) measured calcarenite (mostly Plio-Pleistocene eolianite) density, again by the Archimedean principle, from tsunami boulders (eolianite and limestone) on Mallorca Island. Based on an analysis of 16 samples, the boulders were found to have bulk densities that ranged from 2.363 to 2.412 g/cm^3 with a mean value of 2.388 g/cm^3. This is close to the Bonaire values reported by Engel and May (2012), but about 15 percent less than those measured by Spiske et al. (2008), also from Bonaire.

Frohlich et al. (2009) describe very large boulders in the west of Tongatapu near the village of Fahefa, at an elevation of +10 m asl and 130 m inland. Others can be found up to +20 m asl and 400 m inland. The long axis of the largest boulder measured 15 m with a height of 9 m and volume around 1200 m^3. The material is broken from reef rock that builds the base along an active cliff. As the age of the reef rock is from the last interglacial, the dislocation has clearly taken place during younger Holocene times. This is evident by the stripping of soil cover due to flow erosion around the boulder, as well as (missing) karstification differences between the wider environment of the rock platform and the boulder-resting place. More boulders of a similar size have been found on the Tuamotus, French Polynesia (cf. Bourrouilh-Le Jan and Talandier, 1985; Talandier and Bourrouilh-Le Jan, 1987). In many coastal environments, the sheer size of boulders, their isolation from cliff top sources or barriers and location many meters above sea level, their apparent age, and the lack of other, smaller boulders from frequent tropical cyclones in their vicinity would point to their origin from rare and extreme powerful events, that is, tsunamis (Figures 3.16–3.18).

As the comparison of field facts and models, as well as laboratory wave tests have documented, a lot of unexplained features and processes exist in coastal boulder dislocation, and most questions remain open for continued debate. The main reason is that although storm wave models produce results that match observations rather well, the dimensions of runup, inundation, and flow depth and velocity of tsunamis are not easily translated into

FIGURE 3.16 Boulders on Curacao N (Watamula site): notches at boulders, uplifted for at least 7 m against gravity, weights >50 t (Scheffers, 2002).

FIGURE 3.17 A 170-t overturned boulder on a 13-m cliff top of northeast Barbados (Scheffers and Kelletat, 2006).

laboratory or model studies. The same is true for the local factors of friction, coastal gradients, nearshore bathymetry, etc. Nonetheless, there are increasing numbers of direct observations of extreme events and their capability of boulder transport according to measurements of weight,

FIGURE 3.18 Giant boulders (>100 t) at Quobba (north of Carnarvon, Western Australia) at about 11 m asl that have been broken into several pieces by crashing down from huge waves (compare also Goto et al., 2010a). *Photo credit: A. Scheffers.*

volume, distance, and altitude relative to sea level. The observations have shown that boulders significantly distant from the shoreline (several decameters and more), in altitudes of many meters or decameters (transported this amount against gravitation), and of volumes of >10–20 m^3 or 20–30 t, may occur as extreme singularities associated with the strongest likely cyclone waves at the most suitable locations, but generally belong to the category of tsunami dislocation. From this, we have a reasonable threshold for field research on paleostorms and paleotsunamis (compare also Benner et al., 2010; Goto et al. (2009), 2010).

Studies to discriminate between storm-dislocated boulders and those moved onshore by tsunami have been conducted in regions where both processes occur rather often, for example, in southeastern Japan (cf. Goto and Imamura, 2007; Imamura et al., 2008; Goto et al., 2010a,b,c). Other papers also discuss the problem of boulder transport by storms or tsunamis, but their results and arguments differ in many ways (Goff et al., 2010a; Kortekaas and Dawson, 2007; Nandasena and Tanaka, 2013; Regnauld et al., 2010, and others).

Nott (2003) distinguished the difference between three important pre-transport settings of boulders; submerged, subaerial, or joint bounded. He came to the conclusion that submerged boulders need the least wave energy for transportation relative to the other settings, and several authors followed this argument. However, a submerged boulder needs the longest/highest lift against gravity to reach a new position on a cliff top, and it may be not submerged just in front of a large wave with an extreme drawdown of water in its preceding trough. Also not considered is the mode of vertical transport against gravitation, the altitude of deposition, the obstacle of a cliff in the wave path, or the roughness of the surface. The latter has been discussed by Weiss (2012), who described the importance of surface friction as a factor during boulder transport. This may be true for sliding and pushing processes (cf. Figures 3.14 and 3.15), but as Imamura et al. (2008) stated, it is insignificant if the process of movement is as saltation, or rolling with saltation, or a transportation via a single "jump" over a long distance (Figures 3.17–3.21).

Weiss' (2012) reflecting on "The mystery of boulders moved by tsunamis and storms" discussed the processes of movement of a spherical boulder (although most published reports of boulders that have been moved by tsunamis, or expected to have been so translocated, indicate that the boulders are plates or cuboids). He stated that sliding is the main process of transportation and therefore friction is the main physical parameter of concern. He concluded that a vertical bed roughness >30 percent of the boulder radius (or ~15 percent of the boulder height) will prevent boulder transportation. This means that a spherical boulder of 2 m in diameter, which would have a volume of about 4.2 m^3 and a weight close to 10 t, will not be moved on a bed with a roughness of >0.3 m. This would exclude movement of this

FIGURE 3.19 This boulder (>1.5 m long) from the immediate supratidal with fine sculpturing of rock pools at the east coast of Bonaire has been dislocated without any destruction of the finest bioerosive sculptures and smashed down (and thereby broken) about 400 m from its original place (Scheffers, 2002).

FIGURE 3.20 Imbrication of large, platy boulders in a cliff top position south of Rabat, Morocco. Every second boulder has been turned over (showing rock pools on their lower sides) and weights often exceed 30–50 t (Mhammdi et al., 2008).

boulder size and form on all limestone coastlines with bioerosive rock pools, as those typically have depths >0.5 m. Weiss also concluded that storm waves have a greater total energy and higher frequency compared to those of tsunami waves. Therefore, they would leave boulders in organized patterns of clusters and lines, whereas tsunamis should produce unorganized boulder deposits. If this is the case, it is difficult to explain the sources of the forces that have overturned the giant (tsunami) boulders from the Lisbon event AD 1755 in northwest Morocco (Mhammdi et al., 2008; Figures 3.20 and 3.21), the imbrication and imbrication trains often occurring in tsunami boulder assemblages, or the steep inclination of single boulders against obstacles. Also difficult to explain are deposits of large (>30 t) boulders balancing on top of boulder ridges/piles that are several meters high, in excess of a 100-m distance from the sea and >15–25 m asl (cf. Figures 3.21 and 3.22). Weiss (2012) pointed out that storm and tsunami waves of the same height may move boulders of the same size and that it is even possible that storm waves with an amplitude lower than that of tsunami waves may move a boulder of the same size. So far as we know from the literature and based on our experience observing very strong storm wave impacts (with wave heights of at least 12 m directly at the shoreline), these never moved boulders similar to those transported by tsunamis, but they definitely moved much smaller ones. We agree with Weiss (2012) that a critical need exists for more field observations and measurements, but we disagree that quantification of bed roughness and documentation of boulder arrangements are the most important measurements needed to distinguish storm-moved from

FIGURE 3.21 Overturned boulders with well-preserved rock pools facing downwards, located south of Rabat, Morocco (Mhammdi et al., 2008). Maximum weights of these cliff top boulders are of the order of 100 t.

FIGURE 3.22 On the Aran Islands (Galway Province, western Ireland), large boulders, 4 m long and with masses >18 t, can be found balancing on top of boulder piles well removed from the modern shoreline and at >15 m asl (Scheffers et al., 2009).

tsunami-moved boulders. Bed roughness is of little to no importance if the movement is not predominantly by rolling or sliding across a surface. Further, overturning of boulders >8 m long around their *a*-axis (as in many imbrication trains, see Figures 3.20 and 3.21) demonstrates the strength of lift forces in tsunami flow, as does the settling of very large boulders high on top of boulder clusters and ridges of different forms (Figure 3.22).

Benner et al. (2010) tried to estimate thresholds for storm wave movements by calculations based upon the physics of the processes involved, to be used to accompany field observations for a first step of discrimination. The results show that a cubic boulder weighing >80 t or a flat one >30 t, cannot be moved by storm waves because the moment of impact is too short. On a horizontal surface a boulder of 37 t ceases movement after 1 s at a distance of 1 m when transport is caused by a wave with a horizontal velocity of 10 ms^{-1}. If the wave moves at 12 ms^{-1}, the boulder will stop after 2 s at a distance of 4 m, and at a 16-ms^{-1} wave velocity (not possible even in the strongest cyclones!), after 3 s, a 37-t boulder will move a distance of 10 m (not considering the usually very rough surfaces along limestone coasts, such as those with older coral reef terraces and bioerosional forms (as in Figure 3.15)). In theory, the maximum value caused by a 16-ms^{-1} flow would dislocate a 247-t boulder horizontally for 1.3 m, or 0.13 m vertically. Storm waves >10 m in height directly at the coast on a 10- slope may move a 34-t boulder for a maximum distance of 10 m and altitude of 2 m.

Generally, in an environment with low friction, a small slope and without a cliff, storm waves may move boulders of around 20 m^3 and >30 t just a few meters inland. All larger boulders, or those moved across a longer distance and at greater altitude, need a strong and constant flow to establish a wider movement. These conditions are only given by tsunami inundation with a flow depth exceeding the height of individual boulders. Nevertheless, more tests in wave channels and in particular more direct observations before, during, and after strong wave events (storms and tsunamis) are needed to allow general agreement on the many open questions of the movement of large coastal boulders and to learn what the most important parameters are during each of the very different movement processes.

3.3 FUTURE RESEARCH DIRECTIONS

Based on the foundation from different aspects in paleotsunami research, we can define some subjects that promise interesting results in the near future. Among these are

1. Filling the gaps of local and regional knowledge on tsunami impacts globally.
2. Inspecting and reinterpreting samples from geoarchives, in particular regarding fine sediment characteristics.
3. More precise definition on tsunami criteria in fine sediment stratigraphies.
4. Detecting tsunami histories using archaeological techniques for Mesolithic to Bronze Age times and uncovering the reasons of the shift of settlements into inland. Was it extreme events or just general sea-level rise?
5. Modeling boulder movement thresholds. How can we better replicate natural processes in our laboratories and in numerical simulations?
6. Extending tsunami histories into time spans of past higher sea levels such as those during Interglacial periods (Figures 3.23–3.26).

FIGURE 3.23 Two strata of graded bedding (both ~1 m thick) in a sandy matrix from the southwest coast of Fuerteventura, Canary Islands, Spain. *Photo: D. Kelletat.*

FIGURE 3.24 Chaotic deposit with partial imbrication (moved from the left to the right) and some shell material, from the southern part of St Lucia, southern Caribbean island arc. The largest boulder has a diameter of about 1.6 m, the site is +20 m asl. *Photo: A. Scheffers.*

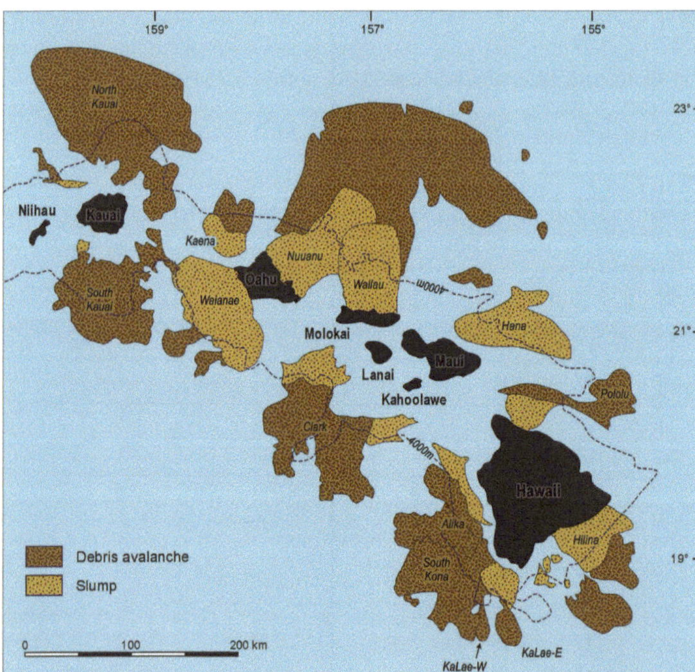

FIGURE 3.25 Giant slides around the volcanoes of the Hawaiian island group. *Modified from Moore et al. (1989); see Whelan and Kelletat (2003).*

FIGURE 3.26 Well-rounded boulder (>22 t) on beach cobbles near Los Vilos, central Chile, on a 25 -m asl terrace from the Last Interglacial. *Photo: G. Seehof.*

With the increasing knowledge and understanding concerning submarine landslides (e.g., Figure 3.25 and summaries in Whelan and Kelletat, 2003), we could ask the question, "Where are the documents of the related megatsunamis (as these slides commonly occurred during the last interglacial with former higher sea levels and dislocated hundreds to thousands of cubic kilometers of sediments)?" The El Golfo Slide off Hierro, Canary islands, that involved 400 km^3 of material, or the Cumbre Nueva Slide off La Palma that moved 200 km^3 of sediment (Carracedo et al., 1999; Masson et al., 2008), or the Alika two slide off Maui (Hawaii) that involved 200 km^3 of material, have been dated to approximately 125,000 years BP (coinciding with the last interglacial sea-level high stand MIS 5e). The Alika 1 and 3 slides off Maui that involved about 2,000 km^3 of mass are dated to about 247,000 years BP (the same timing as the second to last interglacial high stand). Larger slides, such as Hilina off Hawaii of 10,000-km^3 magnitude, or Nuuanu off Oahu with 5000 km^3, have not yet been dated (Moore et al., 1989).

3.4 CONCLUSIONS/OUTLOOK

As we have tried to demonstrate, a lot of knowledge has been gained in 1989–2014 of paleotsunami research, especially as a consequence of field observations. The number of open questions, however, has not been reduced because each new observation and model enforces new interpretations. The lack of process observations is evident, as is the difficulty to find agreements concerning the most relevant physical processes in boulder movement by storm waves or tsunami flow. The same is true for agreements that indicators in fine sediment deposits are significant and indisputably tsunamigenic, as well as what size boulders (by form, mass, slope, altitude, pretransport setting, bed roughness, etc.) are indisputably moved only by tsunamis and never by the highest storm waves. Convincing results by transport tests on smaller scales within water channels are also on the agenda. More and better identification of tsunami-like deposits in older geologic units, back to the Mesozoic age, are published based on studies on the K/T-boundary phenomenon. However, the identification of Quaternary paleotsunami deposits older than the last half of the Holocene are extremely scarce, because of the long time spans of sea levels similar to modern day levels within many interglacial periods. These remarks should not be taken as being pessimistic concerning the present state or the near future for our research subject. They are simply typical for a young and actively growing branch of geoscience.

REFERENCES

Abbott, D., Martos, S., Elkinton, H., Bryant, E.F., Gusiakov, V., Breger, D., 2006. Impact craters as sources of mega-tsunami generated chevron dunes. Geol. Soc. Am. Abstr. Progr. 38 (7), 299.

Altinok, Y., Ersoy, S., 2000. Tsunamis observed on and near the Turkish coast. Nat. Hazards 21, 185–205.

Ambrasey, N., Synolakis, C., 2010. Tsunami catalogs for the eastern Mediterranean revisited. J. Earthquake Eng. 14, 309–330.

Andrade, C., 1992. Tsunami generated forms in the Algarve barrier islands (South Portugal). Sci. Tsunami Hazards 10 (1), 21–34.

Atwater, B.F., 1987. Evidence for great Holocene earthquakes along the outer coast of Washington State. Science 236 (4804), 942–944.

Bahlburg, H., Weiss, R., 2007. Sedimentology of the December 26, 2004, Sumatra tsunami deposits in eastern India (Tamil Nadu) and Kenya. Int. J. Earth Sci. 96 (6), 1195–1209.

Baptista, M.A., Miranda, J.M., 2009. Revision of the Portuguese catalog of tsunamis. Nat. Hazards Earth Syst. Sci. 9, 25–42.

Baptista, M.A., Miranda, P.M.A., Miranda, J.M., Mendes Victor, L., 1999. Constraints on the source of the 1755 Lisbon tsunami inferred from numerical modelling of historical data on the source of the 1755 Lisbon earthquake. J. Geodyn. 25 (2), 159–174.

Bartel, P., Kelletat, D., 2003. Erster Nachweis holozäner Tsunamis im westlichen Mittelmeergebiet (Mallorca, Spanien) mit einem Vergleich von Tsunami- und Sturmwellenwirkung auf Festgesteinsküsten. – Beiträge der 20. Jahrestagung des Arbeitskreises "Geographie der Meere und Küsten. Berichte Forschungs- und Technologiezentrum Westküste der Universität Kiel 28, 93–107. Büsum.

Benner, R., Browne, T., Brückner, H., Kel,letat, D., Scheffers, A., 2010. Boulder transport by waves: progress in physical modelling. Z. Geomorphol. 54, 127–146.

Bishop, P., Sanderson, D., Hansom, J., Chaimanee, N., 2005. Age-dating of tsunami deposits: lessons from the 26 December 2004 tsunami in Thailand. Geogr. J. 171 (4), 379–384.

Bondevik, S., Svendsen, J.-I., Mangerud, J., 1997a. Tsunami sedimentary facies deposited by the Storegga tsunami in shallow marine basins and coastal lakes, western Norway. Sedimentology 44, 1115–1131.

Bondevik, S., Svendsen, J.I., Johnsen, G., Mangerud, J., Kaland, P.E., 1997b. The Storegga tsunami along the Norwegian coast, its age and runup. Boreas 26, 29–53.

Bondevik, S., et al., August 5, 2003. Record-breaking height for 8000-year-old tsunami in the North Atlantic. Eos 84 (31).

Bourrouilh-Le Jan, F.G., Talandier, J., 1985. Major high-energy events in a reef environment: tsunamis, hurricanes and tropical cyclones and their effects on the sedimentology and geomorphology of an atoll: Rangiroa, Tuamotu (SE Pacific). Mar. Geol. 37, 263–272.

Bourrouilh-Le Jan, F., Talandier, J., 1985. Sedimentation et fracturation de haute energie en milieu recifal: tsunamis, ouragans et cyclones, leurs effets sur la sedimentation et la geometrie d'un atoll. Mar. Geol. 67.

Bourgeois, J., Weiss, R., 2009. "Chevrons" are not mega-tsunami deposits; a sedimentologic assessment. Geology 37, 403–406.

Brill, D., Brückner, H., Jankaew, K., Kelletat, D., Scheffers, A., Scheffers, S., 2011. Potential predecessors of the 2004 Indian Ocean Tsunami—sedimentary evidence of extreme wave events at Ban Bang Sak, SW Thai-land. Sediment. Geol. 239, 146–161.

Bruins, H.J., MacGillivray, J.A., Synolakis, C.E., Benjamini, C., Keller, J., Kisch, H.J., Klügel, A., van der Plicht, J., 2008. Geoarchaeological tsunami deposits at Palaikastro (Crete) and the Late Minoan IA eruption of Santorini. J. Archaeol. Sci. 35, 191–212.

Brune, S., Babeyko, A.Y., Ladage, S., Sobolev, S.V., 2010. Landslide tsunami hazard in the Indonesian Sunda Arc. Nat. Hazards Earth Syst. Sci. 10, 589–604.

Bryant, E.A., Young, R.W., Price, D.M., Wheeler, D.J., Pease, M.I., 1997. The impact of tsunami on the coastline of Jervis Bay, southeastern Australia. Phys. Geogr. 18, 441–459.

Bryant, E.A., Nott, J., 2001. Geological indicators of large tsunami in Australia. Nat. Hazards 24, 231–249.

Bryant, E.A., Walsh, G., Abbott, D., 2007. Cosmogenic mega-tsunami in the Australia region: are they supported by Aboriginal and Maori legends?. In: Piccardi, L., Masse, W.B. (Eds.), Myth and Geology, vol. 273. Geological Society, London, Special Publications, pp. 203–214.

Bryant, E., 2008. Tsunami: The Underrated Hazard, second ed. Springer-Praxis Publishing, Chichester.

Bryant, E., Haslett, S.K., Scheffers, S., Scheffers, A., Kelletat, D., 2010. Tsunami chronology supporting late Holocene impacts. J. Sib. Fed. Univ. 1, 63–71.

Bryn, P., Berg, K., Forsberg, C.F., Solheim, A., Kvalstad, T.J., 2005. Explaining the Storegga slide. Mar. Petrol. Geol. 22, 11–19.

Bunch, T.P., Kennett, J., Kennett, D.K., 2008. Impacts, mega-tsunami, and other extraordinary claims. Geol. Soc. Am. 18 (6), 11.

Carracedo, J.C., Day, S.J., Guillou, H., Perez Torrado, F.J., 1999. Giant Quaternary landslides in the evolution of La Palma and El Hierro, Canary Islands. J. Volcanol. Geotherm. Res. 94, 169–190.

Charalampakis, M., Stefatos, A., Mpourdopoulos, K., Ferentinos, G., 2007. Towards the mitigation of the tsunami risk by submarine mass failures in the Gulf of Corinth: the Xylocastro resort town case study. Adv. Nat. Technol. Hazards Res. 27, 367–376.

Choowong, M., Murakoshi, N., Hisada, K., Charusiri, P., Daorerk, V., Charoentitirat, T., Chutakositkanon, V., Jankaew, K., Kanjanapayont, P., 2007. Erosion and deposition by the 2004 Indian Ocean tsunami in Phuket and Phang-nga Provinces, Thailand. J. Coastal Res. 23 (5), 1270–1276.

Cita, M.B., Aloisi, G., 2000. Deep sea tsunami deposits triggered by the explosion of Santorini (3500 yrs BP), eastern Mediterranean. Sediment. Geol. 135 (1–4), 181–203.

Dawson, A.G., Long, D., Smith, D.E., 1988. The Storegga slides: evidence from eastern Scotland for a possible tsunami. Mar. Geol. 82, 271–276.

Dawson, A.G., Shi, S., 2000. Tsunami deposits. Pure Appl. Geophys. 157, 875–897.
Dawson, A.G., Stewart, I., 2007. Tsunami deposits in the geological record. Sediment. Geol. 200, 166–183.
De Martini, P.M., Barbano, M.S., Smedile, A., Gerardi, F., Pantosti, D., Del Carlo, P., Pirrotta, C., 2010. A unique 4000 year long geological record of multiple tsunami inundations in the Augusta Bay (eastern Sicily, Italy). Mar. Geol. 276, 42–57.
Dominey-Howes, D., 2004. A re-analysis of the Late Bronze Age eruption and tsunami of Santorini, Greece, and the implications for the volcano-tsunami hazard. J. Volcanol. Geotherm. Res. 130, 107–132.
Dominey-Howes, D., Humphreys, G., Hesse, P.P., 2006. Tsunami and palaeotsunami depositional signatures and their potential value in understanding the late-Holocene tsunami record. The Holocene 16 (8), 1095–1107.
Eaton, J.P., Richter, D.H., Ault, W.U., 1961. The tsunami of May 23, 1960, on the Island of Hawaii. Seis. Soc. Am. Bull. 51, 135–157.
Engel, M., Brückner, H., Wennrich, V., Scheffers, A., Kelletat, D., Vött, A., Schäbitz, F., Daut, G., Willershäuser, T., May, S.M., 2010. Coastal stratigraphies of eastern Bonaire (Netherlands Antilles): new insights into the palaeotsunami history of the southern Caribbean. Sediment. Geol. 231, 14–30.
Engel, M., May, S.M., 2012. Bonaire's boulder fields revisited: evidence for Holocene tsunami impact on the Leeward Antilles. Quat. Sci. Rev. 54, 126–141.
Etienne, S., Paris, R., 2010. Boulder accumulations related to storms on the south coast of the Reykjanes Peninsula (Iceland). Geomorphology 114, 55–70.
Frohlich, C., Hornbach, M.J., Taylor, F.W., Shen, C., Moala, A., Morton, A.L., Kruger, J., 2009. Huge erratic boulders in Tonga deposited by a prehistoric tsunami. Geology 37, 131–134.
Fujino, S., Naruse, H., Matsumoto, D., Sakakura, N., Suphawajruksakul, A., Jarupongsakul, T., 2010. Detailed measurements of thickness and grain size of a widespread onshore tsunami deposit in Phang-nga Province, southwestern Thailand. Isl. Arc 19, 389–398.
Goff, J.R., 2008. The New Zealand Palaeotsunami Database. NIWA Technical Report 131, Christchurch.
Goff, J., Lane, E.M., Arnold, J., 2009. The tsunami geomorphology of coastal dunes. Nat. Hazards Earth Syst. Sci. 9, 847–854.
Goff, J., Weiss, R., Courtney, C., Dominey-Howes, D., 2010a. Testing the hypothesis for tsunami boulder deposition from suspension. Mar. Geol. 277, 73–77.
Goff, J.R., Chagué-Goff, C., Courtney, C., Dominey-Howes, D., 2010b. Analysis of the Mahuika comet impact tsunami hypothesis. Mar. Geol. 271, 292–296.
Goff, J., Nichol, S.L., Kennedy, D., 2010c. Development of a palaeotsunami database for New Zealand. Nat. Hazards, 193–208.
Goodman-Tchernov, B.N., Dey, H.W., Reinhardt, E.G., McCoy, F., Mart, Y., 2009. Tsunami waves generated by the Santorini eruption reached Eastern Mediterranean shores. Geology 37, 943–946.
Goto, K., Imamura, F., 2007. Numerical models for sediment transport by tsunamis. Quat. Res. 46, 463–475.
Goto, K., Okada, K., Imamura, F., 2009. Characteristics and hydrodynamics of boulders transported by storm waves at Kudaka Island, Japan. Mar. Geol. 262, 14–24.
Goto, K., Miyagi, K., Kawamata, H., Imamura, F., 2010. Discrimination of boulders deposited by tsunamis and storm waves at Ishigaki Island. Japan. Mar. Geol. 269, 34–45.
Goto, K., Chavanich, S.A., Imamura, F., Kunthasap, P., Matsui, T., Minoura, K., Sugawara, D., Yanagisawa, H., 2007. Distribution, origin and transport process of boulders deposited by the 2004 Indian Ocean tsunami at Pakarang Cape, Thailand. Sediment. Geol. 202, 821–837.

Goto, K., 2008. The genesis of oceanic impact craters and impact-generated tsunami deposits. In: Shiki, T., Tsuji, Y., Minoura, K., Yamazaki, T. (Eds.), Tsunamiites—Features and Implications. Elsevier, Amsterdam, 411 pp., pp. 277–298.

Goto, K., Kawana, T., Imamura, F., 2010a. Historical and geological evidence of boulders deposited by tsunamis, southern Ryukyu Islands, Japan. Earth Sci. Rev. 102, 77–99.

Goto, K., Okada, K., Imamura, F., 2010b. Numerical analysis of boulder transport by the 2004 Indian Ocean tsunami at Pakarang Cape, Thailand. Mar. Geol. 268, 97–105.

Goto, K., Miyagi, K., Kawamata, H., Imamura, F., 2010c. Discrimination of boulders deposited by tsunamis and storm waves at Ishigaki Island, Japan. Mar. Geol. 269, 34–45.

Gracia, F.J., Alonso, C., Benavente, J., Anfuso, G., Del Rio, L., 2006. The different coastal records of the 1755 tsunami waves along the south Atlantic Spanish coast. Z. Geomorphol., NF Suppl. 146, 195–220.

Gutscher, M.A., 2005. Destruction of Atlantis by a great earthquake and tsunami? A geological analysis of the Spartel Bank hypothesis. Geology 33, 685–688.

Hadler, H., Vött, A., Koster, B., Mathes-Schmidt, M., Mattern, T., Ntageretzis, K., Reicherter, K., Sakellariou, D., Willershäuser, T., 2011. Lechaion, the ancient harbour of Corinth (Peloponnese, Greece) destroyed by tsunamigenic impact. In: 2nd INQUA-IGCP-567 International Workshop on Active Tectonics, Earthquake Geology, Archaeology and Engineering, Corinth, Greece: 70–73.

Haflidason, H., et al., 2004. The Storegga Slide: architecture, geometry and slide development. Mar. Geol. 213, 201–234.

Hall, A.M., Hansom, J.D., Williams, D.M., 2010. Wave-emplaced coarse debris and megaclasts in Ireland and Scotland: boulder transport in high-energy littoral environment: a discussion. J. Geol. 118, 699–704.

Hall, A.M., Hansom, J.D., Williams, D.M., Jarvis, J., 2006. Distribution, geomorphology and lithofacies of cliff-top storm deposits, examples from high-energy coasts of Scotland and Ireland. Mar. Geol. 232, 131–155.

Hamzah, L., Puspito, N., Imamura, F., 2000. Tsunami catalogue and zones in Indonesia. J. Nat. Disast. Sci. 22, 25–43.

Hansom, J.D., Barltrop, N.D.P., Hall, A.M., 2008. Modelling the processes of cliff-top erosion and deposition underextreme storm waves. Mar. Geol. 253 (1-2), 36–50.

Hansom, J.D., Hall, A.M., 2009. Magnitude and frequency of extra-tropical North-Atlantic cyclones: a chronology from cliff-top storm deposits. Quat. Int. 195, 42–52.

Hieke, W., Werner, F., 2000. The Augias megaturbidite in the central Ionian Sea (central Mediterranean) and its relation to the Holocene Santorini event. Sediment. Geol. 135, 205–218.

Hills, J.G., Mader, C.L., 1997. Tsunami produced by the impacts of small asteroids. Ann. N.Y. Acad. Sci. 822, 381–394.

Hori, K., Kuzumoto, R., Hirouchi, D., Umitsu, M., Janjirawuttikul, N., Patanakanog, B., 2007. Horizontal and vertical variations of 2004 Indian tsunami deposits: an example of two transects along the western coast of Thailand. Mar. Geol. 239, 163–172.

Imamura, F., Goto, K., Ohkobo, S., 2008. A numerical model for the transport of a boulder by tsunami. J. Geophys. Res. 113. http://dx.doi.org/10.1029/2007JC004170.

Kelletat, D., 1991. The 1550 BP tectonic event in the Eastern Mediterranean as a basis for assessing the intensity of shore processes. Z. Geomorphol., NF Suppl. 81, 181–194.

Kelletat, D., Scheffers, A., 2004. Bimodal tsunami deposits—a neglected feature in palaeo-tsunami research. In: Schernewski, G., Dolch, T. (Eds.), Geographie der Meere und Küsten. Coastline Reports, vol. 1, pp. 1–20.

Kelletat, D., Scheffers, A., Scheffers, S., 2007. Field signatures of the SE Asian mega-tsunami along the west coast of Thailand compared to Holocene palaeo-tsunami from the Atlantic region. Pure Appl. Geophys. 164, 1–19.

Kench, P.S., S.L., Nichol, Smithers, S.G., McLean, R.F., Brander, R.W., 2008. Tsunami as agents of geomorphic change in mid-ocean reef islands. Geomorphology 95, 361–383.

Khan, S., Robinson, E., Rowe, D.A., Coutou, R., 2010. Size and mass of shoreline boulders moved and emplaced by recent hurricanes, Jamaica. Z. Geomorphol. 54, 281–299.

Kopp, H., Weinrebe, W., 2009. Ursachen von Tsunamis: ein Überblick. Geogr. Rundschau 61, 20–27.

Kortekaas, S., Dawson, A.G., 2007. Distinguishing tsunami and storm deposits: an example from Martinhal, SW Portugal. Sediment. Geol. 200, 208–221.

Kristan-Tollmann, E., Tollmann, A., 1994. The youngest big impact on earth deduced from geological and historical evidence. Terra Nova 6, 209–217.

Lander, J.F., Whiteside, L.S., Lockridge, P.A., 2002. A brief history of tsunamis in the Caribbean Sea. Sci. Tsunami Hazards 20, 57–94.

Lario, J., Luque, L., Zazo, C., Goy, J.L., Spencer, Ch, Cabero, A., Bardaji, T., Borja, F., Dabrio, C.J., Civis, J., Gonzales-Delgado, J.A., Borja, C., Alonso-Azcarate, J., 2010. Tsunami vs. storm surge deposits: a review of sedimentological and geomorphological records of extreme wave events (EWE) during the Holocene in the Gulf of Cadiz, Spain. Z. Geomorphol., NF 54 (Suppl. 3), 301–316.

Lavigne, F., Paris, R., Wassmer, P., Gomez, C., Brunstein, D., Grancher, D., Vautier, F., Sartohadi, J., Setiawan, A., Syahnan, T.G., Fachrizal, B.W., Mardiatno, D., Widagdo, A., Cahyadi, R., Lespinasse, N., Mahieu, L., 2006. Learning from a major disaster (Banda Aceh, December 26th, 2004): a methodology to calibrate simulation codes for tsunami inundation models. Z. Geomorphol. 146, 253–265.

Lay, T., Kanamori, H., Ammon, C.J., Nettles, M., Ward, S.N., Aster, R.C., Beck, S.L., Bilek, S.L., Brudzinski, M.R., Butler, R., Deshorn, H.R., Ekstrom, G., Satake, K., Sipkin, S., 2005. The great Sumatra–Andaman earthquake of 26 December 2004. Science 308, 1127–1133.

Lorang, M.S., 2000. Predicting threshold entrainment mass for a boulder beach. J. Coastal Res. 16, 432–445.

Martinez-Solares, J.M., 2001. Los efetos en Espana del terremoto de Lisboa (1 de noviembre de 1755). Inst. Geograf. Nc. Monogra. 19, 756 pp.

Masson, D., Le Bas, T.P., Grevemeyer, I., Weinrebe, W., 2008. Flank collapse and largescale landsliding in the Cape Verde Islands, off West Africa. Geochem. Gephys. Geosyst. 9. http://dx.doi.org/10.1029/2008GC001983.

McCoy, F.M., Heiken, G., 2000. Tsunami generated by the late bronze age eruption of Thera (Santorini), Greece. Pure Appl. Geophys. 157, 1227–1256.

Mhammdi, N., Medina, F., Kelletat, D., Amahmou, M., Aloussi, L., 2008. Large boulders along the Rabat coast (Morocco); possible emplacement by the November, 1st, 1755 A.D. Lisbon tsunami. Sci. Tsunami Hazards 27 (1), 1–30.

Moore, J.G., Clague, D.A., Holcomb, R.T., Lipman, P.W., Normark, W.R., Torresan, M.E., 1989. Prodigious submarine landslides on the Hawaiian Ridge. J. Geophys. Res. 94, 17465–17484.

Morton, R.A., Richmond, B.M., Jaffe, B.E., Gelfenbaum, G., 2008. Coarse-clast ridge complexes of the Caribbean: a preliminary basis for distinguishing tsunami and storm-wave origins. Journal of Sedimentary Research 78, 624–637.

Nandasena, N.A.K., Tanaka, N., 2013. Boulder transport by high energy: numerical model-fitting experimental observations. Ocean Eng. 57, 163–179.

National Geophysical Data Center/World Data Center (NGDC/WDC), 2013. Historical Tsunami Database. Boulder, CO, USA. http://www.ngdc.noaa.gov/hazard/tsu_db.shtml (accessed 15.11.13.).

Nichol, S.L., Lian, O.B., Carter, C.H., 2003. Sheet-gravel evidence for a late Holocene tsunami run-up on beach dunes, Great Barrier Island, New Zealand. Sediment. Geol. 155, 129–145.

Noormets, R., Felton, E.A., Crook, K.A.W., 2002. Sedimentology of rocky shorelines: 2. Shoreline megaclasts on the north shore of Oahu, Hawaii – origins and history. Sediment. Geol. 150, 31–45.

Noormets, R., Crook, K.A.W., Felton, E.A., 2004. Sedimentology of rocky shorelines: 3. Hydrodynamics of megaclast emplacement and transport on a shore platform, Oahu, Hawaii. Sediment. Geol. 172, 41–65.

Nott, J., 2000. Records of prehistoric tsunamis from boulder deposits: evidence from Australia. Sci. Tsunami Hazards 18, 3–14.

Nott, J., 2003a. Waves, coastal boulder deposits and the importance of the pre-transport setting. Earth Planet. Sci. Lett. 210, 269–276.

Nott, J., 2003b. Tsunami or storm waves? Determining the origin of a spectacular field of wave emplaced boulders using numerical storm surge and wave models and hydrodynamic transport equations. J. Coastal Res. 19, 348–356.

O'Loughlin, K., Lander, J.F., 2003. Caribbean Tsunamis—A 500-Year History from 1498–1998. Kluwer Academic Publishers, Dordrecht.

Papadopoulos, G.A., Chalkis, B.J., 1984. Tsunami observed in Greece and the surrounding area from antiquity up to present times. Mar. Geol. 56, 309–317.

Papadopoulos, G.A., Fokaefs, A., 2005. Strong tsunamis in the Mediterranean Sea: a re-evaluation. ISET J. Earthquake Tech. 42, 159–170.

Pareschi, M.T., Boschi, E., Mazzarini, F., Favalli, M., 2006. Large submarine landslides offshore Mt. Etna. Geophys. Res. Lett. 33, L13302.

Pirazzoli, P.A., 1986. The early Byzantine tectonic paroxysm. Z. Geomorphol, N Suppl. 62, 31–49.

Plafker, G., Savage, J., 1970. Mechanism of the Chilean earthquakes of May 21 and 22, 1960. Geol. Soc. Am. Bull. 81, 1001–1030.

Powars, D.S., 2000. The Effects of the Chesapeake Bay Impact Crater on the Geologic Framework and the Correlation of Hydrogeologic Units of Southeastern Virginia, South of the James River. U.S. Geological Survey, Professional Paper 1622. http://purl.access.gpo.gov/GPO/LPS21905 (accessed 29.01.12.).

Regnauld, H., Oszwald, J., Planchon, O., Pignatelli, C., Piscitelli, A., Mastronuzzi, G., Audevard, A., 2010. Polygenetic (tsunami and storm) deposits? A case study from Ushant Island, western France. Zeitschrift für Geomorphologie 54 (3), 197–217.

Reinhardt, E.G., Goodman, B.N., Boyce, J.I., Lopez, G., van Hengstum, P., Rink, W.J., Mart, Y., Raban, A., 2006. The tsunami of 13 December A.D. 115 and the destruction of Herodot the Great's harbour at Caesarea Maritima, Israel. Geology 34, 1061–1064.

Richmond, B.M., Jaffe, B.E., Gelfenbaum, G., Morton, R.A., 2006. Geologic impacts of the 2004 Indian ocean tsunami on Indonesia, Sri Lanka, and the Maldives. Z. Geomorphol. 146, 235–251.

Saintilan, N., Rogers, K., 2005. Recent storm Boulder deposits on the Beecroft Peninsula, New South Wales, Australia. Geogr. Res. 43 (4), 429–432.

Satake, K., Rabinovich, A., Kânoğlu, U., Tinti, S., 2011. Introduction to "tsunamis in the world ocean: past, present, and future. Volume I". Pure Appl. Geophys. 168, 6–7.

Sawai, Y., Jankaew, K., Martin, M.E., Prendergast, A., Choowong, M., Charoentitirat, T., 2009. Diatom assemblages in tsunami deposits associated with the 2004 Indian Ocean tsunami at Phra Thong Island, Thailand. Mar. Micropaleontol. 73, 70–79.

Scheffers, A., 2002. Palaeotsunami evidence from boulder deposits on Aruba, Curacao, and Bonaire. Sci. Tsunami Hazards 20, 26–37.

Scheffers, A., 2004. Tsunami imprints on the Leeward Netherlands Antilles (Aruba, Curaçao, Bonaire) and their relation to other coastal problems. Quat. Int. 120, 163–172.

Scheffers, A., Kelletat, D., 2003a. Chevron-shaped accumulations along the coastlines of Australia as potential tsunami evidences? Sci. Tsunami Hazards 21 (3), 174–188.

Scheffers, A., Kelletat, D., 2003b. Sedimentologic and geomorphologic tsunami imprints worldwide—a review. Earth Sci. Rev. 63, 83–92.

Scheffers, A., Kelletat, D., 2005. Tsunami relics on the coastal landscape west of Lisbon, Portugal. Sci. Tsunami Hazards 23 (1), 3–16.

Scheffers, A., 2005. Coastal response to extreme wave events—hurricanes and tsunami on Bonaire. In: Essener Geographische Arbeiten, vol. 37, 100 pp.

Scheffers, A., 2006. Coastal transformation during and after a sudden neotectonic uplift in western Crete. Z. Geomorphol. 146, 97–124.

Scheffers, A., Scheffers, S., Kelletat, D., 2005. Paleo-Tsunami relics on the southern and central antilleanisland arc (Grenada, St. Lucia and Guadeloupe). J. Coastal Res. 21 (2), 263–273.

Scheffers, A., Kelletat, D., 2006b. Recent advances in palaeo-tsunami field research in the intra-Americas-Sea (Barbados, St. Martin and Anguilla). In: Proceedings of the NSF Caribbean Tsunami Workshop. World Scientific Publishers, Puerto Rico, pp. 178–202.

Scheffers, A., Scheffers, S., 2006. Documentation of hurricane Ivan on the coastline of Bonaire. J. Coastal Res. 22, 1437–1450.

Scheffers, A., Scheffers, S., 2007. First significant tsunami evidence from historical times in western Crete (Greece). Earth Planet. Sci. Lett. 259 (3–4), 613–624.

Scheffers, A., 2008. Tsunami boulder deposits. In: Shiki, T., Tsuji, Y., Yamazaki, T., Minoura, K. (Eds.), Tsunamiites—Features and Implications. Elsevier, Berlin, pp. 299–318.

Scheffers, A., Scheffers, S., Kelletat, D., Abbott, D., Bryant, E., 2008. Chevrons—enigmatic sedimentary coastal features. Z. Geomorphologie NF 52 (3), 375–402.

Scheffers, A., Kelletat, D., Engel, M., 2009a. Die Entwicklung der Tsunamiforschung nch der Katastrophe vom 26. Dezember 2004. Geogr. Rundschau 12, 12–18.

Scheffers, A., Scheffers, S., Kelletat, D., Browne, T., 2009b. Wave-emplaced coarse debris and megaclasts in Ireland and Scotland: boulder transport in a high-energy littoral environment. J. Geol. 117, 553–573.

Scheffers, S.R., Scheffers, A., Kelletat, D., Bryant, E.A., 2009c. The Holocene palaeotsunami history of west Australia. Earth Planet. Sci. Lett. 270, 137–146.

Scheffers, A., Kelletat, D., Haslett, S.K., Scheffers, S., Browne, T., 2010a. Coastal boulder deposits in Galway Bay and the Aran Islands, western Ireland. Z. Geomorphol 54, 247–279.

Scheffers, A., Scheffers, S., Kelletat, D., 2010b. Reply to: Wave-emplaced coarse debris and megaclasts in Ireland and Scotland: storm waves or tsunamis? By A. M. Hall, J. D. Hansom and D. M. Williams. J. Geol. 118, 705–709.

Scheffers, A., Engel, M., May, M., Scheffers, S., Joannes-Boyau, R., Haensler, E., Kennedy, K., Kelletat, D., Brueckner, H., Voett, A., Schellmann, G., Schaebitz, F., Radtke, U., Sommer, B., Willershaeuser, T., Felis, T., August 8, 2013. Potential and limits of combining studies of coarse and fine-grained sediments for the coastal event history of a Caribbean carbonate environment. J. Geol. Soc. London. http://dx.doi.org/10.1144/SP388.4. Special Issue. Geological Society, London, Special Publications, 388.

Scicchitano, G., Monaco, C., Tortorici, L., 2007. Large boulder deposits by tsunami waves along the Ionian coast of south-eastern Sicily (Italy). Mar. Geol. 238, 75–91.

Shaw, B., Ambraseys, N.N., England, P.C., Floyd, M.A., Gorman, G.J., Higham, T.F.G., Jackson, J.A., Nocquet, J.-M., Pain, C.C., Piggott, M.D., 2008. Eastern Mediterranean tectonics and tsunami hazard inferred from the AD 365 earthquake. Nature Geoscience 19, 268–276.

Shiki, T., Tsuji, Y., Minoura, K., Yamazaki, T. (Eds.), 2008. Tsunamiites—Features and Implications. Elsevier, Amsterdam, 411 pp.

Sironi, M.B.C., Rimoldi, B., 2005. Prehistoric mega-tsunami in the eastern Mediterranean and its sedimentary response. Atti Della Academina Nazionale Dei Lincei Rendiconti Lincei Scienze Fisiche E Naturali 16 (3), 137–157.

Soloviev, S.L., Solovieva, O.N., Go, C.N., Kim, K.S., Shchetnikov, N.A., 2000. Tsunamis in the Mediterranean Sea 2000 B.C.—2000 A.D. Kluwer Academic Publishers, Dordrecht, 240 pp.
Soloviev, S.L., 1990. Tsunamigenic zones in the Mediterranean Sea. Nat. Hazards 3, 183—202.
Spiske, M., Böröcz, Z., Bahlburg, H., 2008. The role of porosity in discriminating between tsunami and hurricane emplacement of boulders—a case study from the Lesser Antilles, southern Caribbean. Earth Planet. Sci. Lett. 268, 384—396.
Spiske, M., Bahlburg, H., 2011. A quasi-experimental setting of coarse clast transport by the 2010 Chile tsunami (Bucalemu, Central Chile). Mar. Geol. 289 (1-4), 72—85.
Stiros, S.C., 2001. The AD 365 Crete earthquake and possible seismic clustering during the 4th—6th centuries AD in the Eastern Mediterranean: a review of historical and archaeological data. J. Struct. Geol. 23, 545—562.
Stiros, S.C., Drakos, A., 2006. A fault model for the tsunami-associated magnitude $\geq$ 8.5 eastern Mediterranean AD 365 earthquake. Z. Geomorphol. 146, 125—137.
Suanez, S., Fichaut, B., Magne, R., 2009. Cliff-top storm deposits on BanneglsIand, Briggany, France: effects of giant waves in the eastern Atlantic Ocean. Sediment. Geol. 220, 12—28.
Switzer, A.D., Pucillo, K., Haredy, R.A., Jones, B.G., Bryant, E.A., 2005. Sea level, storm, or tsunami: enigmatic sand sheet deposits in a sheltered coastal embayment from southeastern New South Wales. Aust. J. Coastal Res. 21 (4), 655—663.
Switzer, A.D., Jones, B.G., 2008. Large-scale washover sedimentation in a freshwater lagoon from the southeast Australian coast: sea-level change, tsunami or exceptionally large storm? Holocene 18 (5), 787—803.
Synolakis, C.E., Okal, E.A., 2005. 1992—2002: perspective on a decade of post-tsunami surveys. In: Satake, K. (Ed.), Tsunamis: Case Studies and Recent Developments: Adv. Natur. Technol. Hazards, vol. 23, pp. 1—30.
Synolakis, C.E., Bernard, E.N., Titov, V.V., Kâ Noğ LU, U., Gonzalez, F.I., 2008. Validation and verification of tsunami numerical models. Pure Appl. Geophys. 165, 2197—2228.
Talandier, J., Bourrouilh-Le Jan, F., 1987. High energy sedimentation in French Polynesia: cyclone or tsunami? In: El-Sabh, M.I., Murty, T.S. (Eds.), Natural and Manmade Hazards. Reidel Publishing, pp. 181—187.
Sedimentary features of tsunami deposits – their origin, recognition and discrimination: an introduction. In: Tappin, D.R. (Ed.), Sediment. Geol. 200, 151—154.
Tester, E.W., Abbott, D.H., Meyers, C.A., 2007. Evidence for a Late Holocene oblique impact into the Gulf of Carpentaria, Australia. In: Geological Society of America Annual Meeting, Denver, CO. Abstracts with Programs, vol. 39, p. 312.
Tinti, S., Maramai, A., 1996. Catalogue of tsunami generated in Italy and in Cote d'Azur, Francde, a step towards a unified catalogue of tsunami in Europe. Ann. Geofis. 39, 1253—1300.
Tinti, S., Maramai, A., Graziani, L.: The Ital-ian Tsunami Catalogue (ITC), Version 2, available at: http://www.ingv.it/servizi-e-risorse/BD/catalogo-tsunami/catalogo-degli-tsunami-italiani, 2007.
Tsuji, Y., Namegaya, Y., Matsumotu, H., Iwasaki, S., Kanbua, W., Sriwichai, M., Meesuk, V., 2006. The 2004 Indian tsunami in Thailand: surveyed runup heights and tide gauge records. Earth Planets Space 58, 223—232.
Tsunamis, hurricanes and neotectonics as driving mechanisms in coastal evolution. In: Scheffers, A., Kelletat, D. (Eds.), Proceedings of the Bonaire Field Symposium, March 2—6, 2006, Zeitschrift für Geomorphologie, Supplementband 146. Borntraeger, Berlin, 265 pp.
Vött, A., Brückner, H., Schriever, A., Luther, J., Handl, M., van der Borg, K., 2006. Holocene palaeogeographies oft the Palairos coastal plain (Akarnania, Northwest Greece) and their geoarchaeological implications. Geoarchaeology 21, 649—664.

Vött, A., Brückner, H., Brockmüller, S., Handl, M., May, S.M., Gaki-Papanastassiou, K., Herd, R., Lang, F., Maroukian, H., Nelle, O., Papanastassiou, D., 2009. Traces of tsunamis across the Sound of Lefkada, NW Greece. Global Planet. Change 66, 112–128.

Vött, A., Lang, F., Brückner, H., Gaki-Papanastassiou, K., Maroukian, H., Papanastassiou, D., Giannikos, A., Hadler, H., Handl, M., Ntageretzis, K., Willershäuser, T., Zander, A., 2011a. Sedimentological and geoarchaeological evidence for multiple tsunamigenic imprint on the Bay of Palairos-Pogonia (Akarnania, NW Greece). Quat. Int. 242 (1), 213–239.

Vött, A., Bareth, G., Brückner, H., Lang, F., Sakellariou, D., Hadler, H., Ntageretzis, K., Willershäuser, T., 2011b. Olympia's harbour site Pheia (Elis, western Peloponnese, Greece) destroyed by tsunami impact. Die Erde 142 (3), 259–288.

Ward, S.N., Day, S.J., 2001. Cumbre Vieja Volcano; potential collapse and tsunami at La Palma. Canary Isl. Geophys. Res. Lett. 28–17, 3397–3400.

Weiss, R., 2012. The mystery of boulders moved by tsunamis and storms. Mar. Geol. 295–298, 28–33.

Whelan, F., Kelletat, D., 2002. Geomorphic evidence and relative and absolute dating results for tsunami events on Cyprus. Sci. Tsunami Hazards 20 (1), 3–18.

Whelan, F., Kelletat, D., 2003. Submarine slides on volcanic islands—a source for mega-tsunamis in the Quaternary. Prog. Phys. Geogr. 27 (2), 198–216.

Whelan, F., Kelletat, D., 2005. Boulder deposits on the southern Spanish Atlantic coast: possible evidence for the 1755 AD Lisbon tsunami? Sci. Tsunami Hazards 23, 25–38.

Williams, D.M., Hall, A.M., 2004. Cliff-top megaclast deposit of Ireland, a record of extreme waves in the North Atlantic: storms or tsunamis? Mar. Geol. 206, 101–117.

FURTHER READING

Engel, M., Brückner, H., 2011. The identification of palaeo-tsunami deposits—a major challenge in coastal sedimentary research. Coastline Rep. 17, 65–80.

Hansom, J.D., Barltrop, N.D.P., Hall, A.M., 2008. Modelling the process of cliff-top erosion and deposition under extreme waves. Mar. Geol. 253, 36–50.

Nadesna, N., Paris, R., Tanaka, N., 2011. Reassessment of hydrodynamic equations: minimum flow velocity to initiate boulder transport by high energy events (storms, tsunamis). Mar. Geol. 281, 70–84.

Sawai, Y., Fujii, Y., Fujiwara, O., Kamataki, T., Komatsubara, J., Okamura, Y., et al., 2008. Marine incursions of the past 1500 years and evidence of tsunamis at Suijin-numa, a coastal lake facing the Japan trench. Holocene 18 (4), 517–528.

Zhou, Q., Adams, W.M., 1986. Tsunamigenic earthquakes in China, 1831 BC to 1980 AD. Sci. Tsunami Hazards 4, 131–148.

Chapter 4

Tsunami Case Studies

Eugene J. Farrell[1], Jean T. Ellis[2] and Kieran R. Hickey[1]

[1] *Discipline of Geography, National University Ireland Galway, Galway, Ireland,* [2] *Department of Geography and Marine Science Program, University of South Carolina, Columbia, SC, USA*

ABSTRACT

Tsunamis are caused by geological processes, such as earthquakes, landslides, or volcanic eruptions, that displace large volumes of ocean water. Large-magnitude, subduction zone earthquakes, where two plates in the ocean push into each other, are the most common source of the recent large tsunamis. Submarine landslides, sometimes triggered by earthquakes, and coastal or submarine volcanoes also cause tsunamis. This chapter describes 10 modern and historic tsunami events that were significant in terms of their size, impact, extent, and/or triggering mechanisms. Each tsunami event is described using four different categories: (1) tsunami generation; (2) tsunami size, and extent (3) impact of the event at the local, regional and, where applicable, global scales; and (4) lessons learned in the aftermath of the event. The case studies are grouped according to the tsunamigenic source: earthquake (2004 Indian (Sumatra–Andaman) earthquake, 2011 Tōhoku earthquake, 1964 Alaska earthquake, 1960 Valdivia earthquake, 1946 Aleutian Island earthquake, 1908 Messina-Reggio earthquake, 1755 Great Lisbon earthquake), landslide (Storegga Slides 30,000 and 7,200–7,000 YBP, Papua New Guinea, 1998), and volcano (Krakatoa 1883).

4.1 INTRODUCTION

Tsunami, or harbor (*tsun-*) wave (*-ami*) in Japanese, are caused by geological processes, such as earthquakes, landslides or volcanic eruptions, which displace large volumes of ocean water. The displaced sea surface propagates outward from the source as a series of ocean waves with extremely long wavelengths and periods. Wind-generated waves cause water motion to depths of 150 m; a tsunami involves water movement to the sea floor bottom. The Earth-moving event is most typically an earthquake, but tsunamis are also generated from submarine or terrestrial landslides and volcanoes (Table 4.1). Other chapters in this volume review multiple characteristics of tsunamis: Kânoğlu and Synolakis (generation, modeling, and dynamics), Nott (paleotsunamis), and

Coastal and Marine Hazards, Risks, and Disasters. http://dx.doi.org/10.1016/B978-0-12-396483-0.00004-2
Copyright © 2015 Elsevier Inc. All rights reserved.

TABLE 4.1 Summary of Tsunami Case Studies Reviewed in This Chapter (M_w Earthquake Moment Magnitude)

Generating Source (Event Magnitude, if Applicable)	Date	Name	Fatalities	Areas Affected
Earthquake (M_w 9.0)	03/11/2011	Tōhoku earthquake or Great East Japan earthquake and tsunami	15,885; 2,615 missing	Japan, North America, South America, Antarctica, Northern Europe
Earthquake (M_w 9.1–9.3)	12/26/2004	2004 Indian Ocean (Sumatra–Andaman) earthquake and (boxing day tsunami)	230,000 estimated	South and Southeast Asia including, but not limited to, Indonesia, Sri Lanka, Thailand, India, India, Maldives, Bangladesh, and Malaysia
Earthquake (M_w 9.2)	03/27/1964	1964 Great Alaskan earthquake and Good Friday (Crescent city) tsunami	139	Alaska, West coast of the United States, Hawaii, Japan
Earthquake (M_w 9.5)	05/22/1960	1960 Valdivia earthquake (Great Chilean earthquake) and tsunami	2,183–6,000	Chile, Hawaii, Japan, Philippines, New Zealand, and Australia
Earthquake (M_w 8.1)	04/01/1946	1946 Aleutian Islands earthquake and tsunami	165	Aleutian Islands, Hawaii
Earthquake (M_w 7.2)	12/28/1908	1908 Messina-Reggio earthquake and tsunami	60,000–123,000	Italy
Earthquake (M_w 8.5–9.0)	11/01/1755	1755 Great Lisbon earthquake and tsunami	40,000–50,000 estimated	Portugal, Spain, North Africa, Northwest Europe
Landslide	M_w 7.1 07/07/1998	1998 Papua New Guinea tsunami	2,189	Papua New Guinea
Landslide	30,000 YBP; 7,200–7,000 YBP	Storegga slides tsunamis	Unknown	Norway, Scotland, Faroe Islands, Greenland
Volcano	08/27/1883	Krakatoa volcanic eruption and tsunami	36,417–120,000	Indonesia, South Africa, New Zealand, Australia, Japan, Hawaii, Alaska, North and South America, and UK

Hansom (Chapter 11, this volume). Rather, this chapter describes 10 historic and modern tsunami events that were significant in terms of their size, impact, extent, and/or triggering mechanisms. Large magnitude subduction-zone earthquakes, where two plates in the ocean collide, are the most common source of large tsunamis in recent years (California Coastal Commission, 2011). For example, the 2004 Indian Ocean (Sumatra–Andaman) earthquake and the 2011 Tōhoku earthquake were a subduction-zone type of earthquake that generated destructive tsunamis. Submarine landslides, sometimes triggered by earthquakes, are another source of tsunami, as witnessed in Papua New Guinea in 1998. Coastal or submarine volcanoes are a third source of tsunami; Krakatoa is the most famous of this type. Geologic evidence also exists that meteor strikes have generated large tsunamis, but this is beyond the scope of this chapter.

In this chapter, we report on earthquake-generated tsunamis in the Indian (Sumatra–Andaman Earthquake 2006), Pacific (Tōhoku Earthquake 2011; Alaska Earthquake 1964; Valdivia Earthquake 1960; Aleutian Islands 1948, 1946), and Atlantic (The Great Lisbon Earthquake 1755) Oceans and the Mediterranean Sea (Messina-Reggio Earthquake 1908). Tsunamis triggered by submarine landslides (Storegga Slides 30,000 and 7,200–7,000 YBP, Papua New Guinea 1998) and volcanic eruptions (Krakatoa 1883) are also presented. Each tsunami event is described using four different categories: (1) tsunami generation; (2) tsunami size and extent; (3) impact of the event at the local, regional, and, where applicable, global scales; and (4) lessons learned in the aftermath of the event.

The most vivid and popularized depiction of tsunami is likely the *Kanagawa Oki Nami-Ura* (*The Great Wave Off Kanagawa*, Figure 4.1(a)) woodblock print (or woodcut) published between between 1830 and 1833 by Katsushika Hokusai. This piece is the first print in Hokusai's *Fugaku Sanljūrokkei* (*Thirty-six Views of Mount Fuji*) series and illustrates an enormous wave threatening boats off the coast of Kanagawa in the Kantō region of Honshu, which is the largest and most populous island of Japan that also encompasses the Greater Tokyo area. *Kanagawa-oki Honmoku no zu* (*View of Honmoku off Kanagawa*, c.1803, Figure 4.1(b)) and *Oshiokuri Hato Tsusen no Zu* (*Fast Cargo Boat Battling the Waves*, c.1805, Figure 4.1(c)) by the same artist are precursors to *The Great Wave Off Kanagawa*. In 1834, Hokusai created a second series prints, including *Kaijo no Fuji* (*One hundred views of Mount Fuji*, Figure 4.1(d)), which also shows tsunami waves and Mount Fuji, the latter is considered a sacred symbol of Japan's national identity.

The Hokusai illustrations showing the tsunami wave approaching the land with ships depicts the inherent conflict between nature and humans. Tsunami hazards become natural disasters with substantial fatalities and infrastructure damage (Table 4.1). The latter is exacerbated when nuclear plants, such as Fukushima Dai-ichi and Fukushima Dai-ni in Japan, are emplaced along vulnerable and populated coastlines, for example. The extent of a disaster is

FIGURE 4.1 Tsunami artwork by Hokusai: (a) Kanagawa Oki Nami-Ura (The Great Wave off Kanagawa) c. 1830–1833; (b) Kanagawa-oki Honmoku no zu (View of Honmoku off Kanagawa) c. 1803; (c) Oshiokuri Hato Tsusen no Zu (Fast Cargo Boat Battling The Waves) c. 1805; (d) Kaijo no Fuji (100 views of Mount Fuji) c. 1834.

mitigated by increased coastal resilience and improved warning systems. For example, at least 43 percent of Japan's coastline has some type of engineered structure to protect against large typhoon- or tsunami-generated waves. However, the 2011 Tōhoku, Japan, event made it evident that very few, if any, coastal structures are able to prevent large tsunamis from laying waste to low lying coastal regions. Kânoğlu and Synolakis (Chapter 2, this volume) document the recent advances to tsunami modeling; however, in addition to improved models, information dissemination warning inhabitants of the impending tsunami in adequate time for evacuation is critical. The warning in advance to the 2011 Japanese tsunami event saved some lives, but ultimately the tsunami resulted in many fatalities because the warning was not issued with adequate time.

4.2 EARTHQUAKE-GENERATED TSUNAMIS

4.2.1 The 2011 Tōhoku Earthquake (Great East Japan Earthquake) and Tsunami

4.2.1.1 Generation

The Great East Japan earthquake that generated the tsunami occurred at 05:46 UTC (14:46 local time) on 11 March, 2011. Historically, this region has been very seismically active with previous substantial earthquakes and tsunamis occurring in 1611, 1896, and 1933. Large earthquakes, ones with moment magnitudes (M_w) of 7.8–8.0, are predicted to occur in this region with a 99 percent probability by 2040 (ADRC, 2011). The 11 March, 2011, event had a one in 1,000 year return period and greatly exceeded any pre-disaster expectations. The mechanism causing the tsunami consisted of not only of a slipping movement of the deep plate boundaries that lead to normal ocean trench earthquakes in the region, but also a simultaneous slipping movement at the shallow plate boundaries (GFDRR, 2012). The plate-boundary thrust-faulting earthquake that generated the tsunami had a moment magnitude of 9.0. It is the fourth largest earthquake on record since modern records began in 1900 and the largest in Japanese history. Numerous strong aftershocks ($>M_w$ 7.0) were recorded and continue to occur (as of June 2014). According to the Japanese Meteorological Agency, there have been 776 aftershocks greater than M_w 5.0 magnitude (as of June 2014).

The March 2011 earthquake epicenter was located 70 km east of the Oshika Peninsula in the Tōhoku region in the northeast portion of the main island of Honshu. The earthquake had a relatively shallow depth of 32 km (USGS, 2011) and lasted for 6 min, which, historically, is unusually long. The earthquake was powerful enough to shift parts of Japan's main island of Honshu eastward, or 2.5 m closer to the United States mainland. The seismic event also affected the Earth's axis and orbit (Chang, 2011). The earthquake resulted in significant plate boundary movement and had a rupture area that was approximately 500 km long (north to south) and 200 km wide (east to west), which caused massive displacement of several million tons of water.

4.2.1.2 Size and Extent

The maximum tsunami height was approximately 40 m and consisted of a number of main and subsidiary waves. The tsunami struck the Japanese coastline 36 min after the earthquake. Field surveys indicated that the highest runup height was 38.9 m. The tsunami spread over the entire Pacific Ocean; 10 h after the earthquake a 2-m tsunami wave impacted Chile, which is approximately 17,000 km away from the epicenter. The tsunami broke icebergs off the Sulzberger Ice Shelf 13,000 km away. This was the first time the United Sates National Oceanic and Atmospheric Administration (NOAA) reported observational evidence from satellites linking tsunami to ice calving in Antarctica. Reports of impacts in Norwegian fjords also occurred (Brunt et al., 2011). Two-meter waves were observed on tide gauges in Russia, South America, Hawaii, and along the west coast of the United States.

4.2.1.3 Impacts

The tsunami impacts were exacerbated by the geologic (magnitude, depth, duration, and displacement) and geographic (proximity to coastline) characteristics of the generating earthquake. Japan is a leading country regarding tsunami-disaster prevention and has many structural and nonstructural tsunami countermeasures along the coast, especially in Sanriku where the coastline configuration can cause tsunami waves to amplify to heights exceeding 10 m (Suppasri et al., 2013). An earthquake with a magnitude equal to the one that occurred in March 2011 was never predicted, nor planned for. Emplaced sea defenses were overwhelmed by the tsunami. Despite the issuance of the highest-level tsunami warning (i.e., a "major tsunami" with a tsunami of at least 3 m) by the Japan Meteorological Agency, very little chance existed for people to evacuate. The actual tsunami exceeded the warnings and 101 designated tsunami evacuation sites were hit by the waves resulting in >1,000 deaths (Japan Times, 2011). The relative ineffectiveness of these defenses was a result of the sudden 2-m subsidence of >400 km of the east coast of Japan and the wave heights that exceeded the seawall heights of 10 m (Chang, 2011). It is now accepted that the pre-disaster predictions and assumptions greatly underestimated the actual earthquake and tsunami magnitude and devastation, which initiated a review of Japan's strategies of hazard risk and management. One suggested approach is to include earthquakes, such as the Jogan Sanriku earthquake of 869, the Keicho Sanriku earthquake of 1611, the Enpo Boso earthquake of 1677, and the Meiji Sanriku earthquake in 1896 to tsunami-hazard prediction models (GFDRR, 2012).

As of 10 April, 2014, The National Police Academy of Japan confirmed 15,885 deaths; 6,148 injures; and 2,623 missing persons resulting from the 2011 tsunami. In the latter case, the bodies were likely washed out to sea or buried so deep in sediment and/or debris along the coastline that they are unlikely to be discovered. In addition, 340,000 people were displaced because 127,290 buildings collapsed, 272,788 buildings were classified as "half collapsed," and 747,989 buildings were partially damaged. The tsunami also caused one death in Jakarta, Indonesia, and one death in the Klamath River, California, USA.

This event was widely covered by international news outlets. The media portrayed entire coastal towns and villages swept away by one or more major waves and showed many people marginally escaping and thousands losing their lives. The number of fatalities was amplified by the landward penetration of the tsunami, which in some areas was >10 km. The Geospatial Information Authority of Japan reported that the tsunami inundated an area of approximately 561 km^2 in Japan.

Extensive damage occurred to almost all parts of NE Japan. An estimated 4.4 million people were left without electricity because of the damage to electricity-generating stations and power lines and/or the precautionary shutdown of some nuclear power stations. The scale of building devastation was enormous, and it included the complete destruction of 11 hospitals and damage to over 300 more. Three major commercial ports were destroyed, and >300 fishing ports were damaged. One oil refinery and a natural gas processing plant were badly damaged. One dam collapsed, causing some fatalities, and others were damaged. Seven hundred fifty four cultural properties were damaged, which included five properties classified as National Treasures in the affected region (Anon, 2011). Major transport disruption occurred, with damage to critical roads and rail lines that caused large-scale service cancellations and major delays. Telecommunications were also badly affected. An estimated 230,000 vehicles were destroyed or damaged, mostly from the tsunami. The total cost of damage is estimated at US$250 billion, which makes it the most expensive natural disaster in human history. This estimate is likely to continue to rise, particularly because of the nuclear implications of this event. In particular, two tsunami-related events are going to continue to contribute to the total event-related costs, discussed below.

The first is the partial meltdown of the Fukushima nuclear power plant that was partially inundated by the tsunami. This was the worst nuclear disaster since Chernobyl in the Ukraine in 1986. A 20-km exclusion zone was emplaced for protection against radiation, which affected 80,000 residents. Starting in 2014, a limited number of residents have been allowed to return to their homes in this zone, but many are too fearful to do so because of their concern about the radiation levels. In sum, 8 percent of Japan was blanketed by radiation, which results in a huge cleanup effort that will take decades. One current issue is the removal of 23 million tons of contaminated soil that was scraped from the surface (50-mm layer) from hundreds of thousands of hectares of farmland (Kamiya, 2011). The cleanup at Fukushima is ongoing and will continue indefinitely (Kingston, 2012). The Fukushima meltdown has resulted in citizen protest and a reconsideration of Japan's use of nuclear power.

The second effect is the generation of a mass of between one and two million tons of debris, which includes human remains that are being distributed around the Pacific Ocean (Laurent et al., 2013). The first debris to wash up on the coast of North America occurred in Oregon, USA, on 6 February, 2013. In response, the United States has emplaced protocols through NOAA's Marine Debris program, for example, to manage this material (Figure 4.2).

4.2.1.4 Relief Efforts

In response to this major disaster, the Japanese government quickly mobilized the Self-Defense Forces, including all emergency services and the Army. The importance of such a rapid response of personal was an important lesson learned from the Kobe earthquake of 1995 when it took two days to activate relevant emergency services and the Army. Several countries sent search and rescue teams. Up to two weeks after the event, many thousands of people were rescued from the rubble and mud. Japanese and worldwide aid organizations responded, with the Japanese Red Cross reporting USD$1 billion in donations (Nebehay, 2011).

FIGURE 4.2 Tsunami debris-watch placard generated by National Oceanic and Atmospheric Administration (NOAA) and placed in communities along the US Pacific coast.

4.2.1.5 *Aftermath*

Suppasri et al. (2013) provided a comprehensive analysis of the performance of the tsunami countermeasures that were in place in March 2011 and, subsequently, the lessons learned and recommendations for future management strategies. Their findings include the following recommendations: (1) the tsunami countermeasures were not designed to resist an event with the magnitude of the 2011 earthquake event; (2) construction of massive structures to completely protect against 500- to 100-year return-period tsunamis cannot be achieved when budget and time are limited; (3) future structures should have stronger foundations and seawall gates should be remotely controlled; (4) control forests should be planted as secondary barriers at higher elevations behind the seawalls; (5) wooden structures should be replaced by reinforced concrete structures in areas where tsunami inundation is expected; (6) the elevation of railways and roads should be raised to serve as secondary or tertiary tsunami barriers; (7) the design and location of evacuation buildings should be reconsidered; (8) increased awareness and training of citizens will reduce the number of fatalities; (9) land-use policies for future development should avoid tsunami-prone areas; and (10) hard (engineered) and soft (education; evacuation plans) countermeasures should be used to increase tsunami awareness and community readiness to tsunamis.

4.2.2 The 2004 Indian Ocean (Sumatra–Andaman) Earthquake and (Boxing Day) Tsunami

4.2.2.1 Generation

Tsunamis from seismic activity are much more rare in the Indian Ocean compared to the those in the Pacific Ocean (Table 4.1). Nonetheless, the 2004 event in the Indian Ocean occurred at 00:58 UTC (08:58 local time) on 26 December, 2004. The earthquake-generating tsunami had a moment magnitude between 9.1 and 9.3. It was the second most powerful event since modern seismic records began in 1900 and was the largest for the preceding 40 years (McKee, 2005). The event affected the orbit of the Earth and triggered other earthquakes approximately 11,000 km away in Alaska (West et al., 2005). Since 1900, only two earthquakes have been recorded with a similar magnitude; the 1960 Great Chilean Earthquake (also called the Valdivia Earthquake, M_W 9.5) and the 1964 Great Alaskan Earthquake (also called the Good Friday Earthquake, M_W 9.2) both of which generated significant tsunamis and are reported in this chapter.

The epicenter of the Sumatra–Andaman megathrust event was 30 km undersea around 250 km NW of Sumatra along the Indo-Australian plate boundary. It is estimated that this section of the plate had not moved for >200 years, which during that time, accumulated a lot of energy (McKee, 2005). At the time of impact, the earthquake set a new record for the longest duration at between 8 and 10 min (Walton, 2005). The earthquake ruptured the Sumatra and Sunda subduction zones over a length of 1,300 km (Sibuet et al., 2007), which generated a massive tsunami consisting of two or three main waves and numerous smaller ones. Based upon seabed surveys, it is estimated that there was at least 10 and 4–5 m of lateral and vertical movement, respectively, along the fault line (Bagla, 2005).

The main earthquake was followed by a series of aftershocks that were recorded in the Andaman Islands archipelago in the Bay of Bengal between India and Myanmar. The largest aftershock registered a magnitude of 8.7 off the coast of Sumatra, which prompted a debate among seismologists on whether to classify the event as an aftershock or a "triggered earthquake." Indeed, one of these aftershocks is classified here as separate event (Indian Ocean Aftershock 2006, M_W 7.7) because it generated a tsunami that resulted in substantial loss of life.

4.2.2.2 Size and Extent

The tsunami took between 15 min (Sumatra) and 7 h (Somalia) to reach various locations along the Indian Ocean coastline. Locations closest to the epicenter in the northern regions of the Indonesian island of Sumatra were hit very quickly, whereas Sri Lanka and the east coast of India were hit roughly 2 h later, for example. Thailand was also struck about 2 h later, despite being closer to the epicenter, because the tsunami traveled more slowly in the shallow Andaman Sea. The tsunami was recorded by tide gauges on Australia's west coast within 5 h of the event. The tsunami impacted the Pacific Ocean where it produced small but measurable waves (<0.50 m) along the western coast of North and South America. The main tsunami wave was recorded on tide gauges along the Chilean and Peruvian coastlines 27 and 28 h later, respectively. Numerical models, such as ones produced by NOAA and based on quality-controlled source data (e.g., tide gauges), were used to replicate the generation and propagation of the Indian Ocean tsunami and to illustrate how the waves propagated around the world's ocean basins.

The height of the tsunami varied greatly and depended on its distance and direction from the epicenter and other factors such as the local bathymetry and coastal topography. The height ranged from 2 to 3 m at the African coast (Kenya) and up to 10–15 m at Sumatra, the region closest to the earthquake epicenter. The maximum height of the main tsunami wave was between 24 and 30 m and the second main wave was between 10 and 15 m high.

4.2.2.3 Impacts

This event caused the largest loss of life of any known tsunami with 230,00–280,000 estimated fatalities (Diacu, 2009). Based on fatalities, this event outranks almost every natural disaster. This, and the 2010 Haiti earthquake that killed between 100,000 and 316,000 are the deadliest natural disasters of the twenty-first century (Columbian Journalism Review, 2012). The impact of the Boxing Day tsunami, was exacerbated by the scale of seabed displacement, its proximity to the coastline of Banda Aceh Province in Indonesia where most fatalities and damage occurred, and the lack of an Indian Ocean Tsunami Detection and Warning System. Even with a functioning warning system in place, the people of Banda Aceh would have had less than 15 min to evacuate. Ultimately approximately 170,000 of the fatalities occurred in this area. One of the great tragedies of this event is that with a warning system in place, it is believed that all 60,000+ lives lost in Sri Lanka, Thailand, India, Maldives, Bangladesh, Malaysia, Myanmar, and in several countries in East Africa could have been avoided (Table 4.2). This estimate includes around 9,000 tourists, mainly from Europe, vacationing in Thailand; for example, Sweden lost >500 citizens.

In addition to the extensive fatalities, approximately 1.69 million people were displaced. Indonesia and Sri Lanka were the worst impacted with over half million people displaced in each country (Meisl et al., 2006). Many survivors lost their livelihoods due to the destruction of their fishing boats and coastal farms. In the case of Thailand, the tourism sector was impacted; even areas that were not affected by the tsunami experienced a substantial drop in bookings (Jayasuriya and McCawley, 2010).

TABLE 4.2 Estimated Arrival Time of the First Tsunami Wave

Country	Distance from Epicenter (km)	Elapsed Time to Reach Coast
Sumatra	100	<15 min
Thailand	550–600	1 h
Malaysia	550–600	1 h
Myanmar (Burma)	750	1 h 15 min
Andaman Islands	750–800	1 h 20 min
Sri Lanka	1,500	2 h 30 min
India	1,850	2 h 45 min
Maldives	2,350	3 h 30 min
Somalia	4,750	7 h 30 min

Source: After Kowalik et al. (2005)

The scale of destruction and damage was enormous, but in most places, it was limited to within 1 or 2 km of the coastline. Large expanses of the coastal zone were cleared of every building, all vegetation, and soil. The total cost of damage was estimated at around \$15 billion, which is rather low given the scale of the devastation because most of the areas affected were in developing countries where standards are low compared to the global average. Indonesia was the worst affected with an estimated damage of at least \$4.5 billion, followed by Sri Lanka with an estimated \$3.5 billion, and India and Thailand with an estimated damage of >\$1.5 billion each.

4.2.2.4 Relief Efforts

No doubt exists that outside of wartime, the relief effort associated with this tsunami was the most extensive the world had ever seen. Direct involvement occurred by almost every nation around the world. Significant public donations were raised, and many governments contributed to the relief effort. In sum, governments, aid agencies, and individual donations totaled >US\$14 billion (Jayasuriya and McCawley, 2010). Nations that lost their citizens supported the relief effort, possibly more than they would had their country not been affected. Disaster relief teams arrived from all over the world, along with some military assistance. The initial purpose of the relief teams was to rescue individuals trapped in buildings and to help treat the injured. Unfortunately, many still died unnecessarily due to lack of medical supplies, clean water, and the essentials required to live. The UN designated former US President Bill Clinton as a United Nations (UN) Special Tsunami Envoy. His main objective was to ensure that countries follow through on their commitments for disaster relief. Clinton was also involved in mediating the two preexisting conflicts that existed in the region, the Tamil-Sri Lankan issue (separatist activity since 1983–65,000+ victims) and the Aceh–Indonesian issue (separatist activity since 1976–10,000+ victims). Most residents in the Aceh Province are Muslim and did not want foreign military present, especially those from the United States (Barron et al., 2005). The Indian government did not want assistance on some of the remote Andaman and Nicobar Islands where there are strategic Indian military bases and indigenous tribes. On the Indian mainland thousands of homeless and starving "untouchables" were denied access to clean water and food in relief camps for fear of spiritual contamination (Jayasuriya and McCawley, 2010). A substantial concern existed that large portions of the aid were stolen by corrupt individuals and government officials. In cases where this was confirmed, harsh sentences were distributed (Asian Development Bank, 2005).

4.2.2.5 Aftermath

In the aftermath of the tsunami, the UN Intergovernmental Oceanographic Commission (IOC), comprising UN Educational, Scientific, and Cultural Organization and other partners, began coordinating efforts to create an Indian Ocean early warning system and administering evacuation plans. At a 2005 UN Meeting in Kobe, Japan, it was agreed to establish a warning system that would become operational in June, 2006. The warning system consists of 25 seismographic stations reporting to 26 national tsunami information centers and six DART® (Deep-Ocean Assessment and Reporting of Tsunami) buoys. In 2012, Thailand successfully launched their national warning system, which was eight years after the Andaman coast was destroyed and 5,395 people were killed, many of whom were tourists. The Thai National Disaster Warning Center established 136 warning towers and three tsunami-detection buoys in the Andaman Sea that are connected to the United States Geological Survey, the World Meteorological Organization, and other authorized disaster-monitoring agencies. The efficacy of Indonesia's early warning system was tested in April 2012 when an 8.6 M_w earthquake occurred 400 km southwest of Banda Aceh. Despite the initial (and likely justifiable) panic by people in many at-risk coastal locations, according to Thorkild Aarup, Head of the Tsunami Unit of the UN IOC, "The three early warning systems (and evacuation drills) functioned as they should have across the board. The Indonesian early warning was issued at 8.43 UTC—five minutes after the quake happened. The Australian warning was issued 10 min after, while India's was issued eight minutes after the earthquake." The region has certainly become much better equipped in the eight years following the devastating 2004 Boxing Day tsunami. However, it is essential that governments continue to train people to respond to warning systems and prepare the coastal inhabitants for tsunami disasters.

4.2.3 Great Alaskan Earthquake and Good Friday (Crescent City) Tsunami

4.2.3.1 Generation

Alaska is the most seismically active region in the United States; the Alaska Earthquake Information Center locates and reports about 22,000 earthquakes each year. This tectonically active area comprises the Alaska–Aleutian subduction zone where the oceanic Pacific Plate is being thrust under the continental North American plate at a rate of 0.06–0.07 m per year, which produces large and destructive earthquakes. The shallower crust of southern Alaska, as a result of the stresses generated by this plate collision, also aids the production of numerous strong earthquakes. Three of the seven largest earthquakes in the twentieth century occurred in Alaska (1957 Aleutian, 1964 Great Alaskan/Prince William Sound, and 1965 Rat Islands). Also noteworthy, the 10 largest earthquakes recorded in the United States have occurred in Alaska, most of which were megathrust earthquakes along the Alaska–Aleutian subduction zone (USGS, 2014). During the past century, three large and well-documented tsunamis were generated by seismic events in the northern Pacific near the Alaskan coast. These include the 1946 (discussed later in this chapter) and 1957 Aleutian events, and the 1964 Alaskan event (discussed below). Although all three tsunamis were produced by seismic activity, the intensities and areas of affected coastline differ.

On Good Friday, 27 March, 1964, a M_w 9.2 earthquake struck at 02:36 UTC (17:36 local time). The epicenter was 125 km east of Anchorage and 65 km west of Valdez, AK. The earthquake lasted approximately 5 min. It was the largest ever recorded in the United States and the second-most powerful measured by a seismograph worldwide—the strongest being the 1960 Chilean earthquake. The earthquake seismic activity was reported worldwide. Studies following this earthquake revealed evidence of at least nine previous similar-type megathrust earthquakes in south central Alaska in the last 5,500 years. Statistically, events of this size have a recurrence interval of about 600 years (USGS, 2014). The 1964 earthquake resulted in large vertical displacements over an area approximating 260,000-km^2, much of this beneath Prince William Sound. The land subsided up to 3 m west of the fault system. Parts of the Gulf of Alaska, east of the epicenter, were uplifted 11 m and exposed former tidelands of Montague Island on Prince William Sound. Over 180,000 km^2 of Alaska, much of which lies along the coast, subsided over 1 m, thereby making these areas vulnerable to flooding during high spring tides. The large vertical displacement also generated a major tectonic tsunami that destroyed coastal towns along the southeast coast of Alaska and caused damage all the along the Pacific coast as far as California.

4.2.3.2 Size and Extent

The earthquake generated a massive *trans*-oceanic tsunami that propagated across the Pacific Ocean to Hawaii and Japan. Landslides generated from the shaking generated many local tsunamis in Prince William Sound. It is estimated that about one third of the fatalities were due to the open ocean tsunami in Alaska, Oregon, and California. Locally generated waves in Prince William Sound claimed at least 82 lives. The tsunami killed 30 people in Port Valdez and 23 of the 68 people living in nearby Chenaga. One hundred thirty one fatalities occurred from the earthquake and tsunamis, and US$311 million was spent on repairing damages (equivalent to US$2.3 billion in 2014) (USGS, 2014).

4.2.3.3 Impacts

The greatest impacts of the tsunami were along the southeastern coast of Alaska where many coastal fishing communities along Prince William Sound and Kodiak Island were completely devastated. Reports also occurred of tsunami damage in Hawaii and Japan. The tsunami waves generated by the 1964 earthquake killed 119 people. In addition to the primary tectonic tsunami, the violent shaking during the earthquake caused many large rockslides and submarine landslides, which in turn, produced destructive local tsunamis 12–21 m high along Prince William Sound, such as Port Valdez and Chenaga (Bryant, 2001). Survivor accounts and field investigations describe a wave that was >8 m high. Reports also occurred of survivors who reached higher ground by out-running the wave.

All three Pacific west coast states of the United States were impacted by the tsunami. In Washington and Oregon, the reported tsunami damage was mainly to coastal properties and infrastructure (houses, cars, roads, flood defenses, etc.) that were impacted by 2- to 5-m waves. In some cases, the waves were amplified along estuary channels causing extensive inland damage. The most notable damage was in Crescent City, CA, where the waves were amplified, causing 10 fatalities and $15 million damage from a 6.5-m-high tsunami run-up event (relative to Mean Low Water) that flooded most of the town. The town is the second-most western point in California and is where the shape of the sea floor amplifies the tsunami energy and channels it towards the coast. Eyewitness accounts describe the inundation as occurring slowly with water levels rising at approximately 0.3 m/min (Dengler and Magoon, 2005). South of Crescent City, wave heights at Humboldt Bay and Eureka exceeded 4 m. The tsunami caused an estimated US$1 million in damage to San Francisco Bay, where the waves were >1 m high. Wave heights reached >3 m at Half Moon Bay and Santa Cruz, 2.5 m in Monterey, and nearly 2 m in San Diego. In Long Beach, CA, the tsunami caused one fatality when the wave struck the Cerritos Channel. The 1964 tsunami, like those from smaller events in 1946 and 1957, traveled across the Pacific striking the Hawaiian Islands, but caused little damage. Maximum wave heights reached 3.8 m at Hilo, 3.4 m at Kuhului, and 0.3 m a Kana'i. When the tsunami reached Japan, it was <0.25 m high. With the aid of the Pacific Tsunami Warning System, a warning was issued in Honolulu, HI within 46 min of the earthquake. At the time, this warning system had recently been updated following the 1960 Chilean earthquake and tsunami events.

4.2.3.4 Aftermath

A siren-based tsunami warning system was established in Crescent City following the 1964 event. Evacuation maps were designed and posted along evacuation routes, which makes Crescent City one of the designated "Tsunami Ready" cities in the United States according to the National Weather Service (NWS). This program promotes tsunami-hazard preparedness and is a collaboration among federal, state, and local emergency management agencies; the public; and the NWS.

In 2011, the Tōhoku earthquake in Japan resulted in a large *trans*-ocean tsunami that arrived in Crescent City over six and half hours following the earthquake. California deployed a pretsunami field team and collected valuable data during and after the event to help improved tsunami prediction models, hazard maps, and guidance for coastal communities (Wilson et al., 2011). Planners and architects have worked on trying to reduce the tsunami energy that focuses through the narrow Crescent City harbor entrance and reflects off the steep harbor walls. In 2014, a new inner boat basin was completed that comprises a 5-m-wide, 2.5-m-deep dock and a hanging underwater wall of concrete intended to block the energy of an incoming tsunami. The US$33 million inner boat basin is considered the first tsunami-resistant harbor in the United States (Spencer and Grube, 2014).

The 1964 event was a benchmark for the scientific community, as it was the first to provide direct evidence that led to a basic understanding of *trans*-ocean tectonic tsunami generation and inundation mapping over large geographic regions. At the time of publication, Alaska is conducting the Alaska Shield Exercise, which commemorates the 50-year anniversary earthquake by replicating the ground shaking and tsunami impacts associated with the event. This exercise will assess the State's progress toward meeting their preparedness objectives, validate their capabilities, identify shortfalls, and educate the coastal inhabitants. This activity is a centerpiece of FEMA's (Federal Emergency Management Agency) National Exercise Program, which has participants from across the United States and focuses on five mission areas that enable the country to build, sustain, and deliver core capabilities: prevention, protection, mitigation, response, and recovery (FEMA, 2014) (Table 4.3).

4.2.4 The 1960 Valdivia Earthquake (Great Chilean Earthquake) and Tsunami

4.2.4.1 Generation

The Valdivia Earthquake occurred at 19:11 UTC (15:11 local time) on 22 May, 1960, and is the largest earthquake recorded (M_w 9.5). It was initially

TABLE 4.3 Wave RunUp Statistics along the Pacific Ocean Coast from the 1964 Alaskan Tsunami

Region	Average Height (m)	Maximum Height (m)
Source area	11.0	67
Canada	4.8	9.8
Washington State	1.8	4.5
Oregon	2.8	4.3
California	1.6	4.3
Central America	0.4	2.4
South America	1.2	4.0
Hawaii	2.3	4.9
Pacific Islands	0.2	0.6
Japan	0.4	1.6
Australia, New Zealand	0.2	0.2

Source: From Bryant (2001).

assessed in 1960 with a magnitude (M_w) of 8.6, but was recalculated as a 9.5 in 2007 (USGS, 2007). A swarm of foreshocks occurred prior to the main seismic event, measuring as high as M_w 8.0, indicating the severity of the stress at the subduction zone off the coast of Chile. The aftershocks continued until 6 June, 1960.

The rupture was a result of the movement of the western margin of the South American Plate relative to the subducting Nazca Plate off the coast of Chile. The M_w 9.5 event was a megathrust-type earthquake generated from sudden ruptures of a long segment of a subduction zone (Kanamori and Cipar, 1974). The horizontal movement was estimated at 18.3 m and occurred in an area measuring approximately 966 km long by 161 km wide, impacting a total area of 155,526 km^2 (Atwater et al., 2005). This massive movement generated a substantial tsunami consisting of multiple waves propagating across the Pacific Ocean.

4.2.4.2 Size and Extent

Ten to 15 min after the 9.5 magnitude earthquake, the tsunami wave struck land impacting >800 km of the coast of Chile. The maximum height of the main tsunami was approximately 25 m. Tsunami waves as high as 10.7 m were recorded up to 10,000 km away from the epicenter in Hawaii, Japan, and the Philippines. A Hawaiian tide gauge that survived the tsunami recorded eight main waves and numerous smaller ones 15 h after the earthquake (Atwater et al., 2005) (Figure 4.3). The tsunami arrived at Hilo, HI, 15 h after the earthquake event. This was the hardest hit city in the Hawaiian Islands (Lander and Lockridge, 1989a). Locations along the Japanese coast measured the arrival of the tsunami 22 h after the earthquake and experienced considerable coastal flooding and damage. The Philippines also recorded the tsunami and experienced coastal damage. In Crescent City, CA, the first wave arrived 15.5 h after the earthquake and the wave runup reached 1.7 m (Lander and Lockridge, 1989b). The tsunami caused minimal damage to coastal and fishing structures and some vessels. The tsunami was also recorded along the Alaskan coastline and caused minor damage nearly a day after the earthquake (Lander and Lockridge, 1989c).

4.2.4.3 Impacts

It is unclear how many people were killed in the earthquake. Reports have varied from 2,183 to 6,000. At least 3,000 people were injured, and two million people were initially displaced because their homes were destroyed (USGS, 2007). In Japan, 142 people were killed by the tsunami (Figure 4.4). In Hawaii, 61 people were killed as a result of coastal inundation from the tsunami. In Chile, the majority of the fatalities were associated with earthquake damage. Although many of the reported casualties appear to be caused by the tsunami, it is impossible to quantify the exact cause of deaths.

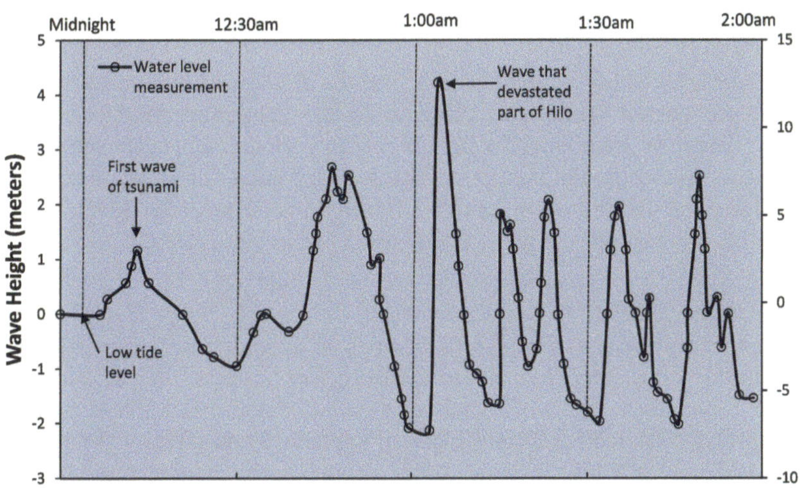

FIGURE 4.3 Water level measurements associated with tsunami waves on the island of Hawaii. Low tide = approximately 0 m; tsunami = approximately 4 m; wall of water = 4 m. *After Atwater et al. (2005).*

FIGURE 4.4 A view from high ground of one of the first large (>5 m) waves during the 1960 Chilean tsunami at Onagawa, Japan. *After Atwater et al. (2005).*

The biggest earthquake and tsunami impact was at around the city of Valdivia city in southern Chile, which is located closest to where the initial plate movement occurred. In Valdivia, it was estimated that 40 percent of the houses were destroyed causing 20,000 people to become homeless. The damage and destruction along the Chilean coast was also felt in the agricultural sector in southern Chile (Weischet and Von Huene, 1963), which had large economic repercussions for the region. The monetary damage from the tsunami varied from $400 to $800 million (USGS, 2007).

The earthquake was responsible for triggering at least one volcanic eruption in the Andes and numerous landslides. One of these landslides generated a lake tsunami that killed two people in Argentina 200 km east of Valdivia. Danger was also associated with the buildup of water behind rivers that were blocked from tsunami-generated landslides (Weischet and Von Huene, 1963).

This disaster caused a ritual human sacrifice of a 5-year-old orphan boy in the isolated coastal village of Collileufu by the indigenous Mapuche community. The local traditional healer and religious leader demanded the sacrifice to calm the Earth and the ocean. Two men served in jail for murder when the judge ruled that those involved had "acted without free will, driven by an irresistible natural force of ancestral tradition" (Tierney, 1990).

4.2.4.4 Relief Efforts

The relief effort was relatively of a small scale given the economic state of Chile at this time. Recovery was a relatively slow process. Outside aid was provided including, but not limited to, a field hospital and a public school donated by the United States and Mexico, respectively. A national committee was established in Chile in 1960 to manage the problems generated by the earthquake. The committee was still in operation in 1974 when it became the National Office of Emergency of the Interior Ministry, responsible with national responsibility for natural disasters.

4.2.5 The 1946 Aleutian Island Earthquake and Tsunami

4.2.5.1 Generation

At 09:28 UTC (00:28 local time) on 1 April, 1946, a magnitude M_w 8.1 earthquake occurred in the North Pacific near the Aleutian Island chain in Alaska. The epicenter was 25 km deep near the Unimak Island, which is the largest and eastward island in the Aleutian Island chain. The 1946 earthquake generated a Pacificwide tsunami, and at least 40 significant aftershocks were recorded (López and Okal, 2006).

4.2.5.2 Size and Extent

The tsunami was unusually large relative to the generating earthquake. The waves traveled southward with an average speed of 800 km/h. It took 4.5–5 h to reach Kauai Island, HI, and an hour more to strike Hilo on the Island of Hawaii, located 3,815 km from the earthquake epicenter. The only reliable record of the arrival time of the tsunami at the Hawaiian Islands is from the tide-gauge record in Honolulu Harbor, which shows the beginning of the first rise 4.5 h after the earthquake. The tsunami spread over the entire Pacific Ocean.

4.2.5.3 Impacts

The most detailed and well-documented account of the 1946 Aleutian tsunami come from the Hawaiian Islands and Scotch Cap that is located on Unimak Island. The tsunami reached the Hawaiian Islands approximately 5 h after the earthquake and caused considerable damage and loss of life in some locations. The height and violence of the wave along the Hawaiian coastline varied greatly. At some locations, the waves were up to 16.7 m high and violently inundated the coast. Elsewhere, they rose slowly without turbulence (Shepard et al., 1950). Tsunami waves across the islands reached estimated maximum heights of 16.7, 11, and 10 m on Hawaii, Oahu, and Maui, respectively. The tsunami killed 159 people on the Island of Hawaii and six people in Alaska. The majority of fatalities occurred in Hilo, the largest settlement on the Big Island (Figure 4.5).

FIGURE 4.5 People running fleeing from the tsunami wave in Hilo, Hawaii. This is probably the first photograph of a large tsunami wave. *Wikimedia commons photograph.*

It is estimated that one-third of Hilo was destroyed by three large waves. Approximately 500 buildings were destroyed, 1,000 buildings were severely damaged, and significant damage occurred to coastal roads, bridges, railway lines, and shore docks. In Laupahoehoe, 60 km north of Hilo, a school house was destroyed by the tsunami, killing a teacher and 25 students.

Within minutes of the earthquake, Unimak Island was impacted by a tsunami. The initial wave height was approximately 30.5 m. The wave destroyed the Scotch Cap Lighthouse and killed five lighthouse keepers. The lighthouse was a steel-reinforced concrete structure standing approximately 30 m above sea level. The 1946 tsunami had little effect on the Alaskan mainland due to the presence of the Aleutian Islands, which absorbed the brunt of the tsunami's power and shielded the mainland. The total estimated cost of tsunami damage in 1946 prices was US$26 million (Hamblyn, 2009).

4.2.5.4 Relief Efforts

The US military led the relief effort, in part because they had a significant presence on Hawaii post-World War II. General George F. Moore, Commanding General of US Army Forces in the Pacific, ordered an immediate large-scale response when the scale of the destruction became known. The US National Guard worked with voluntary groups, such as the Boy Scouts, to search for survivors, recover bodies, and clear the debris in locations where it was causing significant problems for transport and access to key infrastructure (Hamblyn, 2009).

4.2.5.5 Aftermath

A key outcome from the 1946 tsunami was the establishment of the first tsunami warning system in the Pacific Ocean. In 1948, the US Coast Guard and US Geodetic Survey formed the Seismic Sea Wave Warning System, which was later renamed as the Pacific Tsunami Warning System. No loss of life would have occurred if the system had existed prior to this event because the residents would have around 4–5 h advance notice of a possible tsunami. Shepard et al. (1950) made numerous recommendations echoing the need for an early detection and warning system. They also recommended that damage from future tsunamis could be lessened by (1) avoiding construction in known danger areas; (2) constructing suitable seawalls; (3) limiting construction in heavily populated areas; and (4) raising structures in rural areas on stilts. These recommendations are similar to those presented by Suppasri et al. (2013) in the aftermath of the Japanese 2011 tsunami disaster.

4.2.6 1908 Messina–Reggio Earthquake and Tsunami

4.2.6.1 Generation

At 04:20 UTC (05:20 local time) 28 December, 1908, a 7.2 magnitude earthquake occurred in the Messina Straits, which separates Sicily from Calabria in the southern mainland of Italy. This region sits along the Africa–Eurasia convergent plate boundary and has a history of destructive earthquakes.

As the African continental plate pushes against the sea floor underneath Europe at a rate of 25 mm per year, vertical displacements generate earthquakes; however, it is unclear as to which fault in the region caused the earthquake. This was one of the deadliest earthquakes in recent European history and also one of the first to be investigated with modern instruments (seismometers). The ground shook for about 30–40 s, and the area of destruction was approximately 6,000 km^2. Over 90 percent of the structures in Messina were destroyed.

It has been proposed recently that the tsunami was not generated by the 7.2 earthquake, but rather by a large submarine landslide that occurred offshore near Giardini, Sicily, in the Ionian Sea. Different evidence has been used to link the landslide with the tsunami, but it is still a matter of debate and uncertainty. Broken submarine telegraph cables immediately after the earthquake suggest large turbidity currents moved the sediments. Billi et al. (2008) describe tsunami arrival times, bathymetric, and seismological data to explain that a submarine landside triggered the tsunami.

4.2.6.2 Size and Extent

Within minutes after the passage of the seismic waves, a tsunami with maximum runup of about 10 m struck the coastlines of Calabria and Sicily (Billi et al., 2008). Eyewitness accounts revealed that the sea receded up to 200 m along the coastal regions. Three large waves occurred, which according to Tinti and Maramai (1996), lasted many hours along the coast, and ranged in heights from 3 m to in excess of 10 m. In some instances, the first wave was the largest, whereas at others, it was the second. Field surveys reported runup heights for hundreds of kilometers along the coastal regions in the weeks after the event show that wave heights were 1–3 m in the northern part of the Straits and 5–10 m in the southern part.

4.2.6.3 Impacts

The earthquake and subsequent tsunami nearly completely destroyed two major cities of Messina (Sicily) and Reggio Calabria (Calabria) that are located 10 km apart and face the Straits. At least 60,000 people were killed, but estimates of 80,000 (Baratta, 1910) to >100,000 (Omori, 1909) have been made. Guidoboni et al. (2007) contended that most people were killed by the earthquake, but as many as 2,000 died as a result of a 12-m-high tsunami that struck the shorelines south of Messina and south of Reggio Calabria. The tsunami completed the destruction of the collapsed buildings.

4.2.6.4 Aftermath

News of the disaster was carried by Italian torpedo boats to Nicotera, which is located on the Calabria coast. Even though the telegraph lines were still working, news was not transmitted until midnight that day. The Italian military provided immediate emergency management by searching for, treating, and evacuating the wounded. They also provided security by shooting looters. The earthquake and tsunami disasters made worldwide headlines and international relief efforts were delivered by the Red Cross and the Russian and British navy fleets to expedite the search for survivors and assist in postdisaster cleanup.

4.2.7 The 1755 Great Lisbon Earthquake and Tsunami

4.2.7.1 Generation

During the early hours of All Saints Day (November 1) in 1755, many people of Lisbon were attending religious celebrations when a powerful earthquake violently shook the western and southern parts of the country. The earthquake epicenter was located approximately 200 km southwest of the southwestern tip of Portugal and the town of Cape São Vicente. Lisbon is located 200 km north of Cape São Vicente. Anecdotal accounts describe the event as a series of three distinct jolts that started at 09:30 UTC (09:30 local time) and lasted for about 9 min (Pereria, 2006). The earthquake occurred along the Azores-Gibraltar Fault zone (AGFZ), which is the westernmost continuation of the boundary between the Africa and Eurasia plates. The seismicity along this part of AGFZ is diffuse and the plate boundary is poor (DEFRA, 2006). Neither the location, nor the character of the plate boundary, are as well understood as seismically active regions. In fact, the exact source of the 1755 earthquake remains unknown. The tsunami that occurred from a large earthquake in the region (M_w 7.9) on February 28, 1969, has been extensively modeled (e.g., Heinrich et al., 1994; Gjevik et al., 1997) in the hopes of gaining a better understanding of the 1755 event (DEFRA, 2006). The 1969 earthquake generated a small tsunami that was recorded at tidal stations in Portugal, Spain, Morocco, the Azores, and the Canary Islands. Seismicity in the region suggests that compression is occurring. The fact that a tsunami was generated means that a thrust mechanism can be assigned to the 1755 earthquake with relatively high confidence (DEFRA, 2006). The region has a long history of major, but infrequent, large tsunamis and more frequent smaller earthquakes and tsunamis (Zitellini et al., 1999).

4.2.7.2 Size and Extent

The Lisbon earthquake is one of the most violent and longest seismic events on record. In a historical context, only nine subsequent earthquakes have magnitudes either very similar to, or higher than, the Lisbon earthquake in the 259 intervening years. The earthquake generated a significant tsunami with multiple waves. The tsunami that engulfed Lisbon arrived 40 min after the earthquake. A British merchant in Lisbon at the time of the earthquake provided an account of his experiences on the morning when the earthquake struck (Maxwell, 2006):

Not long after...a general Pannic was raised from a Crowd of People's running from the Waterside, all crying out the Sea was pouring in and would certainly overwhelm the City. This new Alarm, created such Horrors in the agitated Minds of the Populace, that vast Numbers of them ran screaming into the ruinated City again, where, a fresh Shock of the Earthquake immediately following, many of them were buried in the Ruins of falling Houses. This Alarm was, however, not entirely without Foundation. For the Water of the River rose at once above twenty Feet perpendicular, and subsided again to its natural Pitch in less than a Minute's time. I was of the Number that continued where we were, but the Horror and Distraction of the Multitude were so increased by this astonishing Phænomena, that I confess they appeared more shocking to me than even the very Operations of the Earthquake.

Maximum wave height was estimated at 6 m in Lisbon, but much larger (up to 30 m) along the west coast of Portugal. At Cape São Vicente, the runup height, evaluated from historical data, was >15 m (Baptista et al., 1998). In the Algarve, there are reports that the sea receded >36 m before the tsunami struck. The wave pushed inland up to distance of 5 km. In Sagres, the sea rose to >6 m, and there are reports of 26 m waves and inland runup over 2.5 km (Tedim and Goncalves, 2007).

The resulting tsunami traveled >1,400 km north and was recorded as far away as Southern England along the west and south coasts of Ireland (Robertson et al., 1755–1756), Belgium, and the Netherlands. In Cornwall, the tsunami arrived at 14:00 and wave heights of 3–4 m were reported (Pereira, 2006). In Ireland, eyewitness accounts in Kinsale, Cork, indicate that the main wave was 4.35 m high (GBCO, 1852). The tsunami crossed the Atlantic Ocean, reaching the Lesser Antilles in the afternoon. Reports from Antigua, Martinique, and Barbados note that the sea first rose by 1.5 m, followed by large waves (Robertson et al., 1755–1756).

4.2.7.3 Impacts

Earthquakes with such a huge release of energy are likely to generate tsunamis. Many of the deaths were directly attributable to buildings collapsing and the fire that subsequently burned for three days in the city. A substantial number of people died drowned as the tsunami swept through the city and along the west coast of Portugal.

Almost all the coastal towns and villages of the Algarve on the south coast of Portugal were heavily damaged, except Faro, which was protected by sandy banks. In this region, >1,000 people died with up to 10 percent of some communities perishing. In the town of Peniche, situated 80 km north of Lisbon, many fatalities were recorded. In Kinsale, Cork, the tsunami damaged parts of the "Spanish Arch" section of the city wall protecting Galway Harbor on the west coast of Ireland. The tsunami killed approximately 2,000 in Cadiz and Huelva in southwest Spain.

Pereria (2006) provides a substantial account of the social and economic costs of the earthquake and tsunamis, based on anecdotal accounts of the event. Estimates of the total casualties from the 1755 earthquake vary tremendously, ranging from 10,000 to almost 100,000 in Lisbon. The lack of reliable estimates of the Portuguese population prior to 1755 contributes to the ambiguity regarding the death toll. The uncertainty surrounding casualties also persists because, fearing the advent of plague and disease, the government ruled that the corpses should be disposed swiftly; many of them were thrown out in the sea. The total death toll outside Lisbon was around 2,500–4,000. The number of fatalities includes fatalities from the earthquake and subsequent fires and tsunamis in Portugal, Spain, and Morocco (Pereira, 2006).

Some 250 aftershocks occurred in the six months after the main earthquake, which led to the collapse of many unstable buildings. The scale of the devastation in Lisbon was described in Mullin (1992), 6,000 of the 20,000 houses were uninhabitable and many of the rest barely so; 35 of the 40 churches were completely destroyed; 75 of the 85 convents were completely destroyed; and 33 palaces were destroyed. In summary, at least 10 percent of the wealth in Portugal was lost because of this event.

In southwestern Spain, the tsunami caused damage to Cadiz and Huelva, and the waves penetrated the Guadalquivir River, reaching Seville. On the coastline of the Madeira Islands, the 15-m tsunami waves caused damage and casualties to the western coast of Morocco, from Tangier, where the waves reached the walled fortifications of the town, to Agadir, where the waters passed over the walls, with many reported fatalities. In Ceuta, on the north coast of Africa, the tsunami was strong, but it decreased rapidly as it entered the Mediterranean Sea (Pereria, 2006).

4.2.7.4 Relief Efforts

A majority of Lisbon was destroyed by the impact of earthquakes, fires, and tsunamis. However, this devastation enabled the city to rebuild and modernize their narrow streets and overhanging buildings. The King of Portugal appointed the Marquês de Pombal to manage the rebuilding efforts. His first duty was to declare martial law that enabled the military to bring order to the city and establish a food system. The dead were buried at sea and looters were hanged (Mullin, 1992). The Marquês de Pombal became a hero to many Portuguese as the city was raised from the ruins with its new street network that was gridded and wide (previously unknown in Portugal), open squares, and a rebuilt port that served as the lifeline for the economic prosperity of the city.

4.2.7.5 Aftermath

The 1755 Lisbon earthquake represents a turning point for how humans viewed natural disasters. It represents the first attempt to shift away from an "act of God" viewpoint to a more scientific approach. A significant debate raged at the time, as people sought ways to understand how a major European Christian city could be destroyed. Many at the time still saw the disaster as punishment by God, and this led to a substantial religious revival in Portugal and around Europe. Others, such as Immanuel Kant (1724–1804) and Jean-Jacques Rousseau (1712–1778), proposed alternative and more rational views, including describing the disasters as a natural event and emphasized the need to avoid building in hazardous locations (Hamblyn, 2009). In many respects, you can trace the foundation of modern natural hazards research and the development of seismology to this event.

4.3 LANDSLIDE-GENERATED TSUNAMI

4.3.1 The 1998 Papua New Guinea Tsunami

4.3.1.1 Generation

New Guinea is a seismically active region that is located on the border of the Pacific and Australian plates that are converging at a rate of 111 mm/year (Pegler et al., 1995). A complex seismic history exists in the region involving different types of faulting, including steep dipping reverse faults with vertical displacements capable of generating large tsunamis. A major earthquake (M_w 7.1) occurred in the Bismarck Sea approximately 19 km off the northern shore of Papua New Guinea at 08:49 UTC (18:49 local time). Two after-shocks (M_w 5.6 and 5.9) occurred within 20 min of the initial earthquake. Minutes after the first earthquake, eyewitness accounts describe a loud noise that sounded like a jet fighter landing. In the following few minutes, three large waves estimated at 4–15 m arrived and destroyed three villages along a 40-km stretch of coastline. These relatively modest earthquake sizes are not typically associated with such destructive tsunamis, which led to scientists to investigate if there were other potential sources of generation. One theory is that the wave motion was focused by the bathymetric pattern along the coast that caused the waves to be more destructive than otherwise expected for an earthquake of M_w 7.1 (McSaveney et al., 2000). Models of tsunami waves indicate that an earthquake with this magnitude could not have caused a tsunami >2 m near the coast. Other examinations of the angle of approach and timing of the waves suggested that the earthquake was not the tsuna-migenic source. More recently, the evidence from offshore marine expeditions and surveys suggests that this was a seminal event because it demonstrated, for the first time, that relatively small deepwater submarine mass failures can trigger large devastating tsunamis (see Tappin et al., 2008 for a review). The seismic marine surveys indicate that the youngest slump in the region that failed on 17 July, 1998, was approximately 4.2 km wide, 4.5 km long, and 750 m for a total volume of 6.4 km^3. It is thought that this slump or submarine landslide occurred 13 min after the initial earthquake and occurred because of the shaking (Synolakis et al., 2002). Subsequently, as survivors provided more stories and reflections on the timing and sequence of the events, the evidence began to suggest movement of submarine sediment after the earthquake (cf., Davies, 1998).

4.3.1.2 Size and Extent

The tsunamis struck the Sissano Lagoon coastline in northern Papua New Guinea. This section of the coast is about 15 km long and is dominated by a long, low lying, narrow sand spit that fronts a lagoon, and was a favored site for residential communities. The tsunami that struck these villages was estimated at 11 m high and was traveling at about 70 km/h (McSaveney et al., 2000). Eyewitness reports describe three waves occurring within a few minutes of the earthquake with very little time between. The entire sequence of events lasted about 35 min and did not extend much beyond this stretch of coastline, which is very unusual for a tsunami. Reports also occurred of tsunami impacts at Aitape, which is located 40 km south of the Sissano Lagoon.

4.3.1.3 Impacts

The Sissano Lagoon coastline was the central disaster area and all structures, regardless of the quality of the construction, were destroyed (McSaveney et al., 2000). Two of the fishing coastal communities at Warapu and Arop, located on the slender spit of land that separated the Sissano lagoon from the sea, were completely swept away. Most of the deaths occurred in these villages. The failure of people to try to evacuate was not surprising as they had no prior history of tsunamis, which made it an unprecedented event. However, there is evidence that the area was impacted by a tsunami from an earthquake that occurred in 1907 (Carey, 1935; Ripper and Letz, 1999). Between 20 percent and 40 percent of the local population perished in the tsunami. At least 2,189 people lost their lives, although the true death toll will never be known.

4.3.1.4 Aftermath

Over 10,000 people were permanently displaced. This caused huge social and economic disruption to the area, such as adaptation problems at sites of new villages in inland locations that were hotter, more insect-infested and had water and sanitation issues (Dengler and Preuss, 2003). The Sissano Lagoon appears to have experienced at least three tsunami since 1907, but may have a longer history of tsunami disasters. Since the 1997 tsunami, the residential locations on the low-lying sand barrier fronting the Sissano Lagoon are now considered unsafe for habitation (McSaveney et al., 2000). The subsequent ocean surveys and field land surveys in the years preceding this event have led to improved models of tsunami generation and also caused a reevaluation of tsunamigenic sources and the risk associated with tsunami hazards.

4.3.2 Storegga Slide Tsunamis

4.3.2.1 Generation

Three substantial submarine landslides took place between 30,000 and 7,200–7,000 years BP in the Norwegian Sea on the northern edge of the Atlantic Ocean. These slides are cumulatively referred to as the Storegga Slides. These events were most likely caused by isostatic uplift west of Norway in an area that is away from plate boundaries, and therefore considered tectonically inactive. It is probable that continental slopes were steepening from increased land-based erosion causing large volumes of sediment to deposit in this area. It is unknown how long this sediment took to accumulate and whether it was over one or multiple glacial cycles. The denudation rates increased during the Quaternary when extensive ice sheets periodically developed and retreated on these uplifted areas (Long and Holmes, 2001). Sediment delivery to the continental shelf is only possible when the coastal waters and adjoining lands are not covered by permanent ice and it is only during glacial retreat and ice melt could this magnitude of sediment erosion could take place. The massive sediment erosion and subsequent deposition eventually resulted in large-scale submarine landslides. This slope failure could have been triggered by small earthquakes related to the isostatic uplift. Three exceptionally large submarine landslides have been mapped on the sea floor. It is likely that all three generated tsunamis (Dawson et al., 1998).

4.3.2.2 Size and Extent

A numerical model simulation predicts that the waves generated from Storrega Slides would have inundated most coastline bordering the Norwegian and North Seas (Harbitz, 1992). Tsunami deposits have been traced in nearshore coastal lake basins in western Norway, Shetland Islands, and Faroe Islands where accumulations of fine grained organic muds were interrupted by marine sands and gravels. Sand layers were found several meters inland in raised estuarine mudflats in Scotland (Long et al., 1989) and peat outcrops in Shetland (Bondevik et al., 2003), which is indicative of tsunami deposits. These field observations have been used to reconstruct runup heights for the tsunamis. Bondevik et al. (2012) and Bryn et al. (2005) report runup heights for the west coast of Norway (3–12 m), Scotland (3–6 m), Shetland (20–30 m), and the Faroe Islands (>15–20 m). The first slide occurred approximately 30,000 years BP and covered an area of 34,000 km^2 with an average deposit thickness of 114 m. This event occurred during the Ice Age, which makes it difficult to find evidence of a tsunami (Dawson et al., 1988). The second and third submarine slides both occurred approximately 7,200–7,000 years BP. One landslide covered an area of 19,200 km^2 with an average thickness of 88 m. The second covered an area of 6,000 km^2 and the average deposit thickness is not known.

4.3.2.3 Impacts

The first of the 7,200-year-old event most likely caused substantial impact along the northeast coastline of Scotland and western Norway. A coastal site in Inverness, Scotland, shows a prehistoric occupation layer that is covered by a layer of tsunami deposits. Little evidence occurs for any occupation on this site for hundreds of years following. It is likely that a substantial loss of life and communities occurred, although it is presumed that populations were sparse during this time (Dawson, 1990). Based on sedimentological evidence, the tsunami extended several hundred meters inland of the former coastline with a runup of 1–2 m in open areas, but much higher in enclosed bays (Long et al., 1989). It has been suggested that the submarine slide 7,000 years BP is responsible for the most recent flooding of Doggerland, which was the name given to the now flooded part of the southern North Sea, and the separation of the land bridge between the United Kingdom and Europe (Weninger et al., 2008).

4.3.2.4 Aftermath

Recent concerns have developed that anthropogenic activities, in particular oil and gas extraction, may generate another large submarine slide (Masson et al., 2006). However, recent research has found this to be unlikely and only realistically possible following an Ice Age (Bondevik et al., 2003). If a repeat tsunami occurs, it would be catastrophic from an economic perspective, but with relatively low numbers of fatalities because the coastal areas are sparsely populated.

4.4 VOLCANO-GENERATED TSUNAMI

4.4.1 Krakatoa Volcanic Eruption and Tsunami

4.4.1.1 Generation

Krakatau is a volcanic island in Indonesia (formerly called the Dutch East Indies) located in the shallow waters of the tectonically active Sunda Strait. This region is located in the transition zone between the Sumatra and Java fault zones where the Indo-Australian plate is subducting under the Southeast Asian plate. The 1883 eruption comprised four large explosions that took place on 27 August, 1833, at 23:30 UTC (05:30 local time), 00:44 (06:44), 04:02 (10:02), and 04:41 (10:41). The 1883 eruption was estimated to be equivalent to 200 megatons of trinitrotoluene or 10,000 Hiroshima atomic bombs (Pararas-Carayannis, 2003). It is estimated that >17 km^3 of debris was ejected, creating a huge plume of gas and ashes that rose 50 km into the atmosphere that was dense enough to block out the midday sun. The debris caused a volcanic winter and worldwide temperatures were reduced by an average of 1.2 °C for 5 years. Interestingly, it has been proposed that the dramatic red sky shown in Edvard Munch's 1893 painting *The Scream* is a depiction of the

Norwegian sky shortly after the 1883 eruption of Krakatoa. Pyroclastic flows and huge landslides of rock, ash, and superheated gas destroyed communities across the islands. Each explosion triggered large tsunamis, some are estimated at 35 m high, which completely submerged the adjacent islands and wiped out all traces of human settlement. The final eruption was large enough that the noise and the shockwaves were recorded as far away as Perth, Western Australia, and near Mauritius, which are located 2,000 and 3,000 miles from Krakatoa, respectively.

4.4.1.2 Size and Extent

Eyewitness accounts reported that inhabitants of the low-lying 1,000 Islands, located 80 km east of Krakatoa between Java and Sumatra, had to climb trees to save themselves. Similarly, eyewitness accounts from the passenger ship S.S. *Loudon*, which was anchored in Lampong Bay 75 km north of Krakatoa, describe "a gigantic wave of prodigious height advancing toward the seashore with considerable speed" (Pararas-Carayannis, 2003). This passenger ship was loaded with sightseers on a volcano trip to tour the volcano activity that had started on 20 May, 1883. The heroic efforts of its Captain Lindemann and his crew are well documented: "immediately, the crew… managed to set sail in face of imminent danger; the ship had just enough time to meet with the wave from the front. The ship met the wave head on and the Loudon was lifted up with dizzying rapidity and made a formidable leap… The ship rode at a high angle over the crest of the weave and down the other side. The wave continued on its journey toward land, and the benumbed crew watched as the sea in a single sweeping motion consumed the town. There, where an instant before had lain the town of Telok Betong, remained nothing but the open sea" (Aleshire, 2007). Other ships such as the steamship P.S. *Berouw* were not as fortunate. The P.S. *Berouw* was lifted over 3 km inland and deposited nearly 10 m above sea level, killing all 28 crew members. The aftermath is described by one witness in de Boer and Sanders (2004) as: "Thousands of human beings and also carcasses of animals still await burial, and make their presence apparent by the indescribable stench. They lie in knots and entangled masses impossible to unravel, and often jammed along with coconut stems among all that had served these thousands as dwellings, furniture, farming implements, and adornments for houses and compounds".

4.4.1.3 Impacts

Simkin and Fiske (1983) provide the most detailed account of the event based on 88 eyewitness accounts from nearby coastal areas or ships and supplementary data (barometric, tide gauge, and geology). Total fatalities were reported by Dutch authorities as 36,417; 90 percent of which were killed by the tsunamis. However, nobody knows how many people were washed out to sea by these enormous waves. For months after the eruption, the Sunda Straits where congested with thick pumice banks, often containing at least 50 corpses. Large portions of coral reef were displaced to the land (Figure 4.6). The impacts of the tsunami were felt worldwide, but the impact on coastlines were insignificant compared with the devastation felt in the region of Krakatoa. Small sea level fluctuations were recorded by tide gauges in South Africa, New Zealand, Australia, Japan, Hawaii, Alaska, North and South America, and the United Kingdom (Pararas-Carayannis, 2003).

Based on comparisons of maps generated before and immediately after the eruptions, it was estimated that two thirds of the island and archipelago were completely destroyed. At the time of the eruptions, Krakatoa consisted of three distinct adjoining cones, Rakata (800 m), Danan (450 m), and Perboewatan (120 m). Two weeks prior to the 1883 eruptions, a relatively detailed map of Krakatoa was generated by a Dutch topographical engineer Captain Ferzenaar who was interested in the presence of several craters and fumaroles that were remnants of smaller eruptions that had occurred and caused damage in May and June the same year. These were correlated with previous sketches in 1748 and 1836 (van Padang, 1955). Captain Ferzenaar's advice to colonial authorities to suspend visits to the island were a precaution to the ongoing volcanic activity he observed on the islands. Two weeks after his visit, the series of eruptions occurred, cumulating in the large explosions on 6 August, 1883. This is one of the deadliest volcanic events in recorded history with the estimated fatalities ranging from 36,000 to 120,000.

4.4.1.4 Aftermath

Since 1927, submarine eruptions have resulted in new land emerging from the ocean caldera and the birth of Anak Krakatau (Child of Krakatau), which

FIGURE 4.6 Photographed in 1885, this massive chunk of coral was scraped off the seabed and deposited on the western coastline of Java during the eruptions of Krakatoa. It is believed that a chunk similar in size to this, smashed into the Fourth Point Lighthouse (near Anjer in Java, east of Krakatoa), ripping it off its foundations. *From: http://commons.wikimedia.org/wiki/File: COLLECTIE_TROPENMUSEUM_Groot_brok_koraal_uit_zee_dat_bij_Anjer_op_land_is_geworpen_na_de_uitbarsting_van_de_Krakatau_in_1883._TMnr_60005541.jpg.*

presently reaches a height of around 300 m. These eruptions generated tsunamis, but their heights were rapidly attenuated away from the source region (Pararas-Carayannis, 2003). Anak Krakatoa is some of the youngest land on Earth and makes headlines with its constant volcanic activity that sends plumes of ash 2 km in the air and sends lava flows back into the ocean. The unique age of this land makes it of interest to ecologists and biogeographers that observe the development and establishment of pioneer plant species and communities in areas with no previous footprint of life. Camus et al. (1987) reported on the sequence of new eruptions that have started on Anak Krakatoa. They state that the possibility of the main threat posed by Anak Krakatau at the present time could be the generation of tsunamis by failure of parts of the SW flank of the island, similar to the failure that caused the 2-m-high tsunami experienced on Rakata during the night of October 19–20, 1981.

4.5 Conclusions

This chapter has demonstrated that when very large tsunamis are generated, whether by earthquake, landslide or volcano, the effects on the coastline and the human populations is devastating, and sometimes results in substantial loss of life. Tsunamis and their potential impacts received considerable attention following the major events in Japan in 2004 and 2011. Preceding these events, only a few black and white photographs of tsunamis existed. The events of 2004 and 2011 were documented with of high quality still and video imagery. Given the rarity of a major tsunami, it is remarkable that two large and devastating events occurred within an eight-year time span; the last one prior to these with a comparable magnitude occurred in 1755. The visual impact of the recordings of these two recent events was extraordinary for the general public and provided new data for researchers worldwide. The Pacific Ocean is most vulnerable to future tsunamis impacts due to the frequency and size of earthquakes that occur along the 'Ring of Fire;' however, no ocean is without risk. Minor and micro-tsunamis occur frequently, but are rarely noticed by the general public and are only detected using high quality tide gauges. The need for all oceans to have a tsunami detection and warning system is emphasized in light of the loss of life the 2004 and 2011 events.

REFERENCES

USGS [U.S. Department of the Interior, U.S. Geological Survey], 2014. The 1964 Great Alaska Earthquake & Tsunami. http://www.usgs.gov/blogs/features/usgs_top_story/the-1964-great-alaska-earthquake-tsunami/ (accessed June 2014).

ADRC, 2011. Great East Japan Earthquake. Update on Damage and Recovery (2nd Report). Asian Disaster Reduction Center.

Anon, 2011. Damage to Cultural Properties in the 'Great East Japan Earthquake'. Agency for Cultural Affairs, Japan.

Aleshire, P., 2007. Ocean Ridges and Trenches. Chelsea House Publications, 148pp.

Asian Development Bank, 2005. Curbing corruption in tsunami relief operations. In: Proceedings of the Jakarta Expert Meeting Organized by the ADB/OECD Anti-Corruption Initiative for Asia and the Pacific and Transparency International. Hosted by the Government of Indonesia Jakarta, Indonesia 7–8 April 2005.

Atwater, B.F., Cisternas, V.M., Bourgeois, J., Dudley, W.C., Hendley II, J.W., Stauffer, P.H., 2005. Surviving a Tsunami—Lessons from Chile, Hawaii and Japan. Circular 1187, Version 1.1, Revised Edition. U.S. Geological Survey.

Bagla, P., January 28, 2005. After the Earth Moved. Science Now.

Baptista, M.A., Heitor, S., Miranda, J.M., Mendes Victor, L., 1998. The 1755 Lisbon Tsunami: Evaluation of the Tsunami Parameters. J. Geodynamics 25 (2), 143–157.

Baratta, M., 1910. La catastrophe sismica calabro-messinese (28 dicembre 1908), relazione. Soc. Geogr It., Rome.

Barron, P., Clark, S., Daud, M., 2005. Conflict and Recovery in Aceh: An Assessment of Conflict Dynamics and Options for Supporting the Peace Process. World Bank, Washington D.C. USA.

Billi, A., Funiciello, R., Minelli, L., Faccenna, C., Neri, G., Orecchio, B., Presti, D., 2008. On the cause of the 1908 Messina tsunami, southern Italy. Geophys. Res. Lett. 35, L06301. http://dx.doi.org/10.1029/2008GL033251.

Bondevik, S., Stormo, S.K., Skjerdal, G., 2012. Green mosses date the Storegga tsunami to the chilliest decades of the 8.2 ka cold event. Quat. Sci. Rev. 45, 1–6.

Bondevik, S., Dawson, S., Dawson, A., Lohne, Ø., 2003. Record-breaking height for 8000-year old tsunami in the North, EOS. Trans. Am. Geophys. Union 84 (31), 289–293.

Bryant, E., 2001. Tsunami: The Underrated Hazard. Cambridge University Press, 320 pp.

Brunt, K.M., Okal, E.A., MacAyeal, D.R., 2011. Antarctic ice-shelf calving triggered by the Honshu (Japan) earthquake and tsunami, March 2011. J. Glaciol. 57 (205), 785–788.

Bryn, P., Berg, K., Forsberg, C.F., Solheim, A., Kvalstad, T.J., 2005. Explaining the Storegga Slide. Marine and Petroleum Geology 22, 11–19.

California Coastal Commission, 2011. The Tōhoku Tsunami of 11 March, 2011: A Preliminary Report on Effects to the California Coast and Planning Implications. State of California, Natural Resource Agency, California Coastal Commission, 40 pp.

Camus, G., Gouraud, A., Vincent, P.M., 1987. Petrologic evolution of Krakatau (In-donesia): implications for a future activity. J. Volcanol. Geotherm. Res. 33, 299–316.

Carey, S.W., 1935. Preliminary notes on a recent earthquake in New Guinea. Aust. Geogr. 2 (8), 8–15.

Chang, K., March 13, 2011. Quake Moves Japan Closer to U.S. and Alters Earth's Spin. The New York Times.

Columbia Journalism Review, January 20, 2012. Two Years Later, Haitian Earthquake Death Toll in Dispute.

Davies, H., 1998. The Sissano Tsunami. University of Papua New Guinea, Port Moresby.

Dawson, A.G., 1990. Evidence for a tsunami from a mesolithic site in inverness, Scotland. J. Archaeol. Sci. 17 (5), 509–512.

Dawson, A.G., Long, D., Smith, D.E., 1988. The Storegga slides: evidence from eastern Scotland for a possible tsunami. Mar. Geol. 82 (3–4), 271–276.

DEFRA, 2006. Tsunamis—Assessing the Hazard for the UK and Irish Coast. Study Commissioned by the Defra Flood Management Division. The Health and Safety Executive and the Geological Survey of Ireland, 103 pp.

Dengler, L.A., Magoon, O., 2005. The 1964 tsunami in Crescent City, California: a 40-year retrospective. In: Proceedings from Solutions to Coastal Disasters 2005. American Society of Civil Engineers, pp. 639–648.

Dengler, L., Preuss, J., 2003. Mitigation lessons from the July 17, 1998 Papua New Guinea tsunami. Pure Appl. Geophys. 160, 2001–2031.

Diacu, F., 2009. Megadisasters: Predicting the Next Catastrophe. Oxford University Press, Oxford.

FEMA, 2014. National Exercise Program (NEP). National Level Exercise—Capstone Exercise 2014. http://www.fema.gov/national-exercise-program-nep-capstone-exercise-2014 (accessed June 2014).

GFDRR, 2012. Lessons learned from the Great East Japan Earthquake: hazard information and damage scenarios to inform effective countermeasures to extreme events. In: Improving the Assessment of Disaster Risks to Strengthen Financial Resilience. A Special Joint G20 Publication by the Government of Mexico and the World Bank, p. 297.

Gjevik, B., Pedersen, G., Dybesland, E., Harbitz, C.B., Miranda, P.M.A., Baptista, M.A., Mendes-Victor, L., Heinrich, P., Roche, R., Guesmia, M., 1997. Modeling tsunamis from earthquake sources near Gorringe Bank southwest of Portugal. J. Geophys. Res. C102, 27931–27949.

Great Britain Census Organisation, 1852. Census of Ireland 1851: Tablet of Deaths. HMSO, London.

Guidoboni, E., Ferrari, G., Mariotti, D., Comastri, A., Tarabusi, G., Valensise, G., 2007. CFTI4Med, Catalogue of Strong Earthquakes in Italy (461 B.C.–1997) and Mediterranean Area (760 B.C.–1500). INGV-SGA.

Hamblyn, R., 2009. Tales of the Earth: Four Events That Changed the World. Picador, London.

Harbitz, C.B., 1992. Model simulation of tsunamis generated by the Storegga slides. Mar. Geol. 105, 1–21.

Heinrich, P., Baptista, M.A., Miranda, P., 1994. Numerical simulation of the 1969 tsunami along the Portuguese coasts. Preliminary results. Sci. Tsunami Hazards 12, 3–23.

Japan Times, April 14, 2011. Tsunami Hit More than 100 Designated Evacuation Sites. Japan Times.

Jayasuriya, S., McCawley, P., 2010. The Asian Tsunami: Aid and Reconstruction after a Disaster. Edward Elgar, Cheltenham UK and Northampton MA USA.

Kamiya, S., June 20, 2011. Debris Removal, Recycling Daunting. Japan Times.

Kanamori, H., Cipar, J.J., 1974. Focal process of the great Chilean earthquake May 22 1960. Phys. Earth Planet. Inter. 9 (2), 128–136.

Kingston, J., 2012. Natural Disaster and Nuclear Crises in Japan: Response and Recovery after Japan's 3/11. Routledge.

Kowalik, Z., Knight, W., Logan, T., Whitmore, P., 2005. Numerical modelling of the global tsunami: Indonesian tsunami of 26 December 2004. Sci. Tsunami Hazard 23 (1), 40–56.

Lander, J.F., Lockridge, P., 1989a. Tsunami in Hawaii in US Department of Commerce. National Oceanic and Atmospheric Administration. United States Tsunami 1690–1988 (including Unites States Possessions 1690–1988).

Lander, J.F., Lockridge, P., 1989b. Tsunami on West Coast of United States in US Department of Commerce. National Oceanic and Atmospheric Administration. United States Tsunami 1690–1988 (including Unites States Possessions 1690–1988).

Lander, J.F., Lockridge, P., 1989c. Tsunami in Alaska in US Department of Commerce. National Oceanic and Atmospheric Administration. United States Tsunami 1690–1988 (including Unites States Possessions 1690–1988).

Laurent, C., Lebreton, M., Borrero, J.C., 2013. Modeling the transport and accumulation floating debris generated by the 11 March 2011 Tōhoku tsunami. Mar. Pollut. Bull. 66, 53–58.

Long, D., Holmes, R., 2001. Submarine landslides and tsunami threat in Scotland. In: ITS Proceedings, pp. 355–366.

Long, D., Smith, D.E., Dawson, A.G., 1989. A Holocene tsunami deposit in eastern Scotland. J. Quatern. Sci. 4, 61–66.

López, A.N., Okal, E.A., 2006. A seismological reassessment of the source of the 1946 Aleutian 'tsunami' earthquake. Geophys. J. Int. 165 (3), 835–849.

Masson, D.G., Harbitz, C.B., Wynn, R.B., Pedersen, G., Levholt, F., 2006. Submarine landslides: processes, triggers and hazard prediction. Phil. Trans. R. Soc. A., 1–31.

Maxwell, K., 2006. Lisbon 1755: the first 'Modern' disaster (but if modern, how is it so?). In: Paper Presented at the Conference the Treaty of Windsor and 620 Years Do Anglo-Portuguese Relations, 18 October 2006, St. Peters College. University of Oxford, UK.

McKee, M., 9/2/2005. Power of tsunami earthquake heavily underestimated. New Sci..

McSaveney, M.J., Goff, J.R., Darby, D.J., Goldsmith, P., Barnett, A., Elliott, S., Nongkas, M., 2000. The 17 July 1998 tsunami, Papua New Guinea: evidence and initial interpretation. Mar. Geol. 170, 81–92.

Meisl, C.S., Safaie, S., Elwood, K.J., Gupta, R., Kowsari, R., 2006. Housing reconstruction in northern Sumatra after the December 2004 great Sumatra earthquake and tsunami. Earthquake Spectra 22 (S3), 777–802.

Mullin, J.R., 1992. The Reconstruction of Lisbon Following the Earthquake of 1755: A Study in Despotic Planning. Landscape Architecture & Regional Planning Faculty Publication Series. Paper 45.

National Police Agency of Japan, April 10, 2014. Damage Situation and Police Countermeasures Report.

Nebehay, S., March 11, 2011. Japan Request Foreign Rescue Teams. Reuters News Agency.

Neumann Van Padang, M., 1955. A Swedish sketch of Krakatoa in 1748. Bull. Volcalog., Series 2 (17), 135–139.

Omori, F., 1909. Preliminary report on the Messina–Reggio earthquake of December 28, 1908. Bull. Imp. Earthquake Invest. Comm. 3-2, 37–46.

Pararas-Carayannis, G., 2003. Near and far-field effects of tsunamis generated by the paraoxysmal eruptions, explosions, caldera collapses and massive slope failures of the Krakatoa volcano in Indonesia on August 26–27, 1883. Int. J. Tsunami Soc. 24 (4).

Pegler, G., Das, S., Woodhouse, J.H., 1995. A seismological study of the eastern New Guinea and the western Solomon sea regions and its tectonic implications. Geophys. J. Int. 122, 961–981.

Pereria, A.S., 2006. The Opportunity of a Disaster: The Economic Impact of the 1755 Lisbon Earthquake. Discussion Paper 06/03. Centre for Historical Economics and Related Research, York University, 37pp.

Ripper, I.D., Letz, H., 1999. The Sissano Lagoon (Aitape) Tsunami: Which Earthquake Was Responsible? Papua New Guinea, Department of Minerals and Energy, Geological Survey Report 99/7, 10 pp.

Robertson, J., Carteret Webb, P., Adee, S., Hodgson, J., Cranbrook, J., Pringle, J., Mills, H., Birch, T., Thomlinson, M., Philips, R., Crusius, L., Blair, J., Viscount Parker, Lord, Huxham, J., Borlase, W., Swanzey, S., Arderon, W., Barber, T., Harrison, J., Cowper, S., Gardener, R., Colquhoun, J., Nicola, L., Brocklesby, R., Tom, H., Steplin, J., De Hondt, M., Allamond, M., 1755–1756. An extraordinary and surprising agitation of the waters, though without any perceptible motion of the earth, having been observed in various parts of this island, both maritime and inland, on the same day, and chiefly about the time, that the more violent comotions of both earth and waters so extensively affected many very distant parts of the globe; the following accounts, relating to the former, have been transmitted to the society, in which are specified the times and places when and where they happened. Philos. Trans. 49, 351–398.

Sibuet, J.-C., Rangin, C., Le Pichon, X., Singh, S., Cattaneo, A., Graindorge, D., Klingelhoefer, F., Lin, J.-Y., Malod, J.-A., Maury, T., Schneider, J., Sultan, N., Umber, M., Yamaguchi, H., the Sumatra Aftershocks Team, 2007. 26th December 2004 great Sumatra–Andaman earthquake: co–seismic and post–seismic motions in northern Sumatra. Earth Planet. Sci. Lett. 263, 88–103.

Shepard, F.P., MacDonald, G.A., Cox, D.C., 1950. The tsunami of April 1, 1946. Bull. Scripps Instit. Oceanogr. 5 (6), 6–33.

Simkin, T., Fiske, R.S., 1983. Krakatau 1883: The Volcanic Eruption and its Effects. Smithsonian Institution Press, 464pp.

Spencer, A., Grube, N., March 23, 2014. Why Are We Tsunami Prone? Del Norte Triplicate. http://www.triplicate.com/News/Local-News/Why-are-we-tsunami-prone (accessed June 2014).

Suppasri, A., Shuto, N., Imamura, F., Koshimura, S., Mas, E., Yalciner, A.C., 2013. Lessons learned from the 2011 great east Japan tsunami: performance of tsuanmi countermeasures, coastal buildings, and tsunami evacuation in Japan. Pure Appl. Geophys. 170, 993–1018.

Synolakis, C.E., Bardet, J.P., Borrero, J., Davies, H., Okal, E., Silver, E., Sweet, J., Tappin, D., 2002. Slump origin of the 1998 Papua New Guinea tsunami. Proceedings of the Royal Society of London A 458, 763–769.

Tappin, D.R., Watts, P., Grilli, S.T., 2008. The Papua New Guinea tsunami of 17 July 1998: anatomy of a catastrophic event. Nat. Hazards Earth Syst. Sci. 8, 243–266.

Tedim, F., Gonçalves, J., 2007. Simulation of the 1755 tsunami flooding area in the Algarve (Southern Portugal): the case-study of Portimão. Territorium 14, 21–31.

Tierney, P., 1990. The Highest Altar: Unveiling the Mystery of Human Sacrifice. Penguin, New York.

Tinti, S., Maramai, A., December 1996. Catalogie of tsunamis generate din Italy and in Cote d'Azur, France: a step towards a unified catalogue of tsunamis in Europe. Annali Di Geofisica XXXIX (N6), 1253–1299.

USGS [United States Geologic Survey], 2007. The Largest Earthquake in the World. USGS Earthquake Hazards Program.

USGS [United States Geologic Survey], 2011. Magnitude 9.03—Near the East Coast of Honshu, Japan. USGS Earthquake Hazards Program.

Walton, M., 20 May 2005. Scientists: Sumatra Quake Longest Ever Recorded. CNN.

West, M., Sanches, J.J., McNutt, S.R., 2005. Periodically triggered seismicity at mount Wrangell, Alaska, after the Sumatra earthquake. Science 308 (5725), 1144–1146.

Weischet, W., Von Huene, R., 1963. Further observations of geologic and geomorphic change resulting from the catastrophic earthquake of May 1960. Bull. Seismol. Soc. Am. 53 (6), 1237–1257.

Weninger, B., Schulting, R., Bradtmöller, M., Collard, M., Edinborough, K., Hilpert, J., Jöris, O., Niekus, M., Rohling, E.J., Wagners, B., 2008. The catastrophic final flooding of Doggerland by the Storegga Slide tsunami. Documenta Praehistorica XXXV, 1–26.

Wilson, R., Dengler, L., Borrero, J., Synolakis, C., Jaffe, B., Barberopoulou, A., Ewing, L., Legg, M., Ritchie, A., Lynett, P., Admire, A., McCrink, T., Falls, J., Treiman, J., Manson, M., Davenport, C., Lancaster, J., Olson, B., Pridmore, C., Real, C., Miller, K., Goltz, J., 2011. The Effect of the 2011 Tōhoku Tsunami on the California Coastline. California Department of Conservation website. http://www.conservation.ca.gov/cgs/geologic_hazards/Tsunami/Documents/ssa_2011_california_tohoku_small.pdf (accessed June 2014).

Zitellini, N., Chierici, F., Sartori, R., Torelli, L., 1999. The tectonic source of the 1755 Lisbon earthquake and tsunami. Annali Di Giofisica 42 (1), 49–55.

FURTHER READING

De Boer, J.Z., Sanders, D.T., Volcanoes in Human History: The Far Reaching Effects of Major Eruptions. Princeton University Press, 295 pp.

Folger, T., February 2012. The calm before the wave. Where and when will the next tsunami hit? Nat. Geogr., 54–77.

Gonzalez, F.I., Milburn, H.M., Bernard, E.N., Newman, J.C., 1998. Deep-ocean assessment and reporting of tsunamis (DART®): brief overview and status report. In: Proceedings of the International Workshop on Tsunami Disaster Mitigation, 19–22 January 1998. Tokyo, Japan.

Meining, C., Eble, M.C., Stalin, S.E., 2001. System development and performance of the deep-ocean assessment and reporting of Tsunamis (DART) system from 1997–2001. In: ITS 2001 Proceedings, NTHMP Review Session, pp. 235–242. PaPer R-24.

Pino, N.A., Piatanesi, A., Valensise, G., Boschi, E., 2009. The 28 December 1908 Messina Straits earthquake (M_w 7.1): a great earthquake throughout a Century of seismology. Seismol. Res. Lett. 80 (2), 243–259.

Williams, S., June 28, 2006. Indian Ocean Tsunami Warning System up and Running. UNESCOPRESS. Press Release No. 2006-69.

Chapter 5

Palaeostorm Surges and Inundations

Jonathan F. Nott
College of Science, Technology and Engineering, James Cook University, Cairns, Queensland, Australia

ABSTRACT
Prehistoric records of tropical-cyclone (TC)-generated storm surges and marine inundations occur in the form of ridges of coral rubble, sand, shell, sand and shell, and pumice and barrier washover deposits. As yet, such records have not been identified beyond approximately 7,000 years of age. Recent studies of palaeomarine inundation deposits in the United States, Japan, and Northern Australia have shown a substantial degree of synchroneity in global intense TC behavior over the past 3,000–5,000 years. One of the most striking aspects of these records is that they all display extended alternating periods (centuries to millennia) of relative quiescence and heightened intense TC activity irrespective of both the resolution and type of long-term, marine inundation record. These deposits have unique sedimentary signatures and they can now be used to determine return intervals of storm marine inundations using robust statistical techniques.

Storm surges and associated marine inundations can be generated by all storms approaching the coast from the sea. A storm surge is an elevation of the sea surface ahead of, and during the passage of, a storm toward the coast. It is a long-gravity wave with a wavelength comparable to the diameter of the generating storm and can be likened to a raised dome of water that inundates the coast. Variations in the height of surges are caused by the lowered atmospheric pressure associated with the storm, the forward or translational speed of the storm, the radius of maximum winds, the angle of approach of the storm track to the coast, the offshore bathymetry, and the shape of the coastline. Surge heights can exceed 7 m under optimal conditions. The surge forms part of the total marine inundation associated with the storm. Other components of the inundation include the tide, wave setup (addition to the height of the water by breaking waves), wave action (waves on top of the surge), and wave run-up (uprush of waves against or over an object or sloping shoreline, respectively).

Coastal and Marine Hazards, Risks, and Disasters. http://dx.doi.org/10.1016/B978-0-12-396483-0.00005-4
Copyright © 2015 Elsevier Inc. All rights reserved.

Storm tide refers to the combined inundation level resulting from the storm surge and tide. Storm surges can be generated by both temperate low-pressure systems and tropical cyclones (TCs). The latter tend to generate the largest surges and inundations because they achieve lower central air pressures and can generate stronger onshore winds. This chapter describes the types of evidence used to reconstruct past or palaeomarine surges and inundations generated by TCs. These long-term records help to reduce risk from this hazard because they allow more robust estimates of magnitude/frequency relationships and return intervals to be made compared to those that can be derived from short historical records alone.

Palaeostorm inundations are recorded principally by sediments deposited inland and above the normal limit of wave-deposited sediments along a coast. The prehistoric record of storm inundations is largely restricted to the latter half of the Holocene, which is approximately the last 5,000–6,000 years, or since termination of the Holocene marine transgression. Although TCs would have formed during the period of lower sea level between the present and last interglacial period, any sediments deposited by these events would likely have been reworked during the Holocene sea level rise. The last interglacial storm deposits are preserved in some locations, but to date, no such deposits have been positively identified.

Sediments deposited during TCs take the form of ridges of coral rubble, cheniers, sand beach ridges, shell ridges, gravel ridges, pumice ridges, sand splays, and washover deposits, being layers of sand within otherwise muddy or organic sediments in back barrier lagoons. All these sediments are deposited by surge and/or waves. Substantial quantities of sediment, mainly sand and rarely isolated shells, can be transported inland and deposited by the high-velocity winds. These deposits are difficult to recognize because they occur in environments where normal aeolian processes have deposited similar sand-sized particles. Storm-wind-blown deposits have also not been recognized as forming distinct sedimentary layers or units in the back beach environment, unlike aeolian deposits blown by lower velocity winds (Nott, 2014). The following sections describe the sedimentary and morphological characteristics of each of these types of palaeomarine inundation deposits.

5.1 CORAL RUBBLE RIDGES

Coral rubble ridges occur in locations where coral reefs occur close to shore. During a storm, coral fragments are eroded from near-shore reefs by wave action and transported either onshore or offshore (Baines et al., 1974; Davies, 1983; Hughes, 1999; Rasser and Riegl, 2002). Fragments can also be transported from existing accumulations of offshore coral rubble. The offshore accumulations result from a number of erosional processes, such as biodegradation and wave action, during storms and fair weather conditions (Hughes, 1999; Rasser and Riegl, 2002). It has been suggested that the angle of the offshore reef slope plays a role in whether the eroded fragments are transported predominantly offshore or onshore. Steep reef foreslopes favor offshore transport of fragments, commonly to depths of greater than 50 m, which is too deep for the fragments to be reworked and transported by storm waves. Shallow and wide reef fronts favor transport onshore and the formation of coral rubble ridges. However, some sites, such as Curacao Island in the central Great Barrier Reef, Australia, (Figures 5.1−5.3) that are fronted by narrow, steep reef slopes, have extensive coral rubble ridge development on land (Hayne and Chappell, 2001; Nott and Hayne, 2001). The sites with minimal accumulation of coral rubble in the shallow waters of the reef and maximum accumulation of rubble in the deeper offshore waters below wave base (depth to which waves will entrain and transport sediment on ocean floor) suggest that the onshore ridges could have formed from predominantly live coral fragments broken off during the storm. At other sites, however, little doubt exists that onshore ridges were formed from the reworking of existing accumulations of rubble in the shallow waters offshore. It is difficult to know whether the rubble ridge is deposited gradually during the storm, or as one, or as a series of sediment units moving landward from existing offshore accumulations. Scoffin (1993) has described coral rubble ridge building as ridges that "have been transported and deposited like large asymmetric waves of sediment; material picked up on the seaward side is rolled up the ridge and dropped down the advancing slope." This suggests that an entire, or substantial part, of an offshore accumulation of rubble is moved onshore as a single unit during the storm. Alternatively, if the ridges accumulate gradually, then it could be assumed that the ridge will increase in height over time during the storm. In this instance, wave run-up may influence their formation.

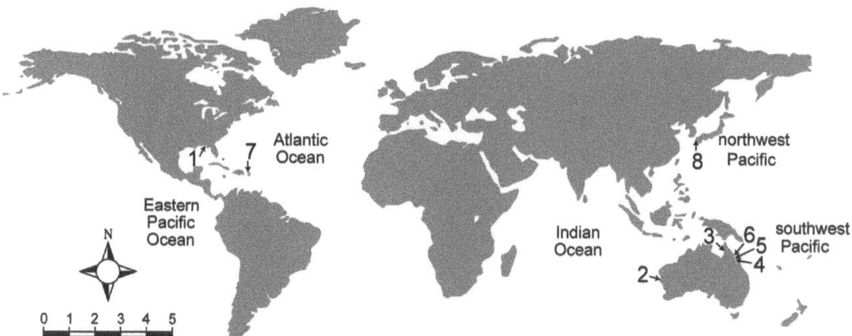

FIGURE 5.1 Map of global late Holocene tropical cyclone sedimentary records. 1, Texas Wallace and Anderson (2010); 2, Alabama/NW Florida Liu and Fearn (1993, 2000); 3, NW Florida 2 Lane et al. (2011); 4, New York City Scileppi and Donnelly (2007); 5, Puerto Rico Woodruff et al. (2008); 6, Belize McCloskey and Keller (2009); 7, Japan Woodruff et al. (2009); 8, Western Australia Nott (2011); 9, Gulf of Carpentaria Rhodes et al. (1980); 10, Wonga Beach Forsyth et al. (2012); 11, Cowley Beach Nott et al. (2009); 12, Rockingham Bay Forsyth et al. (2010).

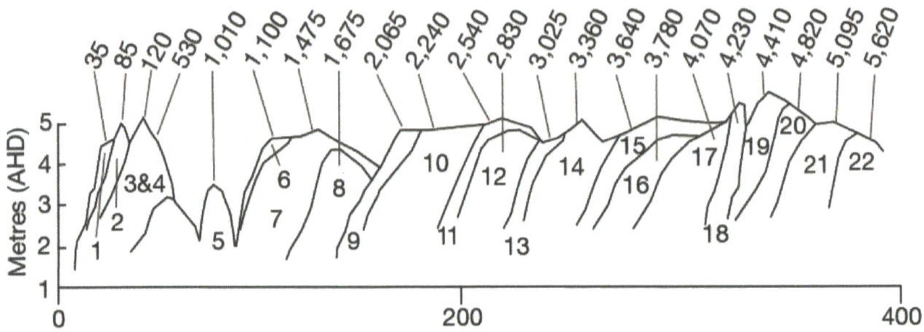

FIGURE 5.2 Stratigraphy and chronology (radiocarbon mean calibrated and reservoir-corrected ages) of coral shingle ridges at Curacao Island, Northeast Australia.

It is likely that the height of an onshore coral rubble ridge is a function of the combined storm surge, tide, and wave setup and wave run-up. Determining to what extent wave run-up is responsible for the height of the resulting ridge is important, as run-up can equal or exceed the height of the storm surge, depending upon various conditions. Wave run-up is a function of significant wave height and wave period/length, wave refraction/diffraction, bathymetry, beach slope angle, and roughness and permeability of beach material (Neilsen and Hanslow, 1991). With very rough, coarse-grained, permeable substrates, Losada and Gimenez-Curto (1981) wave run-up can be 0.30–0.75 times that of run-up on a sandy, largely impermeable beach under the same storm conditions. Observations of historical surge-emplaced rubble ridges suggest that wave run-up may, in some circumstances, play an insignificant role. For example, a 3.5-m-high ridge was deposited on Funafuti Atoll during TC Bebe in 1971. The surge accompanying the storm was 5 m above the level of the

FIGURE 5.3 Oblique aerial photo of Curacao Island showing transect of cross-section shown in Figure 5.2. *Photo taken by D. Hopley.*

reef flat, or mean low tide level, and 1.5 m higher than the elevation of the resulting ridge crest (Maragos et al., 1973). A similar situation occurred at Mission Beach, south of Cairns, North Queensland, Australia, where an intense TC struck in March 1918. There, a 4.5- to 5.1-m-high ridge of pumice was deposited by a surge as the cyclone crossed the coast. Eyewitness observations, results from numerical storm surge and wave models of the event, and knowledge of the tide level at the time show that together the combined storm tide (surge plus tide) and wave setup amounted to an inundation level of 4.7–4.9 m. This suggests that wave run-up could have only contributed 0.2–0.4 m of the ridge height at its highest elevation. Elsewhere, where the ridge is only 4.5 m high, wave run-up does not appear to have contributed to the formation of the pumice ridge.

Coral rubble ridges contain a number of distinct sedimentary facies or units of sediment. These include storm beach face, berm, crest, and washover facies (Hayne and Chappell, 2001). Beach face and berm facies include porous, clast-supported, coarse biogenic shingle deposits that rarely dip seaward but are generally structureless. Crest facies are horizontally bedded and are finer grained than beach face deposits. Washover facies are bedded, dip landward up to 15°, and sometimes contain imbricated clasts (imbrication is a sedimentary feature where particles are arranged in an overlapping shingle-like pattern dipping in one direction). Each of these facies or units combines to make a storm deposit. Storm deposits are commonly separated by "ground surfaces" being lenses of pumice pebbles and a weak sooty or earthy palaeosol (ancient soil). These ground surfaces are really former ground surfaces, or the surface of the feature that was exposed for sufficient time between individual cyclone events so that some soil development was able to take place. Ridges often contain only one storm deposit, but it is possible for two or more storm deposits to occur in one ridge. Careful excavation of the ridge is necessary in order to determine the number of storm deposits comprising the ridge. Samples collected for geological dating from only one storm deposit, when more than one storm deposit is present, may bias the age determination of that ridge.

Coral rubble ridges often accumulate on the sheltered side of islands, presumably because particles are constantly removed by the largest or most intense TCs on the exposed side of the island. The sheltered sides will experience lower wave energy, but perhaps the full effects of the surge. Because the wave energy is reduced, the likelihood of the ridge being removed during subsequent cyclones is lessened. Where the preservation potential for ridges is high, a number of ridges are sometimes able to accumulate over time. Curacao Island on the Great Barrier Reef has 22 consecutive coral rubble ridges paralleling the shore on its northwestern (NW) or sheltered side (Figures 5.2 and 5.3). Individual ridges extend for over 100 m along the shore and rise to over 5 m above the midtide level (the tidal range is approximately 3 m). The ridges were deposited by successive cyclones so that new ridges are deposited seaward of the previously emplaced ridge.

The age of the ridges increases progressively with the distance inland. Curacao Island is typical of many sites that preserve coral rubble ridges; however, not all sites retain as many ridges. Elsewhere, such as on Fitzroy Island in North Queensland (Nott, 2003), only one or two ridges paralleling the shore are preserved. This commonly results from site's exposure and the cyclone frequency. The time interval between cyclones will determine the extent to which compaction and lithification, or induration of the ridge can proceed. Over time, individual coral clasts within a ridge will weather or break down and will provide carbonate that will progressively cement clasts together. Cementation characteristically begins within the core of the ridge. It is common for loose fragments to remain on the crest and sides of the ridge several centuries after deposition. Throughout time, the ridge becomes resistant to further wave attack, although looser fragments on the ridge crest can be removed and replaced by younger fragments if subsequent storm surges are high enough to overtop that ridge. Also, in this fashion, more recent storm deposits can be superimposed on older ones.

Prior to stabilization, and depending upon the geomorphic setting, ridges can migrate inland by waves washing over the ridges and transporting clasts to the landward side of the ridge. The 19-km-long, 3.5-m-high, and 35-m-wide coral rubble ridge deposited by TC Bebe (in 1971) on Funafuti Atoll continued to move inland and along shore for many years (Baines and McLean, 1976). Indeed, in some instances on coral atolls, the ridges will move inland across the reef flat and abut or lap onto existing ridges through normal, noncyclonic wave action. Individual storm or cyclone deposits and ridges will be more difficult to recognize at these locations compared to sites such as Curacao Island where individual ridges often remain distinct.

5.2 CHENIERS AND SHELLY BEACH RIDGES

Cheniers are ridges composed partly, or entirely, of marine shells that have been deposited onto a mud substrate. They are also separated from each other by the substrate (Chappell and Grindrod, 1984). Beach ridges can be composed of sand or sand and shell, and sometimes isolated coral fragments. Unlike cheniers, beach ridges are separated by sand swales. The ridge and swale topography forms a distinct and continuous sand unit that may be deposited onto any type of substrate. This substrate is commonly, but not always, composed of estuarine muds deposited during a lower sea level stand. Cheniers and beach ridges are not restricted to tropical regions, and hence can form independently of TCs.

It is likely that all cheniers are deposited by storm waves; if the cheniers are in the tropics, these waves are likely to be due to TCs. On the other hand, beach ridges have been recognized to form by a number of processes, including deposition by swash during low- or high-wave energy conditions, or aggradation above the mean sea-level by an offshore sand bar (Taylor and

Stone, 1996). Although all these processes, and the associated beach ridges, can occur independently of TCs, the ridges that occur well above mean sea level, and contain layers and/or beds of shell within tropical regions, are likely to have been deposited during cyclones.

Excellent examples of these ridges occur along the shores of the Gulf of Carpentaria, Australia (Figure 5.1) (Rhodes et al., 1980). Up to 80 individual ridges paralleling the shore form a beach ridge plain that extends inland for over 3 km in places. These ridges, along the eastern and southern shores of the gulf, contain shell-rich layers up to 2 m thick, interspersed within medium- to coarse-grained sand (Rhodes et al., 1980). The ridges rise up to 6 m above mean sea level (tidal range of approximately 2 m) and extend along the coast for up to 10 km. A number of factors suggest that these ridges were deposited by storm surge and waves including: the height of these ridges above sea-level, the presence of abundant shell layers in the ridge stratigraphy, that sea levels have not varied by more than 2 m in the region during the Holocene, and that the Gulf of Carpentaria is especially prone to the development of intense TCs because of its warm, shallow waters. Radiocarbon dating of the ridges by Rhodes et al. (1980) showed that they increase in age with distance inland.

5.3 SAND BEACH RIDGE PLAINS

Beach ridges are triangular to convex, swash aligned, swash- and storm-wave-built ridges formed in the backshore, at or above the normal spring high tide level. They are composed of principally or purely marine deposits (Hesp, 2006). Foredunes are convex shore-parallel features that form through aeolian sand deposition within backshore vegetation (Hesp, 2006).

A number of studies have been conducted to extract records of actual TC marine inundations over the past 6,000 years from beach ridges along the coast of North Queensland, Australia (Nott and Hayne, 2001; Nott, 2003; Nott et al., 2009; Forsyth et al., 2010) and Western Australia (Nott, 2011). Nott and Hayne (2001), Hayne and Chappell (2001), and Nott (2003) suggested that entire coral shingle ridges could be deposited during a single storm event. Rhodes et al. (1980) suggested the same for the sand/shell beach ridges along the western coast of the Gulf of Carpentaria, Queensland. However, this is unlikely the case for sand ridges within the coarse-grained beach ridge plains of the wet tropics of Northeast (NE) Australia (Figure 5.1). Rather, it is more likely that units of sand between 0.1 and 1.5 m thick are deposited onto the crests of a ridge during marine inundations; a ridge will grow in height progressively with each marine inundation event. For example, TC Larry (2006) deposited a relatively thin unit of sand (0.02–0.1 m tapering landward) onto the incipient ridge at the back of Cowley Beach, approximately 150 km south of Cairns, Australia, in 2006 (Figure 5.4).

Sand units were deposited on top of the seaward beach ridge along Cairns' northern beaches during several TC-induced inundations between 1996 and

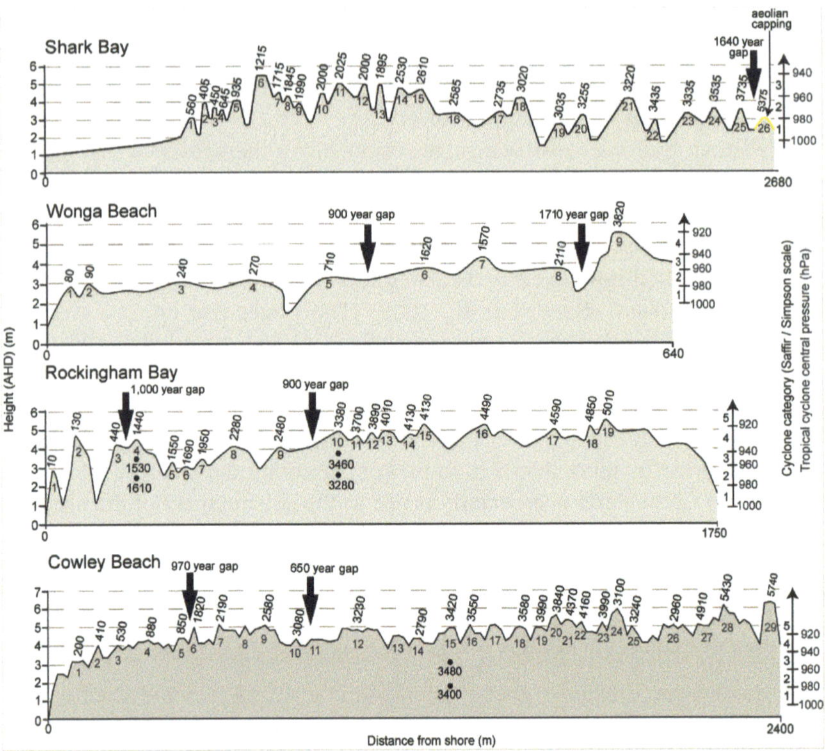

FIGURE 5.4 Beach ridge cross-sections and chronologies (Shark Bay is dated using radiocarbon and remainder using OSL). OSL, optically stimulated luminescence. *From Nott and Forsyth (2012).*

2001. One of these units can be seen in Figure 5.5, which shows deposition during an inundation generated by TC Justin in 1997. The unit varies in thickness from 0.05 to 0.4 m and extends for approximately 100 m along the crest of an existing sand beach ridge that stands 2 m above Australian Height Datum (AHD). TC Justin was a category 2 system (Australian category system) when it made landfall, and the inundation generated by this storm was able to overtop the lower lying section of the beach ridge. Elsewhere, the beach ridge rises to 3 m AHD and the marine inundation did not overtop the ridge in these locations. As a consequence, no sand unit was deposited.

These events and others that have deposited similar sand units provide a glimpse of how these beach ridges develop over the longer term. When a marine inundation is sufficiently large enough to overtop or reach the crest of an existing ridge, the inundation results in the deposition of a unit of sand, causing the ridge to grow in height with each successive event. Different magnitude inundations can be responsible for depositing a ridge until the ridge approaches a height that is attainable by wave run-up generated by only very

FIGURE 5.5 Photograph of sand unit deposited onto the first beach ridge at Clifton Beach (Cairns, North Queensland, Australia) during TC Justin, 1997. The photograph shows the sand unit extending down the rear flank of the beach ridge and extending into the swale behind the ridge. This sand deposit was coarse grained and was over 30 cm thick in places. *From Nott (2010).*

intense TCs. Hence, the most extreme inundations will deposit the final units of sediment on the ridge. The initial units of sediment constituting a ridge (those lowest in the ridge stratigraphy) could have been deposited by a range of inundations starting from noncyclonically induced inundations, such as very high tides and strong trade-wind-generated wave conditions, to the most intense TC-generated inundations (Figure 5.6). Progressively higher run-up is required to deposit sand onto the ridge crest as it grows in height. An elevation point will be reached where the vast majority of inundations can no longer reach the ridge crest and the ridge will cease to grow. The timing of this terminal stage may also be influenced by the growth rate of the next seaward ridge. This ridge was likely initiated before the ridge to its landward side reached its maximum height. Hence, the rate and volume of sediment delivery to the coastal system will play a role in the height that ridges can finally attain. Periods when sediment delivery rates and volumes to the coastal system are high may result in a ridge not attaining its maximum height as a function of the intensity of TCs alone. This is because the ridge on its seaward side has attained sufficient height to diminish the ability of wave run-up to reach this next inland ridge. Hence, most of the sedimentation will occur on the most seaward, likely slightly lower elevation, ridge, and the next inland one will become starved of new sediment. Ridges may have the opportunity to reach their maximum possible height, which is determined by the maximum inundation height for that region, during periods of relatively low sediment delivery rates (and volumes) to the coastal system. In this fashion, there will be interplay between the processes operating to attain the maximum ridge

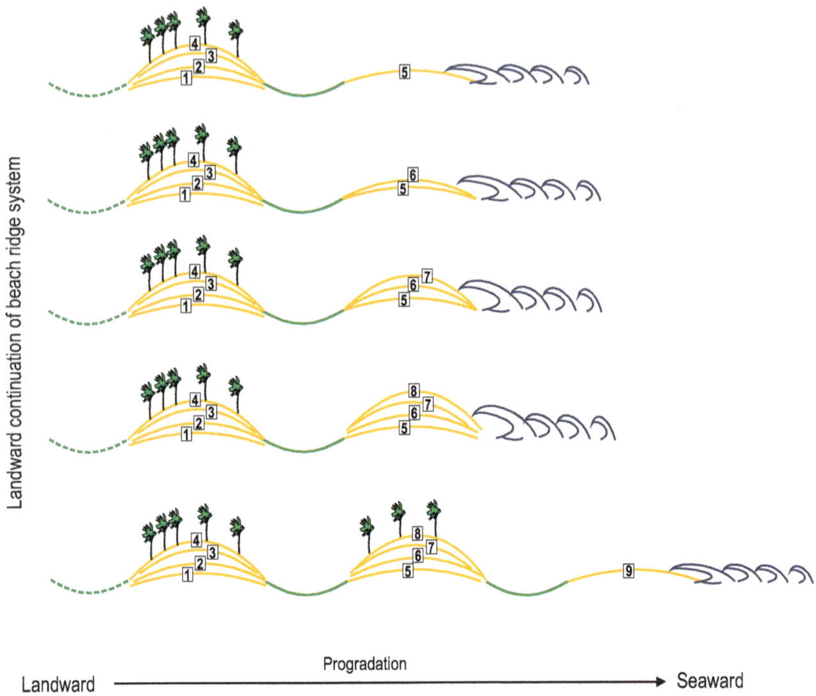

FIGURE 5.6 Schematic sequence of accretion of sand beach ridges in Northeast Queensland. The number and thickness of individual units will vary between ridges. In this sequence, unit 1 is the oldest and unit 9 is the most recently deposited. Lower units can be deposited by low-magnitude inundations but increasingly higher inundations (hence higher magnitude storms) are required to deposit uppermost units on a ridge. *From Nott et al. (2013).*

height—these processes being the rate of sediment delivery to the near-shore environment and the maximum height of the inundations able to be generated by TCs. In the case of the ridge plains along the wet tropical coast of NE Australia, for example, the final inundations responsible for depositing the uppermost units of sediment on the beach ridges were generated by extreme intensity TCs (Nott et al., 2009; Nott, 2010; Forsyth et al., 2010). Hence, using the heights of these ridges and methods developed by Nott (2003) for calculating the intensity of the TCs responsible suggests a 5,000- to 6,000-year record of intense TCs.

Beach ridges constructed by surge and waves display a key sedimentary signature. They have a sudden textural coarsening and coarse-skewed distribution that occurs at the base of each sedimentary unit deposited during an event. Unlike coral shingle ridges, which can be deposited entirely during a single storm event, sand beach ridges appear to accrete progressively over time. Another ridge develops seaward; it progressively increases in height until inundations can no longer reach its crest and a beach ridge plain develops. Each layer deposited during a storm can be identified by the sedimentary characteristics mentioned. Coarse-skewed trends with minimal change in mean grain size characterize the upper levels of these deposits when the sedimentary unit is deposited at a location within the zone of maximum onshore winds. These same trends are not apparent in sediments deposited at locations that experience predominantly offshore winds during the TC event, which in the case of Eastern Australia is north of the eye-crossing location. Hence, it may be possible to identify the relative landfall locations of past TCs using this diagnostic sedimentary signature (Nott et al., 2013). This coarse-skewed sedimentary trend, along with the initial textural coarsening of each event unit, may be useful for identifying TC sedimentary units in older beach ridges where advanced pedogenesis has obscured the visual stratigraphic markers that are evident in more recent storm-built beach ridges. Results from this research have also shown that geochemical signatures and microfossil data (diatoms and foraminifera) can be used to identify palaeocyclone deposits as well as the limit of inundation of older cyclones and thus help distinguish them from other deposits associated with fair weather conditions (Nott et al., 2013).

5.4 SHELL RIDGES

Unlike cheniers, shell ridges do not form as separate features on a mud substrate. Rather, they are much more akin to sand beach ridges, but are composed entirely of marine shells. They are not common around the coast of Northern Australia, although they are extensively preserved at Shark Bay, Western Australia. Here, the shells of the species *Fragum erugatum*, otherwise known as the cardiid cockle, flourish in these hypersaline waters (Figure 5.4). Limited circulation of waters occurs between the open ocean and the deeply indented bays here, particularly Hamlin Pool, because of the formation of the Faure Sill, a sand bar that developed during the mid to late Holocene and stretches across the mouth of Hamlin Pool. The limited number of predators of *F. erugatum* means that this species thrives in the near-shore environment and provides an abundant supply of shells of remarkably uniform size (5−8 mm diameter) for potential transport onshore during marine inundations induced by TCs. The resulting ridges are composed entirely of this shell species. Sand occurs only in the most shoreward ridges. This sand appears to be removed from the ridges with time because the second and third rows of ridges inland are devoid of sand.

Approximately 40 ridges parallel the coast at Hamlin Pool. The ridges increase in age with distance inland. The majority of the ridges were deposited from 5,500 years before present (BP) to the present (Nott, 2011). Pleistocene ridges are also preserved landward of the Holocene sequence. The Pleistocene ridges differ from the Holocene ones, as the former are composed of a much wider array of shell species, which are also larger in size compared to the single species of uniform size in the Holocene ridges. The larger Pleistocene shells are open ocean species and suggest greater water circulation between the open ocean and the indented bays during the Pleistocene compared with that at present. The shell ridges extend in height up to 8 m AHD. Such ridges could only, within this very protected and calm water environment, have been formed by TC-induced waves and surge.

5.5 GRAVEL RIDGES

Ridges composed of gravel and rare coral fragments are common along the Kimberley Coast of NW Western Australia. These ridge sequences form gravel barriers and commonly impound back barrier lagoons in embayments along sections of coast dominated by steep rock cliffs. They are particularly common along the western side of Cambridge Gulf north of Wyndham in the East Kimberley region. The gravel ranges in size up to 1.6 m diameter (A-axis) and 1.4 m (B-axis) and have been deposited into sequences of up to nine ridges paralleling the shore. At La Crosse Island (Figure 5.7) offshore from the mouth of the Ord River, gravel ridges have been deposited in every embayment and form a discontinuous sequence that surrounds the island. The ridges at two sites on opposite sides of the island extend up to 5 m AHD. Radiocarbon samples on coral fragments embedded with the core of ridges from each of seven ridges show that they were deposited from approximately 5,000 years BP until recently (Nott, 2000). The radiocarbon samples do not show a progressive increase in age with distance inland, suggesting that the ridges are regularly overtopped and reworked by marine inundations. Interestingly, the most landward ridges within many of the ridges have concave, circular depressions up to several meters across and 1 m deep. The origin of these features is intriguing because they do not appear to have formed by tree fall and disturbance of the gravels by the uprooting of the tree, as this could be expected to form an asymmetric-shaped depression. Also, no evidence exists that trees ever grew on the ridges. Although large saltwater crocodiles are

FIGURE 5.7 Gravel ridges on La Crosse Island, NW Australia. NW, northwestern.

known to inhabit the island, nesting in ridges composed of such coarse particles is neither a recognized behavior of this animal nor is it known to be a feature of turtle nesting, especially considering that the depressions are located on the most landward ridge and a turtle would have to traverse six or more substantial ridges. High-velocity, onshore flows generated by tsunamis could potentially generate such depressions. However, the size of the particles composing these ridges is within the range of particle sizes able to be transported by TC-generated waves (Nott, 2003). This part of the Western Australian coast is not known to have experienced tsunami impact, unlike the Indian-Ocean-facing coast of Western Australia during historical times. The size of the gravel particles and the height of the ridges above sea level further suggest that these features are likely a product of TC-generated surge.

5.6 PUMICE RIDGES

Pumice ridges, to date, have only been documented from NE Queensland, Australia. Similar to other ridges they are convex in morphology and stand up to 5–6 m in height and 10–20 m in width. They are composed entirely of clasts of pumice varying in size from 0.01 to 0.2 m. A ridge of pumice was deposited at North Mission beach, Queensland, during the marine inundation generated by an intense TC on March 10, 1918 (Taylor, 1982). The inundation occurred at high tide and was reported to have reached at least 3.5 m above normal sea level. The inundation resulted in many deaths and transported and deposited supplies from a shed on Dunk Island approximately 5 km offshore onto the mainland. In addition, one large bag of flour was deposited over 3.5 m high in a tree; the flour in the center of the bag was still sufficiently dry to be able to make damper (a type of bread) the day after the maelstrom (Taylor, 1982). The crest of the pumice ridge here reaches 5 m AHD and it extends alongshore for approximately 500 m.

5.7 SAND SPLAYS

Coastal sand dunes are often eroded, and diminished substantially in height, when surge and waves overtop them. The eroded sand can be transported inland as a splay or sheet that thins landward. Similar deposits can also occur when tsunamis inundate sandy coasts (Goff et al., 1998).

An excellent example of a TC-deposited sand splay occurred when TC Vance, with a central pressure of 910 hPa, crossed the Western Australian coast near Tubridgi Point in March 1999. At the time of occurrence, this was the most powerful cyclone to cross the Australian coast. Winds around the eye were estimated at 300 km/h and it generated the strongest wind gust ever recorded in Australia of 267 km/h at Learmonth, approximately 30 km from the eye. The zone of maximum winds struck a section of coast with a sand barrier composed of three rows of parallel dunes approximately 6 m above midtide level behind a wide sandy beach. In most locations, the first two rows of dunes were completely eroded and the sand was removed, presumably offshore. The third or the most inland dune row was eroded to form a steep scarp. Where the zone of maximum winds struck, however, all three rows of dunes were destroyed and the sand transported inland as an extensive splay approximately 400–500 wide that extended 200–250 m inland from the position of the former first dune row (Figure 5.8). This sand splay decreased in thickness from 1.5 m immediately to the rear of the position of the former third row of dunes to 0.75 m thick at its most inland extent. The splay terminated abruptly at a salt marsh where it was marked by a steep fronted ($\sim 30^{\circ}$ angle) toe slope. Sediments within the splay were deposited as steep ($\sim 30^{\circ}$) tabular cross-beds (Figure 5.8). Medium- to coarse-grained sand occurred at the base of the unit, along with clasts of coral and shells, and graded upward into medium- to fine-grained sand. The tabular cross-beds, along with other field evidence, suggest that the surge struck the coast with considerable force and moved inland as a reasonably high velocity bore. Such conclusions were supported by the presence of scour pits measuring up to 10 m long and 3 m wide on the lee (inland) side of trees, imbricated gravels and small boulders of lithic rock within the splay on its seaward side, and the sheer volume of sand transported inland. Too few studies have been undertaken to develop facies models to allow for distinction between sand splays deposited by TCs and tsunami. The available literature suggests that tsunami sand splays are sometimes structureless, or have horizontal bedding, or have cross-beds indicating bidirectional flow (Goff et al., 1998). However, the Tubridgi Point sand splay from TC Vance differs, as it has tabular cross-beds that suggest unidirectional flow. Whether these different styles of sedimentation bear any specific relationship to tsunami versus storm wave processes, however, remains to be ascertained.

FIGURE 5.8 Toe of sand splay, Tubridgi Point, Western Australia.

5.8 WASHOVER DEPOSITS

The deposition of sand layers, up to 0.5 m thick, in back barrier lagoons and swamps where fine-grained sediments are generally deposited, has been interpreted as evidence of storm washover events. Sediments within back barrier lagoons are normally muddy, organic, or fine grained. Interbedded sand layers within these fine-grained sediments can be due to storm surge and waves overtopping a sand dune barrier, and transporting sand into an environment where it is not normally deposited. By geologically dating the sand layers, the long-term history of cyclones in a region can be ascertained. It is important to demonstrate that these sand layers are indeed from storm surges during TCs and not from other sources such as river flow or tsunami. The marine origin of the sand layers can be ascertained through identification of marine environment micro- and macrofossils such as foraminifera and diatoms within them.

Sand layer stratigraphies from washover events due to TCs have been studied, predominantly along the shores of the Gulf of Mexico and the Southeastern and Eastern United States. Liu and Fearn (1993, 2000) examined sand layers in lakes along the Florida and Alabama coasts, and Donnelly et al. (2001a,b) studied similar deposits in New England and New Jersey along the US Atlantic coast (Figure 5.1). All these studies assume that the height and general nature of the barrier has remained unchanged over the length of the washover record. Such assumptions seem reasonable when separate sites some distance apart show the same chronology of events, or at least clusters of events, suggesting that some regional factor has influenced the behavior of TCs at different times during the past. For example, in Liu and Fearn's (1993, 2000) studies along the Gulf of Mexico of Shelby Lake, Alabama, and Western Lake, Florida (Figure 5.1), which are approximately 150 km apart, they showed a distinct clustering of apparently high-magnitude cyclones approximately 3,200−1,000 years BP. Prior to and following this time there were considerably fewer intense TCs. Scott et al.'s (2003) overwash record from South Carolina revealed alternating phases of TC activity, which were out of phase with the Gulf of Mexico sites. These records suggest that the most intense storm activity occurred in South Carolina over the last 1,000 years when six intense storms occurred (at that specific site), whereas the period 1,000−2,000 years BP had three intense storms and only two intense storms occurred between 2,000 and 3,000 years BP. This out-of-phase activity between the Atlantic and Gulf of Mexico regions has been attributed to meridional shifts in the jet stream, which controls the position of the Bermuda High pressure system. When the jet stream and Bermuda High are further north, TCs are directed toward the Atlantic coast and when further south, TCs tend to be directed toward the Gulf of Mexico (Liu and Fearn, 2000; Scott et al., 2003).

Donnelly and Woodruff (2007) noted periods of activity and inactivity along the Atlantic US seaboard and the Caribbean region (Puerto Rico) over the past 5,000 years. They found that heightened activity occurred between

5,000 and 3,500 years BP and between 2,500 and 1,500 years BP with less activity in the intervening times. Comparisons with a long-term proxy El Niño Southern Oscillation (ENSO) record highlighted that the phases of more frequent TC occurrence were associated with periods dominated by La Niñas and phases of lower frequency TC occurrence, with periods dominated by El Niños. Interestingly, Donnelly and Woodruff (2007) noted that ENSO appeared to have been a stronger control on long-term TC activity compared to sea surface temperatures. A study of overwash deposits by Woodruff et al. (2009) found that TC activity in this region of the NW Pacific was out of phase with the West Atlantic region, so that in Japan extended episodes of La Niñas were associated with less TC activity and the reverse was the case for El Niños. Also using the overwash technique, McCloskey and Keller (2009) found two periods of TC hyperactivity between 4,500 and 2,500 years BP in Belize. Scileppi and Donnelly (2007) found multiple overwash events at Long Island, New York, between 2,200 and 900 years BP, with no TC occurrences between AD 1300 and 1600 and a return to intense events after AD 1600. In Puerto Rico, Woodruff et al. (2008) uncovered considerable variation in the frequency of TCs, as recorded by overwash events, on centennial to millennial scales over the past 5,000 years.

5.9 PREHISTORIC TC INTENSITY

A method to determine the intensity of the prehistoric cyclones responsible for building ridges and eroding terraces in raised gravel beaches was introduced by Nott and Hayne (2001) and Nott (2003). The height of these features is assumed to represent the minimum height of the storm inundation during the event responsible. The elevation of these features is accurately surveyed to datum, and samples of coral and/or shell radiocarbon dated to determine the minimum height and times of inundation, respectively. The height of the inundation is related to the intensity of the palaeocyclone, which is determined through the use of numerical storm surge and shallow water wave models (GCOM2D model, see Nott (2003)). The models are used to determine the relationship between surge height and central pressure for each location containing evidence of palaeocyclones. Also, the relationship between surge height and translational velocity of the cyclone, the radius of maximum winds, and the track angle of the cyclone as it approaches and crosses the coast are determined. Model results are compared to measured surge heights from recent or historical cyclone events near the study sites. The central pressure of the cyclone responsible for formation of the ridge or terrace is determined by modeling the magnitude of the surge plus wave setup, and run-up and tide required to inundate the ridge or terrace.

The tide height at the time of the prehistoric event is unknown but can be estimated (at the 95% confidence level) to have occurred within the 2 sigma probability tidal range of the frequency distribution nodal tide curve for each site. This tidal range forms the uncertainty margins associated with the mean central pressure of the cyclone.

The sand beach ridges are geologically dated using optically stimulated luminescence (OSL), or similar methods, and the coral shingle and shell ridges or sand ridges containing shell beds are dated using radiocarbon. The chronologies allow a frequency of events to be determined over the late Holocene and the numerical modeling allows the magnitude of the events to be determined (Figure 5.4).

5.10 GAPS IN CHRONOLOGY

One of the most notable characteristic of all the ridge plain chronologies examined thus far in Australia is the presence of substantial gaps in ridge plain formation over the late Holocene (Figure 5.4). The sand beach ridge plains of NE and Northern Australia show two prominent chronological gaps. Rockingham Bay has a 900-year gap between 3,380 and 2,480 years BP and a 1,000-year gap between 1,440 and 440 years BP. Cowley Beach has a 650-year gap between 3,230 and 2,580 years BP and a 970-year gap between 1,820 and 850 years BP. Two major gaps of 1,700 years and 900 years occur at Wonga Beach between 3,820 and 2,110 years BP, and 1,620 and 710 years BP, respectively. These gaps could be the result of either erosion of existing ridges or a lack of ridge development because of limited sediment supply via stream channels to the coast and/or a decrease in the frequency of TC-generated inundations capable of causing ridge accretion. Erosion is unlikely to be responsible because the gaps are broadly coeval across all three ridge plains. Also, there is no evidence either in the plan form and ridge morphology or stratigraphy to suggest that the chronological gaps are due to erosion. These gaps are most likely due to periods when fewer high-intensity TCs made landfall. These periods resulted in fewer or no sufficiently large surges capable of depositing enough sediment to promote the development of the highest ridges in the plain.

The coarse-grained sand beach ridge plain in the Gulf of Carpentaria was dated using radiocarbon analyses of marine shells incorporated within the ridge (Rhodes et al., 1980). These radiocarbon analyses were recalibrated by Nott and Forsyth (2012) using the INTCAL09 and MARINE09 radiocarbon age calibration curves (Reimer et al., 2009) (Calib 6.0). The ages of these ridges extends to almost 7,500 years BP and major gaps in the chronology occur between approximately 5,500 and 3,500 years BP; 2,700 and 1,800 years BP; and 1,000 and 500 years BP (Figure 5.9). Only one major gap occurs in the radiocarbon chronology (also marine corrected and calibrated) for the shell beach ridges in Western Australia and this occurs between approximately 5,400 and 3,600 years BP, which is the same approximate time of earliest major gap in the Gulf of Carpentaria record (Figures 5.4 and 5.9). A second possible gap occurs in the Western Australia sequence between 500 years BP

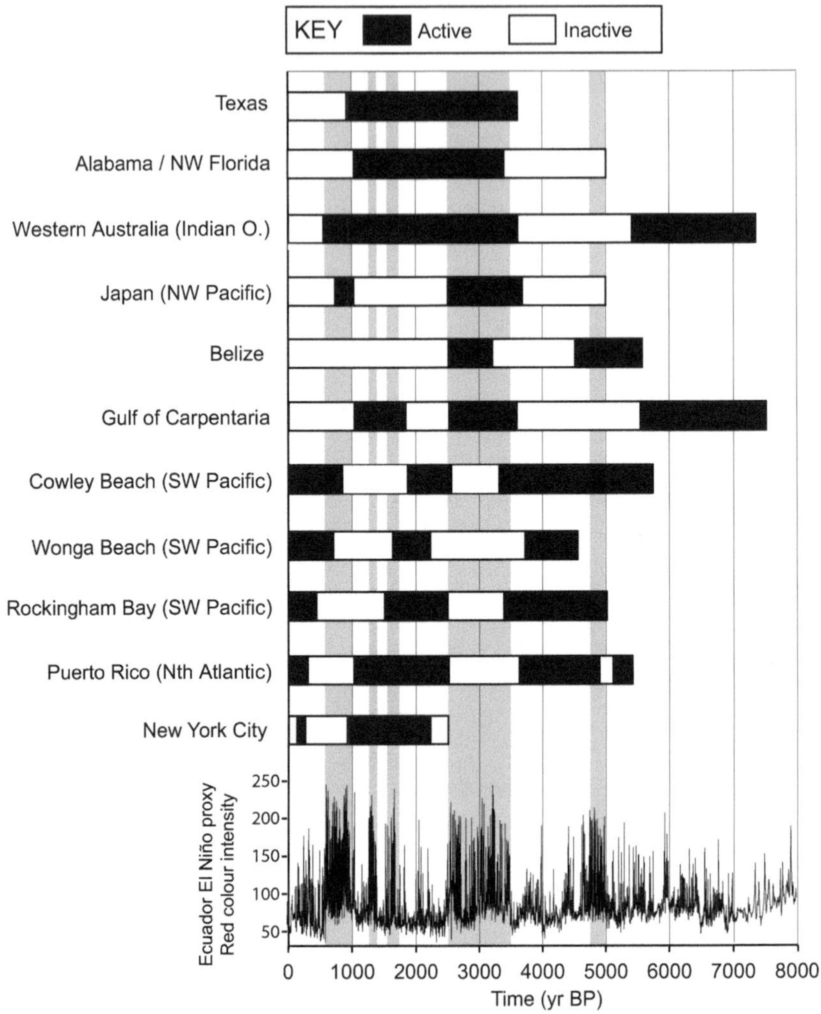

FIGURE 5.9 Global phases of TC activity and inactivity over past 5,000–7,000 years. See caption of Figure 5.1 for references. TC, tropical cyclones. *From Nott and Forsyth (2012).*

and the present, although it is not certain whether this is genuine cessation in TC activity or an erosional gap in the record. The first major gap in the Western Australia sequence is definitely not due to erosion, but rather is climatic in origin as the ridges deposited prior to this gap have been buried by a 0.5- to 0.75-m-thick layer of well-sorted, very fine-grained sediment that is aeolian in origin (Figures 5.4 and 5.9). This wind-blown unit does not occur on the ridges after the gap, suggesting that this was caused by a markedly drier period of climate during which few, if any, TCs made landfall. This does not necessarily suggest there were fewer intense TCs in the Eastern Indian Ocean during this period, simply that there was a substantial reduction in landfalling TCs (Nott, 2011). This same period (5,400–3,600 years BP), known as the Neoglacial, was also marked by aridity in Eastern and Central Australia (McGowan et al., 2008). The gaps in TC activity in the Gulf of Carpentaria and in Western Australia are out of phase with those in NE Australia, hence the periods of heightened TC activity in NE Australia occurred during periods of relative quiescence in Western Australia and the Gulf of Carpentaria (Figure 5.9).

The overwash sedimentary records from the North Atlantic (Puerto Rico) and the Gulf of Mexico also have major gaps suggestive of periods of quieter hurricane activity (Liu and Fearn, 1993, 2000; Donnelly and Woodruff, 2007; Wallace and Anderson, 2010). These two regions also display an out-of-phase relationship with each other in terms of hurricane activity during the late Holocene (Figure 5.9). The Atlantic deposits suggest hurricanes were relatively inactive between approximately 3,500 and 2,500 years BP and again between approximately 1,000–500 years BP (Scileppi and Donnelly, 2007; Woodruff et al., 2008), whereas the Gulf of Mexico deposits show decreased hurricane inactivity from approximately 5,000 to 3,400 years BP and again from approximately 1,000 years BP to the present (Liu and Fearn, 1993; Donnelly and Woodruff, 2007). The NW Florida deposits (Lane et al., 2011) display broadly similar gaps except for a brief period of activity at approximately 800 years BP (Figure 5.9). Comparisons between the NE Australian and Puerto Rico deposits show that both the southwest (SW) Pacific and North Atlantic oceans experienced two major periods of TC quiescence at approximately the same times since 5,000 years BP. Coevally there was heightened TC activity in the NW Pacific (Japan), East Indian Ocean, and the Gulfs of Carpentaria and Mexico (Figure 5.9). When the NW Pacific, Indian Ocean, and Gulfs experienced a major phase of TC quiescence centered on 4,500 to 4,000 years BP, the North Atlantic and SW Pacific experienced heightened TC activity. The same was true of the phase of TC quiescence centered on approximately 2,000 years BP in the NW Pacific and Gulf of Carpentaria, although the Indian Ocean and Gulf of Mexico continued to have high TC activity during this time (Figure 5.9).

The periods of inactivity over the past 5,000 years in the Atlantic basin have been associated with periods dominated by El Niños (Donnelly and Woodruff, 2007). Comparisons with a long-term record of ENSO from Ecuador (Moy et al., 2002) suggest the same may be true of the SW Pacific (Figure 5.9) between approximately 3,500 and 2,500 years BP. The later phase of reduced TC activity in the SW Pacific (1,800–900 years BP), however, appears to begin between 500 and 800 years earlier than the Atlantic record (1,000–500 years BP), suggesting it may not be due to more frequent El Niños. If El Niños are the cause of this punctuated activity globally then it could be expected there would be an antiphase relationship between the Atlantic/SW Pacific and the East Indian Ocean, NW Pacific (Southern Japan), and the Gulf of Carpentaria. Apart from the Gulf of Mexico, where the North

Atlantic Oscillation (NAO) has possibly played an important role (Liu and Fearn, 1993, 2000), modern-day El Niños can cause an increase in TC activity in the Gulf of Carpentaria and Western Australia and the Southern Japan region (Chan and Xu, 2009). A much higher percentage of TCs develop off the Western Australian coast during El Niños (61%) compared to La Niñas (38%) (Broadbridge and Hanstrum, 1998). The same is true of the Gulf of Carpentaria where TC activity appears to be depressed during La Niñas. The monsoon trough, within which most TCs form in this region, tends to move southward of the Gulf of Carpentaria over the Australian landmass during La Niñas and sits over the Gulf during El Niños. TC activity in the Gulf of Carpentaria today is out of phase with TC activity in the nearby SW Pacific, which experiences fewer TCs during El Niños (Basher and Zheng, 1995). The long-term TC records do show an antiphase relationship between these locations and the Atlantic/SW Pacific between 3,500 and 2,500 years BP, but this relationship is not as obvious during the late Holocene. No clear coincidence occurs between the major periods of El Niños suggested by the Ecuadorian record and all the periods of less TC activity in the Atlantic and SW Pacific or the periods of heightened TC activity in the NW Pacific and East Indian Oceans and the Gulf of Carpentaria. This may in part be due to the Ecuadorian ENSO record itself, for it differs from other, albeit much shorter, proxy records of ENSO (Lander and Guard, 1998; Cobb et al., 2003; Mann et al., 2009; Conroy et al., 2009).

The lags in the precise onset and cessation of TC activity between the ocean basins and gulfs globally may also be due to differences in depositional processes between overwash and beach ridge sedimentary environments. Furthermore, these lags may be in part due to the different chronological techniques used between sites—the North Atlantic, NW Pacific and Indian Oceans, and Gulfs of Mexico and Carpentaria have been radiocarbon dated, whereas OSL has been used on the sand beach ridge records of the SW Pacific. It is also possible that regional climatic factors have influenced TC activity in each basin such as the North Atlantic Oscillation (NAO) in the North Atlantic and possibly the Interdecadal Pacific Oscillation (IPO) in the Pacific and Indian Ocean Dipole in the Indian Ocean. Despite these factors, and our inability to attribute definitive climatic causes, it is clear that all the long-term TC records recovered to date display alternating phases of more activity and less activity, which appears to be a genuine characteristic of the long-term climatology of intense TCs.

5.11 DERIVING ROBUST RETURN PERIODS FROM PALAEOSTORM DEPOSITS

Only beach ridge plains have been used to determine a statistically robust measure of the return periods of past storm inundations. Because beach ridges register the most extreme inundation events over many millennia, they can be modeled using a generalized extreme value theory and Bayesian analysis approach (Nott and Jagger, 2013). This method was applied to the beach ridge plain at Rockingham Bay, NE Queensland, which was directly impacted by TC Yasi in 2011, one of the most severe storms to have impacted the east coast of Australia for nearly a century (Nott and Jagger, 2013). The statistical analysis suggested that the storm surge that generated this event had a 1 in 1,000- year return period. Return periods for TC marine inundations are usually derived from synthetic data sets generated from deterministic models or by extrapolating short historical records. Such approaches contain considerable uncertainties because it is difficult to test their veracity until a sufficiently long period has elapsed. These approaches also often only consider storm surges or storm tides and not the total inundation, which includes waves, setup, and run-up, likely to flood a coastal property. The beach ridge records do reflect these additional components of the inundation and hence are a more realistic guide to the extent of flooding of properties. Comparisons of the results from the beach ridge approach with those methods that model only storm surge or storm tide (McInnes et al., 1999; Hardy et al., 2004) suggest that the former produced a shorter return period (more frequent) for the inundation generated by TC Yasi than the synthetically generated TC climate approach for this site, which suggested a return period of between 1 in 5,000 and 1 in 10,000 years.

5.12 CONCLUSION

Palaeostorm surge/inundation deposits offer considerable insight into the nature of storms that have impacted a region. Patterns and periodicities of prehistoric storms can be reconstructed from sediments deposited by surge and waves at elevations or distances inland beyond the reach of normal marine processes. Millennial- and century-scale periodicities have been identified from sedimentary records in the South and Southeastern United States, Japan, and Northern Australia and such patterns highlight synchroneity in TC activity between different regions globally.

Deposits such as sand beach ridges contain distinct sedimentary signatures that allow the frequency and intensity of past surge/inundations events to be reconstructed. Robust statistical techniques can be applied to these records to generate return intervals that can be used to estimate levels of risk and exposure as human populations, urbanization, and tourism grow rapidly along tropical and subtropical coasts.

REFERENCES

Baines, G.B.K., McLean, R.F., 1976. Sequential studies of hurricane deposit evolution at Funafuti atoll. Mar. Geol. 21, M1–M8.

Baines, G.B.K., Beveridge, P.K., Maragos, J.E., 1974. Storms and island building at Funafuti Atoll, Ellice Islands. In: Proc. 2nd Int. Coral Reef Symp, vol. 2, pp. 485–496.

Basher, R.E., Zheng, X., 1995. Tropical cyclones in the southwest Pacific: spatial patterns and relationships to the Southern Oscillation and sea surface temperature. J. Clim. 8, 1249–1260.

Broadbridge, N.W., Hanstrum, B.N., 1998. The relationship between tropical cyclones near Western Australia and the Southern Oscillation index. Aust. Meteorol. Mag. 47, 183–189.

Chan, J.C.L., Xu, M., 2009. Interannual and interdecadal variations of landfalling tropical cyclones in East Asia. Part I: time series analysis. Int. J. Climatol. 29, 1285–1293.

Chappell, J., Grindrod, J., 1984. Chenier plain formation in northern Australia. In: Thom, B. (Ed.), Coastal Geomorphology in Australia. Academic Press, Sydney.

Cobb, K.M., Charles, C.D., Cheng, H., Edwards, R.L., 2003. El Niño/Southern Oscillation and tropical Pacific climate during the last millennium. Nature 424, 271–276.

Conroy, J.L., et al., 2009. Unprecedented warming of surface temperatures in the eastern tropical Pacific Ocean. Nat. Geosci. 2, 46–50.

Davies, P.J., 1983. Reef growth. In: Barnes, D.J. (Ed.), Perspectives on Coral Reefs. Aust. Inst. Mar. Sci. B. Clouston Publishing, Manuka, pp. 69–106.

Donnelly, J.P., Woodruff, J.S., 2007. Intense hurricane activity over the past 5,000 years controlled by El Niño and the West African monsoon. Nature 447, 465–468.

Donnelly, J.P., Roll, S., Wengren, M., Butler, J., Leder, R., Webb III, T., 2001a. Sedimentary evidence of intense hurricane strikes from New Jersey. Geology 29, 615–618.

Donnelly, J.P., Smith-Bryant, S., Butler, J., Dowling, J., Fan, L., Hausmann, N., Newby, P., Shuman, B., Stern, J., Westover, K., Webb III, T., 2001b. 700 yr sedimentary record of intense hurricane landfalls in southern New England. Geol. Soc. Am. Bull. 113, 714–727.

Forsyth, A., Nott, J., Bateman, M., 2010. Beach ridge plain evidence of a variable late-Holocene tropical cyclone climate, North Queensland, Australia. Palaeogeogr. Palaeoclimatol. Palaeoecol. 297, 707–716.

Forsyth, A., Nott, J, Bateman, M., Beaman, R., 2012. Juxtaposed beach ridges and foredunes within a ridge plain - Wonga Beach, northeast Australia. Mar. Geol. 307–310, 111–116.

Goff, J., Crozier, M., Sutherland, V., Cochran, U., Shane, P., 1998. Possible tsunami deposit from the 1855 earthquake, North Island, New Zealand. In: Stewart, I.S., Vita-Finzi, C. (Eds.), Coastal Tectonics, vol. 46. Geological Society, London, Special Publication, pp. 353–374.

Hardy, T., Mason, L., Astorquia, A., July 2004. Surge Plus Tide Statistics for Selected Open Coast Locations along the Queensland East Coast. Queensland Climate Change and Community Vulnerability to Tropical Cyclones. Ocean Hazards Assessment Stage 3. Queensland Government, Report.

Hayne, M., Chappell, J., 2001. Cyclone frequency during the last 5,000 yrs from Curacao Island, Queensland. Palaeogeogr. Palaeoclimatol. Palaeoecol. 168, 201–219.

Hesp, P.A., 2006. Beach ridges – definition and re-definition. J. Coastal 39, 72–75. Research, Special Issue.

Hughes, T.P., 1999. Off-reef transport of coral fragments at Lizard Island, Australia. Mar. Geol. 157, 1–6.

Lander, M.A., Guard, C.P., 1998. A look at global tropical cyclone activity during 1995: contrasting high Atlantic activity with low activity in other basins. Mon. Weather Rev. 126, 1163–1173.

Lane, P., Donnelly, J.P., Woodruffe, J.D., Hawkes, A.D., 2011. A decadally-resolved paleohurricane record archived in the late Holocene sediments of a Florida sinkhole. Mar. Geol. 287, 14–30.

Liu, K., Fearn, M., 1993. Lake sediment record of late Holocene hurricane activities from coastal Alabama. Geology 21, 793–796.

Liu, K., Fearn, M., 2000. Reconstruction of prehistoric landfall frequencies of catastrophic hurricanes in northwestern Florida from lake sediment records. Quat. Res. 54, 238–245.

Losada, M., Gimenez-Curto, L., 1981. Flow characteristics on rough, permeable slopes under wave action. Coastal Eng. 4, 187–206.

Mann, M.E., et al., 2009. Global signatures and dynamical origins of the Little Ice Age and Medieval climate anomaly. Science 326, 1256–1260.

Maragos, J., Baines, G., Beveridge, P., 1973. Tropical cyclone Bebe creates a new land formation on Funafuti Atoll. Science 181, 1161–1164.

McCloskey, T.A., Keller, G., 2009. 5000 year sedimentary record of hurricane strikes on the central coast of Belize. Quat. Int. 195, 53–68.

McGowan, H., Petherick, L., Kamber, B., 2008. Aeolian sedimentation and climate variability during the late Quaternary in southeast Queensland, Australia. Palaeogeogr. Palaeoclimatol. Palaeoecol. 265, 171–181.

McInnes, K., Walsh, K., Pittock, B., 1999. Impact of Sea-level Rise and Storm Surges on Coastal Resorts. A Report for CSIRO Tourism Research. CSIRO Atmospheric Research, Melbourne, Australia.

Moy, C.M., Seltzer, G.O., Rodbell, D.T., Anderson, D.M., 2002. Variability of El Niño/Southern Oscillation activity at millennial timescales during the Holocene epoch. Nature 420, 162–165.

Neilsen, P., Hanslow, D.J., 1991. Wave run-up distributions on natural beaches. J. Coast. Res. 7, 1139–1152.

Nott, J., 2014. Grain size and the origin of wave constructed beach ridges. A Discussion of 'Beach ridges and prograded beach deposits as palaeoenvironment records'. Earth Sci. Rev. 132, 82–84.

Nott, J., Chague-Goff, C., Goff, J., Sloss, C., Riggs, N., 2013. Anatomy of sand beach ridges: evidence from severe tropical cyclone Yasi and its predecessors, northeast Queensland, Australia. J. Geophys. Res.: Earth Surf. 118, 1–10.

Nott, J., Jagger, T., 2013. Deriving robust return periods for tropical cyclone inundations from sediments. Geophys. Res. Lett. 1029/2012GL054455.

Nott, J., Forsyth, A., 2012. Punctuated global tropical cyclone activity over the past 5,000 years. Geophys. Res. Lett. 39. L14703.

Nott, J., 2011. A 6000 year tropical cyclone record from Western Australia. Quat. Sci. Rev. 30, 713–722.

Nott, J.F., 2010. A theory (involving tropical cyclones) on the formation of coarse-grained sand beach ridges in northeast Australia. Geol. Soc. London 346 (6), 7–22. Special Publications.

Nott, J., Smithers, S., Walsh, K., 2009. Sand beach ridges record 6000 year history of extreme tropical cyclone activity in northeastern Australia. Quat. Sci. Rev. 28, 1511–1520.

Nott, J., 2003. Intensity of prehistoric tropical cyclones. J. Geophys. Res. 108 (D7), 4212–4223.

Nott, J., Hayne, M., 2001. High frequency of 'super-cyclones' along the Great Barrier Reef over the past 5,000 years. Nature 413, 508–512.

Nott, J., 2000. Records of prehistoric tsunamis from boulder deposits – evidence from Australia. Sci. Tsunami Hazards 18, 3–14.

Rasser, M.W., Riegl, B., 2002. Holocene reef rubble and its binding agents. Coral Reefs 21, 57–72.

Reimer, P.J., et al., 2009. INTCAL 09 and MARINE09 radiocarbon age calibration curves, 0–50,000 years Cal BP. Radiocarbon 51, 1111–1150.

Rhodes, E.G., Polach, H.A., Thom, B.G., Wilson, S.R., 1980. Age structure of Holocene coastal sediments, Gulf of Carpentaria, Australia. Radiocarbon 22, 718–727.

Scileppi, E., Donnelly, J.P., 2007. Sedimentary evidence of hurricane strikes in western Long Island, New York. Geochem. Geophys. Geosyst. 8, Q06011.

Scoffin, T., 1993. The geological effects of hurricanes on coral reefs and the interpretation of storm deposits. Coral Reefs 12, 203–221.

Scott, D.B., Collins, E.S., Gayes, P.T., 2003. Records of prehistoric hurricanes on the South Carolina coast based on micro-paleontological and sedimentological evidence, with comparison to other Atlantic Coast records. Geol. Soc. Am. Bull. 115, 1027–1039.

Taylor, M., Stone, G.W., 1996. Beach ridges: a review. J. Coastal Res. 12, 612–621.

Taylor, R.J., 1982. The Lost Plantation: A History of the Australian Tea Industry. G.K. Bolton Publishers, Cairns, Australia, 64pp.

Wallace, D.J., Anderson, J.B., 2010. Evidence of similar probability of intense hurricane strikes for the Gulf of Mexico over the late Holocene. Geology 38, 511–514.

Woodruff, J.D., Donnelly, J.P., Emanuel, K., 2008. Assessing sedimentary records of paleohurricane activity using modeled hurricane climatology. Geochem. Geophys. Geosyst. 9, Q09V10.

Woodruff, J.D., Donnelly, J.P., Okusu, A., 2009. Exploring typhoon variability over the mid-to-late Holocene: evidence of extreme coastal flooding from Kamikoshiki, Japan. Quat. Sci. Rev. 28, 1774–1785.

Chapter 6

Storm Surge Warning, Mitigation, and Adaptation

Diane P. Horn
Birkbeck College, University of London, London, UK

ABSTRACT
This chapter provides a systematic overview of measures which play a role in mitigation and adaptation to hazards associated with storm surge. Mitigation and adaptation are interpreted in a manner comparable to the disaster risk reduction literature, referring to the amelioration of storm surge disaster risk through the reduction of existing hazards, exposure, or vulnerability. The chapter discusses major storm surge barriers such as the Dutch Delta Works and the Thames Barrier, and describes more recently built barriers in St Petersburg (Russia), New Orleans, and Venice. The chapter also reviews storm surge early warning systems, with examples from Bangladesh, the Philippines, the United Kingdom, and the United States.

The physical causes of storm surge are well known, and models are increasingly effective at predicting the storm surge associated with particular cyclone conditions. Despite this, we continue to see loss of life from storm surges. For example, the surge from Typhoon Haiyan in November 2013 was not unexpected; the strength of the storm was predicted and understood and over 800,000 people were evacuated, yet it led to 6,111 deaths and over 5 million were people displaced (Chan et al., 2013). In comparison, the 2004 Indian Ocean tsunami and Hurricane Katrina "only" displaced 1–1.5 million each (Ferris, 2013). The greatest loss of life related to a tropical cyclone is from storm surge (Doocy et al., 2013). In the United States, the loss of life in the three deadliest hurricanes (Galveston, Texas, 1900, over 8,000 deaths; Lake Okeechobee, Florida, 1928, 2,500 deaths; Hurricane Katrina 2005, 1,833 deaths) was primarily due to storm surge (Weindl, 2012). Hurricane-surge-induced flooding has killed more people in the United States than all other hurricane threats combined in the twentieth and twenty-first centuries (NOAA, 2007). The extratropical Cyclone Xynthia claimed 47 lives in 2010, Europe's highest storm surge toll since 1962 (Kron, 2013). Why does such loss of life continue despite better understanding of the physics of storm surge and

Coastal and Marine Hazards, Risks, and Disasters. http://dx.doi.org/10.1016/B978-0-12-396483-0.00006-6
Copyright © 2015 Elsevier Inc. All rights reserved.

improved modeling and warning systems? How can we best reduce the negative impacts of storm surges?

6.1 MITIGATION AND ADAPTATION

The terms "mitigation" and "adaptation" are used in many disciplines, which results in a range of interpretations. For example, the term "mitigation" can be used to describe actions taken to reduce the likelihood of an event occurring (e.g., reducing greenhouse gas emissions in order to reduce increases in global temperature and thus reduce the rate of sea level rise) or actions taken to reduce the impact if the event does occur (e.g., building flood defenses). Within a particular discipline the meaning of such terms may be clear. There is no interdisciplinary consensus, however, and until recently, climate scientists and disaster risk reduction researchers had very different definitions of mitigation. In the climate change literature, mitigation refers to the reduction of the rate of climate change via the management of its causal factors (emission of greenhouse gases from fossil fuel combustion, agriculture, land use changes, etc.). However, in the disaster risk reduction literature, mitigation refers to the amelioration of disaster risk through the reduction of existing hazards, exposure, or vulnerability. Recent Intergovernmental Panel on Climate Change (IPCC) reports have revised their definitions to accommodate both interpretations of mitigation. IPCC definitions now distinguish between mitigation of climate change, which is defined as "a human intervention to reduce the sources or enhance the sinks of greenhouse gases," and mitigation of disaster risk and disasters, which is defined as "the lessening of the potential adverse impacts of physical hazards, including those that are human-induced, through actions that reduce hazard, exposure, and vulnerability" (IPCC, 2012).

Wilby and Keenan (2012) and Cooper and Pile (2013) reviewed the different definitions of adaptation, as summarized above, and Hallegatte (2009) identified five categories of practical adaptation strategies. The first is *no-regret measures*, which yield benefits even in the absence of increasing hazards (although they are not cost free). The second category is *reversible strategies*, which are flexible enough to reduce as much as possible the cost of being wrong about future risks. The third category, *safety margin strategies*, reduces vulnerability at low or no cost. The fourth category, *soft strategies* such as land use planning and insurance, influence individual and institutional decisions and therefore can have an effect on hard investments. The final category is *strategies to reduce decision-making time horizons*. Hallegatte (2009) also distinguished between "hard" adaptation (e.g., building sea defenses) and "soft" adaptation such as land use planning, early warning systems, and financial instruments such as insurance. Wilby and Keenan (2012) distinguished between the broader enabling environment for adaptation (information provision, institutional arrangements, and preparedness) and specific implementing measures to reduce flood risk. They classified these implementing measures into three categories: defending against flood risk, living with flood risk, and withdrawing from flood risk. Defense measures typically involve some form of engineering to protect existing land use. Hard engineering structures such as levees and storm surge barriers are probably the most common; however, there is a growing interest in soft engineering approaches, such as wetland creation. Accommodation actions include raising buildings and roads above flood level, establishing evacuation routes and warning systems, the creation or enhancement of stormwater system capacity, and zoning policies aimed at preventing development in high-risk areas. Retreat policies include those aimed at discouraging rebuilding in high-risk areas and the reclamation or abandonment of highly flood-prone lands. Table 6.1 summarizes the range of available approaches to coastal hazards (not necessarily restricted to storm surge). Cooper and Pile (2013) characterized adaptation measures as two types, those that modify the environment and those that are aimed at changing human activities. They noted that the reduction of harm and the realization of benefits to humans are central to all definitions of adaptation. Wilby and Keenan (2012) argued that in the case of flood risk management (floods in general, not storm surge explicitly), much of what is labeled adaptation could just be described as good practice.

In the context of storm surge, mitigation can be thought of acting to reduce the number and severity of storm surges (which would mean reducing the number and severity of tropical, subtropical, and extratropical cyclones), while adaptation is learning to live with the risk of storm surges. Adaptation measures are unlikely to eliminate risk, but should aim to reduce risks to levels that are acceptable within the limits of available resources. The measures discussed in this chapter are adaptive measures only, which may reduce the vulnerability to storm surge rather than the storm surge itself—none of them can stop cyclones from occurring.

6.2 STORM SURGE BARRIERS

The most common hard engineering structure used to protect against surge-induced flooding is a storm surge barrier. These structures provide temporary protection from flooding, generally for a few hours before and after high tide and are often partly open during normal conditions to allow navigation and salt water exchange with estuarine areas landward of the barrier (Jonkman et al., 2013). A storm surge barrier may be only one component of a larger flood protection scheme, which will often include other structures such as seawalls and levees. Storm surge barriers are usually built at a position where the barrier can be closed during times where flooding is predicted. When the barrier is not closed, it allows free passage of water and shipping. The main disadvantages of a storm surge barrier system are the huge construction and maintenance costs. Movable barriers also require simultaneous investment in flood warning systems, which provide information on when to close the barrier (Aerts et al., 2013b).

TABLE 6.1 Coastal Hazard Adaptation Strategies (Not Restricted to Storm Surge)

Adaptation Option	No-Regret Strategy	Reversible/ Flexible	Safety Margins Available	Soft Strategy	Type of Response
Hard defenses (e.g., seawalls, levees)	+	–	+		Defend against flood risk
Storm surge barriers	+	–	+		Defend against flood risk
Restore natural coastal defenses (e.g., salt marsh, mangrove, dunes)	+	+	+		Defend against flood risk
Temporary/demountable defenses		+	+		Live with flood risk
Enhanced drainage systems	+	–	+		Live with flood risk
Land use planning (e.g. rezoning, setback, compulsory purchase, restrictions on development in flood zones)	+	+	+		Live with flood risk
Improved building standards/flood-resilient construction (buildings and infrastructure)	+		+		Live with flood risk
Information and warning	++	+		+	Live with flood risk
Evacuation schemes	++	+		+	Live with flood risk
Flood/storm surge shelters	+	+	+		Live with flood risk
Insurance (household to national level)	++	+		+	Live with flood risk
Relocation and retreat	–	–			Withdraw from flood risk

In the classification used by Hallegatte (2009), ++ indicates options which yield benefits even without climate change, while + indicates options that are "no-regret" only in some cases, depending on local characteristics and – indicates that the strategy would entail significant losses in the current climate.
Source: Adapted from Hallegatte (2009) and Wilby and Keenan (2012).

Most storm surge barriers were implemented after a flood disaster occurred (Aerts et al., 2013b). The best known barriers are the Dutch Delta Works and the Thames Barrier, which were both built in response to the 1953 North Sea storm surge when 305 people died in the United Kingdom and 1835 died in the Netherlands. The Delta Works (Figure 6.1) is the largest flood protection project in the world and includes dams, sluices, locks, bridges, tunnels, dikes, levees, and storm surge barriers that protect southern Holland against a 1:10,000-year storm surge (Zhong et al., 2012). The Oosterscheldekering (Eastern Scheldt storm surge barrier) is the largest of 13 structures which make up the Delta Works. Construction on the Delta Works began in 1958 and was officially completed in 1997 at a cost of $7 billion (Bijker, 2002). However, the Netherlands continues to add infrastructure to the system as needed (Pilarczyk, 2012), with a seawall near Harlingen completed in 2010. The Delta Committee was set up to investigate the impact of climate change and projected sea level rise in the twenty-first century. Their 2008 report (Delta Vision Committee report, 2008) led to the Delta Programme, which addresses future flood risk management and freshwater supplies, and the establishment of the Delta Fund. Significant investments will need to be made after 2050; for

FIGURE 6.1 The Dutch Delta Works. *http://en.wikipedia.org/wiki/File:Deltawerken_na.png*

example, the Maeslant Barrier protecting Rotterdam is due to be replaced after 2070 to build a new closable storm surge barrier. The Delta Fund, which went into effect at the start of 2013, sets aside money for the government investments to works set out in the Delta Programme (Zevenbergen et al., 2013). The central government and the water boards (regional government bodies which levy their own taxes) have agreed to pay an equal share of the costs of current and future flood protection measures (Verduijn et al., 2012). They will each contribute €131 million in 2014 and €181 million annually from 2015, with a total of €16.6 billon available from all funding sources between 2014 and 2028 and €22 billion between 2029 and 2050 (Kabat et al., 2009). All the resources from the Delta Fund are fully allocated until 2019, when funding for investment in new flood risk management measures will be available (Delta Programme, 2014). The impact of the Delta Works has been extensively studied, with research investigating the impact on ecology (e.g., Hostens et al., 2003; Tangelder et al., 2012; Jansen et al., 2013; van Wesenbeeck et al., 2014), geomorphology (e.g., Louters et al., 1998; Roelvink et al., 2001; Tönis et al., 2002; Hudson et al., 2008), and hydrology and hydrodynamics (e.g., Jonkman et al., 2008; Augustijn et al., 2011; Zhong et al., 2012). The references reported here represent only a small number of the studies which have been carried out on the areas affected by the Delta Works. Zhong et al. (2012) evaluated the impact of the Maeslant storm surge barrier on flood frequency under rising sea levels and showed that the operation of the Maeslant Barrier reduces flood frequency and can partly compensate for the effect of future sea level rise, although the return periods of all water levels will decrease as the sea level rises. The Maeslant Barrier is currently closed when water level in Rotterdam reaches 3.0 m. According to the model of Zhong et al. (2012), the return period of this water level in 2010 without the barrier, 10.9 years, is increased to 2,400 years with the barrier. Under the current closing decision water level of 3.0 m, the port of Rotterdam will be closed once every 3.2 years in 2050 and once every 1.1 year in 2100. The design safety level, 4.0 m, will be reached with a return period of 46,948 years in 2010, 16,420 years in 2050, and 3,849 years in 2100.

Although the water level in the 1953 storm surge was the highest ever recorded in London, the height of the storm surge was about 3 h before high tide and the river level was low after a dry spell, and thus only a few locations in east London were flooded. After the lucky escape in 1953, the UK government appointed the Waverley Committee to study flood dangers to the city. The Committee recommended that a structure be constructed across the Thames, backed up with a considered approach to development in the floodplain, and 41 proposals at six sites were put forward following the report (Waverley, 1954). However, no action was taken until the problem was passed to the Greater London Council in 1968 with a request that a full investigation into the problem be carried out as a matter of considerable urgency (Horner, 1979). This led to the construction of the barrier, begun in 1976 and completed in 1984 at a cost of

£560 million (equivalent to about £1.6 billion in 2014 prices), including the cost of strengthened defenses both upstream and downstream of the Thames Barrier and the associated construction of three minor barriers, two major floodgates, and nine minor floodgates on tributaries of the Thames. The Thames Barrier was designed to protect against the 1,000-year flood, with an allowance for an increase in water level up to 2030. In 2010, the UK Environment Agency (EA) published a study (TE2100) on planning for flood risk management in the Thames Estuary, which aimed to determine the appropriate level of flood protection needed for London and the estuary for the next 100 years (Environment Agency, 2010). The TE2100 policy recommendations were divided into three time periods. In the first 25 years (2010–2035), the strategy is to continue to maintain the current flood defense system including planned improvements, ensure effective floodplain management is in place, and monitor change indicators. The strategy for the transition period (2035–2049) is to replace and upgrade current defenses and to make the final decision on building a new barrier or other end-of-century option, and to start planning for this. The agreed end-of-century option is to be planned, designed, and constructed between 2050 and 2100. The extensive benefit–cost analyses carried out in the TE2100 analysis suggested that improving defense standards now is not cost effective, as the extra benefit gained is generally not worth its costs until climate and socioeconomic change begin to create additional risk from 2050 and toward 2070 (Penning-Rowsell et al., 2013a). No costings for the unspecified end-of-century option were given in the TE2100 report but other information from the EA indicated that investment of £1.5 billion will be needed for the first 25 years, with another £1.5 billion needed for the middle 15 years, and £6–7 billion for 2050–2100 (Environment Agency, 2013). The contracts for the first stage of TE2100 went out for bidding in September 2013. The phase 1 program represents the capital-funded work needed to maintain tidal defenses from 2015 to 2025 and includes refurbishment of fixed assets (such as tidal walls and embankments), active assets (including the Thames Barrier gates), and new assets such as pumping stations. The contract for phase 1 will be awarded in September 2014.

As of April 2014, the Thames Barrier has been closed 174 times since it became operational in 1982, with increasingly frequent closures since 2000. Half of these closures were to protect against tidal flooding and half to protect against river flooding. The closure on December 6, 2013, was associated with the biggest storm surge since 1953 and the highest tide in response to which the Thames Barrier has closed, at an elevation of 4.1 m (Atkin, 2014). The EA published a map showing the probable impact on central London if the barrier had not been in place (Figure 6.2). The TE2100 report recommended that the Thames Barrier should not be closed more than 50 times a year to reduce the chance of it failing. The Thames Barrier was expected to reach this design limit from 2135 onward, and the TE2100 report suggested that once this design limit is reached, it may not be possible to close the barrier to protect against fluvial

FIGURE 6.2 The UK Environment Agency's simulation of the flooding which should have resulted from the storm surge in December 2013 without the Thames Barrier. http://www.nce.co.uk/news/nce-live-news-updates-tuesday-10-december-tidal-surge-broke-records-london-will-choke-without-crossrail-2/8656557.article.

flooding in order to maintain the reliability of the barrier against tidal flooding (TE2100). The Thames Barrier came under unprecedented pressure in the first months of 2014, reaching its operational safety limit of 50 closures on 4 March of that year. This is the record for the highest number of times it has been closed in a single season (Figure 6.3), with many of the closures to protect West London against river flooding in the wettest winter since records began in 1766. The EA described these frequent closures as a "blip" and is still forecasting that a replacement is not needed until 2070 (BBC, 2014). However, the EA will carry out an investigation of the robustness of the Thames Barrier in response to a request from the Mayor of London and will report its findings at a Thames Estuary Steering Group meeting in early summer 2014 (London Assembly, 2014). Even with the impetus of the loss of life in 1953, it took years to plan and build the Delta Works and the Thames Barrier. Padron and Forsyth (2013) estimated that the average interval between planning and constructing a major storm surge barrier is 27 years, suggesting that perhaps the timetable for the Thames end-of-century option should be brought forward and that something like the Delta Fund should be set up now to plan for its financing.

Storm surge barriers in other cities are reviewed by Dircke et al. (2013) and summarized in Table 6.2; this chapter will describe only a few more examples. The St Petersburg Flood Prevention Facility Complex is a storm surge barrier

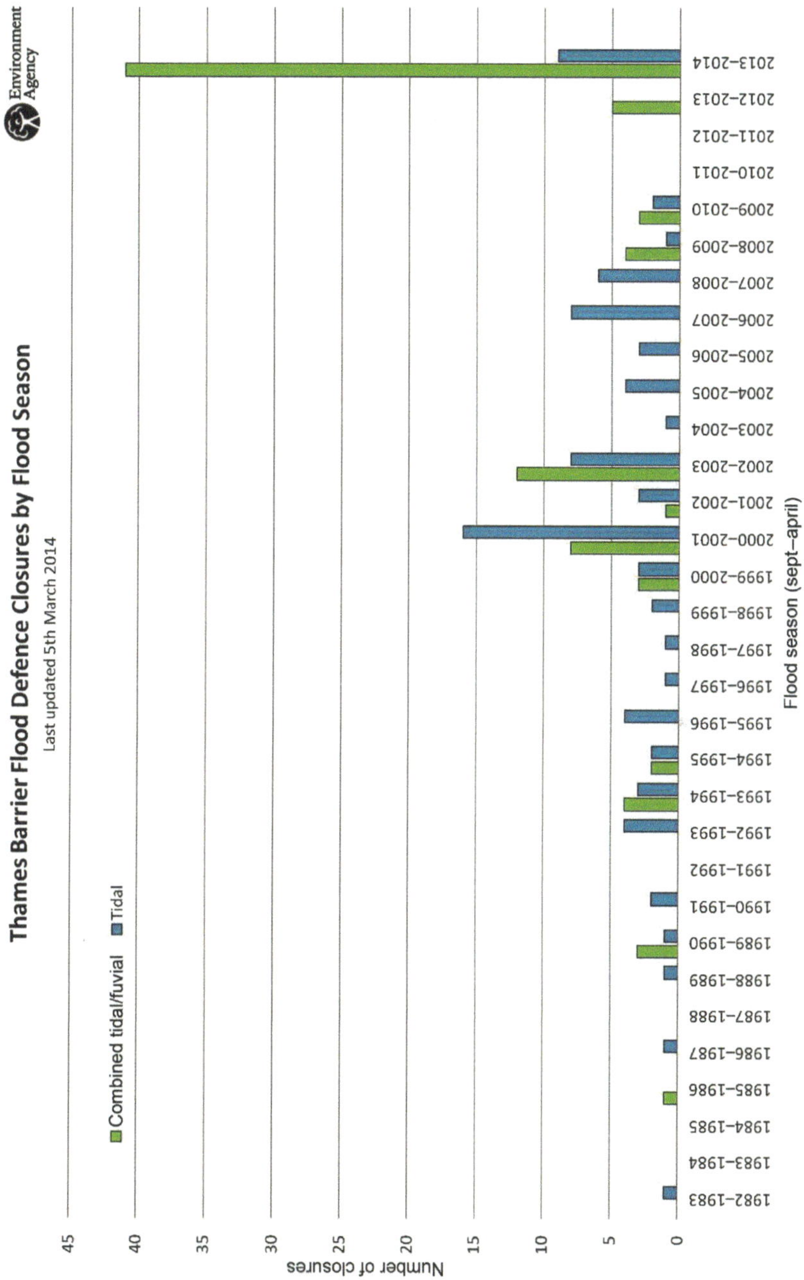

FIGURE 6.3 Closures of the Thames Barrier since it was built. https://www.gov.uk/the-thames-barrier

TABLE 6.2 Overview of Storm Surge Barriers

Barrier	Type	Country	Design and Construction Time (years)	Year of Operation	Approx Cost (Million Dollars)
Hollandse Ijssel storm surge barrier	Lifting gate	Netherlands	4	1958	127
Oosterschelde storm surge barrier	Lifting gate	Netherlands	10	1986	5,005
Maeslant storm surge barrier	Floating sector gate	Netherlands	8	1997	709
Europoort Barrier and Hartel Barrier	Lifting gate	Netherlands	6	1997	329
Ramspol storm surge barrier	Rubber dam	Netherlands	5	2002	88
Venice storm surge barrier	Flap gate	Italy	13 estimated	2016 estimated	7.6
New Bedford storm surge barrier	Rolling sector gate	New Bedford, MA, USA	4	1968	72
Stamford storm surge barrier	Flap gate	Stamford, CT, USA	4	1968	19
Harvey canal flood protection barrier, Gulf Intracoastal Waterway West Closure Complex (GIWWCC)	Sector gate	New Orleans, LA, USA	3	2008	1,014

IHNC New Orleans hurricane protection barrier	Sector gate	New Orleans, LA, USA	4	2011	1,318
Thames Barrier	Segment gate, radial gate	United Kingdom	8	1982	2,730
Fox Point hurricane barrier	Radial gate	Providence, RI, USA	6	1966	75
St Petersburg Barrier	Lifting gate, floating sector gate	Russia	Started in the 1990s, resumed in 2002	2010	6,600
Eider Barrage	Radial gate	Germany	6	1973	113
Ems storm surge barrier	Lifting gate, radial gate, segment gate	Germany	5	2002	286
Marina Barrage	Crest gate	Singapore	3	2008	173

Source: Based on Dircke et al. (2013), Expanded and Updated.

system which includes 11 embankment dams, 67 hydraulic gates, a bridge and tunnel, and two navigation channels with gates that can be closed to protect against storm surge (Hunter, 2012). Construction on the barrier began in 1979, halted at the breakup of the Soviet Union, and was completed in 2011 at a cost of about $3 billion. The scheme is designed to protect against a 1,000-year storm surge of 4.55 m with an extreme limit of 5.15 m for a 1:10,000-year flood (Ivanov et al., 2012).

A storm surge barrier system to protect Venice, known as MOdulo Sperimentale Elettromeccanico (MOSE), is under construction. The MOSE project consists of four tidal defense barriers at the openings in the barrier islands where Adriatic tides enter Venice lagoon (Bajo et al., 2007). The MOSE defenses include 21 gates in a north channel at the Lido inlet, 20 gates in the south channel and an intermediate island linking the two barriers, a barrier with 19 gates at the Malamocco inlet, and a barrier with 18 gates in the Chioggia inlet (Fontini et al., 2010). Construction started in 2003 and is expected to be completed in 2016 with an estimated cost of €5.493 million (about $7.6 million). There has been a long debate about the optimal closing level of the MOSE barriers: increasing the barriers' functioning frequency improves the degree of protection against storm surge; however, this also increases interference with harbor activities (Fontini et al., 2010). Researchers have also debated whether the yet-to-be-completed Venice barrier will protect the city as sea level rises (e.g., Pirazolli, 2002; Bras and Harleman, 2002; Pirazzolli and Umgeisser, 2006; Umgiesser and Matticchio, 2006; Rinaldo et al., 2008). The current expectation is that the MOSE barriers will protect Venice until sea level rise is greater than 0.60 m; at that stage, new options for the future of the Venice lagoon will have to be considered (Munaretto et al., 2012).

Hurricane Katrina highlighted the need for improved defenses for New Orleans and led to the construction of the Hurricane and Storm Damage Risk Reduction System (HSDRRS) for Southeast Louisiana. The storm surge from Katrina breached 50 floodwalls and levees and collapsed a 1,220-m section of the Inner Harbor Navigation Canal (IHNC), a waterway that connects the Mississippi with Lake Borgne and the Gulf of Mexico (Miller et al., 2013). The $14.45 billion HSDRRS consists of 563 km of levees and floodwalls, 73 pumping stations, three canal closure structures with pumps, and four gated outlets (USACE, 2010). Because evacuation is now seen as a major part of the new emphasis on risk reduction, the HSDRRS also includes an elevated road and bridge over a new floodwall at the entrance to the Lake Pontchartrain Causeway Bridge (Eqecat, 2009; Reid, 2013). One component of the HSDRSS, the 3-km IHNC—Lake Borgne Storm surge barrier—was completed in 2013 at a cost of $1.1 billion and is designed to protect the areas which were hit hardest by Katrina. This barrier closes off Lake Borgne, situated at the east side of New Orleans, and closes both the lake and the canals northwest and southwest of the lake. The navigable Seabrook surge barrier protects the Lake Pontchartrain entrance of the IHNC. A movable gate has been built in the canal northwest of Lake Borgne, the Gulf Intracoastal Waterway (GIWW), making navigation possible during normal conditions. The canal southwest of Lake Borgne, the Mississippi River–Gulf Outlet, has been permanently closed by the storm surge barrier. The new system is designed to withstand the 1:100-year storm event in 2057 (USACE, 2010). The HSDRSS system was tested by Hurricane Isaac in 2012 and performed as designed (Reid, 2013; USACE, 2013), although flooding did occur in areas outside the system. This was attributed to an intense and long-duration surge caused by the long duration of tropical force winds, which in some cases were aggravated by extreme local rainfall (USACE, 2013). However, is a 1:100 protection standard adequate to protect New Orleans? Compare this to the 1:10,000 standard of the Delta Works.

Hurricane Sandy provided a demonstration of the vulnerability of New York City to storm surge, reviving discussion of the possibility of building a protective barrier (e.g., Bowman et al., 2005; Colle et al., 2008; Aerts and Botzen, 2012; Coch, 2013; Hill, 2012, 2013). Proceedings of a conference held in 2009 under the sponsorship of the Infrastructure Group of the American Society of Civil Engineers were reissued in 2013 in the aftermath of Sandy (Aerts et al., 2013a). Aerts et al. (2013b) outlined three possible configurations of storm surge barriers. The cost of these barriers is estimated at up to $15 billion, comparable to the amount spent on the Louisiana HSDRSS system (Aerts et al., 2013b). Note that these barriers would also protect against only the 1:100 storm surge. In contrast, other researchers argue that the city should commit to protecting all areas to a 1:500-year standard (Tollefson, 2013). The system envisaged by these researchers, including Bowman et al. (2005), Colle et al. (2008), and Hill (2013), would comprise two or three barriers to protect New York and New Jersey, including an 8-km-wide barrier approximately 6 m high that could be opened and closed at the entrance to New York's harbor, and a second barrier at the entrance to Long Island Sound (Tollefson, 2012). The state panel's cost estimates for such a system range from $7 to $29 billion, depending on the design (Tollefson, 2013). Wagner et al. (2014) argued that historically, financial resources, policies, and public support have coalesced after catastrophic events to act as catalysts for policy change; however, the arrival of Sandy during the economic crisis restricted the availability of funding.

Structural measures can never entirely eliminate flood risk, and may not always be an appropriate response to such risk. Storm surge barriers can protect a large area with relatively small structures, but are not appropriate on an open coastline. In addition, surge barriers are expensive and therefore not economically feasible for all locations. As with all flood defense structures, the presence of the barrier may encourage development in hazardous areas. Many coastal cities will not be able to rely on such major infrastructure, and will need instead to rely on a mix of structural and nonstructural measures.

6.3 STORM SURGE WARNING

If it is not possible or desirable to completely eliminate flood risks, adaptation can be accomplished in the form of adequate provision of emergency procedures (Cooper and Pile, 2013). Improvements in forecasting, early warning systems, and evacuation and shelter procedures have reduced storm-surge-related mortality (Doocy et al., 2013). Real-time observational and predictive modeling techniques make it possible to issue cyclone and related storm surge warnings, which may trigger the implementation of emergency procedures and provide an opportunity for the evacuation of vulnerable areas. To be effective, early warning systems must integrate four elements: (1) knowledge of the risks faced, (2) technical monitoring and warning service, (3) dissemination of meaningful warnings to those at risk, and (4) public awareness and preparedness to act (UNISDR, 2009). Although comprehensive coverage of early warning systems for storms and tropical cyclones is available, disasters such as Hurricane Katrina have highlighted inadequacies in technologies for enabling effective and timely emergency response (Grasso and Singh, 2009). Storm surge is generally treated as a corollary of cyclones (both tropical and extratropical), with few early warning systems specifically for storm surge in existence. For example, Grasso and Singh (2009) presented detailed tables of early warning systems for different types of events, but they did not include storm surge in these tables, with one table for floods (which they interpreted as primarily fluvial and pluvial) and another table for severe weather, storms, cyclones, and hurricanes. The report did show that there are often inadequate flood warning and monitoring systems, especially in developing or least developed countries. In most of the cases that they surveyed, Grasso and Singh (2009) found that communication systems and adequate response plans were lacking. They argued that predictions are not useful unless they are translated into a warning and action plan that the public can understand.

Storm surge warning systems are most often operated at a national scale and are usually linked to predictions of the path and landfall of tropical or extratropical cyclones. Many countries have or are developing storm surge prediction models and warning systems: this review has found references to at least 28 national programs, but only a few will be described here. The Joint Technical Commission for Oceanography and Marine Meteorology (JCOMM) has compiled a comprehensive list of national storm surge products (JCOMM, 2014). Although that list is not exhaustive and does not explicitly address storm surge warning systems, it gives an indication of international activity in this area. JCOMM is currently carrying out a worldwide survey on operational storm surge models and data, with data collection scheduled to be completed by the end of May 2014. JCOMM has also initiated the World Meteorological Organization (WMO) Coastal Inundation Forecasting Demonstration Project to assist countries at risk of coastal flooding to operate and maintain a reliable integrated forecasting and warning system. Four national subprojects have begun since 2011 (Bangladesh, Fiji, Dominican Republic, and Indonesia), with two more planned (Shanghai and South Africa).

In 2007, Cyclone Sidr made landfall in southern Bangladesh, causing 3,406 deaths. Despite advance warnings being issued 5 days before landfall, only slightly over a third of the population complied with cyclone warnings and evacuation orders (Paul, 2012). After Sidr, the Bangladesh government modernized the early warning systems and initiated the construction of 2,000 new cyclone shelters and as a result, in 2009, Cyclone Aila caused only 330 deaths (Haque et al., 2012; Ahamed, 2013). In Myanmar, Cyclone Nargis caused over 146,000 deaths and more than $10 billion in damages in 2008. In contrast, 2 years earlier a slightly stronger storm, Cyclone Mala, hit the Myanmar coast about 150 km north of the Nargis track, but only led to 22 deaths after a well-executed evacuation effort (Fritz et al., 2011). The area hit by Nargis lacked evacuation plans and shelters, with residents having no experience of storm surge flooding (Brakenridge et al., 2013). In February 2011, Cyclone Yasi hit Queensland, Australia. The cyclone was 500 km wide with 285 km/h wind speeds, yet no lives were lost as a result of advance warning and an evacuation that was completed before the cyclone struck (Haque et al., 2012). Early warnings disseminated 4 days before Cyclone Phailin hit Eastern India in October 2013 allowed the evacuation of nearly 1.2 million people, with only 44 deaths. In contrast, a comparable cyclone hit the same area in 1999, with the loss of more than 10,000 lives (Harriman, 2014).

The United Nations Development Program identified 29 developing countries and four developed nations that are at risks from cyclones, but 42% and 27% of cyclone deaths in the past two centuries have occurred in Bangladesh and India, respectively (Doocy et al., 2013). (Note, however, that although the Doocy et al. paper was published in 2013, the data analyzed in the study was from 1980 to 2008 and does not, therefore, include any deaths from storm surges since then, particularly Typhoon Haiyan.) Bangladesh is particularly vulnerable to storm surge because of its location, low-elevation and flat topography, high population density, and limited coastal defense structures. Penning-Rowsell et al. (2013b) estimated that nearly 700,000 deaths have been caused by cyclones affecting Bangladesh since 1960. Two storms accounted for the majority of these deaths, with more than 300,000 deaths from Bhola Cyclone in 1970 and 138,866 from Cyclone Gorky in 1991. Economic losses due to cyclones are not declining significantly, although the reduction in potential gross domestic product associated with major disasters has gradually become smaller as the national economy has grown and become less dominated by agriculture (Haque et al., 2012).

The Bangladesh Cyclone Preparedness Programme began in 1972, in response to the 1970 disaster, when no warnings were received in vulnerable rural areas and coastal villages despite the cyclone being detected by coastal radar and satellite. The loss of life has decreased dramatically under this program, with improved warning systems, infrastructure improvement to facilitate evacuation, access to elevated cyclone shelters, the construction of coastal embankments and more resilient buildings, and the reforestation of mangrove forests (Haque et al., 2012; Paul, 2012). Forecasts from the Storm Warning Centre are used to produce early warnings supported by emergency telecommunications and cyclone preparedness activities in the coastal zone. In particular, a large network of trained local volunteers is equipped with megaphones and public address systems to disseminate the warning. Warning dissemination was originally the weak link in the system and the large death toll in 1991 has been attributed to this, and also to the fact that only two out of every five shelters were useable because of flooding (Haque et al., 2012). Since 2007 there has been a major change in the availability of communication technology with the widespread use of mobile phones. Colorful hot air balloons can also be used to convey cyclone warning messages in remote and coastal areas of Bangladesh (Haque et al., 2012). However, continued challenges of illiteracy, lack of awareness, and communication problems mean that some people do not understand or follow the evacuation warnings. Coastal areas are not well connected by road networks, public bus service is very limited, and few people own a car. Most coastal residents walk from their homes to public cyclone shelters, which are mostly located within 3–5 km of their homes; however, this can be difficult in adverse weather conditions. In addition, many of the at-risk population are unwilling to use public shelters because of the poor condition (e.g., lack of latrines and separate rooms for women, no drinking water, and lighting). The shelters that are in reasonably good condition can only accommodate about 15% of the coastal population of Bangladesh (Paul, 2012). Despite improvements in warning systems, precyclone evacuation remains a challenge. Haque et al. (2012) recommended that a network of small, sturdy, and safe multipurpose buildings should be established within a 2-km walking distance of households and villages instead of developing larger cyclone shelters.

Typhoon Haiyan (known locally as Typhoon Yolanda), which struck the Philippines in November 2013, was the deadliest storm in the country's history, with 6,111 deaths and 1,779 people still missing, and estimated losses of $12.9 billion (Neussner, 2014). Haiyan is also the strongest storm yet recorded at landfall, with sustained winds of up to 315 km/h. The storm surge was not measured, but has been estimated at 2.3–5 m, with some estimates as high as 7–8 m (eSurge, 2014; NASA, 2014; Neussner, 2014). Although warnings were issued and the Philippines National Disaster Risk Reduction and Management Council (NDRRMC), with local authorities, evacuated more than 800,000 people, the official death toll still exceeded 6,000 (Neussner, 2014). In the current system, a warning is issued by Philippine Atmospheric, Geophysical and Astronomical Services Administration (PAGASA), which gives a warning to NDRRMC, which then releases a national statement on what to expect from a storm. The information goes from the national to regional, provincial, municipal, and finally to the village level. Officials from PAGASA said that the public lacked a clear understanding of what was meant by a storm surge (Santos, 2013). There is no Filipino term for storm surge, and ironically, if the warning had been described as a tsunami, people would have had a better idea what to expect (Yates, 2013). Many residents of coastal areas and political decision makers admitted that they were not familiar with the term "storm surge" (Neussner, 2014). Although PAGASA accurately forecast the typhoon's path and strength, public warnings failed to sufficiently explain and clarify the specific dangers, and warnings related to storm surge were issued relatively late (Neussner, 2014). In some locations, however, such as Balankayan, East Samar, no lives were lost because of effective warning and evacuation drills (Yates, 2013). Postdisaster studies have suggested that up to 94% of the victims of Haiyan could have been saved if they had been properly warned and evacuated to safe areas (Neussner, 2014). PAGASA and NDRRMC emphasized rain, flood, and landslide warnings, but did not stress very strongly the storm surge to come. The official storm surge map grossly underestimated the inundation area of the storm surge (Neussner, 2014). In addition, many of the evacuation centers were single-floor buildings without stilts and thus inappropriate as storm surge shelters, and were flooded by the storm surge. Even when a clear evacuation message was given by the government, a large part of the population did not leave their homes in the danger zones. One of the main reasons given by respondents of the Gesellschaft fur Internationale Zusammenarbeit (GIZ) survey was the fear of theft and/or looting. Some also dismissed the evacuation order or underestimated the height and force of the water (Neussner, 2014). In February 2014, the Philippines Science Secretary proposed the adoption of a storm surge advisory protocol and flood advisory system. If the storm surge warning protocol is adopted, Storm Surge Advisory (SSA) No. 1 would indicate a storm surge height of up to 2 m; SSA No. 2, up to 5 m; and SSA No. 3, more than 5 m. Unlike public weather warning signals, which are updated in 24 h, a storm surge advisory will be given 48 h in advance. The new advisory systems for storm surges and floods are part of a disaster preparedness road map that the President wants adopted before the onset of the rainy season in June 2014. The storm surge disaster risk reduction efforts are geared at a reduction of exposure in the coastal areas by implementing "no-build" zones, development of natural barriers and construction of man-made barriers to reduce impact of hazard, and resiliency of home, building, and other infrastructure (Ubac, 2014).

In December 2013, the United Kingdom experienced the largest storm surge since 1953, but suffered relatively limited damage thanks to improved defenses. About 10,000 homes were evacuated in Eastern England after receiving coastal flood warnings from the UK Coastal Monitoring and Forecasting service (UKCMF). The UKCMF is managed by the EA in partnership with the Met Office, the National Oceanography Centre (NOC), the Flood

Forecasting Centre, and the Centre for Environment, Fisheries and Aquaculture Science (CEFAS). The UKCMF represents a move from forecasting alone to forecasting and monitoring. It was originally called the Storm Tide Warning Service and was set up after the 1953 storm. The UKCMF uses a network of 44 tide gauges and a strategic wave monitoring network hosted by CEFAS. The surge model uses inputs from the Met Office mesoscale atmospheric model, the WAM wave prediction model, tide-surge models developed by Proudman Oceanographic Laboratory (part of NOC), and bathymetric data to predict the height and timing of surge events. Surge models run four times a day producing forecasts up to 2 days ahead, and include local models which provide more accurate total water level forecasts (surge plus tide). Model results are transmitted to the EA along with real-time data from the tide gauge network. The EA operates an online flood warning service (for all floods, not just coastal floods or storm surge) which publishes a live flood warning map updated every 15 min, local flood information by postcode, a 3-day flood forecast, flood warnings on Facebook, and flood warnings by phone, text or, e-mail. The Flood Forecasting Centre also sends Flood Guidance Statements to emergency responders.

In the United States, the National Hurricane Center (NHC) monitors cyclone tracks and intensity in the North Atlantic and Eastern North Pacific east of 140° W. The NHC issues a package of advisory products which are updated every 6 h, and give information on areas of disturbed weather and their potential for development into a tropical cyclone during the next 5 days, which are tied to the anticipated arrival time of tropical storm force winds. Each of the advisory categories (tropical storm watch, tropical storm warning, hurricane watch, hurricane warning) has a clearly defined meaning which is included in the NHC's public advisory bulletins and includes a series of parameters which describe the storm. Local Weather Forecast Offices issue Hurricane Local Statements, which include detailed information on weather conditions, evacuation decisions, recommended precautionary actions, storm surge and tide information, and potential for other hazards (e.g., tornado, flash flood, rip currents, beach erosion).

The NHC predicts storm surge using the Sea, Lake and Overland Surges from Hurricanes (SLOSH) model. SLOSH estimates winds and storm surge heights resulting from historical, hypothetical, or predicted hurricanes for a range of input variables: atmospheric pressure, cyclone intensity, size, forward speed, and direction of motion (hurricane track). These inputs are used to create a model of the wind field which drives the storm surge. Predictions for a particular location are calculated for the local bathymetry, including the adjacent continental shelf, the local shoreline plan form, and angle of approach of the cyclone relative to the shoreline. SLOSH does not include waves, normal river flow and rain, or the astronomical tide (although operational runs can be carried out at different tide levels via an initial water level anomaly). SLOSH outputs strongly depend on the accuracy of the meteorological input, and meteorological uncertainty will dominate over storm surge model specifications (Lowry, 2012).

The NHC uses the SLOSH model to make advance calculations of potential surge for 37 basins, which represent sections of coastline that are centered on particularly susceptible features such as inlets, large coastal centers of population, low-lying topography, and ports. These basins include all of the US East and Gulf coasts, the Bahamas, Puerto Rico, the Virgin Islands, and parts of Hawaii and Guam. The NHC normally updates the storm surge data sets for five or six basins a year. These are done after a storm changes the shape of a coastline, the high-resolution land elevation for the area has been updated, or local emergency managers need updated surge data for planning. When a hurricane threatens a particular basin, the NHC runs thousands of simulated hurricanes for all five storm categories, changing various factors such as forward speeds, directions of approach, and landfall locations. Occurrence probabilities of these hurricanes are not considered. The SLOSH model produces three storm surge products: probabilistic surge (P-Surge), Maximum Envelope of Water (MEOW), and Maximum of Maximums (MOM). Storm-specific uncertainties are accounted for in the P-Surge product, which shows the overall chances that the specified storm surge height will occur at each individual location on the forecast map during the indicated forecast period. The probabilities are based on errors during recent years in the official track and intensity forecasts issued by the NHC. Variability in storm size is also incorporated (NOAA, 2014).

Each model run for a category creates a separate MEOW, which provides a worst-case snapshot for a particular storm category, forward speed, trajectory, and initial tide level. All the MEOWs for a given basin are combined to form an MOM, which is an ensemble product of maximum storm surge heights for all hurricanes of a given category. MOMs are created by pooling all the MEOWs for a given basin, separated by category and tide level (zero/high), and selecting the MEOW with the greatest storm surge value for each basin grid cell. This procedure is repeated for each storm category. In total 10 MOMs are made available for each basin: one MOM per storm category and tide level. MOMs represent the worst-case scenario for a given category of storm under "perfect" storm conditions (NOAA, 2014). Neither MEOWs nor MOMs are storm specific and no single hurricane will produce the regional flooding depicted in the MEOWs and MOMs, which are intended to capture the worst-case high water value at a particular location for evacuation planning. Although these acronyms may seem frivolous, the idea is that they are easier to remember—and maybe to show that modelers have a sense of humor. Perhaps to balance the cat names (e.g., Cat 5 storm, MEOWs), animations of storm surge generated by SLOSH have the extension .rex, which reportedly was the name of the developer's dog (Williams, 2013).

The most familiar way of describing the intensity, thus destructive potential, of tropical cyclones is the Saffir–Simpson Hurricane Wind Scale

(SSWS), which is a classification of hurricanes into five categories based on a hurricane's sustained wind speed at a particular time. The scale provides examples of the type of damage and impact in the United States associated with winds of indicated intensity and indicates the maximum sustained wind and measure of the degree of damage possible in areas that experience the maximum wind (typically the eyewall). The SSWS is not an overall indicator of the severity of the storm, as it does not indicate the size of the wind field or the size of the area impacted, the magnitude of the storm surge, or the amount of rain that will fall. Earlier versions of the scale incorporated central pressure and storm surge as components of the categories. However, storm surge is affected by many factors other than wind intensity: the central pressure of the cyclone, the maximum wind speed, the speed of motion of the storm, the size of the cyclone and its wind field, the angle of approach to the coast, the width and slope of the continental shelf, and local features of the coastline (Irish and Resio, 2010). None of these are included in the SSWS. In 2010, after storms such as Katrina (2005) and Ike (2008) demonstrated conclusively that estimates of storm surge based simply on wind intensity are misleading, the NHC removed central pressure and storm surge from the scale.

Extreme storm surges such as those resulting from Katrina and Ike showed that the Saffir–Simpson categories, which were designed to be an index of the potential intensity of wind damage, are not a good public warning scale for storm surge (Irish and Resio, 2010; Kantha, 2013). Most people interpret the hurricane category as an indication of the severity of the storm. Using the SSWS category to make evacuation decisions may leave people at risk from storm surge. For example, Katrina was a category 3 hurricane at landfall, yet caused more damage and loss of life than either Hurricane Camille (1969) or Hurricane Andrew (1992), both category 5 storms. A similar situation occurred with Hurricane Ike (2008), a weak category 2 storm which caused extensive storm surge damage on the Texas coast. Hurricane Andrew had a landfall wind speed of 75 m/s but a radius of only 77 km and a storm surge of 2.4 m. In contrast, Katrina had a landfall speed of 56 m/s and a radius of 217 km, with a storm surge of 7.5–8.5 m. Ike had a landfall speed of 49 m/s and a radius of 195 km, with a storm surge of 4.8–5.9 m. The relatively large sizes of both Katrina and Ike demonstrated that the impact of a tropical cyclone is also a function of its size (Kantha, 2013). The larger the hurricane, the greater is the impact potential. However, because only the SSHS category is widely disseminated, the lay public, and local officials not privy to sophisticated models and other data at the federal level, are generally unaware of the true destructive power of the storm surge (Kantha, 2013).

The need for a change in the method of reporting storm surge hazards became even more obvious after hurricanes Isaac and Sandy in 2012, where despite the very accurate forecasts, many people did not respond to the warnings from forecasters (Baker et al., 2012). At landfall, Isaac was a category 2 hurricane and Sandy was a posttropical cyclone with hurricane-force winds. These storm descriptions were interpreted by the general public as low-risk categories, and the extent of the storm surge surprised many people. After the storm, forecasters talked to focus groups and found that the public interpreted the storm's danger in terms of its hurricane category, and storm surge warnings did not encourage residents of threatened areas to evacuate (Lazo and Morrow, 2013). Many residents were not fully aware of the nature of the warnings for their area, how strong a storm Sandy was, the risk that it posed, or how long impacts would persist. Residents who lived within one block of the water felt that the greatest threat that the storm posed was wind rather than water (Baker et al., 2012). Both National Weather Service (NWS) forecasters and emergency managers reported difficulty determining inundation guidance at specific locations. One of the biggest surprises in Sandy was the impact of the surge and how fast it moved in; the public and National Oceanic and Atmospheric Administration/NWS partners did not clearly understand what storm surge was or how dangerous it could be (NOAA, 2013). A poststorm assessment concluded that the highest priority need was for improved high-resolution storm surge forecasting and communication (NOAA, 2013).

Starting in the 2014 hurricane season, the NHC is issuing potential storm surge flooding maps as one of its products. These maps show land areas where, based on the latest NHC forecast, storm surge could occur and the elevation that the water could reach in those areas. The maps are created from multiple SLOSH runs and allow for uncertainties in the track, landfall location, intensity, and size of the cyclone. However, the maps do not account for wave action, pluvial flooding, or flooding inside levees and overtopping (NOAA, 2014). The potential storm surge flooding maps will be issued when a hurricane or tropical storm watch is first issued for any portion of the Gulf or east coast of the United States, which is approximately 48 h before the anticipated onset of tropical storm force winds. The maps will be updated every 6 h in association with each full advisory package, but because of the extensive computer processing times, they will not be available until 45–60 min after the advisory is released. The new maps will be color coded, much like the radar maps on the local news showing rain and severe weather. The new surge warnings will not include categories; instead, the colors will represent depths above the ground. The new maps will be used on an experimental basis for at least 2 years while the NHC collects feedback from emergency personnel and the public. A package of storm surge watch and warning advisories, similar to existing hurricane products, will be rolled out in 2015.

6.4 STORM SURGE DISASTER RISK REDUCTION

Disaster risk derives from a combination of physical hazards and the vulnerabilities of exposed elements. Hazard refers to the chance and characteristics of the hazardous phenomenon itself (e.g., flood depth, flood extent).

Exposure refers to the presence (location) of people, livelihoods, physical and biological systems, infrastructure, or economic, social, or cultural assets in places that could be affected by physical events and which, thereby, are subject to potential future harm, loss, or damage (IPCC, 2012). Vulnerability is defined as the characteristics and circumstances of a community, system, or asset that makes it susceptible to the damaging effects of a hazard, for example, due to unsafe housing and living conditions (UNISDR, 2009). Both exposure and vulnerability to storm surge are increasing. Jongman et al. (2012) calculated that the global population exposed to 1:100-year coastal flooding was about 271 million in 2010, with exposed assets of $13 trillion. By 2050, Jongman et al. (2014) calculated that 345 million people will be living in 1:100 coastal flood hazard locations, with exposed assets of $43 trillion, an increase of approximately 200% from 2010. Dasgupta et al. (2009) considered the potential impact of a 1:100 storm surge in 84 developing countries along with 577 of their cyclone-vulnerable coastal cities with populations over 100,000, and compared it to a more intense impact later in the century. Their study showed a significant asymmetry in storm surge risk, with three cities (Manila, Alexandria, and Lagos) out of the 577 accounting for 25% of future coastal population exposure and 10 cities accounting for 53% of the future exposure. The increased vulnerability of these cities can be attributed to urban growth combined with physical exposure to storm surge inundation.

The UNISDR 2013 Global Assessment Report on Disaster Risk Reduction reported that as of December 2012, 85 countries had established multisector national platforms for disaster risk reduction. With a few exceptions, these institutional and legal systems have remained focused on disaster preparedness and response rather than encouraging risk reduction. There is no quantitative index to measure the scale of disaster caused by storm surge in different regions. In addition, there is no easily understood scale to indicate the severity of storm surge, with nothing comparable to the earthquake magnitude and intensity scales or the SSWS. Few disaster risk management systems have been able to employ land use planning to encourage disaster risk reduction (UNISDR, 2013). As with any natural hazard, storm surge risk is a function of the physical hazard (storm surge depth and extent), exposure (the location and number of people and economic assets in locations that can be inundated by storm surge), and vulnerability (susceptibility to damage and loss). Although human vulnerability to storm surge is expected to increase in future years due to population growth, urbanization, increased coastal settlement, poverty, and changing weather patterns (Doocy et al., 2013), storm surge risk could be reduced through a range of adaptive strategies. Storm surge barriers can decrease the chance of flooding and thus reduce the physical hazard. Limiting development in the most hazardous locations can reduce exposure. Improved building codes, requiring flood-damaged buildings to be rebuilt, and new developments to be constructed in a resilient fashion can reduce vulnerability, as can early warning systems and the establishment of effective procedures to prepare for and recover from storm surge disasters. Insurance can help to cover the remaining risk to individuals, can assist national governments in financing recovery and reconstruction after a disaster, and can provide price signals to discourage development in hazardous locations. As international efforts move toward developing a post-2015 framework for disaster risk reduction, perhaps progress will be made toward a unified approach to storm surge warning similar to that which has been established for tropical cyclone and tsunami warnings.

ACKNOWLEDGMENTS

Thanks to the Climate Change and Sea Level Rise Initiative at Old Dominion University, who supported my research on flood insurance through a Visiting Scholarship, and where I started thinking seriously about the issues related to adaptation.

REFERENCES

Aerts, J.C.J.H., Botzen, W.J.W., 2012. Hurricane Irene: a wake-up call for New York City? Nat. Hazards Earth Syst. Sci. 12, 1837–1840.

Aerts, J.C.J.H., Boutzen, W.J.W., de Moel, H. (Eds.), 2013a. Cost Estimates for Flood Resilience and Protection Strategies in New York City. Annals of the New York Academy of Sciences, 1294, pp. 1–104.

Aerts, J.C.J.C., Bowman, M., Boutzen, W.J.W., de Moel, H., 2013b. Storm surge barriers for New York city and New Jersey. In: Aerts, J., Boutzen, W.J.W., de Moel, H. (Eds.), Cost Estimates for Flood Resilience and Protection Strategies in New York City, Annals of the New York Academy of Sciences, 1294, pp. 49–68.

Ahamed, M., 2013. Community based approach for reducing vulnerability to natural hazards (cyclones, storm surges) in coastal belt of Bangladesh. Procedia Environ. Sci. 17, 361–371.

Atkin, A., April 2014. The calm after the storms. Environment Agency Blog 14. https://environmentagency.blog.gov.uk/2014/04/14/the-calm-after-the-storms/.

Augustijn, D.C.M., van den Berg, M., de Bruine, E., Korving, H., 2011. Dynamic control of salt intrusion in the Mark-Vliet river system, The Netherlands. Water Resour. Manage. 25, 1005–1020.

Baker, E.J., Broad, K., Czajkowski, J., Meyer, R., Orlov, B., 2012. Risk Perceptions and Preparedness Among Mid-Atlantic Coastal Residents in Advance of Hurricane Sandy. Risk Management and Decision Processes Center, The Wharton School, University of Pennsylvania. Working paper 2012-18.

Bajo, M., Zampato, L., Umgeisser, G., Cucco, A., Canestrelli, P., 2007. A finite element operational model for storm surge prediction in Venice. Est. Coast. Shelf Sci. 75 (1), 236–249.

BBC, March 5, 2014. Thames Barrier Breaks Closure Record. Available from http://www.bbc.co.uk/news/uk-england-london-26453484.

Bijker, W.E., 2002. The Oosterschelde storm surge barrier: a test case for Dutch water technology, management and politics. Technol. Cult. 43 (3), 569–584.

Brakenridge, G.R., Syvitski, J.P.M., Overeem, I., Higgins, S.A., Kettner, A.J., Stewart-Moore, J.A., Westerhof, R., 2013. Global mapping of storm surges and the assessment of coastal vulnerability. Nat. Hazards 66, 1295–1312.

Bras, R.L., Harleman, D.R., 2002. Obsolete? No. Necessary? Yes. The gates will save Venice. Eos 83 (10), 224–226.

Bowman, M.J., Flood, R.D., Hill, D., Wilson, R.E., 2005. Hydrologic Feasibility of Storm Surge Barriers to Protect the Metropolitan New York – New Jersey Region. Marine Sciences Research Center Technical Report. Stony Brook University.

Chan, E.Y.Y., Liu, S., Hung, K.C., 2013. Typhoon Haiyan and beyond. Lancet 382 (9908), 1873. http://dx.doi.org/10.1016/S0140-6736(13)62415-0.

Coch, N.K., 2013. Hurricane flood protection in NYC – now is not the time for multiple surge barriers. Discussion of Hill, D. 2012. The lessons of Katrina: learned and unlearned. Journal of Coastal Research 28(2): 324-331 J. Coastal Res. 28 (5), 1313–1315.

Colle, B.A., Buonaiuto, F., Bowman, M.J., Wilson, R.E., Flood, R., Hunter, R., Mintz, A., Hill, D., June 2008. New York City's vulnerability to coastal flooding: storm surge modeling of past cyclones. Bull. Am. Meteorol. Soc., 829–841.

Cooper, J.A.G., Pile, J., 2013. The adaptation – resistance spectrum: a classification of contemporary adaptation responses to climate-related coastal change. Ocean and Coastal Manage. Available from http://dx.doi10.1016/j.ocecoaman.2013.09.006.

Dasgupta, S., Laplante, B., Murray, S., Wheeler, D., 2009. Climate Change and Future Impacts of Storm-surge Disasters in Developing Countries. CGD Working Paper 182. Center for Global Development, Washington, DC. Available from http://www.cgdev.org.

Delta Committee Report., 2008. Available from http://www.deltacommissie.com/en/advies.

Delta Programme. Netherlands Ministry of Infrastructure and Environment and Ministry of Economic Affairs, 2014. Working on the Delta: Promising Solutions for Tasking and Ambitions. Available from http://www.deltacommissaris.nl.

Dircke, P.T.M., Jongeliong, T.H.G., Jansen, P.L.M., 2013. In: Hill, D., Bowman, M.J., Khinda, J.S. (Eds.), Storm Surge Barriers to Protect New York City: Against the Deluge; 2009 March 30–31. American Society of Civil Engineers, Reston, Virginia, pp. 65–87.

Doocy, S., Dick, S., Daniels, A., Kirsch, T.D., April 16, 2013. The human impact of tropical cyclones: a historical review of events 1980–2009 and systematic literature review. PLOS Curr. Disasters. http://dx.doi.org/10.1371/currents.dis.f4deb457904936b07c09daa98ee8171a. Edition 1.

eSurge, 2014. Tropical Cyclone Haiyan. Available from http://www.storm-surge.info/storm-surge-tropical-cyclone-haiyan.

Environment Agency (UK), 2010. Thames Estuary Plan: Managing Flood Risk through London and Thames Estuary. Crown Copyright, London. Available from http://cdn.environment-agency.gov.uk/LIT7540_43858f.pdf.

Environment Agency, 2013. Thames Estuary Phase 1 Programme – Industry Event. http://a0768b4a8a31e106d8b0-50dc802554eb38a24458b98ff72d550b.r19.cf3.rackcdn.com/TEP1_Industry_Day_Presentation_LIT_CODE_8772_(2)_9bae3c.pdf.

Eqecat, 2009. Technical Report: The Hurricane and Storm Damage Risk Reduction System in New Orleans, Louisiana. Eqecat.

Ferris, E., November 26, 2013. For Typhoon Haiyan Survivors, the Recovery Is Just Beginning. NOVA Next. Available from http://www.pbs.org/wgbh/nova/next/earth/typhoon-haiyan-recovery/.

Fontini, F., Umgeisser, G., Vergano, L., 2010. The role of ambiguity in the evaluation of the net benefits of the MOSE system in the Venice lagoon. Ecol. Econ. 69, 1964–1972.

Fritz, H.M., Blount, C., Thwin, S., Thu, M.K., Chan, N., 2011. Observations and modelling of cyclone Nargis storm surge in Myanmar. In: Wallendorf, L.A., Jones, C., Ewing, L., Battalio, B. (Eds.), Proceedings of Solutions to Coastal Disasters: 2011 June 26–29; Anchorage, Alaska. American Society of Civil Engineers, Reston, Virginia, pp. 1–9.

Grasso, V.F., Singh, A., 2009. Early Warning Systems: State-of-art Analysis and Future Directions. Draft report. United Nations Environment Program.

Hallegatte, S., 2009. Strategies to adapt to an uncertain climate. Global Environ. Change 19, 240–247.

Harriman, L., 2014. Cyclone Phailin in India: early warning and timely action saved lives. Article reproduced from UNEP Global Environmental Alert Service (GEAS). Environ. Dev. 9, 93–100.

Haque, U., Hashizume, M., Kolivras, K.N., Overgaard, H.J., Das, B., Yamamoto, T., 2012. Reduced death rates from cyclones in Bangladesh: what more needs to be done? Bull. W.H.O. 90, 150–156.

Hill, D., 2012. The lessons of Katrina: learned and unlearned. J. Coastal Res. 28 (2), 324–331.

Hill, D., 2013. New York City plans to be flooded. J. Coastal Res. 29 (6A) (iii-iv).

Horner, R.W., 1979. The Thames barrier project. Geogr. J. 145 (2), 242–253.

Hostens, K., Hummel, H., Mees J., Vincx, M., Heip, C.H., 2003. The fish fauna of the Oosterschelde, a decade after completion of the engineering works. In: Hostens, K. The Demersal Fish and Macro-invertebrate Assemblages of the Westerschelde and Oosterschelde Estuaries (Southern Bight of the North Sea) (Ph.D. thesis). 59–77.

Hudson, P.F., Middelkoop, H., Stouthamer, E., 2008. Flood management along the lower Mississippi and Rhine rivers (The Netherlands) and the continuum of geomorphic adjustment. Geomorphology 101, 209–236.

Hunter, P., 2012. The St Petersburg Flood Protection Barrier: Design and Construction. CETMEF PIANC Paris. HR Wallingford, 1–9. http://eprints.hrwallingford.co.uk/603/.

IPCC, 2012. Managing the Risks of Extreme Events and Disasters to Advance Climate Change Adaptation. A special report of Working Groups I and II of the Intergovernmental Panel on Climate Change. Cambridge University Press, Cambridge.

Irish, J.L., Resio, D.T., 2010. A hydrodynamics-based surge scale for hurricanes. Ocean Eng. 37, 69–81.

Ivanov, S.V., Kosukhin, S.S., Kaluzhnaya, A.V., Boukhanovsky, A.V., 2012. Simulation-based collaborative decision support for surge floods prevention in St. Petersburg. J. Comput. Sci. 3, 450–455.

Jansen, O.E., Aarts, G.M., Reijnders, P.J.H., 2013. Harbour porpoises *Phocoena phocoena* in the Eastern Scheldt: a resident stock or trapped by a storm surge barrier? PLoS One 8 (3), e56932. http://dx.doi.org/10.1371/journal.pone.0056932.

JCOMM (Joint WMO/IOC Committee on Oceanography and Marine Meteorology), 2014. Guide to Storm Surge Forecasting Geneva. WMO, Switzerland. Available from http://www.jcomm.info/index.php?option=com_content&view=article&id=175&Itemid=37#04.

Jongman, B., Ward, P.J., Aerts, J.C.H., 2012. Global exposure to river and coastal flooding: long term trends and changes. Global Environ. Change 22, 823–835.

Jonkman, S.N., Bockarjova, M., Kok, M., Bernardini, P., 2008. Integrated hydrodynamic and economic modelling of flood damage in the Netherlands. Ecol. Econ. 66, 77–90.

Jonkman, S.N., Hillen, M.M., Nicholls, R.J., Kanning, W., van Ledden, M.M.van Ledden, 2013. Costs of adapting coastal defences to sea-level rise - new estimates and their implications. J. Coast.l Res. 29 (5), 1212–1226.

Jongman, B., Hochrainer-Stigler, S., Feyen, L., Aerts, C.J.H., Mechler, R., Botzen, W.J.W., Bouwer, L.M., Pflug, G., Rojas, R., Ward, P.J., 2014. Increasing stress on disaster-risk finance due to large floods. Nat. Clim. Change 4, 264–268.

Kabat, P., Fresco, L.O., Stive, M.J.F., Veerman, C.P., van Alphaen, J.S.L.J., Parmet, B.W.A.H., Hazeleger, W., Katsman, C.A., 2009. Dutch coasts in transition. Nat. Geosci. 2, 450–452.

Kantha, L., 2013. Classification of hurricanes: lessons from Katrina, Ike, Irene, Isaac and Sandy. Ocean Eng. 70, 124–128.

Kron, W., 2013. Coasts: the high-risk areas of the world. Nat. Hazards 66, 1363–1382.

Lazo, J.K., Morrow, B.H., 2013. Survey of Coastal U.S. Public's Perspective on Extra Tropical – Tropical Cyclone Storm Surge Information. National Center for Atmospheric Research Societal Impacts Program Technical Note.

London Assembly, February 26, 2014. London's Flood Risk. Response to Questions to the Mayor. http://questions.london.gov.uk/QuestionSearch/searchclient/.

Louters, T., van den Berg, J., Mulder, J., 1998. Geomorphological changes of the Oosterschelde tidal system during and after the implementation of the Delta Project. J. Coastal Res. 14 (3), 1134–1151.

Lowry, M.R., 2012. Storm Surge Forecasting and Products. Presentation of the US National Hurricane Center. Available from http://www.nhc.noaa.gov/outreach/presentations/2012_08nhcL311_stormSurgeForecastingProducts.pdf.

Miller, D., Desoto-Duncan, A., Hertzler, B., 2013. Hurricane Katrina and the Inner Harbor Navigation Canal Lake Borgne surge barrier. La Houille Blanche 2, 5–11.

Munaretto, S., Vellinga, P., Tobi, H., 2012. Flood protection in Venice under conditions of sea-level rise: an analysis of institutional and technical measures. Coastal Manage. 40, 355–380.

NASA, 2014. Haiyan (Northwestern Pacific Ocean), November 20, 2013. Available from http://www.nasa.gov/content/goddard/haiyan-northwestern-pacific-ocean/#.UyAT3xaoXFI.

National Oceanic and Atmospheric Administration (NOAA). Technical Memorandum NWS TPC 5, 2007. http://www.nhc.noaa.gov/pdf/sshws_statement.pdf.

NOAA, 2013. Hurricane/Post-tropical Cyclone Sandy, October 22-29, 2012. Service assessment. NOAA, Washington DC.

NOAA, 2014. Storm Surge and Coastal Inundation: Forecasts and Warnings. Available from http://www.stormsurge.noaa.gov/forecast_warning.html.

Neussner, O., 2014. Assessment of Early Warning Efforts in Leyte for Typhoon Haiyan/Yolanda. Deutsche Gesellschaft für Ineternationale Zusamenerarbeit (GIZ), Manila.

Padron, D.V., Forsyth, G., 2013. NY/NJ Outer harbor Gateway. In: Hill, D., Bowman, M.J., Khinda, J.S. (Eds.), Storm Surge Barriers to Protect New York City: Against the Deluge. American Society of Civil Engineers, Reston, Virginia, pp. 218–249.

Paul, B.K., 2012. Factors affecting evacuation behavior: the case of 2007 cyclone Sidr, Bangladesh. Prof. Geogr. 64 (3), 401–414.

Penning-Rowsell, E.C., Haigh, N., Lavery, S., McFadden, L., 2013a. A threatened world city: the benefits of protecting London from the sea. Nat. Hazards 66, 1383–1404.

Penning-Rowsell, E.C., Sultana, P., Thompson, P.M., 2013b. The 'last resort?' Population movement in response to climate-related hazards in Bangladesh. Environ. Sci. Policy, S44–S59.

Pilarczyk, K.W., 2012. Impact of the Delta Works on the Recent Developments in Coastal Engineering. World Scientific Publishing Company, London.

Pirazzoli, P.A., 2002. Did the Italian government approve an obsolete project to save Venice? Eos 83 (10), 223–224.

Pirazzoli, P.A., Umgeisser, G., 2006. The projected "MOSE" Barriers against flooding and the expected global sea-level rise. J. Mar. Environ. Eng. 8 (3), 247–261.

Reid, R.L., 2013. Special report: defending New Orleans. Civ. Eng. – ASCE 83 (11), 48–67.

Rinaldo, A., Nicótina, L., Alessi Celegon, E., Beraldin, F., Botter, G., Carniello, L., Cecconi, G., Delfina, A., Settin, T., Uccelli, A.,D., Alpaos, L., Marani, M., 2008. Sea level rise, hydrologic runoff, and the flooding of Venice. Water Resour. Res. 44, W12434. http://dx.doi.org/10.1029/2008WR007195.

Roelvink, J.A., Jeuken, M.C.J.L., van Holland, G., Aarninkhof, S.G.J., Stam, J.M.T., 2001. Long-term process-based modelling of complex areas. In: Hansen, H., Larson, M. (Eds.), Proceedings of Coastal Dynamics '01, 2001 June 11–15. Lund, Sweden. American Society of Civil Engineers, New York, pp. 383–392.

Santos, L.A., 2013. Storm Surge: Lost in Translation and Interpretation. Available from https://www.devex.com/news/storm-surge-lost-in-translation-and-interpretation-82311.

Tangelder, M., Troost, K., van den Ende, D., Ysebaert, T.J.W., 2012. Biodiversity in a Changing Oosterschelde: From Past to Present. Wettelijke Onderzoekstaken Natuur & Milieu, Wageningen. WOT-werkdocument 288.

Tollefson, J., 2012. Hurricane sweeps US into climate-adaptation debate. Nature 491 (7423), 167–168.

Tollefson, J., 2013. New York vs. the sea. Nature 494, 162–164.

Tönis, I.E., Stam, J.M.T., van de Graff, J., 2002. Morphological changes of the Haringvliet estuary after closure in 1970. Coastal Eng. 44, 191–203.

Ubac, M.L., February 1, 2014. New Warning Signals Eyed for Storm Surges. Philippine Daily Inquirer. Available from http://newsinfo.inquirer.net/572322/new-warning-signals-eyed-for-storm-surges.

Umgiesser, G., Matticchio, B., 2006. Simulating the mobile barrier (MOSE) operation in the Venice Lagoon, Italy: global sea level rise and its implication for navigation. Ocean Dyn. 56, 320–332.

United Nations international Strategy for Disaster Reduction, 2009. UNISDR Terminology on Disaster Risk Reduction. United Nations International Strategy for Disaster Reduction, United Nations Office for Disaster Risk Reduction, Geneva, Switzerland. Available from http://www.unisdr.org/we/inform/terminology.

United Nations international Strategy for Disaster Reduction, 2013. Global Assessment Report on Disaster Risk Reduction. United Nations International Strategy for Disaster Reduction, United Nations Office for Disaster Risk Reduction, Geneva, Switzerland. Available from http://www.unisdr.org/we/inform/publications/33013.

US Army Corp of Engineers, 2010. IHNC-Lake Borgne Surge Barrier. USACE, New Orleans.

US Army Corps of Engineers, 2013. Hurricane Isaac with and without 2012 100-year HSDRSS Evaluation. Final report. USACE, New Orleans.

van Wesenbeeck, B.K., Mulder, J.P.M., Marchand, M., Reed, D.J., de Vries, M.B., de Vriend, H.J., Herman, P.J., 2014. Damming deltas: a practice of the past? Towards nature-based defenses. Est. Coast. Shelf Sci 140, 1–6.

Verduijn, S.H., Meijerink, S.V., Leroy, P., 2012. How the second delta committee set the agenda for climate adaptation policy: a Dutch case study on framing strategies for policy change. Water Altern. 5 (2), 469–484.

Wagner, M., Chhetri, N., Sturm, M., 2014. Adaptive capacity in light of Hurricane Sandy: the need for policy engagement. Appl. Geogr. 50, 15–23.

Waverley, J.A., 1954. Report of the Departmental Committee on Coastal Flooding. HMSO, London.

Weindl, H., 2012. Tropical cyclones. In: Severe Weather in North America: Perils, Risks, Insurance. Munich Re, Munich, pp. 35–57.

Wilby, R.L., Keenan, R., 2012. Adapting to flood risk under climate change. Prog. Phys. Geogr. 36, 348–378.

Williams, J., August 21, 2013. National Hurricane Center Gives Storm Surge Modeling a Boost. Washington Post. Available from http://www.washingtonpost.com/blogs/capital-weather-gang/wp/2013/08/21/hurricane-center-gives-storm-surge-modeling-a-boost/.

Yates, R., 2013. When Word Save Lives: 'storm Tsunami' Vs. 'storm Surge'. Available from http://plan-international.org/about-plan/resources/news/when-words-save-lives-storm-tsunami-v-storm-surge/.

Zevenbergen, C., van Herk, S., Rijke, J., Kabat, P., Bloemen, P., Ashley, R., Speers, A., Gersonius, B., Verbeek, W., 2013. Taming flood disasters: lessons learned from Dutch experience. Nat. Hazards 65, 1217–1225.

Zhong, H., van Overloop, P.-J., van Gelder, P., Rijcken, T., 2012. Influence on a storm surge barrier's operation on the flood frequency in the Rhine Delta Area. Water 4, 474–493.

FURTHER READING

Luettich, R., 2014. Generation, characteristics and modeling storm surges. In: Sherman, D.J., Ellis, J.T. (Eds.), Coastal and Marine Hazards and Disasters. Elsevier.

Chapter 7

Storm Surge Case Studies

Hans von Storch[1], Wensheng Jiang[2] and Kazimierz K. Furmanczyk[3]
[1] *Institute of Coastal Research, Helmholtz Zentrum Geesthacht, Geesthacht, Germany,*
[2] *Ocean University of China, Qingdao, China,* [3] *University of Szczecin, Szczecin, Poland*

ABSTRACT
This chapter presents details on a number storms surge cases: along the Southern Baltic Sea coast, the estuary of the Elbe in Germany, and the Yellow China Sea coast at Qingdao. These case studies feature storm surge characteristics, specifically, losses of life and property, erosion extent, and relationship to extratropical and tropical storm intensity. These cases demonstrate the severity of the issue and the need of precautionary measures, not only for limiting the possible damages but also for being able to manage for a possible failure of the coastal defense measures.

7.1 INTRODUCTION

Storm surges are the major geophysical risk in coastal regions (von Storch and Woth, 2008; Gönnert et al., 2001); they are often associated with significant losses of life and property (Figure 7.1). Along the Bangladesh coast, tropical storms and their surges in 1876, 1891, 1970, and 1991 went along with a toll of 100,000 and more lives, and it was only in 2008 that the tropical storm Nargis killed more than 100,000 people in Myanmar (Fritz et al., 2009). In midlatitudes, the number of losses is usually several orders of magnitude smaller, namely, up to a few hundred, which is, of course, bad enough.

All coastal regions of the world where strong storms occasionally or regularly pass are affected by storm surges, which comprise most of the world's coasts (Figure 7.2). There are two major types of storms, tropical and extratropical storms. In principle there are more, such as polar lows, cold surges, and medicanes (Mediterranean hurricanes), which regionally play a role with storm surges, but this chapter is limited to the two main types. The different characteristics of these storms and the associated surges are listed in Table 7.1.

The hazard of storm surges is related to high water levels, which may flood low-lying areas with strong near-shore currents and waves. The former threatens life and property; in historical times, large stretches of land were

Coastal and Marine Hazards, Risks, and Disasters. http://dx.doi.org/10.1016/B978-0-12-396483-0.00007-8
Copyright © 2015 Elsevier Inc. All rights reserved.

FIGURE 7.1 Historical engraving of a storm surge with dike failure in The Netherlands in 1673. (von Storch and Woth, 2008).

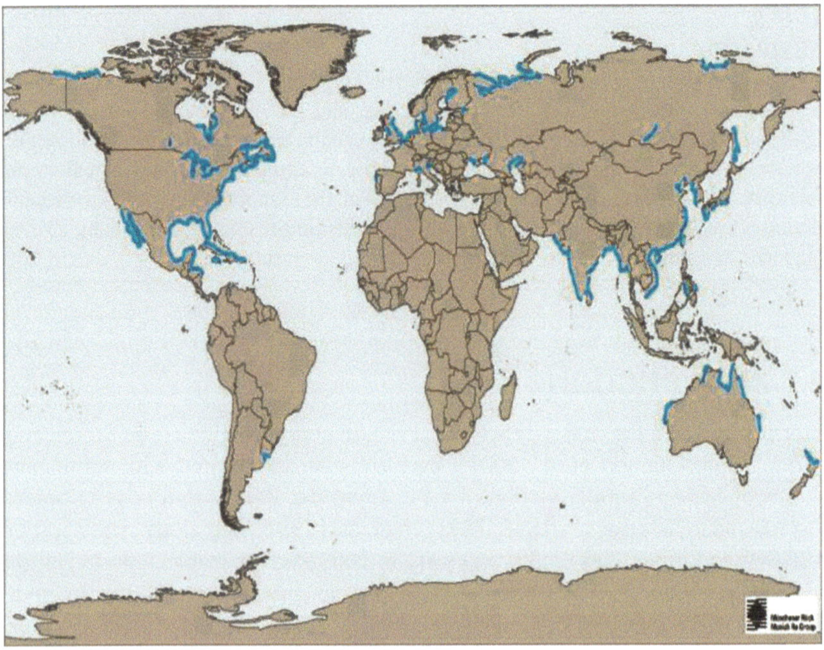

FIGURE 7.2 Coasts endangered by storm surges. *With permission by Munich Re.*

never recovered (such as the Dollart at the German/Dutch border; e.g., de Kraker, 2009). The latter is associated with enhanced erosion and loss of dunes, beaches, and cliffs.

A storm surge is a phenomenon that has always affected coastal inhabitants (e.g., Lu, 1984; Petzelberger, 2000). When assessing the intensity of historical storm surges, the simplest method to determine the maximum water level is to measure markings on buildings (Figure 7.3(a) shows a marking on a house in Schleswig at the German Baltic Sea coast; Figure 7.3(b) shows a modern tide

TABLE 7.1 Characteristics of Storm Surges Caused by Tropical Storms (Hurricanes, Typhoons) and Extra-tropical Storms (von Storch and Woth, 2008; Gönnert et al., 2001).

Parameter	Tropical Cyclone	Extratropical Cyclone
Spatial scale of storm	500 ± 200 km	1,000 ± 500 km
Representation in weather reanalyses of past decades (since 1960)	In earlier decades underrepresented; sometimes cyclones are missed	Mostly well described, in particular in well-monitored Northern Hemisphere regions; some inhomogeneities remain
Amplitude of surges	Larger—Hurricane Camille caused a surge of 7.5 m in Gulfport, MS, USA, in August 1969	Smaller—surges of 5 m and more are infrequent events
Duration of surge	Several hours, up to half a day	2–5 days
Length of coastline affected by the surge	Less, usually <200 km	Several hundred kilometers
Geometry of the storm	Compact and nearly symmetrical	Ill defined and sprawling geometry

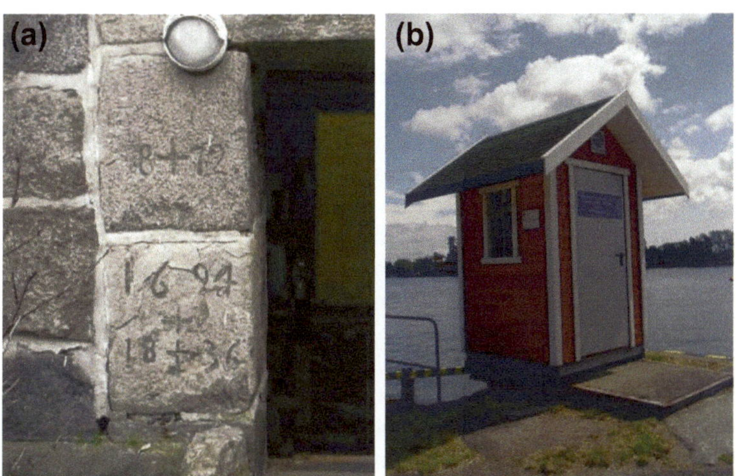

FIGURE 7.3 Recording storm surge heights—historical marks of high water (a) in 1872, 1694, and 1836 in Schleswig, Germany, and a modernized tide gauge in Swinoujście, Poland (b), which has with only short interruptions been operating since 1811.

gauge station in Swinoujście, Poland). The measures for protecting people and property against these extreme and dangerous events have a long history, as explained by Niemeyer et al. (1996).

This chapter introduces case studies from three regions on the globe: the Yellow China Sea coast at Qingdao, the estuary of the Elbe and Hamburg, and the southern Baltic Sea coast. These examples are not representative of all possible issues and problems associated with storm surges; however, the cases we selected highlight some of the major issues and problems related to the global hazard of storm surges.

7.2 THE CASE OF QINGDAO, CHINA

China faces the West Pacific Ocean and has more than 18,000 km of coastline. Storm surges are the major marine disaster in China and are mainly caused by typhoons. According to Hou et al. (2011), the Shanghai to Quanzhou and the Zhujiang Estuary to Northeast Hainan coastal areas are the two regions of China most affected by storm surge disasters. A total of 14 tide gauge stations are operating along the Chinese coast. From 1949 to 2009, storm surge heights exceeded 2.00 m 43 times, with most of them impacting the two aforementioned regions. Hou et al. (2011) also found that the severity of these disasters, which occur predominantly in September, has increased in the past two decades.

China has a 4,000-year history with written documentation of storm surges. Lu (1984) derived records of storm surges from Chinese historical literature and determined that the first storm surge ever recorded in China, and likely in the world, is in a Chinese history book titled the "Book of Han" that was completed in the first century. A storm surge in the Bohai Sea was described in this book: "It rained for a long time. The northeast wind blew and the sea water was overflowed to southwest. The water intruded into the land for more than 100 km and the land of Nine-River area was inundated" (Lu, 1984). Unfortunately, the time of the surge was not mentioned.

This chapter focuses on the storm surge record from Qingdao. Qingdao is an important harbor city in China that faces the Yellow Sea and is located on the west coast of the Jiaozhou Bay (Figure 7.3). The city of Qingdao was a small village, established in 1891. In the past approximately 125 years, it has grown to a city with more than 8 million people. The major cause for storm surges in this area is typhoons, even though extratropical storms can cause surges and damages.

Zhang et al. (2006) examined the tide gauge data in Qingdao from 1949 until 2003, and found 14 cases of high water levels related to typhoons. Nine of these events were associated with sea levels exceeding +5.10 m.[1] The highest absolute water level was +5.48 m in 1997—including tide and storm effect. The highest surge height, related only to the wind effect above the tidal component, was 1.47 m in 1952.

1. Here, and elsewhere in this article, sea level heights are given relative to a local reference.

Peak surge heights in the Qingdao depend on which one of the four typical typhoon paths is followed. The first type is typhoon makes landfall on the Jiangsu coast before continuing north to Qingdao. The second one represents typhoons that make landfall in Fujian Province and then move north until turning east to Qingdao. The third moves north from the East China Sea to Qingdao. And the fourth is for typhoons that move northwest to Qingdao.

The following describes two severe storm surge disasters in Qingdao. The first storm was a (still nameless) typhoon that formed on 22 August 1939 in the Pacific Ocean. On the evening of 30 August the typhoon approached Qingdao and was accompanied by severe precipitation. At 6 am on 31 August, the typhoon center was about 120 km south of Qingdao and the wind speed in Qingdao likely exceeded 150 km/h. At 9 am the typhoon made landfall on the west coast of Jiaozhou Bay, which is the location of the city of Qingdao. In Qingdao city, 17 people were killed by the storm surge. More than 1,000 houses were destroyed and an additional 3,000 houses were damaged. About 460 ha of farmland near the coast was inundated. The loss of grain harvest was estimated at approximately 11,000 tons. The total economic loss was equivalent to US$ 4 million (c.1939). The disaster is still remembered by the local people through oral transmission from the community elders (Cai et al., 2010).

More recently, in August 1985, a storm surge disaster happened when Typhoon Mamie hit Qingdao (Figure 7.4). In this disaster 29 people were killed and 368 people were wounded. More than 8,000 m of sea dikes and other coastal defense measures were destroyed. The economic loss was equivalent to around US$ 200 million.

Along this coast, extratropical meteorological storms can also cause storm surges. In May 2013 a storm surge hit Qingdao. The famous Zhanqiao Pier was damaged for the fourth time in 100 years (Figure 7.5).

Qingdao is an important city to the economy of China, therefore the ability to adapt to the risk of storm surges is crucial. The well-established storm surge forecasting service in China is a prerequisite for managing the storm surge hazard. While an intensification of storm surges due to the general rise in sea level is almost certain, possible changes of storm intensities due to anthropogenic climate change are still under investigation.

7.3 THE CASE OF HAMBURG AND THE ELBE ESTUARY[2]

Hamburg, Germany, is a harbor in the estuary approximately 140 km upstream along the river Elbe. The estuary opens to the northwest into the German Bight, which is prone to storm surges (Figure 7.6). The history of storm surges in Hamburg since 1750 is characterized by three phases: (1) a frequent damage period (prior to 1850), (2) a calm period (1855–1962), and (3) a period of

2. See von Storch and Woth (2008), von Storch et al. (2008).

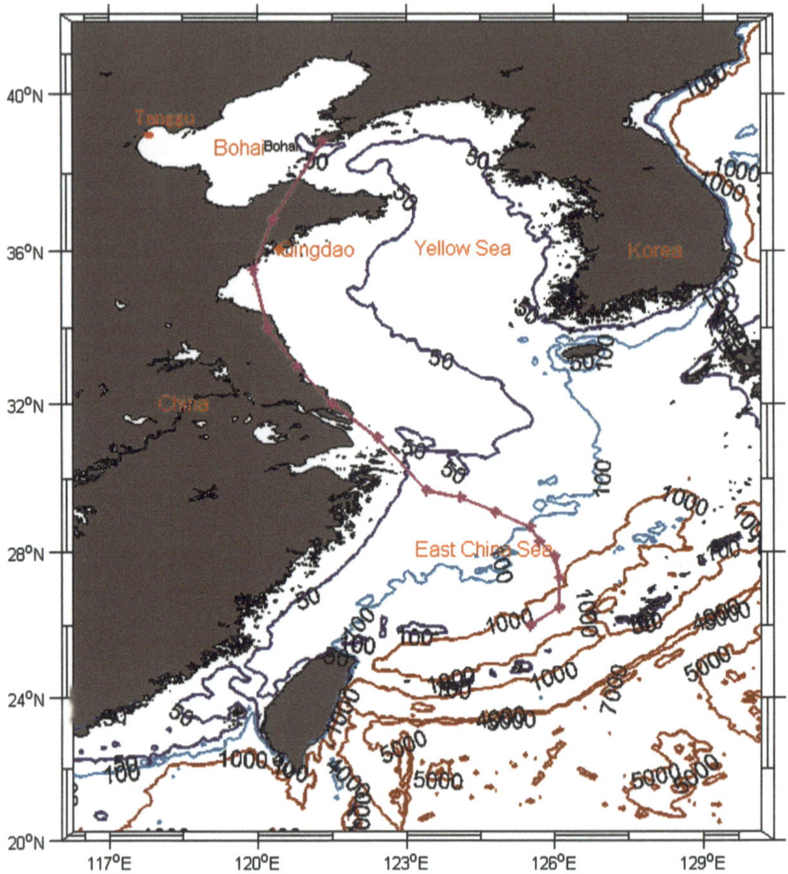

FIGURE 7.4 Bathymetry (in meters) of the East China Sea, with Qingdao at the southern coast of the Shandong peninsula. The pink line denotes the path of the Typhoon Mamie in August 1985.

elevated but well-managed storm surge levels (1962 to present) (Figure 7.7). In 1962, a storm surge disaster killed more than 300 people in Hamburg. This event served as warning to the city and its residents, and coastal defense was massively improved in the following years.

In the eighteenth century, storm surges and breaking dikes were relative frequent in Hamburg. The dike failures when water levels reached approximately +5.20 m. Interestingly, these storm surges occurred in clusters. After the severe storm surge in 1825, dike heights were raised to +5.70 m. After beginning to raise the dikes, and until 1962, only one severe storm surge happened (in 1855). After this storm and for more than 100 years until 1962, the improved dike levels were not really challenged; all gauge readings were well below +5.00 m.

2013年5月 28

青島晚報

买家具 购建材 首选富尔玛

82860085

岛城普降大到暴雨局部特大暴雨，为同期52年来第一次

巨浪逞凶 栈桥险成"断桥"

桥体约50米坍塌，已封闭待修，历史上曾3次被冲断 [3-8版]

纪检监察人员限期清退会员卡

中纪委要求限6月20日前"退卡"，做到"零持有、零报告" [2版]

女骑警花费受质疑

长春盗车杀婴凶手判死刑

苏沪杭新品丝绸服装服饰

FIGURE 7.5 The local newspaper reporting the damage of the Zhanqiao Pier in Qingdao in 2013.

The 1962 event proved to be a major event in the perceptions of storms surges in Hamburg. After the catastrophe of World War II (WWII), which left large parts of Hamburg in ruins, people were focusing on economic progress and had little sense for the risk of natural disasters. The more than 300 drowning victims were from poor quarters of the city, many of whom resettled after having fled from the east at the end of WWII and were unaware of the storm surge risk (Figure 7.8). It became obvious that the coastal defense was insufficient; the badly maintained dikes broke in several locations (Figure 7.9), and the civil defense for the case of a dike failure turned out to be inefficient. After the 1962 catastrophe massive investments into the coastal defense were made; dikes were raised to +7.20 m.

The next substantial flood following the dike improvements occurred in 1976 and it exceed the 1962 surge level and reached +6.45 m. The newly enforced coastal defense held and damages were insignificant in Hamburg. Nevertheless, dikes were raised again to a level between +8.00 m

FIGURE 7.6 Satellite image of the Elbe estuary, with Cuxhaven and Hamburg-St Pauli. *With permission of Schönfeld, GKSS*. The distance between Hamburg-St Pauli and Cuxhaven is about 140 km. (von Storch et al., 2008).

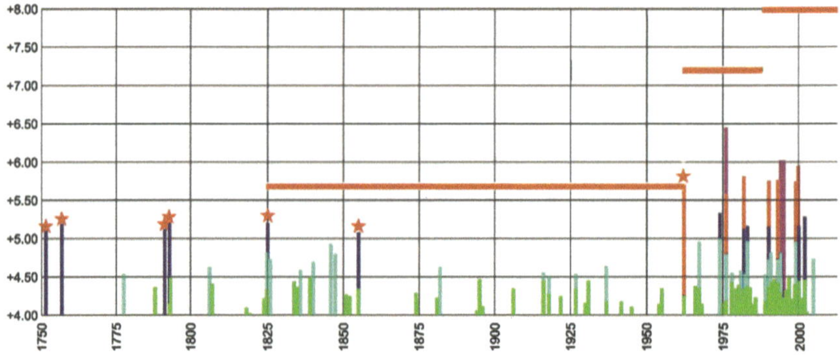

FIGURE 7.7 Storm surge heights (vertical bars) and dike heights (red horizontal lines) as recorded at the tide gauge St Pauli in Hamburg from 1750 to 2004. The color coding represents different surge heights. The red stars indicate dike failures. (von Storch et al., 2008).

and +9.30 m. Since 1962 several very high storm surges took place with heights between +5.50 m and +6.00 m, but only minor damage was reported.

It has been speculated that the increase of storm surge occurrence in Hamburg since 1962 is related to anthropogenic climate change. This

Hamburger Abendblatt

Berichte und Bilder von der Flutkatastrophe auf den Seiten 3, 5, 9, 10, 11, 12, 13, 18

Hamburger, bitte helft! Nehmt Obdachlose auf!

Bisher 119 Tote geborgen

Wasser und Nahrung für die Eingeschlossenen

Die Flut geht zurück

Die Türen auf!

Straßen im Freihafen heute morgen blockiert

Landestrauer für drei Tage

FIGURE 7.8 Reporting in the local press about the 1962 storm surge disaster. The main headline reads: "People of Hamburg, please help!".

speculation is very likely false; instead the main part of the increase is likely due to the improvement of coastal defense and the dredging of the shipping channel. The intensification of the North Atlantic Oscillation during the period between 1960 and 1995 may have contributed to a minor increase (Weisse and Plüß, 2005). A measure of the effect of the former two causes accounts for the difference of storm surge heights in Cuxhaven, at the mouth of the Elbe estuary, and in Hamburg (for locations, see Figure 7.6). Prior to 1962, storm surges in Hamburg were on an average about 30 cm higher than in Cuxhaven. After 1962 this difference rose to about 1 m (Grossmann et al., 2007). Experts estimate that about three-quarters of this increase is related to coastal defense measures and one-quarter to the deepening of the shipping channel from less

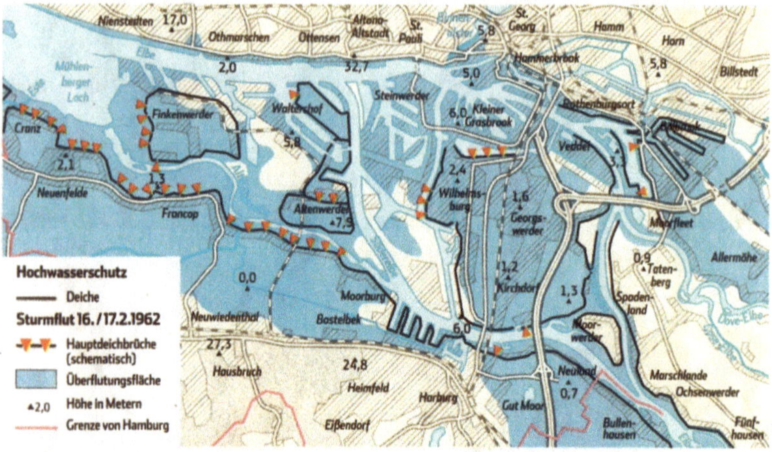

FIGURE 7.9 Flooded areas (blue) during the storm surge of February 16, 1962, in Hamburg. The black line describes the coastal defense line; the red triangles represent dike failures.

than 11 to 14.50 m. Thus, modifications of the river Elbe have significantly increased the storm surge height in Hamburg, whereas climatic effects are being rather minor (cf. Weisse and Plüß, 2005; WASA, 1998; Alexandersson et al., 1998).

7.4 THE CASE OF THE SOUTHERN BALTIC SEA COAST

The Baltic Sea is a semiclosed basin with limited connection to the North Sea via the Danish Straits. The tidal range in the Baltic Sea does not exceed a couple of centimeters and is practically negligible. Every couple of years or so the "filled-basin" phenomena occurs when a very strong wind blowing from (west – northwest) moves water from the western to the eastern parts of the Southern Baltic Sea, pumping water from the North Sea to the Baltic Sea via the Danish Straits at the same time. This causes more water to accumulate in the South Baltic and the level subsequently rises. When the strong wind changes direction to N–NE (north – northeast) in a filled-basin situation, even more water accumulates at the western part of the South Baltic, and the water level increases significantly, reaching levels of +1.5–2.0 m (Furmańczyk, 2013). Such a situation leads to catastrophic effects on the coast. Low dunes are eroded and water overflows, thus causing coastal flooding. Cities located at the river mouths are usually affected by this event.

A filled-basin scenario has occurred many times in the history of the southern and western parts of the Baltic Sea. In historical reports we find that in the past many strong storms at the Southern Baltic took place, which were associated with significant coastal and complete villages being destroyed. For example, a storm at the beginning of thirteenth century separated Ruden Island

FIGURE 7.10 Southern Baltic Sea Storm surge on November 13/14, 1872, as depicted by Illustrirte Zeitung, Leipzig, December 21, 1872.

from Rugia Island. A storm in 1497 created a new connection between Wisła Lagoon and the Baltic Sea. A storm in 1558 and the years following destroyed Łeba village, and the barriers on Rugia Island, Usedom Island, and Hel Peninsula were broken in many locations.

The greatest documented storm event in this region took place on 13–14 November 1872 (Figure 7.10). Many people lost their lives due to the storm. Two years later another great storm occurred with high storm surges in Kołobrzeg (+2.20 m) and in Świnoujście (+1.96 m). Another strong event took place on 30–31 December 1904, at the Pomeranian Bay. Water levels near Rugia Island were about +2.5 m to +2.8 m; in Świnoujście, +1.8 m; and in Warnemünde and Flensburg, about +1.9 m to +2.2 m (Majewski et al., 1983). The highest water levels observed at the tide gauges located along the Polish part of the coast are presented in Table 7.2 (after Wiśniewski and Wolski, 2009).

The storm event on 13–14 November 1872, was successfully reconstructed from pressure maps and surge heights, scattered air pressure data, and numerical modeling by Rosenhagen and Bork (2009). The intensity of 1872 event is put into perspective when plotting annual maximum storm surges from 1926 to 2006 (Figure 7.11). The maximum surge height in 1872 was about +3.30 m, while in all other years the level of +2 m was almost never passed. Indeed, the 1872 event was an outlier. If it had not been documented so well at multiple locations, one may be misled to suspect the recoding as an error. It is suggested that this event was due to the concurrent effect of a basin filling, the wind-induced surge, and a contribution of the major eigen oscillation in the Baltic Sea.

At the end of the nineteenth century along the southern coast of the Baltic Sea, coastal dunes were perceived as a dam to protect against flooding

TABLE 7.2 The Highest Water Levels Observed at the Tide Gauges Located along the Polish Coast, From West to East (Wiśniewski and Wolski, 2009)

Station	Maximum Water Level above reference level (NN) (m)	Date
Świnoujście	1.96	10.02.1874
Dziwnów	1.15	10.02.1874
Kołobrzeg	2.22	13.11.1872
Darłowo	1.59	9.01.1914
Ustka	1.58	15.12.1898
Łeba	1.68	15.12.1898
Władysławowo	1.44	23.11.2004
Hel	1.22	14.01.1993
Gdynia	1.32	23.11.2004
Gdańsk	1.64	16.12.1843

Note that the dates vary across the entire time.

(Furmańczyk, 2013). People began to protect them by constructing fences to increase the deposition and accumulation of sand. These dune restoration strategies took place in Dziwnów, Poland, for example.

Dziwnów is a small town having approximately 3000 residents located at the Dziwna Strait that connects the Kamieński Lagoon with the Pomeranian Bay. Dziwna Strait is one of the conduits transporting water from the Szczecin Lagoon, via the Kamieński Lagoon, into the Baltic Sea. At the end of the first half of the nineteenth century, Dziwnów was originally a small fishing village.

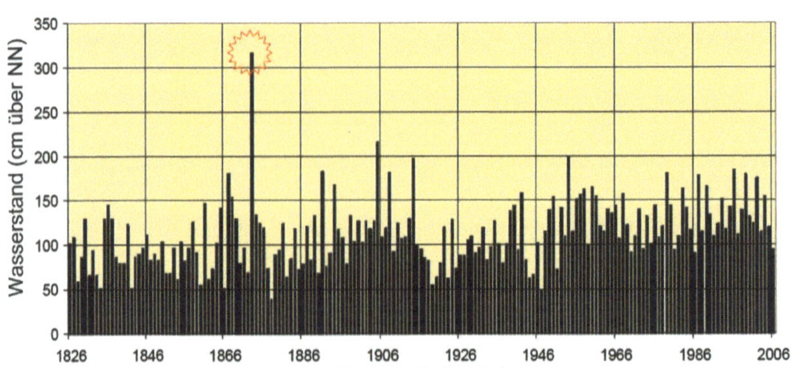

FIGURE 7.11 Annual maximum surge heights in the years 1826–2006 in Travemnde, Germany. Notice the peak for 1872. *From Rosenhagen and Bork (2009).*

Since then, it has grown dramatically and has gained importance as a holiday and spa resort. Beginning in 1890, hotels and health centers, a beach hall, a promenade on the dune, beach baths (on piles over the beach), and a pier were constructed (Figure 7.12). In a short period of time starting after 1890, Dziwnów became the most popular holiday destination on the Baltic coast, and was designated as the "Pearl of the Baltic" (Gerlach-Jósewicz et al., 1996).

Unfortunately, on 30 December 1913, a substantial storm devastated the "Pearl of the Baltic" and the community has not since been rebuilt. The devastation started on 27 December with strong N–NW winds (Beaufort 8–9). On 30 December the winds shifted to N–NE and formed a strong storm surge. Elevated water levels in the region were observed, for example: Kołobrzeg (+1.59 m), Świnoujście (+1.98 m), Travemünde (+1.95 m), and Kiel (+1.82 m) (Majewski et al., 1983). In the afternoon of 30 December water overflowed the low dunes and flooded the low-lying hinterland. Part of the town was affected and several buildings and tourist infrastructure were partly destroyed or filled with water (Figure 7.13).

The first hydrodynamic reconstruction of the 1913 event was completed using the XBeach model (Bugajny et al., 2013). The modern-day coastal profile was adopted and the dune morphology was approximated from old postcards; the

FIGURE 7.12 Postcards from Dziwnów, Poland, before the storm surge on 30 December 1913. (a) The "Kurhaus," beach hall and promenade and (b) beach activity and vegetated foredune.

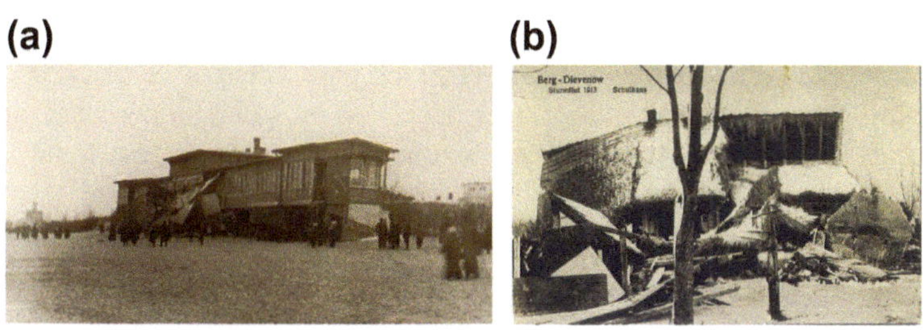

FIGURE 7.13 Postcards showing damages in Dziwnów, Poland, caused by the storm surge on 30 December 1913. (a) Beach hall and (b) schoolhouse.

dune top and width were estimated at +4 m and +16 m, respectively. A water level of +1.0 m at the beginning of the storm was used with a maximum level of +1.14 m (maximum water level in Dziwnów was noted +1.15 m in 1874). Significant wave height was at 4.5 m (0.5 m higher than was noted at the time of storm "Xavier" that took place at the same place in December 2013).

The XBeach simulation demonstrated an overflow effect after a couple of hours from storm initiation (Taranowicz, 2013). The model results corresponded well with descriptions provided by local witnesses shortly after the storm. This same storm impacted other communities along the Southern Baltic coast. In Gdańsk, located along the eastern shore of Gdańsk Bay, water level of +1.56 was measured. There was a lot of damage to the port and tourist infrastructure. In Sopot, a famous resort in the Gdańsk Bay, water overflowed low dunes and caused flooding in the city.

The 30 December 1913, event was one of the greatest (biggest) storm surges in the southern part of the Baltic Sea coast. After the storm 5 tons of amber was deposited. Since then many protection measures have been implemented, including beach nourishment and dune restoration to make this area less vulnerable to future storm surge events.

7.5 CONCLUDING REMARKS

The risk emanating from storm surges is a global, albeit regionalized phenomenon, affecting a large percentage of the world population, many urban conglomerates, and centers of commerce and trade (McGranahan et al., 2007; Hoozemans et al., 1993; Nicholls and Hoozemans, 2000). The scientific community, interested in the dynamics, prediction, and management of the hazard of storm surges, is composed of oceanographers, meteorologists, hydrologists, coastal engineers, geographers, planners, historians, and other earth and social scientists. Because of the breadth and the regionally limited impact of storm surges, this scientific community is highly fragmented, and is separated by regional languages, cultures, and management concepts. More efforts are needed to bring this community together (Kremer et al., 2013).

The specific case studies presented in this chapter demonstrate the multiple processes that result in storm surge and the various associated results. Changing coastal morphology, pumping ground water, and compacting deltas are some examples of the results. While climate change, in particular via changing "the degree of filling" (sea level), is of great importance, storm surges deserve attention because of the challenge they represent today even without the predicted effects on climate change.

REFERENCES

Alexandersson, H., Schmith, T., Iden, K., Tuomenvirta, H., 1998. Long-term trend variations of the storm climate over NW Europe. Global Atmos. Ocean Syst. 6, 97–120.

Bugajny, N., Furmańczyk, K., Dudzińska-Nowak, J., Paplińska-Swerpel, B., 2013. Modelling morphological changes of beach and dune induced by storm on the Southern Baltic coast using XBeach (case study: Dziwnow Spit). Special Issue No. 65. In: Conley, D.C., Masselink, G., Russell, P.E., O'Hare, T.J. (Eds.), J. Coastal Res. pp. 672–677.

Cai, Q., Liu, S.-S., Sun, H., 2010. A study of the storm tide disaster and the relief work in 1939 in Qingdao. Orient. Forum 5, 12–16.

Fritz, H.M., Blount, C.D., Thwin, S., Thu, M.K., Chan, N., 2009. Cyclone Nargis storm surge in Myanmar. Nat. Geosci. 2, 448–449. http://dx.doi.org/10.1038/ngeo558.

Furmańczyk, K., 2013. Poland. In: Pranzini, E., Williams, A. (Eds.), Coastal Erosion and Protection in Europe. Routledge, pp. 81–95.

Gerlach-Jósewicz, A., Walasek, L., Wójcik, K., 1996. Studium Ruralistyczne (Skrócone) Gminy Dziwnów. Biuro Dokumentacji Zabytków, Szczecin.

Gönnert, G., Dube, S.K., Murty, T., Siefert, W., 2001. Global Storm Surges. Die Küste, vol. 63, Kuratorium für das Küsteningenieurwesen, Boyens Medien, Heide, Germany. 623 pp. ISBN:3-8042-1054-6.

Grossmann, I., Woth, K., von Storch, H., 2007. Localization of Global Climate Change: Storm Surge Scenarios for Hamburg in 2030 and 2085. Die Küste, vol. 71, Kuratorium für das Küsteningenieurwesen, Boyens Medien, Heide, Germany. pp. 169–182.

Hoozemans, F.M.J., Marchand, M., Pennekamp, H.A., 1993. A Global Vulnerability Analysis, Vulnerability Assessments for Population, Coastal Wetlands and Rice Production on a Global Scale, second ed. Delft Hydraulics and Rijkswaterstaat, Delft and the Hague.

Hou, J.M., Yu, F.J., Yuan, Y., Fu, X., 2011. Spatial and temporal distribution of red tropical storm surge disasters in China. Mar. Sci. Bull. 30 (5), 535–539.

Lu, R.J., 1984. Historical Records of Disastrous Tide in China. Ocean Press, Beijing, 295 pp.

Majewski, A., Dziadziuszko, Z., Wiśniewska, A., 1983. Monografia Powodzi Sztormowych 1957–1975. Wydawnictwo Komunikacji i Łączności, Warszawa.

McGranahan, G., Balk, D., Anderson, B., 2007. The rising tide: assessing the risks of climate change and human settlements in low elevation coastal zones. Environ. Urban. 19, 17–37.

Nicholls, R.J., Hoozemann, F.M.J., 2000. Global vulnerability analysis. In: Schwartz, M. (Ed.), Encyclopedia of Coastal Science. Kluwer Academic Publishers.

Niemeyer, H.D., Eiben, H., Rohde, H., 1996. History and heritage of German coastal engineering. In: Kraus, N.C. (Ed.), History and Heritage of Coastal Engineering. American Society of Civil Engineers, 45 pp.

Petzelberger, B.E.M., 2000. Coastal development and human activities in NW Germany. In: Pye, K., Allen, J.R.L. (Eds.), Coastal and Estuarine Environments: Sedimentology, Geomorphology and Geoarchaeology, vol. 175. Geological Society, London, Special Publication, pp. 365–376.

Rosenhagen, G., Bork, I., 2009. Rekonstruktion der Sturmflutwetterlage vom 13. November 1872. Küste, vol. 75, Kuratorium für das Küsteningenieurwesen, Boyens Medien, Heide, Germany. pp. 51–70.

Taranowicz, W., 2013. Upadek Perły Bałtyku. http://echokamienia.pl/wiadomosci/powiat-region/5651-sztorm-i-cofka-w-dziwnowie-czy-dziwnow-znajdzie-sie-pod-woda; as of 5 September 2014.

von Storch, H., Woth, K., 2008. Storm surges, perspectives and options. Sustainability Sci. 3, 33–44. http://dx.doi.org/10.1007/s11625-008-0044-2.

von Storch, H., Gönnert, G., Meine, M., 2008. Storm surges – an option for Hamburg, Germany, to mitigate expected future aggravation of risk. Environ. Sci. Policy 11, 735–742.

WASA, 1998. Changing waves and storms in the Northeast Atlantic? Bull. Amer. Met. Soc. 79, 741–760.

Weisse, R., Plüß, A., 2005. Storm related sea level variations along the North Sea coast as simulated by a high-resolution model 1958–2002. Ocean Dyn. 56 (1), 16–25. http://dx.doi.org/10.1007/s10236-005-0037-y.

Wiśniewski, B., Wolski, T., 2009. Katalog Wezbrań I Obniżeń Sztormowych Poziomów Morza Orz Ekstremalne Poziomy Wód Na Polskim Wybrzeżu. Wydawnictwo Naukowe Akademii Morskiej, 156 pp.

Zhang, X.-H., Sheng, C.-Y., Shao, Z.-H., 2006. Analysis of storm surge at coastal area of Qingdao. Mar. Forecasts S1, 42–46.

FURTHER READING

De Kraker, A.M.J., 2009. Stormweer in de Nederlanden tijdens de 15^{e} en 16^{e} eeuw. In: Essink, K. (Ed.), Verdronken Geschiedenis. Mededelingen van de Vereniging voor Terpenonderzoek, Nr. 12 (juni 2009). http://www.verdronkengeschiedenis.nl/nl/stormvloed/docs/3-Adriaan%20de%20Kraker-2-NL-opmaak-def-def-3.pdf.

Kremer, H., Nicholls, R., Ratter, B., Weisse, R., 2013. Interdisciplinary research and perspectives on current and future storm surges. Nat. Hazards 66, 1293–1294.

Chapter 8

Sea-Level Rise: Causes, Impacts, and Scenarios for Change

Robert J.N. Devoy
The Coastal and Marine Research Centre, Beaufort Research and the Department of Geography, University College Cork, Cork, Ireland

ABSTRACT
Studies of sea-level changes (SLCs) are well established, as is reflected in an extensive research and wider publication literature. People globally have long recognized that ancient marine shorelines and beach sediments occur inland, well away from present day coasts. The causes of SLC center upon a series of primary controls resulting from the operation of an integrated atmosphere–ocean system. These include changes in the total water volume of the oceans over time, temperature effects on ocean water expansion, alterations in the shape of the ocean basins from Earth crustal movements, and changes in their accommodation space through sediment accumulation. Satellite surveys also show that the topography of the oceans and shelf seas is influenced by gravitational controls, which define the Earth geoid, with an up to 180 m difference in interregional sea surface heights. In the Quaternary, past glaciations have driven sea levels to approximately −120 m global mean sea level (gmsl). The significance of SLC today lies in establishing an understanding and projections of how sea levels will change in the future. Climate warming during the twenty-first century will result in the melting of the Earth's remaining ice masses, with gmsls likely to rise at rates of 4–5 mm/year by 2050, reaching levels of 0.5–0.9 m by 2100 under differing Intergovernmental Panel on Climate Change (IPCC) global warming scenarios. The impacts of these movements will potentially result in initial losses of up to 30% of coastal wetlands and an increasing "squeeze" of people and biological systems into the reorganizing coastal zone.

8.1 INTRODUCTION

Sea-level and sea surface changes (SLCs) are a primary driver in coastal systems' functioning; their analysis and quantification forms a critical element in the study of coastal environments (e.g., Devoy, 1987; Warrick et al., 1993;

Coastal and Marine Hazards, Risks, and Disasters. http://dx.doi.org/10.1016/B978-0-12-396483-0.00008-X
Copyright © 2015 Elsevier Inc. All rights reserved.

Carter and Woodroffe, 1994; Barthel et al., 1999; Smith et al., 2000; Church et al., 2010; Nicholls, 2010; Nicholls et al., 2007, 2011; Bengtsson et al., 2012; Wong and Losada, 2014; Church et al., 2014; Masselink and Gehrels, 2014; Pugh and Woodworth, 2014). The fact that SLCs have occurred regularly over varying timescales, both in vertical extent (height range) and consequent spatial exposure of the world's continental margins, is well established (e.g., Devoy, 1979, 1987, 1997; Hallam, 1981; Haq et al., 1987; Haq, 1991; Pirazzoli, 1996; Peltier, 1998, 2004; Cronin, 1999; Edwards, 2006; Bindoff et al., 2007; Haq and Schutter, 2008; Muller et al., 2008; Murray-Wallace and Woodroffe, 2014; Church and Clark et al., 2014) (Figure 8.1). Most recently in Earth's history, the Quaternary glacial and interglacial cycles have been the main control in these vertical movements of coastal positions worldwide. Changes in ice-mass volumes (from glaciers and ice sheets) through Croll–Milankovitch forcings of Earth temperature and the linked operation of the hydrological cycle, have caused a wide range of sea-level movements. Ice-mass changes have driven the position of the coastal zone (CZ) regularly across the continental shelves at timescales of 10^{3-4} years over the last 0.5–1.0 M years (Lowe and Walker, 1997; Siegert, 2001) (Figure 8.2). This close correlation of glaciations with SLC and its crustal impacts was recognized early by the first Quaternary and Earth scientists, such as Agassiz, Jamieson, and later Daly and Zeuner (Smith and Dawson, 1984; Devoy, 1987; Dawson, 1992). The erosion associated with these CZ movements may have created the low-angle surfaces ($<1°$) and wide extent of many passive ocean-continental shelf margins (Open University, 1998). Low-stand sea-level positions of −120 m to −130 m global mean sea level (gmsl) (e.g., Fairbanks, 1989; Bard et al., 1990, 1996; Peltier, 1998), during glaciations and high stands in interglacials of +2 to 6 m gmsl have occurred during the Quaternary, with changed coastal positions identified from most world coastal regions (Figure 8.1(b)) (e.g., Shackleton, 1987; Chappell et al., 1996; Church et al., 2010; Murray-Wallace and Woodroffe, 2014). Since the ending of the last glacial stage (10, 000 years ago), the subsequent Holocene rates of sea-level rise (SLR) have fallen from high values of 20–40 mm/year (Bard et al., 1996), following the major deglaciations in the Northern Hemisphere, to relative mean global rates of <1 mm/year during the late Holocene, post-3000 years BP (e.g., Bindoff et al., 2007; Delaney et al., 2012; Church and Clark et al., 2014).

The Earth history connection between continual SLC and CZ movements has been highlighted in the five Intergovernmental Panel on Climate Change reports since 1990 (1990–2014) (e.g., IPCC WGI, 2001, 2007, 2014; IPCC WGII, 2014). People ignore the message of the past scales and cycles of rapid environmental changes at their peril: SLC is a good analog in developing an understanding of the dynamic and complex functioning of the Earth. Detailed studies of former SLCs have occurred since the 1970s, with the development under UNESCO of the International Geological Correlation Programme's projects (IGCP) on "Sea levels." These have continued

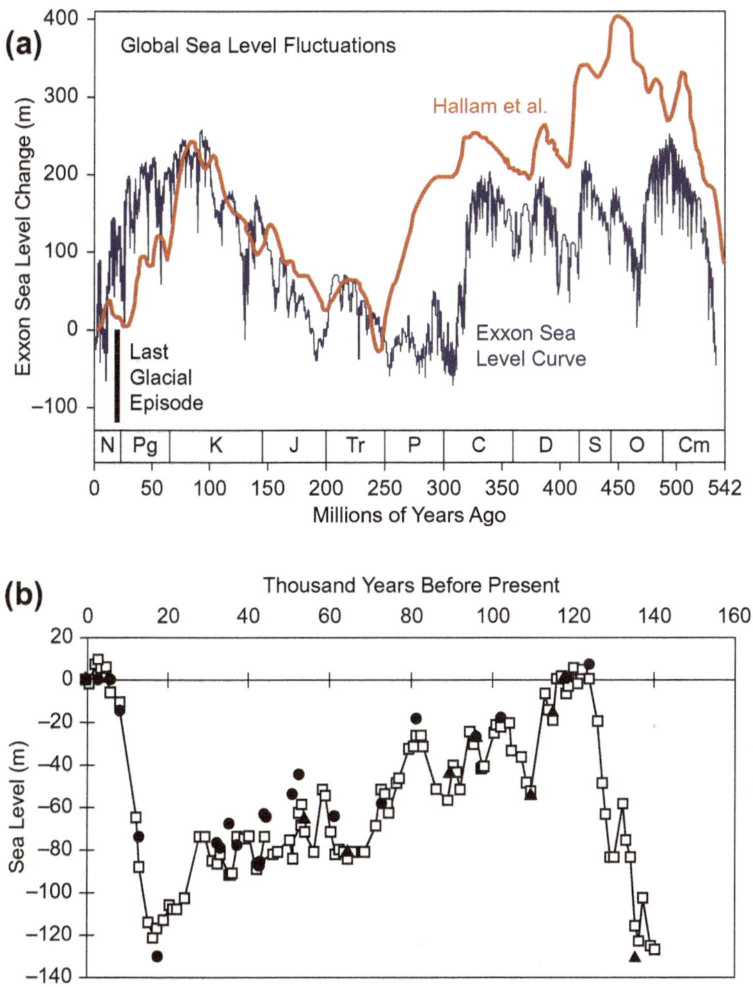

FIGURE 8.1 (a) The pattern of long-term (10^{6-8} years) global-relative, SLCs, interpreted from Exxon-type continental Shelf sedimentary coastal on-lap and off-lap data *(Source: Open web, after Vail et al. (1977), Hallam (1981), and Haq (1991).).* (b) Late Quaternary RSL changes for Marine Isotope Stages 1–5, based upon geomorphological shoreline, coral-terrace data from the Huon Peninsula, Papua New Guinea (dots), and oxygen isotope records from benthic foraminifera (squares) *(Source: Open web source figure, based upon data from Peltier (1998), after Shackleton (1987) and Chappell et al. (1996).).*

unbroken to the present day. Many of the projects, for example, Projects 61, 200, 274 (Greensmith and Tooley, 1982; van de Plassche, 1986; Pirazzoli, 1991, 1996) have concentrated on the proxy records and operations of past SLCs. More recent work, for example, Project 588, “Preparing for Coastal Change” have made links to CZ process and applied research (IGCP 588,

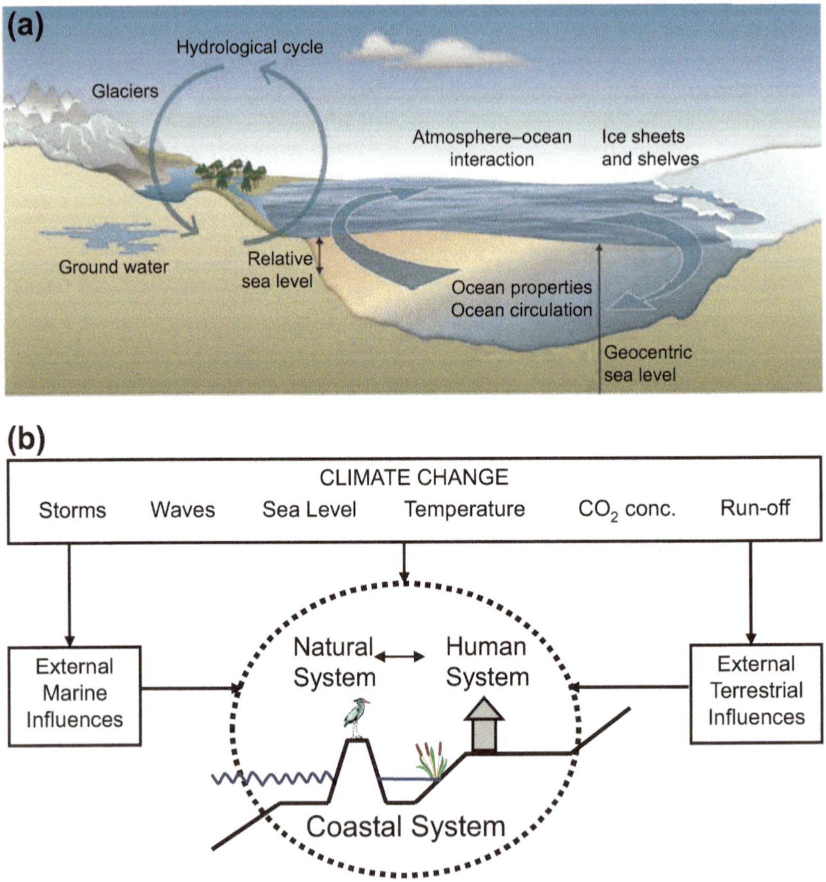

FIGURE 8.2 (a) Operation of the coupled Ocean–Atmosphere systems and their linkage to RSL changes at the land margins, in which the large-scale mass–energy fluxes and the ocean water properties, for example, of temperature, salinity, and density, will influence RSL values and the positioning of the CZ *(Source: Church et al. (2014).)*. (b) Definition and functioning of the CZ in the context of RSL changes at the land–ocean margins (Figure 8.2(a)) and showing the links to climate forced coastal drivers *(Source: IPCC WGII, 2007, Nicholls et al., 2007.)*.

2014). In the late 1990s, the IGCP projects were joined by the International Geosphere and Biosphere Programme (IGBP), with the study of SLC and coastal issues represented by the Land and Ocean Interactions in the Coastal Zone (LOICZ) program (LOICZ, 2014). These international SLC studies and many others worldwide, together with their incorporation into the understanding of coastal processes functioning (e.g., Carter and Woodroffe, 1994; Hardisty, 1990, 1994; Duffy and Devoy, 1999; Sanchez-Arcilla et al., 2000; Davison-Arnott, 2010; Masselink et al., 2011), have produced an extensive sea level and related literature. Examples include, the NATO Advanced

Science Institutes Research Workshop Series (as per the titling in the book Sabadini et al., 1991); the European Science Foundation colloquium series, Interglacial Sea-level Changes in 4D (Devoy, 1999); the European "Environmental Framework Research Programmes III–VII" (EU CORDIS Website, 2014; Barthel et al., 1999; Smith et al., 2000); and in the United States, Sea Grant (federally sponsored by the National Oceanic and Atmospheric Administration), the Environmental Protection Agency and many other Federal and States' funding programs (e.g., Barthel and Titus, 1984; Titus et al., 1991, 2010). It should be noted that although much of this work began in the mid-twentieth century in the Euro-North American "centered world," including Australia and New Zealand, it has only recently been owned and developed by other societies, for example, India, Japan, China and Southeast Asia, Africa, and South America (e.g., viz. IGCP Projects; IGBP–LOICZ; Devoy, 1987; Tooley, 1993; Murray-Wallace and Woodroffe, 2014). Current concerns over future SLR under climate warming, as a cause of coastal change, also now link SLR intimately to the fields of atmosphere–ocean systems modeling (e.g., IPCC, 2001a; IPCC WGI, 2007, 2014). Work in this field is involved in two key areas relevant to coastal vulnerability issues. First, in establishing the ranges of projected SLR to 2100 and beyond, with the past sea-level records for use as analog and in the validation of projected SLC. Second, in helping develop the viable management options for future coasts (e.g., Barthel et al., 1999; Devoy, 2008; Cooper and Cummins, 2009; Titus et al., 1991, 2010; Gray et al., 2014; Cooper and Pilkey, 2012; Cooper et al., 2014; CoastAdapt, 2014).

8.2 SEA-LEVEL DEFINITIONS, MEASUREMENTS, AND CAUSES

Sea level is the primary reference surface in establishing the point of the land–ocean divide in Earth systems and consequently in the definitions of the CZ (Figure 8.2) (Woodroffe, 2002; Cooper and Cummins, 2009; Nicholls et al., 2007, 2010; Wong and Losada, 2014). SLCs determine the immediate position of coastlines and the effective extent and wider spatial movements of the CZ over time. Further, it is the surface level or "platform," as by analogy with a medieval "siege tower," from which hydrodynamic processes operate in developing shorelines (Carter and Devoy, 1987; Carter, 1988; Hardisty, 1990, 1994; Carter and Woodroffe, 1994; Stone and Orford, 2004; Masselink et al., 2011; Tibaldi et al., 2012; Ranasinghe et al., 2013; Kremer et al., 2013; Devoy, in press). The sea surface is constantly perturbed by wind-driven (gravity) waves, tides, storm surges, tsunamis, many "miscellaneous" wave forms, the effects of internal and long-period waves and by ocean currents, steric, density, and meteorological dynamic effects (Warrick et al., 1993; Smith et al., 2000; Open University, 1998; Pugh, 2004; Church et al., 2014). The surface is spatially (local–regional) and temporally complex in its form and functioning, differing in shape and elevation from microscale to macroscale (1 m^2 to 10^3 km and seconds to 10^{6-8} years). In paleoenvironmental reconstructions of SLCs (e.g., the IGCP projects; Peltier, 1998, 2009; Lambeck, 1991, 1995, 2001; Church et al., 2010; Murray-Wallace and Woodroffe, 2014), and from the contemporary monitoring of sea levels, it is clear that no globally uniform sea level exists (van de Plassche, 1986; Warrick et al., 1993; Smith et al., 2000; Pugh, 2004).

The sea surface is commonly defined at local–regional scales in relation to a mean sea level (msl) (i.e., the averaged position of all states of sea level over a period of time, generally greater than one year), as a mean tide level, or referenced to a bathymetric, chart datum on marine maps; in the British Isles taken commonly as the lowest point of astronomical tides (e.g., Woodworth, 1990, 1993, 1999; Woodworth et al., 1991, Woodworth and Blackman, 2004; Zerbini et al., 2000; Pugh, 2004; Church et al., 2010; Cronin and Devoy, 2010; Admiralty Tide Tables, 2014). These levels have been measured and monitored by land-based tide gauges at reference locations that are situated ideally in "stable" locations, free of Earth crustal subsidence and uplift movements. This is often difficult to determine accurately; hence, the relocation of datum points at times (e.g., in Britain, from Liverpool to Newlyn in 1921, or in the Republic of Ireland from Dublin to Malin Head in 1965). Historically, these expressions of sea level (e.g., msl) have been used as the reference point for geodetic land height surveys Ordnance Datums, for example, OD (Britain), NAP (the Netherlands), NGF (France), and will vary between regions, countries, and continents (IOC, 1985, 2009; Woodworth et al., 1991, 2002, 2007, 2009a; Woodworth and Player, 2003). These datums do not allow for easy comparison of sea surface height changes over long distances, or even the within-region recognition of the complexity in the operation of SLC (e.g., at scales of >50–100 km). Consequently, the advent of satellite altimetry and its integration into more extensive and precise land-based tide gauge networks, and whole earth spatial coverage, have led to a much better understanding of sea-level variability (e.g., Global Sea-level Observing System, GLOSS, 2014; Merrifield et al., 2009; IOC, 2009) and the development of geocentric systems of measuring Earth shape (Carter et al., 1993; Zerbini et al., 2000) (Figure 8.3). These survey systems now form the basis for measurements of SLC and particularly for definition of the first-order, long-term "static-semi-static" component of sea surface levels, namely, differences in the Earth geoid, that is, the gravitational equipotential surface shape of the Earth (Warrick et al., 1993; Zerbini et al., 2000; Pugh, 2004; Masselink and Gehrels, 2014). The geoid surface results primarily from the gravitational effects of land masses, rock density differences with Earth geophysical functioning, and extraterrestrial gravitational forces.

Although predicted earlier (e.g., Lisitzin, 1974; Carey, 1981), satellite surveys clearly show the geoid distortions of Earth shape, with the distribution of relative high and low points of the sea surface (Figure 8.4). These evidence

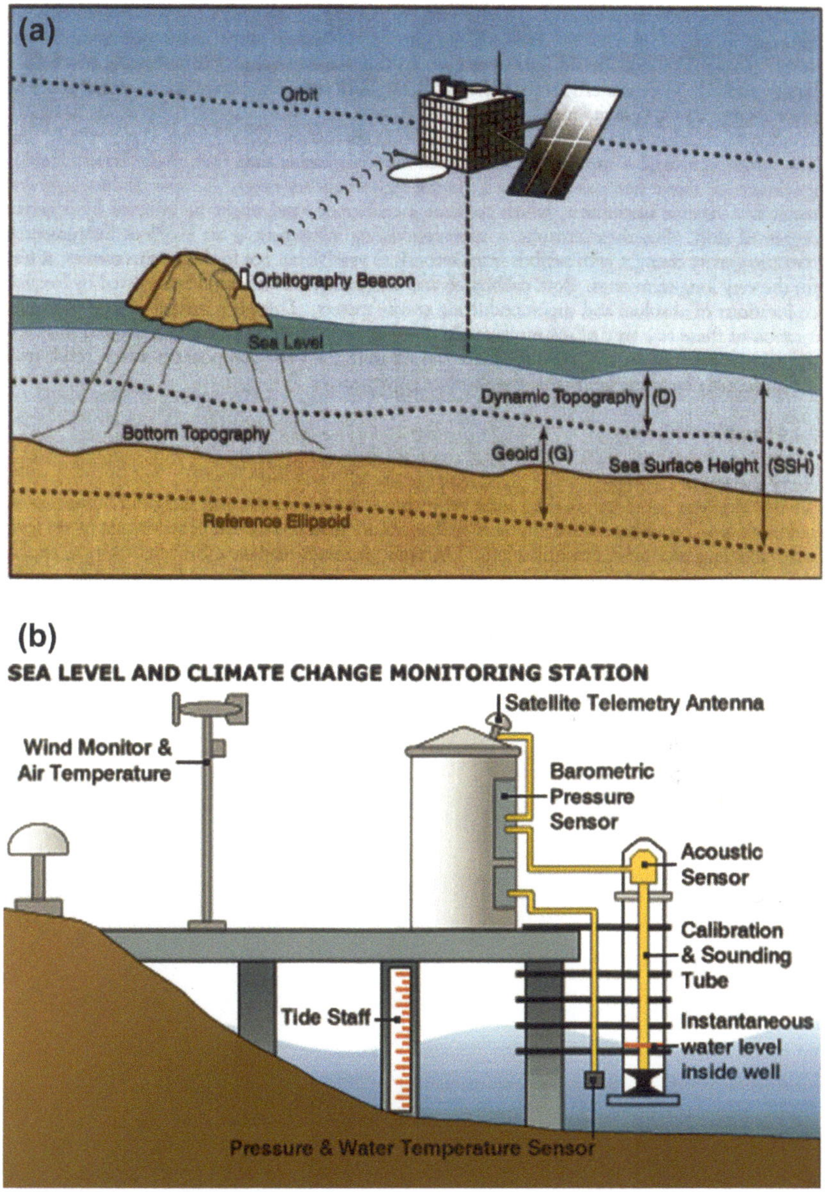

FIGURE 8.3 (a) Definition of sea-level surfaces in relation to the geoid and of other key reference parameters, using satellite altimetry *(Source: Zerbini et al. (2000).)*. (b) Land-based tide gauge and satellite altimetry-linked, monitoring station *(Source: Open Web, National Tidal Centre, Australia.)*.

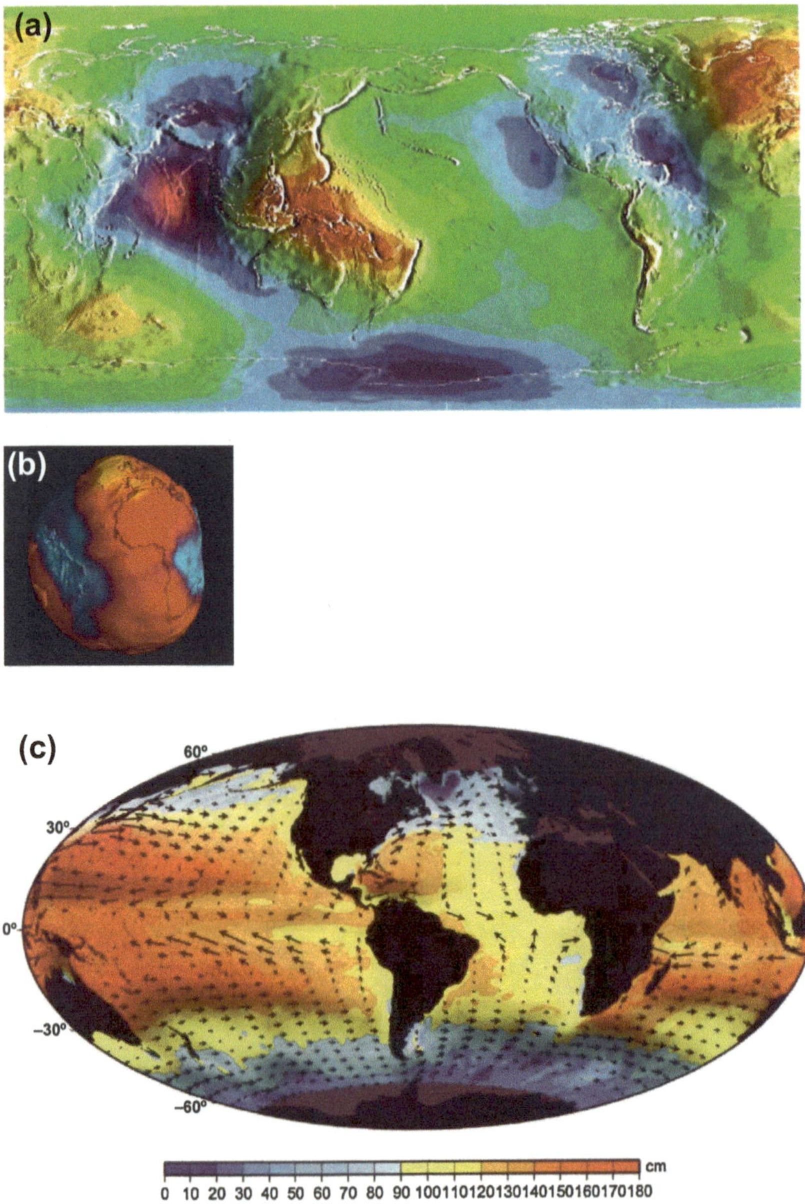

FIGURE 8.4 (a) The Earth's geoid, measured relative to the reference ellipsoid and showing areas of maximum surface elevation (orange–red) of +85 m and minimum levels (blue–purple–black) of approximately 106 m. The differences result primarily from the uneven mass and density distributions within the Earth *(Source: Pugh (2004).)*. (b) Earth geoid shape (*Source: Open web*). (c) Dynamic sea surface as gmsl variations from the geoid, measured in dynamic centimeters, caused by meteorological, ocean density, and current changes with time. The black arrows show the main ocean current pathways and are associated with some of the largest height differences *(Source: From Pugh, Changing Sea Levels, (2004).)*.

significant relative sea-level (RSL) height differentials of up to 180 m between the "lows" and "highs," for example, of southern India compared to the central–eastern North Atlantic. Earlier conjecture that these "static" RSL positions migrate in long-term timescales (>10^6 years) (e.g., Mörner, 1987a,b) lack support (Devoy, 1987; Tooley, 1993; Peltier, 1998). Further, it is recognized that at continental land margins (Clark et al., 1978; Clark and Primus, 1987; Clarke et al., 2005; Church et al., 2014), and also around major ice masses, other mass/gravity distortions of the geoid exist (scales of >1-m sea surface set up). Melting of the present day major ice masses (e.g., for the Greenland and Antarctica ice sheets) will lead to the relaxation of the ice-marginal sea surfaces and to regional and wider redistributions of the gravitationally held water "bulges" (e.g., Peltier, 1998, 2004; Lambeck, 2001; Milne et al., 2009; Mitrovica et al., 2009, 2011; Bindoff et al., 2007; Church et al., 2010; Church et al., 2014).

Additionally, the return of land-based melt water to the oceans under Earth rotation results in centrifugal movements of water and large-scale time and spatial differential SLCs. In the lateglacial deglaciation of the Northern Hemisphere, early rapid rises of sea level occurred in the Southern Hemisphere in response to the ice melt, as reflected in the Holocene "family" of sea-level curves (Thom and Roy, 1983; Devoy et al., 1994; Murray-Wallace and Woodroffe, 2014), with RSL reaching or exceeding present levels by 6,000 BP (Figure 8.5(a) and (b)). Subsequent equilibriation of water levels took place, with return flows northward later in the Holocene and RSL continuing to present levels (Clark and Lingle, 1977; Clark and Primus, 1987; Tooley, 1993; Pirazzoli, 1996; Peltier, 1998, 2004, 2009; Milne and Mitrovica, 1998), with overall rapid early rises of gmsl (Bard et al., 1996). These changes in ice and ocean water masses have feedback effects on changes in Earth tilt, "wobble" and rate of spin, which also undergo other periodic changes over time, and together result in variations in sea surface levels (Church et al., 2014). Under modeled ice melt with future climate warming (to 2100 and beyond), then similar mechanisms of water flows and the recording of SLC in the Northern and Southern Hemispheres will occur. These will be consequent upon the differences in the expected timings of Greenland and Antarctic downwasting, though rapid global SLC responses to ice melting are likely (Church et al., 2014; Mitrovica et al., 2011; Rahmstorf, 2007, 2010, 2012).

More significant at these large-regional spatial scales are the short term (10^1 years) to rapid (seconds to days) movements in sea levels created by the dynamic sea surface (Lisitzin, 1974; Zerbini et al., 2000; Pugh, 2004; Church et al., 2010) (Figure 8.4(c)). These are caused mainly by meteorological and coupled Earth atmosphere–ocean energy drivers (including, e.g., ocean currents and steric changes, atmospheric pressure fields, winds, open ocean rainfall, and large-scale river discharges). Resultant phenomena (e.g., El Nino Southern Oscillation (ENSO), El Niño, and La Niña events;

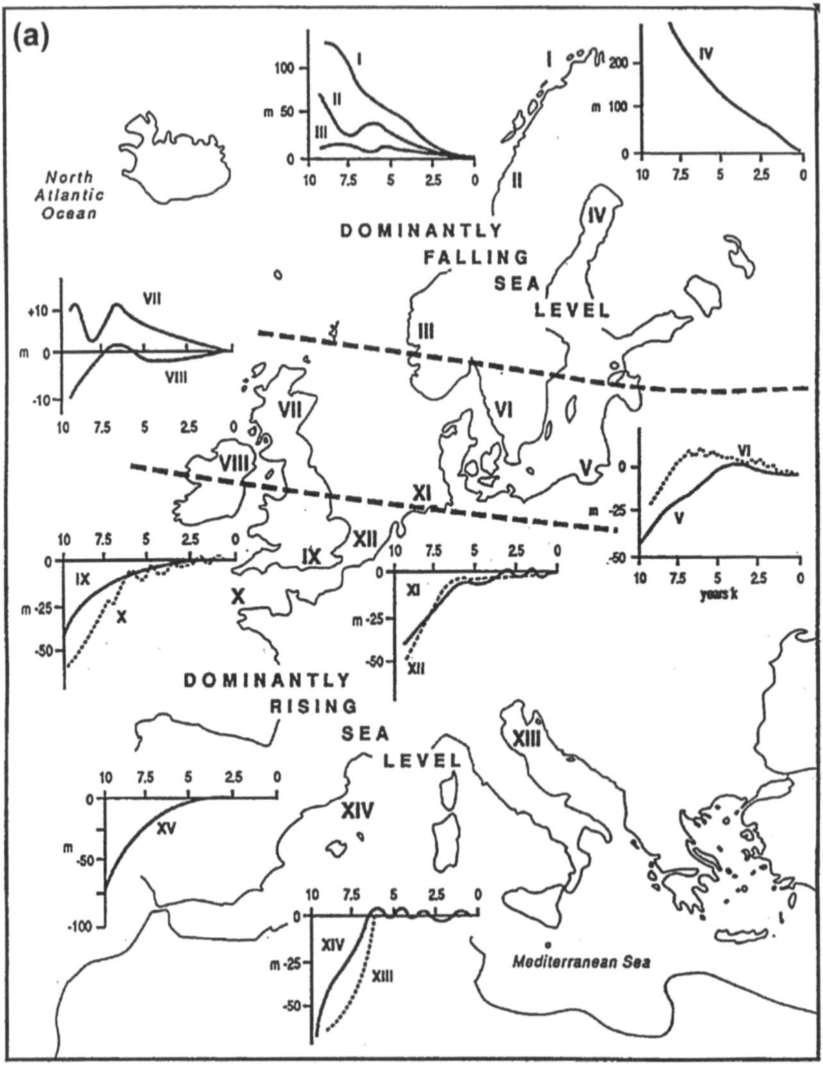

FIGURE 8.5 (a) The Holocene "family" of RSL curves from Northwest Europe, showing the different coastal zonal "signatures" in simplified sea-level trends resulting from Earth crustal isostatic responses to loading changes following deglaciation *(Source: From Carter, 1992, with permissions from the Quaternary Research Association.)* and (b) the Holocene SLC patterns from areas of Australia and New Zealand, indicating RSLs at or above present day levels by 6,000 years BP *(Source: Devoy et al. (1994); Thom and Roy (1983).)*. (c) Approaches in the construction and identification of Holocene sea-level patterns, trends and tendencies. *(Source: From Carter, 1992, with permissions from the Quaternary Research Association.)*.

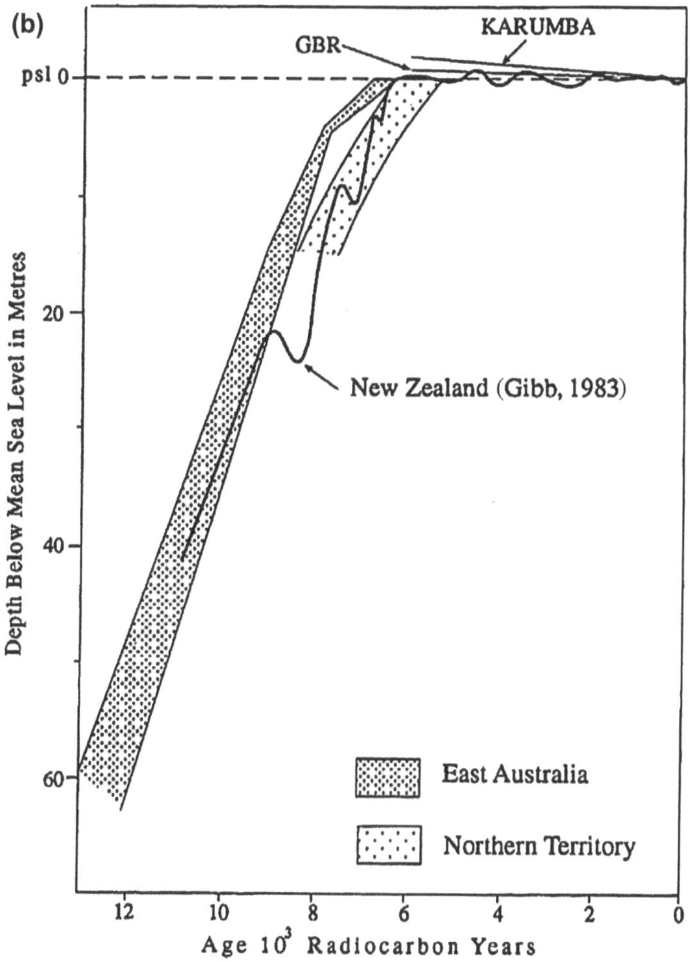

FIGURE 8.5 cont'd

Pacific Decadal Oscillation (PDO); North Atlantic Oscillation (NAO)), produce changes in mean sea surface heights of 1−2 m from the geoid and locally >0.3 m (Figure 8.6). These sea surface variations may be interannual, or longer-term periodic to quasicyclical in occurrence. In the case of ENSO events, return periods of 3−7 years in the nineteenth/twentieth centuries (McGregor, 1992; IPCC WGI, 2014). Paleo-El Niño events from proxy ocean core records show different frequencies during the Holocene (Ortlieb and Maclane, 1993; Fagan, 2000). Most importantly, the warming of ocean water results in the expansion of the water column and thermostatic (steric) SLR. This effect is regionally varied, due to changes in water

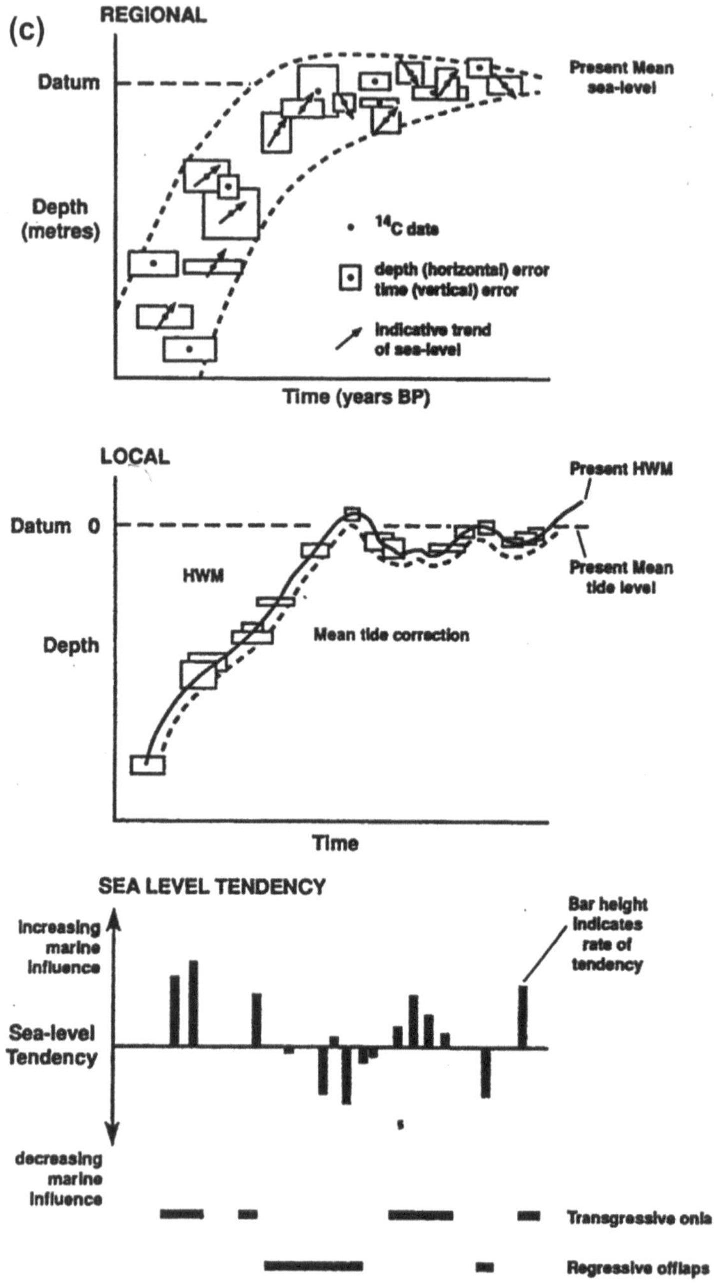

FIGURE 8.5 cont'd

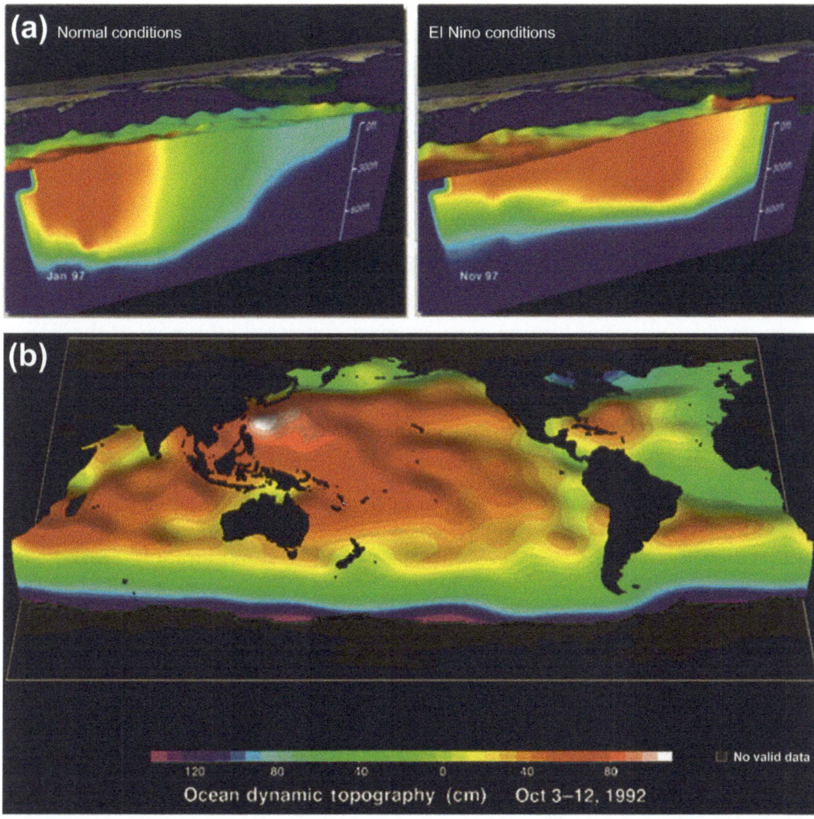

FIGURE 8.6 (a) Equatorial cross-section of the Pacific Ocean, showing the development of an El Nino event, showing normal conditions in January 1997 to the full El Nino development in November 1997 in the eastern Pacific, with raised sea surface levels. Changed sea surface topography is shown linked to the temperature changes, 30 °C as red to 8 °C as dark blue *(Source: Open web, NASA.)*. (b) El Niño event developing in 1992, showing the height detail of the surface topography *(Source: Open Web.)*.

density and linked characteristics. Estimates indicate that during the period from 1954–2014, steric changes have caused averaged rates of SLR of 0.33 mm ± 0.7 mm/year, to as high as 0.52 mm/year (in the top 700 m of the oceans), with rates of 0.4 mm ± 0.9 mm/year indicated for the global oceans to depths of 3000 m. Maximum effects of these changes are concentrated in the subtropical North Atlantic and tropical eastern Pacific regions (IPCC WGI, 2001, 2007, 2014).

Dynamic sea surface changes, particularly where coupled with storminess patterns, have important consequences on coasts, in terms of increased rates of erosion, flooding from coupled marine inundation, and river discharges and through wider impacts in sediment movements and coastal processes functioning (Duffy and Devoy, 1999; Woodroffe, 2002; Nicholls et al., 2007; Saito et al., 2007; Kremer et al., 2013). Along East Pacific coasts, El Niño events result in elevated sea surfaces of >0.1 m and La Niña episodes can cause lowered surfaces (Figure 8.6). When these temporary changes are associated with storminess, significant phases of increased erosion and beach change can occur, as recorded along western coasts of the United States, eastern Australia, and in the wider central Pacific region (Short et al., 1995; Storlazzi and Griggs, 2000; Soloman and Forbes, 1999; USGS, 2011). Widespread impacts can also occur upon offshore marine systems in sea surface temperature (SST) changes, ocean productivity, and famine consequences for people throughout the South American Pacific regions (Open University, 1998; Fagan, 2000; IPCC, 2001a; IPCC WGI and WGII, 2007, 2014). Operation of the NAO and linked changes in the position of the Icelandic Low Pressure system, influenced possibly by variations in Arctic sea-ice distributions, have similar erosion and coastal process impacts in the North Atlantic (Stone and Orford, 2004). Periods of increased storminess in the late Holocene have been identified from this region, associated with barrier breaching, coastal flooding, and other sedimentary–hydrodynamic consequences resulting from these teleconnection types of factors (Gilbertson et al., 1999; Dawson et al., 2004; Delaney et al., 2012).

8.2.1 Larger-Scale Causes for SLCs

The longer-term causes of SLCs have been a topic of study since at least the eighteenth Century and subject to varying changes in paradigms and understanding through the twentieth Century (Smith and Dawson, 1984; Devoy, 1987; Shennan and Tooley, 1987; Smith et al., 2000; Murray-Wallace and Woodroffe, 2014). Much debate has centered on the definition of the term *eustasy*, which formerly implied absolute changes of water level within the oceans and controlling the position of shorelines (e.g., Suess, 1906; Daly, 1934; Visser, 1980; Tooley, 1993). Attached to the concept initially was that SLCs were global in nature and the term was synonymous with the "water volume" of the oceans. The work of the IGCP *Sea Levels* Projects 61, 200 and 274 (e.g., Greensmith and Tooley, 1982; van de Plassche, 1986; Pirazzoli, 1996) particularly, altered this understanding. They established clearly the worldwide variability of Holocene and Quaternary interglacial SLC, knowledge of the component drivers to SLCs and the range and heterogeneity of data used in the reconstruction of past sea levels. Further, this IGCP project series (200, 274, 367, 437…588), among other research programs, raised questions on the value and the approaches needed in using paleosea levels as predictors of future SLCs and of the limitations of these classic geologically based sea-level studies in establishing coastal functioning (Devoy, 1987; Shennan and Tooley, 1987; Pirazzoli, 1996; Smith et al., 2000). What became clear in the

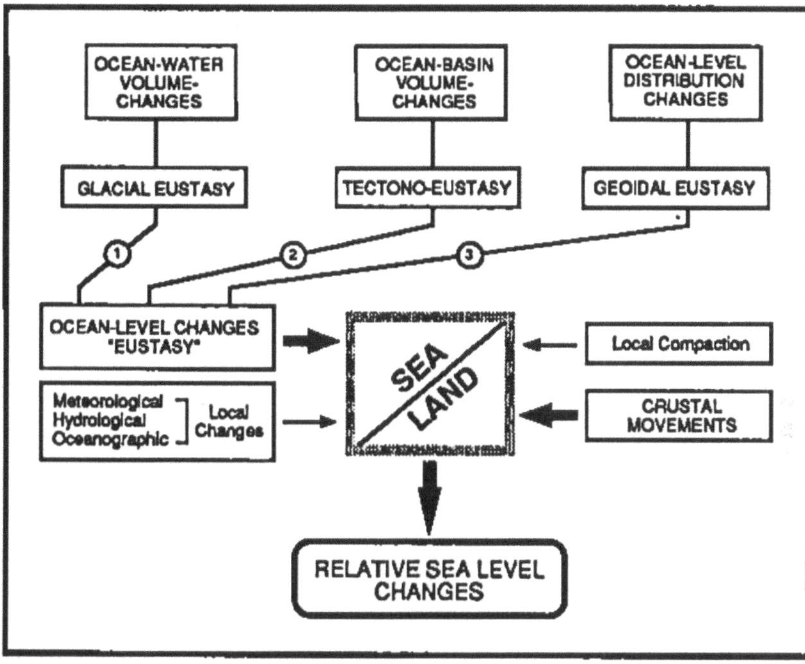

FIGURE 8.7 Primary controls in the development of relative sea-level changes. *Source: From Dawson (1992); after Mörner (1980).*

1980s is that eustasy, and the other linked controls on sea-level positions, are not spatially and temporarily uniform, or worldwide in operation. This understanding is supported by the work on the causes and drivers to longer-term, pre-Quaternary RSL changes (Vail et al., 1977; Hallam, 1981; Haq et al., 1987; Haq, 1991; Haq and Schutter, 2008; Conrad, 2013) (Figure 8.1(a)).

As morphodynamic and hydrodynamic studies of coastal processes emphasize (e.g., Carter and Woodroffe, 1994; Duffy and Devoy, 1999; Short, 1999; Cooper et al., 1995, 2001, 2007; ICS, 2002, 2009, 2011, 2013; Ranasinghe et al., 2013; O'Shea and Murphy, 2013; Cooper and Jackson, 2014) coasts operate at local scales. In turn, the same is true in the deconstruction of the causes of SLCs and their use in determining shoreline positions. SLCs comprise many macroscale to microscale temporal and spatial Earth environmental factors. Understanding of these SLC components helps define the term eustasy as, "absolute sea-level changes regardless of causation and including the main family of vertical and horizontal geoid and dynamic changes" (Mörner, 1987a,b) (Figure 8.7 and Table 8.1). The significance of this approach to SLCs has been in making the linkage to problem solving in the study of coastal processes and associated management issues (e.g., Jennings et al., 1998; Cowell and Thom, 1994; Cooper et al., 2001, 2007) and

TABLE 8.1 Summary of the Components Involved in Defining Eustasy and Relative SLCs

<table>
<tr><td rowspan="19">E U S T A S Y
OCEAN LEVEL CHANGES</td><td rowspan="16">VERTICAL AND HORIZONTAL GEOID CHANGES</td><td rowspan="10">OCEAN BASIN VOLUME</td><td rowspan="10">TECTONO-EUSTASY</td><td colspan="2">EARTH-VOLUME CHANGES</td></tr>
<tr><td rowspan="5">TECTONICS</td><td>OROGENY</td></tr>
<tr><td>MID-OCEANIC RIDGE GROWTH</td></tr>
<tr><td>PLATE TECTONICS</td></tr>
<tr><td>SEA FLOOR SUBSIDENCE</td></tr>
<tr><td>OTHER EARTH MOVEMENTS</td></tr>
<tr><td colspan="2">SEDIMENT IN-FILL</td></tr>
<tr><td rowspan="3">ISOSTASY</td><td>LOCAL ISOSTASV</td></tr>
<tr><td>HYDRO - ISOSTASV</td></tr>
<tr><td>INTERNAL LOADING ADJUSTMENT</td></tr>
<tr><td rowspan="2">OCEAN WATER VOLUME</td><td colspan="3">GLACIAL EUSTASY</td></tr>
<tr><td></td><td colspan="2">WATER IN SEDIMENT, LAKES AND CLOUDS, EVAPORATION, JUVENILE WATER</td></tr>
<tr><td rowspan="4">OCEAN MASS / LEVEL DISTRIBUTION</td><td rowspan="4">GEOIDAL EUSTASY</td><td colspan="2">GRAVITATIONAL WAVES</td></tr>
<tr><td colspan="2">TILTING OF THE EARTH</td></tr>
<tr><td colspan="2">EARTH'S RATE OF ROTATION</td></tr>
<tr><td colspan="2">DEFORMATION OF GEOID RELIEF (DIFFERENT HARMONICS)</td></tr>
<tr><td rowspan="3">DYNAMIC CHANGES</td><td rowspan="3" colspan="2">DYNAMIC SEA LEVEL CHANGES</td><td colspan="2">METEOROLOGICAL</td></tr>
<tr><td colspan="2">HYDROLOGICAL</td></tr>
<tr><td colspan="2">OCEANOGRAPHIC</td></tr>
</table>

Source: After Tooley (1993).

in the development of numerical models in simulations of coastal changes (de Vriend, 1991; de Vriend et al., 1993; Peltier, 1998; Stive, 2004, 2006; Stive and de Vriend, 1995; Stive and Wang, 2003; Wang et al., 2010; Cronin et al., 2009; Cronin, 2010; Dan et al., 2011; Ranasinghe et al., 2013). The approach emphasizes the limitations of concentrating on the short-term characterization of coastal systems and the need to incorporate feedbacks and inheritance from mesoscale controls to macroscale coastal boundary controls, such as SLCs. Operation of these longer-term factors will continually alter the projections of coastal-system outcomes provided by the numerical process models used to establish coastal functioning, such as those of Delft 3-D, Deltares; MIKE 21, Danish Hydraulics Institute (e.g., Stive et al., 1991; Stive and de Vriend, 1995; Stive, 2004, 2006; Cronin, 2010; O'Shea and Murphy, 2013; Devoy, in press). Although varying modeling approaches have been used to help resolve these scale and feedback issues, the problems remain, particularly in the inclusion of SLC factors (IPCC WGI, 2007, 2014).

8.2.2 Shorter-Term SLCs and Related Drivers to Coastal Vulnerability

Assessments and numerical modeling of physical coastal-system changes under climate warming indicate the critical role of SLR as a forcing control (Smith et al., 2000; Church et al., 2014). For cohesionless, "soft" sedimentary coasts (e.g., sand and cobble-sized beachs and dune-barrier systems) the Bruun Rule relationship of SLR as a "proportional" driver in the landward retreat of shorelines, will account for 25–50% of likely coastal changes (Ranasinghe et al., 2013; Nicholls et al., 2007; Devoy, 2008; Church et al., 2010). The remaining 50% of coastal "retreat" will be influenced by climate-driven rainfall and river-discharge factors, with iterative repercussions in linked coastal process operations. In this, SLR will be important in contributing to feedbacks in coastal-sediment flux and particularly in the sedimentary infilling of coastal-accommodation spaces (e.g., Cowell and Thom, 1994; Ranasinghe et al., 2013; de Groot et al., 2012). These types of responses to SLRs are connected closely to the recognized impacts of people in controlling the pathways and volumes of material mass (i.e., sediments and water) moving to coasts from land sources. The storage of water in reservoirs and other artificial water basins accounts for 3% of potential future SLR (Church et al., 2014). The impacts of this disruption to drainage systems, including major modifications of drainage basins, from vegetation clearances, changed land uses, and sediment retention, have had widespread significance for the rates of observed coastal changes (Bindoff et al., 2007; Nicholls et al., 2007). Where river and sediment flows have been interrupted, estuaries, deltas, and other coastal systems have become sediment starved. This has caused feedbacks in with consequent feedbacks in the effectiveness of SLR in driving increased rates of coastal erosion and the observed landward retreat of coastal wetlands—ecosystems. Examples of this phenomenon are common, from the Mississippi River, where this was in part a cause of the 2004 Hurricane Katrina disaster in New Orleans, to the Ganges–Brahmaputra river catchments and the large delta systems of Southeast Asia (Saito et al., 2007). Similarly, the impact of large-scale groundwater pumping, for coastal megacities and in estuary–delta zones, has speeded land subsidence in these areas since the 1950s and added to the acceleration of SLRs (Smith et al., 2000; Nicholls et al., 2007; Church et al., 2010; Church et al., 2014).

8.3 HUMAN LINKS AND DRIVERS: IMPACTS OF SLCs ON PEOPLE

Understanding the components and causes of SLCs has its intrinsic value as part of science and the study of Earth Systems' functions (Smith et al., 2000; Schwartz, 2006). Yet the primary significance of SLCs is in their likely impacts on the risks and vulnerabilities for people living in the CZ and linked marine environments (e.g., ISOS, 1991; Barthel et al., 1999; de Groot and Orford, 2000; Lim et al., 2005; Adger et al., 2005; Oft and Tsuma, 2006; Olsson et al., 2004, 2007; IPCC WGII SPM, 2007; Alcamo et al., 2007; Devoy, 1992, 2008; Nicholls et al., 2007, 2011; Lange et al., 2010; Marchand, 2009; Nicholls, 2010; European Commission, 2011; Gray et al., 2014; Wong and Losada, 2014; CoastAdapt, 2014; Cooper et al., 2014). SLCs have long influenced where and how people live and interact in the CZ (Cooper and Cummins, 2009). Numerous examples exist (e.g., North Sea, The Netherlands, Black Sea and Mediterranean Sea margins, Nile, northeast Arabian Sea—Indus delta) of drowned land surfaces and of forced changes in human and biotic system's living spaces and habitats following the postglacial SLR (e.g., Jelgersma, 1966; Roeleveld, 1974; Jelgersma et al., 1979). Migrations of people, plants, and animals have also been conditioned by SLCs in the operation of landbridges (e.g., Fairbridge, 1961; Kurtén, 1968; Devoy, 1985, 1995; Edwards and Brooks, 2008). The causes of SLC impacts range from the immediate effects of RSLs and linked coastal process operations, for example, tsunami, storms, earthquake, and volcanicity-induced land movements (Figure 8.8), to those of longer-term SLCs from deglaciation and the consequent return of water to the oceans, with marine inundation and earth crustal land uplift or subsidence (e.g., Pirazzoli, 1996; Sabadini et al., 1991; Peltier, 1998; Lambeck, 1995, 2001, Lambeck et al., 1996, 1998, 2012). Expected RSL rise of >1 m by 2100 through climate warming, and to higher levels by 2300, will cause increasing rates of SLR (4–5 mm/year by 2030) and the onshore movement of coastal systems (Bindoff et al., 2007; Devoy, in press; European Space Agency, 2014; Church et al., 2014; Rahmstorf, 2010, 2012; Rahmstorf et al., 2007, 2012a; Rahmstorf and Coumou, 2011; Royal Society & US National Academy of Sciences, 2014).

FIGURE 8.8 (a) Impacts of the Durban tsunami, 2007 on coastal structures *(Photograph Source: Denis Linehan, Department of Geography, UCC.)*. (b) Storm damage in January 2014 on coastal road and infrastructure at Rossbehy, southwest Ireland *(Source: Valerie O'Sullivan, Killarney, Ireland.)*.

Currently, 25–30% of the world's population lives "at the coast" (±1 m of msl), with this figure estimated to rise to 50% by 2100 (Nicholls et al., 2007, 2011; Wong and Losada, 2014). People and coastal systems will be forced into diminishing land spaces, as controlled by the continental land margin gradient, with all the potential effects of "coastal squeeze" established in the coastal management literature (e.g., Barthel et al., 1999; de Groot and Orford, 2000; Nicholls et al., 2007; Cooper and Cummins, 2009, 2014, Cooper and Jackson, 2012; Cooper and Pilkey, 2012). In this future onshore movement of coastal systems, 20–30% of the world's coastal wetlands will be initially "lost" through erosion, or inundated by 2050 (Church et al., 2010; Church et al., 2014; Royal Society & US National Academy of Sciences, 2014). Together with the physical coastal systems, these wetlands and linked biotic environments will have to adjust within the new spaces developed under SLRs in the current back-CZ areas. During this process, both people and biological systems will be stressed and vulnerable to the impacts of the increasing magnitudes, and possibly frequencies, of coastal events, such as storm surges, sediment movements, and erosion (Figure 8.8) (Sanchez-Arcilla et al., 2000; Lozano et al., 2004; Lowe et al., 2009; Jenkins et al., 2009; Church et al., 2014). In the case of storms, although such events are themselves causes of temporary increases in sea surface levels, the effects of storminess will be increasingly magnified on coasts with the immediate background drivers in SLR into the twenty-first Century under climate warming and have an intimate linkage as such to SLCs per se (Wong and Losada, 2014). More widely, the effects of SLRs with other coastal process functions have had significant impacts on the world's coasts. The effects of El Niño-driven coastal erosion for the western USA, eastern Australia, and central Pacific islands have been noted earlier (Section 8.2). In southeast Asia, the effects of direct SLRs, coupled with sediment starvation on coasts from river catchment changes, are having a profound impact on the major river and delta systems of the region (Saito et al., 2007; Hijioka et al., 2014). Globally, large delta and estuaries are similarly recording the impacts of SLRs magnified by the contemporary increased effects of sediment starvation, land subsidence through groundwater abstraction, urban developments, and linked changes in wetland extent and functioning (Nicholls et al., 2007; Wong and Losada, 2014).

Consequently, radical adjustments to these forcings will have to be made through mitigation and adaptation measures (see Nicholls, this volume). This will continue to require active political decision making and governance, in the development of Coastal Zone Management, linked Marine Spatial Planning policies and legislation, and, importantly, in economic-business-industry and lifestyle responses (Cooper and Cummins, 2009; European Commission, 2011). For coastal communities, awareness of and responses to these issues of SLRs and linked coastal changes are often difficult and complicated by a lack of technical knowledge of integrated coastal-system functioning and of the wider science. The extensive research literature now established shows the need for coastal inhabitants, and societies as a whole, to build capacity to respond to future SLRs and coastal changes. This may mean abandoning the traditional approaches of built, shoreline-protection measures (Cooper and Pilkey, 2012) and adopting increasingly soft-engineering techniques in defense strategies and of hazard zoning, together with structured shoreline retreat of the most vulnerable coasts, as in the large delta areas, such as the Mississippi River. Equally, coastal dwellers will need to engage with the realities of where future coasts will be prone to SLR and learn new techniques for adaptation (e.g., Titus et al., 2010; Cooper and Cummins, 2009; Lim et al., 2005; European Commission, 2011, 2013; MCCIP, 2012; Gray et al., 2014; Cooper et al., 2014; Devoy, in press; Wong and Losada, 2014).

8.4 SEA-LEVEL PATTERNS, TRENDS, AND MODELS

Sea-level patterns and trends are a complex subject, but one that underpins the development and reliability of the modeled projections of future SLRs, for example, from natural irradiative forcings of Earth climate, as in Croll–Millankovitch planetry change factors in climate functioning, to anthropogenically induced atmosphere warming and scenarios for change (e.g., IPCC WGIII SRES, 2000; IPCC WGI, 2007, 2014; IPCC WGI SPM, 2014). It is not the intention here to present the detail of the different models used in establishing the patterns and trends in SLCs at the macroscales to microscales, but to provide a brief guide to the different approaches.

The models used in SLCs and related studies range from conceptual–behavioral types to those of varying numerical form (Woldenberg, 1985; Kirkby et al., 1993; Raper, 1993; Raper et al., 2000; India and Bonillo, 2001; Allen, 1997, 2003; Allen and Pye, 2009; Marchand, 2009; IPCC WGI, 2007, 2014). Numerical models include empirical and deterministic, quasi-deterministic types, based upon mathematical–statistical interpretations of data, for example, sea-level index points (van de Plassche, 1986), which assume linear relationships, to stochastic and linked simulation modeling (e.g., de Vriend, 1991; Schlesinger, 1993; Cowell and Thom, 1994; Raper, 1993; Raper et al., 2000; Stive and Wang, 2003; Wang et al., 2010; Dan et al., 2011; Rahmstorf, 2007, 2012; Tebaldi et al., 2012; Meehl et al., 2007; IPCC WGI, 2014). Numerical modeling and simulation techniques are the most relevant in developing projections of SLR under climate warming scenarios. As with all models, these often suffer from a lack quality, environmental-boundary data as inputs for model forcings, and validation. Also, in the direct recording of SLCs, good data may be lacking in defining the different contributors to SLRs, and particularly in helping understand the feedbacks involved and in the validation of model outputs (Raper, 1993; Raper et al., 2000; Zerbini et al., 2000; Lowe et al., 2009; IPCC WGI, 2007, 2014; Church et al., 2010, 2013). However, the quality of these complex models has improved steadily, particularly with the introduction of "ensemble"

Atmosphere and Ocean General Circulation Models (AOGCMs) and the wider use of Regional Circulation Models (RCMs). This modeling and better knowledge of the constraints on systems contributing to SLC, for example, of glacier mass balance and ice melt and ocean mixing, have provided greater confidence in model results (IPCC WGI, 2014). Many would argue that numerical model projections should still be used only as a guide to possible environmental system outcomes (e.g., of SLR), rather than certainties (Cooper and Pilkey, 2007, 2008).

8.4.1 Macroscale to Meso-scale Changes

At the marcro- and meso-scales, the patterns and trends of SLCs, as evidenced in sedimentary and other Earth proxy records, are controlled by domain-dominant, environmental-forcing controls (Devoy, 1982; Carter and Woodroffe, 1994; Jennings et al., 1998). The long-term (10^{6-8} years) SLC drivers of Earth tectonics, such as ocean ridge and plate growth decay, mountain building, ocean-sediment accumulation, and geoid changes together, are shown in Figure 8.7 and Table 8.1 (Devoy, 1987; Ota et al., 1992; Woodroffe, 2002). They are joined by changes in the total volume of water available in the oceans; through additions of juvenile water via volcanicity and other Earth exogenic processes, especially in the early- to mid-Phanerozoic, coupled with water storage and exchanges with epicontinental seas and ocean basins. The formation of the Mediterranean and the Messinian salinity crisis forms a good example of this (e.g., Open University, 1998; Muller et al., 2008).

Understanding long-term SLC patterns (10^{6-8}–10^4 years), as measured against the land "freeboard" margins (Schubert and Reymer, 1985), have been based on seismic and sea-bed drilling data from the continental shelves (e.g., Vail et al., 1977; Hallam, 1981; Haq, 1991; Haq and Schutter, 2008). These have generated "signature" logs of sedimentary cyclical coastal on-laps and off-laps (Figure 8.1(a)). The links of these changes to RSLs, and to their causative geophysical and Earth crustal movements, together with other large-scale environmental drivers, have been developed into conceptual–behavioral models and shown in deterministic numerical expressions. These provide primarily first-order projections of SLCs over geological timescales (Pitman, 1978; Devoy, 1987; Cronin, 1999).

Similar approaches have developed in the study of the Quaternary (Figure 8.1(b)) and particularly in the Holocene SLC. Sedimentary and other data sets of inferred sea-level positions (index points) have been used to develop time–depth sea-level curves (Figure 8.5(c)) (Bloom, 1977; van de Plassche, 1986; Pirazzoli, 1991, 1996). Statistical analyses of these data (Devoy, 1982, 1987; Shennan and Tooley, 1987; Shennan et al., 1983, 2002; Shennan and Andrews, 2000) show that empirical, deterministic models provide only a first-order and generally weak explanation of sea-level records, because of the complexity of controls involved in developing the observed

SLC patterns. Stochastic and environmental feedback mechanisms are shown as becoming increasingly important during the Holocene following reduction in the initial rapid rises of RSLs (Bard et al., 1990, 1996) and in generating local to regional patterns of SLCs (Jennings et al., 1998; Roy et al., 1994; Allen, 1997, 2003; Allen and Pye, 2009). A significant problem in using these "time–depth" data are that they invariably represent RSLs, often with significant error margins (van de Plassche, 1986). Analytical approaches have been developed (e.g., Jelgersma, 1966; Jelgersma et al., 1979; Andrews, 1970; Tooley, 1978; Mörner, 1980', 1987a,b; Shennan et al., 1983; Shennan and Tooley, 1987; Carter, 1992) to address and help reduce these limitations. This has resulted in attempts at isolating the tectonic, crustal isostatic (uplift and subsidence), and "ocean water volume" components involved (Figure 8.7), thereby generating more reliable explanations of the sea-level data as actual SLC and the reconstruction of shoreline positions. Andrews' (1970), and Mörner's (1971, 1976, 1980, 1987a,b) related conceptual–empirical approaches form early examples of this, expressed generically as:

$$\mathrm{U}t = \mathrm{R}t - \mathrm{E}t \tag{8.1}$$

where Ut is the isostatic uplift over time, R is the position of RSL at a specific time t as measured in reference to present sea level, and Et is the position of eustatic sea level over time (after Devoy (1987) and Tooley (1993)).

A related method, which introduced a new approach in isolating "sea-level tendencies," challenged the use of the height component of proxy sea-level data, with their many error problems (Figure 8.5(c)). These included those of measurement errors, the use of inhomogeneous data sets, the compaction and consolidation of cored sedimentary data, the different possible water-level meanings of index points and paleotidal changes at different sites (Shennan, 1982; Hennan & Tooley, 1987; Long et al., 1998; Zong and Tooley, 1996; Shennan and Andrews, 2000; Shennan et al., 2002). In this approach, only the chronological elements of sea-level data are used, emphasizing the dependence on the number of index point dates available and their precision (e.g., of radiometric and other "absolute" dating evidence). Analyses provide a basis for the numerical study and modeling of the patterns of Holocene SLCs at regional scales, but again, not for the projection of future SLCs:

$$\mathrm{E} = 0.5 \sin(\mathrm{T}\pi/500) \quad \text{if } \mathrm{T} < 300 \tag{8.2}$$

$$\mathrm{E} = 0.5 \sin(\mathrm{T}\pi/500) + 50(\cos((\mathrm{T} - 3100)\pi/2000) - 1) \quad \text{if } \mathrm{T} > 3100 \tag{8.3}$$

where E is regional eustasy for time T (after Shennan, 1987).

Further, this methodology is dependent on a number of boundary assumptions, particularly of former tidal regimes and vertical land-level movements (Devoy, 1987; Tooley, 1993). This has led to integration in subsequent analyses and modeling of Holocene sea-level data with independent models of

Earth crustal behavior and tides at regional scales (e.g., Lambeck, 1991, 2001; Lambeck et al., 1996, 1998; Shennan et al., 2002, 2006; Edwards and Brooks, 2008; Lowe et al., 2009).

Understanding the "land movement" components of SLCs and RSLs at varying mesotime and spatial scales has been based upon the geophysical modeling of crustal rheology and behavior (e.g., Walcott, 1972; Clark et al., 1978; Mörner, 1980; Lambeck et al., 1996; Lambeck, 2001; Peltier, 1998, 2004; Brooks et al., 2007; Bradley et al., 2011). The models have been used widely in paleoshoreline and coastal map reconstructions, primarily for the Holocene. However, many of these approaches use sea-level index points in model validation and in tuning the appropriate rheological characteristics. Consequently, the models also suffer to different degrees from built-in error problems, in terms of the of height and time accuracy of the coastal reconstructions (Devoy, 1991; Zerbini et al., 2000). Issues of regional variations inherent in Earth rheology and crustal response time to uncertainties in the calculations of former ice and water loadings also exist (i.e., ice volumes and rates of change) and may be difficult to constrain (e.g., Peltier, 1998; Lambeck, 2001; Ehlers and Gibbard, 2004; Ehlers et al., 2011; Benn and Evans, 2013; BRITICE-CHRONO, 2014). The primary terms of these geophysical models for past SLCs can be expressed most simply as:

$$\begin{aligned}\Delta\zeta(\varphi,\lambda:t) &= \zeta(\varphi,\lambda:t) - \zeta(\varphi,\lambda:t_0)\\ &= \Delta\zeta_e(t) + \Delta\zeta_r(\varphi,\lambda:t) + \Delta\zeta_i(\varphi,\lambda:t) + \Delta\zeta_W(\varphi,\lambda:t)\end{aligned} \quad (8.4)$$

where the positions of past sea levels at latitude φ, longitude λ, and time t are calculated as the sum of four terms in relation to present day sea level (t_0), though the operation of vertical tectonics is not accounted for. The first term in Eqn (8.4) is for the equivalent SLR, defined as the change in ocean volume/(divided by) ocean surface area; the second term is the SLC for the gravitational potential from changed ice and water loadings; the third is the rebound from changes in ice loadings, and includes further gravity effects, and the last term is the Earth response to sea floor loadings by melt water addition or losses as water uptake to ice masses. Where a number of ice sheets contributes to the SLC in time, as is likely, Eqn (8.4) can be expressed as:

$$\Delta\zeta = \Delta\zeta_e + \sum_{n=1}^{N}\left\{\Delta\zeta_r^{(n)}\left(\varphi,\lambda:t\right) + \Delta\zeta_I^{(n)}\left(\varphi,\lambda:t\right) + \Delta\zeta_W^{(n)}\left(\varphi,\lambda:t\right)\right\} \quad (8.5)$$

where $\Delta\zeta_e$ is the equivalent SLC for the combined ice sheet contributions. The other terms represent the components of the position of the site at which SLCs are being calculated relative to a particular ice mass (from Lambeck (1991)).

In spite of the problems in defining time-based changes in the model terms, these geophysical models have been essential in helping resolve crustal responses to ice, water, and sediment loadings and redistributions during glacial–deglacial hemicycles as part of SLC (e.g., IGCP projects 274, 367, 437 and Devoy, 1991). These form the isostatically driven uplift and subsidence components in SLC, including the operation and significance of different regional signatures of ice-margin and far-field crustal forebulge movements (Sabadini et al., 1991; Devoy, 1991, 1995; Peltier, 1998; Lambeck, 1991, 1995, 2001; Lambeck et al., 1996, 1998; Carter, 1992; Brooks et al., 2007; Edwards and Brooks, 2008; Bradley et al., 2011). The different crustal loadings can also affect changes in planetary rotation and "wobble," with feedbacks to the geoid and potentially to the timings and scales of SLCs (Clark and Lingle, 1977; Clark and Primus, 1987; Peltier, 1998; Milne and Mitrovica, 1998; Mitrovica et al., 2009, 2011). Importantly, the geophysical models established the key first order SLC "finger-prints" with ice-melt patterns for world coasts at regional scales (Figure 8.9) (e.g., Clark et al., 1978). Linked to this work comes an increased understanding and quantification of changes in the geoid. Model results provide a valuable basis and analogs for simulating the likely impacts of Greenland and Antarctic ice deglaciation with future climate warming (Milne and Mitrovica, 1998; Mitrovica et al., 2009, 2011; Milne et al., 2009; Church et al., 2014). Changes in the mass of the present major ice sheets will cause gravitational "relaxation/collapse" of the sea surfaces at their margins and related far-field consequences in SLR. For the North Atlantic, this mechanism alone may result in as much as a 0.19 m SLR (Raper, 1993, 2000; IPCC WGI, 2007, 2014).

8.4.2 Meso-scale to Micro-scale Changes

Meso-scale SLCs are determined primarily by factors of tectonic and glacial eustasy (Figure 8.7 and Table 8.1). At this level of resolution, "oscillatory," as cyclical and quasiharmonic type sea-level movements, are evident and to be expected, particularly as correlated with the Quaternary glacial and interglacial stages (Figure 8.1(b)) (e.g., Devoy, 2005; Roe et al., 2009; Murray-Wallace and Woodroffe, 2014). Superimposed on these glacioeustatic trends are shorter-term sea-level movements, driven by meteorological and other dynamic sea surface factors, together with the feedbacks from longer-term controls (Figure 8.4(b)). These are rarely preserved clearly in the geological proxy records of SLC, except perhaps as sea-level pattern variabilities, or "pattern noise" (e.g., Devoy, 1982; Allen, 1997, 2003; Delaney et al., 2012). At these scales, the question arises as to whether these shorter term drivers really do result also in higher resolution "oscillatory" patterns in sea-level behavior. This question was equally part of the causes, form and nature of Holocene sea-levels "debate" in the early IGCP Sea Levels Projects 61 and 274 (Figure 8.5) (Bloom, 1977; Greensmith and Tooley, 1982; Tooley, 1993; Pirazzoli, 1996; Zerbini et al., 2000; Church

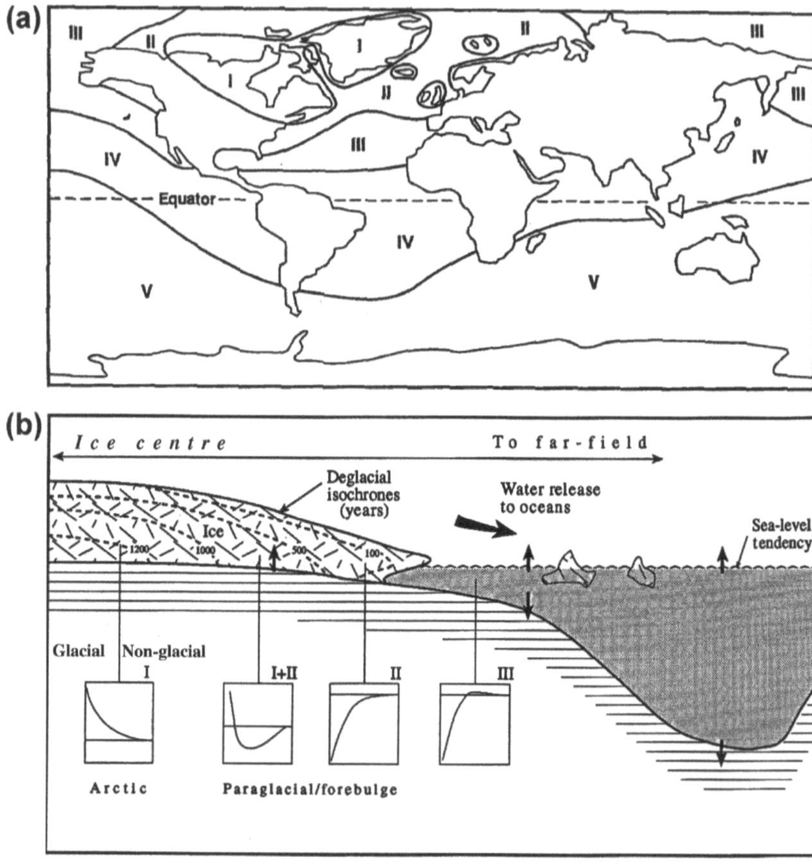

FIGURE 8.9 (a) Modeled major world zones of RSL change, resulting from changes in Earth crustal loading by primarily ice, water, and sediments, linked to Earth geophysical functioning. (b) Near (as former ice center, following Deglaciation), to Far-field principal RSL change signatures ("fingerprints") from Earth crustal modeling predicted different zones of RSL changes, shown in Figure 8.9(a) *(Source: from Carter (1992), with permission from the Quaternary Research Associations, UK).*

et al., 2010). Alternatively, for the meso-scales, the concept of nonoscillatory and "smooth," negative exponential decay in SLRs at local–regional scales was advanced (Jelgersma, 1961, 1966; Shennan and Tooley, 1987). Much has hinged on the sea-level data sets used and the analytical approaches and interpretations taken. Establishment of the principle that no common global sea level exists (van de Plassche, 1986; Pirazzoli, 1996), coupled with more rigorous and numerical approaches to studying the data sets (e.g., Devoy, 1987; Shennan and Tooley, 1987; Allen, 1997, 2003), suggests that the "smooth" pattern in Holocene and other interglacial sea-level recoveries is likely to be an artifact of the data sets used and the study location.

More realistically, current knowledge of atmosphere–ocean–coastal systems' functioning and evidence of changes in the rates of SLR in the nineteenth to twentieth centuries suggests that "oscillations" in sea-level patterns at local–regional scales (e.g., Douglas et al., 2001; Bindoff et al., 2007; Nicholls et al., 2007; Milne et al., 2009; Church et al., 2014) are likely, driven by a wide range of short-term sea-surface change factors, such as salinity–density; temperature–steric effects, local currents, ENSO, PDO, and NAO. What is fundamental in developing an understanding of the likely variability of short-term SLC, is the need for extensive, high-quality observational, non-proxy sea- and land-level measurements, for example, through interlinked land-based tide gauge, ocean wave buoys, gravimetry, and global positioning system (GPS) with satellite altimetry airbourne light detection and ranging (LIDAR) techniques (Devoy et al., 2000; Zerbini et al., 2000; Kandrot, 2013). The deployment of such systems since the 1990s has improved this situation greatly, removing the former tide gauge observational data bias from the northern hemisphere (viz. GLOSS, 2014; Woodworth et al., 2002, 2009a,b; Woodworth and Blackman, 2004; IPCC WGI, 2014). Satellite altimetric and related system measurements currently show a global mean SLR of 3.4 mm/year ± 0.2 mm (Church et al., 2010; Church et al., 2014). This represents an increase on the late twentieth Century mean rates of approximately 1.8 mm/year, though this value has itself varied since the nineteenth Century (Church et al., 1991; Church and White, 2006, 2011; Church et al., 2010) and brings into sharp focus the importance of high-resolution studies of contemporary SLC as a basis for the validation and tuning of appropriate forecasting/projection techniques for future SLCs.

8.4.3 Modeling the Future

The projections of SLCs at all scales are approached through the use of numerical models, requiring high computing power (IPCC WGI, 2007, 2014). Two main approaches are used: (1) physical process-based General Circulation Models (GCM) and RCMs and (2) semiempirical types of models. Both approaches use scenarios of future temperature rises to produce projected outcomes of SLRs (e.g., Schlesinger, 1993; IPCC–SRES, 2000; Raper, 1993; Raper et al., 2000; Church et al., 2014; IPCC WGI, 2001, 2007, 2014; Rahamstorf and Ganopolski, 1999, Rahmstorf et al., 2012a,b).

Physical-process models have been developed as Atmosphere–Ocean General Circulation Models (AOGCM), which numerically integrate the vertical and horizontal exchanges of mass and energy in these environmental systems, using a variable number of vertical thickness layers (multilayered series approach) through the atmosphere and ocean (Church et al., 1991, 2010; Raper, 1993; Raper et al., 2000). The system boundaries are forced by scenarios of changes in the sun's energy inputs (irradiance), volcanic eruption, and Earth heat flows, together with the gas composition of the atmosphere, particularly now as driven by anthropogenic changes in greenhouse gas (GHG) constituents. Formerly, these models were developed by different research groups and run independently (IPCC WGI, 2001; Church et al., 1991, 2001; IPCC WGI, 2007). Now "ensemble" modeling is undertaken, with 21 models currently interlinked and run under the Coupled Model Intercomparison Project Phase 5 (CMIP5) (IPCC WGI, 2014), providing cross-checks and giving greater homogeneity in spatial and temporal model outputs. For higher resolution needs at regional scales, downscaling techniques are applied based on the AOGCM outputs (e.g., Lozano et al., 2004; von Storch et al., 2008). Essential dynamic elements incorporated within these models include changes in ocean density and circulation, as influenced by sea surface winds, flows of air-sea heat and freshwater inputs; Atlantic Meridional Overturning Circulation; and the generation of internal ocean system states causing climate variability (e.g., operation of ENSO, PDO, and NAO).

For SLCs, the outputs from these models, run under different atmosphere–ocean simulation scenarios (Special Report on Emissions Scenarios (SRES), Representative Carbon Pathways (RCPs) (Church et al., 2001; IPCC WGI, 2001, 2007, 2014; Church et al., 2014), primarily provide the changes in the thermal signal driving SLR (i.e., ocean steric effects). Critically, the results from a number of other separate system numerical models are also needed to provide the remaining contributors to future SLRs (Figures 8.7, 8.10, and Table 8.1). Projections of water inputs from ice melt over time are made using Surface Mass Balance (SMB) and ice-dynamics models, again run under the range of climate scenarios (i.e., the different glacier and ice-sheet melting rates) (Braithwaite and Olesen, 1989; Raper, 1993; Raper et al., 2000; Raper and Braithwaite, 2005; Schoof., 2007, 2011; Fausto et al., 2009; IPCC WGI, 2007, 2014). Geophysical models provide quantification of the effects on SLCs and RSLs of crustal loading changes from the redistributions occurring in the remaining ice-cover, returned melt-water, and sediment exchanges (e.g., Lambeck, 1991, 1995, 1998, 2001; Lambeck et al., 1996, 1998; Peltier, 1998, 2004; Clarke et al., 2005; Mitrovika et al., 2009; Tamisiea et al., 2010; Tamisiea, 2011; Riva et al., 2010). Information on wave climate and storm surge impacts, driven primarily by wind stress and atmospheric pressure changes, are provided by dynamic process models, run at higher resolution regional to local scales (Lozano et al., 2004; Wang et al., 2010; Lowe et al., 2010). Simulations of contributions to SLRs from permafrost areas are not well constrained, but estimations are that only small ice–melt contributions from this source will occur before 2100, even with the higher RCP 8.6 scenario (>4.5 °C mean global warming) (IPCC WGI, 2007, 2014; von Storch et al., 2008; Rahmstorf, 2012; Rahmstorf et al., 2012a,b; Schlesinger, 1993; Meehl et al., 2012). Further, until recently, accurate information on small glacier and ice-sheet melt rates, SMB changes and data on the mixing and heat diffusion characteristics of ocean water layers have also been seen as inadequate (IPCC WGI, 2007, 2014). Consequently, the upper-end scenarios, for example, A1F1,

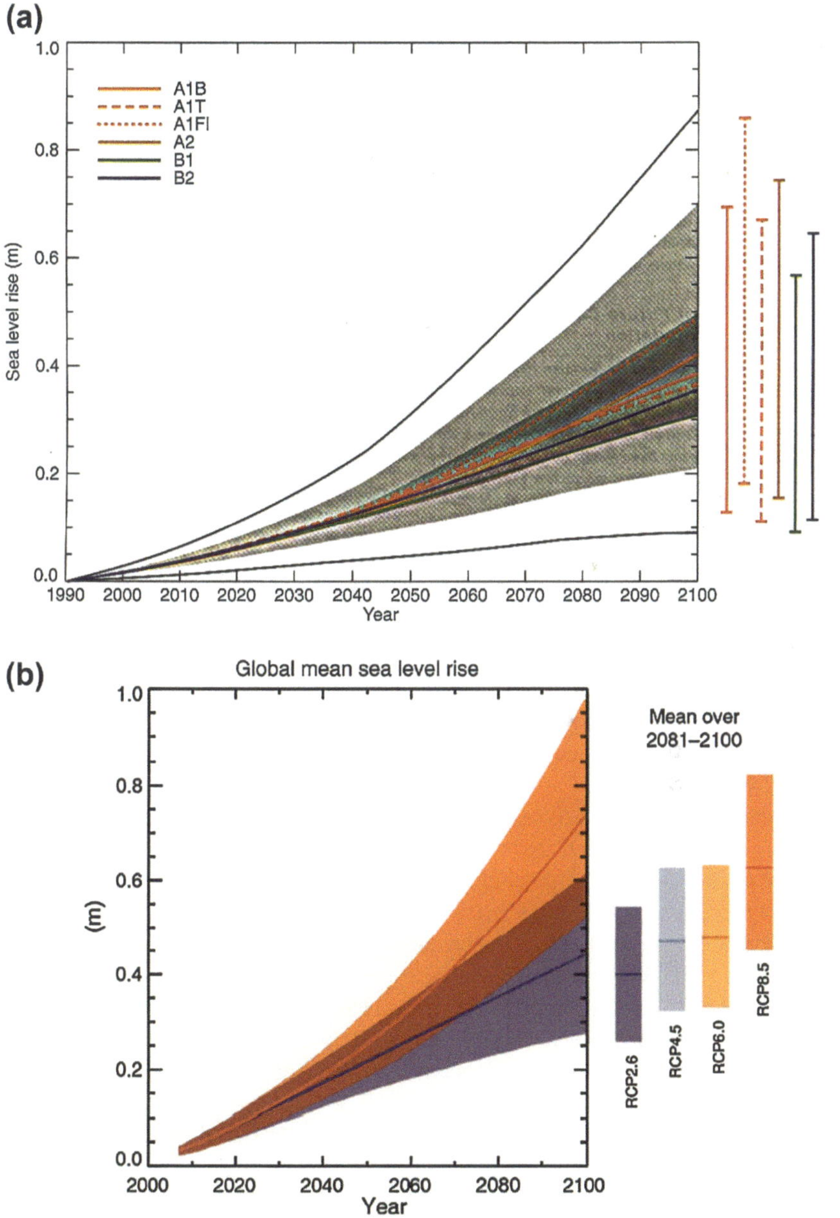

FIGURE 8.10 (a) Process AOGCM projections from the IPCC AR3 of gmsl rise from 1990 to 2100. The colored lines represent the average of the AOGCMs for each of six SRES scenarios. The central area of dark shading showing the AOGCM range of the average gmsl for all 35 SRES scenarios and the light shading shows the gmsl range of all AOGCMs for all scenarios *(Source: IPCC WGI, 2001; IPCC, 2001a,b.)*. (b) IPCC AR5 AOGCM projections of gmsl to 2100 relative to 1986–2005 values using the CMIP5 AOGCMs ensemble. The colored shaded bands show the likely range of gmsl for the RCP 2.6 and RCP 8.5 scenarios *(Source: IPCC AR WGI (2014); IPCC WGI SPM (2014).)*.

A2 (IPCC SRES, 2000) of global mean SLR projections from the IPCC AR4 (2007) of 0.6 m by 2100, were seen as low and not matching observations (Rahmstorf, 2007, 2012; Church et al., 2014; IPCC WGI, 2014) (Figure 8.10(a)). New data on small glacier losses and on Greenland and Antarctic ice-sheet change since 2007, coupled with the CMIP5 ensemble modeling approach, provide it is argued a better fit with current rates of SLRs and information on ice-sheet SMB losses. The IPCC AR5 projections give a maximal global mean SLC of 0.9–1.0 m by 2100 for the high RCP 8.5 scenario of 2.6–4.8 °C mean global warming (IPCC WGI, 2014; Church et al., 2014) (Figure 8.10(b)).

The semiempirical sea-level model approach is simpler than that of dynamic process models (e.g., Schlessinger, 1993; Rahmstorf, 2007, 2010, 2012; Rahmstorf et al., 2012b; Vermeer and Rahmstorf, 2009; Tibaldi et al., 2012; IPCC WGI, 2014). It uses the principle that sea level rises faster with increasing coupled atmosphere–ocean temperature rise and that *sea-level sensitivity* provides a measure (coefficient (Eqn. 8.6)) of the rate of acceleration of SLRs for incremental global temperature changes. The unknown parameters in this model are determined through tuning with observational data (Rahmstorf, 2007, 2012; Rahmstorf et al., 2012b). The method is summarized in the following generic expression:

$$dH/dt = a(T(t) - T_0) \tag{8.6}$$

where H is the sea level, T is the global temperature and T_0 represents a "baseline temperature at which sea level is stable" (after Rahmstorf (2012)).

Scenarios and modeled simulations of future global temperature changes (e.g., IPCC WGI, 2007, 2014) can then be used to produce projections of future SLRs. Results of this semiempirical approach appear to fit well with proxy records for SLR rates from the recent past (e.g., Kemp et al., 2011). However, the projections are necessarily a simplification and although improvements in the approach have been proposed (Vermeer and Rahmstorf, 2009), by limiting the timescale of operation and adding in definition of the short-term response of sea level, the models do not take account of the complex feedback links that occur and that are ideally better represented in process modeling (Church et al., 2014). Projections for SLRs from semiempirical models often exceed those of the process-based techniques, by as much as a factor of x2-3, though in the IPCC AR4 projections, this may in part result from the lack of knowledge and inclusion of ice melt rates and feedbacks (Rahmstorf, 2010; Rahmstorf and Ganopolski, 1999; Church et al., 2014). Overall, a significant weakness in this approach is that it is uncertain that the boundary relationships used to tune the models, linked to the past rates of SLR, will continue for the future. Consequently, the approach is likely to remain more as an approximation for future trends in SLCs (Rahmstorf, 2012; Rahmstorf et al., 2012b).

8.5 CONCLUSIONS, GLOBAL WARMING, AND FUTURE SEA LEVELS

The fact that SLCs have occurred since oceans first formed is well documented in proxy environmental records of varying types (Devoy, 1987; Haq and Schutter, 2008; Murray-Wallace and Woodroffe, 2014). In the long-term, these SLCs, as RSL changes, have created the sedimentary on-lap and off-lap cycles found on passive continental Shelf margins worldwide (Vail et al., 1977; Cronin, 1999). These represent RSL changes against continental free boards of the scale of 100–300 m over timescales of 10^{6-7} years, formed through primarily tectonic and linked ocean shape changing controls. The economic significance of these patterns of environmental change has been the creation of extensive oil, gas and coal deposits under these alternating marine transgressive and regressive marine sedimentary changes. At shorter timescales of 10^{5-6} years, glacial forcings, most recently in the Quaternary (Ehlers et al., 2011), have resulted in drops in RSL of −120 m gmsl, with the postglacial SLR reaching present day levels by 2,000–3,000 BP. At even shorter timescales, the contemporary dynamic sea surface movement, including those from storms, daily wave actions and currents represent the drivers to observable coastal process operations.

In the context of the formation and responsiveness of different coastal systems to changes (Carter and Devoy, 1987; Woodroffe, 2002; Devoy, in press; Cooper and Jackson, 2014), then SLC might appear to be a "passive" control on coasts. However, past SLC records show that rapid accelerations of the rates of SLC can occur as part of integrated Earth system functioning (Bard et al., 1990, 1996). Thus, SLCs can form an "active" driver to fast changes in coastal domains (Cowell and Thom, 1994; Jennings et al., 1998). The significance of this knowledge is that it provides valuable analogs for understanding future alterations in the position, physical form, and likely behavior of the CZ (Cowell and Thom, 1994; Masselink and Gehrels, 2014). Current awareness that the Earth is undergoing a period of major atmosphere–ocean warming, induced by the actions of people (IPCC WGI, 2014), can be linked confidently to future SLR, currently at global rates of >3 mm/year; about a third faster than in the mid–late twentieth Century. The magnitude of future SLR is uncertain. The fact that it will affect most world coastlines to some degree, particularly low-lying islands, "soft" sedimentary coasts, coral reefs, estuaries, and deltas is certain from the knowledge of the past. Locations, such as rock-cliffed coasts and areas of isostatic and tectonically driven uplift (Smith et al., 2000), will be less affected.

The key uncertainties for future SLCs at varying times and spatial scales center on how much global temperature change will occur (Wigley and Raper, 1993; Raper et al., 2000; IPCC, WGI, 2007). Current simulations of GHG forcings of mean surface temperature indicate a rise of 0.5–1.5 °C for the RCP 2.6 scenario and 4–5 °C for RCP 8.5 over 1986–2005 values by

2081–2100, with significantly higher northern polar temperatures and other hemispherical to local scale variations (IPCC WGI, 2014). For future SLCs, projected temperature changes will cause land-based ice masses to return their stores of water to the oceans (Figure 8.2(a)). Temperature alone is likely to account for 50% of the future SLCs, as thermal expansion (steric changes) of the oceans. The critical uncertainty is, again, how fast will these temperature-induced changes occur in the different ice systems and will some of these even continue to gain mass for a time? The earlier IPCC AR4 projections suggested a slow rate of melt response for Greenland and some ice-mass gains for areas of Antarctica (IPCC WGI, 2007). Subsequent observations show a continued rapid melt of small glaciers, a faster retreat for glaciers, and ice sheet downwasting in Greenland, with a similar pattern now emerging for West Antarctica and possibly Eastern Antarctica (ESA, 2014; IPCC WGI, 2007, 2014).

Dependent on coastal distance from the main areas of land ice melting, some delays in SLR responses at local to regional scales may be expected. However, the eventual outcome of continued land ice melting will be gmsl rises of 0.25 m (RCP 2.6) to 0.95 m (RCP 8.5) by 2100 (Figure 8.10(b)) at rates of >4–5 mm/year by 2050. This "elevator style" SLR against coasts will carry with it the other repercussions of ocean–atmosphere warming, in the increased impacts of storminess, coastal erosion, groundwater and sediment flux changes, wetland losses, and "coastal squeeze" on living space for biotic systems and people (Lozano et al., 2004; Kremer et al., 2013; Devoy, 2008; Cooper et al., 2014; Wong and Losada, 2014). How people will deal with these impacts of SLC remains a critical question, particularly in the context of a stressed world economy and the increasing human pressures on coastal living spaces and resources (Jackson, 2011; Eurpoean Comission, 2011, 2013). Approaches to help mitigate SLR focus upon the large scale, for example, of on-land water storages; in lakes, reservoirs, and groundwater; and in water transfer schemes (Nicholls et al., 2007; Wong and Losada, 2014). The value and advisability of building coastal defense structures, a traditional response to SLC in the "Canute syndrome," remains highly questionable (Gray et al., 2014; Cooper and Pilkey, 2012).

REFERENCES

Admiralty Tide Tables, 2014. Admiralty Tide Tables: Volume 1, European Waters. UK Hydrographic Office, Exeter.

Adger, N., Hughes, T.P., Folke, C., Carpenter, S.R., Rockstrom, J., 2005. Social-ecological resilience to coastal disasters. Science 309 (5737), 1036–1039.

Alcamo, J., Moreno, J.M., Nováky, B., Bindi, M., Corobov, R., Devoy, R.J.N., Giannakopoulos, C., Martin, E., Olesen, J.E., Shvidenko, A., 2007. Europe. In: Parry, M.L., Canziani, O.F., Palutikof, J.P., van der Linden, P.J., Hanson, C.E. (Eds.), Climate Change 2007: Impacts, Adaptation and Vulnerability. Contribution of Working Group II to the Fourth Assessment Report of the Intergovernmental Panel on Climate Change. Cambridge University Press, Cambridge, UK, pp. 541–580.

Allen, J.R.L., 1997. Simulation models of saltmarsh morphodynamics: some implications for high-intertidal sediment couplets related to sea-level change. Sediment. Geol. 113, 211–223.

Allen, J.R.L., 2003. An eclectic morphostratigraphic model for the sedimentary response to Holocene sea-level rise in northwest Europe. Sediment. Geol. 161, 31–54.

Allen, J.R.L., Pye, K., 2009. Saltmarshes: Morphodynamics, Conservation, and Engineering Significance, second ed. Cambridge University Press, Cambridge.

Andrews, J.T., 1970. A Geomorphological Study of Postglacial Uplift with Particular Reference to Arctic Canada. Institute of British, Special Publication 2, Alden and Mowbray, Oxford.

Bard, E., Hamelin, B., Fairbanks, R.G., Zindler, A., 1990. Calibration of the timescale over the past 30,000 years using mass spectrometric U-Th ages from Barbados corals. Nature 345, 405–409.

Bard, E., Hamelin, B., Arnold, M., Montaggioni, L., Cabioch, G., Faure, G., Rougerie, F., 1996. Deglacial sea-level record from Tahiti corals and the timing of global meltwater discharge. Nature 382, 214–244.

Barth, M.C., Titus, J.G., 1984. Greenhouse Effect and Sea Level Rise. Van Nostrand Reinhold, New York.

Barthel, K.-G., Suranyi, M., Barth, H., et al., 1999. Conference Proceedings of the Third European Marine Science and Technology Conference Lisbon, 23–27 May 1998. European Commission, Luxembourg.

Bengtsson, L., Koumoutsaris, S., Bonnet, R.-M., Herland, E.-A., et al., 2012. The Earth's Cryosphere and Sea Level Change. Springer, Dordrecht.

Bindoff, N.L., Willebrand, J., Artale, V., et al., 2007. Observations: oceanic climate change and sea level. In: Solomon, S., Qin, D., Manning, M., Chen, Z., Marquis, M., Averyt, K.B., Tignor, M., Miller, H.L. (Eds.), Climate Change 2007: The Physical Science Basis. Contribution of Working Group I to the Fourth Assessment Report of the Intergovernmental Panel on Climate Change. Cambridge University Press, Cambridge.

Bradley, S.L., Milne, G.A., Shennan, I., Edwards, R., 2011. An improved glacial isostatic adjustment model for the British Isles. J. Quat. Sci. 26, 541–552.

Braithwaite, R.J., Olesen, O.B., 1989. Calculation of glacier ablation from air temperature, West Greenland. In: Oerlemans, J. (Ed.), Glacier Fluctuations and Climate Change. Kluwer Academic Publishers, Dordrecht, pp. 219–233.

Brooks, A.J., Bradley, S.L., Edwards, R.J., Milne, G.A., Horton, B., Shennan, I., 2007. Postglacial relative sea-level observations from Ireland and their role in glacial rebound modelling. J. Quat. Sci. 23, 175–192.

Benn, D.I., Evans, D.J.A., 2013. Glaciers and Glaciation, second ed. Routledge, Abingdon, UK.

Bloom, A., 1977. Atlas of sea-level curves: IGCP project 61. Cornell University, Ithaca, New York.

BRITICE-CHRONO Project, 2014. Constraining Rates and Style of Marine-Influenced Ice Sheet Decay. https://www.dur.ac.uk/geography/research/research_projects/?mode=project&id=661.

Carey, W.S., 1981. Causes of sea-level oscillations. Proc. R. Soc. Victoria 92, 13–17.

Carter, R.W.G., 1988. Coastal Environments: An Introduction to the Physical, Ecological and Cultural Systems of Coastlines. Academic Press, London.

Carter, R.W.G., 1992. Sea-level changes: past, present and future. Quat. Proc. 2, 111–132.

Carter, R.W.G., Devoy, R.J.N., 1987. The hydrodynamic and sedimentary consequences of sea-level change. Prog. Oceanogr. 18 (1–4). Pergamon Press, Oxford.

Carter, R.W.G., Woodroffe, C.D. (Eds.), 1994. Coastal Evolution: Late Quaternary Shoreline Morphodynamics. Academic Press, London.

Carter, W.E., Scherer, W., Diamante, J.M., 1993. Global absolute sea level: Hawaiian and US Atlantic coast-Bermuda regional networks. In: Warrick, R.A., Barrow, E.M., Wigley, T.M.L. (Eds.), Climate and Sea Level Change: Observations, Projections and Implications. Cambridge University Press, Cambridge, pp. 72–80.

Chappell, J., Omura, A., Esat, T., McCulloch, M., Pandelfi, J., Ota, Y., Pillans, B., 1996. Reconciliation of late Quaternary sea levels derived from coral terraces at Huon Peninsula with deep sea oxygen isotope records. Earth Planet. Sci. Lett. 141, 227–236.

Church, J.A., White, N.J., 2006. A twentieth century acceleration in global sea-level rise. Geophys. Res. Lett. 33, L01602.

Church, J.A., White, N.J., 2011. Sea-level rise from the late 19th to the early 20th century. Surv. Geophys. 32, 585–602.

Church, J.A., Godfrey, J.S., Jackett, D.R., McDougall, T.J., 1991. A model of sea-level rise caused by ocean thermal expansion. J. Clim. 4, 438–456.

Church, J.A., et al., 2001. Changes in Sea Level. Climate Change 2001: The Scientific Basis. Contribution of Working Group I to the Third Assessment Report of the Intergovernmental Panel on Climate Change. CUP, Cambridge, 639–693.

Church, J.A., Molnselesan, D., Gregory, J.M., Marzeion, B., 2013. Evaluating the ability of process based models to project sea-level change. Environ. Res. Lett. 8, 015051.

Church, J.A., Woodworth, P.L., Aarup, T., Wilson, W.S., 2010. Understanding Sea-Level Rise and Variability. Wiley-Blackwell.

Church, J.A., Clark, P.U., et al., 2014. Sea Level Change. Contribution in Working Group I to the IPCC Fifth Assessment Report on, Climate Change 2014: The Physical Science Basis. Cambridge University Press, Cambridge, United Kingdom.

Clarke, P.J., Lavallee, D.A., Blewitt, G., van Dam, T.M., Wahr, J.M., 2005. Effect of gravitational consistency and mass conservation on seasonal surface mass loading models. Geophys. Res. Lett. 32, L08306.

CoastAdapt, 2014. CoastAdapt: The Sea as Our Neighbour. European Union, Northern Periphery Programme (2007–2013). http://coastadapt.org.

Cooper, J.A.G., Muir, D., Petursdottir, G., 2014. Coastal Climate Change in the Northern Periphery of Europe. Special Issue of Coastal & Ocean Management 94. Elsevier.

Conrad, C.P., 2013. The solid Earth's influence on sea level. Bull. Geol. Soc. Am. 125, 1027–1052.

Clark, J.A., Farrell, W.E., Peltier, W.R., 1978. Global changes in postglacial sea level: a numerical calculation. Quat. Res. 9, 265–287.

Clark, J.A., Lingle, C.S., 1977. Future sea-level changes due to West Antarctic ice sheet fluctuations. Nature 269, 206–209.

Clark, J.A., Primus, J.A., 1987. Sea-level changes resulting from future retreat of ice sheets: an effect of CO_2 warming of the climate. In: Shennan, I., Tooley, M.J. (Eds.), Sea Level Changes. Blackwell, Oxford.

Cooper, J.A.G., Pilkey, O.H., 2007. Rejoinder to Cowell, P.J. and Thom, B.G., 2006. Reply to Pilkey, O.H. and Cooper, J.A.G., 2006. Management of uncertainty in predicting climate change impacts on beaches. J. Coastal Res. 22, 232–245. In: J. Coastal Res. 22, 1577–1579; J. Coastal Res. 23, 277–280.

Cooper, J.A.G. and Pilkey, O.H., 2008. Discussion of Broker et al., 2007. Morphological modelling: a tool for optimisation of coastal structures. J. Coastal Res., 23, 1148–1158. In: J. Coastal Res. 24, 814–816.

Cooper, J.A.G., Cummins, V. (Eds.), 2009. Coastal Management in Northwest Europe. Elsevier, Amsterdam, 869 pp.

Cooper, J.A.G., Jackson, D.W., 2014. Coastal Spits. Springer, Dordrecht.

Cooper, J.A.G., Pilkey, O., 2012. Pitfalls of Shoreline Stabilization. Springer, Dordrecht.

Cooper, A., Orford, J.D., McKenna, J., Jennings, S.C., Scott, B., Malvarez, G., 1995. Mesoscale Behaviour of Atlantic Coastal Systems under Secular Climate and Sea-Level Rise. Final Report, CEC Environment and Climate Programme No. EV5V-CT-0266.

Cooper, J.A.G., Hooke, J.M., Bray, M.J., 2001. Predicting coastal evolution using a sediment budget approach: a case study from southern England. Ocean Coastal Manage. 44, 711–728.

Cooper, J.A.G., McKenna, J., Jackson, D.W.T., O'Connor, M., 2007. Mesoscale coastal behavior related to morphological self-adjustment. Geology 35, 187–190.

Cowell, P.J., Thom, B.G., 1994. Morphodynamics of coastal evolution. In: Carter, R.W.G., Woodroffe, C.D. (Eds.), Coastal Evolution: Late Quaternary Shoreline Morphodynamics. Academic Press, London, pp. 33–86.

Cronin, K.M., 2010. The Value of Different Modelling Approaches to Investigte Estuarine Morphodynamics (Ph.D. thesis). The National University of Ireland, University College Cork.

Cronin, K.M., Devoy, R.J.N., 2010. From 19th century to present: changes in hydrographic surveying techniques and determination of sounding accuracy. Int. Hydrogr. Rev. 4, 15–25.

Cronin, K.M., Devoy, R.J.N., Gault, J., 2009. The value of different modelling approaches to investigate estuarine morphodynamics. J. Coastal Res. SI 56, 932–936. Proceedings of the 10th International Coastal Symposium, Lisbon, Portugal.

Cronin, T.M., 1999. Principles of Paleoclimatology: Perspectives in Paleobiology and Earth History. Columbia University Press, New York.

Daly, R.A., 1934. The Changing World of the Ice Age. Yale University Press, New Haven.

Dan, S., Walstra, R., Stive, M.J.F., Panin, N., 2011. Processes controlling the development of a river mouth spit. Mar. Geol. 280, 116–129.

Davison-Arnott, R., 2010. Introduction to Coastal Processes and Geomorphology. Cambridge University Press, Cambridge.

Dawson, A., 1992. Ice Age Earth: Late Quaternary Geology and Climate. Routledge, London.

Dawson, A.G., Elliott, L., Noone, S., Hickey, K., Holt, T., Wadhams, P., Foster, I.D.L., 2004. Historical storminess and climate 'see-saws' in the North Atlantic region. Mar. Geol. 210, 247–259.

Delaney, C.A., Devoy, R.J.N., Jennings, S.C., 2012. Mid to late Holocene relative sea level and sedimentary changes in southwest Ireland. In: Duffy, P.J., Nolan, W. (Eds.), At the Anvil: Essays in Honour of William J. Smyth. Geography Publications, Dublin, pp. 697–746.

Devoy, R.J.N., 1979. Flandrian sea-level changes and the vegetational history of the lower Thames estuary. Philos. Trans. R. Soc. London B 285 (1010), 355–407.

Devoy, R.J.N., 1982. Analysis of the geological evidence for Holocene sea-level movements in southeast England. Proc. Geol. Assoc. 93, 65–90.

Devoy, R.J.N., 1985. The problem of a late Quaternary landbridge between Britain and Ireland. Quat. Sci. Rev. 4 (1), 43–58.

Devoy, R.J.N. (Ed.), 1987. Sea Surface Studies: A Global View. Croom Helm-Chapman Hall, London.

Devoy, R.J.N., 1991. The study of inferred patterns of Holocene sea-level change from Atlantic and other European coastal margins as a means of testing models of earth crustal behaviour. In: Sabadini, R., Lambeck, K., Boschi, E. (Eds.), Glacial Isostasy, Sea-Level and Mantle Rheology, pp. 213–235, NATO ASI Series C: Mathematical and Physical Sciences, vol. 334. Kluwer Academic Publishers, London, pp. 213–236.

Devoy, R.J.N., 1992. Questions of coastal protection and the human response to sea-level rise in Ireland and Britain. Irish Geogr. 25 (1), 1–22.

Devoy, R.J.N., 1995. Deglaciation, Earth crustal behaviour and sea-level changes in the determination of insularity: a perspective from Ireland. In: Preece, R.C. (Ed.), Island Britain: A Quaternary Perspective, vol. 96. Geological Society of London, Special Publication, pp. 181–208.

Devoy, R.J.N. (Ed.), 1997. Glacial-Interglacial Sea-Level Changes in Four Dimensions. ESF Meeting, Blarney, February 1997. European Science Foundation Colloquium Conference Series, Strasbourg.

Devoy, R.J.N., 1999. Late Quaternary sea levels, ocean circulation and climate in the North Atlantic. In: Dawson Convenor, A.G. (Ed.), Glacial-Interglacial Sea-level Changes in Four Dimensions: Quaternary Sea Levels, Climate Change and Crustal Dynamics. European Research Conferences, Albufeira, Portugal. European Science Foundation, Strasbourg, pp. 13–18. February 1999.

Devoy, R.J.N., 2005. Cork city and the evolution of the Lee Valley. In: Crowley, J., Devoy, R.J.N., Linehan, D., O'Flanagan, P. (Eds.), Atlas of Cork City. Cork University Press, Cork, pp. 7–16.

Devoy, R.J.N., 2008. Coastal vulnerabilty and the implications of sea-level rise for Ireland. J. Coastal Res. 24 (2), 325–341.

Devoy, R.J.N., 2014. The development and management of the Dingle Bay spit-barriers of southwest Ireland. In: Cooper, J.A.G., Jackson, D.W. (Eds.), Coastal Spits. Springer, Dordrecht.

Devoy, R.J.N., de Groot, T., Jelgersma, S., Marson, I., Nicholls, R.J., Paskoff, R., Plag, H.-P., 2000. Sea-level changes and coastal processes: recommendations. In: Smith, D., et al. (Eds.), Sea Level Change and Coastal Processes: Implications for Europe. European Commission, Brussels, pp. 1–8.

Devoy, R.J.N., Dodson, J., Thom, B.G., 1994. Holocene environments in the Hawkesbury valley, New South Wales: a comparison of terrestrial and marine records. Quat. Sci. Rev. 13, 241–256.

Douglas, B.C., Kearney, M.S., Leatherman, S.P. (Eds.), 2001. Sea Level Rise: History and Consequences. Academic Press, San Diego.

Duffy, M.J., Devoy, R.J.N., 1999. Contemporary process controls on the evolution of sedimentary coasts under low to high energy regimes: western Ireland. Geol. Mijnbouw 77, 333–349.

Edwards, R., 2006. Sea levels: change and variability during warm intervals. Prog. Phys. Geogr. 30, 785–796.

Edwards, R., Brooks, A.J., 2008. The island of Ireland: drowning the myth of an Irish landbridge. In: Davenport, J.J., Sleeman, D.P., Woodman, P.C. (Eds.), Mind the Gap: Postglacial Colonisation of Ireland. Special Supplement to the Irish Naturalists' Journal, pp. 19–34.

Ehlers, J., Gibbard, P.L., 2004. Quaternary Glaciations, Extent and Chronology: Part 1, Europe. Elsevier, Amsterdam.

Ehlers, J., Gibbard, P.L., Hughes, P.D., 2011. Quaternary Glaciations—Extent and Chronology: A Closer Look. Elsevier, Amsterdam.

European Space Agency, May 2014. Cryosat results on east Antarctica. http://www.esa.int/Our_Activities/Observing_the_Earth/CryoSat/Antarctica_s_ice_loss_on_the_rise.

European Commission, 2011. Developing a Maritime Strategy for the Atlantic Ocean Area. European Commission, Brussels.

European Commission, 2013. Action Plan for a Maritime Strategy in the Atlantic Area: Delivering Smart, Sustainable and Inclusive Growth. Communication from the Commission to the European Parliament, the Council, the European Economic and Social Committee and the Committee of the Regions, 13.5.2013 COM (2013) 279 Final, Brussels.

Fagan, B., 2000. Floods, Famines and Emperors: El Niño and the Fate of Civilizations. Pimlico, London.

Fairbanks, R.G., 1989. A 17,000-year glacio-eustatic sea-level record: influence of glacial melting rates on the Younger Dryas event and deep-ocean circulation. Nature 342, 637–642.

Fairbridge, R.W., 1961. Eustatic changes in sea level. In: Ahrens, L.H., Press, F., Rankama, K., Runcorn, S.K. (Eds.), Physics and Chemistry of the Earth. Pergamon Press, Oxford, pp. 90–185.

Fausto, R.S., Ahlstrom, A.P., van As, D., Johnsen, S.J., Langen, P.L., Steffen, K., 2009. Improving surface boundary conditions with focus on coupling snow densification and meltwater retention in large-scale ice sheet models of Greenland. J. Glaciol. 55, 869–878.

de Groot, T.A.M., Orford, J.D., 2000. Implications for coastal management. In: Smith, D., Raper, S.B., Zerbini, S., Sanchez-Arcilla, A. (Eds.), Sea-Level Change and Coastal Processes. European Commission, Luxembourg, pp. 214–242.

de Groot, A.V., de Vries, S., Keijsers, J.G.S., Riksen, M.J.P.M., Ye, Q., Poortinga, A., Arens, S.M., Bochev-Van der Burgh, L.M., Wijnberg, K.M., Schretlen, J.L., van Thiel, J.S.M., de Vries, J., 2012. Measuring and modelling coastal dune development in the Netherlands. In: Proceedings of NCK-Days 2012: Crossing Borders in Coastal Research. Enschede, The Netherlands. http://dx.doi.org/10.3990/2.17.

Gilbertson, D.D., Schwenninger, J.L., Kemp, R.A., Rhodes, E.J., 1999. Sand-drift and soil formation along an exposed North Atlantic coastline: 14,000 years of diverse geomorphological, climatic and human impacts. J. Archaeol. Sci. 26, 439–469.

GLOSS, 2014. Information Website. http://www.gloss-sealevel.org/.

Gray, S.R.J., Gagnon, A.S., Gray, S.A., O'Dwyer, B., O'Mahony, C., Muir, D., Devoy, R.J.N., Falaleeva, M., Gault, J., 2014. Are coastal managers detecting the problem? Assessing stakeholder perception of climate vulnerability using fuzzy cognitive mapping. Ocean & Coastal Management 94, 74–89.

Greensmith, T., Tooley, M.J. (Eds.), 1982. IGCP Project 61: Sea-Level Movements during the Last Deglacial Hemicycle (c.15,000 Years). Proceedings of the Geologists' Association 93, 3–125.

Hallam, A., 1981. Facies Interpretation and the Stratigraphic Record. W.H. Freeman and Company, Oxford.

Haq, B.U., 1991. Sequence stratigraphy, sea-level change and significance for the deep sea. In: McDonald, D.I.M. (Ed.), Sedimentation, Tectonics and Eustasy. International Association of Sedimentologists. Special Publication 12, 3–39.

Haq, B.U., Hardenbol, J., Vail, P., 1987. Chronology of fluctuating sea levels since the Triassic. Science 235, 1156–1167.

Haq, B.U., Schutter, S.R., 2008. A chronology of Paleozoic sea-level changes. Science 322, 64–68.

Hardisty, J., 1990. Beaches: Form and Process. Unwin Hyman, London.

Hardisty, J., 1994. Beach and nearshore sediment transport. In: Pye, K. (Ed.), Sediment Transport and Depositional Process. Unwin Hyman, London.

Hijioka, Y., Lin, E., Pereira, J., et al., 2014. Asia, Chapter 24. In: IPCC Fifth Assessment Report, Climate Change 2014: Impacts, Adaptation and Vulnerability, Volume II. Regional Aspects. Cambridge University Press.

ICS (International Coastal Symposia), 2002. In: Cooper, J.A.G., Jackson, D.W.T. (Eds.), Proceedings of the 7th International Coastal Symposium. Journal of Coastal Research, SI 36.

ICS (International Coastal Symposia), 2007. In: Klein, A.F.H., Beaumord, A.C., Finkl, C.W., Diehl, F.L., Calliari, L.J. (Eds.), Proceedings of the 8th International Coastal Symposium 2004, Itajaí – SC – Brazil. Journal of Coastal Research, SI 39, 2 Vols.

ICS (International Coastal Symposia), 2009. In: da Silva, C.P., Vaz, B., Abrantes, P., Estanqueiro, R. (Eds.), Proceedings of the 10th International Coastal Symposium 2009, Lisbon, Portugal. Journal of Coastal Research, SI 56, 2 Vols.

ICS (Internat. Coastal Symposium), 2011. In: Furmanczyk, K., Giza, A., Terefenko, P. (Eds.), Proceedings of the 11th International Coastal Symposium, Szczecin, Poland 9–13. Coastal Education & Research Foundation, pp. 2083.

IOC (Intergovernmental Oceanographic Commission), 1985. Manual on Sea Level Measurement and Interpretation. UNESCO.

IOC (Intergovernmental Oceanographic Commission), 2009. Papers, Presentations and Reports from the Eleventh Session of the GLOSS Group of Experts (GLOSS GE XI) and Associated Workshop, 11–15 May 2009, Paris, France.

IPCC Special Report on Emissions Scenarios (SRES), Nakicenovic, N., Alcamo, J., Davis, G., de Vries, B., Fenhann, J., Gaffin, S., Gregory, K., Grübler, A., et al., 2000. Special Report on Emissions Scenarios, Working Group III, Intergovernmental Panel on Climate Change (IPCC). Cambridge University Press, Cambridge, 595. http://www.grida.no/climate/ipcc/emission/index.htm.

IPCC WGIII, 2000. Special Report on Emissions Scenarios (SRES): Summary for Policymakers. Intergovernmental Panel on Climate Change.

IPCC, 2001a. Climate change 2001: synthesis report. In: Watson, R.T., Core Writing Team (Eds.), A Contribution of Working Groups I, II and III to the Third Assessment Report of the Intergovernmental Panel on Climate Change. Cambridge University Press, Cambridge.

IPCC, 2001b. Technical Summary of Climate Change 2001: The Scientific Basis. Contribution of Working Group I to the Third Assessment Report of the Intergovernmental Panel on Climate Change. Cambridge University Press.

IPCC WGI, 2001. Contribution of Working Group I to the Intergovernmental Panel on Climate Change Third Assessment Report on, Climate Change 2001: The Physical Science Basis. Cambridge University Press, Cambridge.

IPCC WGI, 2007. Contribution of Working Group I to the Intergovernmental Panel on Climate Change Fourth Assessment Report on, Climate Change 2014: The Physical Science Basis. Cambridge University Press, Cambridge.

IPCC WGII, 2007. In: Parry, M.L., Canziani, O.F., Palutikof, J.P., van der Linden, P.J., Hanson, C.E. (Eds.), Climate Change 2007: Impacts, Adaptation and Vulnerability. Contribution of Working Group II to the Fourth Assessment Report of the Intergovernmental Panel on Climate Change. Cambridge University Press, Cambridge.

IPCC WGII SPM, 2007. Summary for policymakers. In: Parry, M.L., Canziani, O.F., Palutkof, J.P., van der Linden, P.J., Hansen, C.E. (Eds.), Climate Change 2007: Impacts, Adaptation and Vulnerability. Contribution of Working Group II to the Fourth Assessment Report of the Intergovernmental Panel on Climate Change. Cambridge University Press, Cambridge.

IPCC WGI, 2014. Contribution of Working Group I to the Intergovernmental Panel on Climate Change Fifth Assessment Report on, Climate Change 2014: The Physical Science Basis. Cambridge University Press, Cambridge, United Kingdom.

IPCC WGII, 2014. Contribution of Working Group II to the Fifth Assessment Report of the Intergovernmental Panel on Climate Change 2014: Impacts, Adaptation and Vulnerability. Cambridge University Press, Cambridge, United Kingdom.

IGCP 588, 2014. International Geological Correlation Programme Project 588. http://www.coastal-change.org/.

India, M.B., Bonillo, D.L., 2001. Detecting and Modelling Regional Climate Change. Springer, Berlin.

ISOS, 1991. Impact of Sea Level Rise on Society: A Case Study of the Netherlands. Delft Hydraulics, Rijkswaterstaat, Amsterdam.

Jackson, T., 2011. Prosperity without Growth: Economics for a Finite Planet. Earthscan, London.

Jelgersma, S., 1961. Holocene sea level changes in the Netherlands. Meded. van de Geol. Sticht. Ser. C. VI, 1–100.

Jelgersma, S., 1966. Sea-level changes during the last 10,000 years. In: Sawyer, J.S. (Ed.), World Climate from 8,000 to 0 BC. Royal Meteorological Society, London, pp. 54–69.

Jelgersma, S., et al., 1979. Sea-level changes in the North Sea basin. In: Oele, E., Schuttenhelm, R.T.E., Wiggers, A.J. (Eds.), The Quaternary History of the North Sea. Uppsala, pp. 233–248. Acta Universitatis Upsaliensis, Symposia Universitatis Upsaliensis Annum Quingentesimum Celebrantis, 2. Almqvist & Wiksell Intl., Stockholm, Sweden.

Jenkins, G.J., Murphy, J.M., Sexton, D.M.H., Lowe, J.A., Jones, P., Kilsby, C.G., 2009. UK Climate Projections: Briefing Report. Meteorological Office Hadley Centre, Exeter, UK.

Jennings, S., Orford, J.D., Canti, M., Devoy, R.J.N., Straker, V., 1998. The role of relative sea-level rise and changing sediment supply on Holocene gravel barrier development: the example of Porlock, Somerset, UK. The Holocene 8 (2), 165–181.

Kandrot, S., 2013. Quantifying Post Storm Dune Recovery Using Terrestrial Laser Scanning in Dingle Bay, County Kerry (Ph.D. submission). National University of Ireland, University College Cork and in Chimera, 28, Department of Geography, UCC.

Kemp, A.C., Horton, B.P., Donnelly, J.P., Mann, M.E., Vermeer, M., Rahmstorf, S., 2011. Climate related sea-level variations over the last two millennia. Proc. Natl. Acad. Sci. U.S.A. 108, 11017–11022.

Kirkby, M.J., Naden, P.S., Burt, T.P., Butcher, D.P., 1993. Computer Simulation in Physical Geography. Wiley, Chichester.

Kremer, H., Nicholls, R., Ratter, B.M.W., Weisse, R., 2013. Risk and Management of Current and Future Storm Surges. Springer, Dordrecht.

Kurtén, B., 1968. Pleistocene Mammals of Europe. Aldine Publishing Company, Chicago. Re landbridges.

Lambeck, K., 1991. A model for Devensian and Flandrian glacial rebound and sea-level change in Scotland. In: Sabadini, R., Lambeck, K., Boschi, E. (Eds.), Glacial Isostasy, Sea-Level and Mantle Rheology. Kluwer Academic Publishers, Dordrecht, pp. 33–62.

Lambeck, K., 1993. Glacial rebound of the British Isles, I: Preliminary results and II: A high-resolution, high precision model. Geophys. J. Int 115, 941–990.

Lambeck, K., 1995. Late Devensian and Holocene shorelines of the British Isles and North Sea from models of glacio-hydro-isostatic rebound. J. Geol. Soc. London 152, 437–448.

Lambeck, K., 2001. Glacial crustal rebound, sea levels and shorelines. In: Steele, J.H., Thorpe, S.A., Turekian, K.K. (Eds.), Encyclopedia of Ocean Sciences. Academic Press, San Diego, pp. 1157–1167.

Lambeck, K., Johnston, P., Smither, C., Nakada, M., 1996. Glacial rebound of the British Isles—III. Constraints on mantle viscosity. Geophys. J. Int. 125, 340–354.

Lambeck, K., Smither, C., Ekman, M., 1998. Tests of glacial rebound models for Fennoscandinavia based on instrumented sea- and lake-level records. Geophys. J. Int. 135, 375–387.

Lambeck, K., Purcell, A., Dutton, A., 2012. The anatomy of interglacial sea levels: the relationship between sea levels and ice volumes during the last interglacial. Earth Planet. Sci. Lett. 315, 4–11. Cambridge University Press, Cambridge, United Kingdom.

Lange, M., Burkhard, B., Garthe, S., Gee, K., Kannen, A., Lenhart, H., Windhorst, W., 2010. Analysing Coastal and Marine Changes: Offshore Wind Farming as a Case Study. Zukunft Kuste—Coastal Futures Synthesis Report. LOICZ Research and Studies No. 36. GKSS Research Centre, Geesthacht, Germany.

Lim, B., Burton, I., Malone, E., Huq, S. (Eds.), 2005. Adaptation Policy Frameworks for Climate Change: Developing Strategies, Policies and Measures. Cambridge University Press, Cambridge.

Lisitzin, E., 1974. Sea-Level Changes. In: Oceanography Series, vol. 8. Elsevier, Amsterdam.

LOICZ, 2014. Land and Ocean Interactions in the Coastal Zone. http://www.loicz.org/.

Long, A.J., Innes, J.B., Kirby, J.R., Lloyd, J.M., Rutherford, M.M., Shennan, I., Tooley, M.J., 1998. Holocene sea-level change and coastal evolution in the Humber estuary, eastern England: an assessment of rapid coastal change. The Holocene 8, 229–247.

Lozano, I., Devoy, R.J.N., May, W., Andersen, U., 2004. Storminess and vulnerability along the Atlantic coastlines of Europe: analysis of storm records and of a greenhouse gases induced climate scenario. Mar. Geol. 210, 205–225.

Lowe, J., Walker, M., 1997. Reconstructing Quaternary Environments, second ed. Addison Wesley Longman, Harlow, UK.

Lowe, J.A., Howard, T.P., Pardaens, A., Tinker, J., Holt, J., Wakelin, S., Milne, G., Leake, J., Wolf, J., Horsburgh, K., Reeder, T., Jenkins, G., Ridley, J., Dye, S., Bradley, S., 2009. UK Climate Projections Science Report: Marine and Coastal Projections. Meteorological Office Hadley Centre, Exeter, UK.

Lowe, J.A., et al., 2010. Past and future changes in extreme sea levels and waves. In: Understanding Sea-Level Rise and Variability. Wiley-Blackwell, Chichester, pp. 326–375.

Merrifield, M., et al., 2009. The Global Sea Level Observing System (GLOSS). Community White Paper for the OceanObs'09 Conference, Venice, September 2009 check in text re. GLOSS.

Masselink, G., Hughes, M.G., Knight, J., 2011. Coastal Processes and Geomorphology. Hodder Education, London.

Masselink, G., Gehrels, R. (Eds.), 2014. Coastal Environments and Global Change. Wiley-Blackwell, Oxford.

Marchand, M. (Ed.), 2009. Modelling Coastal Vulnerability: Design and Evaluation of a Vulnerability Model for Tropical Storms and Floods. IOS Press BV, The Netherlands.

MCCIP, 2012. In: Buckley, P.J., Dye, S.D., Frost, M., Wallace, C.J. (Eds.), Marine Climate Change Impacts Knowledge Gaps. Summary Paper, MCCIP, Lowestoft.

McGregor, G.R., 1992. Temporal and spatial characteristics of coastal rainfall anomalies in Papua New Guinea and their relationship to the Southern Oscillation. Int. J. Climatol. 12, 449–468.

Meehl, G.A., et al., 2007. Global climate projections. In: Solomon, et al. (Eds.), The Physical Science Basis. Contribution of Working Group I to the Fourth Assessment Report, Intergovernmental Panel on Climate Change. Cambridge University Press, pp. 755–828.

Meehl, G.A., et al., 2012. Relative outcomes of climate change mitigation related to global temperature versus sea-level rise. Nat. Clim. Change 2, 576–580.

Milne, G.A., Mitrovica, J.X., 1998. Postglacial sea-level change on a rotating Earth. Geophys. Int. 133, 1–19.

Milne, G.A.W., Gehrels, R., Hughes, C.W., Tamisiea, M.E., 2009. Identifying the causes of sea-level change. Nat. Geosci. 2, 471–478.

Mitrovica, J.X., Gomez, N., Clark, P.U., 2009. The sea-level fingerprint of West Antarctic collapse. Science 323, 753.

Mitrovica, J.X., Gomez, N., Morrow, E., Hay, C., Latychev, K., Tamisiea, M.E., 2011. On the robustness of predictions of se-level fingerprints. Geophys. J. Int. 187, 729–742.

Mörner, N.-A., 1971. Eustatic changes during the last 20,000 years and a method of separating the isostatic and eustatic factors in an uplifted area. Palaeogeogr. Palaeoclimatol. Palaeoecol. 19, 63–85.

Morner, N.-A., 1976. Eustasy and geoid changes. J. Geol. 84, 123–151.

Mörner, N.-A., 1980. Earth Rheology, Isostasy and Eustasy. Wiley, Chichester.

Mörner, N.-A., 1987a. Pre-quaternary long term changes in sea level. In: Devoy, R.J.N. (Ed.), Sea Surface Studies: A Global View. Croom Helm – Chapman Hall, London, pp. 233–241.

Mörner, N.-A., 1987b. Models of global sea-level changes. In: Tooley, M.J., Shennan, I. (Eds.), Sea Level Changes. Blackwell, Oxford, pp. 332–355.

Muller, R.D., et al., 2008. Long term sea-level fluctuations driven by ocean basin dynamics. Science 319, 1357–1362.

Murray-Wallace, C.V., Woodroffe, C.D., 2014. Quaternary Sea-Level Changes: A Global Perspective. C.U.P, Cambridge.

Nicholls, R.J., 2010. Impacts of and responses to sea-level rise. In: Church, J.A., Woodworth, P.L., Aaruo, T., Stanley Wilson, W. (Eds.), Understanding Sea-level Rise and Variability. Chapter 2. Wiley-Blackwell, pp. 17–43.

Nicholls, R.J., Marinova, N., Lowe, J.A., Brown, S., Vellinga, P., de Gusmao, D., Hinkel, J., Tol, R.S.J., 2011. Sea-level rise and its possible impacts given a beyond 4°C world in the twenty-first century. Phil. Trans R. Soc. Lond. A: Maths, Phys & Eng. Sci's 369, 161–181.

Nicholls, R.J., Wong, P.P., et al., 2007. In: Parry, M.L., Canziani, O.F., Palutkof, J.P., van der Linden, P.J., Hansen, C.E. (Eds.), Climate Change 2007: Impacts, Adaptation and Vulnerability. Contribution of Working Group II to the Fourth Assessment Report of the Intergovernmental Panel on Climate Change. Cambridge University Press, Cambridge, pp. 315–356.

Oft, P., Tsuma, W., 2006. 'Mobilising Resilience from below', Linking Institutions to Actors and Knowledge to Decisions: A Case Study from the Netherlands. Centre for Development Research, University of Bonn.

Olsson, P., Folke, C., Berkes, F., 2004. Adaptive co-management for building resilience in social-ecological systems. Environ. Manage. 34, 75–90.

Olsson, Carl Folke, P., Galaz, Victor, Hahn, T., Schultz, L., 2007. Enhancing the fit through adaptive co-management: creating and maintaining bridging functions for matching scales in the Kristianstads Vattenrike Biosphere Reserve, Sweden. Ecol. Soc. 12, 17.

Open University, 1998. The Ocean Basins: Their Structure and Evolution, second ed. Pergamon, Milton Keynes.

Ortlieb, L., Maclane, J., 1993. Former El Niño events: records from western South America. Planet. Change 7, 181–202.

O'Shea, M., Murphy, J., 2013. Predicting and monitoring the evolution of a coastal barrier dune system post breaching. J. Coastal Res. 29, 38–55.

Ota, Y., Nelson, A.R., Berryman, K.R. (Eds.), 1992. Impacts of Tectonics on Quaternary Coastal Evolution. Special Issue of Quaternary International 15/16, 184 pp.

Peltier, W.R., 1998. Postglacial variations in the level of the sea: implications for climate dynamics and solid-earth geophysics. Rev. Geophys. 36, 603–689.

Peltier, W.R., 2004. Global glacial isostasy and the surface of the ice-age earth: the ICE-5G (VM2) model and GRACE. Annu. Rev. Earth Planet. Sci. 32, 111–149.

Peltier, W.R., 2009. Closure of the budget of global sea-level rise over the GRACE era: the importance and magnitudes of the required corrections for global glacial isostatic adjustment. Quat. Sci. Rev. 28, 1658–1674.

Pirazzoli, P.A., 1991. World Atlas of Holocene Sea Level Changes. Elsevier, Amsterdam.

Pirazzoli, P.A., 1996. Sea Level Changes: The Last 20,000 Years. Wiley, Chichester.

Pitman, W.J., 1978. Relationship between eustasy and stratigraphic sequences of passive margins. Bull. Geol. Soc. Am. 89, 1289–1403.

Pugh, D., 2004. Changing Sea Levels: Effects of Tides, Weather and Climate. Cambridge University Press, Cambridge.

Pugh, D., Woodworth, P.L., 2014. Sea-Level Science: Understanding Tides, Surges, Tsunamis and Mean Sea-Level Changes. Cambridge University Press, Cambridge.

van de Plassche, O., 1986. Sea Level Research: A Manual for the Collection and Evaluation of Data. Geo Books, Norwich.

Rahmstorf, S., 2007. A semi-empirical approach to projecting future sea-level rise. Science 315, 368–370.

Rahmstorf, S., 2010. A new view on sea-level rise. Nat. Rep. Clim. Change 4, 44–45.

Rahmstorf, S., Coumou, D., 2011. Increase of extreme events in a warming world. Proc. Nat. Acad. Sci's of USA 108, 17905–17909.

Rahmstorf, S., 2012. Modelling sea-level rise. Nat. Educ. Knowl. 3 (10), 4.

Rahmstorf, S., Ganopolski, A., 1999. Long-term global warming scenarios computed with an efficient coupled climate model. Clim. Change 43, 353–367.

Rahmstorf, S., Cazenave, A., Church, J.A., Hansen, J.E., Keeling, R.F., Parker, D.E., Somerville, R.C.J., 2007. Recent climate observations compared to projections. Science 316, 709.

Rahmstorf, S., Foster, G., Cazenave, A., 2012a. Comparing climate projections to observations up to 2011. Environ. Res. Lett. 7, 044035.

Rahmstorf, S., Perrette, M., Vermeer, M., 2012b. Testing the robustness of semi-empirical sea-level projections. Clim. Dyn. 39, 861–875.

Ranasinghe, R., Duong, T.M., Uhlenbrook, S., Roelvink, D., Stive, M., 2013. Climate-change impacts assessment for inlet-interrupted coastlines. Nat. Clim. Change 3, 83–87.

Raper, S.C.B., 1993. Observational data on the relationships between climatic change and the frequency and magnitude of severe tropical storms. In: Warrick, R.A., Barrow, E.M., Wigley, T.M.L. (Eds.), Climate and Sea Level Change: Observations, Projections and Implications. Cambridge University Press, Cambridge, pp. 192–212.

Raper, S., 2000. Global changes in the volume and mass of the ocean. In: Smith, D., et al. (Eds.), Sea-level Changes and Coastal Processes: Implications for Europe. European Commission, Luxembourg, pp. 9–80.

Raper, S.C.B., et al., 2000. Global changes in the volume and mass of the ocean. In: Smith, D., Raper, S.B., Zerbini, S., Sanchez-Arcilla, A. (Eds.), Sea-Level Change and Coastal Processes. European Commission, Luxembourg, pp. 9–80.

Raper, S.C.B., Braithwaite, R.J., 2005. The potential for sea-level rise: new estimates from glacier and ice cap area and volume distributions. Geophys. Res. Lett. 32, 4.

Riva, R.E.M., Bamber, J.L., Lavillee, D.A., Wouters, B., 2010. Sea-level fingerprint of continental water and ice mass change from GRACE. Geophys. Res. Lett. 37, L19605.

Roe, H.M., Coope, G.R., Devoy, R.J.N., Harrison, C.J.O., Penkman, K.E.H., Preece, R.C., Schreve, D.C., 2009. Differentiation of MIS 9 and MIS 11 in the continental record: vegetational, faunal, aminostratigraphic and sea-level evidence from coastal sites in Essex. U.K. Quat. Sci. Rev. 28, 2342–2373. http://dx.doi.org/10.1016/j.quascirev.2009.04.017.

Roeleveld, W., 1974. The Holocene evolution of the Groningen Marine-Clay District. Ber. Rijkdienst Oudheidkd. Bodemonderz. 24, 1–132.

Roy, P.S., Cowell, P.J., Ferland, M.A., Thom, B.G., 1994. Wave dominated coasts. In: Carter, R.W.G., Woodroffe, C.D. (Eds.), Coastal Evolution: Late Quaternary Shoreline Morphodynamics. Cambridge University Press, Cambridge, pp. 121–186.

Royal Society & US National Academy of Sciences, 2014. Climate Change Evidence & Causes: An Overview. Royal Society, London.

Sabadini, R., Lambeck, K., Boschi, E. (Eds.), 1991. Glacial Isostasy, Sea-Level and Mantle Rheology. Kluwer Academic Publishers, Dordrecht.

Saito, Y., Chaimanee, N., Jarupongsakul, T., Syvitski, J.P.M., 2007. Sinking megadeltas in Asia: sea-level rise and sediment reduction impacts from case study of the Chao Phraya delta. LOICZ Newsl. 2007 (2), 3–9.

Sánchez-Arcilla, A., Hoekstra, P., Jimenez, J.A., Kaas, E., Maldono, A., 2000. Climate change implications for coastal processes. In: Smith, D., Raper, S.B., Zerbini, S., Sanchez-Arcilla, A. (Eds.), Sea-Level Change and Coastal Processes. European Commission, Luxembourg, pp. 173–213.

Schoof, C., 2007. Marine ice sheet dynamics. Part 1. The case of rapid sliding. J. Fluid Mech. 573, 27–55.

Schoof, C., 2011. Marine ice sheet dynamics. Part 2. A Stokes flow contact problem. J. Fluid Mech. 679, 122–155.

Schlesinger, M.E., 1993. Model projections of CO_2 – induced equilibrium climate change. In: Warrick, R.A., Barrow, E.M., Wigley, T.M.L. (Eds.), Climate and Sea-Level Change. Cambridge University Press, Cambridge, pp. 169–191.

Schubert, G., Reymer, A.P.S., 1985. Continental volume and freeboard through geological time. Nature 316, 336–338.

Schwartz, M., 2006. Encyclopedia of Coastal Science. Springer, Dordrecht.

Shackleton, N.J., 1987. Oxygen isotopes, ice volume and sea level. Quat. Sci. Rev. 6, 183–190.

Shennan, I., 1982. Imterpretation of Flandrian sea-level data from the Fenland, England. In: Greensmith, J.T., Tooley, M.J. (Eds.), IGCP Project 61 Sea Level Movements During the Last Deglacial Hemicycle (about 15,000 years), 93. Proc. Geol. Ass, pp. 53–63.

Shennan, I., Andrews, J. (Eds.), 2000. Holocene Land–Ocean Interaction and Environmental Change around the Western North Sea. Geological Society: Special Publications, London.

Shennan, I., Tooley, M.J., 1987. Sea Level Changes. Blackwell, Oxford.

Shennan, I., Tooley, M.J., Davies, M.J., Haggart, B.A., 1983. Analysis and interpretation of Holocene sea-level data. Nature 302, 404–406.

Shennan, I., Peltier, W.R., Drummond, R., Horton, B.P., 2002. Global to local scale parameters determining relative sea-level changes and the postglacial isostatic adjustment of Great Britain. Quat. Sci. Rev. 21, 397–408.

Shennan, I., Bradley, S., Milne, G., Brooks, A.J., Bassett, S., Hamilton, S., Hiller, C., Hunter, A., Woodal, R., 2006. Relative sea-level changes, glacial isostatic modelling and ice sheet reconstructions from the British Isles. J. Quat. Sci. 21, 585–599.

Short, A.D. (Ed.), 1999. Handbook of Beach and Shoreface Morphodynamics. John Wiley & Sons, Chichester.

Short, A.D., Cowell, P.J., Cadee, M., Hall, W., van Dijk, B., 1995. Beach rotation and possible relation to the Southern Oscillation. In: Aung, T.H. (Ed.), Ocean Atmosphere Pacific Conference. National Tidal Facility, Adelaide, pp. 329–334.

Siegert, M.J., 2001. Ice Sheets and Late Quaternary Environmental Change. Wiley, Chichester.

Smith, D., Dawson, A., 1984. Shorelines and Isostasy. Academic Press, London.

Smith, D., Raper, S.B., Zerbini, S., Sanchez-Arcilla, A., 2000. Sea-Level Change and Coastal Processes. European Commission, Luxembourg.

Soloman, S.M., Forbes, D.L., 1999. Coastal hazards and associated management issues on South Pacific islands. Ocean Coastal Manage. 42, 523–554.

Stive, M.J.F., 2004. How important is global warming for coastal erosion. Clim. Change 64, 27–39.

Stive, M.J.F., 2006. Morphodynamics of coastal inlets and tidal lagoons. J. Coastal Res. SI 39, 28–34.

Stive, M.J.F., de Vriend, H.J., 1995. Modelling shoreface profile evolution. Mar. Geol. 126, 235–248.

Stive, M.J.F., Wang, Z.,B., 2003. Morphodynamic modelling of tidal basins and coastal inlets. In: Lakhan, V.C. (Ed.), Advances in Coastal Modeling. Elsevier, Amsterdam, pp. 367–392.

Stive, M.J.F., R.J., de Vriend, H.J., 1991. Sea-level rise and shore nourishment: a discussion. Coastal Eng. 16, 147–163.

Storlazzi, C.D., Griggs, G.B., 2000. Influence of El Nino Southern Oscillation (ENSO) events on the evolution of central California's shoreline. Geol. Soc. Am. Bull. 112, 236–249.

Stone, G.W., Orford, J.D., 2004. Storms and their significance in coastal morpho-sedimentary dynamics. Mar. Geol. (Spec. Iss.) 210. Elsevier, Amsterdam.

Suess, 1906. The Face of the Earth, vol. 2. Clarendon Press, Oxford.

von Storch, H., Zorita, E., Gonzalez-Rouco, J.F., 2008. Relationship between global mean sea level and global mean temperature in a climate simulation of the past millennium. Ocean Dyn. 58, 227–236.

Tamisiea, M.E., 2011. Ongoing glacial isostatic contributions to observations of sea-level change. Geophys. J. Int. 186, 1036–1044 check in text re.

Tamisiea, M.E., Hill, E.M., Ponte, R.M., Davis, J.L., Velicogna, I., Vinogradova, N.T., 2010. Impact of self attraction and loading on the annual cycle in sea level. J. Geophys. Res. 115, C07004.

Tebaldi, C., Strauss, B.H., Zervas, C.E., 2012. Modelling sea-level rise impacts on storm surges along US coasts. Environ. Res. Lett. 7, 1–11. http://dx.doi.org/10.1088/1748-9326/7/1/014032.

Thom, B.G., Roy, P.S., 1983. Sea-level change in New South Wales over the past 15,000 years. In: HopleSy, D. (Ed.), Australian Sea Levels in the Last 15,000 Years: A Review. Department of Geography. James Cook University of North Queensland Monograph Series, Occasional Paper No. 3, pp. 64–84.

Titus, J.G., Park, R.A., Leatherman, S.P., et al., 1991. Greenhouse-effect and sea-level rise: the cost of holding back the sea. Coastal Manage. 19, 171–204.

Titus, J.G., Trescott, D.L., Hudgens, D.E. (Eds.), 2010. The Likelihood of Shore Protection along the Atlantic Coast of the United States. Volume 2: New England and the Southeast. Report to the U.S. Environmental Protection Agency, Washington, DC. http://risingsea.net/ERL/FL.html.

Tooley, M.J., 1978. Sea Level Changes in Northwest England During the Flandrian Stage. Oxford Research Studies in Geography. Clarendon Press, Oxford.

Tooley, M.J., 1993. Long term changes in eustatic sea level. In: Warrick, R.A., Barrow, E.M., Wigley, T.M.L. (Eds.), Climate and Sea Level Change: Observations, Projections and Implications. Cambridge University Press, Cambridge, pp. 81–107.

United States Geological Survey (USGS), July 9, 2011. El Niño Modoki, 2009–10. Am. Geophys. Union Res. Lett.

de Vriend, J.H., 1991. Mathematical modelling and large-scale coastal behaviour: parts 1 and 2. J. Hydraul. Res. 29, 727–740 and 741–753.

de Vriend, H.J., Capobianco, M., Chesher, T., De Swart, H.E., Latteux, B., Stive, M.J.F. (Eds.), 1993. Approaches to Long Term Modelling of Coastal Morphology: A Review. Coastal Eng., vol. 21.

Vail, P.R., Mitchum, R.M., Thompson, S., 1977. Seismic stratigraphy and global changes of sea level. In: Payton, C.E. (Ed.), Seismic Stratigraphy: Applications to Hydrocarbon Exploration. Am. Assoc. Petrol. Geol. Oklahoma, Tulsa, pp. 83–98.

Vermeer, M., Rahmstorf, S., 2009. Global sea level linked to global temperature. Proc. Natl. Acad. Sci. U.S.A. 106, 21527–21532.

Visser, W.A., 1980. Geological Nomenclature. Bonn, Scheltema and Holkema, Utrecht.

Walcott, R.I., 1972. Past sea levels, eustasy and deformation of the Earth. Quat. Res. 2, 1–14.

Wang, X.L., Swail, V.R., Cox, A., 2010. Dynamical versus statistical downscaling methods for ocean wave heights. Int. J. Climatol. 30, 317–332.

Warrick, R.A., Barrow, E.M., Wigley, M.L., 1993. Climate and Sea Level Change: Observations, Projections and Implications. Cambridge University Press, Cambridge.

Wigley, T.M.L., Raper, S.C.B., 1993. Future changes in global mean temperature and sea level. In: Warrick, R.A., Barrow, E.M., Wigley, T.M.L. (Eds.), Climate and Sea Level Change: Observations, Projections and Implications. Cambridge University Press, Cambridge, pp. 111–133.

Woldenberg, M. (Ed.), 1985. Models in Geomorphology. Allen & Unwin, London.

Wong, P.P., Losada, I.J., 2014. Coastal systems and low-lying areas. In: Contribution of Working Group II to the Fifth Assessment Report of the Intergovernmental Panel on Climate Change on, Climate Change 2014: Impacts, Adaptation and Vulnerability. Cambridge University Press, Cambridge, United Kingdom.

Woodroffe, C.A., 2002. Coasts: Form, Process and Evolution. Cambridge University Press, Cambridge.

Woodworth, P.L., 1990. A search for accelerations in records of European mean sea level. International J. of Climatology 10, 129–143.

Woodworth, P.L., 1993. A review of recent sea-level research. Oceanography and Marine Biology Annu. Rev. 31, 87–109.

Woodworth, P.L., Shaw, S.M., Blackman, D.L., 1991. Secular trends in mean tidal range around the British Isles and along the adjacent European coastline. Geophysical J. International 104, 593–609.

Woodworth, P.L., Blackman, D.L., 2004. Evidence for systematic changes in extreme high waters since the mid-1970s. J. Clim. 17, 1190–1197.

Woodworth, P.L., Player, R., 2003. The permanent service for mean sea level: an update to the 21st century. J. Coastal Res. 19, 287–295.

Woodworth, P.L., Aman, A., Aarup, T., 2007. Sea level monitoring in Africa. Afr. J. Mar. Sci. 29, 321–330. http://dx.doi.org/10.2989/AJMS.2007.29.3.2.332.

Woodworth, P.L., Le Provost, C., Rickards, L.J., Mitchum, G.T., Merifield, M., 2002. A review of sea-level research from tide gauges during the World Ocean Circulation Experiment. Oceanogr. Mar. Biol. Annu. Rev. 40, 1–35.

Woodworth, P.L., Rickards, L.J., Perez, B., 2009a. A survey of European sea level infrastructure. Nat. Hazards Earth Syst. Sci. 9, 927–934.

Woodworth, P.L., Foden, P., Pugh, J., Mathews, A., Aarup, T., Aman, A., Nkebi, E., Odametey, J., Facey, R., Esmail, M.Y.A., Ashraf, M., 2009b. Insight into long term sea level change based on new tide gauge installations at Takoradi, Aden and Karachi. Int. Hydrogr. Rev. 1, 18–23 (May 2009).

Zerbini, S., et al., 2000. Regional and local sea-level variations. In: Smith, D., Raper, S.B., Zerbini, S., Sanchez-Arcilla, A. (Eds.), Sea-Level Change and Coastal Processes. European Commission, Luxembourg, pp. 81–133.

Zong, Y., Tooley, M.J., 1996. Holocene sea-level changes and crustal movements in Morecombe Bay, northwest England. J. Quat. Sci 11, 43–58.

Chapter 9

Adapting to Sea Level Rise

Robert J. Nicholls
Faculty of Engineering and the Environment, University of Southampton, Southampton, UK

ABSTRACT

Coasts contain a large and growing population and economy including world cities such as London, New York, Tokyo, Shanghai, Mumbai, Lagos, and Rio de Janeiro, as well as important habitats and their ecosystem services. Global sea levels are rising due to climate change and this will accelerate through this century: a rise of more than 1 m is possible. In some locations these changes may be exacerbated by (1) increases in storminess due to climate change, although this is less certain, and (2) human-induced subsidence due to ground fluid withdrawal from and drainage of susceptible soils, especially in deltas and alluvial plains. Sea-level rise has a range of potential impacts including higher extreme sea levels (and flooding), coastal erosion, and salinization of surface and ground waters. This threatens the loss of large areas of land and associated assets and economic activity, the displacement of millions of people, and significant coastal habitat degradation. However, adaptation can greatly reduce these impacts and promote prosperous and desirable coasts. Adaptation measures can be characterized as (1) protect, (2) accommodate, or (3) retreat approaches. The provision of information measures such as warnings is improving significantly, while novel methods such as ecosystem-based approaches are attracting interest. Adaptation to sea-level rise should be viewed as a process that requires an integrated coastal management philosophy to be consistent with wider coastal activities and other stresses. Hence, in addition to technical skills, adaptation requires an appropriate institutional capacity. The success or failure of measures of adaptation, especially protection, is contested and this influences our view of sea-level rise as a problem. Adaptation can be best analyzed in the context of understanding the coastal system that includes the effects of all drivers, including sea-level rise, their interactions, and feedbacks: these types of analyses are only just beginning. Some proactive adaptation plans are already being formulated, such as around London and in the Netherlands. Coastal cities will be a major focus for adaptation efforts due to their concentrations of people and assets. However, there are important challenges for adaptation in developing countries, most especially in deltaic areas and small islands.

Coastal and Marine Hazards, Risks, and Disasters. http://dx.doi.org/10.1016/B978-0-12-396483-0.00009-1
Copyright © 2015 Elsevier Inc. All rights reserved.

9.1 INTRODUCTION

Sea-level rise has been recognized as a global major threat to low-lying coastal areas since the 1980s (e.g., Barth and Titus, 1984; Milliman et al., 1989; Tsyban et al., 1990). There is a growing literature demonstrating that the potential impacts of sea-level rise are large. To respond to this challenge interest in adaptation is increasing, even though it is recognized as a difficult and challenging problem (Moser et al., 2012; Wong et al., 2014). Although sea-level rise only directly impacts coastal areas, these are the most densely populated and economically active land areas on Earth. More than 600 million people live below 10 m elevation in the Low Elevation Coastal Zone (McGranahan et al., 2007), and the population is growing rapidly in coastal urban areas (CIESIN, 2013). Coastal areas also support important and productive ecosystems that are sensitive to sea-level rise (Crossland et al., 2005). Coasts are already "risky places" exposed to multiple meteorological and geophysical hazards, including storms and storm-induced flooding (Kron, 2013). Threatened low-lying areas already depend on various flood risk adaptation strategies, be it natural and/or artificial flood defences and drainage or construction methods. Recent flood events such as New Orleans and environs on the US Gulf Coast (Hurricane Katrina, 2005), Irrawaddy delta, Myanmar (Cyclone Nagris, 2008), New York and environs (Cyclone Sandy, 2012), or the Philippines (Typhoon Haiyan, 2013) demonstrate what can happen in low-lying areas during extreme flood events. Rising mean sea level and more intense storms are expected to exacerbate these risks significantly (Wong et al., 2014).

The main focus of this chapter is adaptation, although some brief remarks on the possible role of climate mitigation and other source control responses are included for completeness. To distinguish these responses, mitigation and adaptation are defined as follows:

- Mitigation (or source control of sea level)—reducing the magnitude of human-induced climate change and sea-level rise at the global scale, or reducing the magnitude of human-induced subsidence at the local level; and
- Adaptation (to sea level)—reducing the impacts of sea-level rise via behavioral changes. This includes a range of changes from individual actions to collective coastal management policy, such as upgraded defence systems, warning systems, and land management approaches.

Coastal adaptation to sea-level rise has been considered for the last 25 years (Barth and Titus, 1984; IPCC CZMS, 1990), building on the extensive experience in adapting to climate variability and other stresses. Despite this, the uncertainties about the success or failure of adaptation remain large, contributing significant uncertainty to the overall consequences of sea-level rise for society (Nicholls et al., 2014a).

The chapter is structured as follows. First the coast is considered as a system comprising natural and socioeconomic components, to provide an appropriate framework to analyze coasts, sea-level rise, and adaptation. Second, climate change and sea-level rise are considered in more detail, including the important distinction between global-mean and relative sea-level rise (RSLR). Then the impacts of sea-level rise are briefly considered from a physical and a socioeconomic perspective, including drawing on experience from subsiding cities. This is followed by a brief review of mitigation approaches for sea-level rise and a more detailed consideration of adaptation. This demonstrates the complexity of adaptation and the multiple factors that need to be considered. This is followed by a discussion/conclusion, including research needs.

9.2 COASTAL SYSTEMS

Sea-level rise and the need to adapt to it does not happen in isolation: coasts are changing significantly due to more local factors such as urbanization and changing water/sediment inputs due to river regulation and watershed land use and land cover change (Crossland et al., 2005; Valiela, 2006; Syvitski et al., 2009). These types of problems require a systems approach to analyze the full range of interacting drivers, including feedbacks such as adaptation. Figure 9.1 presents a simplified systems model of the impacts of sea-level rise on the coastal zone (Klein and Nicholls, 1999; Nicholls and Klein, 2005). This model highlights the varying implicit and explicit assumptions and simplifications that are necessary within all the available assessments of coastal impacts in general, including their limitations. It characterizes the overall coastal system as interacting natural and socioeconomic systems, which have the potential to constrain each other's evolution. Both systems can be characterized by their exposure, sensitivity and adaptive capacity to change, both from sea-level rise, related climate change, and nonclimate stresses. Collectively, sensitivity and adaptive capacity, combined with exposure, determine the vulnerability to sea-level rise and other changes.

A range of drivers may influence the boundary conditions (Figure 9.1). Sea-level rise is only one aspect of climate change for coastal areas, and all climate change drivers interact with other nonclimate stresses, often exacerbating impacts (see Table 9.1). Lastly, the socio-economic system is not passive as it influences the natural system through deliberate changes such as construction of sea dykes, destruction of wetlands, and building of port and harbor works, as well as unintended changes such as reductions of sediment and water fluxes due to the building of dams. Hence, the socioeconomic system is shaping the future of the coastal system as much, if not more than, the natural system and issues such as sea-level rise are shaping the socioeconomic system. This raises the prospect of the coast as a coevolving system where the natural system shapes the socioeconomic system and vice

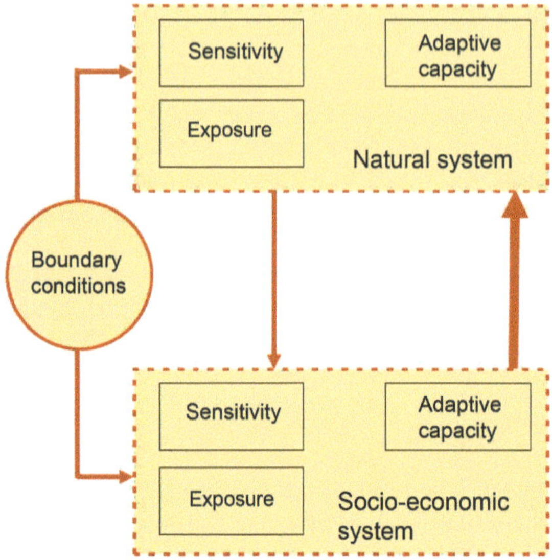

FIGURE 9.1 The coastal system comprises interacting natural and socioeconomic sub-systems which in turn are influenced by changing boundary conditions, such as sea-level rise, climate change, and large-scale nonclimatic stresses.

versa, with adaptation playing an important role in this aspect. It raises a new way of thinking about the future of coasts, which requires further investigation (Lazarus et al., 2014).

9.3 GLOBAL-MEAN AND RSLR

Human-induced climate change is expected to cause a profound series of changes including rising sea level, rising sea-surface temperatures, and changing storm, wave, and run-off characteristics (Wong et al., 2014). Here we will focus on climate-induced sea-level rise, which is mainly produced by (1) thermal expansion of seawater as it warms and (2) the melting of land-based ice, comprising components from (a) small glaciers, (b) the Greenland ice sheet, and (c) the West Antarctic ice sheet (Church et al., 2010; Gornitz, 2013; Pugh and Woodworth, 2014). A global rise in sea level of 17 cm was observed through the twentieth century (i.e., 1.7 mm/year). This observed rise is almost certain to continue and will very likely accelerate through the twenty-first century with a rise of 1 m or more being plausible if the large ice sheets make a large positive contribution (Church et al., 2013). From an impact and adaptation perspective, coastal policymakers are especially concerned about the high end of possible changes (Nicholls et al., 2014b). While the probability of high-end rise is unknown, the large potential impacts make them highly significant in terms of climate risks and policy. There is also concern about

TABLE 9.1 The Main Natural System Effects of Relative Sea-Level Rise and Examples of Adaptation Options. Potential Interacting Factors Which Could Offset or Exacerbate These Impacts Are Also Shown. Some Interacting Factors (e.g., Sediment Supply) Appear Twice as They Can Be Influenced both by Climate and Nonclimate Factors. Adaptation Options are Coded: P—Protection (Hard or Soft); A—Accommodation; R—Retreat

Natural System Effect		Possible Interacting Factors		Possible Adaptation Options
		Climate	Nonclimate	
1. Inundation/flooding	**a.** Surge (flooding from the sea)	Wave/storm climate, erosion, sediment supply	Sediment supply, flood management, erosion, land reclaim	Dikes/surge barriers/closure dams [P—hard], nourishment, including dune construction [P—soft], ecosystem-based barriers (e.g., mangrove afforestation) [P—soft], building codes/flood-proof buildings [A], land use planning/hazard mapping [A/R], planned migration [R].
	b. Backwater effect (flooding from rivers)	Run-off	Catchment management and land use	
2. Wetland loss (and change)		CO_2 fertilization, sediment supply, migration space	Sediment supply, migration space, land reclaim (i.e., direct destruction)	Gabions/breakwaters [P—hard], nourishment/sediment management [P—soft], land use planning [A/R], managed realignment/forbid hard defences [R].
3. Erosion (of "soft" morphology)		Sediment supply, wave/storm climate	Sediment supply	Coastal defences/seawalls/land claim [P—hard], ecosystem-based barriers (e.g., mangroves) [P—soft], nourishment [P—soft], building setbacks/rolling easements [R].

Continued

TABLE 9.1 The Main Natural System Effects of Relative Sea-Level Rise and Examples of Adaptation Options. Potential Interacting Factors Which Could Offset or Exacerbate These Impacts Are Also Shown. Some Interacting Factors (e.g., Sediment Supply) Appear Twice as They Can Be Influenced both by Climate and Nonclimate Factors. Adaptation Options are Coded: P—Protection (Hard or Soft); A—Accommodation; R—Retreat—cont'd

Natural System Effect		Possible Interacting Factors		Possible Adaptation Options
		Climate	Nonclimate	
4. Saltwater intrusion	a. Surface waters	Run-off	Catchment management (over-extraction), land use	Saltwater intrusion barriers [P], desalination [A], move water abstraction upstream [R].
	b. Ground-water	Rainfall	Land use, aquifer utilization	Insert impermeable barriers [P], freshwater injection [P], change water abstraction [A/R].
5. Impeded drainage/higher water tables		Rainfall, run-off	Land use, aquifer utilization, catchment management	Drainage systems/polders [P—hard], change land use/crop type [A], land use planning/hazard delineation [A/R].

Adapted from Nicholls (2010), see also Linham and Nicholls (2010).

higher extreme sea levels due to more intense storms superimposed on mean rise in sea level, but this is much less certain (Church et al., 2013).

When analyzing sea-level rise impacts and responses, it is fundamental that impacts are a product of *relative* (or local) sea-level rise (RSLR) rather than global changes alone (Nicholls et al., 2014b). Relative sea-level change considers the sum of global, regional, and local components of sea-level change: the underlying drivers of these components are (1) climate change, as already discussed, and changing ocean dynamics and (2) nonclimate land level change (i.e., uplift/subsidence) processes such as tectonics, glacial isostatic adjustment (GIA), and natural and anthropogenic-induced subsidence. For large ice sheet changes, gravitational effects due to mass redistribution of melting ice also need to be considered. Hence, RSLR is only partly a response to climate change and varies from place to place (Figure 9.2). Where coasts are subsiding, such as Grand Isle in the Mississippi Delta, Louisiana, RSLR exceeds the global rise. Most populated deltaic areas and alluvial plains are threatened by enhanced subsidence (Ericson et al., 2006; Syvitski et al., 2009; Chaussard et al., 2013). Most dramatically, subsidence can be enhanced by human activity on susceptible soils due to drainage and withdrawal of groundwater as shown in Bangkok (Figures 9.2 and 9.3). Dramatic RSLR has occurred in many coastal cities built on deltas and alluvial plains due to this cause. Over the twentieth century, the parts of Tokyo and Osaka built on deltaic areas subsided up to 4 m and 3 m, respectively, a large part of Shanghai subsided up to 3 m, and the center of Bangkok subsided up to 2 m. Human-induced subsidence can be mitigated by stopping shallow sub-surface fluid withdrawals and managing water levels, but natural "background" rates of

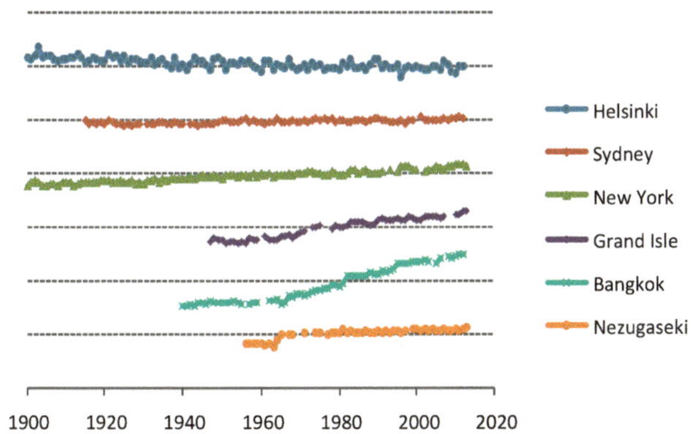

FIGURE 9.2 Selected relative sea-level observations since 1900, illustrating different trends (offset for display purposes). Helsinki shows a falling trend (−2.0 mm/year) as the land is rising, Sydney shows a gradual rise (0.9 mm/year), New York is subsiding slowly (3.1 mm/year), Grand Isle is on a subsiding delta (9.1 mm/year), Bangkok (Station: Fort Phrachula Chomklao) is also on a delta and includes the additional effects of human-induced subsidence (18.9 mm/year from 1962 to 2012), and Nezugaseki shows an abrupt 150–2000 mm rise due to an earthquake. *Data from Holgate et al. (2013); PSMSL, 2014.*

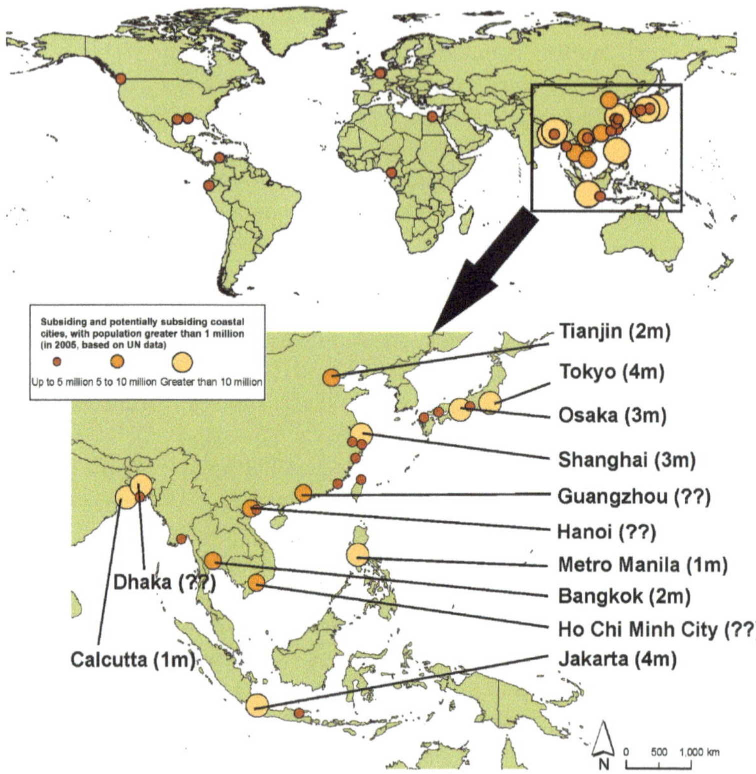

FIGURE 9.3 Subsiding and potentially subsiding coastal cities *(adapted and updated from Nicholls (2010), Hallegatte et al. (2013), with additional data from Kaneko and Toyota (2011), Dang et al. (2014))*. The maximum observed subsidence (in meters) is shown for cities with populations exceeding 5 million people, where known. Maximum subsidence is reported as data on average subsidence is not available.

subsidence that are typical of deltas (1−5 mm/year and maybe more) will continue and RSLR will still exceed global trends in these areas. The four cities mentioned above have all implemented mitigation policies to varying degrees of success, combined with the provision of improved flood defence and pumped drainage systems to avoid submergence and/or frequent flooding. (In Bangkok, subsidence has been greatly reduced in the center of the city, but at the site of the measurements shown in Figure 9.2, which is 20 km to the south, no reduction is evident.) In contrast, other cities such as Jakarta and Metro Manila continue to subside substantially, with maximum subsidence of 4 and 1 m over the last few decades, respectively (Kaneko and Toyota, 2011). Flooding and waterlogging are common and growing problems. There is little systematic policy response to date despite these direct impacts, or the experience described in other cities. This suggests that the problems of enhanced subsidence are likely to be widely repeated in susceptible coastal cities through the twenty-first century. It is important to emphasize that only some cities are prone to this problem: of the 136 coastal cities with a population above 1 million considered by Hallegatte et al. (2013), only 32 have an appropriate geological setting to experience enhanced subsidence as these are cities wholly or partly built in deltaic or alluvial settings (Figure 9.3). Note the concentration of large cities in south, South-East or East Asia.

Greater appreciation of the importance of subsidence is urgently needed to promote responses, including planning and adaptation for RSLR. In much of the developed world quality data is limited. However, new measurement systems permit analysis and quantification, including satellite measurements (Chatterjee et al., 2006) and differential Global Positioning System (DGPS) (Teatini et al., 2005). Beyond this, the political will to tackle these issues is also necessary as discussed by Rodolfo and Siringan (2006) for Manila, the Philippines.

9.4 SEA-LEVEL RISE AND RESULTING IMPACTS

Relative sea-level rise causes more effects than simple submergence (the "bath-tub" effect); the five main effects are summarized in Table 9.1. Flooding/submergence, ecosystem change, and erosion have received significantly more attention than salinization and rising water tables. Along with rising sea levels, there are changes to all processes that operate around the coast. The immediate effect is submergence and increased flooding of coastal lands, as well as saltwater intrusion into surface waters. Longer term effects also occur as the coast adjusts to the new environmental conditions, including wetland loss and change, erosion of beaches and soft cliffs, and saltwater intrusion into groundwater. These lagged changes interact with the immediate effects of sea-level rise and generally exacerbate them. For instance, erosion of saltmarshes, mangroves, sand dunes, and coral reefs degrades or removes natural protection and increases the likelihood of coastal flooding.

A rise in mean sea level also raises extreme water levels. Changes in storm characteristics could also influence extreme water levels. For example, an increase in the intensity of tropical cyclones will generally raise extreme water levels in the areas affected (Church et al., 2013). Extratropical storms may also intensify in some regions, although this effect is uncertain. An improved understanding of these changes is an important research topic to support impact and adaptation assessments.

Changes in natural systems resulting from sea-level rise have many important direct socio-economic impacts on a range of sectors, with these impacts being overwhelmingly negative (Table 9.2). For instance, flooding can damage coastal infrastructure, ports and industry, the built environment, and agricultural areas. In the worst case, flooding leads to significant mortality, as recently demonstrated by Hurricane Katrina (USA) in 2005, Cyclone Nargis (Myanamar) in 2008, Storm Xynthia (France) in 2010, and Cyclone Sandy

TABLE 9.2 Summary of Sea-Level Rise Impacts on Socioeconomic Sectors in Coastal Zones (© *Reprinted with permission from* Nicholls, 2010). These Impacts Are Overwhelmingly Negative

	Sea-Level Rise Natural System Effect (Table 9.1)				
Coastal Socioeconomic Sector	**Inundation/ Flooding**	**Wetland Loss**	**Erosion**	**Saltwater Intrusion**	**Impeded Drainage**
Freshwater resources	**X**	x	–	**X**	**X**
Agriculture and forestry	**X**	x	–	**X**	**X**
Fisheries and aquaculture	**X**	**X**	x	**X**	–
Health	**X**	**X**	–	**X**	x
Recreation and tourism	**X**	**X**	**X**	–	–
Biodiversity	**X**	**X**	**X**	**X**	**X**
Settlements/infrastructure	**X**	**X**	**X**	**X**	**X**

X, Strong; x, Weak; –, Negligible or not established.

(USA) in 2012. Erosion can lead to the loss of beachfront/cliff-top buildings and other infrastructure, and have adverse consequences for sectors such as tourism and recreation. In addition to these direct impacts, there are potential indirect impacts such as mental health problems triggered by floods, or economic effects that cascade through the whole economy. These indirect impacts are poorly understood, but will have economic consequences in terms of the damages caused (and/or the diversion of investment to fund the adaptation to avoid them). Thus, sea-level rise has the potential to trigger a cascade of direct and indirect human impacts.

Importantly, sea-level rise does not occur in isolation and coasts are changing significantly due to nonclimate-induced drivers (Crossland et al., 2005; Valiela, 2006; Wong et al., 2014). Potential interactions of such changes with sea-level rise are indicated in Table 9.1 (column entitled "Potential Interacting Factors") and need to be considered when assessing sea-level rise impacts and adaptation responses. For instance, a coast with a positive sediment budget may not erode given sea-level rise and vice versa. Hence, coastal change ideally requires an integrated assessment approach to analyze the full range of interacting drivers, including the feedback of policy interventions (i.e., adaptation).

9.5 RECENT IMPACTS OF SEA-LEVEL RISE

Over the twentieth century, global sea level rose about 17 cm (or 1.7 mm/year). While this change may seem small, it has had many significant effects, such as reducing the return periods of extreme water levels (Zhang et al., 2000; Menéndez and Woodworth, 2010) and promoting an erosive tendency for coasts (Bird, 1985, 2000). However, linking sea-level rise quantitatively to impacts is difficult as the coastal zone has been subjected to multiple drivers of change over the twentieth century (Nicholls et al., 2009; Wong et al., 2014). Good data on rising sea levels has only been measured in a few locations, and growing coastal populations and infrastructure have increased the exposure available to damage. In addition, adaptation has occurred and, for example, flood defences have often been upgraded substantially, especially in those (wealthy) places where there are long-term sea-level measurements (e.g., Ruocco et al., 2011). Most of these defence upgrades coincide with the expanding populations and wealth in the coastal flood plain and changing attitudes to risk. Hence, RSLR may not have been considered in the design. Equally, impacts can be promoted by processes other than sea-level rise (Table 9.1). For example, widespread human reduction in sediment supply to the coast must contribute to the observed erosional changes around the world and this probably dominates erosion in many locations (Bird, 1985; Syvitski et al., 2009). Hence, while global sea-level rise is a pervasive process, other processes obscure its link to impacts, except in some special cases; most coastal change in the twentieth century was a response to multiple drivers of change.

There have certainly been impacts from RSLR resulting from subsidence (Nicholls, 2010). Notable sites include the iconic city of Venice, which will shortly be protected by the MOSES storm surge barriers (see Horn, in this volume), and the Mississippi delta where thousands of square kilometers of intertidal coastal marshes and adjacent lands were converted to open water in the last 50 years. There are also significant impacts of RSLR in and around subsiding coastal cities (e.g., Figure 9.3), in terms of increased waterlogging, flooding and submergence, and the resulting need for adaptation/management responses.

These empirical observations also provide lessons for adaptation. Subsiding areas with a low population density were often abandoned, such as around Galveston Bay, Texas, and south of Bangkok, Thailand. However, most of the major developed areas that were impacted by RSLR have been defended and continue to experience population and economic growth (Nicholls, 2010). This includes areas where the change in RSLR was rapid—several meters over several decades. However, there are exceptions such as New Orleans (Grossi and Muir-Wood, 2006); its population peaked in 1965 at more than 625,000 immediately before Hurricane Betsy flooded part of the city. Before Hurricane Katrina in 2005, its population was about 500,000. Subsequently, the population has yet to recover to pre-Katrina levels, even though US$15 billion has been invested to significantly upgrade defences (completed 2011). The future of New Orleans will be instructive: Does it prosper or continue to decline behind the new defences and what are the reasons?

Observations since 1900 reinforce the importance of understanding the impacts of sea-level rise in the context of multiple drivers of change; this will remain true under more rapid rises in sea level. RSLR due to human-induced subsidence is of particular interest, but this remains relatively unstudied using a systems approach. Observations also emphasize the ability to protect against RSLR, especially for the most densely populated areas, such as the subsiding Asian megacities or urban areas around the southern North Sea.

9.6 FUTURE IMPACTS OF SEA-LEVEL RISE

The future impacts of sea-level rise will depend on a range of factors, including: (1) the magnitude of sea-level rise, (2) the coastal physiography, (3) the level and manner of coastal development, and (4) the success (or failure) of adaptation. Assessments of the future impacts of sea-level rise have taken place on a range of scales from local to global. They all suggest large potential impacts consistent with Table 9.1, especially increases in inundation, flooding, and storm damage (e.g. Nicholls et al., 2008). Recent studies of flood risk (i.e., expected annual damages) under sea-level rise all emphasize that impacts could be catastrophic when assuming no adaptation (Hallegatte et al., 2013; Hinkel et al., 2014). However, if defences are upgraded and other adaptation takes place, problems are more limited and possibly almost totally avoided. This stresses the importance of understanding adaptation.

In absolute numbers, East, South-East and East Asia, and Africa appear to be most threatened by sea-level rise (Figure 9.4). Vietnam and Bangladesh appear especially threatened due to large absolute and relative populations in low-lying deltaic plains. There are also large absolute threatened populations in India and China. In Africa, Egypt (the Nile Delta) and Mozambique are two potential hotspots for impacts due to sea-level rise. Hotspots also exist outside these regions, such as Guyana, Suriname, and French Guiana in South America. There will be significant residual risk in other coastal areas of the world, such as around the southern North Sea, and major flood disasters are possible in many coastal regions. Small island regions in the Pacific, Indian Ocean, and Caribbean stand out as being especially vulnerable to sea-level rise impacts (Nurse et al., 2014). The populations of low-lying island nations, such as the Maldives, Kiribati, or Tuvalu, face the real prospect of increased flooding, submergence, salinization, and forced abandonment.

9.7 MITIGATION OF SEA-LEVEL RISE

Climate mitigation can slow the global rise in sea level and reduce its impacts. Given the strong inertia of sea-level rise, mitigation stabilizes the *rate* of sea-level rise (rather than stabilizing sea level itself) (Wong et al., 2014). Hence, even with mitigation, sea-level rise will continue and remain a challenge for generations to come (Church et al., 2013). This has been termed

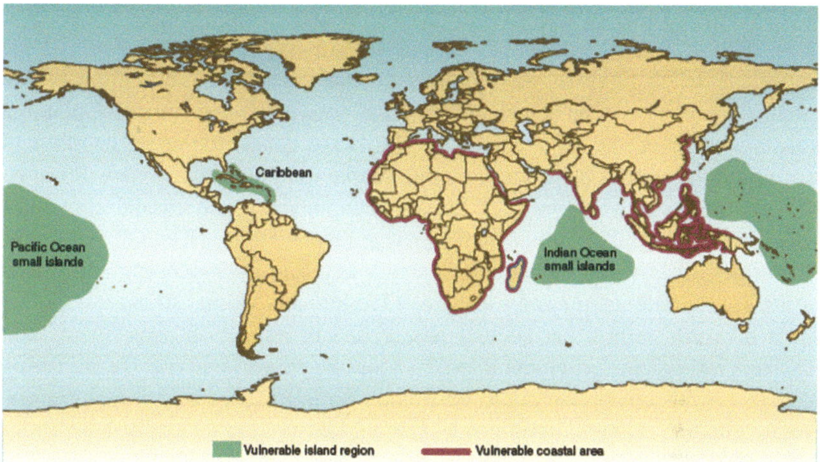

FIGURE 9.4 Regions most vulnerable to coastal flooding and sea-level rise. At highest risk are coastal zones with dense populations, low elevations, appreciable rates of subsidence, and/or inadequate adaptive capacity. *From Nicholls and Cazenave (2010).*

the "commitment to sea-level rise," which leads to a "commitment to adapt to sea-level rise" with fundamental implications for long-term human use of the coastal zone (Nicholls et al., 2007). Hence, adaptation and mitigation are complimentary policy responses for climate change in coastal areas. The fundamental goal of mitigation in the context of coastal areas is to reduce the risk of passing irreversible thresholds concerning the breakdown of the two major ice sheets of Greenland and Antarctica, thus constraining the commitment to sea-level rise to a rate and ultimate rise which can be adapted to at a reasonable economic and social cost. This requires consideration of sea-level rise beyond 2100, which is increasingly being investigated (Church et al., 2013). However, appropriate mixtures of mitigation and adaptation remains to be assessed.

More locally, mitigation of human-induced subsidence also needs to be considered in susceptible areas, as already discussed. This translates into measures to control/reduce groundwater extraction and manage water levels, which have been successfully implemented in a number of cities to date, such as Shanghai, Osaka, and Tokyo (Kaneko and Toyota, 2011). This expertise needs to be transferred more widely to actively subsiding areas such as Jakarta, or sites where these issues are likely to emerge. However, while subsidence can be largely reduced as with climate-induced sea-level rise, it cannot be entirely avoided, and some adaptation to subsidence will be required.

9.8 ADAPTATION TO SEA-LEVEL RISE

Adaptation to sea-level rise involves responding to both mean and extreme rise. It is a complex process with multiple dimensions that different authors characterize differently. The overall field of climate adaptation is also evolving rapidly (e.g., Klein et al., 2014), and this is influencing coastal adaptation, even though coastal adaptation is one of the more mature sectors. It is important to distinguish autonomous (or spontaneous) adaptation versus planned adaptation. One can also distinguish proactive versus reactive planned adaptation. Given the large and rapidly growing concentration of people and activity in the coastal zone, autonomous adaptation processes alone will not be able to cope with sea-level rise. Further, adaptation in the coastal context is widely seen as a public rather than a private responsibility (Klein et al., 2000). Therefore, all levels of government have a key role in developing and facilitating appropriate adaptation measures (Tribbia and Moser, 2008).

It is worth noting that society has tended to react to coastal events and disasters rather than anticipate them. In adapting to sea-level rise we are trying to promote proactive adaptation, where appropriate. There is significant scope for anticipatory adaptation on coasts as many adaptation decisions have long-term (10–100 years) implications (e.g., Hallegatte, 2009). Examples of anticipatory adaptation in coastal zones include upgraded flood defences and drainage systems, higher elevation designs for new coastal infrastructure such as fill levels for land claim and coastal bridges, building standards/regulations to promote flood proofing and resilience, and building setbacks to prevent development in areas threatened by erosion and flooding.

The following section considers adaptation strategies and options, adaptation frameworks, adaptation selection, and adaptation experience.

9.8.1 Adaptation Strategies and Options

Adaptation can be classified in a variety of ways: one of the most widely followed approaches is the Intergovernmental Panel on Climate Change (IPCC) typology of planned adaptation strategies (IPCC CZMS, 1990; Bijlsma et al., 1996; Linham and Nicholls, 2010) (Figure 9.5):

- *(Planned) Retreat*—all natural system effects are allowed to occur and human impacts are minimized by pulling back from the coast via land use planning, development controls, planned migration, etc. (e.g., Figure 9.6);
- *Accommodation*—all natural system effects are allowed to occur and human impacts are minimized by adjusting human use of the coastal zone via changing land use/crop types, flood resilience measures, warning systems, insurance, etc. (e.g., Figure 9.7);
- *Protection*—natural system effects are controlled by soft or hard barriers (e.g., nourished beaches and dunes, or seawalls), reducing human impacts in the zone that would be impacted without protection (e.g., Figure 9.8).

Individually, there are a huge number of potential adaptation options. Examples linked to each natural system impact are provided in Table 9.1. The concept of "attack" has been suggested as a strategy against sea-level rise (e.g., RIBA and ICE, 2010). This is consistent with land claim and

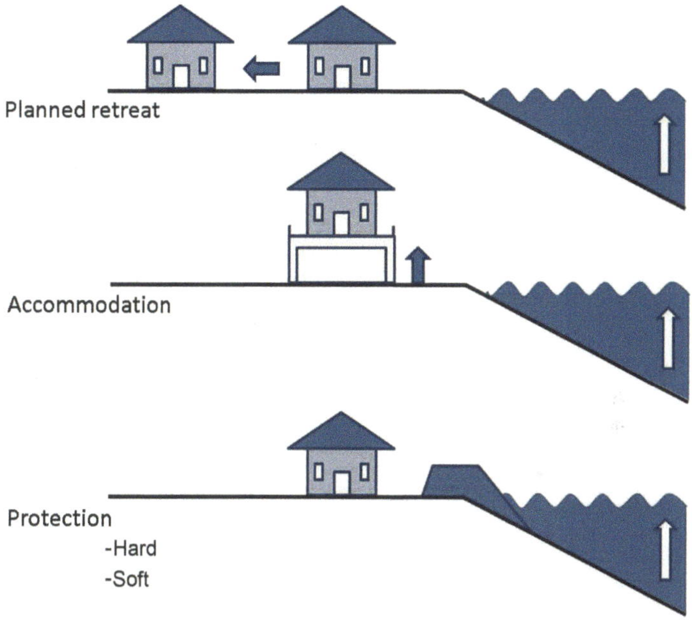

FIGURE 9.5 Generic adaptation approaches for sea-level rise. *Reprinted with permission from Nicholls (2010).*

protection (Linham and Nicholls, 2010). This has a long history in Northwest Europe and East Asia and has been practised in most coastal cities due to space constraints (e.g., Seasholes, 2003). Land claim is an active strategy in many coastal countries such as Singapore, Hong Kong, Dubai, and the Maldives to expand land area for coastal activities: sea-level rise is increasingly being considered in planning of land claim.

Information measures such as disaster preparedness, hazard mapping, and flood warning/evacuation are also increasingly important, and in many ways are cross-cutting and complementary of the three approaches above. There is also increasing interest in ecosystem-based approaches, which have the advantage of being self-sustaining and providing multiple benefits (Borsje et al., 2011; Temmerman et al., 2013). However, the uncertainties about their future state and function are much higher than engineered defences. This emphasizes that many real-world adaptation responses will be hybrid and combine options, possibly from more than one approach. For example, flood protection could use ecosystem buffers in front of artificial defences, reducing the required defence size. In addition, we need to consider the residual risk that remains for all protected areas: this suggests that adaptation needs to be combined with flood forecast and warning systems. Adaptation for one sector may however exacerbate impacts elsewhere: a good example is coastal squeeze of intertidal and shallow coastal habitats where onshore

FIGURE 9.6 An example of a retreat option: managed realignment at Medmerry, West Sussex, UK. The defence line (a shingle barrier beach) was breached allowing the low-lying flood plain behind to be inundated creating new intertidal habitats. To landward, a new (longer) defence line was constructed. *(© Environment Agency)*

FIGURE 9.7 An example of accommodation in a coastal flood plain in the UK. The central property is built at grade while the two adjoining properties have been raised to enhance flood resilience—the design elevation considers extreme water levels plus an allowance for sea-level rise. © *Robert Nicholls*

FIGURE 9.8 An example of a protection option: The Thames (storm surge) Barrier, Greenwich, London. © *Environment Agency*

migration of habitats due to rising sea levels is prevented by hard protection (Jones et al., 2011). In contrast, retreat and accommodation options allow habitat migration. Coastal management needs to consider the balance between protecting socioeconomic activity/human safety and the habitats and ecological functioning of the coastal zone under rising sea levels (Nicholls and Klein, 2005). While the twentieth century saw large losses of coastal habitats due to direct and indirect destruction, most coastal countries now aspire to protect these areas and their ecosystem services: sea-level rise significantly threatens such initiatives.

9.8.2 Adaptation Processes and Frameworks

While adaptation to sea-level rise is relatively new, there is considerable experience of adapting to climate and sea-level variability and other coastal problems. This experience informs decision making under a changing climate. Importantly, adaptation to coastal problems is a multi-stage *process*, including stages such as (1) information and awareness building, (2) planning and design, (3) evaluation, and (4) monitoring and evaluation operating within multiple policy cycles (e.g., Klein et al., 2000; Hay, 2009). The constraints on approaches to adaptation due to broader policy and development goals should also be carefully considered. Once implemented, monitoring and evaluation of adaptation measures is critical and yet easily ignored. This is essential given the large uncertainties associated with sea-level rise and other future conditions, adaptation performance, and coastal management in general.

A range of adaptation frameworks are apparent in the literature, with a diverse range of experience. For example, Integrated Coastal Zone

Management (ICZM) was strongly advocated as the response to sea-level rise in the early 1990s (e.g., IPCC CZMS, 1990; Bijlsma et al., 1996), recognizing that sea-level rise and climate change occur in a multistressed situation. However, it remains unproven as an effective response approach (Wong et al., 2014). Adaptive management where interventions are treated as experiments is also advocated, but again remains largely unproven. The risks of experimenting are raised as concerns. Community-based adaptation (CBA) is widely advocated as a bottom-up development focused approach (e.g., Huq and Reid, 2004; Rawlani and Sovacool, 2011). These provide important benefits for the participants, but in a coastal context there is concern that while they might adapt to small extremes, are the expectations of events such as the 1 in 100 year event appropriate? In the worse case, could CBA promote people to live in hazardous locations? This could be avoided by CBA being placed in a broader framework, including warning systems. Merging climate adaptation with Disaster Risk Reduction (Smith, 2013), which can be seen in part as adapting to climate variability, is also receiving increasing interest (Figure 9.9).

Shoreline management planning (SMP) has emerged in England and Wales over the last two decades as a response to coastal erosion and flood risk management (Nicholls et al., 2013). It provides a framework for thinking about the future of the entire coast over long timescales based on geomorphic principles, including the non-local effects of management. Four generic responses are considered without considering the technical detail: (1) advance the existing defence line; (2) hold the existing defence line; (3) managed realignment; and (4) no active intervention. Options 1 and 2 can be considered as generic protection, while options 3 and 4 can be considered retreat.

FIGURE 9.9 An example of disaster risk reduction: A cyclone shelter in Bangladesh. © *International Federation of Red Cross and Red Crescent Societies. Reprinted with permission from IFRC.*

Note that accommodation is also being implemented in the UK for flood management purposes, as demonstrated by Figure 9.7, but this is implemented at the property level, and hence at a much smaller scale. Supporting the SMP process are national monitoring systems. This high-level approach could be applied widely around the world's coast, including all risks including sea-level rise.

In parallel with this, there has also been recognition that while we need to adapt to sea-level rise, there is great uncertainty about timing and an opportunity to learn. Hence, while we can see different qualitative directions of travel (or possible adaptation pathways), we are not sure how fast we need to travel along the pathway as the magnitude of future sea-level rise is uncertain. Hence, we can define adaptation pathways and even select one and take actions that preserve the option without spending large sums that are needed to realize it, until required (Figure 9.10). Adaptation pathways, combined with monitoring and learning, are an attractive approach for coastal adaptation, especially in cities where large changes will be needed. This approach has been adopted in the Thames Estuary 2100 Project in London (Ranger et al., 2013; Tarrant and Sayers, 2013).

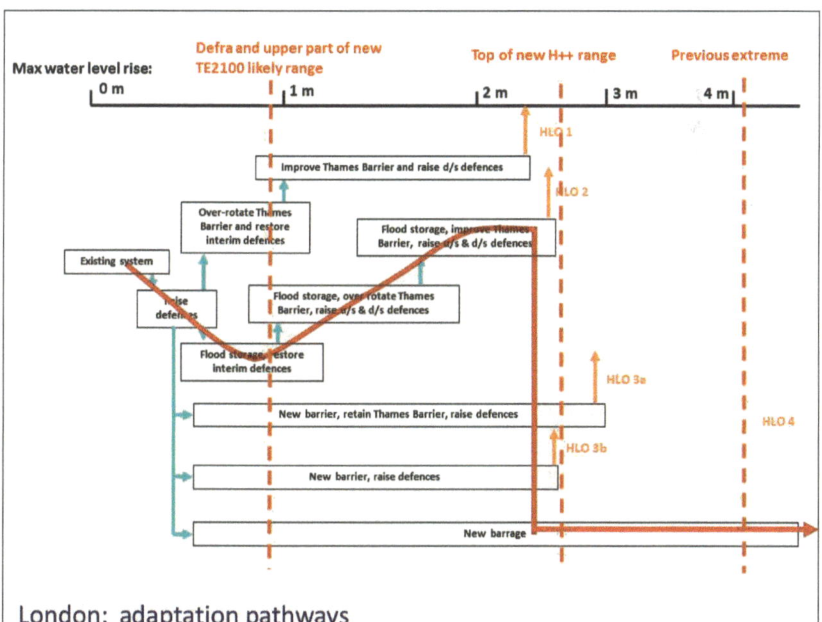

FIGURE 9.10 An example of adaptation pathways for protecting London from coastal flooding: a number of options are shown where effective versus maximum water-level rise, and one possible adaptation pathway through these choices. The number of choices decline with increasing water level and ultimately there is only one option: a new (downstream barrage). *Ranger et al., 2013.*

9.8.3 Choosing between Adaptation Measures/Options

Retreat is often argued as the best response to sea-level rise (e.g., Pilkey and Young, 2009). However, benefit–cost models that compare protection with retreat generally suggest that it is worth investing in widespread protection as populated coastal areas have high economic value (e.g., Fankhauser, 1995; Anthoff et al., 2010; Nicholls et al., 2014a). This does not mean that we *should* protect. Rather the main insight is that these results suggest that significant resources should be available for adapting to sea-level rise, and further that protection can be expected to be a significant part of the portfolio of responses. With or without protection, small island and deltaic areas stand out as relatively more vulnerable in most of these analyses and the impacts fall disproportionately on poorer countries. Even though optimal in a benefit–cost sense, protection costs may overwhelm the capacity of local economies to fund, especially when they are small such as islands (Fankhauser and Tol, 2005; Nicholls and Tol, 2006). While adaptation is essentially a local activity, these funding challenges should be an issue of international concern due to the shared responsibility for climate-induced sea-level rise.

The coastal "adaptation deficit" is an important consideration. This is the cost of adapting coasts to today's climate, before considering adapting to sea-level rise and other climate changes (Burton, 2004; Parry et al., 2009). For example, Hallegatte et al. (2013) identified that US coastal cities have much higher expected damage costs than European coastal cities under current conditions. Equally, less developed and rapidly growing regions such as Africa and Asia are likely have a significant adaptation deficit (e.g., Hinkel et al., 2011), but this requires much more systematic evaluation.

Global cost estimates normally focus on the incremental costs of upgrading defence infrastructure, assuming no adaptation deficit, as this is consistent with the United Nations Framework Convention on Climate Change. Various cost estimates have been produced (Nicholls et al., 2014a), and they are generally smaller than expected, reflecting the high benefit–cost estimates already mentioned. For example, recent global protection costs for flooding were estimated to rise to US$20 and $70 billion/year over the twenty-first century (Hinkel et al., 2014). For 136 coastal cities, Hallegatte et al. (2013) argued that adaptation to sea-level rise would cost about US$50 billion/year. Considering beach erosion, global adaptation costs for sea-level rise only via nourishment estimated costs in 2100 of US$1.5–5 billion/year (Hinkel et al., 2013). These incremental costs seem affordable, but as the cost of the adaptation deficit is not addressed there are uncertainties. At the least, it raises the costs of adaptation in general and protection in particular by a substantial and unknown amount, and it may have the potential to radically change the adaptation pathway we select. Hence the choice between retreat, accommodate, and protect options continues to have significant uncertainties which require further investigation.

It should be noted that in many countries there is limited capacity to address today's coastal problems, let alone consider tomorrow's problems, including sea-level rise. Therefore, promoting coastal adaptation should include developing coastal management capacity and institutions, as already widely recommended (USAID, 2009; Moser et al., 2012).

9.8.4 Adaptation Experience

Through human history, developing technology has increased the range of adaptation options in the face of coastal hazards, and there has been a move from retreat and accommodation to hard protection and active seaward advance via land claim as exemplified by the Netherlands (Van Koningsveld et al., 2008). Rising sea level is one factor calling widespread reliance on protection into question, and the appropriate mixture of protection, accommodation, and retreat, and the whole philosophy of coastal adaptation is being seriously debated as already discussed (Wong et al., 2014).

While there is growing awareness of the need to adapt to sea-level rise, only a few countries or locations are comprehensively preparing for this challenge. Examples include London (Tarrant and Sayers, 2013) and the Netherlands (Kabat et al., 2009; Stive et al., 2011). Both these studies considered a wide range of sea-level rise scenarios, including scenarios of up to 5 m and 4 m, respectively, implicitly thinking beyond 2100. Importantly, they considered adaptation as a process and focused on adaptation pathways as a function of sea-level rise rather than time. This analysis demonstrates that there are options available for large rises in sea level, and in these cases protection upgrade seems feasible for the long-term. This is an effective way to deal with the uncertainty of future sea-level rise (Ranger et al., 2013; Haasnoot et al., 2013). It is worth noting that these activities were more about process and capacity than actual responses at the present time. For instance, the Netherlands has a Delta Commissioner to manage the Delta Plan and develop strategic policy processes and model tools to support this process.

In other locations such as New York City, adaptation is also being carefully considered (Rosenzweig and Solecki, 2010), but the timing of implementation is less clear. In this case, the major event of Cyclone Sandy has accelerated consideration of action, but the outcome is uncertain. In Singapore, new land claim will be raised by approximately a meter to allow for sea-level rise. In general coastal cities are expected to be a major focus for these efforts given the concentration of people and assets, and their ability to fund large investments (Hallegatte et al., 2013; Aerts et al., 2014).

9.9 DISCUSSION/CONCLUSIONS

The chapter illustrates that responding to sea-level rise is a multidimensional problem that crosses many disciplines and embraces natural, social, and engineering sciences, as well as engaging stakeholders, policy, and governance. Sea-level rise has important implications for the world's coast, but the actual outcome will depend on our responses, both in terms of mitigation and adaptation and their success or failure. For adaptation in general, and protection in particular, the likely success or failure is an important uncertainty that deserves more attention, as there are widely divergent views, and this strongly influences how the issue of sea-level rise is considered (e.g., Nicholls and Tol, 2006; Anthoff et al., 2010; Nicholls et al., 2014a). Pessimist and optimist camps exist who have quite different interpretations of the future, especially concerning adaptation. "Pessimists" tend to focus on high rises in sea level, extreme events like Hurricanes Katrina and Sandy and Typhoon Haiyan, and view our ability to adapt to sea-level rise as rather limited, resulting in widespread human migration away from coastal areas, reversing current trends. In contrast, "optimists" tend to focus on uncertainty and lower rises in sea level and stress a high technical ability to protect and the high benefit–cost ratios in developed areas leading to widespread protection. Hence to optimists a major consequence of sea-level rise is the diversion of investment to coastal adaptation in general, and protection in particular.

Optimists have empirical evidence to support their views that we can adapt to sea-level rise in terms of subsiding megacities that are also thriving. Importantly, these analyses suggest that improved protection under rising sea levels is more likely and rational than is widely assumed. The common assumption of a widespread, if not universal, retreat from the shore in the coming decades is not inevitable, and coastal societies will have more choice in their response to rising sea level. However, the pessimists also have evidence to support their view. First, the published protection costs are incremental costs of adapting to sea-level rise, assuming the existence of well-adapted protection infrastructure. The adaptation deficit needs to be considered in the context of adaptation and sea-level rise. Second, assumptions of substantial future population and especially economic growth in coastal areas reinforce the conclusion that protection is worthwhile, lower growth and greater inequalities of wealth may mean less damage in monetary terms, but it will also reduce our adaptive capacity. Third, taking a benefit–cost approach implies a proactive attitude to protection, while historical experience shows most protection has been a reaction to actual or near disaster. Therefore, more frequent coastal disasters and damage are likely in the near- to medium-term, even if the ultimate reactive response is upgraded protection (Hallegatte et al., 2013). Fourth, such disasters could trigger a cycle of decline and abandonment of coastal areas with a profound influence on society's choices concerning coastal protection (cf. Barnett and Adger, 2003). Lastly, maximizing the benefits of retreat and accommodation responses require that implementation occurs soon, which may not happen. Hence, adaptation may not be as successful as some assume, especially for larger rises in sea level. Hence there is much work remaining to understand these diverse issues and shape our coastal future!

Sea-level rise is clearly a threat that demands a response. Climate mitigation can reduce the commitment to sea-level rise, most particularly the potential Greenland and West Antarctic contributions. Local mitigation of human-induced subsidence also needs to be considered where appropriate as it will minimize RSLR in susceptible areas. However, a commitment to sea-level rise remains due to climate effects, supplemented by subsidence in many locations: this requires an adaptation response. There is a need to better understand these threats, including the implications of different mixtures of mitigation and adaptation, and different portfolios of adaptation and adaptation pathways. As the coast is a coupled system, it will be important to examine different scenarios of sea-level rise and climate change, socioeconomic changes, and how adaptation may coevolve with the wider coastal system. Better understanding the long-term effects of adaptation will require such a perspective. There is also a need to engage with and inform the coastal and climate policy process. Research is required at all scales from local to global, but much will be learnt about adaptation in practice (Wong et al., 2014). This will promote more appropriate adaptation options, as well as provide the opportunity to learn from experience.

ACKNOWLEDGMENTS

Dr. Sally Brown is thanked for producing Figure 9.3, and reviewing an earlier version of this chapter as are two anonymous reviewers.

REFERENCES

Aerts, J.C.J.H., Botzen, W.J.W., Emanuel, K., Lin, N., de Moel, H., Michel-Kerjan, E.O., 2014. Evaluating flood resilience strategies for coastal megacities. Science 344, 473–475.

Anthoff, D., Nicholls, R.J., Tol, R.S.J., 2010. The economic impact of substantial sea-level rise. Mitigation Adapt. Strateg. Global Change 15, 321–335.

Barnett, J., Adger, W.N., 2003. Climate dangers and atoll countries. Clim. Change 61, 321–337.

Barth, M.C., Titus, J.G. (Eds.), 1984. Greenhouse Effect and Sea Level Rise: A Challenge for This Generation. Van Nostrand Reinhold, New York, USA.

Bijlsma, L., Ehler, C.N., Klein, R.J.T., Kulshrestha, S.M., McLean, R.F., Mimura, N., Nicholls, R.J., Nurse, L.A., Pérez Nieto, H., Stakhiv, E.Z., Turner, R.K., Warrick, R.A., 1996. Coastal zones and small islands. In: Watson, R.T., Zinyowera, M.C., Moss, R.H. (Eds.), Climate Change 1995: Impacts, Adaptations, and Mitigation of Climate Change: Scientific-technical Analyses. Contribution of Working Group II to the Second Assessment Report of the Intergovernmental Panel on Climate Change. Cambridge University Press, Cambridge, United Kingdom and New York, NY, USA, pp. 289–324.

Bird, E.C.F., 1985. Coastline Changes: A Global Review. Wiley and Sons, New York, USA, 219.

Bird, E.C.F., 2000. Coastal Geomorphology: An Introduction. Wiley and Sons, Chichester, UK.

Borsje, B.W., van Wesenbeeck, B.K., Dekker, F., Paalvast, P., Bouma, T.J., van Katwijk, M.M., de Vries, M.B., 2011. How ecological engineering can serve in coastal protection. Ecol. Eng. 37, 113–122.

Brad Murray, A., Gopalakrishnan, S., McNamara, D.E., Smith, M.D., 2013. Progress in coupling models of human and coastal landscape change. Comput. Geosci. 53, 30–38.

Burton, I., 2004. Climate change and the adaptation deficit. In: French, A., et al. (Eds.), Climate Change: Building the Adaptive Capacity, Meteorological Service of Canada. Environment Canada, pp. 25–33.

Chatterjee, R.S., Fruneau, B., Rudant, J.P., Roy, P.S., Frison, P., Lakhera, R.C., Dadhwal, V.K., Saha, R., 2006. Subsidence of Kolkata (Calcutta) city, India during the 1990s as observed from space by differential Synthetic Aperture Radar Interferometry (D-InSAR) technique. Remote Sens. Environ. 102, 176–185.

Chaussard, E., Amelung, F., Abidin, H., Hong, S.-H., 2013. Sinking cities in Indonesia: ALOS PALSAR detects rapid subsidence due to groundwater and gas extraction. Remote Sens. Environ. 128, 150–161.

Church, J.A., Woodworth, P.L., Aarup, T., Wilson, W.S. (Eds.), 2010. Understanding Sea-level Rise and Variability. Wiley-Blackwell, Hoboken, NJ, USA, 428 pp.

Church, J.A., Clark, P.U., Cazenave, A., Gregory, J.M., Jevrejeva, S., Levermann, A., Merrifield, M.A., Milne, G.A.R., Steven Nerem, R.S., Nunn, P.D., Payne, A.J., Pfeffer, W.T., Stammer, D., Unnikrishnan, A.S., 2013. Sea level change. In: Climate Change 2013: The Physical Science Basis IPCC Working Group I, Fifth Assessment Report. Intergovernmental Panel on Climate Change.

Center for International Earth Science Information Network - CIESIN - Columbia University, 2013. Low Elevation Coastal Zone (LECZ) Urban-rural Population and Land Area Estimates, Version 2. NASA Socioeconomic Data and Applications Center (SEDAC), Palisades, NY. http://sedac.ciesin.columbia.edu/data/set/lecz-urban-rural-population-land-area-estimates-v2 (accessed 21.04.14.).

Crossland, C.J., Kremer, H.H., Lindeboom, H.J., Marshall Crossland, J.I., Le Tissier, M.D.A. (Eds.), 2005. Coastal Fluxes in the Anthropocene. Springer, Berlin, Germany.

Dang, V.K., Doubre, C., Weber, C., Gourmelen, N., Masson, F., 2014. Recent land subsidence caused by rapid urban development in the Hanoi region (Vietnam) using ALOS InSAR data. Nat. Hazards Earth Syst. Sci. 14, 657–674.

Ericson, J.P., Vorosmarty, C.J., Dingman, S.L., Ward, L.G., Meybeck, M., 2006. Effective sea-level rise and deltas: causes of change and human dimension implications. Global Planet. Change 50, 63–82.

Fankhauser, S., 1995. Protection versus retreat: estimating the costs of sea-level rise. Environ. Plann. A 27, 299–319.

Fankhauser, S., Tol, R.S.J., 2005. On climate change and economic growth. Res. Energy Econ. 27, 1–17.

Gornitz, V., 2013. Rising Seas: Past, Present and Future. Columbia University Press, New York, USA, 344 pp.

Grossi, P., Muir-Wood, R., 2006. Flood Risk in New Orleans: Implications for Future Management and Insurability. Risk Management Solutions (RMS), London, UK.

Haasnoot, M., Kwakkel, J.H., Walker, W.E., ter Maat, J., 2013. Dynamic adaptive policy pathways: a method for crafting robust decisions for a deeply uncertain world. Global Environ. Change 23, 485–498.

Hallegatte, S., 2009. Strategies to adapt to an uncertain climate change. Global Environ. Change 19, 240–247.

Hallegatte, S., Green, C., Nicholls, R.J., Corfee-Morlot, J., 2013. Future flood losses in major coastal cities. Nat. Clim. Change 3, 802–806. http://dx.doi.org/10.1038/nclimate1979.

Hay, J.E., 2009. Institutional and Policy Analysis of Disaster Risk Reduction and Climate Change Adaptation in Pacific Island Countries. United Nations International System for Disaster Reduction (UNISDR) and the United Nations Development Programme (UNDP), Suva, Fiji.

Hinkel, J., Brown, S., Exner, L., Nicholls, R.J., Vafeidis, A.T., Kebede, A.S., 2011. Sea-level rise impacts on Africa and the effects of mitigation and adaptation: an application of DIVA. Regi. Environ. Change. http://dx.doi.org/10.1007/s10113-011-0249-2.

Hinkel, J., Nicholls, R.J., Tol, R.S.J., Wang, Z.B., Hamilton, J.B., Boot, G., Vafeidis, A.T., McFadden, Loraine, Ganopolski, A., Klein, R.J.T., 2013. A global analysis of erosion of sandy beaches and sea-level rise: an application of DIVA. Global Planet. Change 111, 150–158. http://dx.doi.org/10.1016/j.gloplacha.2013.09.002.

Hinkel, J., Lincke, D., Vafeidis, A.T., Perrette, M., Nicholls, R.J., Tol, R.S.J., Marzeion, B., Fettweis, X., Ionescu, C., Levermann, A., 2014. Coastal flood damage and adaptation costs under 21st century sea-level rise. Proc. Natl. Acad. of Sci.. http://dx.doi.org/10.1073/pnas.1222469111.

Holgate, S.J., Matthews, A., Woodworth, P.L., Rickards, L.J., Tamisiea, M.E., Bradshaw, E., Foden, P.R., Gordon, K.M., Jevrejeva, S., Pugh, J., 2013. New data systems and products at the Permanent Service for Mean Sea Level. J. Coastal Res. 29, 493–504. http://dx.doi.org/10.2112/JCOASTRES-D-12-00175.1.

Huq, S., Reid, H., 2004. Mainstreaming adaptation in development. IDS Bull. 35, 15–21. http://dx.doi.org/10.1111/j.1759-5436.2004.tb00129.x.

IPCC CZMS, 1990. Strategies for Adaption to Sea Level Rise. Report of the Coastal Zone Management Subgroup, Response Strategies Working Group of the Intergovernmental Panel on Climate Change. Ministry of Transport, Public Works and Water Management, The Hague, The Netherlands.

Jones, L., Angus, S., Cooper, A., Doody, P., Everard, M., Garbutt, A., Gilchrist, P., Hansom, J., Nicholls, R.J., Pye, K., Ravenscroft, N., Rees, S., Rhind, P., Whitehouse, A., 2011. Coastal margins. In: UK National Ecosystem Assessment Technical Report. United Nations Environment Programme World Conservation Monitoring Centre (UNEP-WCMC), Cambridge, UK.

Kabat, P., Fresco, L.O., Stive, M.J.F., Veerman, C.P., van Alphen, J., Parmet, B., Hazeleger, W., Katsman, C.A., 2009. Dutch coasts in transition. Nat. Geosci. 2, 450–452.

Kaneko, S., Toyota, T., 2011. Long-term urbanization and land subsidence in asian megacities: an indicators system approach. In: Taniguchi, M. (Ed.), Groundwater and Subsurface Environments: Human Impacts in Asian Coastal Cities, pp. 249–270. http://dx.doi.org/10.1007/978-4-431-53904-9_13.

Klein, R.J.T., Nicholls, R.J., 1999. Assessment of coastal vulnerability to climate change. Ambio 28, 182–187.

Klein, R.J.T., Aston, J., Buckley, E.N., Capobianco, M., Mizutani, N., Nicholls, R.J., Nunn, P.D., Ragoonaden, S., 2000. Coastal adaptation. In: Metz, B., Davidson, O.R., Martens, J.W., Van Rooijen, S.N.M., Van Wie McGrory, L.L. (Eds.), IPCC Special Report on Methodological and Technological Issues in Technology Transfer. Cambridge University Press, Cambridge, UK.

Klein, R.J.T., Midgley, G.F., Preston, B.L., Alam, M., Berkhout, F.G.H., Dow, K., Shaw, M.R., 2014. Adaptation Opportunities, constraints, and limits. In: Climate Change 2014: Impacts Adaptation and Vulnerability IPCC Working Group II, Fifth Assessment Report. Intergovernmental panel on Climate Change.

Kron, W., 2013. Coasts: the high-risk areas of the world. Nat. Hazards 66, 1363–1382.

Linham, M.M., Nicholls, R.J., 2010. Technologies for Climate Change Adaptation: Coastal Erosion and Flooding. TNA Guidebook Series. UNEP Risø Centre on Energy, Climate and Sustainable Development, Roskilde, Denmark, 150pp.

McGranahan, G., Balk, D., Anderson, B., 2007. The rising tide: assessing the risks of climate change and human settlements in low elevation coastal zones. Environ. Urban. 19, 17–37.

Menéndez, M., Woodworth, P.L., 2010. Changes in extreme high water levels based on a quasi-global tide-gauge dataset. J. Geophys. Res. 115, C10011.

Milliman, J.D., Broadus, J.M., Gable, F., 1989. Environmental and economic implications of rising sea level and subsiding deltas: the Nile and Bengal examples. Ambio 18, 340–345.

Moser, S.C., Williams, S.J., Boesch, D.F., 2012. Wicked challenges at Land's end: managing coastal vulnerability under climate change. Annu. Rev. Environ. Resour. 37, 51–78.

Nicholls, R.J., 2010. Impacts of and responses to sea-level rise. In: Church, J.A., Woodworth, P.L., Aarup, T., Wilson, S. (Eds.), Understanding Sea-level Rise and Variability. Wiley-Blackwell, pp. 17–51.

Nicholls, R.J., Cazenave, A., 2010. Sea level rise and its impact on coastal zones. Science 328, 1517–1520.

Nicholls, R.J., Hanson, S., Herweijer, C., Patmore, N., Hallegatte, S., Corfee-Morlot, J., Chateau, J., Muir-Wood, R., 2008. Ranking Port Cities with High Exposure and Vulnerability to Climate Extremes—Exposure Estimates. Environmental Working Paper No. 1. Paris Organisation for Economic Co-operation and Development (OECD).

Nicholls, R.J., Klein, R.J.T., 2005. Climate change and coastal management on Europe's coast. In: Vermaat, J.E., Ledoux, L., Turner, K., Salomons, W., Bouwer, L. (Eds.), Managing European Coasts: Past, Present and Future. Springer, London, UK (Environmental Science Monograph Series).

Nicholls, R.J., Tol, R.S.J., 2006. Impacts and responses to sea-level rise: a global analysis of the SRES scenarios over the twenty-first century. Philos. Trans. R. Soc. A Math. Phys. Eng. Sci. 364, 1073–1095.

Nicholls, R.J., Wong, P.P., Burkett, V.R., Codignotto, J.O., Hay, J.E., McLean, R.F., Ragoonaden, S., Woodroffe, C.D., 2007. Coastal systems and low-lying areas. In: Parry, M.L., Canziani, O.F., Palutikof, J.P., Van Der Linden, P., Hanson, C.E. (Eds.), Climate Change 2007: Impacts, Adaptation and Vulnerability. Cambridge University Press, Cambridge, UK.

Nicholls, R.J., Woodroffe, C.D., Burkett, V.R., 2009. Coastal degradation as an Indicator of global change. In: Letcher, T.M. (Ed.), Climate Change: Observed Impacts on Planet Earth. Elsevier, Oxford, pp. 409–424.

Nicholls, R.J., Townend, I.H., Bradbury, A., Ramsbottom, D., Day, S., 2013. Planning for long-term coastal change: experiences from England and Wales. Ocean Eng. 71, 3–16. http://dx.doi.org/10.1016/j.oceaneng.2013.01.025.

Nicholls, R.J., Stive, M.J.F., Tol, R.S.J., 2014a. Coping with coastal change. In: Masselink, G., Gehrels, R. (Eds.), Coastal Environments and Global Change. Wiley, Chichester, UK.

Nicholls, R.J., Hanson, S.E., Lowe, J.A., Warrick, R.A., Lu, X., Long, A.J., 2014b. Sea-level scenarios for evaluating coastal impacts. Wiley Interdiscip. Rev.: Clim. Change 5 (1), 129–150.

Nurse, L., McLean, R., Agard, J., Briguglio, L.P., Duvat, V., Pelesikoti, N., Tompkins, E., Webb, A., 2014. Small islands. In: Climate Change 2014: Impacts, Adaptation and Vulnerability IPCC Working Group II, Fifth Assessment Report. Intergovernmental Panel on Climate Change.

Parry, M.L., Arnell, N.W., Berry, P.M., Dodman, D., Fankhauser, S., Hope, C., Kovats, S., Nicholls, R.J., Satterthwaite, D., Tiffin, R., Wheeler, T., 2009. Assessing the costs of adaptation to climate change: a review of the UNFCCC and other recent estimates. International Institute for Environment and Development and Grantham Institute for Climate Change, London, 111pp. Permanent Service for Mean Sea Level (PSMSL), 2014, "Tide Gauge Data", Retrieved 30 Jun 2014 from. http://www.psmsl.org/data/obtaining/.

Pilkey, O.H., Young, R., 2009. The Rising Sea. Island Press/Shearwater Books, Washington, D.C., 203pp.

PSMSL, 2014. Data from the Permanent Service for Mean Sea Level (PSMSL). http://www.psmsl.org/ (accessed 08.05.14.).

Pugh, D., Woodworth, P., 2014. Sea-level Science: Understanding Tides, Surges, Tsunamis and Mean Sea-level Change. Cambridge University Press, Cambridge, 395pp.

Ranger, N., Reeder, T., Lowe, J.A., 2013. Addressing 'deep' uncertainty over long-term climate in major infrastructure projects: four innovations of the Thames Estuary 2100 Project. Eur. J. Decis. Processes 1, 233–262.

Rawlani, A.K., Sovacool, B.J., 2011. Building responsiveness to climate change through community based adaptation in Bangladesh. Mitigation Adapt. Strateg. Global Change 16, 845–863.

RIBA and ICE, 2010. Facing up to Rising Sea Levels. Retreat? Defend? Attack? RIBA (Royal Institute of British Architects) and ICE (Institution of Civil Engineers), London. http://www.buildingfutures.org.uk/assets/downloads/Facing_Up_To_Rising_Sea_Levels.pdf.

Rodolfo, K.S., Siringan, F.P., 2006. Global sea-level rise is recognised, but flooding from anthropogenic land subsidence is ignored around northern Manila Bay. Philipp. Disaster Manage. 30, 118–139.

Rosenzweig, C., Solecki, W. (Eds.), 2010. Climate Change Adaptation in New York City: Building a Risk Management Response, vol. 1196. Annals of the New York Academy of Sciences, pp. 1–354.

Ruocco, A.C., Nicholls, R.J., Haigh, I.D., Wadey, M.P., 2011. Reconstructing coastal flood occurrence combining sea level and media sources: a case study of the Solent, UK since 1935. Nat. Hazards 59 (3), 1773–1796. http://dx.doi.org/10.1007/s11069-011-9868-7.

Seasholes, N.C., 2003. Gaining Ground: A History of Landmaking in Boston. MIT Press, Cambridge, MA, USA.

Smith, K., 2013. Environmental Hazards: Assessing Risk and Reducing Disaster. Routledge, London, UK.

Stive, M.J.C., Fresco, L.O., Kabat, P., Parmet, B.W.A.H., Veerman, C.P., 2011. How the Dutch plan to stay dry over the next century. Proc. Inst. Civ. Eng. 164, 114–121.

Syvitski, J.P.M., Kettner, A.J., Overeem, I., Hutton, E.W.H., Hannon, M.T., Brakenridge, G.R., Day, J., Vörösmarty, C., Saito, Y., Giosan, L., Nicholls, R.J., 2009. Sinking deltas. Nat. Geosci. 2, 681–689.

Tarrant, O., Sayers, P.B., 2013. Managing flood risk in the Thames Estuary – the development of a long-term robust and flexible strategy. In: Sayers, P.B. (Ed.), Flood Risk: Planning, Design and Management of Flood Defence Infrastructure. ICE Publishing, London, UK, pp. 303–326.

Teatini, P., Tosi, L., Strozzi, T., Carbognin, L., Wegmüller, U., Rizzetto, F., 2005. Mapping regional land displacements in the Venice coastland by an integrated monitoring system. Remote Sens. Environ. 98 (4), 403–413.

Temmerman, S., Meire, P., Bouma, T.J., Herman, P.M.J., Ysebaert, T., De Vriend, H.J., 2013. Ecosystem-based coastal defence in the face of global change. Nature 504, 79–83.

Tribbia, J., Moser, S.C., 2008. More than information: what coastal managers need to plan for climate change. Environ. Sci. Policy 11, 315–328.

Tsyban, A., Everett, J., Titus, J., 1990. World oceans and coastal zones. In: McG. Tegart, W.J., Sheldon, G.W., Griffiths, D.C. (Eds.), Climate Change: The IPCC Impacts Assessment. Australian Government Publishing Service, Canberra, pp. 6-1–6-28.

USAID, 2009. Adapting to Coastal Climate Change: A Guidebook for Development Planners. USAID, Rhode Island. Available from. www.crc.uri.edu/download/CoastalAdaptationGuide.pdf (accessed 07.10.10.).

Valiela, I., 2006. Global Coastal Change. Blackwell, Malden, MA, USA.

Van Koningsveld, M., Mulder, J.P.M., Stive, M.J.F., Van Der Valk, L., Van Der Weck, A.W., 2008. Living with sea-level rise and climate change: a case study of the Netherlands. J. Coastal Res. 24, 367–379.

Wong, P.P., Losada, I.J., Gattuso, J., Hinkel, J., Khattabi, A., McInnes, K., Saito, Y., Sallenger, A., 2014. Coastal systems and low-lying areas. In: Climate Change 2014: Impacts, Adaptation and Vulnerability IPCC Working Group II, Fifth Assessment Report Intergovernmental Panel on Climate Change.

Zhang, K.Q., Douglas, B.C., Leatherman, S.P., 2000. Twentieth-century storm activity along the US east coast. J. Clim. 13, 1748–1761.

Chapter 10

Storm-Induced Morphology Changes along Barrier Islands and Poststorm Recovery

Ping Wang[1] **and Tiffany M. Roberts Briggs**[2]

[1] *School of Geosciences, University of South Florida, Tampa, FL, USA,*

[2] *Department of Geosciences, Florida Atlantic University, Boca Raton, FL, USA*

ABSTRACT

Barrier islands, or narrow strips of sand islands in the sea, have the distinction of being among the most vulnerable, yet most desirable sites for human habitation. Vulnerabilities of barrier islands include risks associated with sea-level rise, as well as energetic ocean events, such as tsunamis and storms, the latter of which are crucial in reshaping barrier islands. This chapter discusses barrier-island morphology and subenvironments, storm impacts to barrier-island morphology, and short-term, poststorm recovery of barrier islands, focusing particularly on tropical storms. Sallenger (2000) identified four levels of storm impacts to barrier-island morphology. From the weakest to the strongest, they are swash regime, collision regime, overwash regime, and inundation regime. This chapter describes various examples of each impact scale in terms of morphology changes in each subenvironment. In addition, morphology changes caused by seaward-directed flows associated with ebbing storm surge are reviewed. Beach recovery initiates as the storm energy subsides, generally in the morphologic form of ridge and runnel development. Continued beach recovery includes increased elevation of the ridge crest, that is, growth of beach berm, and overwash deposits in the runnel, eventually welding the ridge.

10.1 INTRODUCTION

Managing the risks of extreme events and adapting to global climate change are major challenges to humankind in the twenty-second century (IPCC, 2012). Densely populated coastal zones are particularly vulnerable to extreme events, especially tropical and extratropical cyclones, coupled with the anticipated (and predicted possibility of accelerating) sea-level rise. Understanding and improving predictive capabilities of storm impacts to coastal regions are

Coastal and Marine Hazards, Risks, and Disasters. http://dx.doi.org/10.1016/B978-0-12-396483-0.00010-8
Copyright © 2015 Elsevier Inc. All rights reserved.

crucial to effective risk assessment management, hazard mitigation, and coastal-zone adaptation. The extreme energy associated with storms often induces large and rapid morphology changes. Along heavily developed coasts, these impacts to homes and infrastructure can be catastrophic and costly.

Although all types of coastal environments are vulnerable in some capacity to extreme storm events, the specific vulnerability and storm impact depend on the antecedent morphology, sediment characteristics, and oceanographic conditions of a particular coast. Especially vulnerable to energetic events are sandy coasts and barrier islands, where large storm-induced morphologic changes are often observed. Barrier islands, narrow strips of sand islands in the sea, have the distinction of being among the most vulnerable, yet most desirable sites for human habitation. Barrier islands are composed of unconsolidated sediment, and are quite mobile with the dynamic interactions between the land and sea. Consequentially, barrier islands are highly susceptible to risks associated with sea-level rise, tsunamis, and storms. In particular, storms play a significant role in reshaping barrier islands at shorter time scales. Sustainability of the natural and human environments on densely populated barrier islands continues to be an immense challenge. In this chapter, we review beach and barrier-island system responses to storms and the poststorm recovery.

Barrier islands comprise approximately 15 percent of the world's coast (Glaeser, 1978) occurring on all continents (except Antarctica) and at nearly all latitudes (Davis, 1994). In general, barrier islands, particularly extensive barrier-island chains, such as those along the US Gulf of Mexico and Atlantic coasts, tend to develop along passive margins (Inman and Nordstrom, 1971). Beyond tectonic controls at a global scale, barrier islands can vary substantially in size, shape, and sediment characteristics. For example, eastern North America (including the US Gulf of Mexico and Atlantic coasts) hosts an extensive distribution of barrier islands, which vary in morphology and sediment composition.

The barrier-island system consists of several subenvironments that are influenced by different coastal processes and interactions (Figure 10.1). The crossshore distribution of subenvironments, from seaward to landward, in a barrier-island system may include the nearshore zone, subaerial beach, dune field, overwash platform, interior wetland, back-barrier beach or wetland, and back-barrier bay (Figure 10.1(a)); however, not all subenvironments are necessarily always present. In the longshore direction, the barrier-island system is composed of the barrier islands separated by tidal inlets (Figure 10.1(b)). The tidal inlet complex may include tidal channels, a flood tidal delta, and an ebb-tidal delta.

Based on relative dominance of wave and tidal forcing, Hayes (1979) proposed a morphodynamic classification of coast, including wave-dominated, mixed-energy, and tide-dominated coasts. Barrier islands do not typically develop along tide-dominated coasts. The morphodynamics of barrier-island systems are strongly influenced by the interaction between the barrier-island

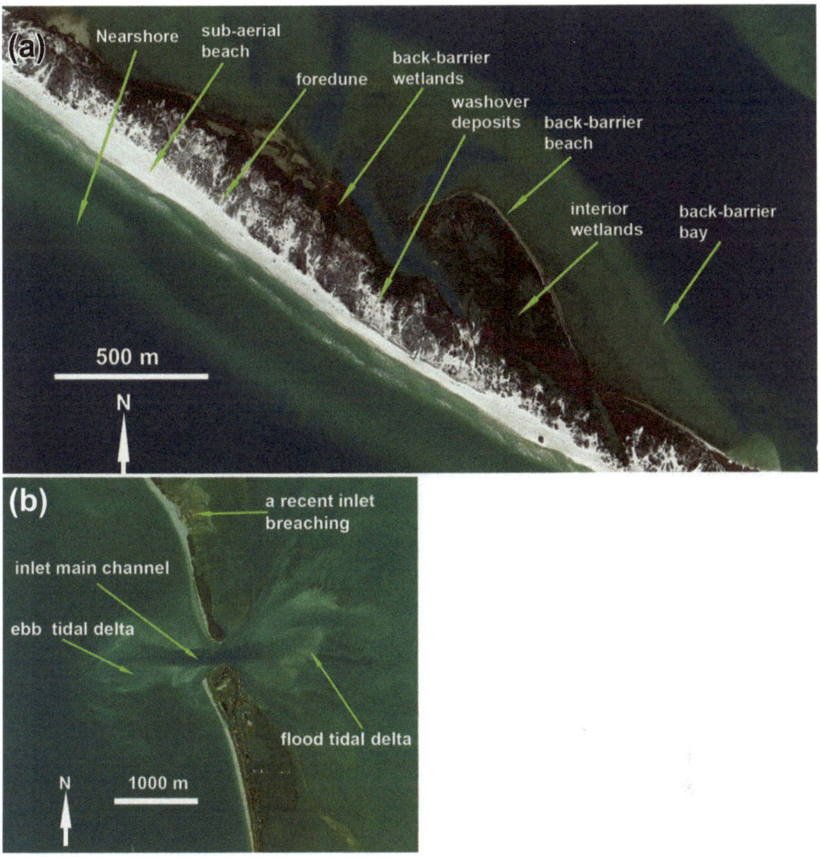

FIGURE 10.1 Subenvironments in a barrier-island system: (a) different barrier-island subenvironments in the crossshore direction; Shell Island, Florida, USA; (b) different barrier-island subenvironments in the longshore direction; Redfish Pass and adjacent barrier islands, Florida, USA.

beach/nearshore environments (controlled mostly by wave forcing) and the adjacent tidal inlets (strongly influenced by tidal forcing). Hayes (1979) and Davis and Hayes (1984) further classified barrier-island systems, based on relative dominance of wave and tidal forcing (Figure 10.2). Wave-dominated barrier islands tend to be long and narrow, with tidal inlets spaced far apart (Figure 10.2(a)). The associated ebb-tidal deltas tend to be small to nonexistent. Wave-dominated barrier islands are typical along the northern and western Gulf of Mexico coast. Mixed-energy barrier islands often are wider on one end compared to the other (i.e., are drumstick shaped), are influenced by wave refraction over the relatively large, ebb-tidal delta, and the resultant sediment transport pattern is along the adjacent barrier-island beach (Figure 10.2(b)). Drumstick barrier islands are common along the Georgia and South Carolina coasts.

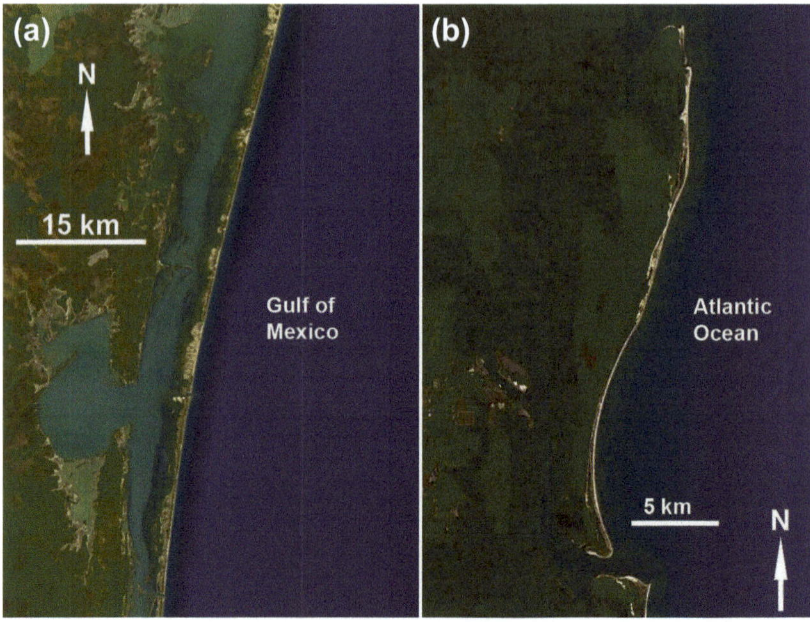

FIGURE 10.2 Morphodynamic classification of barrier-island systems: (a) wave-dominated barrier island, note the long barrier island with relatively narrow tidal inlets that are far apart; Laguna Madre, Tamanlipas, Mexico; (b) mixed-energy barrier island, note the drumstick shape; Cumberland Island, Georgia, USA.

The morphodynamics of barrier-island systems are strongly influenced by high-energy events, or storms. Storms generate large waves, often on top of an elevated water level or storm surge, which attack barrier islands. This causes substantial morphology changes, not only in areas that constantly interact with the ocean (e.g., the nearshore zone), but also in areas that do not normally interact with the ocean (e.g., the dunes and interior wetlands). Further, storms can induce barrier-island breaching and the formation of new tidal inlets. The development of new tidal inlets often induces significant morphology responses along several adjacent barrier islands at a time scale of years to tens of years, as the new inlet captures the tidal prism that is served by existing inlets (Wang et al., 2011; Wang and Beck, 2012).

10.1.1 Subenvironments of the Barrier-Island System

Barrier-island subenvironments (Figure 10.1) are distinguished according to the crossshore or longshore morphology variations within the barrier-island system. Each subenvironment is dominated by different processes, and consequently responds differently to storm impacts. The nearshore zone, within the transition from ocean to land, is the zone where incident ocean wave energy dissipates under both normal and storm conditions. This region, particularly the zone where waves break, is dynamic with active sediment transport and constant morphology change. Varying both temporally and spatially, the nearshore zone may exhibit various morphologies ranging from a monotonic seaward-sloping profile to a profile with one or multiple bars. The nearshore morphology responds to variations in hydrodynamics at various temporal scales including, for example, seasonal, annual, and event scales. The result is a complex feedback exchange between nearshore waves and currents with the nearshore morphology.

The subaerial beach is located above the spring high-tide line. Under normal conditions, sediment transport is typically eolian, as wave action does not directly reach this part of the barrier island. However, during energetic conditions, storm-generated waves are superimposed on elevated water levels due to storm surge, resulting in active wave-induced sediment transport and morphology change across a larger crossshore region of the barrier island than during normal conditions. The subaerial beach change can also be caused by erosion and accretion of the intertidal zone, resulting in landward and seaward movement of the shoreline and subsequent widening and narrowing of the beach.

The highest elevations within the subaerial section of barrier islands are dunes. The dunes directly landward of the beach are referred to as foredunes. They often develop a ridge morphology. Vegetation can play a key role in dune dynamics as a stabilizing agent. Dune development and stability can be a major factor in the morphologic response of the barrier island to large storm waves superimposed on high water levels (discussed in the following sections). Figure 10.3 illustrates the zonation of the dune–beach–nearshore system (USACE, 2002).

Within the low-lying interior of barrier islands, wetlands can be present (Figure 10.1(a)). The relative low elevation of the interior wetlands provides accommodation space for sediment deposition associated with onshore eolian sediment transport or storm overwash. The typically wet surface and vegetation coverage retards sediment transport, making this a predominantly depositional environment. When filled by wind-blown sand and/or storm washover, the interior wetlands can eventually evolve into a dune field.

The shoreline along the bayside of a barrier island is quite different from that along the open coast, and can be composed of sandy beaches, marshes, or mangrove swamps in tropical latitudes (Figure 10.1(a)). Wave energy along the bayside shoreline is typically low due to the limited wind fetch of most back-barrier water bodies. However, given persistent and high-velocity winds, the choppy bay waves can be erosive to back-barrier shorelines. Thus, the morphodynamics of the bayside shoreline are strongly influenced by the wind fetch over the back-barrier bay (Stone et al., 2004). Within low-lying barrier islands, the bayside shoreline tends to be irregularly shaped from overwash deposition.

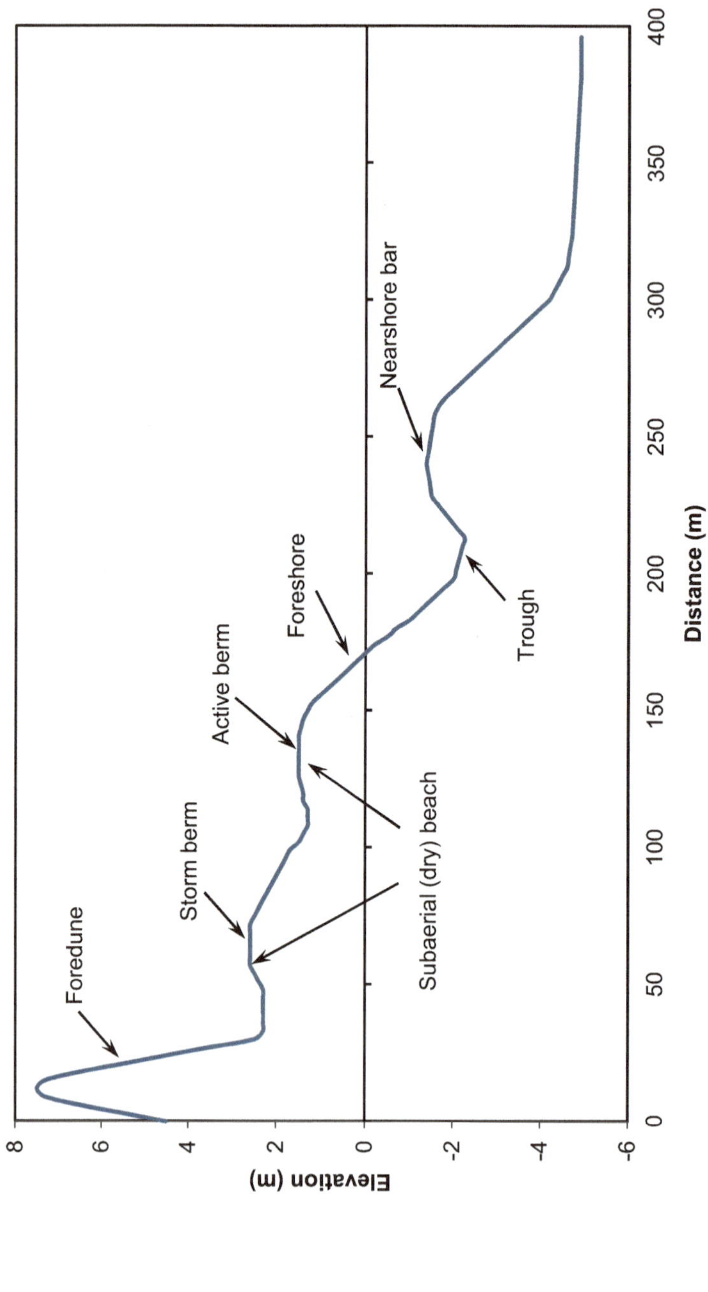

FIGURE 10.3 Detailed morphological zonation of the seaward side of a barrier island.

The back-barrier bay is landward of the subaerial barrier island. The width of the back-barrier bay (the distance from the barrier-island bayside shoreline to the mainland shoreline) varies substantially from a few hundred meters to many kilometers. The back-barrier bay typically does not provide sediment directly to the bayside shoreline. Instead, it ultimately provides accommodation space for storm overwash and inundation. Landward propagation of the bayside shoreline results from overwash and inundation by energetic storms and potential eolian transport.

Tidal inlets interrupt longshore sediment transport along an otherwise continuous shoreline. Interactions between barrier islands and tidal inlets have a strong influence on the adjacent shorelines and on overall morphodynamics of barrier-island systems (Figure 10.1(b)). Barrier islands breaching during storms often form tidal inlets. A flood tidal delta initially forms from storm deposits during breaching, which may be reshaped by flood tides and currents. The ebb-tidal delta, controlled by the combined effects of the ebbing tide and nearshore hydrodynamics, is typically a sediment sink within the barrier-island depositional system. The relatively shallow water over the ebb-tidal delta can have a significant influence on the pattern of wave propagation, and therefore affect the patterns of sediment transport and barrier-island morphology.

In summary, a barrier-island system contains several subenvironments that influence barrier-island morphodynamics. Storms have different impacts on the different barrier-island subenvironments. The nearshore, subaerial beach, and dunes are the general regions that experience storm-induced erosion, whereas interior wetlands, bayside shoreline regions, and the back-barrier bay are likely sites of storm-induced deposition of sediments eroded from the other nearby subenvironments. Morphology changes also occur in the offshore region, typically receiving eroded sediments from the beach and dunes.

10.1.2 The Intensifying and Subsiding Phases of a Storm

The passage of a storm can be divided into two phases: the intensifying phase and the subsiding phase. Morphologic response of the various barrier-island subenvironments can be quite different during the different storm phases, although detailed morphology changes during the storm are not well understood due to difficulties of measurements. Figure 10.4 illustrates the variations of water level and wave height associated with the passage of Superstorm Sandy along the coast of Long Island, New York in October 2012. The storm surge is the difference between the predicted water level and the measured water level (Figure 10.4(a)). Storm surge is an important component of the elevated water levels measured during storms, in addition to wave setup and swash runup.

During the intensifying phase of the storm, the energy at the coast increases rapidly in the form of rising water levels (i.e., storm surge) and increasing wave heights and wind speeds. The barrier-island subenvironments that are

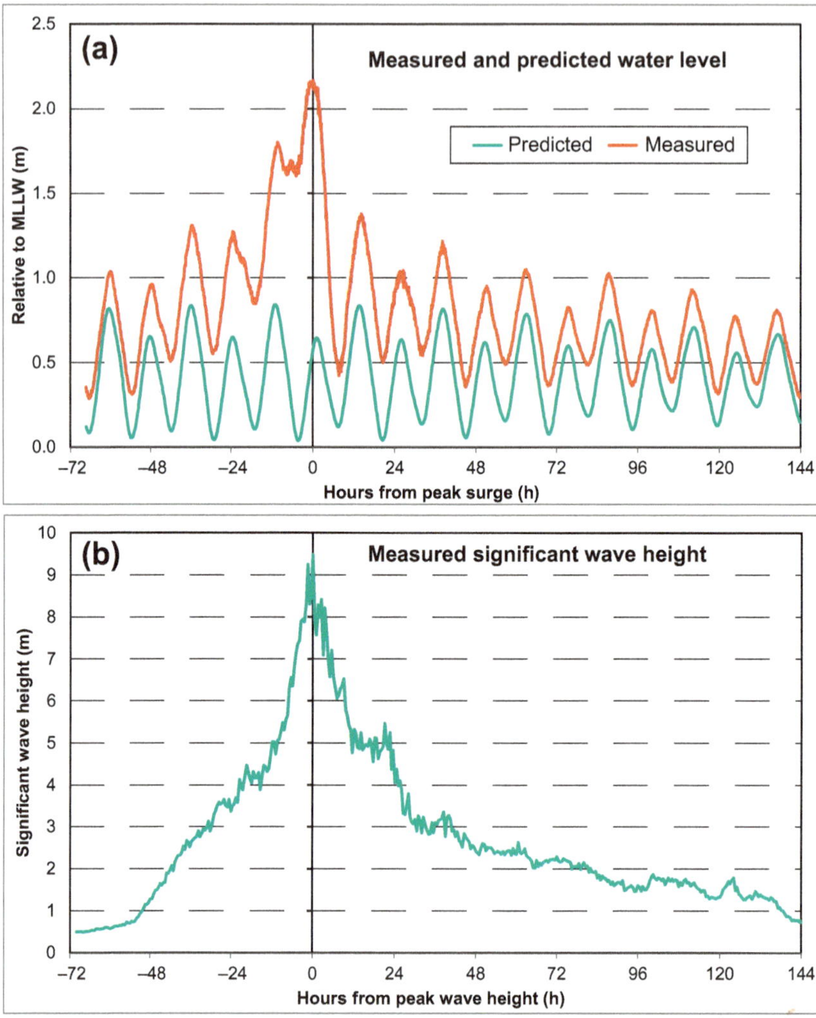

FIGURE 10.4 Measured storm surge and wave height during the passage of Superstorm Sandy. The water level was measured at NOAA station 8510560 at the east tip of Long Island, New York. The wave height was measured at NOAA station 44097, approximately 60 km to the west of Long Island. (a) Measured and predicted water level, note the rising of the water level followed by rapid subsidence. (b) Measured significant wave height, note the rapid increasing of the wave height and the slightly slower decreasing.

typically subaerial and not in constant contact with ocean forcing, for example, backshore and dunes (Figure 10.3), can become submerged and undergo energetic wave actions. The morphology of these subenvironments adjusts rapidly to the rapidly increasing forcing and evolves toward an equilibrium state with energetic storm waves and elevated water levels.

After the storm energy peaks and it moves away from the coast, the system enters the subsiding phase, with receding water levels and decreasing wave height and wind speed. Although the energy is subsiding, it is still much more energetic than normal conditions, and active sediment transport continues. Within hours to days of the storm's passage, the beach and nearshore zones start to recover. Short-term poststorm recovery is discussed in detail in the following sections. It should be kept in mind that barrier-island morphology responses during the intensifying phase can be significantly different from those during the subsiding phase.

10.2 FACTORS CONTROLLING STORM IMPACT ON BARRIER-ISLAND MORPHOLOGY

Barrier-island response to storm impact depends on the hydrodynamic characteristics of the forcing mechanism and the morphological properties of the responding environments. The forcing mechanisms from the storm are controlled by its meteorological and oceanographic characteristics, such as storm intensity, size, track and forward speed, and bathymetric characteristics of the continental shelf. The responding environmental factors include maximum elevation and width of the barrier-island, nearshore morphology (e.g., barred or nonbarred coast), continuity and width of the dune field, beach width, and offshore bathymetric characteristics.

10.2.1 Driving Factors Controlled by the Tropical Storm

This section focuses mostly on driving mechanisms associated with tropical storms. Although extratropical (or winter) storms can generate high waves and storm surges, detailed meteorological conditions are different. The most commonly used tropical storm (particularly hurricanes) classification is the Saffir-Simpson wind scale. This scale is based on the maximum sustained surface wind speed (peak 1-min wind at the standard meteorological observation height of 10 m over an unobstructed exposure) associated with the cyclone (NOAA, 2012). However, in addition to wind speed, storm-induced morphologic impacts along a barrier-island coast are controlled by a combination of several additional factors.

The size, track, and landfall location to the right or left side of the storm (relative to the storm's eye) play important roles determining the magnitude of impact. Tropical cyclones rotate anticlockwise in the northern hemisphere. Thus, the coastline to the right side of the approaching storm experiences onshore winds that generate high waves and storm surge. On the left side of the storm, the offshore-directed winds suppress the storm surge and waves, which reduces morphologic impact compared to the coastline on the right side of the storm. The storm track will determine which coastal stretches will experience stronger onshore forcing (right side of the storm) or reduced wave and surge conditions (left side of the storm). The actual size of the storm determines the area of influence by the storm forcing; the larger the storm, the greater the area of impact.

The forward moving speed of the storm system is another crucial factor determining the degree of impact. For example, although the maximum wind speeds are comparable, Hurricane Ivan (2004) moved much slower than Hurricane Denise (2005), 13 versus 22 km/h at landfall. Additional meteorological and oceanographic information on Hurricane Ivan and Hurricane Denise can be found in Claudino-Sales et al. (2008). Hurricane Ivan induced much greater morphology changes than Hurricane Denise (Wang and Horwitz, 2007; Claudino-Sales et al., 2008). A slow-moving system provides more time for the growth of storm surge and waves, as compared to a fast-moving system. Further, a slow-moving system means more time of elevated water levels and a longer duration for the energetic storm waves to reshape the barrier islands. Therefore, a large, slow-moving storm system is particularly effective in causing substantial barrier-island morphology changes.

The bathymetric characteristics of the continental shelf, especially those of the inner continental shelf, influence the development and propagation of storm surge and waves. A steep continental shelf, such as those along the west coast of the United States, may limit the development of storm surge, whereas a very gentle and wide continental shelf, such as those along the northern Gulf of Mexico coast, may lead to the development of higher storm surge. On the other hand, the gentle continental shelf may limit wave heights due to depth-limited breaking and wave energy loss from bottom friction. Houser et al. (2008) found that the transverse ridges on the inner continental shelf of northwest Florida coast had significant control on the alongshore pattern of morphology changes of dunes and beach-nearshore during the impact of Hurricane Ivan. Haerens et al. (2012) documented the considerable influences of numerous offshore sand banks on storm-induced beach changes along Belgian coast. Lentz et al. (2013) concluded that the offshore bathymetry variations and geological framework exert significant control on both short-term and medium-long morphological evolution of Fire Island, New York.

In summary, impacts of a storm to barrier-island morphology are influenced by the combination of a suite of meteorological and oceanographic characteristics associated with the storm, rather than simply wind speed or any other individual factor. The combination of a large, slow-moving storm over a broad and gentle continental shelf with high wind speeds (e.g., conditions associated with Hurricane Katrina), will likely lead to a higher storm surge and waves, with the capacity for inducing greater morphology change. Fritz et al. (2007) found that Hurricane Katrina, which was a large, slow-moving, category 3 hurricane, generated higher storm surge at all measurement locations as compared to Hurricane Camille, which was a smaller and faster category 5 storm that impacted the same region.

The above discussion focuses on impacts from a single storm. Morphological impacts by multiple closely spaced (in time) storms, often referred to as storm clusters, can be greater than the simple sum of individual storms or one stronger storm. Studies on the morphological effects of storm clusters are limited and are beyond the scope of this paper. Two recent examples, both from the Gold Coast of Australia, include Karunarathna et al. (2014) and Splinter et al. (2014). The short interval (on the order of a few days) between consecutive storms does not allow the beach to recover, and therefore, can enhance the continued erosion by the subsequent storms (Karunarathna et al., 2014).

10.2.2 Responding Factors Associated with Barrier-Island Subenvironments

The overall morphology of barrier islands varies substantially in terms of the characteristics and composition of its subenvironments. The nearshore and dry beach provide the first line of defense as a storm approaches. A gently sloping nearshore with one or multiple bars can effectively dissipate wave energy through a wide zone of wave breaking, reducing the energy arriving at the shoreline. As a storm approaches with increasing surge and wave height, the normally supratidal beach comes in contact with energetic wave forcing and the flat back beach is reshaped to a gentle seaward dipping shape. The wider the beach, the more effective it is at dissipating storm waves and protecting the dunes landward.

As the storm surge and wave height continue to increase, the dunes may come into direct contact with wave forcing. The elevation and degree of connectivity of dunes play crucial roles in barrier-island response to storm impact. Little morphologic change within the barrier-island interior, bayside shoreline, and back-barrier bay will result from storm impacts if the dune ridge maintains its connectivity (i.e., it does not allow water to reach landward of the dunes). The barrier-island storm impact scale, or the Sallenger Impact Scale, developed by Sallenger (2000), is based on the interaction between the storm and dunes, discussed in the following. Claudino-Sales et al. (2008) developed a conceptual model describing the complex factors controlling the survival of dunes during storm impact. In addition to dune height, continuity and width of the dune field, width of the barrier island, distance of the dune field to the ocean, and type and density of vegetation all play important roles in the dunes' ability to survive storm impacts. Houser et al. (2008) quantified the influences of some of the factors through statistical analyses. Along developed barrier-island coasts, continuous dunes also provide crucial protection to infrastructure landward.

The interior wetlands of barrier islands come into contact with storm waves only after the dunes are eroded, overtopped, or breached. The interior wetlands are typically densely vegetated. The vegetation can be effective in preventing sediment erosion. In addition, storm-wave energy may be substantially dissipated by friction by the time the waves reach the barrier interior. Therefore, the low-lying interior wetlands are mainly sediment sinks receiving sediments eroded from the dunes, beach, and nearshore. Depending on how the dunes are breached or overtopped, the overwash deposits can adopt a tongue shape, a fan shape, or an extensive washover terrace with a much greater spatial scale than a single tongue or fan. Wang and Horwitz (2007) found that little to no erosion occurs at the bottom of the overwash deposits in the interior wetlands, suggesting that the storm deposits are mostly depositional, instead of an event started with erosion and followed by deposition. In contrast, the process of erosion followed by deposition is typical of other barrier-island subenvironments such as nearshore, beach, and dunes. Wolner et al. (2013) examined the barrier-island response to storm-induced disturbance incorporating both ecological and physical processes, referred to as "ecomorphodynamic feedbacks." They developed a conceptual model linking the dominance of dune-building grasses and overwash-adapted maintainer species to barrier-island response to storm impacts. They suggest that on low-relief barrier islands that are often overwashed, ecological processes, in this case the dominance of maintainer species can play a significant role in the overall barrier-island morphodynamics.

Under certain circumstances, such as extreme storm energy and relatively narrow barrier islands, the overwash deposition may extend to the bayside shoreline and into the back-barrier bay. Erosion along the ocean side and deposition along the bayside results in landward rollover of barrier islands. Barrier-island rollover, largely driven by episodic storm impacts, is also considered a response of barrier islands to sea-level rise (McBride et al., 1989).

10.2.3 The Sallenger Impact Scale

Recognizing the complex interactive factors controlling the morphologic response of barrier islands to storm impacts, Sallenger (2000) developed a storm impact scale specifically for barrier islands: the Sallenger Impact Scale, referred to as "the Sallenger Scale" in the following. The Sallenger Scale differs fundamentally from existing storm-related scales, such as the Saffir-Simpson wind scale, in that the coupling between storm-forcing processes and the geometry of the responding barrier-island coast is explicitly included. Four regimes, representing different levels of impact, are defined by the Sallenger Scale.

The Sallenger Scale is defined by four vertical levels, R_{HIGH}, R_{LOW}, D_{HIGH}, and D_{LOW} (Figure 10.5). R_{LOW} represents the elevation below which the beach is continuously subaqueous and is not directly related to storms. R_{HIGH}

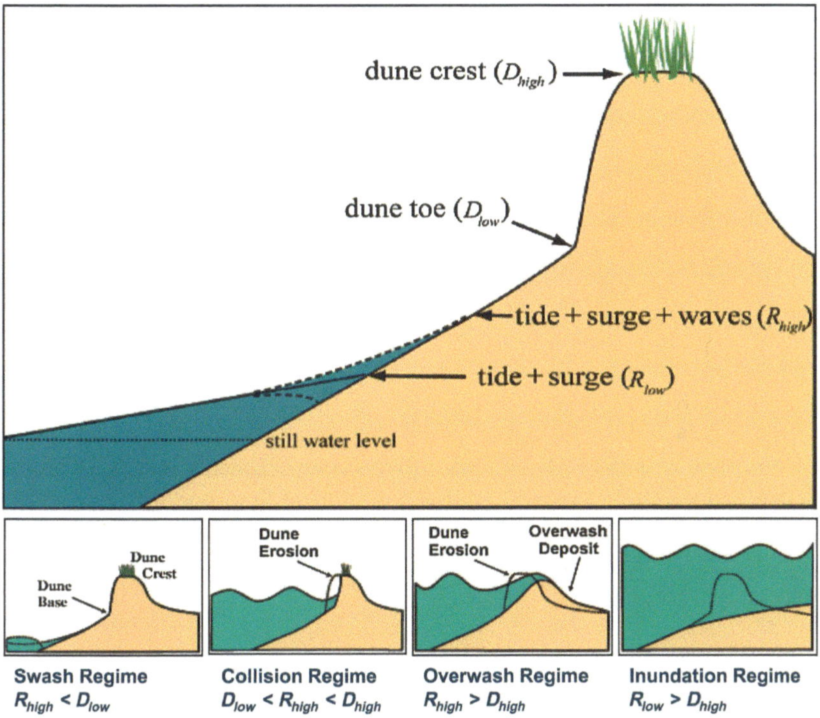

FIGURE 10.5 The Sallenger Scale, including from level 1 to level 4: swash regime, collision regime, overwash regime, and inundation regime. *From United States Geological Survey (USGS) Coastal and Marine Science Center http://coastal.er.usgs.gov/hurricanes/impact-scale/.*

represents a maximum level the water reaches during a storm and is the elevated water level caused by storm surge plus runup of the storm waves:

$$R_{HIGH} = R_{2\%} + \eta_{mean} \tag{10.1}$$

where η_{mean} is the mean measured sea level, for example, from a tide gauge, and includes both astronomical tides and storm surge (Sallenger, 2000). $R_{2\%}$ is the 2 percent exceedance of wave runup on a beach that includes both wave setup and swash runup. The limit of wave runup has been the subject of numerous studies (e.g., Guza and Thornton, 1982; Holman, 1986; Ruggiero et al., 2001; Roberts et al., 2010). Guza and Thornton (1982), for example, suggested that significant wave runup, R_s that includes wave setup and swash runup, is linearly proportional to the significant deepwater wave height (H_0):

$$R_s = 3.48 + 0.71H_0 \quad \text{(units of centimeters)} \tag{10.2}$$

Based on field measurements, several studies (Holman and Sallenger, 1985; Holman, 1986; Ruggiero et al., 2004; Stockdon et al., 2006) argued that more accurate predictions for intermediate beaches, based on the morphodynamics classification of beaches by Wright and Short (1984), can be obtained by including the surf similarity parameter, ξ:

$$\xi = \frac{\tan \beta}{\sqrt{H_0/L_0}} \tag{10.3}$$

Holman (1986) found a dependence of the 2 percent exceedance of runup $R_{2\%}$ on the deepwater significant wave height and the (offshore) surf similarity parameter as

$$R_2 = (0.83\xi + 0.2)H_0 \tag{10.4}$$

Based on a series of large-scale laboratory experiments at SUPERTANK (Kraus and Smith, 1994) and Large-scale Sediment Transport Facility (LSTF) (Wang et al., 2002a, 2002b), Roberts et al. (2010) developed a simple empirical formula linking maximum wave runup (R_{tw}) to significant breaking wave height (H_{bs}) as

$$R_{\text{tw}} = 1.0H_{\text{bs}} \tag{10.5}$$

Two representative elevations are used to define the morphologic characteristic of the barrier island, D_{HIGH} and D_{LOW} (Sallenger, 2000). D_{HIGH} is the elevation of the highest part of the "first line of defense," or the crest elevation of the foredune ridge (Figure 10.5). For a barrier island that does not have a foredune, D_{HIGH} is the elevation of the beach berm crest. D_{LOW} is the elevation of the base of the foredune. If a foredune is absent, $D_{HIGH} = D_{LOW}$.

Thus, the four barrier island impact regimes of the Sallenger Scale are defined based on the relationship among the four parameters, R_{HIGH}, R_{LOW}, D_{HIGH}, and D_{LOW}. The least severe impact (impact level 1), the swash regime, occurs when

$$R_{HIGH} < D_{LOW} \tag{10.6}$$

To satisfy Eqn (10.6), the elevated water level associated with the storm is relatively low and does not substantially exceed the height of the beach berm crest. Swash regime involves morphology changes in the nearshore zone, while dry beach and dunes are not impacted. The collision regime (impact level 2) occurs when the storm-induced, elevated water level exceeds D_{LOW}, but is lower than D_{HIGH} (Eqn (10.7)).

$$D_{LOW} < R_{HIGH} < D_{HIGH} \tag{10.7}$$

For impact level 2, the subaerial beach is overtopped, and the dunes come into direct contact with the storm wave forcing. The collision regime often involves severe beach and nearshore erosion and the foredune ridge can suffer substantial erosion due to direct wave attack. When the storm-induced elevated water level exceeds the dune crest, that is,

$$R_{HIGH} > D_{HIGH} \text{ and } R_{LOW} < D_{HIGH} \tag{10.8}$$

overwash regime (impact level 3) occurs. Landward directed flow overtopping the dune crest, which can exceed 2 m/s (Holland et al., 1991), along with storm waves propagate into the interior of the barrier island. These overwash processes can transport a large amount of sediment landward into the barrier-island interior and back-barrier bay, forming various washover morphologies. The most severe storm impact to a barrier island is the impact level 4, inundation regime. It occurs when

$$R_{LOW} > D_{HIGH} \tag{10.9}$$

During this regime, the entire cross-section of the barrier island is submerged with a depth equal to the difference between R_{HIGH} and R_{LOW} and all of the subenvironments are experiencing intense sediment transport associated with breaking waves. An inundated barrier island exhibits a low, typically at or even below the elevation of the prestorm back beach, and flat topography shaped by the energetic breaking waves. As indicated by Sallenger (2000), this regime is not as well studied as the other regimes. Case studies on each regime impact are discussed in the following:

The Sallenger Scales, combining various storm properties and the morphological characteristics of the barrier island to assess the degree of storm impact to barrier-island morphology, is a complex multivariable scale and does not respond simply to a single parameter. For example, the increasing Sallenger impact levels from one to four do not directly or necessarily relate to the increase of storm intensity, such as that defined by the Saffir-Simpson scale. A Saffir-Simpson category 1 hurricane may cause a Sallenger level 4 impact on a barrier island. Although a category 4 hurricane may only cause a level 1 impact (Sallenger et al., 2006). Stockdon et al. (2007) developed a procedure to model regional-scale barrier-island response to storm impact by applying the Sallenger Scale using R_{HIGH} and R_{LOW} values obtained from numerical modeling of storm surge and wave, and D_{HIGH} and D_{LOW} values extracted from airborne LIDAR (LIght Detection And Ranging) survey data. This model allows a semi-quantitative assessment or prediction of potential storm impact to barrier islands.

Plant and Stockdon (2012) amended the Sallenger (2000) impact scale by applying a Bayesian Network to quantitatively predict beach and dune morphology changes, specifically, the dune elevation, dune-line position, and shoreline position changes. Plant and Stockdon (2012) further suggested that, in addition to the four parameters (Eqns (10.6)–(10.9)) used in the Sallenger Scale, initial beach, and dune widths are crucial in improving the prediction skills. The statistical approach of the Plant and Stockdon (2012) model confirms the qualitative observations by Claudino-Sales et al. (2010).

The various factors controlling storm impacts on barrier islands were also classified by Morton (2002). Although largely qualitative, this classification incorporated more complicated interactions between storm and barrier-island factors compared to the four elevation levels employed by Sallenger (2000). Morton (2002) also provided detailed morphological and sedimentological descriptions of each type of barrier-island response and suggested that near real-time forecasting of expected storm-induced morphology impact, in the form of a geographic information system map, is possible based on his system. Compared to the Sallenger (2000) Scale, the Morton (2002) classification involves more complicated, and somewhat subjective, decision making and is not as readily able to be incorporated in a computer model. Also, the Morton (2002) classification does not provide a probability estimate.

The Sallenger Scale, and subsequent improvements by Stockdon et al. (2007) and Plant and Stockdon (2012), provides a valuable tool and guideline for assessing vulnerability of barrier islands to storm impacts and predicted future sea-level rise. The Sallenger Scale has been applied worldwide for regional-scale assessment of storm impact and for coastal planning and management. Idier et al. (2013) applied the Sallenger Scale to evaluate the vulnerability of several sandy beaches along the French coast to climate variations. Jimenes et al. (2012) analyzed the longer-term (1958–2008) storm impact to Catalan coast (Spain) using the Sallenger Scale. In a large multidisciplinary European project (MICORE and ConHaz), Ciavola et al. (2011a,b) and Harley and Ciavola (2013) applied the impact scale as a main guideline in the development of online predictions of storm-related hydrodynamic and morphodynamic changes at nine study sites along European coast. Armaroli et al. (2012, 2013) developed a dune-erosion threshold, referred to as Dune Stability Factor, based on the concept of the Sallenger Scale. Rodrigues et al. (2012) applied the Sallenger Scale to assess the overwash hazard along several barrier islands in southern Portugal. In all the above studies, the Sallenger Scale was used to guide the numerical modeling efforts for the assessment and classification of different morphological impacts by storms. Also, the aforementioned studies recognized the importance of the storm duration in morphological impacts.

In the case of multiple storm impacts, the Sallenger Scale applied to subsequent storms may be significantly influenced by the previous storms. For example, if a barrier island is severely inundated by one storm, before the barrier-island elevation recovers significantly, which may take years to decades, subsequent storms may also cause inundation, or level 4 impact, even for much weaker storms with slower wind speeds, lower surge, and smaller waves. A similar situation may occur for overwash regime after the foredune is overtopped and eroded by an initial storm. Therefore, for multiple storm impacts, the Sallenger Scale can be strongly influenced by the most energetic storm, allowing subsequent weaker storms to result in a similar impact scale.

Under these circumstances, the Sallenger Scale is dominated by the morphology resulting from the most recent strong storm and would not be able to distinguish the meteorological and oceanographic differences among the storms.

Another important barrier-island morphologic response to storm impacts, not included in the Sallenger Scale, is breaching. Where a barrier island is breached, a new tidal inlet is formed and the barrier island is cut into two islands. If a sufficient tidal prism exists for the new inlet, it will evolve and become a semipermanent tidal inlet, with the formation of flood- and ebb-tidal deltas. The formation and evolution of a new tidal inlet have significant influence to the morphology of the breached barrier island as well as the morphology of nearby barrier islands and existing tidal inlets. For example, a new inlet may capture the tidal prism of the existing nearby inlet. This will cause the existing inlet to become unstable, leading to longshore migration or closure of the existing inlet, and associated regional-scale morphology change of adjacent barrier islands (Wang et al., 2011; Wang and Beck, 2012).

10.3 MORPHOLOGIC RESPONSE OF BARRIER ISLANDS TO STORM IMPACT

The Sallenger Scale provides a conceptual guide for systematic documentation of storm impacts on barrier islands. In the following, various recent case studies, organized based on the Sallenger Scale, are presented. The following discussion presents case studies for each level of the scale and describes the associated morphological changes. It is worth noting that higher impact levels include all impacts associated with lower levels. For example, the collision regime includes swash regime impacts because changes in the nearshore zone also occur as the foredune is under attack by storm waves. In the following, discussion on collision regime focuses on morphology changes in the dune fields, although changes in the nearshore zone also have occurred. Similarly, discussion on overwash regime focuses on the morphology of the overwash deposits, instead of beach and dune erosion, although they are parts of the overwash regime.

10.3.1 Swash Regime

Swash regime occurs when the elevated water level and high storm waves do not reach the dunes. A majority of the morphology changes occur in the nearshore and foreshore areas. Significant beach erosion, with large landward shoreline retreat, takes place due to the active sediment transport induced by high waves on top of the elevated water level. Generally, the high storm waves result in a gentler and wider foreshore and offshore migration of the sand bar (if present). However, as demonstrated by the impact from Tropical Storm

Debby, substantial spatial variations can occur depending on local morphologic and sedimentologic characteristics.

Tropical Storm Debby struck the west-central, barrier-island coast of Florida in late June 2012 (Figure 10.6). Although a rather weak tropical storm with a peak surge of 0.9 m and a peak offshore wave height of 5.5 m, Tropical Storm Debby moved slowly and generated high-energy conditions for nearly three days that resulted in widespread beach erosion. At most locations, beach erosion and offshore-bar migration were measured (Figure 10.7(a)). This offshore-bar migration is similar to the storm-induced, morphology changes documented by numerous studies (Lee et al., 1998; Larsen and Kraus, 1994; Birkemeier et al., 1999). Although beach and nearshore erosion occurred along the entire west-central Florida barrier-island coast, behavior of the nearshore bar varied spatially. Offshore-bar migration was measured along most sections of the beach; however, both upward accretion (Figure 10.7(b)) and onshore bar migration (Figure 10.7(c)) were also measured. The southerly approaching storm and shoreline-orientation change may be attributable to the different bar behavior, although the exact cause of the different bar behavior is not clear.

Not all barrier-island beaches have a sand bar, and a storm bar is not always developed. Roberts et al. (2013) studied the impact of storm Nor'Ida that impacted a mixed sand and gravel beach with a characteristic morphology of an absent nearshore bar during both normal and storm conditions along the Delaware coast from November 11 to 14, 2009 (lasting eight tidal cycles).

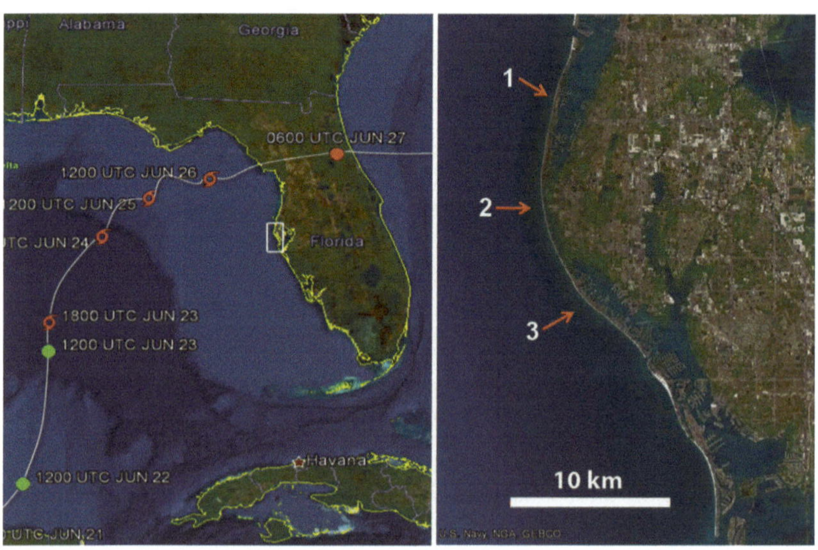

FIGURE 10.6 Left panel: track of Tropical Storm Debby in late June 2012. Each symbol represents the distance traveled by the storm in one day. Right: barrier islands along west-central Florida coast, indicating the locations of example beach profiles shown in Figure 10.7.

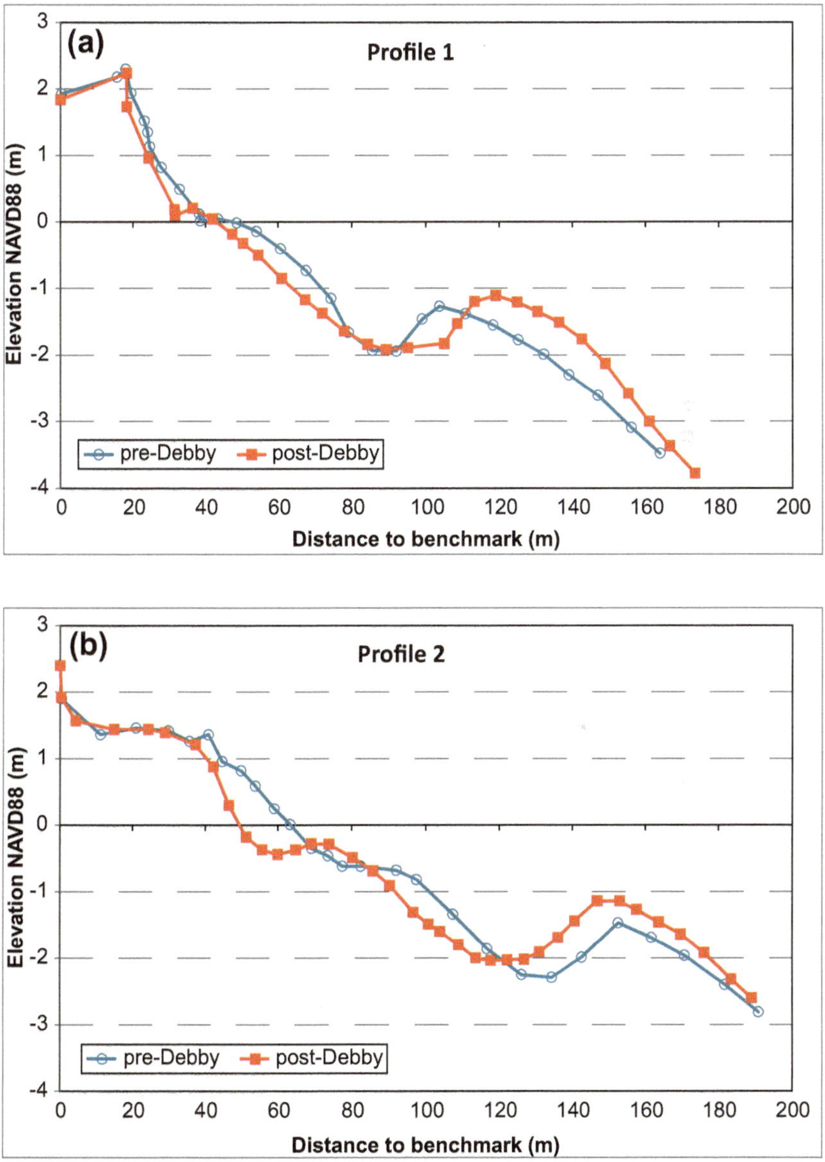

FIGURE 10.7 Beach profiles surveyed immediately before and after the passage of Tropical Storm Debby. Erosion was measured on the beach and in the nearshore. Deposition was measured on the sand bar. Locations of the beach profiles are shown in Figure 10.6. (a) Profile 1 (location shown in Figure 10.6) showing offshore sand bar migration; (b) Profile 2 showing upward accretion of sand bar; (c) Profile 3 showing onshore migration of the sand bar.

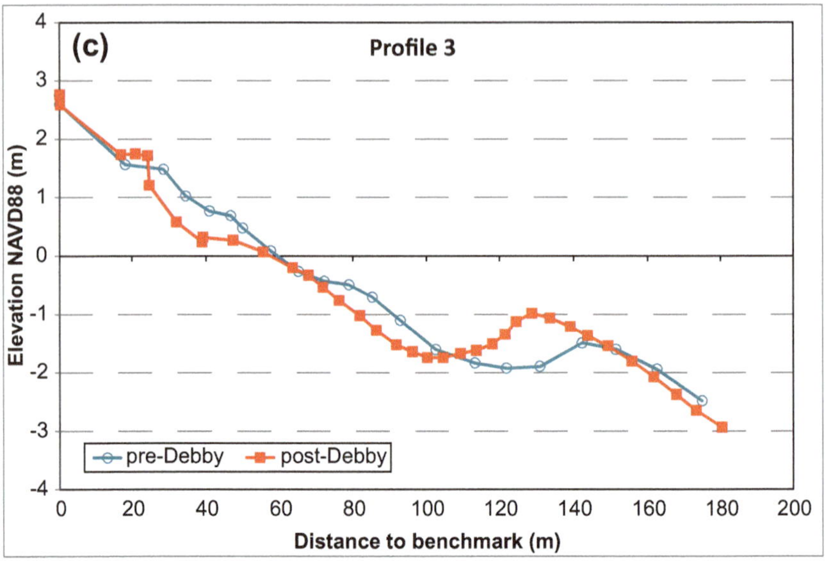

FIGURE 10.7 —Cont'd

At the time, Nor'Ida was considered one of the most energetic storm events on record to impact the northeastern US coast (Grosskopf and Bass, 2010; Herrington and Miller, 2010). Munger and Kraus (2010) determined the return interval for this storm was 45-90 years based on beach erosion potential (prior to the occurrences of Hurricane Irene in 2011 and Superstorm Sandy in 2012). Similar to Superstorm Sandy (2012), Nor'Ida resulted from remnants of Hurricane Ida and a nor'easter that collided off the Atlantic coast (also colloquially referred to as the "Friday the 13th storm"). The maximum significant wave height was 8.11 m and maximum dominant wave period was nearly 14 s; over a five-day interval, Nor'Ida's energetic wave conditions exceeded a 4-m wave height for approximately 52 h (Roberts et al., 2013). Winds were persistently from the northeast during the five-day period, with maximum velocities of 21 m/s. The onshore winds generated a storm surge of nearly 1 m, which lasted over the eight tidal cycles.

Nor'Ida induced widespread, collision regime impacts along the US mid-Atlantic coast. Here, we focus on the characteristic changes in the nearshore area, as an example of swash regime. Sand was eroded from the beach during various storms (studied by Roberts et al., 2013), including Nor'Ida, and was deposited in the offshore area in a nearly planar layer with the extent determined by the storm energy (Figure 10.8). The more energetic storm Nor'Ida deposited, the eroded sand from the beach further offshore than the two previous weaker storms (Roberts et al., 2013). The absence of a sandbar under a

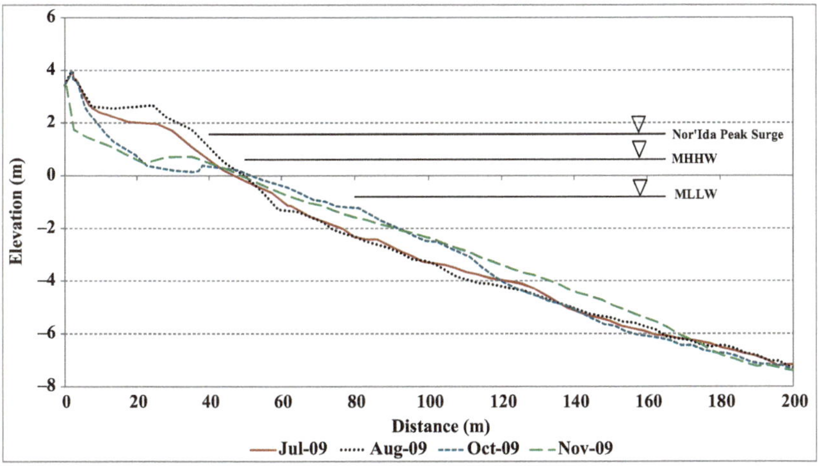

FIGURE 10.8 Beach profiles surveyed immediately before (Jul-09) and after the passage of three storms. The distal passage of a hurricane resulted in beach accretion (Aug-09). The passage of the first significant winter storm resulted in substantial beach erosion (Oct-09). The passage of Nor'Ida resulted in substantial beach and dune erosion (Nov-09). The profile is located along the Delaware coast. Note the absence of a sand bar on all the profiles. *From Roberts et al. (2013).*

variety of wave conditions, both accretionary and erosive, was not exactly clear, but persisted following even a substantial storm impact (i.e., Nor'Ida). Roberts et al. (2013) attributed the constant lack of a sandbar to the mixed sand and gravel nature of beach sediment and a persistent gradient in longshore transport.

10.3.2 Collision Regime

Collision regime occurs when the high storm waves superimposed on the storm surge come into direct contact with dunes, typically the foredune ridge. Along the coast where foredunes are well developed and continuous, extensive dune scarping can occur. Wang et al. (2006) documented dune scarping caused by Hurricane Ivan in 2004 that extended nearly continuously for about 50 km along the northwest Florida coast (Figure 10.9). As illustrated in Figure 10.9, this stretch of the coast is heavily developed with residential structures. The seaward portion of many of the dune walkovers and access stairs to the beach were washed away, leaving the houses standing close to the scarp. The vulnerability of these houses to future storms was dramatically increased by the erosive impacts from Hurricane Ivan.

Dune scarping is caused by the upper portion of the dune collapsing due to the rapid erosion at the dune toe, for example as occurred from the impact of Nor'Ida (Figure 10.10). The collapsing of the upper portion of the dune can also be seen in the photograph from northwest Florida (Figure 10.9). The patches of grass on the steep slope of the scarp in Figure 10.9 apparently

FIGURE 10.9 Extensive dune scarping caused by Hurricane Ivan in 2004 along northwest Florida coast. Note the blocks of sand with grass on the steep slope, which resulted from the collapsing of the upper portion of the dune.

FIGURE 10.10 High waves superimposed on the elevated storm water breaking at the dune, as a result of Nor'Ida. The dune scarp is formed due to the collapsing of the upper portion.

resulted from the mass wasting due to the scour near the base. Therefore, the height of the dune scarp is largely controlled by the morphology of the prestorm dunes, and may not be directly related to the combined level of the storm surge plus wave runup. The magnitude of dune scarping in terms of distance of the landward retreat of the foredune line and the height of the dune scarp should be controlled by the duration of the impact and the prestorm dune morphology, instead of only the elevated water (surge plus wave runup) level. As discussed earlier, a limitation of the Sallenger-impact regimes is its lack of temporal scales. By incorporating beach and dune widths, Plant and Stockdon (2012) indirectly incorporated a temporal scale in the Sallenger (2000) Scale.

Type and density of dune vegetation and width of the beach seaward of the dune play significant roles on the survival of the dunes during storm impact (Claudino-Sales et al., 2008, 2010). Denser bush-type vegetation improves the strength of the sediment and increases the dunes' ability to resist erosion. In addition, a wider beach seaward of the dunes dissipates more wave energy and results in reduced wave height at the dunes. Wider dunes can endure longer attack by storm waves. Further, Houser et al. (2008) found that offshore bathymetry variations also have significant influence on the intensity of the collision regime.

10.3.3 Overwash Regime

Overwash regime occurs when the storm surge plus wave runup overtops the foredune ridge or when the foredune ridge is eroded by prolonged storm wave attack. In the latter case, the Sallenger criterion (Eqn (10.8)) does not need to be satisfied for overwash regime to occur, thus exposing a potential weakness of the Sallenger Scale in that temporal scale is not considered. The Sallenger criteria (Eqns (10.6)–(10.9)) are based solely on (peak) levels of storm surge plus waves relative to elevations of the beach-dune system. The storm duration is considered in regards to generating higher storm surge and waves, but not in regards to the beach-dune system's ability to sustain prolonged erosion.

Morton and Sallenger (2003) provided a detailed description of various morphological characteristics of overwash deposits along the US Gulf of Mexico and Atlantic coasts and associated controlling factors. A general difference in overwash penetration was identified between the microtidal portion of Gulf of Mexico coast and the mesotidal portion of Atlantic coast. Donnelly et al. (2006) compiled an extensive review on existing studies of overwash processes with a goal of developing a numerical model to capture overwash deposits.

Local-scale overwash occurs when one or several gaps exist in an otherwise continuous foredune ridge or a weak (e.g., low and narrow) section of the foredune is overtopped or eroded through. Schwartz (1975, 1982) and Leatherman and Williams (1977, 1983) described detailed morphological characteristics and sedimentary structures associated with small-scale overwash deposits. The overwash deposits are often referred to as washover fans due to their fan shape, resulting from flow spreading out from an "overwash channel." Accommodation space landward of the breached foredune ridge, for example, barrier-island interior wetland, is necessary for the development of overwash fans. Two basic bedding structures were described including horizontal to very low angle, landward-dipping stratification and steep delta-foreset bedding (Schwartz, 1982). Morton and Sallenger (2003) examined the factors controlling the landward penetration of this type of overwash deposits.

Regional-scale overwash occurs when the Sallenger-overwash criterion (Eqn (10.8)) is satisfied over a long stretch of a barrier island, or extended sections of foredune are eroded by prolonged storm impact. Wang and Horwitz (2007) and Claudino-Sales et al. (2010) documented the morphological and sedimentological characteristics of regional-scale overwash, or overwash terrace, along the northwest Florida barrier islands caused by Hurricane Ivan in 2004. Figure 10.11 illustrates the prestorm and poststorm LIDAR topography of a section of Santa Rosa Island. Nearly the entire foredune ridge was eroded or overtopped by Hurricane Ivan, with a few exceptions where the prestorm dunes were wide. The small hummocky dunes landward of the foredune ridge were also eroded and replaced by an extensive overwash terrace, with an elevation of approximately 2 m above mean sea level (Claudino-Sales et al., 2010). Densely vegetated dunes farther landward survived the storm impact (Figure 10.11), likely due to their ability to substantially dissipate the storm wave energy.

The interior wetlands of barrier islands provide accommodation space for overwash deposits, with thickness ranging from a few tens of centimeters to $>$1 m (Figure 10.12). The buried upright marsh grasses suggest that overwash into the barrier-island interior wetlands is mostly a depositional event with minimal erosion. Claudino-Sales et al. (2010) found that the sand volume gained from the washover deposits is smaller than the sand volume eroded from the dune and the beach, suggesting a net sediment loss, likely to the offshore during an overwash regime. Stockdon et al. (2007) concluded that sediment deposited in the overwash fans (or terraces) will not likely return to the beach environment to contribute to poststorm beach recovery by natural processes.

Most existing overwash studies, such as those discussed above, were conducted along sandy coasts. However, overwash also occurs on gravel barrier beaches that are common along the coastlines of northern Europe (Buscombe and Masselink, 2006). Matias et al. (2012) conducted a series of laboratory experiments to quantify overwash threshold for gravel beaches. They found that, in general, findings from sandy beaches are also applicable along gravel barriers.

10.3.4 Inundation Regime

Inundation regime occurs when Eqn (10.9) is satisfied, or at places where the dunes are narrow and discontinuous, and are completely eroded by prolonged storm waves. Inundation often results in landward propagation of the bayside

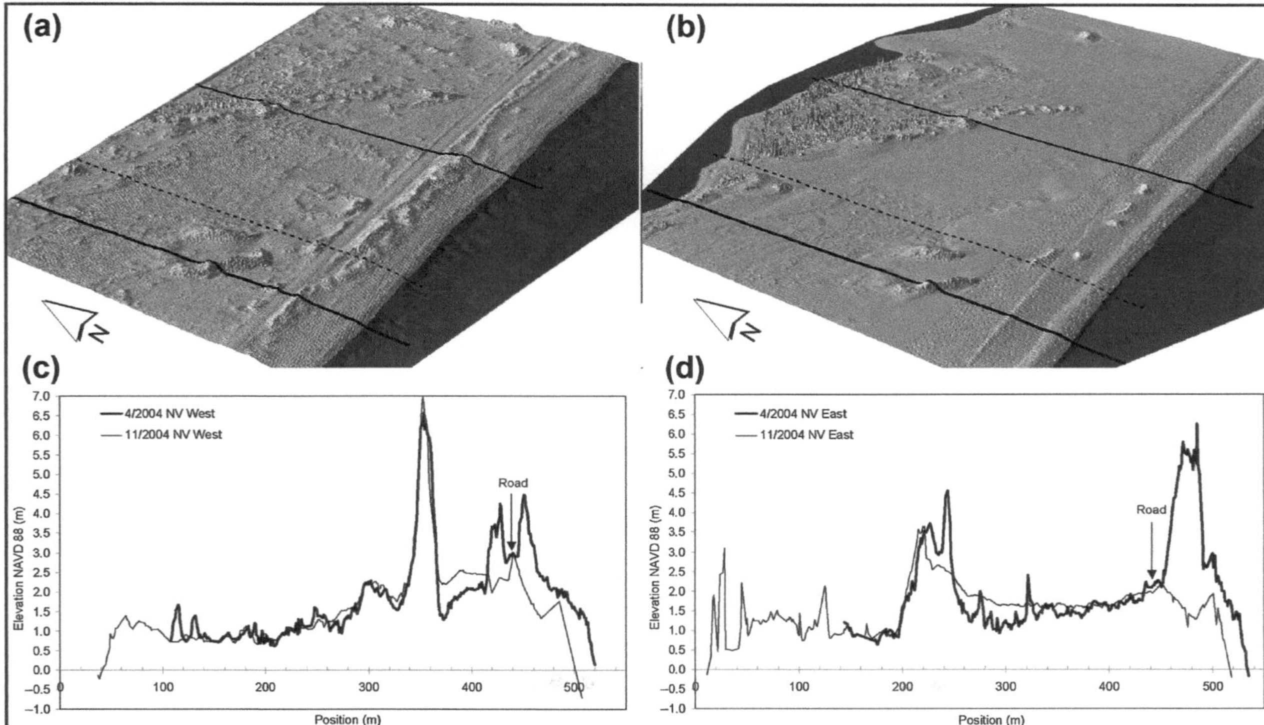

FIGURE 10.11 Overwash induced by Hurricane Ivan (2004) and locations of the two crossisland profiles (solid lines). (a) Pre-Ivan LIght Detection And Ranging (LIDAR) image with 1.5 times vertical exaggeration. (b) Post-Ivan LIDAR image with 1.5 times vertical exaggeration. Note the nearly complete destruction of all the semicontinuous dunes. The orientation of the washover lobes tends to be parallel to the orientation of the preserved dune field. (c) NV-West profile showing the erosion of a relatively low (2-m) dune. (d) NV-East profile showing the complete erosion of a tall (5-m) dune. *From Claudino-Sales et al. (2010).*

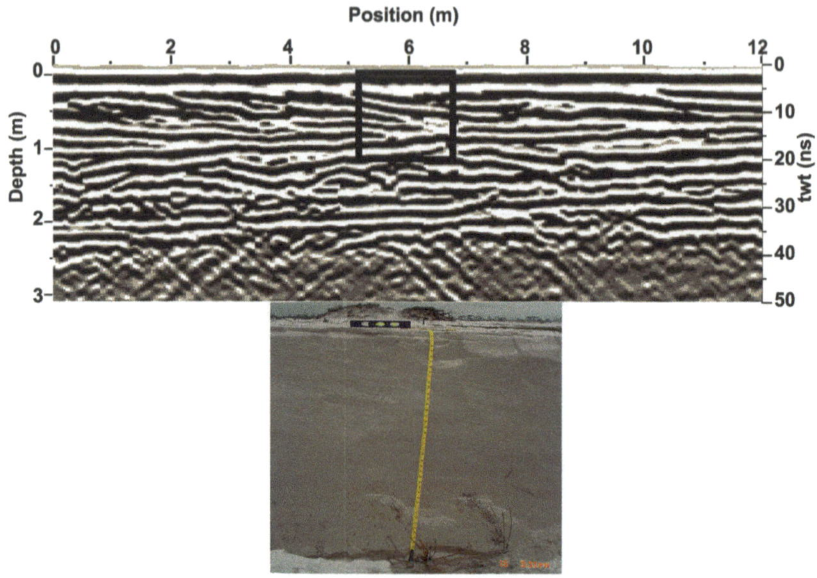

FIGURE 10.12 Upper: 250-MHz Ground-Penetrating Radar (GPR) transect scanned at the terminus of an overwash lobe deposited by Hurricane Ivan along Santa Rosa Island, northwest Florida. Lower: trench excavated through the Ivan washover lobe. The base of Ivan washover layer is marked by freshly buried upright grass, and an abrupt change in sediment color. Little evidence of erosion can be identified at the contact. Stratification within the Ivan washover is inclined to the east at an apparent angle of 16°. The inclined stratification and basal contact exposed in the trench correlate well with reflective patterns in the GPR profile. *Modified from Wang and Horwitz (2007).*

shoreline, or barrier-island rollover. Figure 10.13 illustrates the inundation caused by Hurricane Ivan, a category 5 hurricane at its peak in the Gulf of Mexico, near the western end of Santa Rosa Island (Claudino-Sales et al., 2010). The prestorm barrier-island morphology is characteristic of a low foredune ridge with numerous gaps. The foredune ridge, along with small hummocky dunes in the barrier-island interior, was completely eroded by the storm, resulting in a flat inundation surface. The elevation of the inundation surface is quite low (~1 m above sea level) and is much lower than the overwash terrace elevation (~2 m above sea level) generated by the same storm.

Hurricane Ivan also caused the bayside shoreline to propagate landward for up to 100 m (Figure 10.13). Similar to the overwash case, the sand volume eroded from the nearshore–beach–dune was greater than the sand volume deposited in the back-barrier bay, suggesting a net sediment loss to the offshore region. Morton (1978) documented large-scale rhomboid bedforms associated with active sediment transport by breaking waves superimposed on strong currents during hurricane overwash and inundation.

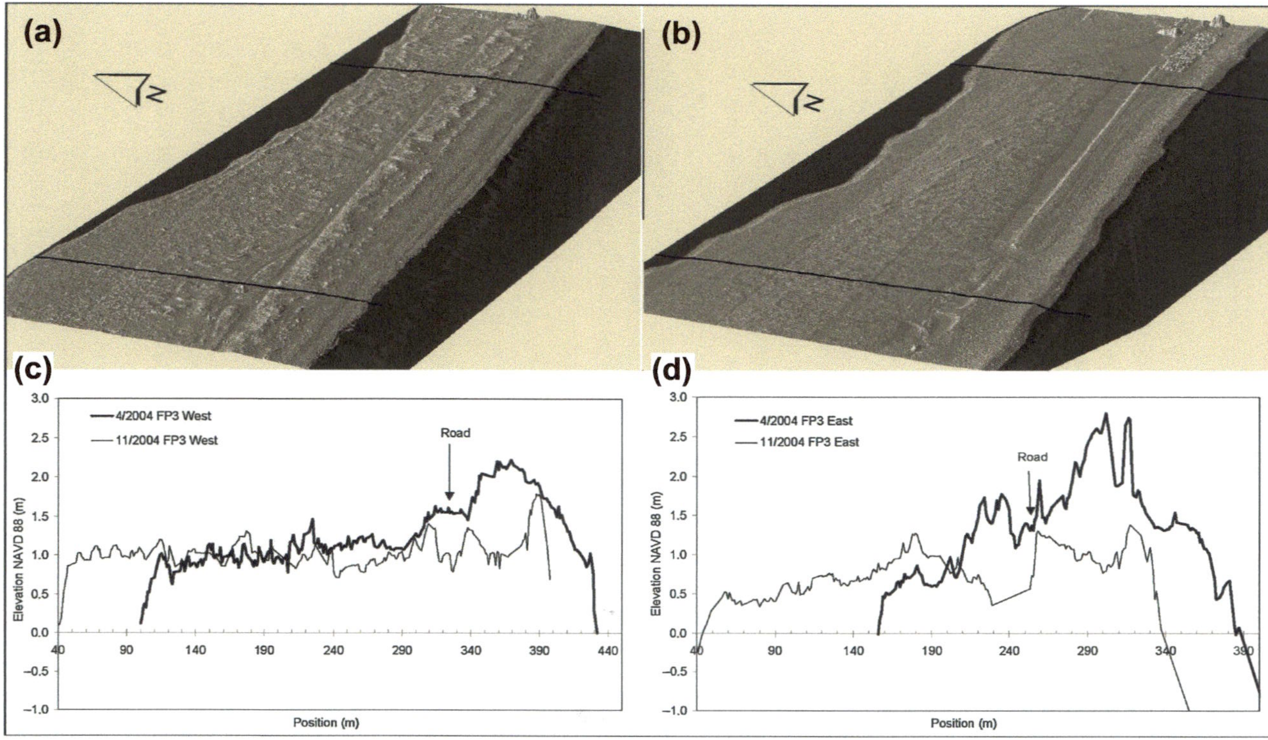

FIGURE 10.13 Inundation induced by Hurricane Ivan (2004) and the locations of the two crossisland profiles. (a) Pre-Ivan LIght Detection And Ranging (LIDAR) image with 1.5 times vertical exaggeration. (b) Post-Ivan LIDAR image with 1.5 times vertical exaggeration. Note the complete destruction of all dunes. (c) FP3-West profile. (d) FP3-East profile. Note the scour hole along the destroyed road at both profile locations and the landward propagation of the bayside shoreline. The rectangular topographic features at the top of panel 6B are piles of road debris. *From Claudino-Sales et al. (2010).*

10.3.5 Other Storm-Induced Morphological Impacts on Barrier Islands

Documented by Sherman et al. (2013), but not included in the Sallenger Scale, are a series of morphological impacts along Bolivar Peninsula in Texas induced by Hurricane Ike, a category 2 at the time of landfall near Galveston, Texas. Substantial beach erosion occurred after the peak of the storm passed during the subsiding phase of the storm and was caused by strong offshore-directed flow associated with the subsidence of the approximately 5-m storm surge. Five types of scour features were distinguished by Sherman et al. (2013) based on the depth, duration, and velocity of the ebbing surge, modulated by local conditions such as vegetation, the presence of streets, or width and depth of beach deposits. Some of the scour features persisted for several months to three years after the storm. Significant seaward sediment transport and deposition induced by ebbing flow associated with subsiding storm surge were also observed by Hayes (1967) after Hurricane Carla (category 5) in 1961 and Hurricane Cindy (category 1) in 1963.

Another significant morphological impact of storms on barrier islands is breaching. Barrier-island breaching is also not included in the Sallenger (2000) Scale. As a matter of fact, many tidal inlets between barrier islands were originally formed by storm breaching. Substantial regional-scale morphological "adjustments" typically occur along several adjacent barrier islands after a new inlet is formed (Wang et al., 2011; Wang and Beck, 2012). Barrier-island breaching and subsequent regional-scale morphological evolution of impacted barrier islands is beyond the scope of this paper.

10.4 POSTSTORM RECOVERY BY NATURAL PROCESSES

10.4.1 Formation of a Ridge and Runnel System as Storms Subside

The poststorm morphologic recovery process depends on the geologic controls and hydrodynamic forcing, although these controls at various temporal scales are not well-understood. A common morphologic feature associated with the natural poststorm recovery is the formation of a ridge and runnel system (King and Williams, 1949; Hayes, 1972; Davis and Fox, 1975; Hine, 1979; Nakashima, 1989; Apoluceno et al., 2002; Aagaard et al., 2004; Figlus et al., 2012; Roberts et al., 2013). The onshore migrating ridge is separated from the subaerial beach by a trough, or runnel (Figure 10.14); the shape and alongshore continuity of the ridge and runnel system varies. The ridge and runnel morphology is an indicator of onshore sediment transport conditions during storm subsidence and recovery. Along the sand and gravel beaches of Delaware, a ridge and runnel system formed in just hours following the peak of Nor'Ida (Roberts et al., 2013). The observed rapid initiation of morphologic

FIGURE 10.14 An onshore migrating ridge (left) is separated from the subaerial beach by a trough, or runnel (right). *From Roberts et al. (2013).*

recovery following a storm has also been observed at other locations (Davis et al., 1972; Owens and Frobel, 1977; Morton et al., 1994).

Morphologic recovery following storm impact does not always occur, nor does it necessarily occur rapidly. Often, recovery mechanisms operate much slower than the initial erosional impact of storms. The temporal rate of full recovery also depends on the size and magnitude of the storm. Although recovery following the passage of a storm may initiate almost immediately, full or even partial morphologic recovery is not always possible. This was observed along the Chandeleur Islands, Louisiana, after Hurricane Frederic in 1979 (category 3 at the time of landfall) where morphologic recovery ceased after the first three months subsequent to the passage of the storm (Kahn, 1986).

10.4.2 Onshore Migration of the Ridge and Attachment to the Beach

After the formation of a ridge and runnel, the ridge typically migrates onshore and welds to the upper foreshore or beach. The process can lead to full recovery, or partial recovery, of the eroded poststorm beach. The onshore migration and eventual welding of the ridge to the beach has been observed as the recovery mechanism along a number of coasts, including the low-energy sandy barrier island of Mustang Island, Texas (Davis and Fox, 1975), the moderately high-energy sandy barriers on the North Danish coast (Aagaard et al., 2004), and the moderate-energy sand and gravel beaches along the mid-Atlantic coast of the United States (Roberts et al., 2013). In each of these studies, the combination of onshore sediment transport during subsiding storm wave energy and the diminished undertow currents, and overwash-like transport across the ridge at high tide, allowed for onshore transport of sediment along the landward edge of the ridge, ultimately depositing in the runnel and effectively welding the ridge and runnel system to the beach. This process is the primary mechanism for the initial recovery of a beach following a storm.

However, the rate of onshore ridge and runnel migration and beach welding is often somewhat proportional to the degree of storm erosion. For example, along the Delaware coast the recovery morphology of a ridge and runnel is frequently recognized. However, the rate of migration and beach welding varies from several weeks to months. Roberts et al. (2013) observed rapid welding of the ridge and runnel following two smaller-scale winter storm erosion events (swash regime), but a slower recovery (several months) following the severe storm erosion by Nor'Ida (collision regime). After the ridge is welded to the back beach, the elevation of the beach berm continued to "recover," or increase. Rapid growth of the post-storm beach berm, in terms of weeks to months, was also measured by Wang et al. (2006) along the northwest Florida coast after Hurricane Ivan.

10.4.3 Recovery of Coastal Dunes

Coastal dune recovery is a slow process, generally following the recovery of the subaerial beach. Dune accretion and aggradation are by eolian processes that require a substantially wide beach so that dry sand can accumulate around nuclei, such as seaweed or remnants of dunes. Short and Hesp (1982) attribute eolian transport from swash deposition to the feedback between actual beach topography and flow regime across the beach. Morton et al. (1994) determined that a minimum width of dry beach, in this case 50 m, is necessary for dunes to form and recover following a category 3 hurricane that impacted the southeastern Texas coast. Depending on the magnitude of dune erosion, the recovery rate can vary. If dune scarping occurred, the dune core remains and serves as the accumulation nuclei. The seaward slope of the foredune can recover within months with adequate sand supply and wind (Suanez et al., 2012). However, with total dune inundation, the process of nebkha formation (isolated dune mounds around individual plants) through incipient and fully vegetated, established dunes may require decades. Priestas and Fagherazzi (2010) found that the presence or absence of vegetation had a significant role in the recovery of the dune field along St George Island, Florida, following the impact of Hurricane Dennis in 2005, a category 3 at the time of landfall. For areas in which the secondary dunes survived, vegetation deterred migration and allowed for recovery. In the flat, low-lying portion of the barrier island, such as overwash fans and aprons, dunes were not observed forming or recovering. In contrast, Weymer et al. (2013) found that despite the complete inundation of intermediate dunes, they exhibited the greatest recovery, with the migration of accretionary mounds and development of embryo dunes along Padre Island in Texas. The smallest dunes often exhibited the least net erosion, but the least recovery and the largest dunes exhibited little poststorm recovery despite scouring at the base. Thus, the impact degree of the storm may have a significant control not only on the extent of erosion but also perhaps the sediment supply for beach-dune recovery.

10.5 SUMMARY

Storm impacts on barrier-island morphology are influenced by the combination of a suite of meteorological and oceanographic characteristics associated with the storm, rather than any other individual factor, such as solely wind speed. The combination of a large, slow-moving, storm over a broad and gentle continental shelf with high wind speeds will likely lead to higher storm surge and waves, and therefore the capacity for inducing greater morphologic change.

Variations in the morphology of barrier-island subenvironments influence the overall storm impact response of the barrier-island system. Generally, the nearshore, subaerial beach, and dunes are regions experiencing storm-induced erosion. However, interior wetlands, the bayside shoreline region, and the back-barrier bay are likely sites of storm-induced deposition of sediments eroded from the other nearby subenvironments. Morphologic changes can also occur in the offshore region, with deposition of eroded beach and dune sediments.

Recognizing the complex interactive factors controlling the morphologic response of barrier islands to storm impacts, Sallenger (2000) developed a storm impact scale specifically for barrier islands. The Sallenger Impact Scale, or Sallenger Scale, considers the interaction between storm-forcing processes and the geometry of the responding barrier-island coast. Four regimes, representing different levels of impact are defined, including swash regime, collision regime, overwash regime, and inundation regime. The morphologic characteristics of each impact scale are reviewed here.

In most cases, the poststorm beach recovery initiates as the storm energy begins to subside. Initial beach recovery often takes the morphological form of ridge and runnel development. Full beach recovery requires subsequent onshore migration and eventual ridge welding to the beach, followed by aggradation of the beach berm to pre-storm elevations. The extent of dune recovery is highly variable, controlled by many factors including sand availability, vegetation, and post-storm morphologic characteristics.

Due to the complex nature of storms and storm-induced morphology change, major current research challenges include the development of realistic predictive capabilities, at multiple spatio-temporal scales. Storm impacts occur at the relative short time scale of days to weeks. Storm recovery occurs at longer time scales of months to years. Qualitatively, storm impacts and poststorm recovery, as summarized here, are reasonably well understood. However, future research should be directed toward a multidisciplinary approach to the development of quantitative predictions of storm impacts and post-storm recovery. This will require improved ability of computing sediment transport under extremely energetic conditions and quantifying the drivers of medium-term (months to years) morphology evolution. Continued development in predictive capabilities of storm impacts to coastal areas will significantly improve risk assessment management, hazard mitigation and coastal-zone adaptation.

REFERENCES

Aagaard, T., Davidson-Arnott, R., Greenwood, B., Nielsen, J., 2004. Sediment supply from shoreface to dunes: linking sediment transport measurements and long-term morphological evolution. Geomorphology 60, 205–224.

Apoluceneo, D.D.M., Howa, H., Dubuis, H., Oggian, G., 2002. Morphodynamics of ridge and runnel systems during summer. J. Coastal Res. SI 36, 222–230.

Armaroli, C., Ciavola, P., Perini, L., Calabrese, L., Lorito, S., Valentini, A., Masina, M., 2012. Critical storm thresholds for significant morphological changes and damage along the Emilia-Romagna coastline, Italy. Geomorphology 143–144, 34–51.

Armaroli, C., Grottoli, E., Harley, M.D., Ciavola, P., 2013. Beach morphodynamics and types of foredune erosion generated by storms along the Emilia-Romagna coastline, Italy. Geomorphology 199, 22–35.

Birkemeier, W.A., Nicholls, R.J., Lee, G., 1999. Storms, storm groups and nearshore morphology change. In: Proceedings of Coastal Sediments'99. ASCE Press, pp. 1109–1122.

Buscombe, D., Masselink, G., 2006. Concepts in gravel beach dynamics. Earth-Sci. Rev. 79, 33–52.

Ciavola, P., Ferreira, O., Haerens, P., Van Koningsveld, M., Armaroli, C., Lequeux, Q., 2011a. Storm impacts along European coastlines. Part 1: the joint effort of the MICORE and ConHaz projects. Environ. Sci. Policy 14, 912–923.

Ciavola, P., Ferreira, O., Haerens, P., Van Koningsveld, M., Armaroli, C., 2011b. Storm impacts along European coastlines. Part 2: lessons learned from the MICORE project. Environ. Sci. Policy 14, 924–933.

Claudino-Sales, V., Wang, P., Horwitz, M.H., 2008. Factors controlling the survival of coastal dunes during multiple hurricane impacts in 2004 and 2005: Santa Rosa barrier island, Florida. Geomorphology 95, 295–315.

Claudino-Sales, V., Wang, P., Horwitz, M.H., 2010. Effect of Hurricane Ivan on coastal dunes of Santa Rosa barrier island, Florida: characterized on the basis of pre- and poststorm LIDAR surveys. J. Coastal Res. 26, 470–484.

Davis, R.A., 1994. Barrier island system—a geologic overview. In: Davis, R.A. (Ed.), Geology of Holocene Barrier Island Systems. Springer-Verlag, Berlin–Heidelberg–New York, pp. 1–46.

Davis, R.A., Fox, W.T., 1975. Process-response patterns in beach and nearshore sedimentation: I. Mustang Island, Texas. J. Sediment. Petrol. 4 (4), 852–865.

Davis, R.A., Fox, W.T., Hayes, M.O., Boothroyd, J.C., 1972. Comparison of ridge and runnel systems in tidal and non-tidal environments. J. Sediment. Petrol. 42, 413–421.

Davis, R.A., Hayes, M.O., 1984. What is a wave-dominated coast? Marine Geol. 60, 313–329.

Donnelly, C., Kraus, N.C., Larson, M., 2006. State of knowledge on measurement and modeling of coastal overwash. J. Coastal Res. 22, 965–991.

Figlus, J., Nobuhisa, K., Gralher, C., 2012. Ridge-Runnel migration. Coastal Eng. Proc. 2012.

Fritz, H.M., Blount, C., Sokoloski, R., Singleton, J., Fuggle, A., McAdoo, B.G., Moore, A., Grass, C., Tate, B., 2007. Hurricane Katrina storm surge distribution and field observations on the Mississippi barrier islands. Estuarine Coastal Shelf Sci. 74, 12–20.

Glaeser, J.D., 1978. Global distribution of barrier islands in terms of tectonics setting. J. Geol. 86, 283–297.

Grosskopf, P.E., Bass, G.P., 2010. The Great Mid-Atlantic storm of 2009: 'Friday the 13th' storm impacts on Ocean City, Maryland. Shore Beach 78 (2), 12–19.

Guza, R.T., Thornton, E.B., 1982. Swash oscillations on a natural beach. J. Geophys. Res. 87 (C1), 483–490.

Haerens, P., Bolle, A., Trouw, K., Houthuys, R., 2012. Definition of storm thresholds for significant morphological changes of the sandy beaches along the Belgian coastline. Geomorphology 143–144, 104–117.

Harley, M.D., Ciavola, P., 2013. Managing local coastal inundation risk using real-time forecasts and artificial dune placements. Coastal Eng. 77, 77–90.

Hayes, M.O., 1967. Hurricanes as Geological Agents; Case Studies of Hurricane Carla, 1961 and Cindy, 1963. Report 61. University of Texas Bureau of Economic Geology, Austin, Texas.

Hayes, M.O., 1972. Forms of sediment accumulation in the beach zone. In: Meyer, R.E. (Ed.), Waves on Beaches. Academic Press, N.Y, pp. 297–356.

Hayes, M.O., 1979. Barrier island morphology as a function of tidal and wave regime. In: Leatherman, S.P. (Ed.), Barrier Islands: From the Gulf of St. Lawrence to the Gulf of Mexico. Academic Press, New York, pp. 1–28.

Herrington, T.O., Miller, J.K., 2010. A comparison of methods used to calculate northeaster damage potential. Shore Beach 78 (2), 20–25.

Hine, A.C., 1979. Mechanisms of berm development and resulting beach growth along a barrier spit complex. Sedimentology 26, 333–351.

Holman, R.A., Sallenger, A.H., 1985. Setup and swash on a natural beach. J. Geophys. Res. 90 (C1), 945–953.

Holland, K.T., Holman, R., Sallenger, A., 1991. Estimation of overwash bore velocities using video techniques. In: Proceedings of Coastal Sediment'91. ASCE Press, pp. 489–497.

Holman, R.A., 1986. Extreme value statistics for wave run-up on a natural beach. Coastal Eng. 9, 527–544.

Houser, C., Hapke, C., Hamilton, S., 2008. Controls on coastal dune morphology, shoreline erosion and barrier island response to extreme storms. Geomorphology 100, 223–240.

Idier, D., Cateslle, B., 33 others, 2013. Vulnerability of sandy coasts to climate variability. Clim. Res. 57, 19–44.

Inman, D.L., Nordstrom, C.E., 1971. On the tectonic and morphologic classification of coasts. J. Geol. 79, 1–21.

IPCC, 2012. Managing the Risks of Extreme Events and Disasters to Advance Climate Change Adaptation. In: Field, C.B., Barros, V., Stocker, T.F., Qin, D., Dokken, D.J., Ebi, K.L., Mastrandrea, M.D., Mach, K.J., Plattner, G.-K., Allen, S.K., Tignor, M., Midgley, P.M. (Eds.), A Special Report of Working Groups I and II of the Intergovernmental Panel on Climate Change. Cambridge University Press, Cambridge, UK, New York, NY, USA, 582 pp.

Jimenes, J.A., Sancho-Garcia, A., Bosom, E., Valedmoro, H.I., Guillen, J., 2012. Storm-induced damages along the Catalan coast (NW Mediterranean) during the period 1958–2008. Geomorphology 143–144, 24–33.

Kahn, J.H., 1986. Geomorphic recovery of the Chandeleur Islands, Louisiana, after a major hurricane. J. Coastal Res. 2 (3), 337–344.

Karunarathna, H., Pender, D., Ranasinghe, R., Short, A.D., Reeve, D.E., 2014. The effects of storm clustering on beach profile variability. Marine Geol. 348, 103–112.

King, C.A.M., Williams, W.W., 1949. The formation and movement of sand bars by wave action. Geogr. J. 113, 70–85.

Kraus, N.C., Smith, J.M. (Eds.), 1994. SUPERTANK Laboratory Data Collection Project. Main Text. Technical Report CERC-94-3, U.S. Army Eng, vol. 1. Waterways Experiment Station, Coastal Engineering Research Center, Vicksburg, MS.

Larson, M., Kraus, N.C., 1994. Temporal and spatial scales of beach profile change, Duck, North Carolina. Marine Geol. 117, 75–94.

Leatherman, S.P., Williams, A.T., 1977. Lateral textural grading in overwash sediments. Earth Surf. Processes Landforms 2, 333–341.

Leatherman, S.P., Williams, A.T., 1983. Vertical sedimentation units in a barrier island washover fan. Earth Surf. Processes Landforms 8, 141–150.

Lee, G., Nicholls, R.J., Birkemeier, W.A., 1998. Storm-driven variability of the beach-nearshore profile at Duck, North Carolina, USA, 1981–1991. Marine Geol. 148, 163–177.

Lentz, E.E., Hapke, C.J., Stockdon, H.F., Hehre, R.E., 2013. Improving understanding of near-term barrier island evolution through multi-decadal assessment of morphologic change. Marine Geol. 337, 125–139.

Matias, A., Williams, J.J., Masselink, G., Ferreira, O., 2012. Overwash threshold for gravel barriers. Coastal Eng. 63, 48–61.

McBride, R.A., Penland, S., Jaffe, B.E., Williams, S.J., Sallenger, A.H., Westphal, K.A., 1989. Erosion and deterioration of the Isles Dernieres barrier island arc - Louisiana, USA: 1853-1988. Transactions of the Gulf Coast Association of Geological Societies 39, 431–444.

Morton, R.A., 1978. Large-scale rhomboid bed forms and sedimentary structures associated with hurricane washover. Sedimentology 25, 183–204.

Morton, R.A., 2002. Factors controlling storm impacts on coastal barriers and beaches—a preliminary basis for near real-time forecasting. J. Coastal Res. 18, 486–501.

Morton, R.A., Paine, J.G., Gibeaut, J.C., 1994. Stages and durations of post-storm beach recovery, Southeastern Texas coast, U.S.A. J. Coastal Res. 10 (4), 884–908.

Morton, R.A., Sallenger, A.H., 2003. Morphological impact of extreme storms on sandy beaches and barriers. J.Coastal Res. 19, 560–573.

Munger, S., Kraus, N.C., 2010. Frequency of extreme storms based on beach erosion at northern Assateague Island, Maryland. Shore Beach 78 (2), 3–11.

Nakashima, L.D., 1989. Shoreline responses to Hurricane Bonnie in southwestern Louisiana. J. Coastal Res. 5 (10), 127–136.

National Oceanographic and Atmospheric Administration (NOAA), 2012. The Saffir-Simpson Hurricane Wind Scale. National Weather Service, National Hurricane Service. http://www.nhc.noaa.gov/pdf/sshws.pdf.

Owens, E.H., Frobel, D.H., 1977. Ridge and runnel systems in the Magdalen Islands, Quebec. J. Sediment. Petrol. 47 (1), 191–198.

Plant, N.G., Stockdon, H.F., 2012. Probabilistic prediction of barrier-island response to hurricanes. J. Geophys. Res. 117 (F03015), 1–17.

Prestas, A.M., Fagherazzi, S., 2010. Morphological barrier island changes and recovery of dunes after Hurricane Dennis, St. George Island, Florida. Geomorphology 114, 614–626.

Roberts, T.M., Wang, P., Kraus, N.C., 2010. Limits of beach and dune erosion in response to wave runup from large-scale laboratory data. J. Coastal Res. 26, 184–198.

Roberts, T.M., Wang, P., Puleo, J.A., 2013. Storm-driven cyclic beach morphodynamics of a mixed sand and gravel beach along the Mid-Atlantic coast, USA. Marine Geol. 346, 403–421. http://dx.doi.org/10.1016/j.margeo.2013.08.001.

Rodrigues, B.A., Matias, A., Ferrerra, O., 2012. Overwash hazard assessment. Geol. Acta 10, 427–437.

Ruggiero, P., Komar, P.D., McDougal, W.G., Marra, J.J., Beach, R.A., 2001. Wave runup, extreme water levels and the erosion of properties backing beaches. J. Coastal Res. 17 (2), 407–419.

Ruggiero, P., Holman, R.A., Beach, R.A., 2004. Wave run-up on a high-energy dissipative beach. J. Geophys. Res. 109 (C06025), 1–12.

Sallenger, A.H., 2000. Storm impact scale for barriers islands. J. Coastal Res. 16, 890–895.

Sallenger, A.H., Stockdon, H.F., Fauver, L., Hanse, M., Thompson, D., Wright, C.W., Lillycrop, J., 2006. Hurricanes 2004: an overview of their characteristics and coastal change. Estuaries Coasts 29, 880–888.

Schwartz, R.K., 1975. Nature and Genesis of Some Storm Washover Deposits. CERC Technical Memo 61. U.S. Army Corps of Engineers, 69pp.

Schwartz, R.K., 1982. Bedform and stratification characteristics of some modern small-scale washover sand bodies. Sedimentology 29, 835–849.

Sherman, D.J., Hales, B.U., Potts, M.K., Ellis, J.T., Liu, H., Houser, C., 2013. Impacts of Hurricane Ike on the beaches of the Bolivar Peninsula, TX, USA. Geomorphology 199, 62–81.

Short, A.D., Hesp, P.A., 1982. Wave, beach and dune interactions in southeastern Australia. Marine Geol. 48 (3–4), 259–284.

Splinter, K.D., Carley, J.T., Golshani, A., Tomlinson, R., 2014. A relationship to describe the cumulative impact of storm clusters on beach erosion. Coastal Eng. 83, 49–55.

Stockdon, H.F., Holman, R.A., Howd, P.A., Sallenger, A.H., 2006. Empirical parameterization of setup, swash, and runup. Coastal Eng. 53, 573–588.

Stockdon, H.F., Sallenger, A.H., Holman, R.A., Howd, P.A., 2007. A simple model for the spatially-variable coastal response to hurricanes. Marine Geol. 238, 1–20.

Stone, G.W., Liu, B., Pepper, D.A., Wang, P., 2004. The importance of extratropical and tropical cyclones on the short-term evolution of barrier islands along the northern Gulf of Mexico, USA. Marine Geol. 210, 63–78.

Suanez, S., Cariolet, J., Cancouët, R., Ardhuin, F., Delacourt, C., 2012. Dune recovery after storm erosion on a high-energy beach: vougot beach, Brittany (France). Geomorphology 139–140, 16–33.

USACE (U.S. Army Corps of Engineers), 2002. Coastal Engineering Manual. Engineer Manual 1110-2-1100. U.S. Army Corps of Engineers, Washington, D.C (in 6 volumes).

Wang, P., Beck, T.M., 2012. Morphodynamics of an anthropogenically altered dual-inlet system: John's pass and blind pass, west-central Florida, USA. Marine Geol. 291–294, 162–175. http://dx.doi.org/10.1016/j.margeo.2011.06.001.

Wang, P., Beck, T.M., Roberts, T.M., 2011. Modeling regional-scale sediment transport and medium-term morphology change at a dual inlet system examined with the coastal modeling system (CMS): a case study at Johns pass and blind pass, west-central Florida. J. Coastal Res. Special Issue 59, 49–60.

Wang, P., Smith, E.R., Ebersole, B.A., 2002a. Large-scale laboratory measurements of longshore sediment transport under spilling and plunging breakers. J. Coastal Res. 18, 118–135.

Wang, P., Ebersole, B.A., Smith, E.R., Johnson, B.D., 2002b. Temporal and special variations of surf-zone currents and suspended-sediment concentration. Coastal Eng. 46, 175–211.

Wang, P., Horwitz, M.H., 2007. Erosional and depositional characteristics of regional overwash deposits caused by multiple hurricanes. Sedimentology 54, 545–564.

Wang, P., Kirby, J.H., Haber, J.D., Horwitz, M.H., Knorr, P.O., Krock, J.R., 2006. Morphological and sedimentological impacts of Hurricane Ivan and immediate post-storm beach recovery along the Northwestern Florida barrier-island coasts. J. Coastal Res. 22 (6), 1382–1402.

Weymer, B.A., Houser, C., Giardino, J.R., 2013. Poststorm evolution of beach-dune morphology: Padre Island national seashore, Texas. J. Coastal Res., 1–12 (online pre-prints).

Wolner, C.W.V., Moore, L.J., Young, D.R., Brantley, S.T., Bissett, S.N., Mcbride, R.A., 2013. Ecomorphodynamic feedbacks and barrier island response to disturbance: insights from the Virginia barrier islands, Mid-Atlantic Bight, USA. Geomorphology 199, 115–128.

Wright, L.D., Short, A.D., 1984. Morphodynamic variability of surf zone and beaches. Marine Geol. 56, 93–118.

Chapter 11

Extreme Waves: Causes, Characteristics, and Impact on Coastal Environments and Society

Jim D. Hansom [1,2], Adam D. Switzer [3,4] and Jeremy Pile [3]

[1] *School of Geographical and Earth Sciences, University of Glasgow, Glasgow, UK,* [2] *Department of Geography, University of Canterbury, Christchurch, New Zealand,* [3] *Earth Observatory of Singapore, Nanyang Technological University, Singapore,* [4] *Division of Earth Sciences, Nanyang Technological University, Singapore*

ABSTRACT

The existence of extreme waves, as observed by seafarers, has been confirmed by data recording and modeling to be more common than previously assumed. Extreme waves mainly occur during major storms at sea by means of constructive interference of wave trains or by nonlinear wave interaction, but extreme waves may also be associated with tsunami or meteotsunami events. If they arrive at the coast, most extreme waves have the potential to cause extensive remodeling and repositioning of the shoreline environment and landforms as well as causing significant damage to human infrastructure and threat to life. The impact of extreme waves on both sedimentary and rocky coasts can be substantial with sediments or rocky boulders eroded from the coastal edge being transported and deposited some distance inland. This characteristic provides clues to the nature of the extreme event and, if recorded within the recent sedimentary record, information about the periodicity of similar events. The impact of extreme waves on coastal communities and environment has prompted a range of mitigation and adaptation strategies to cope with these hazards. These include more robust coast defences, better modeling, prediction and warning systems, improved interagency liaison, improved technical assistance, and storm impact management information for the general public, as well as provision of clear evacuation routes during a wave-related emergency. However, since climate change seems likely to result in increased rates of both sea-level rise and storm-related impacts, there is an emerging consensus that adaptive management of the coastal zone may prove to be a more sustainable strategy than the alternatives.

Coastal and Marine Hazards, Risks, and Disasters. http://dx.doi.org/10.1016/B978-0-12-396483-0.00011-X
Copyright © 2015 Elsevier Inc. All rights reserved.

11.1 WHAT ARE "EXTREME" WAVES?

Even when the sea surface is not particularly stormy, interactions between waves may result in locally higher wave heights that can lead to fatalities at sea and at the coast and thus may become labeled in the media as extreme, giant, freak, or rogue waves. Heavy sea states and severe weather conditions caused the loss of more than 200 large cargo vessels between 1981 and 2000, with over 30 percent of the casualties due to severe weather and an additional 25 percent unexplained (Rosenthal and Lehner, 2008). In many cases single "rogue waves," as well as groups of extreme storm waves, have been reported by the crew members of such ships. The existence of rogue waves had been questioned for a long time, but remote sensing and continuous observation at oil platforms, for example in the North Sea, has demonstrated their occurrence. Rogue waves are defined as a maximum wave height of more than two times the significant wave height or Hs, which is statistically defined as the average of the highest third of all waves over a 20 min period. Using satellite radar and in situ sensors to investigate ship and platform accidents at sea, the European project MaxWave showed that extreme storm waves with heights of 25 m were not uncommon during storms (Rosenthal and Lehner, 2008). In shallow-water coastal settings recent work has clarified the conditions under which exceptionally large and steep waves may form (e.g., Didenkulova and Anderson, 2006; Didenkulova, 2011). Using data of regular deaths in Taiwan resulting from such "mad-dog" waves, Tsai et al. (2004) have shown that they occur mostly along cliff coasts, or at breakwaters fronting waters 110 m deep, with steep offshore slopes.

Although extreme waves can be produced in several ways, they are most commonly generated during major storms at sea and may then propagate toward land with only limited attenuation as a result of diffraction or refraction. A second category of extreme wave is a product of constructive interference produced when different wave fields come together during storms to produce waves that may be several times the height of the original waves, a condition that also occurs in areas where storm waves propagate against a strong opposing tidal or marine current.

Extreme waves can also be produced by Earth movements resulting from tectonic or volcanic activity at or near the coast, or by landslide activity into or beneath the ocean or by extraterrestrial bolide impact. Any of these mechanisms can generate tsunamis (Switzer, 2014). Formerly called tidal waves or seismic sea waves, tsunamis are not tidally produced. Extreme waves can also be produced by meteorological tsunamis or meteotsunamis. Propagating in the same way and with similar coastal dynamics as tsunamis (Monserrat et al., 2006), a meteotsunami is effectively an atmospherically generated large amplitude oscillation or seiche caused by moving air pressure disturbances that can result in waves up to 6 m (Vilibić and Šepić, 2009). Meteotsunamis occur worldwide and can be locally destructive (Bryant, 2001).

11.2 TYPES OF EXTREME WAVES AND THE CONDITIONS THAT PRODUCE THEM

11.2.1 Storm Waves

Extreme waves are commonly produced by very strong winds blowing for lengthy periods over long ocean fetches. However, when does a normal storm become an extreme storm likely to produce extreme waves capable of substantial damage? There are clear criteria for the classification of hurricanes, cyclones, and typhoons and several scales exist to assess the winds generated by these systems, the best known of which is the Saffir–Simpson Hurricane Scale for the North Atlantic and Northeastern Pacific Oceans. Occurring both in the open ocean and along coastlines, the recorded or estimated height reached by extreme waves generated by storms varies considerably with local effects and sampling interval. For example, in a short duration 3-h storm the most probable highest individual wave (Hmpm) is statistically about 1.86 Hs. However, on January 1, 1995, at Statoil's Draupner gas platform (16/11-E) in the North Sea, the Hmpm of the so-called "January" or "New Year Wave" was measured by laser to be 25.6 m (2.4 Hs) (Figure 11.1). This event clearly demonstrated the existence of giant individual waves well in excess of the 10.8 m Hs at that time (Guedes-Soares et al., 2003).

Also in 1995, the BP Amoco platform Schiehallion, sited in deep water in the Northeast Atlantic 160 km west of Scotland, was struck by an extreme wave that ruptured the superstructure 18 m above the waterline (Lawton, 2001). Analysis of extreme waves at Schiehallion produced a 1-year maximum

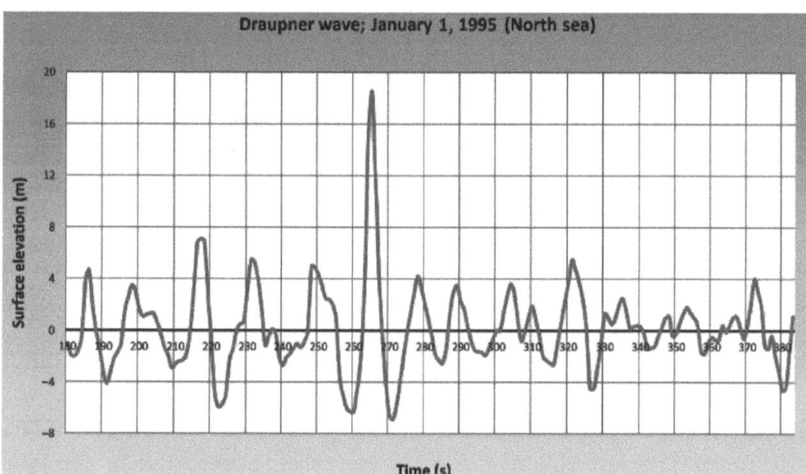

FIGURE 11.1 The "New Year Wave" of January 1, 1995 as measured at Draupner in the North Sea, with time in seconds on the X-axis against height in meters on the Y-axis. *Dysthe et al., 2005.*

individual wave height (Hmax) of 24.3 m (Hansom et al., 2008). Wave buoy K7 (60° 42′ N, 4° 30′ W) nearby, recorded the highest individual waves (Hmax) reaching 28 and 21 m during winter storms in 1992 and 1993, respectively (Hansom et al., 2008). Such waves were of the same order of magnitude as those recorded in 2004 in the Gulf of Mexico under Hurricane Ivan, where an individual Hmax reached 27.7 m (Wang et al., 2005) and in February 2000 near Rockall, 250 km west of Scotland, where an individual Hmax reached 29.1 m and is the highest individual wave ever instrumentally recorded (Holliday et al., 2006).

In the North Atlantic, Hansom and Hall (2009) suggest that the maximum height of extreme waves may have been increasing over recent decades. The Waves and Storms in the North Atlantic project reported increases in annual significant wave height (Hs) of 2.5–7.5 mma^{-1} over the period 1955–1994 (Gunther et al., 1998). This trend may be supported by the observational data of Gulev and Hasse (1999), indicating a 1–3 mma^{-1} increase in annual significant wave height (Hs) in the North Atlantic over the period 1964–1993, a trend that may be linked to intensification of the North Atlantic Oscillation (NAO) (Woolf et al., 2002). Komar and Allan (2008) report similar increases in both annual and winter Hs between 1976 and 2006 in the Northwest Atlantic. Keim et al. (2004) note North Atlantic winter sea conditions to have become rougher over the past 50–100 years, along with an increase in the frequency of very powerful storms. However, geographical variability exists in the Northeast Atlantic with higher rates of increase off the British Isles but less significant changes off Scandinavia (Wang and Swail, 2002). Indeed Feng et al. (2012) detected no increase in Hs in the Norwegian Sea between 2000 and 2009 but found an increase in the annual *mean* of extreme wave height that correlated with the winter NAO as well as an increase in the number of extreme storm waves (Hmax reaching 25.6 m in November 2001).

11.2.2 Giant or "Rogue" Waves

It is evident that very large waves occur regularly at sea during storms. It follows that any interaction or focusing of wave energy that allows these waves to grow larger than their neighbors will result in waves of extreme height. Seafarers commonly describe giant waves during major storms to have a steeper forward face preceded by a deep trough or "hole in the sea," as described by Mallory (1974). There are three known possible mechanisms, the first two of which are described by linear theory and have been understood for some time: (1) time-space focusing; (2) current focusing; and (3) nonlinear focusing.

1. Time-space focusing is the well-known product of longer ocean waves traveling faster than shorter waves to create constructive interference

where the crests coincide. Similar effects occur when wave trains from different directions cross in the ocean or lee of an island. The satellite images of the 100 × 100 km area around Draupner during the January 1 storm in 1995 showed two peaks in the directional spectra, indicating that crossing seas were generating extreme waves (Rosenthal and Lehner, 2008). As a result of MaxWave, Meteo France now uses a Cross Sea Index that uses crossing wave trains as a daily warning criterion of extreme wave conditions in the Mediterranean (Toffoli et al., 2003). A similar effect occurs at coastlines where reflected waves may constructively interact with incident waves to produce abnormally high waves, which may then cause beach and dune erosion, overtopping, and flooding as occurred along the coast of Northeast Scotland in the exceptional storms of December 2012 (Figure 11.2).

2. Current focusing occurs where waves traverse an area of variable currents and, acting like an optical lens, the currents focus wave action into a caustic region to produce freak, rogue, or giant waves (Figure 11.3) (White and Fornberg, 1998). Such focusing of wave energy increases the probability of encountering large waves in these areas. For example, off the east coast of South Africa extreme waves are produced by the strong south-going Agulhas current where it interacts with north-going waves emanating from storms in the Southern Ocean. Many large ships have foundered in this notorious area (Lavrenov, 1998).

FIGURE 11.2 Large peaked wave (solid arrow) at Golspie, Northern Scotland, December15, 2012, likely produced by the interaction of incident high energy storm waves (dashed arrow) with reflected waves from the abrupt end of rock armor to the north of the photograph, although infragravity wave interaction cannot be ruled out. The waves removed the dune face and crest and flooded houses and mobile homes. *Photo: Neil Cameron.*

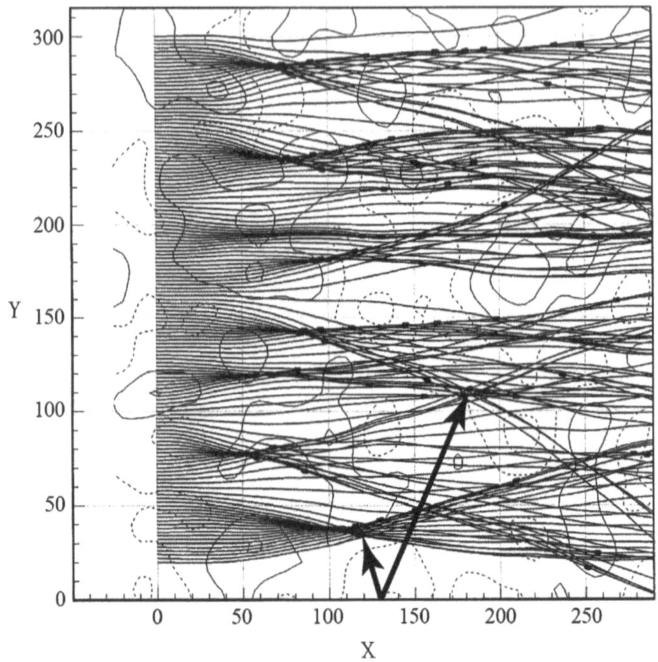

FIGURE 11.3 Modeled wave trajectories (solid dark lines) through an area of variable current (lighter solid and dashed lines). X- and Y-axes represent units of distance from the origin. The parallel wave directions entering from the Y-axis are subject to deflection by the background current to produce areas of increased wave height examples which are shown by the arrows. *Adapted from White and Fornberg (1998).*

3. Nonlinear focusing may occur after the waves are first generated, since they tend to separate into groups, which then become more prominent as they propagate. Linear theory suggests these waves should remain uniform and periodic but, in nature, some waves "grow" at the expense of adjacent waves. Modifications of the nonlinear Schrödinger equation capable of explaining this behavior (Dysthe, 1979; Dysthe and Trulsen, 1999; Liu et al., 2005) have been used to explain freak waves. Called a "breather," over time the wave develops a strong focusing of wave energy so that a small part of the wave train extracts energy from its neighbors and "breathes" itself up to reach enhanced dimensions at their expense (Figure 11.4). A second and distinct type of nonlinear effect, demonstrated to occur on the rocky Atlantic coast of Banneg Island, France, involves large infragravity waves of low-frequency (300 s period) that become trapped against the shore during storms (Sheremet et al., 2014). Such low-frequency infragravity effects produce standing waves that may interact with incident waves during storms, contribute to the generation of abnormally high wave run-up close to the shore, and enhance erosion and inundation levels.

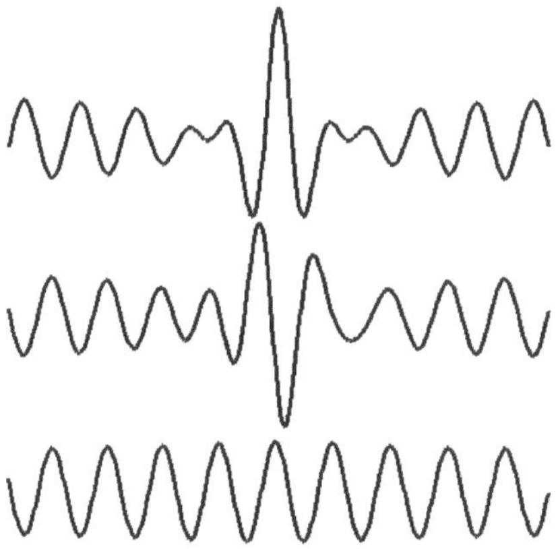

FIGURE 11.4 Development of a "breather" wave in three different stages: in the upper diagram the central waves grow by progressively extracting energy from its neighbors. *Dysthe et al. (2005).*

11.2.3 Tsunami

Tsunamis in the deep ocean typically have very long wavelengths but small wave heights and travel at speeds of more than 700 km h^{-1}. As they slow in shallow water they rapidly gain height and can be more than 30 m high, as occurred in the Indian Ocean in 2004 (Synolakis and Kong, 2006) and in Japan in 2011 (Mori et al., 2011). The highest extreme wave that has ever been reliably observed was produced by a tsunami on July 9, 1958. It was caused by an earthquake-triggered rockslide which released 30.6 million m^3 of rock to plunge from a height of 914 m into the Alaskan fjord of Gilbert Inlet (Figure 11.5). The ensuing tsunami wave swept over the 524 m promontory between Gilbert Inlet and Lituya Bay, uprooting millions of trees that were then swept into the Gulf of Alaska along with three fishing boats, two of which survived (Miller, 1960). Whilst it is clear that the Lituya Bay tsunami caused significant environmental impacts to a remote and largely uninhabited coastline, had it occurred close to a lowland coastal plain with a large resident population and villages, towns, and cities, its impact would have been greatly magnified. Unfortunately, the historical and recent record is littered with tsunamis that have impacted on many such low and densely populated coastlines and have caused widespread destruction of lives, homes, and infrastructure.

11.2.4 Meteotsunami

Meteotsunamis are atmospherically induced ocean waves that are within the tsunami frequency band (Bryant, 2001). Caused by steep pressure jumps along fronts and squalls in the open ocean, these barotropic or pressure-dependent ocean waves are amplified near the coast, particularly within enclosed bays and harbors. Meteotsunami may undergo generation over several hours as the waves move with the storm, in contrast to the instantaneous generation of a

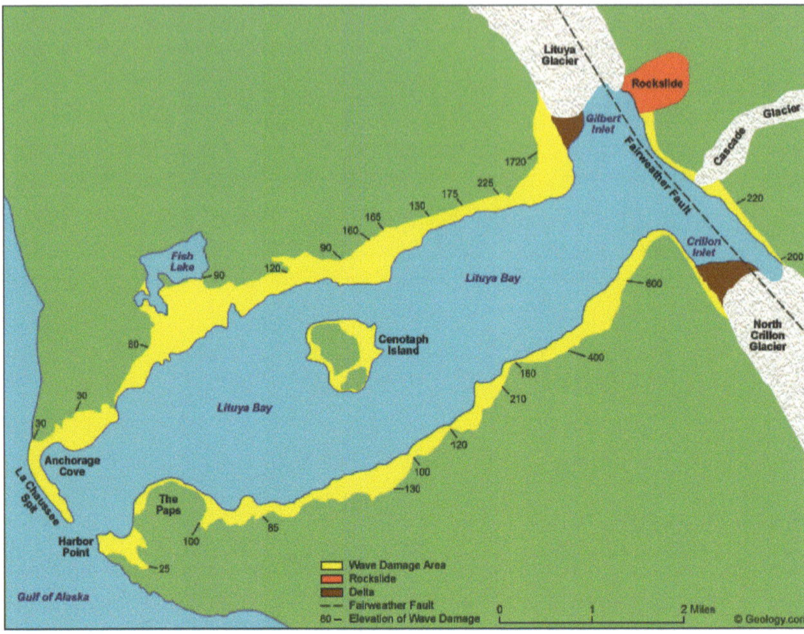

FIGURE 11.5 The glacial fjord of Lituya Bay in the Gulf of Alaska is 11.3 km long and 3.2 km wide with a maximum depth of 219 m. A 9.7 m deep sill separates it from the Gulf of Alaska at La Chaussee Spit. Heights reached by the tsunami are marked in feet. *Geology.com (2014).*

tsunami. Their occurrence has been acknowledged for some time and they are known in the Mediterranean as *rissaga*, *marubbio,* or *stigazzi,* as *Seebär* in the Baltic and as *abiki* or *yota* in Japan. Their destructive effects on coasts and great lakes are now recognized across the world (Monserrat et al., 2006). In July 1992 and October 2008, respectively, Daytona Beach, Florida, and Boothbay Harbor, Maine in the U.S., were hit by waves several meters high that appeared without warning during calm conditions. Similarly, on a sunny day in June 2011, the Yealm River mouth in southwest England reported waves up to 0.8 m high and thought to be linked to a storm 482 km away in the Bay of Biscay (Haslett et al., 2009). As a result of such events and to form the basis of a meteotsunami warning system, the U.S. National Oceanographic and Atmospheric Administration (NOAA) funded the TMEWS project (Towards a MEteotsunami Warning System) along the U.S. coastline to document past potential meteotsunami events and match them with the source, generation, and dynamics of atmospheric disturbances (Šepic et al., 2009; Vilibic et al., 2012).

11.3 IMPACT OF EXTREME WAVES ON THE COASTAL ENVIRONMENT

The impacts of extreme waves may vary from minor erosion and overwash of beaches and dunes (Figure 11.2) to the complete devastation of coastal villages by widespread erosion and flooding events (Figure 11.6). Extreme waves commonly cause localized or regional flooding and these can substantially affect the socioeconomics of coastal communities. In the short term coastal flooding can devastate crops, destroy infrastructure, and take the lives of humans and livestock. In the long term there may be erosional loss of land and loss of agricultural production due to salt water flooding. Recent technological advances have allowed better understanding of the dynamics of coastal flooding from extreme waves. Video footage of extreme wave events including footage of rogue waves, cyclone-induced storm surges, tsunamis, and meteotsunamis now provide a substantial dataset from which researchers can investigate the nature and impact of coastal flooding from extreme wave events. Recent examples include the observations of tsunamis in Indonesia (Fritz et al., 2006) and Japan (Fritz et al., 2012) and meteotsunamis in Korea (Ha et al., 2014).

11.3.1 Extreme Wave Impacts on Sedimentary Coasts (Sand/Gravel/Mud)

When extreme waves strike a coast composed of unconsolidated sediments they commonly cause substantial erosional reconfiguration of the coastal edge and may overwash water and sediment landward of the beach crest that does not return directly to the ocean (Figure 11.7). Anthony (2009) presents a comprehensive review of shore processes and deposits associated with

FIGURE 11.6 The remains of a fishing village near Tacloban, Phillipines after a 5 m + typhoon-produced storm surge in November 2013 killed more than 6,000 people and destroyed more than 100,000 dwellings. Much of the damage at the coast was done by extreme waves and the storm surge. See also Figure 11.17. *Photo: Switzer.*

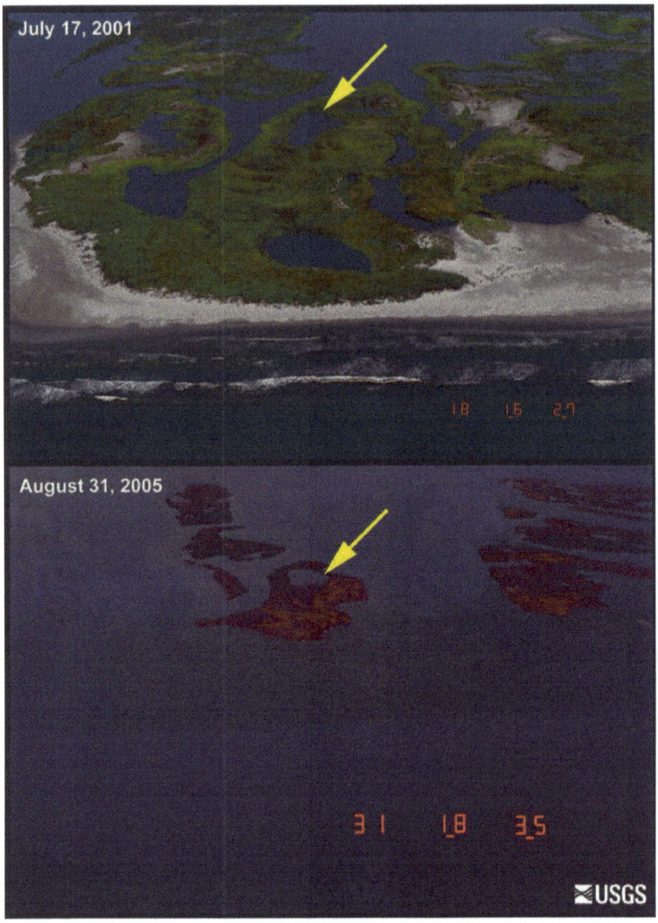

FIGURE 11.7 The Chandeleur Island chain on the Louisiana and Mississippi Coast of the Gulf of Mexico before (top) and after (bottom) the storm surge and large waves from Hurricane Katrina (landfall on August 29, 2005) submerged the islands and eroded large sections of marsh. *Credit: USGS.*

exceptional events. Overwash begins when the run-up level of extreme waves exceeds the local beach or dune crest height inundating the area behind the beach or dune (Switzer, 2014). Although severe overwash is commonly associated with storms or cyclones, it is also caused by tsunami events. It is particularly common on barrier island coasts and is also referred to as marine inundation, marine flooding, or catastrophic saltwater inundation events (Goff et al., 2001). The sediment and coastal flooding deposits that are deposited inland of a beach crest by overwash flow are manifest in a series of landforms including overwash sandsheets, storm fans, and tsunami deposits (Anthony, 2009). The inundation, landward sediment transport, and erosion as a result of

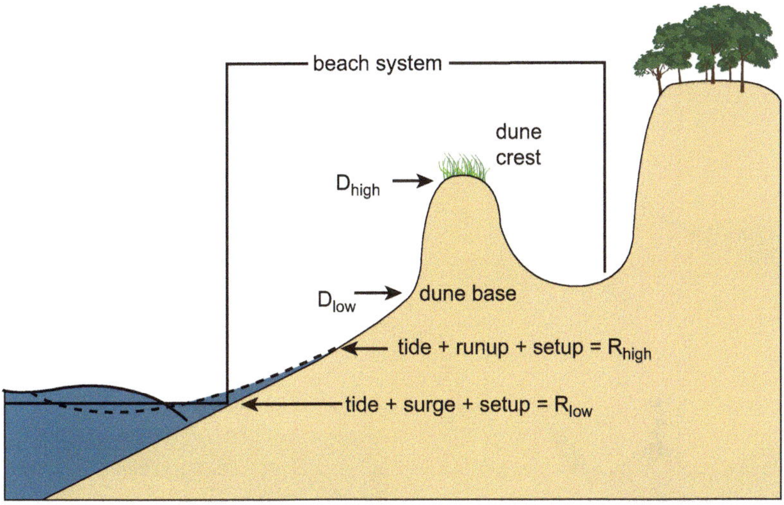

FIGURE 11.8 Diagram showing R_{high} R_{low}, D_{high}, and D_{low} where the dashed line represents the swash excursion deviation from wave setup (solid line). R_{high} is maximum water-level elevation expected during a hurricane and R_{low} is the effective still-water level during a storm, D_{high} is the elevation of the dune crest and D_{low} is the elevation of the dune toe. *Credit: USGS.*

overwash can affect coastal management, primarily through the loss or damage of property as well as damage to, or removal of, infrastructure such as railtrack, roads, recreational areas, car parks, and amenities. Some locations are more susceptible than others to the occurrence of overwash, and the management of any coastal communities affected needs to be aware of the recurrence interval of past events to assess the level of risk from future events. Barrier island coasts are particularly susceptible to the impact of severe storms and associated extreme waves and water levels yet many such low sedimentary coasts are the first line of defence against storm erosion and inundation (Figure 11.7).

In order to achieve a nationally consistent assessment of the hazard associated with storms and hurricanes and improve its capability to assess the vulnerability of the U.S. coast to extreme storm waves, the United States Geological Service (USGS) devised the Storm-Impact Scale model shown in Figure 11.8. The USGS model is based on Sallenger (2000) and comprises four wave energy conditions or regimes of increasing hazard of swash, collision, overwash, and inundation, the impact of each depending on the tide + wave + run-up + storm setup heights (R) relative to the height of the dune toe and crest (D) (Figure 11.8). These four wave-driven regimes range from no net change to the system during normal wave conditions when waves simply reorganize the intertidal sediments (Figure 11.9(a)), through steps of increasing wave impact and erosional change (Figure 11.9(b) and (c)), to major net change involving inundation and erosion of the fronting beach and

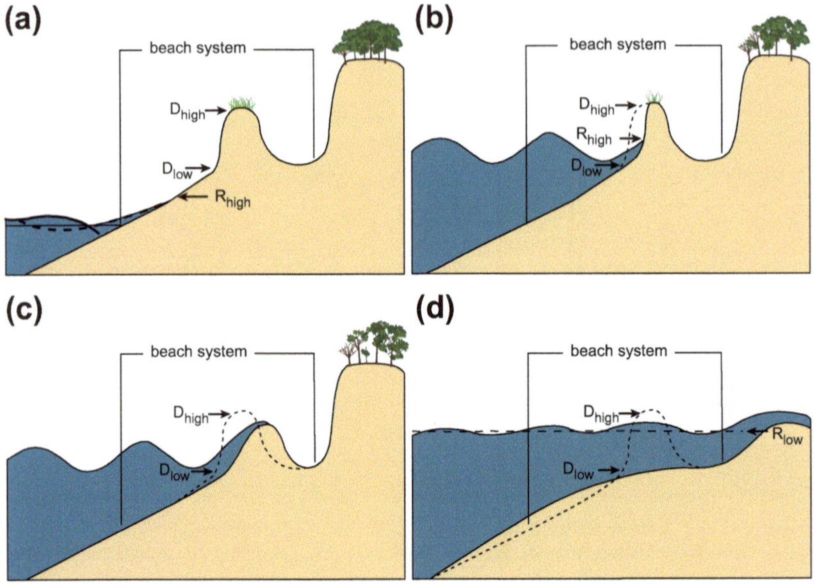

FIGURE 11.9 (a) *Swash regime*: wave run-up confined to beach, temporary erosion, offshore sediment transport followed by nonstorm return and no net change to the system. (b) *Collision regime*: wave run-up exceeds the elevation of the base of the dune, run-up collision causes erosion and net, or (semi-) permanent dune retreat. (c) *Overwash regime*: wave run-up exceeds the dune elevation or beach berm, overtopping transports sand landward to produce net change and landward barrier migration of the order of 100 m. (d) *Inundation regime*: combined wave run-up and storm surge exceeds elevation of low seaward dune to inundate entire beach; if barrier is also narrow, entire island may be inundated; results in net change and barrier migration landward of the order of 1,000 m. *Credit: USGS.*

dune system leading to wholesale landward migration of the system under the impact of severe storms and associated extreme waves (Figure 11.9(d)). The model is useful in providing an estimation of the magnitude and nature of change that can occur when typhoon or hurricane waves make their landfall. The approach compares the known heights of coastal landforms (e.g., derived from LiDAR survey) with modeled elevations of storm-induced water levels (e.g., forecast from probabilistic surge and wave models) in order to define coastal change regimes. These regimes describe the dominant interactions between beach morphology and storm processes along beaches and dunes that serve as the "first line of defence" for many coasts that are exposed to tropical storms and hurricanes, such as the east coast of the U.S. and the land bordering the Gulf of Mexico (Figure 11.7).

Nevertheless, for many such coasts the low frequency of extreme wave events makes it extremely difficult to adequately assess both their recurrence interval and impact, an issue shared in common with all high-magnitude and low-frequency natural events. Both the Tōhoku tsunami (2011) and Haiyan

Typhoon (2013) poignantly showed that catastrophic extreme wave events may be too infrequent for their hazard to be characterized by historical records alone and so the geological and geomorphological records may be used to address this gap. The characteristics and distribution of sediments transported and deposited by extreme waves can help clarify the key processes associated with particular events, including estimates of the generation and track of storms, the source locations of tsunamigenic events, the water depths and velocities of incident waves, and the mapping of past inundation distances. For example, Figure 11.10(a) shows the extent of overwash deposition in southeastern India following Cyclone Thane in 2011. Figure 11.10(b) shows post-event data gathering by field surveyors that will provide geomorphological information for use to better estimate the likely inundation extent and impact of future events.

Wave-derived overwash deposits are also preserved in the geological record and can be used to infer the extent and establish the frequency of past extreme wave events. For example, where stacked sequences of overwash deposits are found, chronologies can be developed via radiocarbon, optically stimulated luminescence, and other dating techniques to help constrain the recurrence interval of past events (Anthony, 2009). Such information may be used to guide adaptation and mitigation efforts aimed at reducing the impact of future extreme wave events. For example, on the coast of Hokkaido, Japan, Nanayama et al. (2003) show large tsunamis to have inundated the coast on average every 500 years between 2,000 and 7,000 years ago, with extreme waves depositing sand sheets (some extending several kms inland). Where preserved and accessible, such records of extreme wave events can be used to help guide land-use management and planning controls to help mitigate the extent and impact of future extreme wave events.

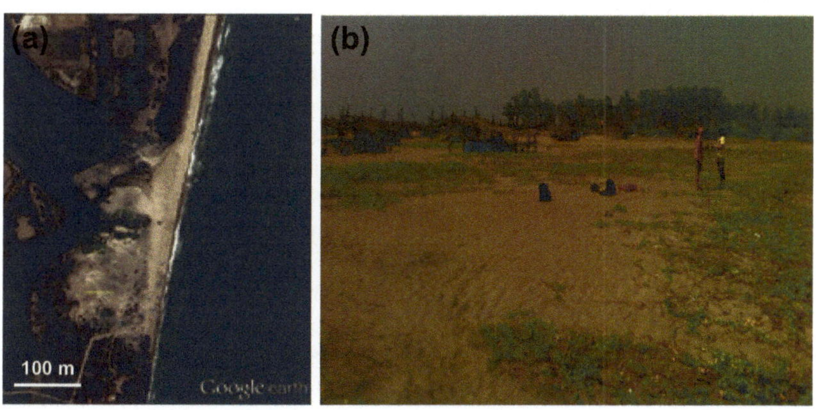

FIGURE 11.10 (a) Cyclone Thane in 2011 produced overwash deposits on the southeast Indian coast near Cuddalore. *(Image from Google Earth).* (b) Students surveying and mapping the recently deposited overwash near Cuddalore. *Photo: Switzer.*

To correctly interpret the local history of extreme waves, investigators often attempt to distinguish between the type of overwash deposit (tsunami or storm) found in the geological record. Comparative studies of both types of deposit reveal some differences in sedimentology, stratigraphy, faunal composition, and inland height and extent. Although researchers have proposed criteria to distinguish these two types of deposits, it is apparent that each must be carefully considered in the context of its regional setting. A variety of physical, sedimentological, and geochemical techniques are available, which, when carefully considered in a local geomorphic and stratigraphic context, may allow a positive discrimination between a storm or a tsumanigenic provenance. Generally, overwash deposits from extreme waves cause a decrease in the total organic matter within coastal plain sediments. Other notable changes include changes in salinity indicators and salts (Chague-Goff et al., 2012), and in the assemblage composition of ostracods, diatoms, foraminifera (often derived from deeper water offshore), pollen, and aquatic plants. The internal sedimentology (e.g., bedding), height above sea level, landward extent, and changes in thickness and regional continuity can also assist in determining the provenance of overwash deposits. In some cases, the existence of deep water heavy mineral and microfaunal assemblages now sited in a terrestrial environment may indicate transport via a tsunami as opposed to a storm (e.g., Switzer et al., 2005). It is clear that overwash deposits from extreme waves can provide an opportunity to evaluate the recurrence of storms and tsunami that are large enough to leave lasting sedimentary signatures. However, at present there exists no single analytical technique that can unambiguously differentiate between tsunami and storm deposits in the geological record.

11.3.2 Extreme Wave Impacts on Rock Coasts

On rocky coasts (herein including reefs and carbonate coasts), coarse clast deposits (boulders) emplaced by extreme waves provide graphic evidence of extreme wave events and can produce a valuable archive of their frequency and magnitude (see Anthony, 2009). However, like the clastic deposits of sand, gravel, and mud mentioned in Section 11.3.1, difficulty often surrounds the interpretation of whether extreme storm or tsunami waves have been responsible for the deposition of coarser sediments (e.g., Switzer and Burston, 2010), and this has led to misinterpretation of some deposits and generated debate in the literature (e.g., Scheffers et al., 2009, 2010; Hall et al., 2010). That said, some general characteristics can be used as broad indicators of either storm or tsunami deposition. Storm deposited clasts close to sea level commonly form ridges, ridge complexes, and clusters composed of a range of sizes (Morton et al., 2007; Etienne and Paris, 2010). This tendency is mirrored at altitude on cliff-top sites where the cliffs are fronted by deep water. This context allows storm waves to gain access to the coast with minimal attenuation and thus retain the capacity for substantial erosion and landward transport of a surprising range of clast sizes (Williams and Hall, 2004; Cox et al., 2012). For example, Hall et al. (2006, 2008) and Hansom et al. (2008) report that Cliff Top Storm Deposits (CTSDs) have accumulated at distances of up to 50 m inland of 15–20 m high cliffs in the northernmost British Isles as a result of multiple extreme wave events. These CTSDs are arranged both as discrete suites of boulder ridges of several meters high and tens of meters wide and, depending on the depositional context, also as individuals or clusters of boulders (Figure 11.11). Composed of large boulders, many of which are over 1 m long axis, the CTSDs show strong imbrication with long axis orientations conforming to the storm wave approach direction (Figure 11.12(a) and (b)) (Williams and Hall, 2004; Hall et al., 2006; Cox et al., 2012). In the Shetland Islands the boulder ridges at altitude have been subject to several episodes of erosional reworking over recent years that can be directly correlated with the impact of extreme waves during identified major storm events with the changes being captured and quantified by time-series Terrestrial Laser Scanning. Dating of the phases of deposition of CTSDs allowed Hansom and Hall (2009) to link the overall storminess of the Eastern Atlantic to the NAO and the sea salt record of the GISP2 Greenland ice core.

A key point is that these deposits have not yet been found to extend more than a few hundred meters inland from the shore (Scoffin, 1993; Scheffers, 2005; Morton et al., 2007; Etienne and Paris, 2010; Cox et al., 2012) even where islands can be demonstrated to have been overwashed at altitude by

FIGURE 11.11 A 3.5 m high cliff-top boulder ridge at 17 m asl facing normal to the wave approach direction and located 50 m from the cliff edge at Grind of the Navir in the Shetland Islands. A range of sizes of freshly quarried boulders can be seen in 2013, some in excess of 1.8 m long axis. *Photo: Hansom.*

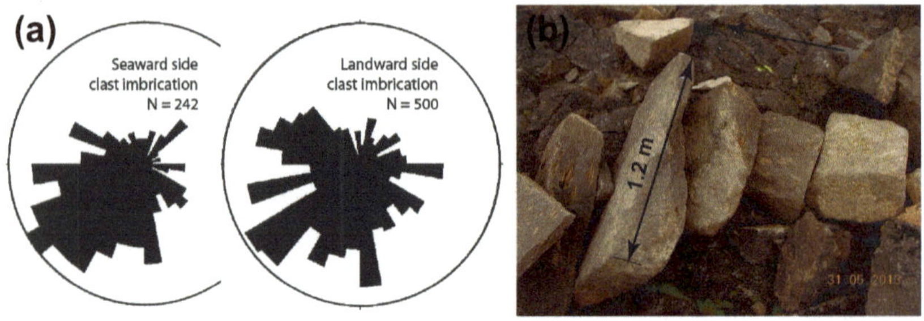

FIGURE 11.12 (a) Imbrication directions for boulder ridges on Inishmore, Aran Islands, Ireland. The dominant imbrication direction in the seaward faces (190–260°) matches the dominant wave approach direction (Williams and Hall, 2004) and reflects emplacement by incident waves (Cox et al., 2012). (b) Imbricate cluster of boulders facing the dominant wave approach direction (arrowed ca. 300°) at 15 m asl and 50 m from the cliff edge at Grind of the Navir in the Shetland Islands, Scotland. Marked long axis is 1.2 m. *Photo: Hansom.*

well-constrained wave heights during specific storms, such as at Banneg Island in Brittany (Suanez et al., 2009; Fichaut and Suanez, 2011). The impact of extreme wave events on rocky shores does not normally impact greatly upon human activity, since the characteristics and relative altitude of rocky shores means that they mostly remain undeveloped and sparsely populated.

In contrast, cobbles and boulders transported inland by tsunamis are typically found as individuals and in expansive fields as a single layer and, relatively speaking, are more dispersed than storm ridges (Watt et al., 2011) and found at substantial distances inland over long stretches of coastline. Modern examples of dispersed tsunami boulder fields have been reported from many parts of the world including Indonesia (Paris et al., 2009), and on elevated platforms in Hawaii (e.g., Richmond et al., 2011).

11.4 IMPACT OF EXTREME WAVES ON SOCIETY AND INFRASTRUCTURE

Over the last decade, the natural hazard presented by extreme wave impact during storms (tropical and extratropical cyclones) and tsunamis has resulted in numerous high-profile coastal disasters and loss of life, property, and infrastructure. The Indian Ocean (2004), Samoan (2009), Chilean (2010), and Japanese (2011) tsunamis, as well as tropical cyclones including Hurricane Katrina in New Orleans, USA (2005), Cyclone Nargis in Myanmar (2008), Hurricane Irene in Northeast USA (2011), Cyclones Yasi (2011) and Ita (2014) in Australia, and Typhoon Haiyan in the Philippines (2013) have highlighted the vulnerability of the coastal zone and its infrastructure, social and ecological systems. In the UK there have been significant increases in the number of severe extratropical cyclones since the 1950s (Alexander et al.,

2005), which appear to be related to changes in the NAO to a more positive phase, but it remains uncertain whether the multidecadal NAO variability is related to climate change (Tsimplis et al., 2005). More recently, the impact of extreme storm waves during the multiple deep extratropical cyclones that affected the British Isles over the winters of 2012/2013/2014 resulted in widespread flooding of low-lying coastal areas and severe erosion of beaches, dunes, and cliffs and the failure of long-lived and substantial coastal defences. The areas affected stretched from the Somerset Levels and Welsh coast in the Atlantic southwest, along much of the South and North Sea coasts of England and extended as far north as Northern Scotland (Figure 11.2). In 2013, the substantial concrete sea wall at Aberystwyth, Wales was partially destroyed and houses evacuated whilst storm waves on the south coast severed the main rail link between London and the southwest of England (Figure 11.13(a) and (b)).

Increasing socioeconomic pressures, human alteration of the landscape, and the continued overexploitation of coastal resources have progressively reduced the resilience of many coastal systems to both short and long-term environmental change, a classic case being the regrading or removal of protective dunes to increase the size of the beach or provide views of the sea by shorefront residents (Nordstrom, 1994). Coastlines worldwide now also face warmer global temperatures and accelerating sea-level rise (IPCC, 2014).When combined with the potential for an increase in storminess and hence extreme wave events, it seems inevitable that these factors will further exacerbate the potential for coastal erosion and flooding of coastal habitats (Jones et al., 2013; Angus, 2014), communities, and infrastructure (Angus and Rennie, 2014; Young et al., 2014; IPCC, 2014), highlighting a need for coastal management responses that are sustainable in the long term.

FIGURE 11.13 (a) Extreme wave impact on the sea wall at Chesilton at the east end of Chesil beach, England, 2013. *(Photo: Poate)*. (b) The main rail line between London and the southwest of England was severed by extreme storm waves in February 2013 and remained closed for 3 months for engineering repairs. *Photo: Geoff Sheppard.*

11.4.1 Mitigation and Adaptation

Despite an increase in the understanding and awareness of extreme waves associated with storms and tsunamis and a clear global increase in efforts to estimate and manage coastal risks, disasters associated with extreme wave events continue to occur and cause significant socioeconomic damage (e.g., Hurricanes Katrina, August 2005 and Sandy, October 2012, the Tōhoku Tsunami, March 2011 and Typhoon Haiyan, November 2013). In order to prepare for future coastal change including the hazard of extreme wave conditions at the coast, stakeholders have been forced to rethink many aspects of hazard response and adaptation strategies including modelling the predicted impact of extreme wave events, engineering the environment through the construction of larger and more robust sea defences and adopting supplementary or alternative responses to such events, e.g., devising evacuation plans and signposting evacuation routes, establishing development set-back zones and investing in adaptation strategies such as managed coastal realignment (Cooper and Pile, 2014). The last of these strategies has been widely used in estuaries and sheltered coastal settings but is yet to find widespread popularity as an adaptation strategy for the outer coast.

The USGS use their Storm Impact Scale model (Figures 11.8 and 11.9) as the basis to predict the impact of wave events of increasing severity along coasts whose characteristics make them physically susceptible to erosion and flooding. For example Figure 11.14 shows the predicted spatial impact of Hurricane Sandy (October 2012) on the Long Island coast in terms of the probability of collision, overwash, and inundation. However, the impact varies depending on the severity of the storm and the nature of the coastline. Overall the low dune elevations along the Gulf of Mexico (2.4 m high) and Southeast Atlantic coasts make 71 percent of their beaches very likely to experience extreme erosion due to overwash on the landfall of a category 1 storm. By contrast, on the mid-Atlantic U.S. coast only 25 percent of the beaches are very likely to experience overwash on landfall of a category 1 storm because the dunes are an average of 2 m higher (Doran et al., 2012). The method is successful and allows before and after impact estimates. For example along the mid-Atlantic US coast, Hurricane Sandy remodeled the beaches and dunes to the extent that the percentage of beach coast very likely to experience dune erosion for a category 1 hurricane landfall, fell from 89 percent prestorm to 75 percent poststorm. Yet for a category 4 hurricane landfall, the percentage of mid-Atlantic beaches that are very likely to experience overwash increased from 92 percent pre-Sandy to 95 percent post-Sandy and the percentage of coastline very likely to experience inundation increased from 66 percent to 68 percent post-Sandy. Along the Gulf coasts a post-Sandy category 4 hurricane landfall is likely to produce overwash along 97 percent and inundation along 83 percent of the coast (Stockdon et al., 2012).

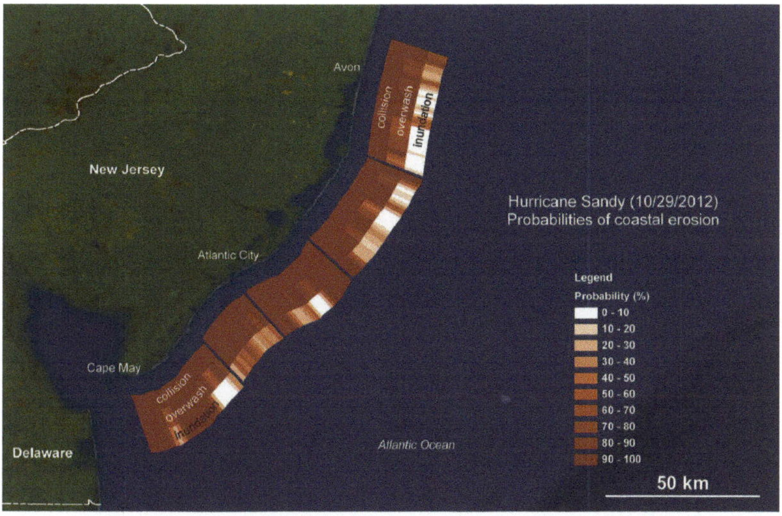

FIGURE 11.14 Color-coded probabilities of the occurrence of collision, overwash, and inundation regimes for Hurricane Sandy on the New Jersey coast, U.S. The predictions suggested beach and dune erosion was likely for the bulk of this coast with significant lengths of the coast likely to be also subject to overwash and net landward movement. More severe inundation of the beach and dune system was expected to be spatially limited during this event. *Credit: USGS.*

Extreme wave events have increasingly demonstrated that coastal states and communities need to improve their resilience by means of effective strategies, tools and resources to conserve, protect, and restore coastal habitats and economies at risk from current impacts as well as those expected due to a changing climate. For example, in a recent detailed report for the U.S. National Climate Assessment, Burkett and Davidson (2012) regarded all US coasts as highly vulnerable to the effects of storms, erosion, flooding and sea-level rise particularly in the more populated low-lying parts of the US coast with the financial risks associated with both private and public hazard insurance expected to increase dramatically (Figure 11.15). The report calls for increased coordination and planning to ensure that US coastal communities are resilient against the effects of climate change. One such initiative was launched in 2008 by the Massachusetts Office of Coastal Zone Management. StormSmart Coasts is targeted to help local officials in coastal communities address the challenges arising from extreme storm wave and flood events, such as occurred in April 2007, by providing direct technical assistance and a Web-based menu of tools for successful coastal floodplain management. This now includes much of the New England and the Gulf Coast states and laid the foundation for a National StormSmart Coasts Network (http://stormsmartcoasts.org/).

Extreme waves generated by midlatitude and tropical cyclones are some of the most economically disruptive natural disasters that occur on Earth

FIGURE 11.15 Developed shoreline of Seaside Heights, New Jersey before (top) and after (bottom) the landfall of Hurricane Sandy in October 2012, showing the extensive beach erosion that destroyed seafront defences, roadways, and properties and the overwash and inundation damage to the low-lying housing coastal strip behind the beach. *Credit: USGS.*

(Switzer, 2014). For example, the category 5 Hurricane Katrina caused an estimated US$ 38,000 million in insured losses alone (Jagger et al., 2008). Modeling by the same authors predicted losses of US$ 10,000 million for a 50-year return period extreme event under unfavorable conditions for the generation of U.S. hurricanes, these losses escalating under favorable conditions for the generation of U.S. hurricanes for the same return period to approximately US$630bn. The U.S. National Oceanic and Atmospheric Administration (NOAA) currently estimate the average annual losses from hurricanes in the USA alone is just under $US 6,000 million with a few extreme events, such as Katrina (US$ 125,000 million, 2005) and Sandy (US$ 68,000 million, 2012), increasing the total annual cost substantially (Figure 11.15).

11.4.2 Prediction and Warning

The catastrophic nature of extreme wave events has helped drive the science of prediction and the deployment of warning systems. Most of the world's oceans are now covered by a network of tsunami warning buoys which are linked to seismic stations and can provide warnings of earthquake-derived tsunamis. Messages are relayed via radio, the internet, or via SMS to subscribed users, generally government agencies and scientists, who then relay the information to the general public through the use of emergency broadcast systems and sirens (Figure 11.16). Similarly, tropical and extratropical storms can also be tracked and modeled, almost from inception, by weather satellites and meteorological agencies throughout the world. Computing power and modeling capabilities have seen a dramatic increase over the past few years, so that it is now possible to run iterative simulations to generate the probable hazard zones and likely wave impacts many days before the landfall of large tropical or extratropical storms of varying levels of intensity. This enables emergency planning organizations to put in place damage mitigation measures and trigger evacuation plans.

However, warning systems and emergency procedures are rarely 100 percent accurate and even the best warning systems may struggle to cope with the sudden onset of extreme events, such as an earthquake triggered tsunami with an epicenter very close to the coast (e.g., Hokkaido earthquake in 1993) or the rapidly changing nature, course, and predicted landfall of a hurricane as it develops in severity to produce different levels of impact on different coastal configurations. Emergency plans and procedures can also suffer from institutional inertia. Even with sufficient warning and accurate weather forecasts, it appeared to take several weeks for the UK authorities to mount effective responses to the winter storms of 2013/2014. A similar situation occurred in New Orleans during and after Hurricane Katrina. Institutional inertia can also be exacerbated by difficulties faced by, and between, the authorities and the general public in communicating the uncertainties attached to predicting a hazard that may be extreme but infrequent, as well as the varying risk involved when vulnerable coastal communities are exposed to hazard.

11.4.3 Vulnerability and Resilience

Coastal vulnerability assessment has emerged as a key concept for understanding the impacts of climate change and natural hazards such as those

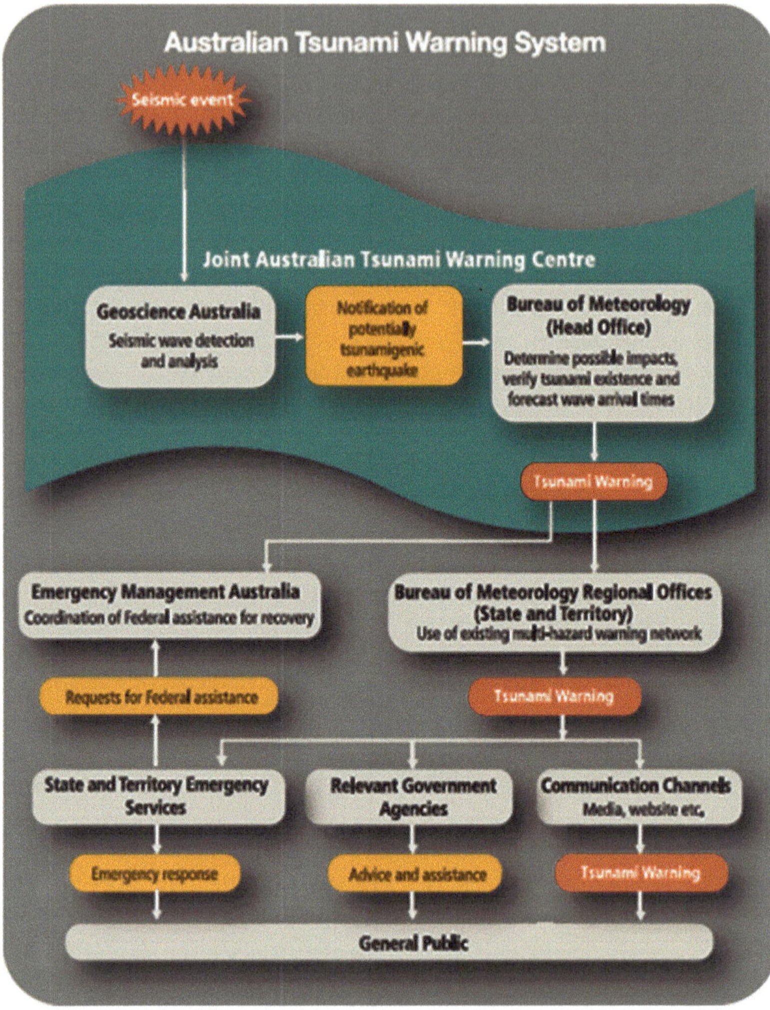

FIGURE 11.16 Australian tsunami warning system flowchart. Notice that, in this model, the initiation of a tsunami alert or warning appears to be dependent on the detection and analysis of a seismic event. *http://www.bom.gov.au/tsunami/about/atws.shtml.*

posed by extreme waves (Switzer, 2014). Depending on the nature and characteristics of the coast, both physical and socioeconomic systems comprise varying levels of sensitivity and resilience to extreme wave events, with coastal vulnerability being the "in-combination" outcome of the interaction of both the physical and socioeconomic effects. Vulnerability of coastal communities to extreme wave events can be regarded as the exposure to hazard of its social, economic, and physical properties, and this may be offset by developing adaptive capacity and coping strategies to reduce adverse impacts and enhance society's resilience. For extreme wave events, assessing physical coastal susceptibility to erosion and flooding in the context of socioeconomic vulnerability has emerged as a key planning consideration. Planners of large coastal infrastructure projects such as road and rail links, ports, airports, and nuclear power plants must now account for the "where," "why" and "how" questions related to potential coastal erosion and flooding and increasingly accommodate the projections of future changes in sea level and storm intensity into their plans. In many cases these effects are intimately linked in that partial or complete loss of a landform or structure that currently provides coastal erosion and/or flood protection may subsequently place the hinterland at risk of more widespread erosion and flood impact than would be the case if only overtopping had occurred. Such "in-combination" effects are difficult to integrate into models of coastal vulnerability since they are dependent on the level of any remedial or adaptive intervention over the intervening time period. In particular, most vulnerability assessment methodologies are not designed to cope with the typical recurrence intervals of extreme events (e.g., 1 in 200 or 1 in 500 year timeframes). There are a variety of coastal vulnerability assessment methods that encompass a broad and diverse range of applications at a variety of spatial scales. Yet despite huge advances in the quality of data for assessing the impact of extreme waves (e.g., tidal data, wave models, satellite imagery, digital terrain models, sophisticated flooding models, and the availability of socioeconomic data), the usefulness of integrated assessment methods at the high end of the scale remains limited.

11.5 CONCLUSION

The impacts of extreme waves have reshaped the history of many coastal cities and communities, with wave erosion and flooding from tsunamis and extreme storms causing widespread devastation. In some cases where extreme events are known to have affected particular coasts in the past, the absence of similar events in the intervening years has resulted in either complacency, an over-reliance on engineered defences (Figure 11.13) or a loss of folk-memory (Figure 11.17). In addition to loss of life and infrastructure, coastal ecosystems and economies are also seriously impacted by the landfall of storm waves and tsunami that have devastating effects on coastal food plantations, fisheries (e.g., seafood farms), and harvesting infrastructure (e.g., boats, processing, and storage facilities). It is clear that whilst engineering structures will remain part of the toolkit to mitigate the effects of extreme storm waves, there is an increasing acceptance that adaptive management (such as land-use zoning, development set-back lines, and managed realignment) may prove a more sustainable alternative to cope with future increases in sea level and extreme storm wave impacts. Assessing the potential hazard posed by extreme waves on any coast and its communities has three main elements: (1) the potential for mortality or social disruption (the social risk) (2) the potential for economic

FIGURE 11.17 An unnamed typhoon struck Tacloban in the Phillipines in 1897, producing impacts identical to those produced on the same coast by Super Typhoon Haiyan (2013). Such recurrence suggests that a major hurdle in preparing coasts for extreme wave events is overcoming the tendency for humans to forget, or ignore, the past record. See also Figure 11.6. *Algue and Jose (1898).*

damage (the economic risk); and (3) the potential for morphological and natural habitat disturbance or destruction (the geoecological risk). Whilst quantifying the last of these elements is perhaps the most achievable, accurate quantification of the in-combination effects of extreme waves on the others remains problematic. Nevertheless, the quantification and prediction of these impacts on the coast and its communities remains an important goal for coastal science in the context of both present and future climate change.

REFERENCES

Algue, P., Jose, S.J., 1898. El Baguio de Samar y Leyte, 12–13 de Octubre de 1897. Observatorio de Manila. Dirigido por los Padres de la Compañia de Jesus. Foto-tipografia de J. Marty. Manila.

Alexander, L.V., Tett, S.F.B., Jonsson, T., 2005. Recent observed changes in severe storms over the United Kingdom and Iceland. Geophys. Res. Lett. 32, L13704. http://dx.doi.org/10.1029/2005GL022371.

Angus, S., 2014. The implications of climate change for coastal habitats in the Uists, Outer Hebrides. Ocean and Coastal Management. 94, 38–43. http://dx.doi.org/10.1016/j.ocecoaman.2014.02.012.

Angus, S., Rennie, A., 2014. An Ataireachd Aird: the storm of january 2005 in the Uists, Scotland. Ocean and Coastal Management. 94, 22–29. http://dx.doi.org/10.1016/j.ocecoaman.2014.02.013.

Anthony, E.J., 2009. Shore Processes and Their Palaeoenvironmental Applications. Developments in Marine Geology, vol. 4. Elsevier, 519pp.

Bryant, E., 2001. Tsunami, the Underrated Hazard. Cambridge University Press, Cambridge, 320pp.

Burkett, V.R., Davidson, M.A. (Eds.), 2012. Coastal Impacts, Adaptation and Vulnerability: A Technical Input to the 2012 National Climate Assessment, p. 150. Cooperative Report to the 2013 National Climate Assessment. http://downloads.usgcrp.gov/NCA/technicalinputreports/Burkett_Davidson_Coasts_Final_.pdf.

Cooper, J.A.G., Pile, J., 2014. The adaptation-resistance spectrum: a classification of contemporary adaptation approaches to climate-related coastal change. Ocean and Coastal Management. 94, 90–98. http://dx.doi.org/10.1016/j.ocecoaman.2013.09.006.

Cox, R., Zentner, D.B., Kirchner, B.J., Cook, M.S., 2012. Boulder ridges on the Aran islands (Ireland): recent movements caused by storm waves, not tsunamis. J. Geol. 120 (3), 249–272.

Chagué-Goff, C., Andrew, A., Szczucinski, W., Goff, J., Nishimura, Y., 2012. Geochemical signatures up to the maximum inundation of the 2011 Tohoku-oki tsunami implications for the 869 AD Jogan and other palaeotsunamis. Sediment. Geol. 282, 5–77. http://dx.doi.org/10.1016/j.sedgeo.2012.05.021.

Didenkulova, I., 2011. Shapes of freak waves in the coastal zone of the Baltic Sea (Tallinn Bay). Boreal Environ.Res. 16, 138–148.

Didenkulova, I., Anderson, C., 2006. Freak waves in 2005. Nat. Hazards Earth Syst. Sci. 6, 1007–1015.

Doran, K.S., Stockdon, H.F., Sopkin, K.L., Thompson, D.M., Plant, N.G., 2012. National Assessment of Hurricane-Induced Coastal Erosion Hazards: Mid-Atlantic Coast. U.S. Geological Survey. Open-File Report 2013–1131, 28 pp. http://pubs.usgs.gov/of/2013/1131.

Dysthe, K.B., 1979. Note on a modification to the nonlinear Schrodinger equation for application to deep water waves. Proc. R. Soc. London, Ser. A. 369, 105–114.

Dysthe, K.B., Trulsen, K., 1999. Note on breather type solutions of the NLS as models for freak-waves. Phys. Scr. T82, 48–52.

Dysthe, K.B., Krogstad, H.E., Socquet-Juglard, H., Trulsen, K., 2005. Freak Waves, Rogue Waves, Extreme Waves and Ocean Wave Climate. (Online) Available from: http://folk.uio.no/karstent/waves/index_en.html. (accessed 20.01.14).

Etienne, S., Paris, R., 2010. Boulder accumulations related to storms on the south coast of the Reykjanes Peninsula (Iceland). Geomorphology 114, 55–70.

Feng, X., Tsimplis, M., Yelland, M., 2012. Extreme waves at the polar front of North Atlantic from 2000 to 2009. Geophys. Res. Abstr. 14, 5719–5721. EGU2012.

Fichaut, B., Suanez, S., 2011. Quarrying, transport and deposition of cliff-top storm deposits during extreme events: Banneg Island Brittany. Mar. Geol. 283, 36–55.

Fritz, H.M., Borrero, J.C., Synolakis, C.E., Yoo, J., 2006. 2004 Indian Ocean tsunami flow velocity measurements from survivor videos. Geophys. Res. Lett. 33 (24).

Fritz, H.M., Phillips, D.A., Okayasu, A., Shimozono, T., Liu, H., Mohammed, F., Skanavis, V., Costas, E., Synolakis, C.E., Takahashi, T., 2012. The 2011 Japan tsunami current velocity measurements from survivor videos at Kesennuma Bay using LiDAR. Geophys. Res. Lett. 39 (7).

Geology.com, 2014. World's Biggest Tsunami. (Online) Available from: http://geology.com/records/biggest-tsunami.shtml. (accessed 22.01.14).

Goff, J., Chagué-Goff, C., Nichol, S., 2001. Palaeotsunami deposits: a New Zealand perspective. Sediment. Geol. 143 (1–2), 1–6. http://dx.doi.org/10.1016/S0037-0738(01)00121-X.

Guedes-Soares, C., Cherneva, Z., Antão, E.M., 2003. Characteristics of abnormal waves in North Sea storm sea states. Appl. Ocean Res. 25 (6), 337–344.

Gulev, S.G., Hasse, L., 1999. Changes of wind waves in the North Atlantic over the last 30 years. Int. J. Climatol. 19, 1091–1117.

Gunther, H., Rosenthal, W., Stawartz, M., Carretero, J.C., Gomez, M., Lozano, I., Serrano, O., Reistad, M., 1998. The wave climate of the northeast Atlantic over the period 1955–1994: the WASA wave hindcast. Global Atmos. Ocean Syst. 6, 121–163.

Ha, T., Choi, J.-Y., Yoo, J., Chun, I., Shim, J., 2014. Transformation of small-scale meteorological tsunami due to terrain complexity on western coast of Korea. In: Green, A.N., Cooper, J.A.G. (Eds.), Proceedings 13th International Coastal Symposium (Durban, South Africa), Journal of Coastal Research, 70, pp. 284–289. Special Issue No.

Hall, A.M., Hansom, J.D., Williams, D.M., Jarvis, J., 2006. Distribution, geomorphology and lithofacies of cliff-top storm deposits: examples from the high energy coasts of Scotland and Ireland. Mar. Geol. 232, 131–155.

Hall, A.M., Hansom, J.D., Jarvis, J., 2008. Patterns and rates of erosion produced by high energy wave processes on hard rock headlands: the Grind of the Navir, Shetland, Scotland. Mar. Geol. 248, 28–46.

Hall, A.M., Hansom, J.D., Williams, D.M., 2010. Wave-emplaced coarse debris and megaclasts in Ireland and Scotland: boulder transport in a high-energy littoral environment: a discussion. J. Geol. 118, 699–704.

Hansom, J.D., Barltrop, N.D.P., Hall, A.M., 2008. Modelling the processes of cliff-top erosion and deposition under extreme storm waves. Mar. Geol. 253, 36–50.

Hansom, J.D., Hall, A.M., 2009. Magnitude and frequency of extra-tropical North Atlantic cyclones: a chronology from cliff-top storm deposits. Quat. Int. 195, 42–52.

Haslett, S.K., Mellor, H.E., Bryant, E.A., 2009. Meteo-tsunami hazard associated with summer thunderstorms in the United Kingdom. Phys. Chem. Earth 34, 1016–1022.

Holliday, N.P., Yelland, M.J., Pascal, R., Swail, V.R., Taylor, P.K., Griffiths, C.R., Kent, E., 2006. Were extreme waves in the Rockall Trough the largest ever recorded? Geophys. Res. Lett. 33, L05613.

IPCC, 2014. Climate Change 2014: Mitigation of Climate Change. Working Group III Contribution to the IPCC 5th Assessment Report. Intergovernmental Panel on Climate Change. http://www.ipcc.ch/report/ar5/wg3/.

Jagger, T.H., Elsner, J.B., Saunders, M., 2008. Forecasting US insured hurricane losses. In: Diaz, H.F., Murnane, R.J. (Eds.), Climate Extremes and Society. Cambridge University Press, pp. 189–208.

Jones, L., Garbutt, A., Hansom, J., Angus, S., 2013. Impacts of climate change on coastal habitats. MCCIP Sci. Rev. 2013, 167–179.

Keim, B.D., Muller, R.A., Stone, G.W., 2004. Spatial and temporal variability of coastal storms in the North Atlantic Basin. Mar. Geol. 210, 7–15.

Komar, P.D., Allan, J.C., 2008. Increasing hurricane-generated wave heights along the U.S. Atlantic Coast and their climate controls. J. Coastal Res. 24 (2), 479–488. http://dx.doi.org/10.2112/07-0894.1.

Lavrenov, I.V., 1998. The wave energy concentration at the Agulhas current of South Africa. Nat. Hazards 17, 117–127.

Lawton, G., June 30, 2001. Monsters of the Deep. New Sci., 28.

Liu, J., Krogstad, H.E., Trulsen, K., Dysthe, K., Socquet-Juglard, H., 2005. The statistical distribution of a nonlinear ocean surface. Int. J. Offshore Polar Eng. 15, 168–174.

Mallory, J.K., 1974. Abnormal waves on the south east coast of South Africa. Intl. Hydrog. 51, 91–29.

Miller, D.J., 1960. Giant waves in Lituya Bay, Alaska. In: United States Geological Survey Professional Paper 354-C. http://earthquake.usgs.gov/earthquakes/states/events/1958_07_10.php.

Monserrat, S., Vilibíc, I., Rabinovich, A.B., 2006. Meteotsunamis: atmospherically induced destructive ocean waves in the tsunami frequency band. Nat. Hazards Earth Syst. Sci. 6 (6), 1035–1051. http://dx.doi.org/10.5194/nhess-6-1035-2006.

Mori, N., Takahashi, T., Yasuda, T., Yanagisawa, H., 2011. Survey of 2011 Tohoku earthquake tsunami inundation and run-up. Geophys. Res. Lett. 38 (7).

Morton, R.A., Gelfenbaum, G., Jaffe, B.E., 2007. Physical criteria for distinguishing sandy tsunami and storm deposits using modern examples. Sediment. Geol. 2008, 184–207.

Nanayama, F., Satake, K., Furukawa, R., Shimokawa, K., Atwater, B.F., Shigeno, K., Yamaki, S., 2003. Unusually large earthquakes inferred from tsunami deposits along the Kuril trench. Nature 424, 660–663.

Nordstrom, K.F., 1994. Beaches and dunes of human-altered coasts. Prog. Phys. Geogr. 18 (4), 497–516.

Paris, R., Wassmer, P., Sartohadi, J., Lavigne, F., Barthomeuf, B., Desgages, E., Grancher, D., Baumert, P., Vautier, F., Brunstein, D., Gomez, C., 2009. Tsunamis as geomorphic crises: Lessons from the december 26, 2004 tsunami in Lhok Nga, west Banda Aceh (Sumatra, Indonesia). Geomorphology 104 (1–2), 59–72. http://dx.doi.org/10.1016/j.geomorph.2008.05.040.

Richmond, B.M., Watt, S., Buckley, M., Jaffe, B.E., Gelfenbaum, G., Morton, R.A., 2011. Recent storm and tsunami coarse-clast deposit characteristics, southeast Hawaii. Mar. Geol. 283 (1–4), 79–89. http://dx.doi.org/10.1016/j.margeo.2010.08.001.

Rosenthal, W., Lehner, S., 2008. Rogue waves: results of the MaxWave project. J. Offshore Mech. Arct. Eng. 130 (2), 8. http://dx.doi.org/10.1115/1.2918126.

Sallenger Jnr, A.H., 2000. Storm impact scale for barrier islands. J. Coastal Res. 16 (3), 890–895.

Scheffers, A., 2005. Coastal response to extreme wave events: hurricanes and tsunami on Bonaire. Essener Geogr. Arbeiten 37, 96.

Scheffers, A., Scheffers, S., Kelletat, D., Browne, T., 2009. Wave-emplaced coarse debris and megaclasts in Ireland and Scotland: boulder transport in a high energy littoral environment. J. Geol. 117, 553–573.

Scheffers, A., Kelletat, D., Scheffers, S., 2010. Wave-emplaced coarse debris and megaclasts in Ireland and Scotland: Boulder transport in a High-Energy littoral environment: a reply. J Geol. 118, 705–709.

Scoffin, T.P., 1993. The geological effects of hurricanes on coral reefs and the interpretation of storm deposits. Coral Reefs 12, 203–221. http://dx.doi.org/10.1007/BF00334480.

Šepic, J., Denis, L., Vilibic, L., 2009. Real-time procedure for detection of a meteotsunami within an early tsunamiwarning system. Phys. Chem. Earth 34, 1023–1031.

Sheremet, A., Staples, T., Ardhuin, F., Suanez, S., Fichaut, B., 2014. Observations of large infragravity wave runup at Banneg Island, France. Geophys. Res. Lett. 41. http://dx.doi.org/10.1002/2013GL058880.

Stockdon, H.F., Doran, K.J., Thompson, D.M., Sopkin, K.L., Plant, N.G., Sallenger, A.H., 2012. National Assessment of Hurricane-Induced Coastal Erosion Hazards—Gulf of Mexico. U.S. Geological Survey. Open-File Report 2012–1084, 51 pp. http://pubs.usgs.gov/of/2012/1084/.

Suanez, S., Fichaut, B., Magne, R., 2009. Cliff-top storm deposits on Banneg Island, Brittany, France: effects of giant waves in the Eastern Atlantic Ocean. Sediment. Geol. 220, 12–28.

Synolakis, C.E., Kong, L., 2006. Runup measurements of the december 2004 Indian Ocean tsunami. Earthquake Spectra 22 (S3), 67–91.

Switzer, A.D., 2014. Chapter 5-Coastal hazards: storms and tsunamis. In: Masselink, G., Gehrels, R.W. (Eds.), Coastal Environments and Global Change. AGU/Wiley, pp. 104–126.

Switzer, A.D., Burston, J.M., 2010. Competing mechanisms for boulder deposition on the southeast Australian coast. Geomorphology 114, 42–54. http://dx.doi.org/10.1016/j.geomorph.2009.02.009.

Switzer, A.D., Pucillo, K., Haredy, R.A., Jones, B.G., Bryant, E.A., 2005. Sea level, storm, or tsunami: enigmatic sand sheet deposits in a sheltered coastal embayment from Southeastern new South Wales, Australia. J. Coastal Res. 21 (4), 655–663. http://dx.doi.org/10.2112/04-0177.1.

Toffoli, A., Lefevre, J.M., Monbaliu, J., Savina, H., Bitner-Gregersen, E., 2003. Freak waves: clues for prediction in ship accidents?. In: Proc of the ISOPE 2003, Hawaii. http://www.isope.org/publications/proceedings/ISOPE/ISOPE%202013/.

Tsai, C.-H., Su, M.-Y., Huang, S.-J., 2004. Observations and conditions for occurrence of dangerous coastal waves. Ocean Eng. 31, 745–760.

Tsimplis, M.N., Álvarez-Fanjul, E., Gomis, D., Fenoglio-Marc, L., Pérez, B., 2005. Mediterranean Sea level trends: atmospheric pressure and wind contribution. Geophys. Res. Lett. 32, L20602. http://dx.doi.org/10.1029/2005GL023867.

Vilibić, I., Šepić, J., 2009. Destructive meteotsunamis along the eastern Adriatic coast: Overview. Phys. Chem. Earth, Parts A/B/C 34 (17), 904–917.

Vilibic, I., Monserrat, S., Amores, A., Dadic, V., Fine, I., Horvath, K., Ivankovic, D., Marcos, M., Mihanovic, H., Pasquet, S., Rabinovich, A.B., Sepic, J., Strelec Mahovic, N., Whitmore, P., 2012. Meteorological tsunamis along the U.S. coastline. Geophys. Res. Abstr. 14, EGU2012–2601.

Wang, D.W., Mitchell, D.A., Teague, W.J., Jarosz, E., Hulbert, M.S., 2005. Extreme waves under Hurricane ivan. Science 309, 896.

Wang, X.L., Swail, V.R., 2002. Trends of Atlantic wave extremes as simulated in a 40-year wave hindcast using kinematically reanalyzed wind fields. J. Clim. 15, 1020–1035.

Watt, S., Buckley, M., Jaffe, B., 2011. Inland fields of dispersed cobbles and boulders as evidence for a tsunami on Anegada, British Virgin Islands. Nat. Hazards 63 (1), 119–131.

White, B.S., Fornberg, B., 1998. On the chance of freak waves at sea. J. Fluid Mech. 355, 113–138.

Williams, D.M., Hall, A.M., 2004. Cliff-top megaclast deposits of Ireland, a record of extreme waves in the North Atlantic - storms or tsunamis? Mar. Geol. 206, 101–117.

Woolf, D.K., Challenor, P.G., Cotton, P.D., 2002. Variability and predictability of the North Atlantic wave climate. J. Geophys. Res.: Oceans (1978–2012) 107 (C10), 9–1 – 9–14.

Young, E., Muir, D., Dawson, A., Dawson, S., 2014. Community driven coastal management: an example of the implementation of a coastal defence bund on South Uist, Scottish Outer Hebrides. Ocean and Coastal Management. 94, 30–37. http://dx.doi.org/10.1016/j.ocecoaman.2014.01.001.

Chapter 12

Rip Currents

Robert W. Brander
School of Biological, Earth and Environmental Sciences, UNSW Australia, Sydney, NSW, Australia

ABSTRACT
Rip currents are strong, narrow seaward flows found on many global beaches where waves break across a surf zone. They represent a major hazard to bathers contributing to hundreds of drownings and tens of thousands of rescues annually. Most rip currents occupy deep channels situated between shallow sandbars or against structures. Rip current flow is unsteady with mean velocities typically on the order of 0.3–0.5 m/s, but with instantaneous flows in excess of 2 m/s. Recent field, laboratory, and numerical modeling studies have challenged traditional views of rip current behavior and, in combination with lifeguard reports, have helped improve understanding of the physical environmental factors leading to maximum rip current risk. Although collaborative research efforts between rip current scientists and beach safety organizations are increasing, significant challenges remain for improving and communicating education and awareness of the rip current hazard to the beachgoing public.

12.1 INTRODUCTION

Rip currents are strong, narrow offshore-directed flows of water that originate in the surf zones of many ocean, inland sea, or lacustrine beaches. They are most commonly associated with breaking waves across three-dimensional surf-zone morphology (Sonu, 1972; Short, 1985; Brander, 1999; MacMahan et al., 2006; Short, 2007; Scott et al., 2009; Dalrymple et al., 2011; Houser et al., 2013; Austin et al., 2014). Where they exist, rip currents are a major coastal hazard to recreational beachgoers as they can quickly carry unsuspecting bathers of all swimming abilities offshore into deeper water often leading to exhaustion, panic, and in many cases, drowning (Klein et al., 2003; Carey and Rogers, 2005; Hartmann, 2006; McCool et al., 2008; Drozdzewski et al., 2012; Arun Kumar and Prasad, 2014; Scott et al., 2014). Consequently, they represent a serious global public hazard and health issue with significant personal, societal, and economic costs related to drowning deaths, near-miss drownings, injuries, trauma, and the associated provision of lifeguard

Coastal and Marine Hazards, Risks, and Disasters. http://dx.doi.org/10.1016/B978-0-12-396483-0.00012-1
Copyright © 2015 Elsevier Inc. All rights reserved.

services and public outreach efforts (Sherker et al., 2008; Brander and MacMahan, 2011).

However, due to inherent logistical difficulties involved in accurate incident reporting, particularly in developing countries (Hammerton et al., 2013; Arun Kumar and Prasad, 2014; Castelle et al., 2014), the amount of people who drown annually in rip currents globally is unknown, but is likely significant. The United States Lifesaving Association (USLA) estimates that rip currents account for 80 percent of surf rescues in the United States and approximately 100 drowning fatalities each year (Brewster, 2010), although values ranging from 35 (Gensini and Ashley, 2010) to 150 (Lushine, 1991) have been reported. In Australia, where over 17,000 rip currents are estimated to occur at any given time (Short, 2007), 89 percent of rescues on surf beaches have been attributed to rip currents (Short and Hogan, 1994) with an average of 21 confirmed drownings per year (Brighton et al., 2013). The latter value exceeds the annual average number of fatalities in Australia caused by bushfires, floods, cyclones, and sharks combined (Brander et al., 2013). In the United States, the number of rip current drownings exceeds fatalities by floods, hurricanes, and tornados (Leatherman, 2013). In this context, the severity of the rip current hazard in terms of human loss of life should not be underestimated. However, while coastal scientists and beach safety organizations are well aware of the hazards that rip currents represent to beachgoers, it is of significant concern that numerous studies have shown that awareness and understanding of rip currents by beachgoers is poor (Wilks et al., 2007; Sherker et al., 2010; Drozdzewski et al., 2012; Hatfield et al., 2012; Williamson et al., 2012; Caldwell et al., 2013; Brannstrom et al., 2014).

Brander and MacMahan (2011) suggest the overall lack of public awareness and understanding of the rip current hazard is partly attributable to several information disconnects that have traditionally hindered the ability to effectively translate basic scientific knowledge of rip currents to beachgoers (Figure 12.1). The first disconnect exists because, until relatively recently (Carey and Rogers, 2005; Scott et al., 2011b; McCarroll et al., 2014a), there have been few direct collaborations between rip current scientists and beach safety practitioners and organizations that have provided concerted two-way communication and development of targeted rip current science and science-based educational programs in relation to the hazard. The second disconnect relates to a lack of consistent and appropriate rip current education strategies aimed at a range of demographic beachgoers, a lack of understanding of their effectiveness, and the challenges involved in making the general public aware of the strategies, or motivating them to seek out this information. Finally, the general public has little direct access to scientific rip current information and few scientific studies have examined social aspects of the rip current hazard to learn what beachgoers can tell us about their knowledge and experiences of rip currents (Drozdzewski et al., 2012; Caldwell et al., 2013; Brannstrom et al.,

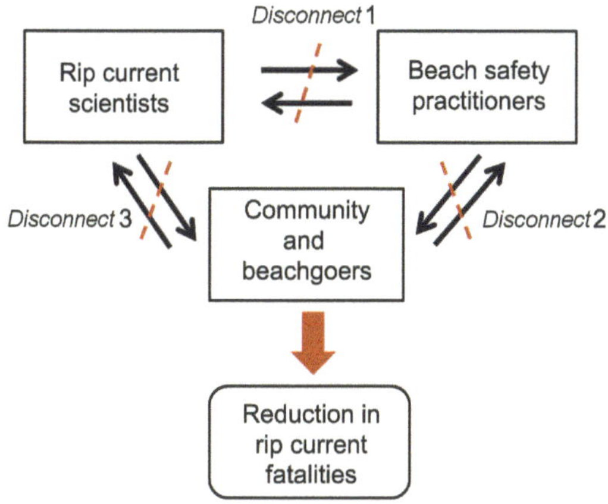

FIGURE 12.1 Information disconnects that have potentially hindered the communication of rip current science, understanding, and awareness between rip current scientists, beach safety organizations and the beachgoing public. *Modified from Brander and MacMahan (2011).*

2014). Brander and MacMahan (2011) argued that addressing these disconnects is critical to reducing rip current fatalities (Figure 12.1).

This chapter uses the information disconnects shown in Figure 12.1 as a platform to review our present understanding of the rip current hazard. It provides an overview of the definitions, formation, types, flow characteristics, and morphodynamic behavior of rip currents in a hazard context with a focus on public beach safety. Recent advances in measuring, monitoring, modeling, and forecasting rip currents, as well as new socio-physical approaches involving beachgoer and bather factors, are described that have both challenged and enhanced our understanding of the rip current hazard. Identification of existing knowledge gaps and directions for future research are identified that will assist both coastal scientists and beach safety organizations in improving understanding, communication, mitigation, and overall awareness of the global rip current hazard.

12.2 RIP CURRENT DEFINITION AND FORMATION

Rip currents have long been the focus of scientific interest as they are an integral component of nearshore cellular circulation involving the exchange of water, sediments, nutrients, biological species, and pollutants between the surf zone and inner continental shelf (Shepard and Inman, 1950; Cook, 1970; Inman and Brush, 1973; Basco, 1983; Talbot and Bate, 1987; Smith and Largier, 1995; Brown et al., 2009; Shanks et al., 2010) and are a key feature of morphodynamic beach state evolution (Wright and Short, 1984; Masselink and Short, 1993; Brander, 1999; Loureiro et al., 2011; Barrett and Houser, 2012). The term "rip current" was first used by Shepard (1936) to describe and distinguish concentrated, commonly channelized, offshore surface flows of water on beaches from the potentially conceptually misleading terms "rip tide" and "undertow," which were gaining popularity in the public vernacular at the time (Davis, 1925, 1931). Both terms are still sometimes used incorrectly in reference to rip currents today. "Rip tide" is a misnomer as rip currents are not tides, and tidal rips are different currents associated with tidal inlets during the ebbing and flooding tide (Leatherman, 2013). Rip currents are also not "undertow," which is a laterally homogenous offshore current near the seabed on beaches (or sections of beaches) with relatively minimal alongshore bathymetric variability at velocities much lower than rip currents (Garcia-Faria et al., 2000; Aagaard and Vinther, 2008).

Our conventional understanding of rip current systems originated from early observations and experiments made at the Scripps Institution of Oceanography in La Jolla, California (Shepard et al., 1941; Shepard and Inman, 1950, 1951; Inman and Quinn, 1952). These studies described rip currents as part of an idealized nearshore circulation cell (Figure 12.2) consisting of alongshore *feeder currents* fully contained within the surf zone, but restricted close to shore, that carry water toward a narrow and fast-flowing shore-normal *rip neck*, which extends through the surf zone. Rip current flow extends well beyond the surf zone where it decelerates as an expanding *rip*

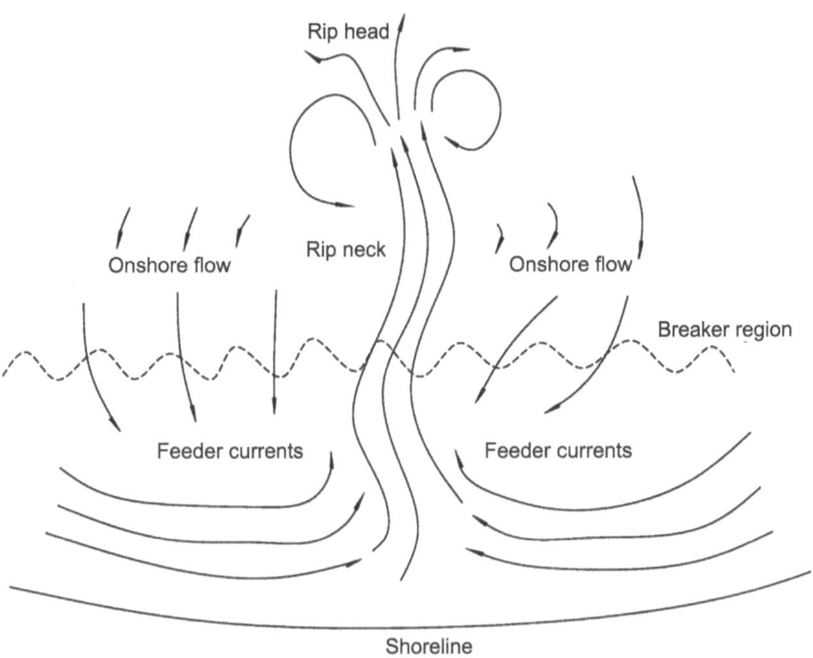

FIGURE 12.2 Idealized schematic of the traditional view of rip current circulation showing the dominance of offshore flow beyond the surf zone. *Source: MacMahan et al. (2006).*

head. This water is able to return shorewards through the action of waves completing the cell (Figure 12.2).

Idealized nearshore circulation cells involving continuous interchange of water between the surf zone and areas offshore (Inman and Brush, 1973) appear in most coastal geomorphology textbooks (e.g., Komar, 1998; Woodroffe, 2002; Davis and Fitzgerald, 2004; Davidson-Arnott, 2010). However, these observations were made on beaches where rip currents were strongly influenced by piers or submarine canyons (Shepard and Inman, 1950; Long and Özkan-Haller, 2005), which force rip flow considerable distances seaward of the surf zone. As such, these rip currents can be considered anomalous relative to more common rip currents that have a strong surf zone morphodynamic association, are modified by directional waves, and are not controlled by structures or offshore topography (Brander and MacMahan, 2011). As discussed in Section 12.4.4, this traditional view of rip current circulation is being challenged by recent studies with important implications for beach safety.

Nearshore cell circulation (Figure 12.2) oversimplifies the formation of rip currents by using a basic mass balance description where breaking waves transport water toward the shoreline and rip currents transport it offshore. Actual rip current formation is more complex involving spatial gradients in wave momentum, referred to as radiation stresses (Longuet-Higgins and Stewart, 1964), that occur in the cross-shore and alongshore directions. Cross-shore momentum gradients are caused by wave shoaling and breaking (Bowen, 1969; Deigaard, 1993) with resulting water level set-up gradients creating offshore-directed, pressure-driven currents (Bowen and Inman, 1969; Haas and Svendsen, 2002). Alongshore set-up gradients occur due to spatial variations in wave breaking and intensity over beaches with non-uniform alongshore, surf-zone bathymetry. Wave breaking is intensified in shallow regions and minimized, or restricted to larger waves, in deeper regions. Currents are formed as water flow moves from regions of intense wave breaking (e.g., shoals) to regions with fewer waves breaking (e.g., channels) that are commonly relatively deeper (Bowen, 1969; Dalrymple, 1978). Changes in water level during a tidal cycle will affect the intensity of wave breaking across three-dimensional surf-zone morphology that results in temporal variability of rip current flow, particularly in meso- and macro-tidal environments (Castelle et al., 2006; Bruneau et al., 2009; Austin et al., 2014).

On beaches with no bathymetric variation, a number of predominantly theoretical, laboratory, and numerical modeling investigations have identified other forcing mechanisms that can result in momentum gradients leading to rip current formation. Most notable are wave–wave and wave–current interactions (Dalrymple et al., 2011). The superposition of incident waves and shoreline-trapped edge waves of the same frequency has been shown theoretically and in the laboratory to create periodic alongshore variations in height of breaking waves (Bowen and Inman, 1969), which leads to offshore flowing rip currents. However, Guza and Inman (1975) suggested that synchronous edge waves are restricted to steep, reflective beaches and cuspate shoreline morphology. Alongshore gradients in wave breaking and resultant rip current flow have also been related to intersecting wave groups from different directions (Dalrymple, 1975; Tang and Dalrymple, 1989; Haller and Dalrymple, 2001). Lastly, shear instability in strong alongshore currents created by waves with large angles of incidence can result in a meandering current that can form eddies that may be ejected from the surf zone (Bowen and Holman, 1989; Oltman-Shay et al., 1989; Özkan-Haller and Kirby, 1999; Reniers, 2007) as transient whirlpool-like rip current flows.

12.3 RIP CURRENT OCCURRENCE, TYPE AND IDENTIFICATION

Given the variability of rip current formation, different types of rip currents occur. Although several classification schemes exist (Short, 1985, 2007; Dalrymple et al., 2011; Leatherman, 2013), the terminology used to describe what are commonly the same type of rip current is inconsistent (Table 12.1). This poses a potential problem for public rip current education efforts in terms of consistency of terminology and generic global messaging. Rip currents may flow through gaps in reefs and tidal inlets, but are generally associated with sandy beaches. Recently, rip currents have been differentiated based on whether they occur on "open coast" or embayed beaches, with the former being far away from permanent topographic features such as headlands or coastal structures (MacMahan et al., 2010a; Dalrymple et al., 2011; Scott et al., 2014). In reality, this distinction is misleading as many embayed beaches support rip currents that are not adjacent to physical boundaries and exhibit the same characteristics as "open coast" beach rip currents. Rather, there are two broad categories: (1) morphologically controlled rip currents that are relatively fixed, or quasi-stationary, in location along a beach over variable temporal periods; and (2) hydrodynamically controlled rip currents that are transient and temporally and/or spatially variable in occurrence (Table 12.1; Figure 12.3).

12.3.1 Morphologically Controlled Rip Currents

Although the location of some rip currents may be determined by offshore morphologic controls such as submarine canyons, offshore shoals, transverse ridges and reefs (Shepard and Inman, 1950; Long and Özkan-Haller, 2005; Houser et al., 2011; Castelle et al., 2014), and associated processes of wave refraction and focusing, it is widely accepted that rip currents occupying deep channels between adjacent shallow sand bars are the most commonly occurring type of rip current (Short, 2007; Brander and MacMahan, 2011; Dalrymple et al., 2011). Their shapes and dimensions are largely dictated by beach type and associated surf-zone morphology (Figure 12.3(a) and (b)), and

TABLE 12.1 Rip Current Types, Key Characteristics and Synonymous Terminology

Rip Current Type	Characteristics	Terminology	Reference
Morphologically Controlled			
Beach	• Occupy deeper channels between adjacent sand bars • Relatively stationary in position for various periods of time • Associated with decreasing or lower wave energy conditions • Most common type	Accretion Low-energy Linear bar-trough Semienclosed Fixed Bar-gap Cusped-shore	Short (1985, 2007) Brander (1999) Dalrymple et al. (2011) Dalyrymple et al. (2011) Brander and MacMahan (2011) Leatherman (2013) Leatherman (2013)
Boundary	• Fixed in location adjacent to natural (headland, rock reef, platform) or anthropogenic (groin, jetty, pier) structures; usually channelized	Topographic Structural Headland Permanent	Short (1985, 2007) Leatherman (2013) Colloq. Colloq.
Megarip	• Large scale boundary rip current associated with large wave conditions • Most common on embayed beaches adjacent to headlands or mid-beach • Erosional with high flow velocities and offshore extent	Megarip	Short (1985, 2007), Dalrymple et al. (2011), Leatherman (2013)
Swash	• Strong backwash in embayments of shoreline beach cusps • Limited seaward extent	Undertow	Colloq.

Continued

TABLE 12.1 Rip Current Types, Key Characteristics and Synonymous Terminology—cont'd

Rip Current Type	Characteristics	Terminology	Reference
Hydrodynamically Controlled			
Transient	• Associated with rising or higher wave energy conditions, confused sea states, wave groups • Episodic/quasi-periodic and can migrate along beach; channels mostly absent • Generally limited offshore flow extent and velocity	Erosional Flash Traveling	Short (1985, 2007) Leatherman (2013) Colloq.

Literature references associated with various terminologies are indicated. See Figure 12.3 for photographic examples.

they have been referred to synonymously in the scientific literature as accretion, low-energy, linear bar-trough, semienclosed, bar-gap, and cusped-shore rips (Table 12.1). Here they are referred to simply as beach rip currents.

Microtidal beach classification models by Wright and Short (1984) and Lippmann and Holman (1990) suggest many sandy beaches follow a progression of stages during post-storm recovery accretionary sequences with offshore/longshore bars migrating landward, becoming more rhythmic and eventually welding to the beachface as transverse bars before reaching a low tide terrace (ridge and runnel) state (Figure 12.4). Rip current channels are present during all of these evolving beach states, with the transverse bar and rip state (Figures 12.3(a) and 12.4(c)) being more prevalent in terms of both occurrence and number of rip channels (Ranasinghe et al., 2000, 2004; Holman et al., 2006; Short, 2007). However, uncertainty still remains regarding the location and persistence of rip current channels after major storm events, the existence of "reset" events, and the role of antecedent morphology and wave energy in driving temporal scales of beach state transition and evolution (Gallop et al., 2011).

Although individual beach rip currents may persist for several months, their location and orientation can vary as channels may migrate alongshore at rates ranging from 2 to 50 m/day, with maximums occurring during storm-wave conditions with oblique angles of wave incidence (Ranasinghe et al., 2000; Bogle et al., 2001; Holman et al., 2006; Turner et al., 2007; Orzech et al., 2010). Rip current channels are not necessarily oriented transverse to the shoreline, but are commonly oblique or sinuous in the offshore direction. Offshore length and alongshore spacing of rip channels are largely dependent on prevailing wave climate. Rip current channels typically extend at least to the modal surf-zone width and rip current spacing has been shown to be more irregular than regular (Turner et al., 2007), ranging from 50 m to more than 500 m as wave energy increases (Short and Brander, 1999; Thornton et al., 2007; Gallop et al., 2011). Rip current channel width and relief can range from 1 to 10's of m's and 1−10 m respectively. Rip current channels are also present on low-tide bar/rip and low-tide terrace rip beach states in meso- and macro-tidal environments (Figure 12.5; Masselink and Short, 1993; Castelle et al., 2006; Bruneau et al., 2009; Scott et al., 2011a) as well as on some multibarred beaches (Short, 1992; Short and Aagaard, 1993; Castelle et al., 2007, 2010a,b). One common characteristic of all channelized rip currents is that they exhibit distinct visual indicators to assist in determining their location.

As wave breaking is reduced over deeper rip channels, beach rip currents characteristically appear as darker green areas between areas of whitewater over adjacent shallow sand bars (Figure 12.3(b)). Other visual indictors include agitated surface water due to the interaction between outgoing rip flow and incoming waves, pronounced shoreline rip embayments, and clouds of suspended sediment ejected just beyond the surf zone (Figure 12.3(a) and (b)).

FIGURE 12.3 Rip current types: (a) beach rip currents associated with transverse bar and rip morphology at Tallows Beach, NSW, Australia *(photo R. Brander)*; (b) beach rip current at Marina, California, illustrating dark gap visual identifier *(photo R. Brander)*; (c) boundary rip current adjacent to a headland at Bondi Beach, Australia *(photo R. Brander)*; (d) boundary rip current adjacent to a groin at Cape Hatteras, North Carolina *(photo S.P. Leatherman)*; (e) megarips forming under high-energy wave conditions adjacent to headland in South Australia *(photo A. Short)*; (f) swash rips formed by backwash from beach-cusp embayments *(photo A. Short)*;

Anecdotally, this type of rip current is considered by lifeguards to be the most dangerous in terms of drowning and rescues, as they typically exist during ideal weather conditions, which promotes large beach crowds, and low-to-moderate wave-energy conditions, which are more attractive for swimming (Gensini and Ashley, 2010; Leatherman, 2013; Brannstrom et al., 2014). Their appearance can also be deceptive to inexperienced beachgoers who may choose to swim in the "calmer" water away from breaking waves (Short, 2007; Sherker et al., 2010).

Many rip currents also occur adjacent to natural boundaries such as headlands, rock platforms, outcrops and reefs (Figure 12.3(c)), or rigid artificial structures (Figure 12.3(d)) such as groins, jetties, and piers (Short, 1985, 1992, 2007; Pattiaratchi et al., 2009; Dalrymple et al., 2011). Various combinations of wave shadowing and refraction processes (Castelle and Coco, 2012), and the deflection of longshore currents (Castelle and Coco, 2013) can act to create quasi-permanent rip current channels against the boundaries that become active during periods of wave breaking. These rip currents have been described in the literature as topographic, structurally controlled, headland, and permanent rip currents (Table 12.1), but are referred to here as boundary rip currents. Boundary rip currents can also occupy deep channels, so can be identified as darker gaps of water, and therefore have the same inherent risk of appearing as safer swimming areas, but with additional risk posed by collision with solid features (Figure 12.3(c) and (d)). Although field measurements of this type of rip current are limited, Pattiaratchi et al. (2009) found persistent recirculation in the lee of a groin, whereas Short and Hogan (1994) and McCarroll et al. (2014a,b) suggested that flow against headlands and rock platforms may be stronger, more confined, and extend further distances offshore than beach rip currents, representing further potential risk to bathers.

Two other types of morphologically controlled rip currents represent extremes in the offshore extent of rip current flow (Table 12.1). Megarips occur during high wave conditions and are more common on embayed beaches, either adjacent to headlands or in the center of longer embayed beaches (Figure 12.3(e)), and can extend 3–4 surf zone widths offshore (Short and Masselink, 1999; Short, 2007; Dalrymple et al., 2011). Short (2007) suggests megarips develop when waves exceed 3 m in height. Although not considered a hazard to bathers due to inclement beach conditions, they can contribute to beach erosion, particularly after storms, which may lead to shoreline recession and property damage (Thornton et al., 2007; Loureiro et al., 2012). They can

(g) small transient rip current at Rehoboth Beach, Delaware, evident by the turbulent water and sediment plume extending seaward from the surf zone *(photo Wendy Carey and Delaware Sea Grant)*; (h) transient rip current south of Perth, Australia, generated by shear instabilities of alongshore currents. Center of the slow moving whirlpool-like plume is approximately 100 m from the shoreline *(photo Scott Patman)*.

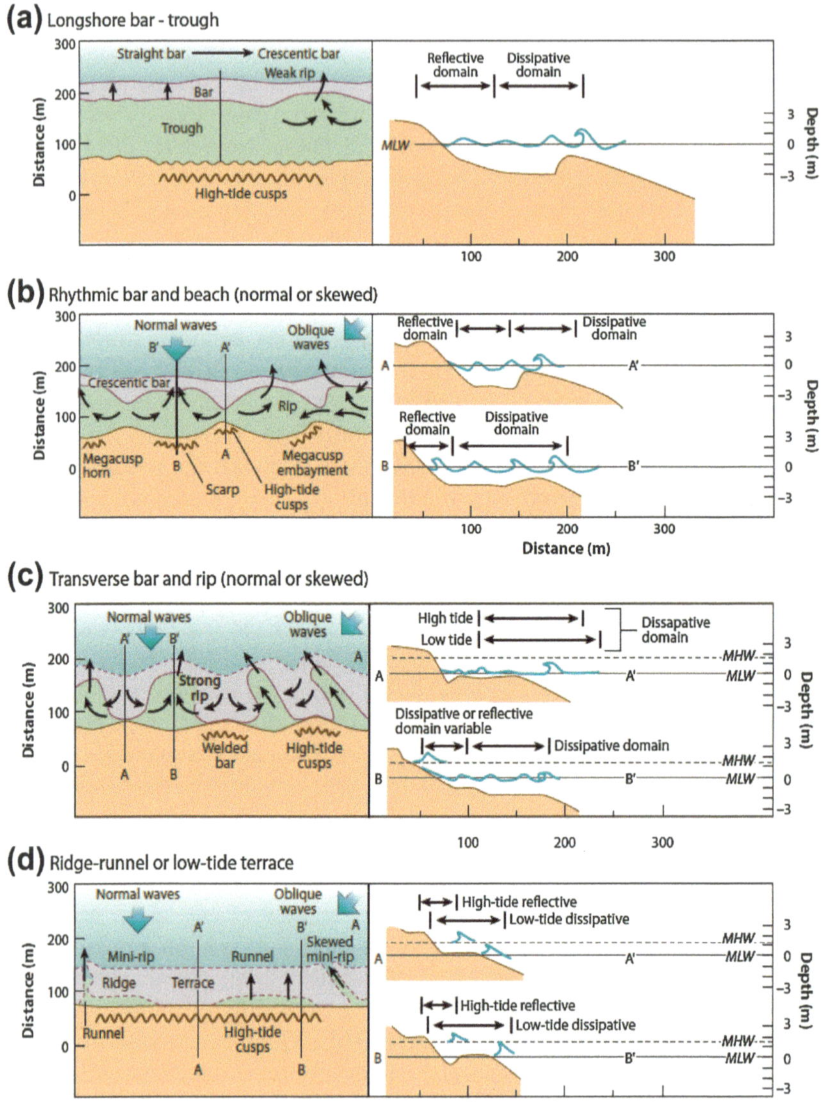

FIGURE 12.4 Intermediate beach states associated with sand-bar and rip channel morphology based on the Wright and Short (1984) micro-tidal beach classification with post-storm down-state beach transition occurring from top (a) to bottom (d). *Modified from Dalrymple et al. (2011).*

be identified by plumes of sediment in the expanding and decelerating rip head (Figure 12.3(g)). In contrast, on steep beaches characterized by shoreline beach-cusp morphology, swash rips may form in the middle of the cusp embayments due to backwash associated with wave run-up processes (Masselink and Pattiaratchi, 1998; Dalrymple et al., 2011). They represent a

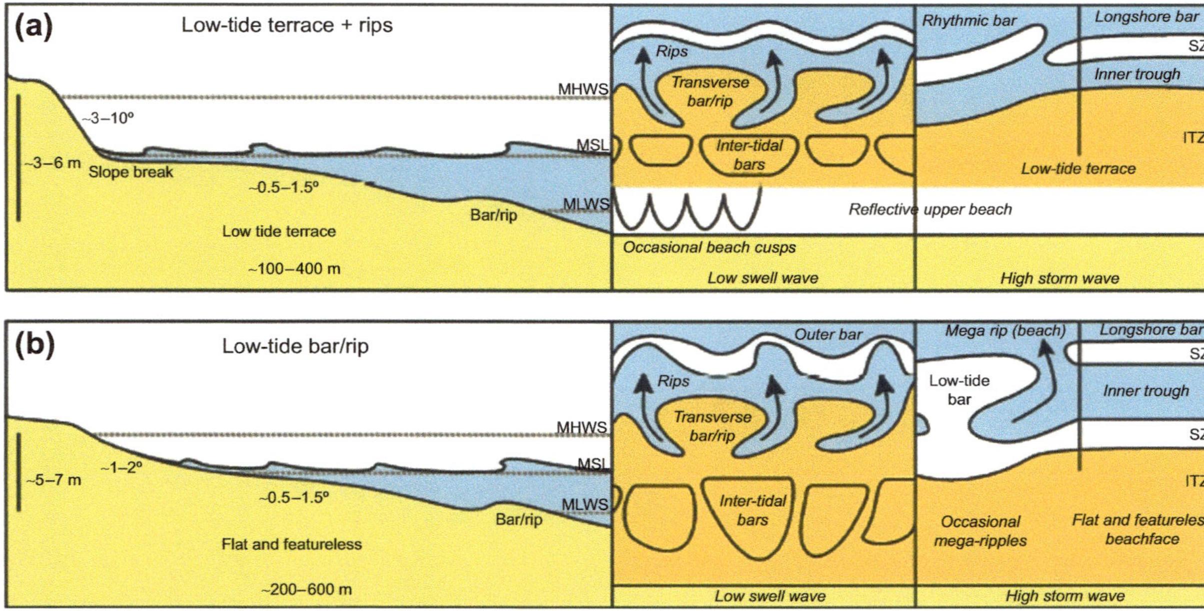

FIGURE 12.5 Meso- and macro-tidal beach states associated with rip currents according to the Masselink and Short (1993) macro-tidal beach classification. Rip currents are denoted by black arrows. Both normal and storm wave conditions are shown. SZ = surf zone; ITZ = intertidal zone. *Modified from Scott et al. (2011b).*

hazard to bathers who can quickly be washed into deeper water just seaward of the shoreline (Figure 12.3(f)). Anecdotal reports suggest it is these rip currents that are often referred to by the general public as "undertow" (Leatherman, 2013). Swash rips can be identified by quasi-regular alongshore spacing of cusp embayments and associated plumes of turbulent water and sediment (Figure 12.3(f)).

12.3.2 Hydrodynamically Controlled Rip Currents

Rip currents may also be quasi-episodic in nature, both temporally and spatially, lasting on the order of 1–10's of minutes at a particular location, or migrating slowly alongshore. These rips have been referred to as transient, flash, erosional, and traveling rips (Table 12.1) and are difficult to measure due to their unpredictable nature. These transient rip currents are not channelized and occur on beaches, or sections of beaches (e.g., sand bars), with planar bathymetric morphology. Their formation is associated with variable wave set-up gradients caused by incident waves of different frequencies and direction (Slattery et al., 2011), dissipative surf-zone conditions (Johnson and Pattiaratchi, 2004; Clark et al., 2012), wave groups (Tang and Dalrymple, 1989; Fowler and Dalrymple, 1990), rising wave conditions (Short, 2007), and shear instabilities in alongshore currents (Reniers et al., 2009; MacMahan et al., 2010a; Murray et al., 2013). Transient rip currents can initiate at any location within a surf zone, typically appearing as narrow bands of choppy churning water with streaky, turbulent, and sediment-laden rip-head plumes extending seaward (Figure 12.3(g) and (h)). They generally have smaller flow velocities than morphologically controlled rip currents, but represent a hazard to bathers due to their unpredictability (Dalrymple et al., 2011).

12.3.3 Rip Current Identification by Beachgoers

Although no study has assessed the use, understanding, or potential uptake of different rip current terminologies by the general public, it is likely that scientific differences between rip current types are largely meaningless to most beachgoers. Instead, the risk posed by rip currents primarily depends on the ability of beachgoers to identify and avoid them, particularly if they choose to swim away from lifeguarded areas or on unpatrolled beaches (Brander, 2013; Brannstrom et al., 2014). However, despite the visual indicators described previously, rip current identification by beachgoers is poor. In a study of domestic and international students on beaches in Queensland, Australia, Ballantyne et al. (2005) found that two-thirds of those who claimed to know what a rip current was were unable to recognize one. Only 18 percent said that rip currents could be identified by the presence of calm water between breaking waves. When shown a photograph of a beach with two channelized beach rip currents, 61 percent of all students selected the rip currents as places they would swim. Sherker et al. (2010) found that 93 percent of surveyed Australian beachgoers thought they could identify a rip current, but only 33 percent could when shown a picture of a beach rip current. Half the survey respondents selected the rip current as the safest place to swim.

Recent studies have shown that rip identification ability by beachgoers in the United States is lower than in Australia. Caldwell et al. (2013) conducted surveys of beachgoers at Pensacola Beach, Florida, using pictures. While the majority of respondents said they could identify rip currents, less than 20 percent were able to do so. On three popular beaches in Texas, Brannstrom et al. (2014) found that when shown a picture of a groin with a boundary rip current on one side and breaking waves (with no rip current) on the other, only 13 percent of beachgoers correctly identified the location of the rip current. In both studies, the majority of respondents identified regions of breaking waves as representing the most hazardous surf conditions and potential for rip current development.

12.4 RIP CURRENT FLOW AND BEACH SAFETY

A major determinant in the level of risk presented by rip currents is the nature of rip current flow itself, which is extremely variable spatially and temporally within and between different rip current types. The following section reviews scientific understanding of rip current flow behavior and describes implications for beach safety in terms of hydrodynamic behavior that may lead to bathers being caught in rip currents and the efficacy of different types of bather self-escape strategies. More detailed information on the physical understanding of rip current flow behavior can be found in the excellent reviews by MacMahan et al. (2006) and Dalrymple et al. (2011).

12.4.1 Spatial Circulation Patterns

Rip currents do not necessarily flow alongshore and then offshore as illustrated by the conventional view shown in Figure 12.2. The recent development and use of low-cost global positioning system (GPS) devices attached to Lagrangian drifters (Johnson et al., 2003; Schmidt et al., 2003; Johnson and Pattiaratchi, 2004; MacMahan et al., 2009) and deployed en masse in surf zones has provided a more holistic understanding (Figure 12.6) of the flow behavior of rip current systems (Austin et al., 2010; MacMahan et al., 2010a; Houser et al., 2013; Austin et al., 2014). Drifter experiments have shown that shore-normal waves approaching nonuniform, alongshore morphology can force a dominantly closed rip circulation (Figure 12.7) characterized by counter- and clockwise rotating eddies that vary in velocity magnitude, shape, and orientation (Reniers et al., 2009; MacMahan et al., 2010a) rather than simply an offshore-directed jet (Figure 12.2).

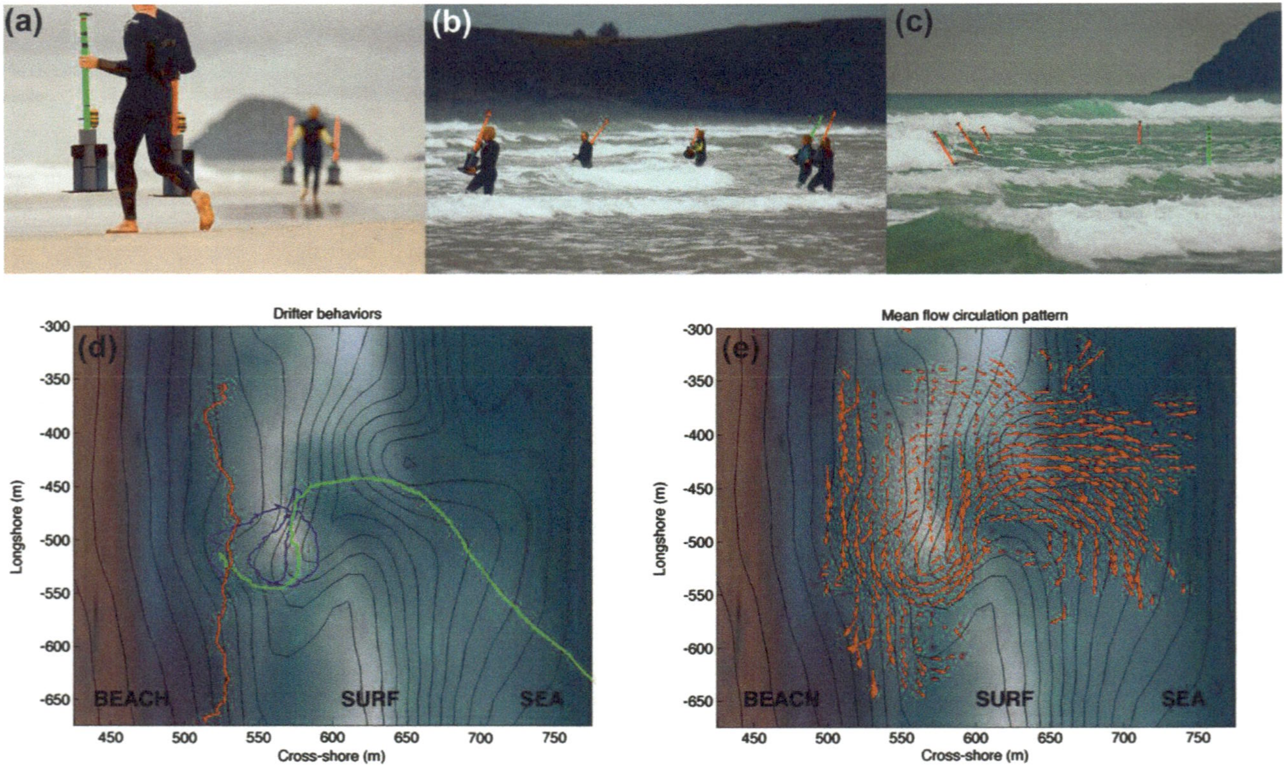

FIGURE 12.6 Example of Lagrangian low-cost global positioning system drifters at Perranporth Beach, UK, showing: (a) design; (b) deployment; (c) drifters in action; (d) individual tracks displaying exiting circulation (green track), enclosed circular rotation within the surf zone (blue track), and alongshore current (red track); and (e) mean flow field of a rip current system in the surf zone based on 164 separate drifter deployments over a 4-h period around low tide. Red arrows show direction and speed (arrow length) of currents. *Modified from Masselink et al. (2014).*

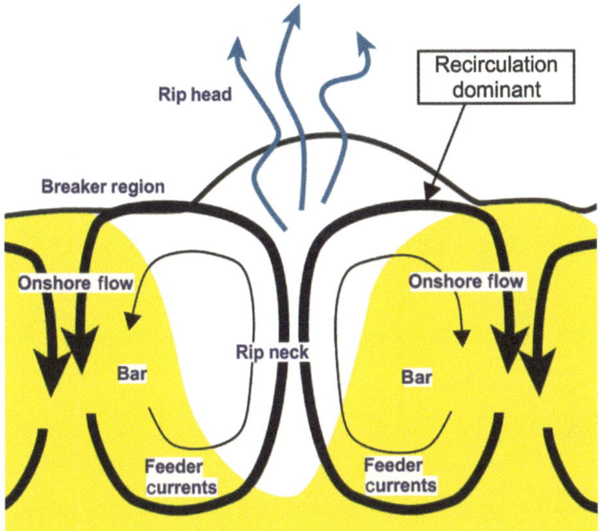

FIGURE 12.7 Modified view of rip current circulation based on recent field measurements by MacMahan et al. (2010) at Sand City, California, indicating that most beach rip currents are dominated by circulation within the surf zone with only episodic exits. *Source: McCarroll et al. (2014a) as modified from Dalrymple et al. (2011).*

As described by MacMahan et al. (2010a), closed rip current eddies may be symmetrical, with two opposing circulation cells that occupy half of the shore-connected shoal and rip channel (Figure 12.8(a)), or asymmetrical, with the current occupying one rip channel and one shore-connected shoal (Figure 12.8(b)). As wave incidence becomes more oblique (Sonu et al., 1966; Sonu, 1972; Houser et al., 2013), or wave height and/or breaking decreases (Austin et al., 2010; MacMahan et al., 2010a,b), a sinuous alongshore current forms that follows the rip channel bathymetry flowing landward over the shoals and seaward in the rip channels (Figure 12.8(c)). MacMahan et al. (2010a) simultaneously observed a combination of rip currents and sinuous alongshore currents, termed a meandering current, where drifters roamed along the beach and migrated between rip current cells and sometimes became temporarily trapped. MacMahan et al. (2010a,b) found rip circulation dominates with larger offshore significant wave heights ($H_S > 1$ m) and more dissipative conditions. Houser et al. (2013) suggested that the transition between sinuous alongshore current to rip current circulation occurs when the relative wave height ratio (H_S/h) exceeds 0.5 (where $h =$ water depth). Other studies (Brander and Short, 2001; Austin et al., 2010, 2014) have identified that the lateral flow of water across shallow sand bars may contribute more water to the rip neck than the idealized alongshore feeder currents shown in Figure 12.2.

Despite evidence of dominant recirculation, exits of water and drifters beyond the surf zone do occur, often in episodic bursts (Figure 12.7).

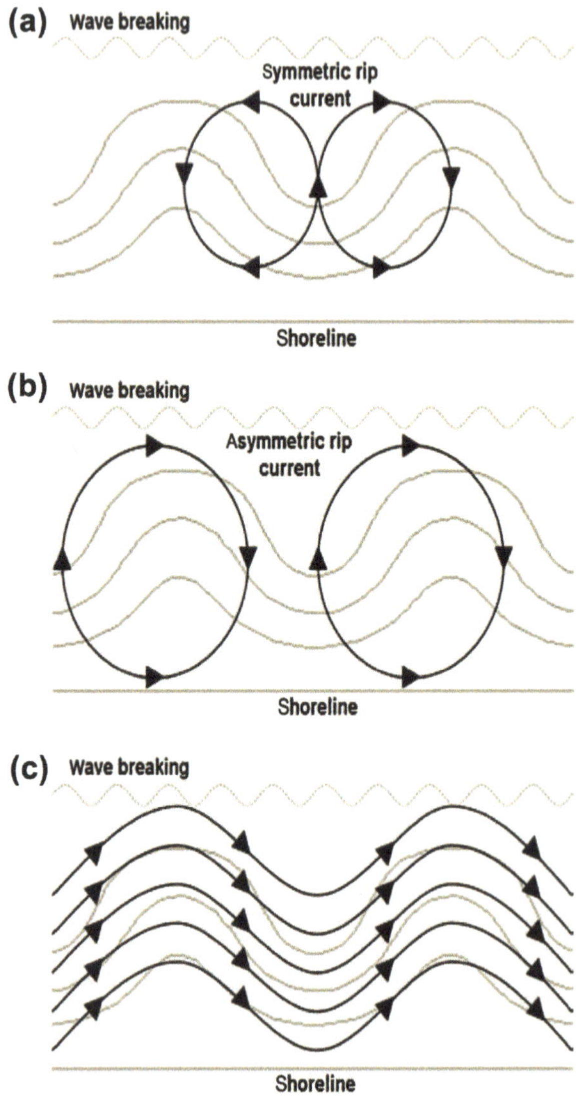

FIGURE 12.8 Idealized circulation patterns for beach rip currents based on global positioning system drifter measurements for: (a) a symmetric rip current; (b) an asymmetric rip current; and (c) sinuous alongshore current. Black lines represent velocity flow field, gray lines are bathymetric contours, and the gray sine wave represents offshore extent of surf zone. *Source: MacMahan et al. (2006).*

Measurements on beaches in California, England, and France (Austin et al., 2010; MacMahan et al., 2010a), with similar transverse-bar surf zone rip channel morphologies and different wave and tidal regimes, found that approximately 15–20 percent (range 0–34 percent) of drifters per hour exited offshore. Similar findings were reported through numerical simulations (Reniers et al., 2009) and in laboratory wave basins by Castelle et al. (2010c), where asymmetric rip currents resulted in more surf-zone exits. Laboratory experiments by Kennedy and Thomas (2004) found higher exit rates with linear bar-trough and rip channel surf-zone morphologies. In field experiments, higher exit rates have been reported in channelized beach rip currents on macro-tidal beaches (up to 73 percent; Scott et al., 2014) and in boundary rip currents adjacent to headlands and rock platforms (up to 90 percent per hour; McCarroll et al., 2014a,b), the latter of which supported modeling results of Castelle and Coco (2012). Houser et al. (2013) reported exit rates of 55 percent of drifters per hour over transverse bar and rip current morphology associated with unsteady nearshore circulation, oblique wave angles, and increased wave dissipation. The degree of surf-zone exits associated with rip current circulation is therefore dependent on a complex series of factors, including the type of surf-zone morphology and rip current, wave energy dissipation, and angle-of-wave approach.

12.4.2 Temporal Flow Behavior and Characteristics

Rip current flow is inherently unsteady. Early qualitative observations recognized that flow velocities were coupled with surf-zone morphology, tidally modulated, increased with increasing wave height, and exhibited pulsating behavior (Shepard, 1936; Shepard et al., 1941; Shepard and Inman, 1950; McKenzie, 1958). Due to the difficulty of deploying and retrieving in situ instruments in a logistically demanding environment, the number of field experiments involving Eulerian measurements of rip current flow is few and primarily associated with beach rip currents (MacMahan et al., 2006; Dalrymple et al., 2011). Nevertheless, measurements from a range of wave and tidal environments on beaches around the world have shown consistent patterns of flow behavior that has helped quantify the early observations.

Hourly mean flow velocities in the necks of channelized beach rip currents under conditions of low-to-moderate wave energy are typically on the order of 0.3–0.5 m/s, with maximum 1-min averaged velocities being much higher, approaching 2 m/s (MacMahan et al., 2006). Measurements of vertical flow structure are few (Sonu, 1972; Brander and Short, 2001; Haas and Svendsen, 2002), but show that velocity increases away from the bed, reaching a maximum just below the surface. In general, rip current flow velocities are tidally modulated (Figure 12.9) with maximums occurring around low tide, decreasing and sometimes ceasing altogether at high tide (Sonu, 1972; Bowman et al., 1988; Aagaard et al., 1997; Brander and Short, 2000; MacMahan et al., 2005). Rip current flow is maximized at lower tidal levels due to a combination of increased wave breaking and dissipation across a wider surf zone. The more shallow (and sometimes exposed) sand bars enhance wave set-up gradients and morphologically constrict the channelized rip current flow

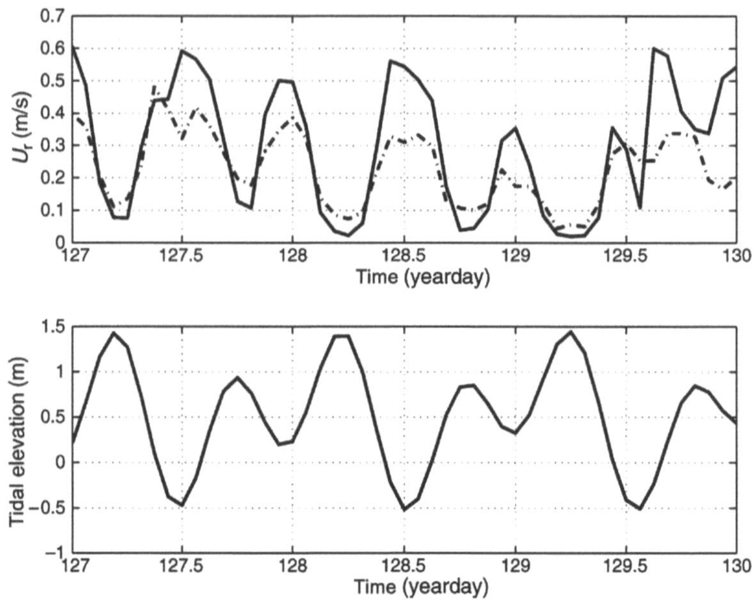

FIGURE 12.9 Tidal modulation of rip current flow velocity based on colocated current meter (top: solid and dashed lines) and pressure sensor (bottom: water levels relative to Mean Sea Level (MSL)) at depths of 1 m in adjacent beach rip channels at Sand City, California. U_r = mean flow velocity. *Source: MacMahan et al. (2006).*

(Brander, 1999; Austin et al., 2010). However, alongshore variations in wave breaking leading to wave set-up gradients are not always associated with lower tidal stages. For example, Truc Vert, France, is a meso-tidal beach where the nearshore bathymetry enhances rip current flow toward high tide (Castelle et al., 2006; Bruneau et al., 2009). On tideless beaches such as inland seas and lakes, rip current flow is solely dictated by episodic storm-generated waves. Flow velocities in other types of rip currents are less well known due to a lack of measurements. Transient rips exhibit lower flow velocities, on the order of 0.2–0.3 m/s (MacMahan et al., 2010a), whereas megarips and beach rips under higher energy wave conditions may experience mean flows in excess of 2 m/s (Brander and Short, 2000; Short, 2007).

Rip current flow has also been observed to pulsate at infragravity frequencies (0.004–0.04 Hz) associated with the arrival and breaking of sets of larger waves (wave groups). These pulsations can accelerate instantaneous rip current flow to over 1 m/s for periods of 25–250 s (Shepard and Inman, 1950; Sonu, 1972; Brander and Short, 2001; MacMahan et al., 2004a). Greater mass transport and wave set-up gradients produced by larger waves in the groups build up significant amounts of water in the surf zone that is transported offshore through rip channels during subsequent, wave height minima (MacMahan et al., 2004a). MacMahan et al. (2004a) also showed that pulsations can result from infragravity cross-shore standing waves associated with the wave groups. Wave basin and numerical modeling studies have shown that the larger the waves in the group, the faster the pulsation response of rip current flow (Haller et al., 1999,2002; Kennedy and Dalrymple, 2001).

Longer pulsations at very low frequencies (VLFs), less than 0.004 Hz, have been observed in both field (Smith and Largier, 1995; Callaghan et al., 2004; MacMahan et al., 2004b) and laboratory studies (Haller and Dalrymple, 2001; Castelle et al., 2010c) and are associated with shear instabilities in the rip current flow. Their intensity has been correlated with underlying bathymetric variations and increasing wave energy. They have also been observed on planar beaches leading to the formation of transient rips (Johnson and Patttiaratchi, 2004; MacMahan et al., 2010a; Dalrymple et al., 2011). Reniers et al. (2010) suggest that VLFs form the dominant exchange mechanism for transporting water from the surf zone to the inner shelf (Figure 12.3(h)).

12.4.3 Implications for Bather Safety

Spatial and temporal rip current flow characteristics have significant implications for bathers and lifeguards as well as beach safety organizations in their efforts to communicate the rip current hazard to the public. Although it is not well documented as to how bathers find themselves caught in rip currents (Drozdzewski et al., 2012), it is not surprising that so many do, given that sudden increases in rip current flow speeds can quickly change conditions from being relatively safe and/or manageable to being very dangerous. Under these circumstances, individual bathers, or groups of bathers, can lose their footing and become caught in a strong rip current flow where their only options are to be rescued by lifeguards or bystanders, or to self-escape.

This has clearly been illustrated by the project on Dynamics of Rips and Implications for Beach Safety (DRIBS), a recent partnership in the United Kingdom between Plymouth University and the Royal National Lifeboat Institution (RNLI). A series of Eulerian and Lagrangian rip current flow measurements on macro-tidal beaches with bar/rip channel surf-zone morphology combined with lifeguard rescue statistics have shown that most individual and mass rescues of bathers are linked to water levels 1–2 h either side of low tide (Scott et al., 2009, 2011b). This is due to enhanced wave breaking across shoals causing rip current flow to suddenly “turn on and off” (Austin et al., 2010, 2014), particularly during spring tides when rip flow activation is maximized (Scott et al., 2009). However, it is important to note that on many beaches across all tidal ranges, subaerial exposure of sand bars at spring low tide can effectively cause ponding of water toward the shoreline, and rip current flow can be strongly reduced or stopped for a short period of time around low tide (MacMahan et al., 2006; Austin et al., 2010, 2014; Scott et al., 2014).

As noted by Scott et al. (2009, 2014), the occurrence of mass rescues is also related to many other factors, the least of which are attractive weather and wave conditions and large number of bathers. Wave groups also play a significant role, particularly on micro-tidal beaches with similar bar/rip channel morphology, where the increase in rip current flow velocities toward low tide is more gradual (Aagaard et al., 1997; Brander and Short, 2001), but mass rescues still occur. These incidents are often reported incorrectly as being caused by "collapsing sand bars." This myth likely began at Australia's Bondi Beach on Sunday, February 6, 1938, when many of the 30,000 people at the beach were bathing on shallow sand bars, when they suddenly found themselves in deep water being dragged out to sea. Within 30 min, 250 bathers required assistance, 35 were rescued unconscious and revived and 5 drowned. Although still widely attributed to a "collapsing sand bar," the real cause was likely the arrival of a large wave group. Local increases in wave set-up due to the larger breaking waves in the group caused bathers to lose their footing and the subsequent combination of rip current and infragravity pulsations carried them offshore. This is supported by eyewitness reports of the arrival of three large waves and bathers being swept into a deep channel (Brander, 2010).

Rip current flow pulsations, at any frequency, are clearly hazardous to bathers, as these rapid flows can quickly take them significant distances offshore in a short period of time (Short, 2007). Although it is often described in rip current educational material and the media that rip currents flow faster than people can swim and/or faster than Olympic swimmers, these statements are somewhat misleading. As described by Brander and MacMahan (2011), competitive times for the 50 m freestyle equate to swimming speeds of 2.3 m/s and average swimmers can easily swim at speeds of 0.5 m/s. In both cases, these swimmers could hypothetically swim against low- to-moderate energy beach rip currents, or to adjacent shallow sand bars, over short distances to reach safety before being taken further offshore. Theoretically, it is only during short-lived rip pulsations of 2 m/s or more that swimmers could be carried seaward of the surf zone without being able to swim out of the rip. The implication here is that non-swimmers and poor swimmers are at greatest risk, but the above argument does not take into consideration swimmer behavior and response when presented with an unfamiliar situation. Even strong swimmers can panic when caught in a rip current (Brander et al., 2011; Drozdzewski et al., 2012).

12.4.4 Rip Current Escape Strategies for Bathers

A review of websites for global beach safety invariably leads to a list of advice on how bathers should respond when caught in a rip current. There are generally variations of four common points: (1) don't panic; (2) don't swim against the rip; (3) escape the rip by swimming parallel to the beach; or (4) stay afloat and signal for help (Brander and MacMahan, 2011). Although agreement is common over the importance of (1) and (2), there has recently been significant debate amongst rip current scientists and beach safety practitioners over the appropriate order of (3) and (4). The long adopted advice promoted by beach safety practitioners globally has been to "swim parallel to the beach" to escape a rip current. This is largely based on the idealized traditional view of rip currents shown in Figure 12.2 where swimming parallel to the beach is logical as rip currents are narrow and flow transverse to the shoreline. Theoretically, swimming parallel will get a swimmer out of the rip current, without swimming directly against the strong rip neck flow, before being carried significant distances offshore. However, this option may be problematic for weak or non-swimmers and, as many rip currents flow obliquely to shore, for competent and experienced swimmers.

Recent findings of rip current circulation patterns using GPS drifters has shown that rip current circulation is more complex than the traditional view with recirculation within the surf zone being dominant at times (Figure 12.7). MacMahan et al. (2010a) found that 80 percent of drifters per hour recirculated within the surf zone and therefore suggested that floating might be a more appropriate escape strategy as bathers would have a high probability of recirculating onto the relative safety of shallow sand bars within minutes while conserving energy. Also, as this re-circulation is not always symmetrical (Figure 12.8(b)), a bather caught in a "circular" rip current flow pattern would find it difficult to choose the correct direction to swim parallel, and may mistakenly swim against an alongshore component of the circulation cell, particularly as the eddy-like nature of rip current flow in the surf zone is often masked by breaking waves (Brander and MacMahan, 2011; Dalrymple et al., 2011). This finding was supported by Miloshis and Stephenson (2011) who used human rip current floaters (Short and Hogan, 1994; Brander and Short, 2000) in a beach rip current in Australia to show that 3 of 11 floaters exited the surf zone. They postulated that alongshore currents were a possible hindrance to swimming parallel.

On a larger scale, the research project on Rip Current Swimmer and Floater Experiments (RIPSAFE), a collaboration between UNSW Australia and Surf Life Saving Australia (SLSA), was initiated to physically test rip current escape strategies in different rip current settings. McCarroll et al. (2014a) described results from an initial field study involving GPS drifter deployments and 185 GPS-equipped swimmer actions in two different beach rip currents and a boundary rip current adjacent to a rock platform. Floating was found to be a longer duration, but more variable escape strategy than swimming parallel. Both strategies failed in some scenarios (McCarroll et al., 2014a). Swim parallel failures were related to swimming against the alongshore component of cellular rip circulation. Float failures were related to surf-zone exits (higher rates for boundary rip current) and instances where the person recirculated without being able to attain safe footing on the shallower sand bar (Figure 12.10). McCarroll et al. (2014a) showed that no single escape strategy

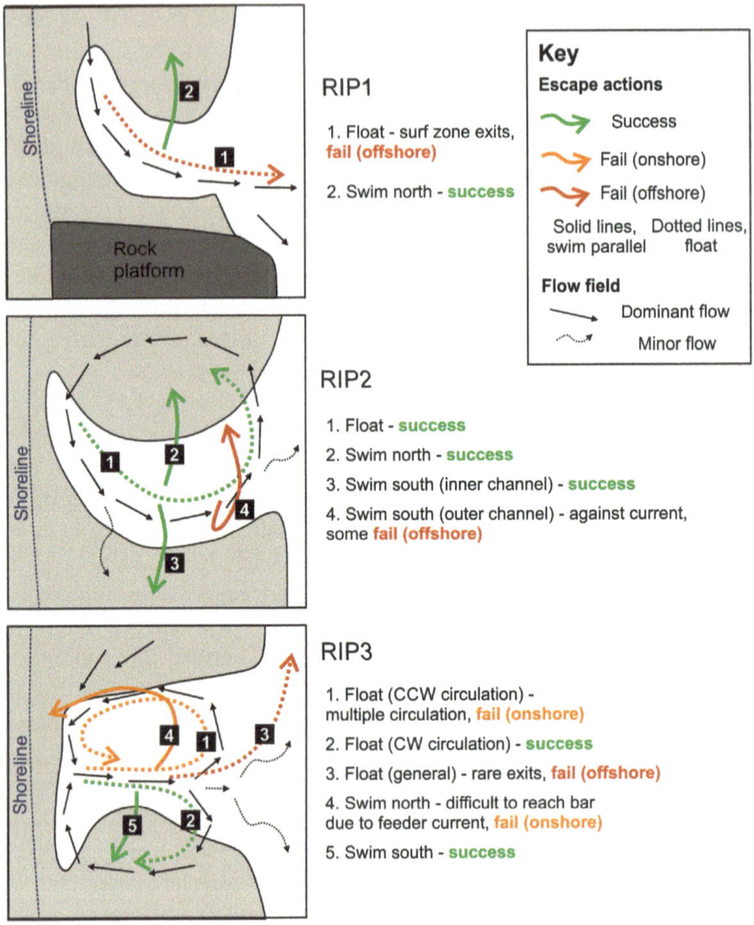

FIGURE 12.10 Idealized representation of global positioning system measurements of rip current flow and swimmer escape strategies at Shelly Beach, NSW, Australia: RIP1 = boundary rip current; RIP2 = adjacent beach rip current; RIP3 = beach rip current further up the beach; success and fail refer to outcomes of swimmer strategy. North is to top of Figure. CW and CCW refer to clockwise and counter-clockwise circulation. *Source: McCarroll et al. (2014a).*

action is appropriate in all rip current flow scenarios, making safety recommendations difficult. They suggested that a combined approach involving elements of both floating and swimming and constant assessment of the situation should be adopted by bathers caught in rip currents.

The RIPSAFE project also includes a social science component using online surveys and interviews of rip current survivors to learn more about their experience being caught in rip currents. Drozdzewski et al. (2012) focused on an informed group and inquired about rip current knowledge, beach experience, and a high, self-rated, swimming ability. The most recalled message when caught in a rip current was to swim parallel (40 percent of respondents) compared to staying afloat (10 percent). However, in terms of their actual response, more respondents reacted by swimming against the rip current back to shore rather than swimming parallel, which reflected the dominant reported emotional response of panicking. Of those that self-escaped, 33 percent did so by swimming parallel compared to 18 percent who floated. Across the whole study demographic, preliminary results suggest that swimming parallel to the shore is the most recalled message by the general public compared to stay afloat. However, regardless of swimming ability and understanding of rip currents, panic tends to override most recalled rip current safety advice suggesting that more efforts should be made to reinforce the "don't panic" message to beachgoers.

12.5 RIP CURRENT MONITORING, MODELING, PREDICTION, AND FORECASTING

Understanding when and where rip currents will form and the conditions leading to particularly dangerous rip current flow behavior in terms of velocity and trajectory is critical for the development of predictive risk indices for the rip currents. Such indices are important as they can potentially provide government agencies with the ability to issue public rip current warnings and enable lifeguarding organizations to facilitate necessary preventative measures (Engle et al., 2002; Arun Kumar and Prasad, 2014). Predictive risk indices for rip currents should correlate meteorological, bathymetric, and offshore wave data with the presence of rip currents, lifeguard rescues, and the number of beachgoers and bathers (Table 12.2). However, for a variety of reasons, these data are not always available and no generic rip current risk forecasting techniques exist that have proven applicable to all beaches with rip currents. Similarly, despite substantial efforts at monitoring and modeling rip currents, we still cannot accurately predict their occurrence or flow behavior.

12.5.1 Application of Remote Video Imaging

The application of relatively inexpensive imagery from remote video cameras and associated image processing techniques in the late 1980s provided a potentially valuable technique for monitoring channel behavior of rip currents. The ARGUS video camera imaging system (Holman and Stanley, 2007) was the first to use time averaging of video images to quantify the presence of sand bars and rip current channels based on the areal extent of breaking waves (Lippmann and Holman, 1989). Video camera imagery from numerous sites has subsequently shown that beach rip currents can persist for many months (Figure 12.10). However, existing studies have not adequately linked instantaneous wave conditions with the number of rip currents, or their alongshore rip current spacing (Ranasinghe et al., 1999, 2004; Holman et al., 2006; Turner

TABLE 12.2 Main Factors Involved when Assessing the Level of Rip Current Risk and Hazard to Bathers

Human/Social	Physical Environment	Rip Current Characteristics
Bather Load	**Marine**	**Type**
Weather conditions (sun, temperature, wind)	Wind (speed, direction)	Beach (channelized)
Surf conditions (calm/stormy)	Waves (height, direction)	Boundary
Water temperature	Waves (period, groupiness)	Megarips
Day of week/holidays	Tides (range, timing)	Swash
Marine flora/fauna (seaweed, jellyfish)	Storms (magnitude, frequency, timing)	Transient
Safety Considerations	**Morphology**	**Morphodynamics**
Lifeguards (location, timing)	Offshore bathymetry	Channel spacing, density
Beach warning flags	Surf-zone bathymetry	Channel width, relief
Rip current signage	Beach type and morphology	Channel offshore extent and orientation
Public awareness/ knowledge/identification	Seasonal beach change	Channel migration
Beachgoer demographics	Antecedent morphology	Circulation patterns (symmetry, alongshore)
Beachgoer swimming ability	Headlands, rock reefs and platforms	Exit rates
Personal behavior	Tidal inlets	Temporal behavior (tidal modulation, pulsing)
Available rip current outreach	Anthropogenic structures (groins, jetties, piers)	Mean and maximum flow velocities

Source: Modified from Leatherman (2013), Scott et al. (2014).

et al., 2007). Instead, Gallop et al. (2011) found that average wave energy conditions over a period of 10 days, combined with storm durations, provided better predictions of rip current morphological patterns. Using a 3-year remote video camera dataset on an embayed beach in New Zealand, they found time lags between wave forcing and morphological response. Rip current channels with short cross-shore length and narrow alongshore spacing responded faster

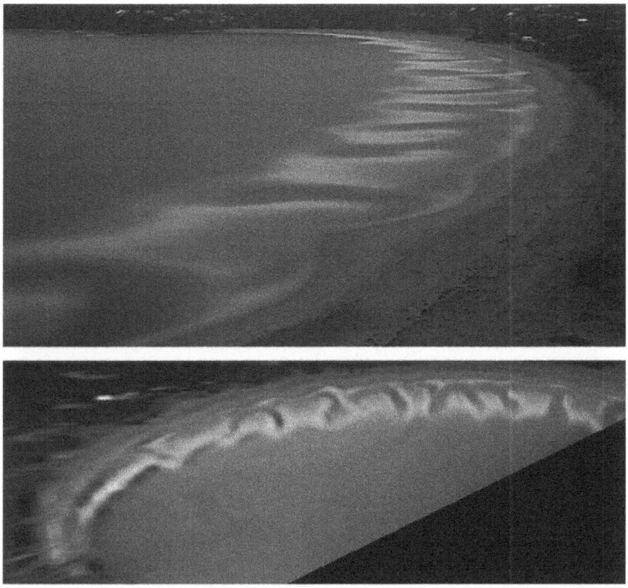

FIGURE 12.11 Oblique time-exposure image from a video camera on Palm Beach, Sydney, Australia (top) and rectification of the same image (bottom). Rip channels are indicated by dark gaps through white areas of wave dissipation. *Modified from Holman et al. (2006).*

to changing wave conditions than rip channels with a long cross-shore length and wider alongshore spacing (Gallop et al., 2011). In general, the application of remote video imaging has illustrated that once formed, beach rip currents are strongly morphologically controlled (Figure 12.11). However, monitoring other types of rip currents, such as transient and megarips, using this methodology is problematic as their short-lived and spatially variable nature does not lend itself well to time-averaged exposure analysis of video imagery, which typically sample images at 1 Hz, but require intervals of 10 min or more to discern distinct hydrodynamic signatures.

12.5.2 Laboratory and Numerical Modeling

Laboratory measurements in wave basins with fixed beds (e.g., Haas and Svendsen, 2002; Kennedy and Dalrymple, 2001; Kennedy and Thomas, 2004) or moveable beds (Castelle et al., 2010c) allow rip current circulation to be tested and replicated under controlled wave climate and tidal variables on a range of beach types (Dalrymple et al., 2011). Castelle et al. (2010a) used drifter observations to assess rip current circulation across eight different barred beach morphologies consistent with the intermediate beach-state classification of Wright and Short (1984; Figure 12.4). For both low- and moderate-energy wave conditions, rip current intensity was found to increase linearly as the degree of alongshore non-uniformity of the surf-zone morphology increased. Peak intensity occurred when rhythmic bars were about to weld to the beach, subsequently decreasing from the transverse bar and rip state (Figure 12.4(c)) to the low-tide terrace beach state (Figure 12.4(d)). Although these results contrast the field measurements of Brander (1999), the rates of drifter exits corresponded well to measurements described in MacMahan et al. (2010).

Morphodynamic process models have the ability to modify surf-zone morphology and input wave conditions in order to replicate complex rip current circulation patterns (Figure 12.12). As such, they have also helped expand our theoretical understanding of the formation, location, evolution, migration, and persistence of rip current channels (e.g., Chen et al., 1999; Reniers et al., 2009; Castelle et al., 2011; Bruneau et al., 2011; Castelle et al., 2013). Calvete et al. (2007) showed that growth and geometry of rip currents is critically controlled by the pre-existing, cross-shore profile, and Castelle and Ruessink (2011) illustrated the importance of the time-varying angle of wave incidence in driving rip channel behavior. Castelle et al. (2012) demonstrated that alongshore perturbations in wave forcing, due to offshore bathymetric anomalies, can cause rip current channels to self-

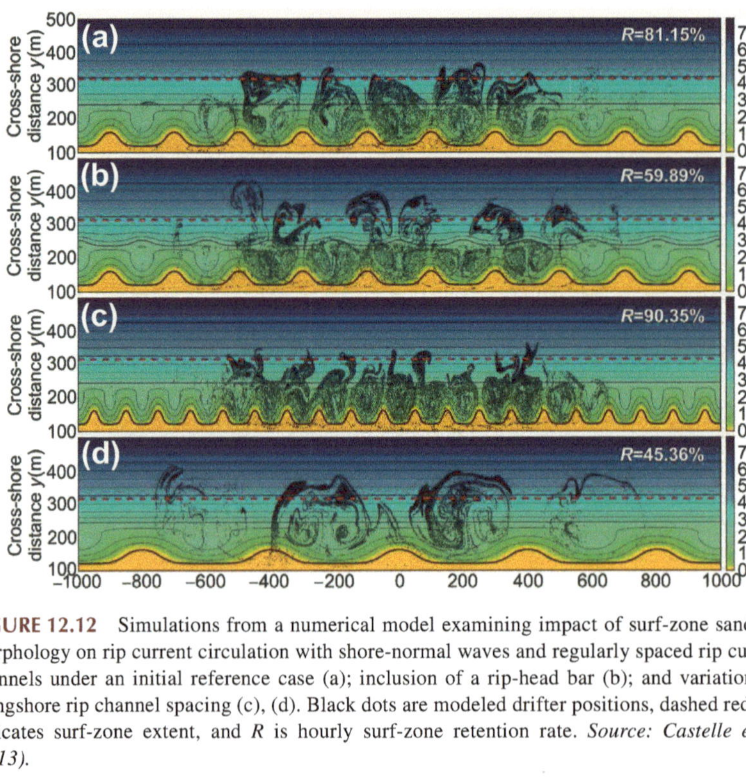

FIGURE 12.12 Simulations from a numerical model examining impact of surf-zone sand bar morphology on rip current circulation with shore-normal waves and regularly spaced rip current channels under an initial reference case (a); inclusion of a rip-head bar (b); and variations of alongshore rip channel spacing (c), (d). Black dots are modeled drifter positions, dashed red line indicates surf-zone extent, and R is hourly surf-zone retention rate. *Source: Castelle et al. (2013).*

organize with different patterns of alongshore length scales and migration rates. Similarly, Castelle and Coco (2012) showed the strong control of headlands and oblique wave forcing on rip current channel behavior on embayed beaches. However, despite their obvious value, numerical models require field validation with bathymetric and directional wave data offshore, which is generally not available, making their application difficult and commonly qualitative.

12.5.3 Correlative Predictions of Rip Current Risk

Existing efforts at forecasting the risk associated with rip currents have primarily relied on correlative predictions that attempt to relate the occurrence of rip current rescues with environmental, meteorological, and oceanographic conditions. Lushine (1991) originally correlated wind direction, wind velocity, swell-wave height, and time of low tide to rescues at a beach in Southeast Florida and developed the Lushine Rip Current Scale (LURCS). Each factor is assigned a score and higher linear summations of these scores equate to a higher risk of rip currents (Dalrymple et al., 2011). Lascody (1998) subsequently modified LURCS for East Central Florida beaches (ECFL) based on the observation that the local wave climate was more severe and wave height should be assigned a greater weighting. These studies largely formed the basis for surf-zone forecasts and rip current outlooks now issued by the US National Weather Service (NWS) directly to the public as part of daily weather forecasts on television news and weather stations (Carey and Rogers, 2005). NWS offices along the Eastern Seaboard, Gulf Coast, California, Great Lakes, Hawaii, and Guam issue rip current outlooks based on a low/moderate/high risk scale. However, although the incorporation of rip current warnings in a public forum is a positive step in increasing awareness of the rip current hazard, their effectiveness has yet to be assessed, and uncertainty remains regarding the accuracy and efficacy of these predictions (Brander and MacMahan, 2011).

Existing rip current risk forecasts are typically not normalized with numbers of beach visitations and can be misleading. For example, high risk levels of rip currents may occur when nobody is on the beach due to inclement weather and swimming conditions, resulting in zero rescues, whereas a low rip current risk level during crowded periods may result in a high number of rescues. In a study on Florida beaches, Engle et al. (2002) found the number of rip current rescues increases with increasing beach visitation. Additionally, days when the number of rip current rescues is significantly large (termed outbreaks) tend to be when drownings occur. Engle et al. (2002) found directional wave spreading is the most important determinant on the occurrence of rip current rescues with shore-normal waves enhancing rip current activity. Unfortunately, many beaches in the United States and elsewhere do not have directional wave sensors, which would dramatically improve rip current predictions.

Nelko and Dalrymple (2008) used rescue data and time-averaged video from Ocean City, Maryland, to show that ECFL and LURCS were not successful and that morphological beach state was also important for rip current predictions. On the Outer Banks of North Carolina, Dusek and Seim (2013b) developed a probabilistic model for rip current forecast using rip current observations made by lifeguards and rip current rescues as a proxy to indicate hazardous conditions. Their model outperformed the NWS rip-forecasting models. Arun Kumar and Prasad (2014) developed an empirical forecasting technique for prediction of rip current risk along the entire extent of the Indian coastline and also showed that the ECFL and LURCS scales were not applicable to the Indian context. The implication of all of these studies is that since rip current activity and wave climate varies between and within different regions and beaches, empirical predictive indices should be calibrated locally and applied elsewhere with caution.

12.5.4 Lifeguard Observations

At a local level, experienced lifeguards (where present) provide a potentially powerful observational tool as they are skilled at recognizing and identifying the location of rip currents. Dusek and Seim (2013a) used lifeguard observations of rip-flow intensity as an alternative to rip current rescue data to determine the influence of wave conditions on rip current activity. Their findings compared well with previous studies using rescue data and measurements and modeling of rip current dynamics (Lascody, 1998; Brander, 1999; Engle et al., 2002; Haller et al., 2002; MacMahan et al., 2005; Johnson and Pattiaratchi, 2006). However, using lifeguards as detailed and rigorous data collectors has also been shown to be difficult and unreliable due to time constraints of lifeguards on patrol (Williamson, 2006).

Perhaps the best example of utilizing lifeguard observations and incident records with rip current bathymetric and flow measurements is the DRIBS study in the United Kingdom (Austin et al., 2013; Scott et al., 2011b, 2014). Analysis of 5-year records of lifeguard incidents from 20 beaches in Southwest England showed that high-risk, high-exposure scenarios for bathers occurred disproportionately around mean low water on days with low wave height ($H_s < 1$ m), long wave period ($T_p > 10$ s), shore-normal wave approach, and light winds (<5 m/s). Extensive Lagrangian and Eulerian field measurements showed that maximum rip current risk was associated with the combination of strong rip-flow speeds and offshore exits that occurred when wave conditions were at or just below average values during periods of strong morphologic control around mean low water (Scott et al., 2014). An operational, rip current prediction tool was developed that showed 64 percent of recorded lifeguard incidents occurred under predicted high-risk conditions, and 36 percent occurred during medium-risk conditions. In general, the DRIBS study, and other results described previously, strongly emphasize the importance of surf-zone morphology of bar-and-rip channels in determining periods of maximum rip risk and that these periods are commonly associated with seemingly benign conditions to beachgoers.

12.6 RIP CURRENT MITIGATION AND OUTREACH

The potential vulnerability of beachgoers to rip currents is strongly related to interactions between waves and surf-zone morphology, but is also influenced by social factors (Table 12.2) such as beachgoer knowledge, awareness and experience of rip currents, choice of swim location, and behavior (Brander, 2013; Brannstrom et al., 2014). As rip currents have been shown to be dangerous during wave and weather conditions that promote large beach crowds and favorable bathing conditions, significant challenges exist for beach safety organization to mitigate the rip current hazard (Table 12.2). As described in Section 3.3, public awareness of rip currents, at least in terms of identification, is poor. Furthermore, the severity of the rip current hazard is commonly not perceived by governments and the general public in the same context as more catastrophic and episodic natural hazards, such as hurricanes, which have the capacity to cause large loss of life. Several reasons exist for this. Rip currents are a high-frequency, low-magnitude hazard that are almost always present on beaches where they occur, but only occasionally result in more than one simultaneous drowning fatality (Brander et al., 2013). The impact of rip current drowning and injury is largely restricted to the immediate family and authorities (lifeguards, emergency services, hospitals, and coroners) dealing with the incidents, and rip currents also do not generally pose an obvious threat to structures or property and cannot be managed using engineering devices (Short and Hogan, 1994). Therefore, despite causing greater overall loss of life over the long term than many other natural hazards, the apparent complacency regarding the rip current hazard has likely had impacts on the level of appropriate funding allocated to rip current risk management and community outreach programs.

12.6.1 Lifeguards, Flags and Signage

Traditionally, the primary form of rip current hazard mitigation has taken the form of beach lifeguard services and warning signs. The value of lifeguards in reducing the incidence of rip current drownings is significant. The chance of death by drowning on beaches in the United States patrolled by lifeguards is 1 in 18 million (Branche and Stewart, 2001). In Australia, almost all rip current-related drownings occur on unpatrolled beaches or outside of supervised areas and lifeguard patrol times (SLSA, 2013). However, the vast majority of global beaches characterized by rip currents are either unpatrolled by lifeguards or only patrolled seasonally during warmer swimming months. It is logistically and economically impossible to provide lifeguarding services on all beaches all of the time (Brander and MacMahan, 2011).

Lifeguard presence is acknowledged visually through towers, vehicles, and beach flags, which can vary globally. The United States uses a colored flag and sign warning system (Figure 12.13) developed by the USLA and the International Lifesaving Federation (ILS) corresponding to the relative safety level of the surf conditions and presence of rip currents (Caldwell et al., 2013). Other countries, such as Australia, New Zealand, and the United Kingdom use a pair of international safety standard red and yellow beach flags (Wilks et al., 2007) with an associated message of "swim between the red and yellow flags." Although such systems may be effective regionally, numerous studies have shown that beachgoers of all demographic groups choose to swim away from the proximity of lifeguards and flags, even if they are aware of the potential hazards (Mitchell and Hadrill, 2004; Ballantyne et al., 2005; McCool et al., 2009; Sherker et al., 2010; Williamson et al., 2012). An unpublished news poll

FIGURE 12.13 Beach warning signage, colored flag system, and definitions used on some beaches in the United States. *Source: National Weather Service.*

survey conducted by Surf Life Saving Australia (SLSA) found that while 92 percent of Australians felt it was important to swim between the red and yellow flags, only 61 percent actually chose to do so. Klein et al. (2003) found that most surf accidents in a popular beach region in Santa Catarina State, Southern Brazil, occurred where warning flags were clearly displayed, suggesting a lack of knowledge or respect of the warning system. The reasons why people choose to swim away from lifeguards and flagged areas remains unclear, although crowded beaches, distance to nearest lifeguard service, narrow patrolled areas, and uncertainty about what the flags mean are all potential factors (White and Hyde, 2010).

The presence and type of rip current warning signs on beaches can vary greatly regionally between different local governments, states, and countries. Often erected for legislative and liability purposes, they are typically placed permanently near public beach access points on both patrolled and unpatrolled beaches, or in front of rip currents by lifeguards during patrols. In general, rip current signs consist of text-based, hazard-warning messages which may or may not use the term "rip current" or include information about how to identify them and react if caught in one (Figure 12.14). However, beach warning signs are rarely evaluated in terms of their use and effectiveness. In a study of Australian beachgoers, Matthews et al. (2014) found less than half of those surveyed observed beach warning signs when present and those that did were more likely to notice hazard symbols rather than textual or diagrammatic information. They cautioned that beach signage may have less of an immediate effect on beachgoers than the responsible authorities may assume and that it was unclear whether recognition of hazard symbols translated into specific knowledge of the hazard.

12.6.2 Education and Awareness Strategies

Most regions with popular recreational beaches characterized by rip currents engage in various forms of public strategies for rip current education and awareness. The primary challenge facing beach safety practitioners engaged in these outreach programs is targeting a range of demographic groups, including school children, adults, domestic and international tourists, with variable surf skills and knowledge, beach visitation rates, languages, and (dis)interest (Brander and MacMahan, 2011). The majority of rip current drownings are associated with adolescent and adult males (Morgan et al., 2008; Gensini and Ashley, 2010; McCool et al., 2009; Woodward et al., 2013; SLSA, 2013), but in practice it is extremely difficult to specifically target this demographic age group. Thus, rip current outreach has traditionally involved the distribution of generic brochures, stickers, refrigerator magnets, and posters, commonly with associated slogans, to tourist information centers, holiday accommodations, and other public tourist amenities near beaches with rip currents. The uptake rate and effectiveness of these types of intervention material is not generally

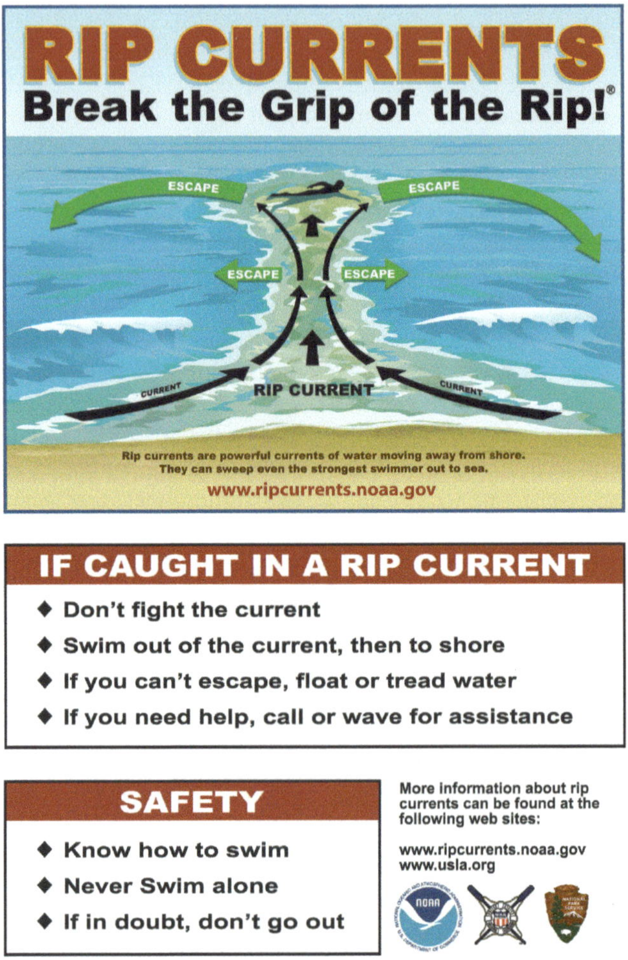

FIGURE 12.14 The standardized "Break the Grip of the Rip" slogan, graphic and rip current educational information occurs as beach warning signage at most public beach access points in the United States. *Source: National Weather Service.*

known. However, recent studies in Australia (Hatfield et al., 2012; Williamson et al., 2012) have shown improved understanding; awareness and identification of rip currents by beachgoers can be achieved through dedicated campaigns involving distribution of posters, postcards, and brochures, the latter of which achieved the highest message recall.

Although beach and surf education is rarely compulsory for schools, numerous beach safety programs exist that deliver rip current education directly to school students. Based on USLA statistics, between 2008 and 2012 an average of 313,000 students per year received beach safety education across the United States by USLA lifeguards (USLA, 2012). In Australia, Surf Life

Saving Australia estimates they provide beach safety education to 240,000 students per year. Providing rip current education to the adult population remains more of a challenge. Aside from generic outreach material, other approaches have ranged from public service announcements that play on community or hotel television channels to highway billboard advertising on approaches to popular destinations for beach holidays. However, as yet, there is little dedicated rip current information provided to inbound airline passengers, either in-flight, or upon arrival at destinations where rip currents are considered to be a problem.

More recently, the growth and increased access to the internet has allowed information about rip currents to reach a larger and global audience through popular social communication networks such as YouTube, Facebook, and Twitter. Dedicated beach safety and rip current internet sites are proliferating globally and numerous interactive websites exist that contain regional databases of beaches with real-time updates of wave, tide, weather, and hazard conditions. As of February 2014, there were more than 200 YouTube videos related to rip currents with over 2 million total views. In particular, a 4-min video "How to Survive Beach Rip Currents," created and uploaded to YouTube in December 2008 by The University of New South Wales in Australia, surpassed 1 million views in January 2014. The use of social media will no doubt increase and has the additional advantage of quickly conveying visual information about the rip current hazard to a large global audience, which may help improve identification of rip currents by beachgoers (Figure 12.15).

Nevertheless, global rip current education and awareness programs have been limited in effectiveness by a lack of consistent information delivery, content, and messaging. To overcome this problem, several countries have developed national rip-education programs that focus outreach efforts around core standardized content so that the general public receives the same consistent safety information. In the United States, a collaborative development effort between the USLA, the National Oceanic and Atmospheric Administration (NOAA), National Weather Service (NWS), and the National Sea Grant Program resulted in the implementation of the "Break the Grip of the Rip" campaign in 2005. This campaign is manifest by a standardized slogan and graphic (Figure 12.14) that appears on all associated forms of

FIGURE 12.15 The use of harmless dye releases in rip currents provides a potentially powerful visual method of communicating the rip current hazard. *Photos: R. Brander.*

generic outreach material and interventions including beach signage. In addition, the first full week of June, which precedes their summer season, has been designated by NOAA as "Rip Current Awareness Week." More recently, SLSA launched a 3-year campaign in 2009 to educate all Australians about rip currents. The campaign was based on the slogan "To Escape a Rip, Swim Parallel to the Beach" with an accompanying Website, widespread advertising, public service announcements on major radio and television networks, and the advent of a National Rip Current Awareness Day (Brander and MacMahan, 2011). However, it is difficult to assess the effectiveness of these national programs due to a lack of rigorous evaluations relating to both awareness of the programs and how the related information has translated into increased awareness and understanding of the rip current hazard by beachgoers.

The collective efforts of rip current mitigation strategies have clearly been successful. For example, since 1960 in the United States, both beach attendance and the incidence of surf rescues has risen, but the total number of drownings has remained relatively stable (Branche and Stewart, 2001). In Brazil, a regional safety education campaign using video, newspapers, leaflets, and signs was developed and an 80 percent reduction in the number of fatal accidents in the region was reported several years after implementation (Klein et al., 2003). The campaign materials were based on social, economic, and cultural data collected on the beach by lifeguards. Nevertheless, the incidence of global rip current drownings remains high, and it remains difficult to assess long-term drowning trends due to the lack of accurate incident reporting, impacts of weather and economic conditions on beach-visitation numbers, and the inherent variability of wave conditions and rip current intensity. Rip current drownings can fluctuate significantly from year to year and over longer cycles, both within and between regions. As such, it is not known whether the incidence of rip current drowning is presently decreasing.

12.7 SUMMARY

Rip currents are common features on many global beaches and represent a significant hazard to beachgoers being responsible for an unacceptably high number of drownings and rescues each year. However, recent advances in field measurements and research collaborations between rip current scientists and beach safety organizations have both challenged our traditional understanding of rip current flow behavior and helped address the information disconnects identified in Figure 12.1. The emerging field of rip current social science has also helped improve our understanding of the human factors involved in rip current incidents and have significant implications toward improving outreach programs targeting the rip current hazard. However, more efforts are needed to examine the awareness and understanding of rip currents in different regions with varying demographic and cultural factors related to beachgoers and usage.

Although our scientific understanding of flow behavior of rip currents is relatively sound, existing measurements have largely focused on channelized beach rip currents. More work is required to examine the complex relationships between rip current flow behavior and nearshore bathymetry in different types of rip currents, particularly boundary rip currents, transient rip current, and megarips, under different wave and tidal conditions. The rapid development of robust techniques of numerical modeling relating to rip current formation and evolution, in combination with remote video imaging, field measurements, and lifeguard incident records, provide potentially powerful tools in improving forecasting and prediction of the rip current hazard.

Global interest in rip current science and improving rip current outreach programs has never been higher, as evident from the recent establishment of the Biennial International Rip Current Symposium in 2010. Ongoing development and expansion of regional and national rip-education campaigns is a major step forward to mitigating the rip current hazard, but a major challenge for beach safety practitioners must be to evaluate the effectiveness of these programs in a meaningful way. The ultimate challenge for rip current outreach is to improve beachgoers' awareness and understanding of the rip current hazard so that they are increasingly motivated to seek out and only swim at beaches supervised and patrolled by lifeguards. However, given that people will always swim outside of patrolled areas, the best alternative may be to promote rip avoidance by improving rip current identification skills, which comes with its own communication challenges. Reducing the number of global fatalities due to rip currents can only be achieved through continued collaborative efforts between rip current scientists and beach safety practitioners, which will take time, but more importantly will require significantly more global recognition and dedicated funding toward this important coastal hazard.

REFERENCES

Aagaard, T., Vinther, N., 2008. Cross-shore currents in the surf zone: rips or undertow? J. Coastal Res. 24 (3), 561–570.

Aagaard, T., Greenwood, B., Nielsen, J., 1997. Mean currents and sediment transport in a rip channel. Mar. Geol. 140, 25–45.

Arun Kumar, S.V.V., Prasad, K.V.S.R., 2014. Rip current-related fatalities in India: a new predictive risk scale for forecasting rip currents. Nat. Hazards 70, 313–335.

Austin, M., Scott, T., Brown, J., Brown, J., MacMahan, J., Masselink, G., Russell, P., 2010. Temporal observations of rip current circulation on a macro-tidal beach. Cont. Shelf Res. 30, 1149–1165.

Austin, M.J., Scott, T.M., Russell, P.E., Masselink, G., 2013. Rip current prediction: development, validation, and evaluation of an operational tool. J. Coastal Res. 29 (2), 283–300.

Austin, M.J., Masselink, G., Scott, T.M., Russell, P.E., 2014. Water level controls on macro-tidal rip currents. Cont. Shelf Res. 75, 28–40.

Ballantyne, R., Carr, N., Hughes, K., 2005. Between the flags: an assessment of domestic and international university students' knowledge of beach safety in Australia. Tourism Manage. 26 (4), 617–622.

Barrett, G., Houser, C., 2012. Identifying hotspots of rip current activity using wavelet analysis at Pensacola Beach, Florida. Phys. Geogr. 33 (1), 32–49.

Basco, D.R., 1983. Surfzone currents. Coastal Eng. 7, 331–355.

Bogle, J.A., Bryan, K.R., Black, K.P., Hume, T.M., Healy, T.R., 2001. Video observations of rip formation and evolution. J. Coastal Res. Spec. Issue 34, 117–127.

Bowen, A.J., 1969. Rip currents. 1. Theoretical investigations. J. Geophys. Res. 74, 5467–5478.

Bowen, A.J., Inman, D.L., 1969. Rip currents. 2. Laboratory and field observations. J. Geophys. Res. 74, 5479–5490.

Bowen, A.J., Holman, R.A., 1989. Shear instabilities of the longshore current, 1. theory. J. Geophys. Res. 94, 18023–18030.

Bowman, D., Rosen, D.S., Kit, E., Arad, D., Slavicz, A., 1988. Flow characteristics at the rip current neck under low energy conditions. Mar. Geol. 79, 41–54.

Branche, C.M., Stewart, S. (Eds.), 2001. Lifeguard Effectiveness: a Report of the Working Group. Centers for Disease Control and Prevention, National Center for Injury Prevention and Control, Atlanta.

Brander, R.W., 1999. Field observations on the morphodynamic evolution of a low-energy rip current system. Mar. Geol. 157, 199–217.

Brander, R.W., 2010. Dr Rip's Essential Beach Book; Everything You Need to Know about Surf, Sand and Rips. UNSW Press, p. 258.

Brander, R.W., 2013. Thinking space: can a synthesis of geography save lives in the surf? Aust. Geogr. 44 (2), 123–127.

Brander, R.W., Short, A.D., 2000. Morphodynamics of a large-scale rip current system at Muriwai Beach, New Zealand. Mar. Geol. 165, 27–39.

Brander, R.W., Short, A.D., 2001. Flow kinematics of low-energy rip current systems. J. Coastal Res. 17 (2), 468–481.

Brander, R.W., MacMahan, J.H., 2011. Future challenges for rip current research and outreach. In: Leatherman, S., Fletemeyer, J. (Eds.), Rip Currents: Beach Safety, Physical Oceanography and Wave Modeling. CRC Press, pp. 1–29.

Brander, R.W., Bradstreet, A., Sherker, S., MacMahan, J., 2011. The behavioural responses of swimmers caught in rip currents: new perspectives on mitigating the global rip current hazard. Int. J. Aquat. Res. Educ. 5, 476–482.

Brander, R.W., Dominey-Howes, D., Champion, C., Del Vecchio, O., Brighton, B., 2013. A new perspective on the Australian rip current hazard. Nat. Hazards Earth Syst. Sci. 13, 1687–1690.

Brannstrom, C., Trimble, S., Santos, A., Brown, H.L., Houser, C., 2014. Perception of the rip current hazard on Galveston Island and North Padre Island, Texas. Nat. Hazards 72 (2), 1123–1138. http://dx.doi.org/10.1007/s11069-014-1061-3.

Brewster, B.C., 2010. Rip current misunderstandings. Nat. Hazards 55 (2), 161–162.

Brighton, B., Sherker, S., Brander, R., Thompson, M., Bradstreet, A., 2013. Rip current related drowning deaths and rescues in Australia 2004-2011. Nat. Hazards Earth Syst. Sci. 13, 1069–1075.

Brown, J., MacMahan, J.H., Reniers, A., Thornton, E., 2009. Surf zone diffusivity on a rip channeled beach. J. Geophys. Res. 114, C11015. http://dx.doi.org/10.1029/2008JC005158.

Bruneau, N., Castelle, B., Bonneton, P., Pedreros, R., Almar, R., Bonneton, N., Bretel, P., Parisot, J.-P., Senechal, N., 2009. Field observations of an evolving rip current on a meso-macrotidal well-developed inner bar and rip morphology. Cont. Shelf Res. 29, 1650–1662.

Bruneau, N., Bonneton, P., Castelle, B., Pedreros, R., 2011. Modeling rip current circulation and vorticity in a high energy meso-macrotidal environment. J. Geophys. Res. 116, C07026, http://dx.doi.org/10.1029/2010JC006693.

Caldwell, N., Houser, C., Meyer-Arendt, K., 2013. Ability of beach users to identify rip currents at Pensacola Beach, Florida. Nat. Hazards 68, 1041–1056.

Callaghan, D., Baldock, T.E., Nielsen, P., Hanes, D., Haas, K., MacMahan, J., 2004. Pulsing and circulation in rip current system. In: Proceedings 26th International Conference on Coastal Engineering. ASCE, pp. 1493–1505.

Calvete, D., Coco, G., Falques, A., Dodd, N., 2007. (Un)predictability in rip channel systems. Geophys. Res. Lett. 34 (5). http://dx.doi.org/10.1029/2006GL028162.

Carey, W., Rogers, S., 2005. Rip currents – coordinating coastal research, outreach and forecast methodologies to improve public safety. In: Proceedings of Coastal Disasters 2005. ASCE, pp. 285–296.

Castelle, B., Ruessink, B.G., 2011. Modeling formation subsequent nonlinear evolution of rip channels: time-varying versus time-invariant wave forcing. J. Geophys. Res. 116, F04008. http://dx.doi.org/10.1029/2011JF001997.

Castelle, B., Coco, G., 2012. The morphodynamics of rip channels on embayed beaches. Cont. Shelf Res. 43, 10–23.

Castelle, B., Coco, G., 2013. Surf zone flushing on embayed beaches. Geophys. Res. Lett. 40 (1–5). http://dx.doi.org/10.1002/grl.50485.

Castelle, B., Bonneton, P., Senechal, N., Dupuis, H., Butel, R., Michel, D., 2006. Dynamics of wave-induced currents over an alongshore non-uniform multiple-barred sandy beach on the Aquitanian Coast, France. Cont. Shelf Res. 26 (1), 113–131.

Castelle, B., Bonneton, P., Dupuis, H., Senechal, N., 2007. Double bar beach dynamics on the high-energy meso-macrotidal French Aquitanian Coast: a review. Mar. Geol. 245, 141–159.

Castelle, B., Ruessink, B.G., Bonneton, P., Marieu, V., Bruneau, N., Price, T.D., 2010a. Coupling mechanisms in double sandbar systems, Part 1: patterns and physical explanation. Earth Surf. Processes Landforms 35, 476–486.

Castelle, B., Ruessink, B.G., Bonneton, P., Marieu, V., Bruneau, N., Price, T.D., 2010b. Coupling mechanisms in double sandbar systems, Part 2: impact on alongshore variability of inner-bar rip channels. Earth Surf. Processes Landforms 35, 771–781.

Castelle, B., Michallet, H., Marieu, V., Leckler, F., Dubarbier, B., Lambert, A., Berni, C., Bonneton, P., Barthélemy, E., Bouchette, F., 2010c. Laboratory experiment on rip current circulations over a moveable bed: drifter measurements. J. Geophys. Res. 115, C12008. http://dx.doi.org/10.1029/2010JC006343.

Castelle, B., Michallet, H., Marieu, V., Bonneton, P., 2011. Surfzone retention in a laboratory rip current. J. Coastal Res. Spec. Issue 64, 50–54.

Castelle, B., Marieu, V., Coco, G., Bonneton, P., Bruneau, N., Ruessink, B.G., 2012. On the impact of an offshore bathymetric anomaly on surf zone rip channels. J. Geophys. Res. 117, F01038. http://dx.doi.org/10.1029/2011JF002141.

Castelle, B., Reniers, A.J.H.M., MacMahan, J., 2013. Numerical modeling of surfzone retention in rip current systems: on the impact of the surfzone sandbar morphology. Proc. Coastal Dyn. 2013, 294–304.

Castelle, B., Almar, R., Dorel, M., Lefebvre, J-P., Sénéchal, N., Anthony, E., Laibi, R., Chuchla, R., du Penhoat, Y., 2014. Flash rip dynamics on a high-energy low-tide-terraced beach (Grand Popo, Benin, West Africa). In: Proceedings 13th International Coastal Symposium (Durban, South Africa). J. Coastal Res., Special Issue No. 66, pp. 633–638.

Chen, Q., Dalrymple, R.A., Kirby, J.T., Kennedy, A., Haller, M.C., 1999. Boussinesq modeling of a rip current system. J. Geophys. Res. 104 (C9), 20617–20638.

Clark, D.B., Elgar, S., Raubenheimer, B., 2012. Vorticity generation by short-crested breaking waves. Geophys. Res. Lett. 39 (24). http://dx.doi.org/10.1029/2012GL054034.

Cook, D.O., 1970. The occurrence and geologic work of rip currents off southern California. Mar. Geol. 9 (3), 173–186.

Dalrymple, R.A., 1975. A mechanism for rip current generation on an open coast. J. Geophys. Res. 80, 3485–3487.

Dalrymple, R.A., 1978. Rip currents and their causes. In: Proceedings 16th International Conference on Coastal Engineering. ASCE, 1414–1427.

Dalrymple, R.A., MacMahan, J.H., Reniers, A.J.H.M., Nelko, V., 2011. Rip currents. Ann. Rev. Fluid Mech. 43, 551–581.

Davidson-Arnott, R., 2010. Introduction to Coastal Processes and Geomorphology. Cambridge University Press, p. 442.

Davis, W.M., 1925. The 'Undertow'. Science 62, 33.

Davis, W.M., 1931. Undertow and rip tides. Science 73, 526–527.

Davis Jr., R.A., Fitzgerald, D.M., 2004. Beaches and Coasts. Blackwell Science Ltd, p. 419.

Deigaard, R., 1993. A note on the three-dimensional shear stress distribution in a surf zone. Coastal Eng. 20, 157–171.

Drozdzewski, D., Shaw, W., Dominey-Howes, D., Brander, R., Walton, T., Gero, A., Sherker, S., Goff, J., Edwick, B., 2012. Surveying rip current survivors: preliminary insights into the experiences of being caught in rip currents. Nat. Hazards Earth Syst. Sci. 12, 1201–1211.

Dusek, G., Seim, H., 2013a. Rip current intensity estimates from lifeguard observations. J. Coastal Res. 29 (3), 505–518.

Dusek, G., Seim, H., 2013b. A probabilistic rip current forecast model. J. Coastal Res. 29 (4), 909–925.

Engle, J., MacMahan, J., Thieke, R.J., Hanes, D.M., Dean, R.G., 2002. Formulation of a rip current predictive index using rescue data. In: Proceedings National Conference on Beach Preservation Technology. FSBPA, Biloxi, MS.

Fowler, R.E., Dalrymple, 1990. Wave group forced nearshore circulation. In: Proceedings 22nd International Conference on Coastal Engineering. ASCE, pp. 729–742.

Gallop, S.L., Bryan, K.R., Coco, G., Stephens, S.A., 2011. Storm-driven changes in rip channel patterns on an embayed beach. Geomorphology 127, 179–188.

Garcia-Faria, A.F., Thornton, E.B., Lippmann, T.C., Stanton, T.P., 2000. Undertow over a barred beach. J. Geophys. Res. 105 (C7), 16999–17010.

Gensini, V.A., Ashley, W.S., 2010. An examination of rip current fatalities in the United States. Nat. Hazards 54 (1), 159–175.

Guza, R.T., Inman, D.L., 1975. Edge waves and beach cusps. J. Geophys. Res. 80, 2997–3012.

Haas, K.A., Svendsen, I.A., 2002. Laboratory measurements of the vertical structure of rip currents. J. Geophys. Res. 107 (C5). http://dx.doi.org/10.1029/2001JC000911.

Haas, K.A., Svendsen, I.A., Zhao, Q., 2003. Quasi-three dimensional modeling of rip current systems. J. Geophys. Res. C Oceans 108 (C7), 3217. http://dx.doi.org/10.1029/2002JC001355.

Haller, M.C., Dalrymple, R.A., 2001. Rip current instabilities. J. Fluid Mech. 433, 161–192.

Haller, M.C., Putrevu, U., Oltman-Shay, J., Dalrymple, R.A., 1999. Wave group forcing of low frequency surf zone motion. Coastal Eng. 41, 121–136.

Haller, M.C., Dalrymple, R.A., Svendsen, I.A., 2002. Experimental study of nearshore dynamics on a barred beach with rip channels. J. Geophys. Res. 107 (C6). http://dx.doi.org/10.1029/2001JC000955.

Hammerton, C.E., Brander, R.W., Dawe, N., Riddington, C., Engel, R., 2013. Approaches for beach safety and education in Ghana: a case study for developing countries with a surf coast. Int. J. Aquat. Res. Educ. 7 (3), 254–265.

Hartmann, D., 2006. Drowning and Beach Safety Management (BSM) along the Mediterranean Beaches of Israel – a long-term perspective. J. Coastal Res. 22 (6), 1505–1514.

Hatfield, J., Williamson, A., Sherker, S., Brander, R., Hayen, A., 2012. Development and evaluation of an intervention to reduce rip current related beach drowning. Accid. Anal. Prev. 46, 45–51.

Holman, R.A., Stanley, J., 2007. The history and technical capabilities of Argus. Coastal Eng. 54, 477–491.

Holman, R.A., Symonds, G., Thornton, E.B., Ranasinghe, R., 2006. Rip spacing on an embayed beach. J. Geophys. Res. 111, C01006. http://dx.doi.org/10.1029/2005/JC002965.

Houser, C., Barrett, G., Labude, D., 2011. Alongshore variation in the rip current hazard at Pensacola Beach, Florida. Nat. Hazards 57, 501–523.

Houser, C., Arnott, R., Ulzhofer, S., Barrett, G., 2013. Nearshore circulation over transverse bar and rip morphology with oblique wave forcing. Earth Surf. Processes Landforms 38, 1269–1279.

Inman, D.L., Quinn, W.H., 1952. Currents in the surf zone. In: Proceedings 2nd International Conference on Coastal Engineering. ASCE, pp. 24–36.

Inman, D.L., Brush, B.M., 1973. Coastal challenge. Science 181, 20–32.

Johnson, D., Stocker, R., Head, R., Imberger, J., Pattiaratchi, C., 2003. A compact, low-cost GPS drifter for use in the oceanic nearshore zone, lakes and estuaries. J. Atmos. Oceanic Tech. 18, 1880–1884.

Johnson, D., Pattiaratchi, C., 2004. Transient rip currents and nearshore circulation on a swell-dominated beach. J. Geophys. Res. 109, C02026. http://dx.doi.org/10.1029/2003JC001798.

Johnson, D., Pattiaratchi, C., 2006. Boussinesq modeling of transient rip currents. Coastal Eng. 53, 419–439.

Kennedy, A., Dalrymple, R.A., 2001. Wave group forcing of rip currents. In: Proceedings of Waves 2001. ASCE, pp. 1426–1435.

Kennedy, A., Thomas, D., 2004. Drifter measurements in a laboratory rip current. J. Geophys. Res. 109, C08005. http://dx.doi.org/10.1029/2003JC001927.

Klein, A.H. da F., Santana, G.G., Diehl, F.L., Menezes, J.T., 2003. Analysis of hazards associated with sea bathing: results of five years work in oceanic beaches of Santa Catarina State, Southern Brazil. J. Coastal Res. Spec. Issue 35, 107–116.

Komar, P.D., 1998. Beach Processes and Sedimentation, second ed. Prentice Hall. p. 544.

Lascody, R.L., 1998. East Central Florida rip current program. Natl. Weather Dig. 22 (2), 25–30.

Leatherman, S.P., 2013. Rip currents. In: Finkl, C.W. (Ed.), Coastal Hazards, Coastal Research Library, vol. 6. Springer, pp. 811–831.

Lippmann, T.C., Holman, R.A., 1989. Quantification of sand bar morphology: a video technique based on wave dissipation. J. Geophys. Res. 94, 995–1011.

Lippmann, T.C., Holman, R.A., 1990. The spatial and temporal variability of sand bar morphology. J. Geophys. Res. 95, 11575–11590.

Long, J., Özkan-Haller, H., 2005. Offshore controls on nearshore rip currents. J. Geophys. Res. 110, C12. http://dx.doi.org/10.1029/2005JC003018.

Longuet-Higgins, M.S., Stewart, R.W., 1964. Radiation stress in water waves, a physical discussion with applications. Deep-Sea Res. 11 (4), 529–563.

Loureiro, C., Ferreira, O., Cooper, J.A.G., 2011. Morphologic change and morphodynamics at high-energy embayed beaches in southwestern Portugal. In: Proceedings of Coastal Sediments 2011. World Scientific, pp. 1375–1389.

Loureiro, C., Ferreira, O., Cooper, J.A.G., 2012. Extreme erosion on high-energy embayed beaches: influence of megarips and storm grouping. Geomorphology 139–140, 155–171.

Lushine, J.B., 1991. A study of rip current drownings and weather related factors. Natl. Weather Dig. 16, 13–19.

MacMahan, J.H., Reniers, A.J.H.M., Thornton, E.B., Stanton, T.P., 2004a. Infragravity rip current pulsations. J. Geophys. Res. 109, C01033. http://dx.doi.org/10.1029/2003JC002068.

MacMahan, J.H., Reniers, A.J.H.M., Thornton, E.B., Stanton, T.P., 2004b. Surf zone eddies coupled with rip current morphology. J. Geophys. Res. 109, C07004. http://dx.doi.org/10.1029/2003JC002083.

MacMahan, J.H., Thornton, E.B., Stanton, T.P., Reniers, A.J.H.M., 2005. RIPEX: observations of a rip current system. Mar. Geol. 218, 118–134.

MacMahan, J.H., Thornton, E.B., Reniers, A.J.H.M., 2006. Rip current review. Coastal Eng. 53, 191–208.

MacMahan, J.H., Brown, J., Thornton, E., 2009. Low-cost handheld global positioning system for measuring surf-zone currents. J. Coastal Res. 25 (3), 744–754.

MacMahan, J., Brown, J., Brown, J., Thornton, E., Reniers, A., Stanton, T., Henriquez, M., Gallagher, E., Morrison, J., Austin, M.J., Scott, T.M., Senechal, N., 2010a. Mean lagrangian flow behavior on an open coast rip channeled beach: a new perspective. Mar. Geol. 268, 1–15.

MacMahan, J.H., Reniers, A.J.H.M., Thornton, E.B., 2010b. Vortical surf zone velocity fluctuations with *O*(10) minute period. J. Geophys. Res. 115, C06007. http://dx.doi.org/10.1029/2009JC005383.

Masselink, G., Short, A.D., 1993. The effect of tide range on beach morphodynamics and morphology: a conceptual beach model. J. Coastal Res. 9 (3), 785–800.

Masselink, G., Pattiaratchi, C.B., 1998. Morphological evolution of beach cusps and associated swash circulation patterns. Mar. Geol. 146, 93–113.

Masselink, G., Austin, M., Scott, T., Russell, P.E., 2014. Rip currents: researching a natural hazard. Geogr. Rev. 27 (3), 37–41.

Matthews, B., Andronaco, R., Adams, A., 2014. Warning signs at beaches: do they work? Saf. Sci. 62, 312–318.

McCarroll, R.J., Brander, R.W., MacMahan, J.H., Turner, I.L., Reniers, A.J.H.M., Brown, J.A., Bradstreet, A., Sherker, S., 2014a. Evaluation of swimmer-based rip current escape strategies. Nat. Hazards 71, 1821–1846.

McCarroll, R.J., Brander, R.W., Turner, I.L., Power, H.E., Mortlock, T.R., 2014b. Lagrangian observations of circulation on an embayed beach with headland rip currents. Mar. Geol. 355, 173–188.

McCool, J.P., Moran, K., Ameratunga, S., Robinson, E., 2008. New Zealand beachgoers' swimming bheaviours, swimming abilities, and perception of drowning risk. Int. J. Aquat. Res. Educ. 1, 7–15.

McCool, J.P., Ameratunga, S., Moran, K., Robinson, E., 2009. Taking a risk perception approach to improving beach swimming safety. Int. J. Behav. Med. 16, 260–366.

McKenzie, P., 1958. Rip current systems. J. Geol. 66 (2), 103–113.

Miloshis, M., Stephenson, W.J., 2011. Rip current escape strategies: lessons for swimmers and coastal rescue authorities. Nat. Hazards 59, 823–832.

Mitchell, R., Haddrill, K., 2004. From the bush to the beach: water safety in rural and remote New South Wales. Aust. J. Rural Health 12 (6), 246–250.

Morgan, D., Ozanne-Smith, J., Triggs, T., 2008. Descriptive epidemiology of drowning deaths in a surf beach swimmer and surfer population. Inj. Prev. 14, 62–65.

Murray, T., Cartwright, N., Tomlinson, R., 2013. Video-imaging of transient rip currents on the Gold Coast open beaches. J. Coastal Res. Spec. Issue 65, 1809–1814.

Nelko, V., Dalrymple, R.A., 2008. Rip currents: mechanisms and observations. In: Proceedings 31st International Conference on Coastal Engineering. World Scientific, pp. 888–900.

Oltman-Shay, J., Howd, P., Birkemeier, W., 1989. Shear instabilities of the mean longshore current, 2. J. Geophys. Res. 94, 18031–18042.

Özkan-Haller, T., Kirby, J., 1999. Nonlinear evolution of shear instabilities of the longshore current: a comparison of observation and computations. J. Geophys. Res. 104, 25953–25984.

Orzech, M.D., Thornton, E.B., MacMahan, J.H., O'Reilly, W.C., Stanton, T.P., 2010. Alongshore rip channel migration and sediment transport. Mar. Geol. 271, 278–291.

Pattiaratchi, C., Olsson, D., Hetzel, Y., Lowe, R., 2009. Wave-driven circulation patterns in the lee of groynes. Cont. Shelf Res. 29, 1961–1974.

Ranasinghe, R., Symonds, G., Holman, R.A., 1999. Quantitative characterisation of rip currents via video imaging. In: Proceedings of Coastal Sediments '99. ASCE, pp. 987–1002.

Ranasinghe, R., Symonds, G., Black, K., Holman, R., 2000. Processes governing rip spacing, persistence and strength in a swell dominated, microtidal environment. In: Proceedings 27th International Conference on Coastal Engineering. ASCE, pp. 455–467.

Ranasinghe, R., Symonds, G., Holman, R.A., 2004. Morphodynamics of intermediate beaches: a video imaging and numerical modelling study. Coastal Eng. 51, 629–655.

Reniers, A.J.H.M., 2007. Modeling of very low frequency motions during RIPEX. J. Geophys. Res. 112, C07013. http://dx.doi.org/10.1029/2005JC003122.

Reniers, A.J.H.M., MacMahan, J., Thornton, E.B., Stanton, T.P., Henriquez, M., Brown, J.W., Brown, J.A., Gallagher, E., 2009. Surfzone retention on a rip channelled beach. J. Geophys. Res. 114, C10010. http://dx.doi.org/10.1029/2008JC005153.

Reniers, A.J.H.M., MacMahan, J.H., Beron-Vera, F.J., Olascoaga, M.J., 2010. Rip current pulses tied to Lagrangian coherent structures. Geophys. Res. Lett. 37 (5). http://dx.doi.org/10.1029/2009GL041443.

Schmidt, W., Woodward, B., Millikan, K., Guza, R., Raubenheimer, B., Elgar, S., 2003. A GPS-tracked surf zone drifter. J. Atmos. Oceanic Tech. 20, 1069–1075.

Scott, T.M., Russell, P.E., Masselink, G., Wooler, A., 2009. Rip current variability and hazard along a macro-tidal coast. J. Coastal Res. Spec. Issue 50, 1–6.

Scott, T.M., Masselink, G., Russell, P., 2011a. Morphodynamic characteristics and classification of beaches in England and Wales. Mar. Geol. 286, 1–20.

Scott, T.M., Russell, P.E., Masselink, G., Austin, M.J., Wills, S., Wooler, A., 2011b. Rip current hazards on large-tidal beaches in the United Kingdom. In: Leatherman, S., Fletemeyer, J. (Eds.), Rip Currents: Beach Safety, Physical Oceanography and Wave Modeling. CRC Press, pp. 225–242.

Scott, T.M., Masselink, G., Austin, M.J., Russell, P., 2014. Controls on macrotidal rip current circulation and hazard. Geomorphology 214, 198–215.

Shanks, A.L., Morgan, S.G., MacMahan, J.H., Reniers, A.J.H.M., 2010. Surf zone physical and morphologica regime as determinants of temporal and spatial variation in larval recruitment. J. Exp. Mar. Biol. Ecol. 392, 140–150.

Shepard, F.P., 1936. Undertow, rip tide or rip current. Science 84, 181–182.
Shepard, F.P., Inman, D.L., 1950. Nearshore circulation. In: Proceedings of the 1st Conference on Coastal Engineering. ASCE, pp. 50–59.
Shepard, F.P., Inman, D.L., 1951. Nearshore circulation related to bottom topography and wave refraction. Trans. Am. Geophys. Union 31 (4), 196–213.
Shepard, F.P., Emery, K.O., Lafond, E.C., 1941. Rip currents: a process of geological importance. J. Geol. 49, 338–369.
Sherker, S., Brander, R.W., Finch, C., Hatfield, J., 2008. Why Australia needs an effective national campaign to reduce coastal drowning. J. Sci. Med. Sport 11, 81–83.
Sherker, S., Williamson, A., Hatfield, J., Brander, R., Hayen, A., 2010. Beachgoers' beliefs and behaviours in relation to beach flags and rip currents. Accid. Anal. Prev. 42, 1785–1804.
Short, A.D., 1985. Rip current type, spacing and persistence, Narrabeen Beach, Australia. Mar. Geol. 65, 47–71.
Short, A.D., 1992. Beach systems of the central Netherlands coast: processes, morphology and structural impacts in a storm driven multi-bar system. Mar. Geol. 107, 103–132.
Short, A.D., 2007. Australian rip systems – friend or foe? J. Coastal Res. Spec. Issue 50, 7–11.
Short, A.D., Aagaard, T., 1993. Single and multi-bar beach change models. J. Coastal Res. Spec. Issue 15, 141–157.
Short, A.D., Hogan, C.L., 1994. Rip currents and beach hazards: their impact on public safety and implications for coastal management. J. Coastal Res. Spec. Issue 12, 197–209.
Short, A.D., Brander, R.W., 1999. Regional variations in rip density. J. Coastal Res. 15 (3), 813–822.
Short, A.D., Masselink, G., 1999. Embayed and structurally controlled beaches. In: Short, A.D. (Ed.), Handbook of Beach and Shoreface Morphodynamics. John Wiley & Sons, pp. 230–250.
Slattery, M.P., Bokuniewicz, H., Gayes, P., 2011. Flash rip currents on ocean shoreline of Long Island, New York. In: Leatherman, S., Fletemeyer, J. (Eds.), Rip Currents: Beach Safety, Physical Oceanography and Wave Modeling. CRC Press, pp. 31–43.
Smith, J.A., Largier, J.L., 1995. Observations of nearshore circulation: rip currents. J. Geophys. Res. 100, 10967–10975.
Sonu, C.J., 1972. Field observations of a nearshore circulation and meandering currents. J. Geophys. Res. 77, 3232–3247.
Sonu, C.J., McCloy, J.M., McArthur, D.S., 1966. Longshore currents and nearshore topographies. In: Proceedings 10th International Conference on Coastal Engineering. ASCE, pp. 502–524.
Surf Life Saving Australia (SLSA), 2013. National Coastal Safety Report 2013. SLSA, Sydney.
Talbot, M.M., Bate, G.C., 1987. Rip current characteristics and their role in the exchange of water and surf diatoms between the surf zone and nearshore. Estuarine Coastal Shelf Sci. 25 (6), 707–720.
Tang, E.C.S., Dalrymple, R.A., 1989. Nearshore circulation: rip currents and wave groups. In: Seymour, R.J. (Ed.), Nearshore Sediment Transport. Plenum Press, pp. 205–230.
Thornton, E.B., Sallenger, A.H., MacMahan, J.H., 2007. Rip currents, cuspate shorelines and eroding dunes. Mar. Geol. 240, 151–167.
Turner, I.L., Whyte, D., Ruessink, B.G., Ranasinghe, R., 2007. Observations of rip spacing, persistence and mobility at a long, straight coastline. Mar. Geol. 236, 209–221.

United States Lifesaving Association (USLA), 2012. National Lifesaving Statistics (accessed 17.03.14.). http://arc.usla.org/Statistics/current.asp?Statistics=5.

White, K.M., Hyde, M.K., 2010. Swimming between the flags: a preliminary exploration of the influences on Australians' intentions to swim between the flags at patrolled beaches. Accid. Anal. Prev. 42, 1831–1838.

Wilks, J., DeNardi, M., Wodarski, R., 2007. Close is not enough: drowning and rescues outside flagged beach patrol areas in Australia. Tourism Mar. Environ. 4 (1), 57–62.

Williamson, A., 2006. Feasibility study of a water safety data collection for beaches. J. Sci. Med. Sport 9, 243–248.

Williamson, A., Hatfield, J., Sherker, S., Brander, R., Hayen, A., 2012. A comparison of attitudes and knowledge of beach safety in Australia for beachgoers, rural residents and international tourists. Aust. N.Z. J. Public Health 36, 385–391.

Woodroffe, C.D., 2002. Coasts; Form, Process and Evolution. Cambridge University Press, p. 623.

Woodward, E., Beaumont, E., Russell, P., Wooler, A., Macleod, R., 2013. Analysis of rip current incidents and victim demographics in the UK. J. Coastal Res. Spec. Issue 65, 850–855.

Wright, L.D., Short, A.D., 1984. Morphodynamic variability of surf zones and beaches: a synthesis. Mar. Geol. 56, 93–118.

Chapter 13

Sea Ice: Hazards, Risks, and Implications for Disasters

Hajo Eicken and Andrew R. Mahoney
Geophysical Institute, University of Alaska Fairbanks, Fairbanks, AK, USA

ABSTRACT

The role of sea ice as a natural hazard is discussed with a focus on Arctic and sub-Arctic regions where risks associated with human activities and ice processes are the greatest. Hazard assessment and emergency response need to consider a range of controlling factors that can lead to events initiating an accident, failure, or full-scale disaster. These factors include environmental hazards, equipment, procedures and settings, and people. Quantifying risks associated with the presence of sea ice requires the joint consideration of the probability of specific hazards and the magnitude of their impacts. Both of these also depend on the type and level of human activity, such that disaster risks are substantially higher in the Arctic than in the Antarctic. We identify three types of sea-ice hazards: (1) broad, long-term hazards and associated risks associated with a rapid reduction in summer ice extent; (2) near-term hazards resulting from changes in sea-ice extent and dynamics such as increased coastal erosion and threats to coastal infrastructure; and (3) immediate risks and the potential for disasters derived from the combination of sea-ice hazards and human activities such as shipping or offshore resource development. A review of key properties and processes governing the role of sea ice as a hazard focuses on recent rapid changes in ice extent and concentration in the Arctic and resulting threats to coastal systems. Other key factors include the distribution of old perennial ice that has a greater thickness and higher mechanical strength than seasonal ice, patterns of ice movement that determine advection of ice hazards, and the degree of ice deformation that can generate thick, rough ice and represent a hazard in its own right. These factors are examined in the context of a case study for the Beaufort and Chukchi Seas in the North American Arctic. Linking specific environmental hazards to the geospatial distribution of human activities and vulnerable ecosystems allows for an integrated Arctic hazards assessment, currently still in its infancy. The need for coordinated environmental observations in informing hazard assessments and emergency response is discussed in the context of recent increases in maritime activities in the Arctic.

Coastal and Marine Hazards, Risks, and Disasters. http://dx.doi.org/10.1016/B978-0-12-396483-0.00013-3
Copyright © 2015 Elsevier Inc. All rights reserved.

13.1 INTRODUCTION: ENVIRONMENTAL HAZARDS, DISASTERS AND SEA ICE

The presence of sea ice in polar and subpolar oceans is a defining environmental factor that governs weather and climate, ecology, and human activities in these regions. Sea ice is also a significant natural hazard, both through direct interaction with assets and infrastructure and through the indirect impacts of variability and rapid changes in its distribution, in particular in Arctic and sub-Arctic regions. Although the former type of threat is clearly defined and falls well within the scope of classic hazard analysis (e.g., Ogorodov et al., 2005; Zhang et al., 2013), the latter aspect relates to the broader problem of threats to benefits or services provided by sea ice. For example, delayed formation of coastal ice as a result of climate change can expose shorelines to erosive action of fall storms and enhance thermal subsidence of coastal permafrost, over time greatly accelerating rates of coastal retreat (Overeem et al., 2011). Nevertheless, both types of threats or hazards, rapid and slow onset, can be understood in the context of sea-ice system services, which describe and potentially quantify the benefits and threats that social–environmental systems derive from sea ice over a range of scales (Eicken et al., 2009). This contribution considers both types of hazards and hence direct and indirect effects of sea ice that can lead to harm or disasters.

Sea ice is an integral part of social–environmental systems, in particular in the northern hemisphere—hence, ice hazards and their potential role in disasters cannot be considered in isolation. This broader perspective is summarized in Figure 13.1, which outlines the combination of factors initiating a sequence of events that may culminate in failure of structures, loss of life and property, or escalate into a large-scale disaster. Hazard mitigation and disaster prevention aim to curb or eliminate initiating events that can potentially escalate with negative consequences for human activities or ecosystems (Vinnem, 2007). It is important to recognize that in such a context, natural or environmental hazards, including sea ice, are part of a combination of factors that can lead to disasters. Other factors include the design or hardware of infrastructure implicated in an event, human judgment, or error as well as procedures in place to ensure safety and reliability of an organizational structure (Figure 13.1). The entire complex of factors, initiating events, prevention, escalation, and disasters can be understood in terms of risks associated with specific consequences (Schneider et al., 2007). Typically, risk is evaluated in terms of a convolution of the probability of occurrence of a particular event and the magnitude of its consequences.

The concept of risk provides some guidance to the scope and thrust of this contribution. By mapping the distribution of sea-ice processes and properties that represent threats and hazards, we can help constrain the probability of occurrence of a disaster or loss event. By evaluating harmful impacts through

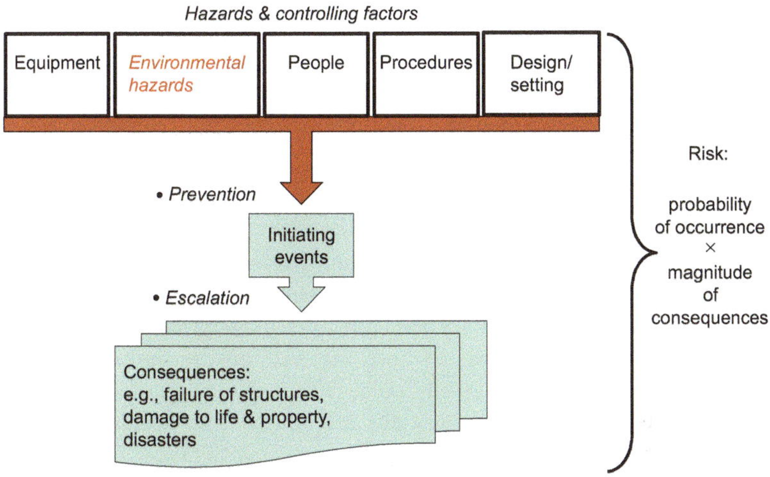

FIGURE 13.1 Key concepts relevant to the study of sea ice as a natural or environmental hazard, building on research in maritime risk assessment (Vinnem, 2007). Environmental hazards are one of the several controlling factors that may lead to a sequence of events culminating in damage to life and property or a potential disaster. The risk associated with such a chain of events is characterized by the convolution of probability of occurrence and the magnitude of the resulting consequences.

direct (e.g., damage to offshore structures or vessels along shipping lanes) and indirect actions (e.g., enhanced coastal erosion or loss of climate regulation through reduced sea ice), we can assess both the probability and the magnitude of consequences in a geospatial context. In combination, such an evaluation provides guidance on both the relevant processes that need to be considered and the specific regions to focus on.

The following section concentrates on a limited range of sea-ice processes and properties relevant for our understanding of sea ice as a natural hazard in both polar regions. At the same time, for geospatially explicit discussions of consequences, the focus is on the Arctic where both the nature of sea-ice hazards and the probability of occurrence of events impacting human activities are more substantial than in the Antarctic. In Antarctica, human activity and vessel traffic are concentrated into a small subregion of the Antarctic Peninsula region, where research bases and tourism account for roughly 100 vessel passages per year (based on 2007/2008 season; Lynch et al., 2010). In the Arctic, such vessel densities are typical of the entire circum-Arctic with, for example, roughly 140 vessels passing through the Canadian Arctic maritime region in 2012 with close to half of these vessels consisting of bulk carriers, tankers, and cargo ships (Pizzolato et al., 2014). Moreover, in contrast with the Antarctic, (sub-)Arctic sea-ice hazards play a prominent role in offshore resource development and associated coastal infrastructure.

13.2 GEOGRAPHIC DISTRIBUTION OF SEA ICE AND KEY PROCESSES AND PROPERTIES

Polar sea ice waxes and wanes with the seasons. Ice fills the Arctic basin and occupies roughly $15 \times 10^6\ km^2$ at its maximum winter extent in March, shrinking to a seasonal minimum in September of $<5 \times 10^6\ km^2$ in recent years (Figures 13.2 and 13.3; Comiso, 2010). As is evident from Figure 13.2, reductions in ice concentration, that is, the areal fraction of ocean surface covered by sea ice, in particular during summer, have exposed large parts of the Arctic shelf seas to open water. In the Pacific Arctic sector, this trend toward reduced summer ice is particularly pronounced and coincides with increases in ship traffic and offshore oil and gas resource exploration (Brigham, 2010; Lovecraft and Eicken, 2011). Note that in contrast the Bering Sea has experienced more extensive, though thinner and typically less concentrated ice cover in recent years (Figure 13.2, left).

During 1979–2014, Arctic sea ice has been found to form later in the fall and break-up earlier in the year. Consequently, the open water season has been increasing in length, in some regions such as the Pacific–Arctic sector (Figure 13.2) by as much as a month or more. However, satellite imagery used to obtain such sea-ice climate data records typically tracks ice only at concentrations >10 percent and may not detect ice during summer months at low concentrations (Meier and Stroeve, 2008). Hence, Arctic shelf seas and coastal regions, such as those highlighted in Figure 13.2, undergoing longer open water periods may still see lingering ice that can represent a hazard but is not captured at a sufficient spatial and temporal resolution (Eicken et al., 2011).

Reductions in ice concentration over the course of summer in the North American and Siberian Arctic (Figure 13.2, right) and a longer open water period have resulted in substantial increases in the amount of solar heating of these waters (Perovich et al., 2007). Further, increased fetch during the fall storm season has resulted in increases in wave amplitudes in fall (Overeem et al., 2011). Thinner, more dynamic ice also appears to be associated with increases in wave heights and deeper propagation of swell into the ice pack in spring (Francis and Atkinson, 2012). These changes work in concert to increase erosive action, thermal subrosion of coastal permafrost, and thereby accelerate rates of coastal retreat with potentially negative impacts on coastal communities and infrastructure.

A key property of relevance from a natural-hazard perspective is the mechanical strength of sea ice. Typically, ice strength varies disproportionately with age since desalination of sea ice during the first and subsequent melt seasons greatly reduces the porosity of the bulk of the ice cover, thereby increasing bending and compressive ice strength by between one half to as much as a factor of three to four (Timco and Weeks, 2010). Similarly, brackish or freshwater ice formed in coastal lagoons or estuaries as well as glacial ice in the form of icebergs will exhibit greater strength than sea ice. Mapping the

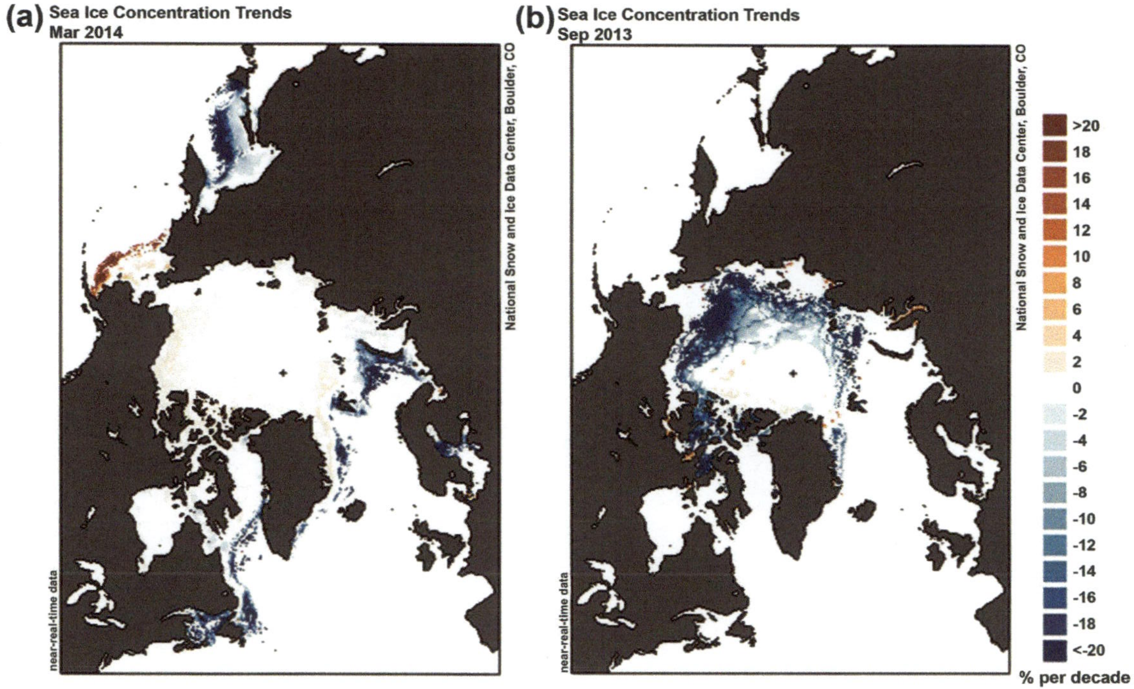

FIGURE 13.2 Monthly Arctic sea-ice concentration trends derived from passive microwave satellite data for the time period 1979–2013 (left: March; right: September). *National Snow and Ice Data Center Sea Ice Index; Fetterer et al., 2013.*

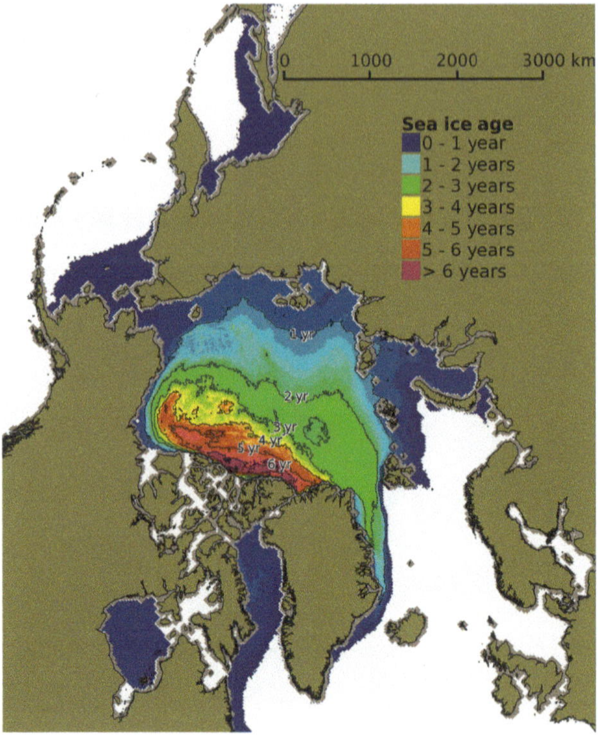

FIGURE 13.3 Mean sea-ice age for the time period 2005–2010. The map is based on ice age fields derived by tracking regions of ice as they drift around the Arctic (Maslanik et al., 2007).

distribution of sea-ice age, using a combination of remote-sensing and drifting buoy data, provides a large-scale measure of the probability of encountering older, stronger ice (Maslanik et al., 2007). For the period 2005–2010, the North American half of the Arctic basin was filled with ice three years or older (Figure 13.3), though the mean age of Arctic sea ice continues to drop substantially as a result of reductions in the extent of perennial ice surviving at least one summer's melt. Here, ice age can also serve as a proxy for level-ice thickness (Maslanik et al., 2007). First-year ice rarely exceeds 2 m and typically ranges between 1 and 1.5 m in thickness across the (sub)Arctic shelf seas in late spring. In the Arctic basin, second-year ice ranges in thickness between 2 and 3 m, and only the oldest level ice is likely to exceed a thickness of 4–5 m. However, the Canadian Archipelago generates much older ice that can exceed a 10-m thickness even in level, undeformed areas and which may enter the Arctic Ocean through some of the straits exiting the Archipelago (Johnston et al., 2009; Barber et al., 2014).

The prevailing patterns of ice motion, largely driven by surface wind forcing and—on time scales of months to years—by sea surface tilt, transport

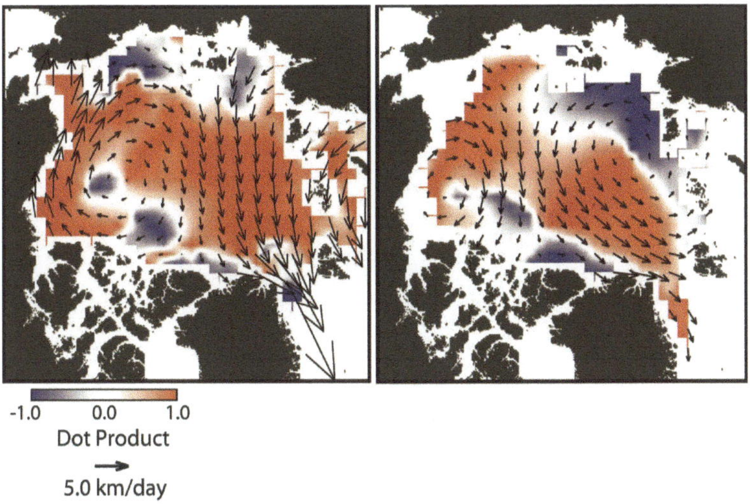

FIGURE 13.4 Mean Arctic sea-ice velocity field for winter (October–May; left) and summer (June–September; right) for the time period 1982–2009 as derived by Kwok et al. (2013). The color scale indicates whether ice velocities are exhibiting a trend toward higher (red) or lower (blue) velocities.

older ice from the high Canadian Arctic into the Pacific Arctic sector and waters off Alaska (Figures 13.3 and 13.4). Along the North American and East Siberian coastlines, ice motion is mostly shore parallel, directed toward the West with the anticyclonic Beaufort Gyre. In central and western Siberia, ice motion is directed away from the shore, whereas north of Greenland and the High Canadian Arctic, it is toward the shore. As a result, some of the oldest, thickest ice is found in the latter regions, whereas some of the youngest ice prevails in the former (Figures 13.3 and 13.4). Recent studies have found an increase in both ice speed and deformation rate (the latter by roughly 50 percent per decade since 1979), likely a result of changes in surface forcing and reduced ice thickness that foster increased ice mobility (Rampal et al., 2009). As is evident in Figure 13.4, regions with the highest mean ice velocities also appear to experience the greatest increase in ice speed. Moreover, recent work by Barber et al. (2014) demonstrates that as a result of the combination of these changes, the complexity of ice motion patterns and hence the challenge to predict ice movement on short time scales has greatly increased.

A measure of the geographic distribution of ice deformation processes can be obtained from an analysis of winter ice movement obtained from radar remote-sensing imagery over the time period 1996–2008 (Figure 13.5; Herman and Glowacki, 2012). Here, the distribution of cells that fall into the top 5-percentile of the mean total deformation rate provides an indication of areas where shear or repeated divergent/convergent deformation events have the potential to produce highly deformed ice. As is apparent from Figure 13.5,

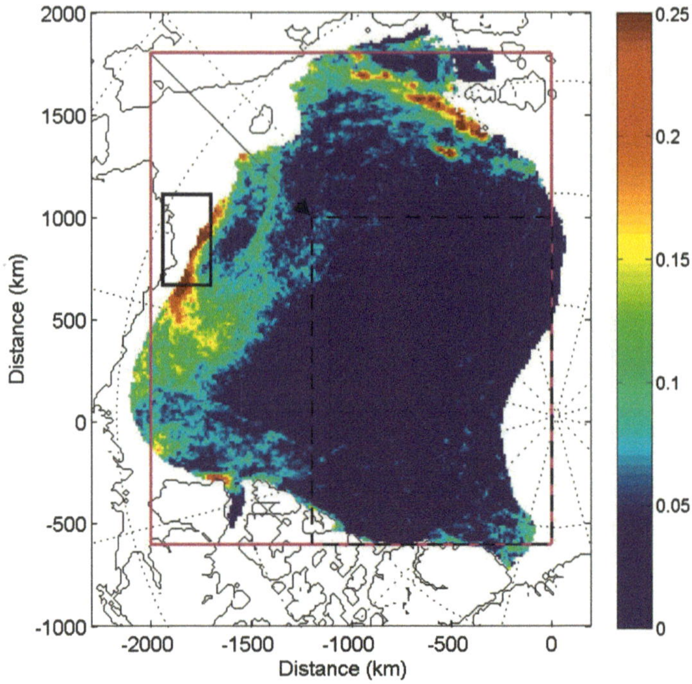

FIGURE 13.5 Occurrence probability of very strong Arctic sea-ice deformation events, corresponding to the top 5-percentile of total deformation rates for the entire Arctic, as derived by Herman and Glowacki (2012). Data cover the winter (November–April) from 1996 to 2008. Rectangle (solid black line) delineates the approximate extent of the data shown in Figure 13.8.

the high deformation regions are confined to the Arctic Ocean marginal seas and are likely the result of ice–land interaction in combination with strong surface forcing by winds (reflected in the mean fields shown in Figure 13.4) and tidal currents (Lyard, 1997). North of the Canadian Archipelago, where some of the oldest ice in the Arctic resides, deformation rates are not as high. This is explained in part by the high strength and compactness of multiyear ice in this region that does not accommodate significant strain. Ice deformation can result in ice thickness multiple times that of thick-level ice, by piling ice floe fragments into ridges or rubble piles. An example of such deformation processes and the resulting features is shown in Figure 13.6 for rafted and ridged ice forced onshore by motion converging on the coastline at Barrow, Alaska. A case study below provides further insight into these processes. The coastal topography and prevailing ice motion (Figures 13.4 and 13.5) favor such ice–coast interaction with potentially significant implications for coastal processes and hazards.

In the Southern Ocean, sea ice girdles the Antarctic continent with a winter maximum extent of roughly 18×10^6 km^2, decreasing to 3×10^6 km^2 in the

FIGURE 13.6 Sea ice forced onshore at Barrow, Alaska in an ice-push event in spring of 2001 (Mahoney et al., 2004). Such events require a compact ice cover offshore and wind or ocean forcing acting over a sufficient fetch of ice. Note how the parent ice sheet of a roughly 1.5-m thickness has been rafted onto the beach at right, while failing in a buckling mode, resulting in an ice pressure ridge at left. Coastal infrastructure, such as powerlines, harbor installations, or buildings (visible in the distance at the right) can be threatened by such ice-push events.

austral summer (Comiso, 2010). Due to higher ocean-to-ice heat fluxes and lack of old multiyear ice, Antarctic sea ice is typically much thinner (maximum level-ice thickness well below 1 m in most regions) than Arctic sea ice. Also, multiyear ice does not undergo the same degree of desalination as in the Arctic, such that the bulk strength of Antarctic multiyear ice is typically of the same magnitude as that of first-year ice. Deformation processes and sea-ice circulation are driven by a similar combination of factors as in the Arctic, creating similar hazards for vessel entrapment in regions of persistent deformation and convergence. Interaction between the sea-ice cover and the Antarctic coastline is substantially different, however, from that of the Arctic because of the prevalence of either ice shelves or rocky shores throughout Antarctica.

13.3 SEA ICE AS NATURAL HAZARD

Building on an understanding of the services sea ice provides to people and ecosystems (Eicken et al., 2009), and taking into consideration in particular the Arctic ice cover's recent transformation, we can evaluate the role of sea ice as a natural hazard. As illustrated in Figure 13.1, this problem cannot be treated as an assessment of, for example, geophysical processes in isolation from the broader setting. Therefore, we recognize three principal aspects of the problem: (1) broad, long-term hazards and associated risks associated with a rapid reduction in summer ice extent, such as geographic shifts in marine ecosystems and warming of submarine permafrost and adjacent land; (2) near-term hazards resulting from changes in sea-ice extent and dynamics such as increased coastal erosion and broad threats to coastal infrastructure; and (3) immediate risks and the potential for disasters derived from the combination of sea-ice hazards and human activities such as shipping or offshore resource development that involve specific, localized assets. The first type of hazard can be thought of in terms of slow-onset events, typically at regional to hemispheric scale, that require a response in the form of mitigation and adaptation (e.g., IPCC, 2012). Hazards of type 3 are associated with rapid-onset events at the local scale (though with regional to global repercussions, e.g., in the case of the threat to major infrastructure or the potential of an oil spill of national or international significance), while type 2 hazards can fall somewhere in between.

Type 1 hazards can be considered in the context of services provided by the sea-ice cover, such as the regulation of global climate and its central role in ice-albedo feedback (Eicken et al., 2009). Rapid changes in ice extent and feedback processes may then have significant impacts, ranging from changes in the Earth's heat budget, to impacts on midlatitude weather to threats to key species of global significance. With the science, for example, on potential linkages between Arctic summer sea-ice reduction and extreme weather events still emerging (e.g., Francis and Vavrus, 2012; Screen and Simmonds, 2013), causal attribution and identification of specific hazards are challenging at this broader scale. Nevertheless, initial model-based attempts have been made to estimate the economic impacts of such hazards (Euskirchen et al., 2013). Future research will have to establish whether hazard mitigation can be directly translated into specific reductions in the risk of, for example, extreme weather events or associated disasters.

For type 2 hazards, the link between sea-ice processes and their variability and change is better established. Thus, the potential for greater significant wave heights and a longer open water period with potential for wave action and coastal flooding has been identified in a number of regions, with major impacts in particular in the Siberian and Pacific Arctic sector (Atkinson, 2005; Overeem et al., 2011; Asplin et al., 2012; Francis and Atkinson, 2012). Under conditions of a more mobile ice cover (Rampal et al., 2009; Barber et al., 2014), this recent development also increases the probability for ice–coastal interaction during the open water and shoulder seasons. For example, in coastal Alaska, revetments put in place to protect the shoreline from wave action have been damaged by storm-induced ice action during the fall freeze-up period. It is not clear whether to expect more frequent occurrence of ice-push and similar events (Figure 13.6) as a result of a changing ice cover. Anecdotal evidence suggests that reductions in the amount of well-anchored, stable shorefast ice have led to an increase in such events in northern and Western Alaska.

Impacts of reduced sea ice on coastal permafrost exacerbate the vulnerability to flooding and coastal retreat. Thus, wave action and warming of surface waters promote the thermal subrosion of both terrestrial and marine permafrost along the coast (Ravens et al., 2012). As a result, rates of coastal retreat and subsidence increase.

Sea-ice hazards of type 3 involve the direct interaction between ice and artificial structures or vessels at the local scale. An assessment of the distribution of human activities that are vulnerable to sea-ice hazards is provided in Figure 13.7. Of particular relevance are the distribution of major shipping lanes, regions of intensifying offshore oil and gas exploration and production, mines accessible only by sea as well as areas frequented by tourist cruise ships. At present, the vast majority of vessel traffic in the Arctic is destinational traffic in support of coastal communities or resource development. With a

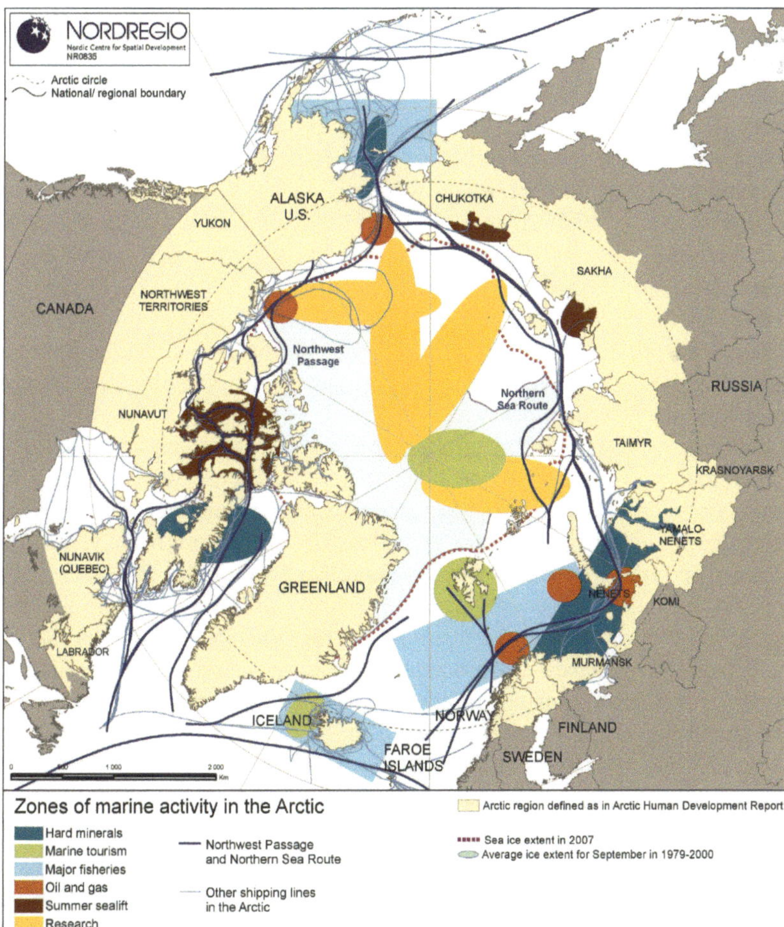

FIGURE 13.7 Map of marine activity in the Arctic. *Sources: Grid-Arendal, Arctic Council, 2009; Arctic Marine Shipping Assessment 2009 Report. Original by H. Ahlenius. Shipping Lines from Arctic Data Portal. Adapted by NordRegio (2011). Design by J. Sterling.*

typical lifetime of mines or offshore oil and gas fields on the order of one to three decades, and with offshore resource exploration efforts underway in all Arctic coastal states, the picture of potential areas of increased risk is likely to persist and evolve gradually over the coming years. Transarctic shipping, along the Northern Sea Route or through the Northwest Arctic Passage (Figure 13.7) is currently of little economic importance, but may increase in significance over coming decades (Brigham, 2010).

In terms of risks emanating from human activities in the Arctic, the infrastructure and activity map shown in Figure 13.7 represents a coarse measure of both the probability of occurrence and the magnitude of the consequences of failure or disaster initiating and escalating events, following the concepts laid out in the Introduction and in Figure 13.1. However, it is only in combination with the key sea-ice processes and properties governing its role as a hazard that a full measure of the regional distribution of hazards and risks can be obtained. The distributions of these latter controlling factors are represented by the maps shown in Figures 13.2−13.5. Most importantly, they include a measure of the probability of encountering sea ice, implicit in the ice concentration trend and ice age distribution maps. Even traces of ice can constitute a major shipping hazard if conditions prevent visual or ship-radar detection. However, as discussed above, the satellite data entering into Figures 13.2 and 13.3 are not of a sufficient resolution and accuracy to provide information on the distribution of such sparse ice (Meier and Stroeve, 2008; Eicken et al., 2011). Thus, areas shown in blue in Figure 13.2 on the right can be considered to contain significant amounts of potential ice hazards throughout much of the open water season.

Secondary factors that determine the severity of a sea-ice hazard as encountered by a vessel or impacting offshore structures relate to the distribution of old ice of high mechanical strength and greater thickness (Figure 13.3). A further measure of the distribution or origin of very thick, deformed ice is the deformation map shown in Figure 13.5. Thick ice, both level old ice and deformed younger ice, is likely to linger long into the melt season. In conjunction with the ice velocity maps shown in Figure 13.4, this accounts for the increased risk associated with massive ice floes or remnants of ice ridges drifting into regions of active offshore oil and gas exploration, such as in the northwestern Alaska Arctic. In the summer of 2012, such older ice remnants required suspension of drilling operations at a borehole late in the season (U.S. Department of Interior, 2013).

The western Canadian and Alaskan Arctic emerges as a region of higher risk with respect to sea-ice hazards, as illustrated by the band of high deformation comparatively close to the coast (Figure 13.5), the potential for advection of strong, thick multiyear ice even in recent years (Figure 13.3), and the high ice velocities observed over the shelf (Figure 13.4). To be sure, these risks are tempered by greatly reduced ice concentration during the open water season, diminished ice velocities in the summer months, and an overall reduction in the thickness and age of ice in the region. Outside of the Arctic proper, the Bohai Sea, a semienclosed body of water along China's northeastern coast emerges as another region of substantial risk. There, a large fishing fleet, offshore oil and gas production platforms, and substantial maritime traffic operate in a region with seasonal sea ice that can reach level-ice thicknesses of up to 1 m and experiences substantial deformation as the result of strong tidal currents and wind forcing (Yue and Li, 2003; Zhang et al., 2013). Moreover, in recent years ice conditions have become more severe, possibly as a result of summer sea-ice loss in the Arctic and its impact on midlatitude weather patterns.

Another type of human activity that is potentially threatened by sea-ice hazards is the use of the coastal ice cover by the indigenous peoples of the Arctic and other northern residents (Eicken et al., 2009). In particular, the shorefast ice serves as a platform and natural extension of the land to provide access to marine waters for subsistence hunting and fishing, as well as an effective transportation corridor between different communities (Krupnik et al., 2010). In addition, ice roads across shorefast ice represent seasonal infrastructure used by the oil industry for nearshore operations (Masterson, 2009). A less stable, less predictable and less trafficable ice cover presents coastal communities with substantial challenges and risks, including the potential to drift out to sea with poorly anchored shorefast ice or for increased dangers of falls through ice (George et al., 2004; Ford et al., 2008). Northern communities have adapted to such changes, and their responses may hold lessons that are relevant in the broader context of addressing sea-ice hazards.

In the Antarctic, at present, the most substantial risk associated with sea ice as a natural hazard is its impact on tourist and research vessel activity. The sinking of a Canadian cruise ship in 2007 as a result of ice damage to the hull illustrates the level of risk associated with high-traffic areas near the Antarctic Peninsula where most of the tourist traffic converges, and where sea ice and icebergs can linger into late summer (Lynch et al., 2010). The temporary trapping of a charter vessel and a research icebreaker in East Antarctica (in late December 2013; J. Turner, theGuardian.com, January 3, 2014) occurred in a region where prevailing patterns of ice movement and deformation (similar to the deformation zones highlighted in Figure 13.5) had resulted in besetting of vessels by ice in the past.

Although thick ice and icebergs represent the greatest risk for structural damage to vessels and other marine assets, it is important to recognize that any ice can be problematic for systems that are designed only for open water. For example, although slush ice may represent no hazard for ice-strengthened structures, it can swamp boats or clog intake ports leading to mechanical failures. Indeed the earliest onset of freeze-up is what defines the end of the offshore drilling season in the Alaska Arctic. Capabilities for predicting and tracking the onset of freeze-up are not currently well developed, which reduces our ability to mitigate the associated risks.

13.4 CASE STUDY: ICE HAZARDS IN THE BEAUFORT AND CHUKCHI SEAS

The Beaufort and Chukchi Seas, comprising Canadian and US waters, provide important sea-ice habitat to a range of organisms, from ice algae to marine mammals. The indigenous population of coastal communities relies on the ice cover as a platform for transportation and subsistence hunting. The region has also seen substantial increases in maritime traffic and offshore oil and gas exploration in recent years, in particular north of the Northwest Territories and on the Northwest Alaskan shelf (areas shown in red in Figure 13.7). Industry may rely on coastal landfast ice as a platform for wintertime exploration and supply of nearshore production platforms. In the summer months and shoulder seasons, sea ice presents a potential hazard to offshore activities.

In evaluating marine ice hazards in this region, it is important to recognize that the use of landfast ice by coastal communities and industry as well as the ecological significance of drifting and landfast sea ice as a habitat benefit from the very types of processes and ice features that represent a hazard to coastal and offshore infrastructure and shipping (see also the summary in Table 13.1). Specifically, the advection of thick multiyear ice from the High Canadian Arctic

TABLE 13.1 Ice-Associated Threats to Maritime Activities in Beaufort and Chukchi Seas (modified from Eicken et al., 2011)

Activity or Asset	Hazard or Threat	Relevant Variable	Setting/Scale
Shipping	Ice contact & damage	Ice concentration, ice type	Marginal ice zone, 10–100 s km
Use of ice as platform	Ice breakout or breaking through of personnel & equipment	Landfast ice stability/ anchoring strength, thickness, morphology	Landfast ice, <1–10 s km
Coastal & offshore infrastructure	Loading, impact & damage by drifting ice, ice push & gouging	Ice velocity, floe size, thickness, ice type/ strength	Coastal & offshore drift ice, <1–10 s km
Emergency/spill response	Oil spill or vessel sinking in ice, dispersal of contaminants by ice	Ice velocity, trajectory of contaminated ice, morphology	Landfast & drift ice, <1–100 s km
Maritime activities	Slush ice build-up, clogging of intake valves etc.	Freeze-up onset, frazil ice formation	Coastal ocean during freeze-up, <1–100 s km

into the region with the prevailing patterns of ice drift (Figures 13.3 and 13.4) as well as the local production of highly deformed ice greatly exceeding level-ice thickness (Figure 13.5) provide potential anchor points that ground the ice in coastal waters and help stabilize the shorefast ice (Mahoney et al., 2014). Uses of the shorefast ice greatly benefit from increased stability, in particular since overall reductions in the extent of multiyear ice in recent years (Figure 13.2) have led to increases in early-season landfast ice breakout events that have required rescue operations to retrieve people and equipment (George et al., 2004).

At the same time, thick multiyear ice and deep-draft pressure ridges or rubbled ice take longer to melt in the summer and present a lingering obstacle to vessel traffic. Although the latter hazard is easily identified and avoided, a potentially bigger challenge involves the threat of such massive ice to drilling platforms and buried pipelines. In the Beaufort and Chukchi Seas, both local and distal sources of such ice occur. The latter include multiyear ice, some of which originates from within the Canadian Archipelago where level ice can reach >10 m in thickness (Johnston et al., 2009), as well as fragments of deteriorating ice shelves and tidewater glaciers in the High Canadian Arctic with thicknesses of several tens of meters (Mueller et al., 2008; Barber et al., 2014). Such high-draft features are driven by a combination of ocean currents, motion of the surrounding pack, and surface winds. With high ice velocities and current speeds (tens of centimeters per second to well above 1 m/s) observed both along parts of the coastline and on the outer shelf (Eicken et al., 2011; Barber et al., 2014) substantial damage could result from such features impacting offshore structures or gouging into the seafloor. Moreover, prediction of movement of deep-draft ice features on time scales of hours to days can be challenging because of the complexity of motion patterns observed (Barber et al., 2014). The prevailing circulation patterns in the region bring such potential hazards within reach of offshore exploration areas in both the eastern Beaufort and eastern Chukchi Seas (Figures 13.4 and 13.7). In summer of 2012, an iceberg several hundreds of meters across with accumulations of boulders and gravel at the surface grounded in water of an approximately 26-m depth north of Barrow, Alaska. That same season, a smaller but taller glacial ice fragment grounded in waters roughly 100 km to the Southwest of Barrow (R. Glenn and R. Shears, personal communication, 2012).

The most prominent local source of deep-draft ice features is Hanna Shoal, which rises to water depths slightly >20 m, and is located approximately 300 km to the Northwest of Barrow. This location is in close proximity to oil and gas exploration leases. An aircraft campaign to measure ice thickness using electromagnetic induction techniques (Haas et al., 2010) revealed highly deformed grounded ice feature up to 26 m thick and more than a kilometer across (Figure 13.8). Satellite imagery analyzed for the time period 1993–2010 indicates that such grounding occurs every year, though somewhat later in the season in more recent years (Mahoney et al., 2012). As evident from the thickness profiles shown in Figure 13.8, deformed ice with a thickness

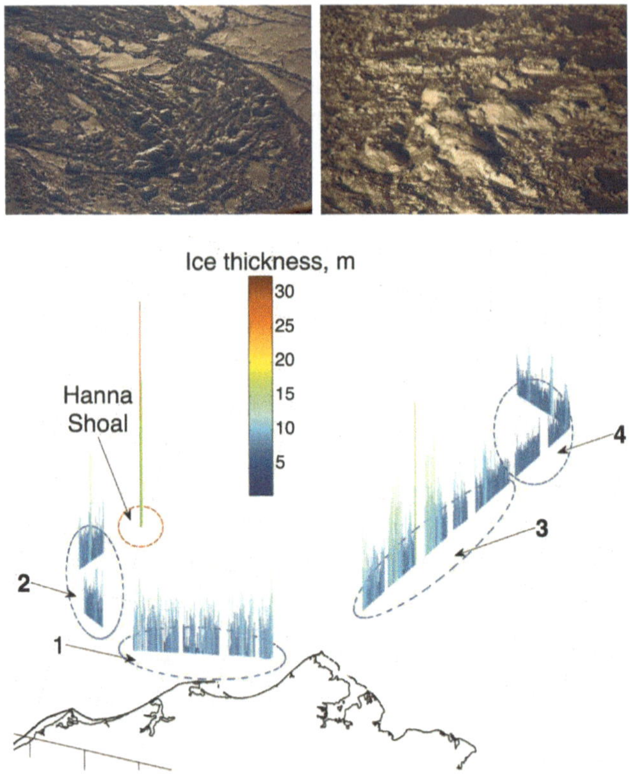

FIGURE 13.8 Ice thickness profiles obtained from airborne electromagnetic sounding (Haas et al., 2010) north of Alaska in April 2011 (location indicated by black rectangle shown in Figure 13.5). Note ice >15 m in thickness in profile section 3 and thicknesses >20 m along a short profile section of a few kilometers' length flown across Hanna Shoal. Much of this ice is heavily deformed and grounded at water depths of <30 m. The photograph at the top left (several hundreds of meters wide) shows a region of highly deformed, grounded ice that is likely helping pin a pan of highly deformed ice roughly 3 km wide in place. The photograph at the top right shows some of the grounded ice features in more detail (roughly hundreds of meters across).

of >15 m can also be found north of Barrow in a zone that corresponds roughly to the band of strong deformation evident in Figure 13.5. This band corresponds to strong shear deformation resulting in the ridging and rubbling of ice as the main body of the pack ice moves past the northernmost tip of the Alaska landmass (Figure 13.4). We interpret the annual grounding of ice at Hanna Shoal to imply that the probability of producing deep-draft ice in this region of strong shear is high enough to result in initial grounding of a pressure ridge at Hanna Shoal every year. This grounded piece of ice then forms the nucleus for further deformation and subduction of strongly deformed ice, resulting in the scalloped, deep-draft features shown in Figure 13.8. Such a local source of deep-draft ice that is known to linger in the region into summer is a potential threat to summer operations as well as buried pipelines that are part of one scenario for potential development of offshore Chukchi oil and gas fields.

The probability of encountering lingering deep-draft and multiyear ice later in summer is still significant, even with observed reductions in summer ice concentration. In 2012, during exploratory drilling near Hanna Shoal in the first half of September, a field of sea ice several tens of kilometers in size drifted into the region and threatened operations. Based on ice forecasting and remote-sensing data analysis by the operator, a potential threat was identified and the drill vessel pulled offsite for a period of approximately two weeks (U.S. Department of Interior, 2013). It is likely that such lingering ice included thick multiyear ice entering the region from the East (Figure 13.3) and persisting well into summer due to greater thickness and reduced melt rates compared to darker, thinner first-year ice. Changes in ice conditions upstream of the region are projected to continue and local deformation patterns seem likely to persist in coming decades. Thus, the Beaufort and Chukchi Seas are likely to retain a full array of ice hazards (Table 13.1). Moreover, factors such as substantial swell or fine-scale ice drift patterns that are not captured by remote-sensing products are likely to increase the overall risk to maritime operations in the region. At the same time, ice use by marine mammals and subsistence hunters will be concentrated into smaller areas and shorter time periods, further affecting key risk factors as outlined in Section 13.1.

13.5 SEA-ICE HAZARDS AND DISASTERS: PREVENTION, MITIGATION, AND RESPONSE

With a variety of controlling factors playing into the potential occurrence and escalation of initiating events (Figure 13.1), prevention of disasters and mitigation of hazards require comprehensive approaches that address a multitude of factors. Here, we briefly discuss the importance of collecting and providing sea-ice information relevant to disaster prevention and response. As a first step, the compilation of hazard and risk maps may help in long-term planning and optimal coordination of emergency response assets. At present, we are not aware of such products being available for broader sectors of the Arctic, although national ice forecasting services may generate local maps that indicate, for example, ice severity as a hazard indicator (e.g., the Chinese Marine Environmental Forecasting Center's Ice Severity Zones, Zhang et al., 2013; or the Barnett Ice Severity Index used in the United States, Eicken et al., 2009). The information summarized in Figures 13.2–13.5, however, would likely figure prominently in the development of such an index. Since such maps can be assembled based largely on satellite-remote sensing and some surface-based measurements, they are of value both as indicators of "normal" hazard conditions and to track specific conditions in a particular year or vulnerable region. At the same time, the type of human activity, infrastructure, or ecosystem service that is potentially impacted also figures into the evaluation of hazard magnitudes. Thus, while old, thick ice constitutes a major hazard for ship traffic and offshore structures, it actually helps stabilize shorefast ice in nearshore regions and thus reduces the risk of catastrophic ice breakout events. Hence, the intertwining of the different controlling factors shown in Figure 13.1 needs to be taken into account when developing hazard maps or ice severity indices.

At present, there is the potential for converging interests and activities related to tracking of environmental variables relevant from a hazards perspective and action by response agencies to address emergency response in the maritime sector across the Arctic. Thus, the International Polar Year 2007–2008 resulted in a major push toward distributed polar environmental observing networks (Krupnik et al., 2011). In the Arctic, this effort has culminated in a number of sustained national and international observing programs, such as the US interagency Arctic Observing Network (IARPC, 2007). While these networks typically comprise a strong academic research component as well as agency–driven activities, they represent a repository of data and information that is critical in the context of environmental security and emergency response. The Arctic coastal states and other nations engaged in Arctic research recognized the importance of such observing efforts and have taken steps toward sustaining such activities at the Arctic Council working group level. At the same time, through the Arctic Council, the coastal states recently have reached agreements on Cooperation on Aeronautical and Maritime Search and Rescue in the Arctic as well as on Cooperation on Marine Oil Pollution Preparedness in the Arctic (www.arctic-council.org/eppr/). Further work is required to ensure that these efforts result in coordinated hazard tracking and response capabilities that are effective in preventing accidents or disasters.

The scientific community has an important role to play in this context, since at present much of the observing capacity in the marine sector of the Arctic is driven by research projects as much as by operational activities. While ocean observing systems developed for lower latitudes have addressed a number of issues central to this challenge (e.g., Schofield et al., 2002), the Arctic—more so than any other region—requires a hybrid approach that meets information needs of the scientific community, Arctic residents and other stakeholders. In this context, two aspects of the problem are deserving of particular notice. First, as discussed in the context of the satellite data shown in Figures 13.2 and 13.3, climate data records and other scientific data products often fail to meet the requirements of operational products meant to inform hazard assessments and emergency response. These issues can be addressed by combining in situ and remote-sensing measurements at the appropriate spatial and temporal scale, such as explored by Barber et al. (2014) in the Canadian Arctic. In a demonstration project for a coastal ice observatory at Barrow, Alaska, we have explored how the combination of ice-based autonomous sensors, marine radar, and remote sensing can serve the needs of multiple ice users and stakeholders (Druckenmiller et al., 2009; Eicken et al., 2011). An important component of this approach is the inclusion of local and indigenous knowledge, in particular since local experts often have considerable expertise and insight into the evaluation of environmental hazards. For example, local observations by Inupiaq ice experts have demonstrated how certain types of coastal ice formations, so-called ice berms, can serve as a natural buffer that can help mitigate some of the wave and ice action on the shoreline and coastal structures. In summary, we see a need for collaborative approaches in hazard identification and management that build on existing and new partnerships emerging in a rapidly changing Arctic.

REFERENCES

Asplin, M.G., Galley, R., Barber, D.G., Prinsenberg, S., 2012. Fracture of summer perennial sea ice by ocean swell as a result of Arctic storms. J. Geophys. Res. 117 http://dx.doi.org/10.1029/2011jc007221.

Atkinson, D.E., 2005. Observed storminess patterns and trends in the circum-Arctic coastal regime. Geo-Mar. Lett. 25, 98–109.

Barber, D.G., McCulloch, G., Babb, D., Komarov, A.S., Candlish, L.M., Lukovich, J.V., Asplin, M., Prinsenberg, S., Dmitrenko, I., Rysgaard, S., 2014. Climate change and ice hazards in the Beaufort Sea. Elementa 2, 10.12952/journal.elementa.000025.

Brigham, L.W., 2010. The fast-changing maritime Arctic. Proc. U.S. Nav. Inst. 136, 55–59.

Comiso, J.C., 2010. Variability and trends of the global sea ice cover. In: Thomas, D.N., Dieckmann, G.S. (Eds.), Sea Ice – An Introduction to Its Physics, Biology, Chemistry and Geology, second ed. Blackwell Science, London, pp. 205–246.

Druckenmiller, M.L., Eicken, H., Pringle, D.J., Williams, C.C., Johnson, M.A., 2009. Towards an integrated coastal sea-ice observatory: system components and a case study at Barrow, Alaska. Cold Reg. Sci. Tech. 56, 61–72.

Eicken, H., Jones, J., MV, R., Kambhamettu, C., Meyer, F., Mahoney, A., Druckenmiller, M.L., 2011. Environmental security in Arctic ice-covered seas: from strategy to tactics of hazard identification and emergency response. Mar. Technol. Soc. J. 45, 37–48.

Eicken, H., Lovecraft, A.L., Druckenmiller, M., 2009. Sea-ice system services: a framework to help identify and meet information needs relevant for Arctic observing networks. Arctic 62, 119–136.

Euskirchen, E.S., Goodstein, E.S., Huntington, H.P., 2013. An estimated cost of lost climate regulation services caused by thawing of the Arctic cryosphere. Ecol. Appl. 23, 1869–1880.

Fetterer, F., Knowles, K., Meier, W., Savoie, M., 2013. Sea Ice Index. Colorado USA, Boulder. National Snow and Ice Data Center. http://dx.doi.org/10.7265/N5QJ7F7W.

Ford, J.D., Pearce, T., Gilligan, J., Smit, B., Oakes, J., 2008. Climate change and hazards associated with ice use in Northern Canada. Arct. Antarct. Alp. Res. 40, 647–659.

Francis, O.P., Atkinson, D.E., 2012. Synoptic forcing of wave states in the southeast Chukchi Sea, Alaska, at nearshore locations. Nat. Hazards 62, 1273–1300.

Francis, J.A., Vavrus, S.J., 2012. Evidence linking Arctic amplification to extreme weather in mid-latitudes. Geophys. Res. Lett. 39, L06801. http://dx.doi.org/10.1029/2012GL051000.

George, J.C., Huntington, H.P., Brewster, K., Eicken, H., Norton, D.W., Glenn, R., 2004. Observations on shorefast ice dynamics in Arctic Alaska and the responses of the Iñupiat hunting community. Arctic 57, 363–374.

Haas, C., Hendricks, S., Eicken, H., Herber, A., 2010. Synoptic airborne thickness surveys reveal state of Arctic sea ice cover. Geophys. Res. Lett. 37, L09501. http://dx.doi.org/10.1029/2010GL042652.

Herman, A., Glowacki, O., 2012. Variability of sea ice deformation rates in the Arctic and their relationship with basin-scale wind forcing. Cryosphere 6, 1553–1559.

Interagency Arctic Research Policy Committee (IARPC), 2007. Arctic observing network: toward a U.S. contribution to pan-Arctic observing. Arct. Res. U.S. 21, 1–94.

IPCC, 2012. Managing the risks of extreme events and disasters to advance climate change adaptation. In: Field, C.B., Barros, V., Stocker, T.F., Qin, D., Dokken, D.J., Ebi, K.L., Mastrandrea, M.D., Mach, K.J., Plattner, G.-K., Allen, S.K., Tignor, M., Midgley, P.M. (Eds.), A Special Report of Working Groups I and II of the Intergovernmental Panel on Climate Change. Cambridge University Press, Cambridge, UK.

Johnston, M., Masterson, D., Wright, B., 2009. Multi-year ice thickness: knowns and unknowns. In: Conference on Port and Ocean Engineering under Arctic Conditions (POAC), Luleå, Sweden, POAC09-120.

Krupnik, I., Allison, I., Bell, R., Cutler, P., Hik, D., López-Martínez, J., Rachold, V., Sarukhanian, E., Summerhayes, C., 2011. Understanding Earth's Polar Challenges: International Polar Year 2007–2008. University of the Arctic, Rovaniemi, Finland.

Krupnik, I., Aporta, C., Gearheard, S., Kielsen Holm, L., Laidler, G., 2010. SIKU: Knowing Our Ice – Documenting Inuit Sea Ice Knowledge and Use. Springer, New York.

Kwok, R., Spreen, G., Pang, S., 2013. Arctic sea ice circulation and drift speed: decadal trends and ocean currents. J. Geophys. Res. Oceans 118, 2408–2425.

Lovecraft, A.L., Eicken, H., 2011. North by 2020: Perspectives on Alaska's Changing Social–ecological Systems. University of Alaska Press, Fairbanks.

Lyard, F.H., 1997. The tides in the Arctic Ocean from a finite element model. J. Geophys. Res. 102, 15611–15638.

Lynch, H.J., Crosbie, K., Fagan, W.F., Naveen, R., 2010. Spatial patterns of tour ship traffic in the Antarctic Peninsula region. Antarct. Sci. 22, 123–130.

Mahoney, A.R., Eicken, H., Shapiro, L.H., Gens, R., Heinrichs, T., Meyer, F.J., Gaylord, A., 2012. Mapping and Characterization of Recurring Spring Leads and Landfast Ice in the Beaufort and Chukchi Seas. Final report, OCS Study BOEM 2012-067.

Mahoney, A.R., Eicken, H., Gaylord, A.G., Gens, R., 2014. Landfast sea ice extent in the Chukchi and Beaufort Seas: the annual cycle and decadal variability. Cold Reg. Sci. Tech. 103, 41–56. http://dx.doi.org/10.1016/j.coldregions.2014.03.003.

Mahoney, A., Eicken, H., Shapiro, L.H., Grenfell, T.C., 2004. Ice motion and driving forces during a spring ice shove on the Alaskan Chukchi coast. J. Glaciol. 50, 195–207.

Maslanik, J.A., Fowler, C., Stroeve, J., Drobot, S., Zwally, J., Yi, D., Emery, W., 2007. A younger, thinner Arctic ice cover: increased potential for rapid, extensive sea-ice loss. Geophys. Res. Lett. 34 (24). http://dx.doi.org/10.1029/2007gl032043.

Masterson, D.M., 2009. State of the art of ice bearing capacity and ice construction. Cold Reg. Sci. Technol. 58, 99–112.

Meier, W.N., Stroeve, J., 2008. Comparison of sea-ice extent and ice-edge location estimates from passive microwave and enhanced-resolution scatterometer data. Ann. Glaciol. 48, 65–70.

Mueller, D.R., Copland, L., Hamilton, A., Stern, D., 2008. Examining Arctic ice shelves prior to the 2008 breakup. EOS Trans. Am. Geophys. Union 89, 502–503.

NordRegio, 2011. Megatrends. Nordic Council of Ministers, Copenhagen.

Ogorodov, S.A., Kamalov, A.M., Zubakin, G.K., Gudoshnikov, Y.P., 2005. The role of sea ice in coastal and bottom dynamics in the Pechora Sea. Geo-Mar. Lett. 25, 146–152.

Overeem, I., Anderson, R.S., Wobus, C.W., Clow, G.D., Urban, F.E., Matell, N., 2011. Sea ice loss enhances wave action at the Arctic coast. Geophys. Res. Lett. 38, L17503. http://dx.doi.org/10.1029/2011GL048681.

Perovich, D.K., Light, B., Eicken, H., Jones, K.F., Runciman, K., Nghiem, S.V., 2007. Increasing solar heating of the Arctic Ocean and adjacent seas, 1979–2005: attribution and role in the ice-albedo feedback. Geophys. Res. Lett. 34, L19505. http://dx.doi.org/10.1029/2007GL031480.

Pizzolato, L., Howell, S.E.L., Derksen, C., Dawson, J., Copland, L., 2014. Changing sea ice conditions and marine transportation activity in Canadian Arctic waters between 1990 and 2012. Clim. Change 122. http://dx.doi.org/10.1007/s10584-10013-11038-10583.

Rampal, P., Weiss, J., Marsan, D., 2009. Positive trend in the mean speed and deformation rate of Arctic sea ice, 1979–2007. J. Geophys. Res. Oceans 114, C05013. http://dx.doi.org/10.1029/2008JC005066.

Ravens, T.M., Jones, B.M., Zhang, J.L., Arp, C.D., Schmutz, J.A., 2012. Process-based coastal erosion modeling for Drew Point, North Slope, Alaska. J. Waterw. Port Coastal Eng. ASCE 138, 122–130.

Schneider, S.H., Semenov, S., Patwardhan, A., Burton, I., Magadza, C.H.D., Oppenheimer, M., Pittock, A.B., Rahman, A., Smith, J.B., Suarez, A., Yamin, F., 2007. Assessing key vulnerabilities and the risk from climate change. In: Parry, M.L., Canziani, O.F., Palutikof, J.P., van der Linden, P.J., Hanson, C.E. (Eds.), Climate Change 2007: Impacts, Adaptation, and Vulnerability. Contribution of Working Group II to the Fourth Assessment Report of the Intergovernmental Panel on Climate Change. Cambridge University Press, Cambridge, pp. 779–810.

Schofield, O., Bergmann, T., Bissett, P., Grassle, J.F., Haidvogel, D.B., Kohut, J., Moline, M., Glenn, S.M., 2002. The long-term ecosystem observatory: an integrated coastal observatory. IEEE J. Oceanic Eng. 27, 146–154.

Screen, J.A., Simmonds, I., 2013. Exploring links between Arctic amplification and mid-latitude weather. Geophys. Res. Lett. 40, 959–964.

Timco, G.W., Weeks, W.F., 2010. A review of the engineering properties of sea ice. Cold Reg. Sci. Technol. 60, 107–129.

U.S. Department of Interior, 2013. Review of Shell's 2012 Alaska Offshore Oil and Gas Exploration Program. http://www.doi.gov/news/pressreleases/upload/Shell-report-3-8-13-Final.pdf. retrieved March 2014.

Vinnem, J.E., 2007. Offshore Risk Assessment – Principles, Modelling and Applications of QRA Studies, second ed. Springer-Verlag, London.

Yue, Q.J., Li, L., 2003. Ice problems in Bohai Sea oil exploitation. In: 17th International Conference on Port and Ocean Engineering under Arctic Conditions (POAC), Trondheim, Norway, June 16–19, 2003, pp. 151–163.

Zhang, X.L., Zhang, Z.H., Xu, Z.J., Li, G., Sun, Q., Hou, X.J., 2013. Sea ice disasters and their impacts since 2000 in Laizhou Bay of Bohai Sea, China. Nat. Hazards 65, 27–40.

Chapter 14

Mangroves, Tropical Cyclones, and Coastal Hazard Risk Reduction

Anna McIvor[1], Thomas Spencer[1], Mark Spalding[2], Carmen Lacambra[3] and Iris Möller[1]

[1] *Cambridge Coastal Research Unit, Department of Geography, University of Cambridge, Cambridge, UK,* [2] *The Nature Conservancy, Department of Zoology, University of Cambridge, Cambridge, UK,* [3] *Grupo Laera, Bogotá, Colombia*

ABSTRACT

Risks from coastal hazards to people and property are expected to increase with near-future sea level rise, changes in storminess, and increasing coastal populations. Evidence from empirical and modeling studies suggests that mangrove forest vegetation can reduce storm surge peak waters levels where mangroves are present over sufficiently large areas. Mangroves are best used alongside other risk reduction measures (embankments, early warning systems) to ensure the lowest possible level of residual risk.

14.1 INTRODUCTION

Risks to lives and livelihoods at the coast, and coastal flood damages, are expected to increase significantly during the twenty-first century with sea level rise (Jevrejeva et al., 2012; Church et al., 2013), possible changes in storminess and potential increases in cyclone intensity (Khairoutdinov and Emanuel, 2013; Woodruff et al., 2013), and increasing population and asset values in the world's coastal lowlands (Seto, 2011; Mendelsohn et al., 2012). A recent modeling study has predicted that, given the maintenance of current sea defenses, but depending on the near-future sea level rise projection used, 0.2–4.6 percent of the global population will be flooded annually by 2100 under 0.25–1.23 m of global mean sea level rise, with expected annual losses of 0.3–9.3 percent of the global gross domestic product (Hinkel et al., 2014). The model suggests that the global costs of protecting the coast with dikes

Coastal and Marine Hazards, Risks, and Disasters. http://dx.doi.org/10.1016/B978-0-12-396483-0.00014-5
Copyright © 2015 Elsevier Inc. All rights reserved.

FIGURE 14.1 (a) Upper delta mangrove forest, Berau River, East Kalimantan, Indonesia; highly diverse associations including *Heritiera littoralis*, *Xylocarpus mollucensis*, *Sonneratia caseolaris*, the mangrove trumpet tree *Dolichandrone spathacea*, the spiny holly mangrove *Acanthus ilicifolius*, the palm *Nypa fruticans*, and the fern *Acrostichum aureum (photograph: M. Spalding)* (b) *Sonneratia alba (photograph: M. Spalding)*; (c) *Rhizophora* mangrove species, Sungei Buloh Wetland Reserve, Singapore Island *(photograph: T. Spencer)*; and (d) *Avicennia germinans*, Salamanca National Park, Colombia *(photograph: C. Lacambra)*.

would require annual investment and maintenance costs of US$12–71 billion while at the same time increasing the risk of catastrophic consequences in the case of the failure of these new defenses (Hinkel et al., 2014). These scenarios point to the need for alternative, long-term coastal adaptation strategies which go beyond traditional engineering solutions. If the goal of coastal management

FIGURE 14.1 (*Contiuned*)

is to reduce risk to acceptable levels of residual risk, then a much wider range of risk reduction methods should be considered, beyond only considering structural scenarios. In this context, the role of coastal ecosystems in natural coastal protection should be pursued more vigorously (Spalding et al., 2014). In this chapter we contribute to this debate by reviewing the role of mangrove forests—assemblages of trees and shrubs typical of saline, waterlogged coastal habitats in the tropics and subtropics (Figures 14.1 and 14.2)—in reducing the risks posed to coastal communities by tropical cyclones (also called hurricanes and typhoons).

Recent estimates of the global coverage of mangrove forests, based on the analysis of Landsat satellite imagery, range from 138×10^3 km^2 (Giri et al., 2011) to 152×10^3 km^2 (Spalding et al., 2010). However, these estimations are based on 1993–2003 data and are unlikely to accurately reflect current coverage. Mangrove loss rates have been estimated at 0.66 percent per year, with 20–35 percent of the world's mangrove area disappearing since the 1980s (FAO, 2007). Although 25 percent of all mangroves occur in protected areas, rates of loss appear highest in less developed countries where mangroves are being cleared for coastal development, aquaculture, timber, and fuel production (Spalding et al., 2014). The case for the importance of retaining mangrove forests has focused on the multiple social and economic benefits that are likely to be derived from the range of ecosystem services that they provide: fisheries, carbon cycling and sequestration, water purification, and high biodiversity (e.g., Sathirathai and Barbier, 2001; Gunawardena and Rowan, 2005; Barbier et al., 2011; Hutchison et al., 2014). A particularly strong argument, however, has been made for mangrove protection and management through their potential role as dissipaters of incident wave energy (e.g., Badola and Husain, 2005), in relation to storm surges, and in response to tsunami impacts, the last of these three being brought into sharp focus by the Asian tsunami of December 2004.[1]

The mixed messages from the attempts to assess the role of mangroves in mitigating the impact of this tsunami event (e.g., Cochard et al., 2008) provide one example of the underpinning lack of basic information regarding the level of coastal protection that mangrove forests can provide in the face of coastal hazards. Recently, a small number of studies have started to address this need.

1. We do not consider the role of mangroves in reducing the long-period wave trains associated with tsunamis in this chapter. The nature of these impacts has been extensively described elsewhere (e.g., Alongi, 2008; Tanaka et al., 2006; Tanaka, 2009). Following the 2004 Asian tsunami, numerous publications (e.g., Wells and Kapos, 2006; Chatenoux and Peduzzi, 2007; Spencer, 2007; Cochard et al., 2008) attempted to make sense of localized reports of reduced impacts behind vegetation (e.g., Kathiresan and Rajendra, 2005; Danielsen et al., 2005). Much controversy has ensued over the nature of such linkages (e.g., Kerr et al., 2006; Kerr and Baird, 2007; Baird et al., 2009; Feagin et al., 2010), and no consensus has, as yet, been reached. More recently, a large-scale study employing a spatial statistical analysis in Aceh, Sumatra, found that coastal vegetation in front of settlements reduced the number of casualties, whereas coastal vegetation behind settlements had the opposite effect (Laso Bayas et al., 2011). Recent modeling studies have explored the effect of coastal vegetation on various tsunami characteristics (run-up height, flow velocity, inundation extent) using both physical and numerical models (e.g., Apotsos et al., 2011; Ohira et al., 2012; Strusińska-Correia et al., 2013). These studies indicate that coastal vegetation (included in numerical models as an increase in surface roughness) can reduce tsunami run-up height and flow velocities but this depends on tsunami characteristics and local bathymetry (e.g., Apotsos et al., 2011). Furthermore, assessments based on data sets derived from moderate-spatial-resolution (>10 m) satellite sensors (e.g., Landsat TM, Landsat MSS and SPOT XS) (e.g., Iverson and Prasad, 2007) fail to register the finer variation in species composition and tree density that often control extreme event impacts (Dahdouh-Guebas and Koedam, 2006).

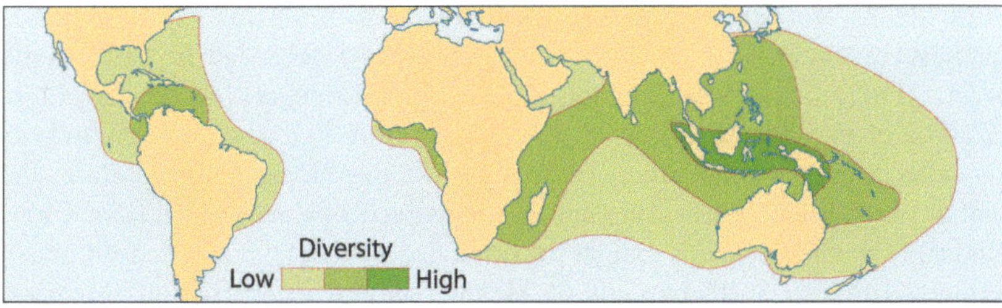

FIGURE 14.2 Global distribution of mangroves *(modified from Veron (1995))*, showing mangrove species diversity. Scale of diversity ranges from 0 to 10 genera (low), 10 to 25 genera (medium), and >50 genera (high). *Adapted from Figure 1.7, Slaymaker, O., Spencer, T., Embleton-Hamann, C., (Eds.), 2009. Geomorphology and Global Environmental Change. Cambridge University Press, Cambridge.*

In particular, Gedan et al. (2011) conducted a broad review of the role of salt marshes and mangroves in coastal protection. They concluded that mangroves and salt marshes can play an important role in reducing risk from coastal hazards. However, they do not address how ecosystems are best incorporated into the design of coastal defense strategies and their implementation. For example, planners and engineers need to know the required mangrove width to reduce a storm surge of a certain height by a certain amount. A review of the evidence for the capacity of mangroves to reduce wave height and storm surge water levels is urgently needed. Here we review studies on the physical processes underlying storm surge reduction, identify important gaps in knowledge, and make some suggestions about the most appropriate ways in which mangroves can be included in coastal defense strategies.

Finally, it is important to note that there are limits to the "biological buffering" capacity of coastal mangroves in relation to storm surge impacts, although at the present time the exact position of these limits in environmental space are poorly known. Cyclones impact mangroves directly through defoliation, branch breakage, toppling, and uprooting (reviewed in Lacambra et al., 2008; Spencer and Möller, 2013) and indirectly, through changes in both tidal and freshwater flushing dynamics and sediment supply (e.g., Paling et al., 2008), processes that disrupt nutrient cycling, and, critically for mangroves, gas exchange between the rhizosphere and the water column/atmosphere (Lugo et al., 1981). Cyclones with typical wind speeds of 120–150 km h^{-1} result in a mosaic of impacted and nonimpacted areas. Damage patterns appear to be related to forest structure, with larger trees more likely to suffer stem breakage or toppling in the path of a cyclone (Roth, 1992; McCoy et al., 1996). However, severe storms, with wind speeds in excess of 200 km h^{-1}, can reduce some areas of mangrove forest cover to little more than residual canopy patches for 50 years or more (Spencer and Möller, 2013). This is partly because such events may lower mangrove surfaces to levels that prevent mangrove seedling reestablishment (Cahoon et al., 2003). However, individual storm tracks are generally narrow (<30 km) and thus damage is invariably spatially restricted. The most intense cyclone ever recorded in the Atlantic basin, Hurricane Wilma (October 2005; up to category 5), destroyed c. 1,250 ha of mangroves (Smith et al., 2009) but this area accounted for only c. 0.4 percent of the total area of mangrove in the Florida. Furthermore, the chance of a particular location being hit in any one cyclone season is very low. Studies of overwash sands and storm surge deposits in lake and coastal marsh sediments in the Northern Gulf of Mexico suggest return periods for catastrophic hurricanes of 300–600 years at particular locations, equating to annual at-a-point landfall probabilities of 0.33–0.6 percent (Liu and Fearn, 2000). Nevertheless, these probabilities rise when one considers that mangrove trees are relatively long-lived in comparison to the time interval between storms. Both Lugo et al. (1976) and Jimenez et al. (1985) have argued that Caribbean mangrove forests can retain a record of past storm impacts in their vegetation canopy structure. Regional maps of historic cyclone tracks across mangrove-populated coasts show a pattern of widespread coverage[2] and when historic frequencies are extrapolated over geological timescales then the total number of landfalls runs into the tens of thousands. Thus, in discussing the Holocene history of the Northern Gulf of Mexico, Conner et al. (1989, p. 46) state "from a long term, regional perspective, hurricanes are not unusual or rare and coastal ecosystems have developed with hurricanes as normal aperiodic events. It is impossible to assess how these systems would have developed without hurricanes, but we believe that they would be different, morphologically and ecologically."

14.2 STORM SURGES

The storm belts of the tropics lie between 5° and 20° north and south of the Equator on the western sides of the ocean basins (Figure 14.3).

In these regions, the high wind stresses experienced over the sea surface and low atmospheric pressures associated with such systems can generate storm surges, i.e., raised water levels, over timescales of hours to days, often well in excess of predicted tidal levels (>1–3 m higher, and >10 m in some extreme cases) (Pugh, 1996; Garrison, 1999; Storch and Woth, 2008) (Figure 14.4). Such events can be particularly marked on coasts with a microtidal range or on coasts with a meso- to macrotidal range when this meteorological forcing coincides with spring high tides (Flather, 2001). Surge events are often enhanced by wind waves, and associated wave run-up, generated by strong onshore winds (Dean and Bender, 2006, Figure 14.4).

2. In a comparable analysis of tropical cyclone impacts on the coral reefs of Australia, Done (1993, p. 126) memorably commented that "maps such as this one (of tropical cyclone paths, 1908–1981) suggests that corals... should have about as much future as a ball of butter in hot spaghetti."

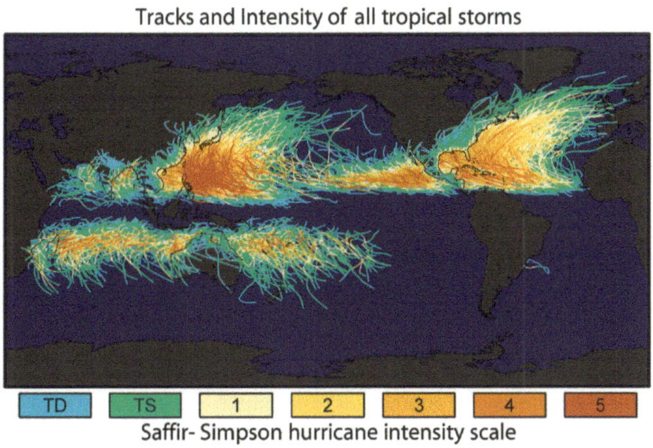

FIGURE 14.3 The tracks of tropical cyclones that formed between 1985 and 2005. The colors represent the strength of the cyclone according to the Saffir–Simpson hurricane wind scale. *Image created by Robert A. Rohde, Global Warming Art; http://www.globalwarmingart.com/wiki/File:Tropical_Storm_Map.png.*

The main atmospheric controls of storm surge height and flood extent include storm intensity, storm size (measured as the radius of maximum wind speed), forward speed of the disturbance, and storm track. Other controls include near-shore bathymetry, coastline geometry (e.g., concave vs convex planform) and orientation, the degree of interconnectivity of coastal water bodies, and the frictional resistance of the land surface (i.e., surface roughness) (Flather, 2001; Dean and Bender, 2006; Resio and Westerink, 2008; Rego and Li, 2009; Spencer and Möller, 2013). Mangroves influence some of these factors; as with

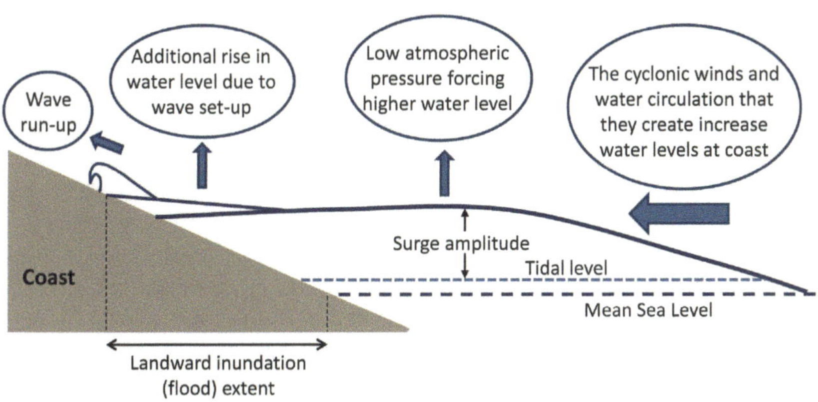

FIGURE 14.4 Schematic diagram showing how storm surges consist of raised water levels at the coast driven by cyclonic winds and low atmospheric pressure. The raised water levels interact with the coastal slope to influence flood extent.

all vegetation, they increase surface roughness (Chow, 1959), reduce the height of surface wind waves (Mazda et al., 2006; Quartel et al., 2007), and reduce the speed of the wind directly over the water surface (Chen et al., 2012) (so long as the vegetation reaches above the water level). Over the longer term (decades to centuries), mangroves can alter the surface elevation of the shore (influencing the bathymetry and topography), the local geometry (e.g., through progradation, the expansion of wetland areas toward the sea), and the location of channels (reviewed in McIvor et al, 2013).

14.3 EVIDENCE FOR REDUCTION OF STORM SURGE IMPACTS BY MANGROVES

Evidence for the ability of mangroves to reduce the impacts of storm surge flooding comes from two sources: (1) direct observations of water levels, and (2) the use of numerical models (with varying degrees of validation) that simulate storm surge behavior in the presence or absence of mangroves.

14.3.1 Observations of Water Level Change

The measurement of storm surge water levels during storms and cyclones presents enormous practical challenges, not least because sensitive (and expensive) equipment may be destroyed or lost during a surge event (Granek and Ruttenberg, 2007). Consequently, very few studies have measured storm surge water levels within mangrove areas. All available measurements are from Southern Florida and hence a rather restricted range of mangrove species compared to the floristically diverse mangrove forests of, for example, Southeast Asia. Here we describe the study by Krauss et al. (2009). Zhang et al. (2012) recorded water level data that they used to validate numerical models of storm surges, and this research is described in Section 14.3.2.

Krauss et al. (2009) analyzed water level measurements collected from a network of water level recorders placed in two wetland areas in Florida during the severe storms of Hurricanes Charley (2004, category 4) and Wilma (2005, up to category 5); Table 14.1 provides detailed information on the recording sites and hurricane characteristics.

As the storm surge from Hurricane Charley passed through the Ten Thousand Islands National Wildlife Refuge, the peak water level reduction was 94 mm km^{-1} through an area that included both mangroves and salt marsh. The following calculations based on data given in Krauss et al. (2009; Figure 14.2 and p. 145) show how the reduction in peak water level through the mangrove area may have been higher. At the first recording point 2.3 km from Faka Union Bay, the peak water level was 786 mm above ground level and 436 mm above the expected high tide level. At the second recording point 3.2 km further inland, at the transition between the mangrove and the marsh, the peak water level was 400 mm above ground level and 296 mm higher than

TABLE 14.1 The Characteristics of Cyclones (in Alphabetical Order), Associated Storm Surges, and the Vegetation they Passed through, which are Discussed in this Review

Location and Source	Cause of the Storm Surge	Wetland type and Width	Water Level Height Reduction if known
Biscayne Bay, east coast of Florida, USA (Xu et al., 2010)	Hurricane Andrew, August 24, 1992, peak wind speed 227 km/h, maximum storm tide 5.2 m	Coastal mangrove zone 1–4 km wide with tree heights of 1–20 m. Species present: *Rhizophora mangle* and *Avicennia germinans* (Smith et al., 1994)	
Ten Thousand Islands National Wildlife Refuge, Florida, USA (Krauss et al., 2009)	Hurricane Charley, August 13, 2004, max winds 240 km/h, peak water level traveled at 0.4 km/h	Mangrove/interior marsh community; dominant species *R. mangle*. Mangrove width 3.2 km	9.4 cm/km across whole area (15.8 cm/km in mangrove area)
Mathbaria, south-western coast of Bangladesh (Tanaka, 2008; ITJSCE, 2008)	Cyclone Sidr, November 15, 2007, maximum wind speed 250 km/h, water levels raised by about 4 m (ITJSCE, 2008)	Forested area, approximately 150 m in width, non-mangrove species *Casuarina equisetifolia* (Tanaka, 2008)	
Along Shark River (Everglades National Park) in Florida, USA (Krauss et al., 2009)	Hurricane Wilma, October 24, 2005, with maximum winds of 195 km/h, peak water traveled at 1.4 km/h, peak water level 5 m	Riverine mangrove, dominant species *R. mangle* (Chen and Twilley, 1999). Distance through mangroves: 14.1 km measured along the Shark River	4.2 cm/km (lower stretch: −0.2 cm/km; upper stretch: 6.9 cm/km)
Gulf Coast, Florida, from Sanibel West to Key West, USA (Zhang et al., 2012)	Hurricane Wilma October 24, 2005, with maximum winds of 195 km/h, peak water level 5 m	Dominant species *R. mangle*, *Laguncularia racemosa*, *A. germinans*. Trees 4–18 m high, stem diameters 5–60 cm. Mangrove width 6–30 km	Models suggest 23–48 cm/km through mangrove area (validated with recorded water levels)

the water level prior to the arrival of the storm surge. This implies a decrease in peak water level of 140 mm (reduction in water level relative to high tide/antecedent water levels) over 0.9 km, equivalent to a reduction in peak water level through mangrove forest of 158 mm km^{-1}.

As the storm surge from Hurricane Wilma passed through the mangrove forest along the Shark River in the Everglades National Park, three recording stations set back from the river by 50–80 m measured a 42 mm km^{-1} peak water level reduction (Krauss et al., 2009). The highest water level reduction was between the two inland stations that were located 9.9 and 18.2 km from the mouth of the river: peak water level fell from 1.040 to 0.462 m, equivalent to a peak water level reduction of 69 mm km^{-1}. Between the seaward recording stations located 4.1 and 9.9 km from the river mouth, there was a slight increase in water level, presumably because of river water backing up behind the surge (Krauss et al., 2009).

Krauss et al. (2009, pp. 147–148) concluded by pointing out that "while our observations indicate that water levels were reduced as storm surge moved though coastal mangrove ecosystems, uncertainty remains over the relative contribution of mangroves over other wetland types, open water or micro-topographic relief along the Gulf Coast over similar distances." It is unclear, therefore, what the exact contribution of mangroves was to the reduction in peak water level, as it is impossible to control for the other factors that may also have affected water level changes. Because of this difficulty, numerical models that include a greater range of controlling factors have an essential role to play in improving storm surge reduction understanding.

14.3.2 Numerical Modeling of Storm Surge Characteristics in the Presence of Mangroves

Numerical models of storm surges offer a complementary approach to exploring the role of mangroves in reducing storm surge water levels (Resio and Westerink, 2008). When such models can be shown to accurately represent storm surge behavior in the presence of mangroves, they can be used to look at the effect of varying parameters such as the width of the mangrove forest (as described in Section 14.4.1 below). Such models are also needed to predict water level reductions due to existing mangrove forests or planned mangrove restorations, where these are intended for use as part of a coastal defense strategy.

To date, few studies have used numerical modeling approaches to better understand the factors affecting storm surge inundation in mangroves. Here we describe available studies relating specifically to mangroves (the first two studies), as well as two other studies based on models including vegetation that can be regarded as broadly similar to mangrove species. Only in one case is the application related to a less developed country. In such settings, local information is crucially needed to effectively validate models, yet such information is typically very scarce or of doubtful quality.

14.3.2.1 The Eulerian–Lagrangian Circulation Model

Xu et al. (2010) used an unstructured Eulerian–Lagrangian Circulation Model (Zhang et al., 2004; using a multiscale grid with cell sizes increasing nearer the coast, reaching 30 m by 30 m in overland cells) to model the surge resulting from Hurricane Andrew (1992; category 5) at Biscayne Bay, Florida east coast, USA (Table 14.1). They found that their model overestimated peak water levels and flooding extent in the southern part of the bay, an area containing mangrove zones with widths of 1–4 km and tree heights of 1–20 m. This suggested that certain land cover types, in particular the large areas of mangroves, had produced significant effects on flood heights and extent.

Xu et al. (2010) explored the effect of land cover type on flood extent by incorporating a measure of surface roughness into their model. They performed sensitivity experiments using the Manning's roughness coefficient ("Manning's n"), which is a measure of surface roughness (Chow, 1959; Chow et al., 1988; Mattocks and Forbes, 2008). Manning's n was set at 0.05, 0.1, and 0.15 within spatial cells containing mangroves (these values were chosen as being representatives of shrubs, woody wetlands, and dense woods, respectively; Chow, 1959; Mattocks and Forbes, 2008). They found that surge inundation extents (Figure 14.5) varied greatly with Manning's n, and most closely matched the observed debris line when a coefficient of $n = 0.15$ was used. They concluded that changes in roughness due to vegetation can significantly influence the local inundation patterns during storm surges.

14.3.2.2 The Coastal and Estuarine Storm Tide Model

Zhang et al. (2012) used the Coastal and Estuarine Storm Tide (CEST) Model to simulate the passage of Hurricane Wilma (2005, up to category 5; Table 14.1) as it passed over a 200-km stretch of the Gulf Coast of South Florida, USA (the relevant mangrove characteristics are given in Table 14.1). With a maximum storm surge of 5 m, Hurricane Wilma resulted in extensive coastal flooding (Smith et al., 2009).

The geographical extent of mangroves to be used in the model was taken from the National Land Cover Database created by the United States Geological Survey (USGS) in 2001. Following Xu et al. (2010), the drag force from mangroves was included in the model by adjusting Manning's n. While Xu et al. (2010) found that Manning's $n = 0.15$ provided the best fit to observed data, Zhang et al. (2012) reduced the coefficient to $n = 0.14$ because of the large number of lakes, rivers, and creeks inside the mangrove zone in this area.

Simulations were undertaken using: (1) Manning's $n = 0.02$ for all spatial cells (this value is typical of the seabed and thus reflects the "no mangrove" case) and (2) Manning's $n = 0.14$ in cells with mangroves, with other land cover types also included using appropriate values of Manning's n (based on values given in Mattocks and Forbes (2008)). Model outputs were compared with water level data collected by various US Federal Government agencies,

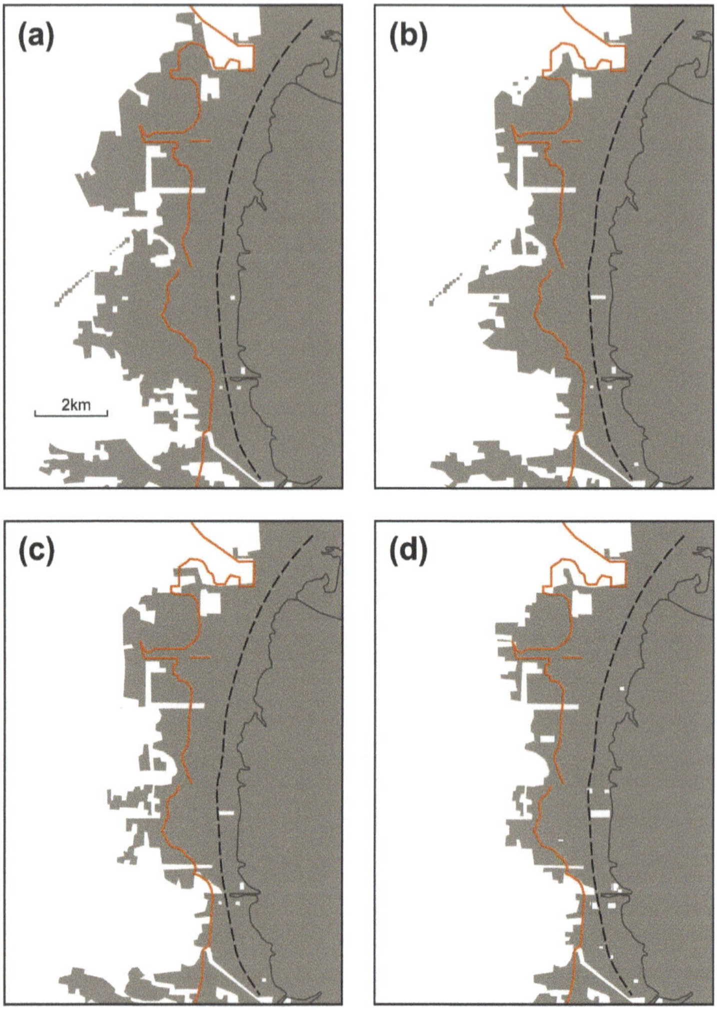

FIGURE 14.5 Flood extents predicted by the Eulerian–Lagrangian Circulation Model *(Xu et al., 2010)* from Hurricane Andrew (1992; category 5) in Biscayne Bay, FL, USA, showing the actual flood extent (red line) and predicted extent from the model, using different values of Manning's n for mangrove areas within the model. (a) $n = 0.02$, (b) $n = 0.05$, (c) $n = 0.1$, and (d) $n = 0.15$. The dashed line shows mangrove extent, and the gray line shows the coastline. *Modified from Xu et al. (2010).*

including the National Oceanic and Atmospheric Administration, USGS, the Federal Emergency Management Agency, and academic researchers. The best match between the simulated water levels and observed water levels was seen when mangroves were included in the model. The root mean square error of computed peak surge heights versus observed ones decreased from 0.60 m (no mangroves) to 0.39 m (with mangroves). Zhang et al. (2012) concluded that including mangroves improved the model's ability to predict storm surge water levels.

The inundation areas predicted by the model were 4,220 km^2 without mangroves, and 2,450 km^2 with mangroves, suggesting that mangroves had a large effect on the inundation extent. Flooding was restricted to within the mangrove zone when mangroves were included in the model (Figure 14.6), and this matched the measured inundation extent taken from surge-induced sediment deposits, which were limited to a zone less than 14 km from the Gulf of Mexico. Storm surge height reduction rates were estimated to have been between 230 and 480 mm km^{-1} across the mangrove areas (Zhang et al., 2012). The simulations indicated that without the presence of a mangrove zone, surge amplitudes would have decreased by 60–100 mm km^{-1}.

However, two further modeling results are noteworthy. First, while the modeled peak water level height was reduced as the storm surge passed through the mangroves, the simulations showed a 10–30 percent increase in water levels in front of the mangrove zone, compared to simulations without mangroves. This is because mangroves can act as an obstruction to the flow of water, causing water levels to build up in front of them. Increased friction within mangroves may also lead to a steeper surge front as the surge moves inland (Resio and Westerink, 2008). Second, Zhang et al.'s (2012) simulations suggested that storm surge reduction was nonlinear across the mangrove width, and this point is discussed in more detail below.

14.3.2.3 Modeling of Wind Wave Set-Up and Set-Down during Surge Events and the Role of Vegetation

In addition to the long-period wave that describes the storm surge itself, short-period wind waves on the water surface often accompany surge events. These wind waves increase the damage caused by the storm surge, and can increase the area that is flooded, through wave set-up and wave run-up. Dean and Bender (2006) used a numerical modeling approach to explore the effect of vegetation (modeled as an array of cylinders) on wave set-up. Their model, based on Airy wave theory (Komar, 1998), predicted that vegetation in shallow water should reduce wave set-up by one-third. Conversely, vegetation in deeper water produces a modeled set-down (i.e., a reduction in water level). The water depth at which this change from a set-up to a set-down occurred can be defined by kh, where k is the wave number ($=2\pi/$wavelength) and h is the still water depth. When waves were modeled using nonlinear wave equations (based on third-order equations, which no longer assume that wave height is small relative to water depth; Stive and Wind, 1982), the presence of vegetation also resulted in a set-down (Dean and Bender, 2006). Dean and Bender's results have not been validated in situ, but they suggest that vegetation, such as mangroves, could have a very large effect on storm surge water levels in those areas where wave set-up makes a

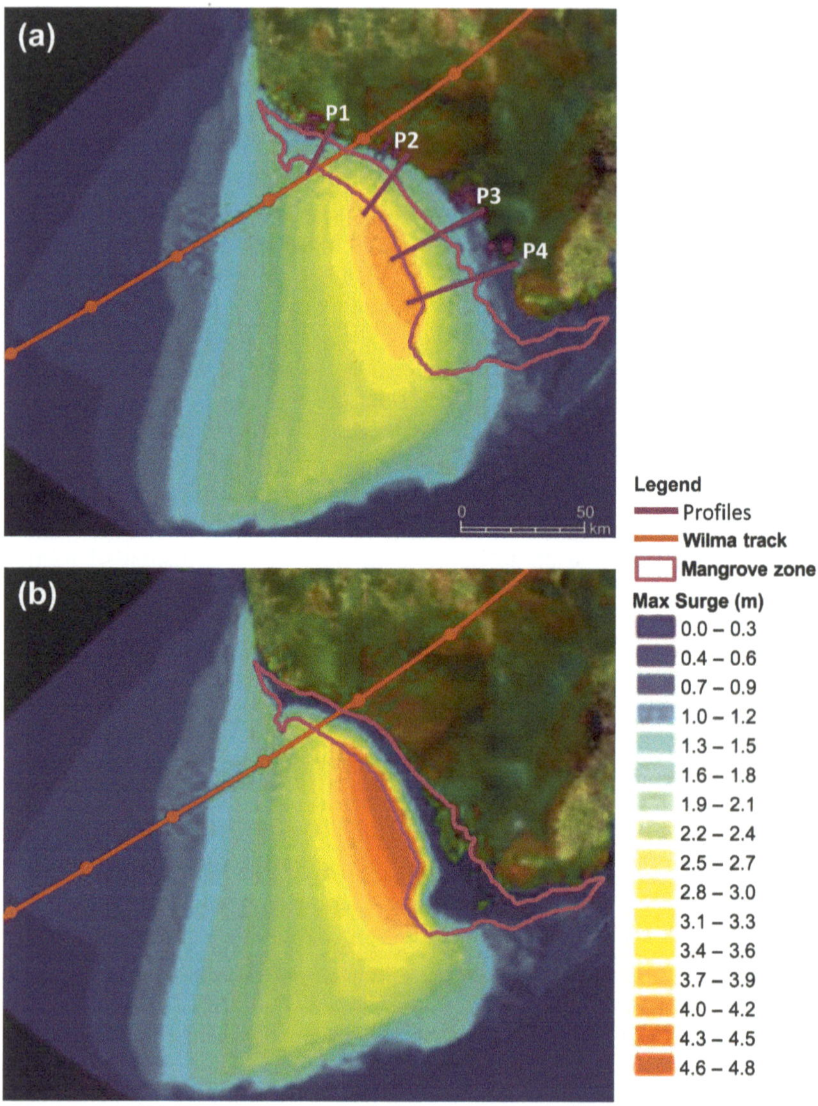

FIGURE 14.6 Peak surge heights over the Gulf Coast of South Florida, USA, from Hurricane Wilma (2005, category 3 on impact) modeled using Coastal and Estuarine Storm Tide *(Zhang et al., 2012)*. (a) Flood extent assuming Manning's $n = 0.02$ across the whole area and (b) flood extent assuming Manning's $n = 0.14$ in mangrove areas (and using appropriate Manning's n values in other land cover types, with land cover taken from the National Land Cover Database). *Adapted from Zhang et al. (2012).*

large contribution to the raised water levels (i.e., areas with relatively narrow continental shelves).

14.3.2.4 Modeling the Storm Surge and Wind Waves from Cyclone Sidr

Tanaka (2008) numerically modeled the storm surge from Cyclone Sidr (Bangladesh, 2007; Table 14.1). A one-dimensional, nonlinear, long wave differential equation was used to explore the effect of a 150-m-wide band of vegetation on a long-period storm surge (wave period 1 or 2 h) and shorter period wind waves (wave period 1 or 2 min). Trees were modeled as cylinders, 10 m high and 0.16 m in diameter, with 0.35 trees m^{-2} in a triangular arrangement. The underlying topography and vegetation measurements matched those seen in transects in Mathbaria, Bangladesh, where the non-mangrove tree *Casuarina equisetifolia* is present.

The model suggested that vegetation had no effect on the water depth ascribed to the long wave component (i.e., the storm surge) over this short distance, although the vegetation slightly decreased the velocity of water inside the vegetation zone and the arrival time of the peak of the storm surge. When wind waves with a period of 1 or 2 min were included on top of the storm surge, the water depth behind the vegetation was reduced by 120 or 280 mm, respectively (compared to no vegetation being present). When the ground slope in the model was reduced from a 1 in 100 gradient to a 1 in 500 gradient, the presence of vegetation reduced maximum water levels by 0.8 m (long wave period 2 h and short wave period 2 min), suggesting that the vegetation reduced flood levels more when present on a shallower slope.

Model results broadly matched field observations: a 150-m-wide band of riverside vegetation near Mathbaria appears to have caused a 0.5–1.0 m difference in water level behind the trees, based on local interviews. This suggests that a relatively narrow band of trees can lower the maximum height reached by storm surge inundation when the effect of vegetation on wind waves is also taken into account, but that further model refinement is required.

14.4 FACTORS AFFECTING STORM SURGE WATER LEVEL REDUCTION IN MANGROVES

Storm surge height reduction through mangroves is likely to depend upon a number of factors, including mangrove characteristics, such as forest width, tree density, and structural complexity (aerial roots, stems, branches, and foliage) of the dominant species; landscape characteristics, such as mangrove surface topography and the presence of channels, ponds, and pools; and storm characteristics, such as the size and forward speed of the cyclone (which may interact with mangroves to influence storm surge reduction). Few quantitative data are available (Krauss et al., 2009); where data do exist, they are generally derived from numerical models rather than observations (Zhang et al., 2012; Liu et al., 2013).

14.4.1 Mangrove Width

Measurements of storm surge reduction rates through coastal wetlands are often quoted in terms of centimeters of water level reduction per kilometer of inland distance, generally measured in the direction of travel of the surge (e.g., Wamsley et al., 2010; Engle, 2011). However, such constant attenuation rates imply a linear reduction in water level with distance into the mangroves (i.e., mangrove width). This is rarely true, both because the landscape is usually heterogeneous (i.e., it is usually a mixture of channels, pools, and vegetation with a varied topography) and the underlying rate of reduction might not be linear even if the environment were homogeneous. Consequently, such linear attenuation rates should be regarded with caution as they can be misleading (Resio and Westerink, 2008; Wamsley et al., 2010). At best, they may serve as rules of thumb around which there is generally a high degree of scatter (Resio and Westerink, 2008; Wamsley et al., 2010). Taking this into account and based on the studies described above, the rate of reduction of surges through mangroves appears to range between 40 and 160 mm km^{-1} (observed water level reduction rates; Krauss et al., 2009), and may be as high as 480 mm km^{-1} (validated numerical model; Zhang et al., 2012). Figure 14.7 compares these attenuation rates with similar measures from within salt marshes.

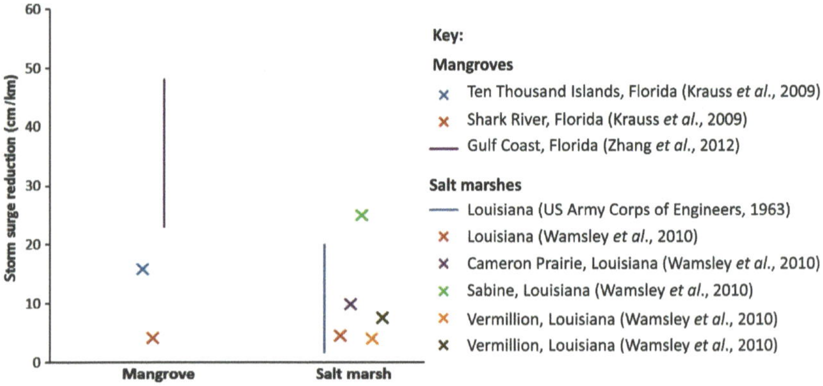

FIGURE 14.7 Summary of storm surge peak water level reduction rates as the surge passes through mangrove and salt marsh vegetation, based on empirical measures or validated numerical models. *(Data sources shown in key; note that Wamsley et al. (2010) combined data from a number of sources.)* Points have been artificially staggered along the *x*-axis to make them more easily visible.

Zhang et al. (2012) used simulations to explore the effects of different widths of mangroves being present. They found that surge attenuation through mangroves was nonlinear, with the largest reduction in peak water levels occurring close to the seaward edge of the mangroves (i.e., along the western shore, Figure 14.6), whereas further inland the water level changed more slowly (Figure 14.8). They suggest that this might explain the relatively low rates of peak water level reduction measured by Krauss et al. (2009), whose field measurements started some distance into the mangroves. The water level reduction rate in the most seaward mangroves might have been higher.

14.4.2 Mangrove Vegetation Characteristics

The density of mangrove vegetation and the diameter of aerial roots and stems are expected to affect the ability of mangroves to reduce storm surge water levels (Krauss et al., 2009; Alongi, 2008). It might be expected that a dense forest would have a greater effect on water levels than a sparse forest, with trees widely spaced. These properties are dynamic as tree densities typically decline with increasing mangrove forest age: forests become less dense, but individual trees larger, due to self-thinning (e.g., Malaysia: Putz and Chan, 1986; Puerto Rico: Jimenez et al., 1985). This characteristic might be quantified through the measurement of projected area, the silhouette of the vegetation as seen from the direction of oncoming waves (Quartel et al., 2007). Projected area varies with height above the ground, and in the case of

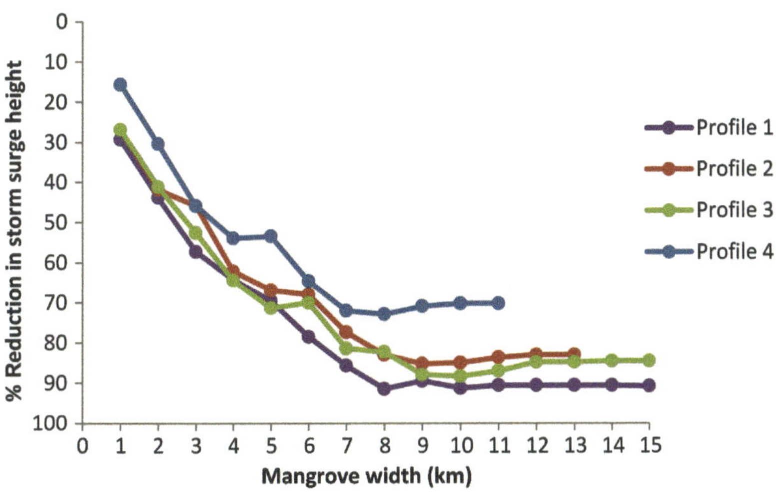

FIGURE 14.8 The reduction in storm surge height as the surge passes through the mangroves along four different shore profiles (shown in Figure 14.6) on the Gulf Coast of South Florida, USA, flooded by a storm surge from Hurricane Wilma (2005, category 3 on impact), and modeled using Coastal and Estuarine Storm Tide *(Zhang et al., 2012). Adapted from Zhang et al. (2012).*

mangroves, the presence (or absence) of aerial roots and low branches is likely to be critical (reviewed in McIvor et al., 2012). However, we know of no data that tests these assumptions under storm surge conditions.

14.4.3 Local Topography and the Presence of Channels and Pools within Mangrove Areas

Mangrove surface topography, and the degree of dissection of such surfaces, is likely to be an important local factor affecting flooding patterns from storm surges, interacting with the peak water level to influence the extent of inundation. However, we know of only two studies which have attempted to assess the importance of these factors. Krauss et al. (2009) and Zhang et al. (2012) consider the presence of channels and pools as likely to decrease the ability of mangroves to reduce peak water levels during surges, because the water is able to pass more easily along the channels and penetrate further inland.

Krauss et al. (2009) recorded higher rates of reduction of peak water levels in intact, relatively unchannelized expanses of mangroves in the Ten Thousand Islands National Wildlife Refuge than through riverine areas along the Shark River (94 mm km^{-1} vs 42 mm km^{-1} respectively). However, they point out that such differences may have been due to differences in the storm characteristics or other factors. The peak water level also propagated more rapidly up the Shark River mangrove area (1.4 km h^{-1} in the Shark River area, compared to 0.4 km h^{-1} in the Ten Thousand Islands National Wildlife Refuge).

Based on their validated numerical model, Zhang et al. (2012) found that surge height decreased at a rate of 230 mm km^{-1} through an area with a mixture of mangrove islands and open water, whereas in areas with less open water, surge height reduction rates ranged from 400 to 480 mm km^{-1}.

14.4.4 Storm Surge Reduction and the Nature of the Generating Mechanism

Reduction in surge heights and flood extent resulting from the presence of mangroves depends in part on the characteristics of the cyclone producing the storm surge (Spencer et al., 2009). CEST model simulations of the passage of storm surges crossing the southern tip of Florida predict that inundation extents associated with a hurricane with high forward speed, small radius of maximum wind speed, and low hurricane category show a greater surge reduction by mangroves than surges created by hurricanes with slow forward speed, large radius of maximum wind speed, and high hurricane category (Zhang et al., 2012; Liu et al., 2013). Thus, for example, the mangroves of south Florida are expected to protect the area behind them against flooding from a category 5 hurricane with a fast forward speed of 11.2 m s^{-1}, but not from a category 5 hurricane with a slow forward speed of 2.2 m s^{-1}. A mangrove forest with a width of tens of kilometers would be needed to attenuate a 2–3 m storm surge from a slow-moving category 5 hurricane (Zhang et al., 2012). Furthermore, the track direction of a hurricane interacts with local bathymetry to determine surge characteristics. Thus, for example, although the south to north track of Hurricane Donna (1960; category 5) produced a 4-m storm surge in the middle Florida Keys as water was driven into shallow Florida Bay, the east to west track of Hurricane Betsy (1965; category 4) across the bay resulted in a lower storm surge in the same region (Perkins and Enos, 1968). Local coastal setting and mangrove area then additionally influences the reduction in flooded area occurring where mangroves are present: Liu et al. (2013) found that the mangrove-induced reduction in flood extent varied between 31.3 and 37.7 percent as the hurricane approach angle varied over 180° in Southern Florida.

14.5 REDUCTION OF SURFACE WINDS BY MANGROVES

It is well known that forest canopies modify near-surface wind speeds (e.g., Raupach and Thom, 1981). Chen et al. (2012) measured wind speed and direction close to mangrove plantations (*Sonneratia apetala* and *Kandelia obovata*) in Sanjiang Bay, Haitian Province, South China. Mean wind speeds of up to 5 m s^{-1}, 2 m above the ground surface and 50 m from the mangrove area, were reduced by more than 85 percent by the mangrove forests. The *Sonneratia* plantation reduced wind speeds more effectively during the warm season, presumably as a result of the denser foliage on the trees at this time of year. During typhoons, mean wind speed and extreme wind speed were reduced by 59.4 percent and 53.2 percent, respectively (the latter in *Sonneratia* forest only; Chen et al., 2012).

It is not possible to directly measure the effect of vegetation-reduced wind speeds on storm surge heights because the effects of reduced wind speeds would never occur independently of other effects, such as the increased drag on the water flow from the vegetation. Using the Advanced Circulation Numerical Model, Westerink et al. (2008) explored how peak water levels varied in hindcasts of Hurricanes Betsy and Andrew when surface wind speeds were modified to reflect the differences in land cover (e.g., dense forested canopies, marshland, or buildings) and the level of local inundation (once a land feature was underwater, it no longer affected wind speeds). Wind direction was also taken into account. Peak water levels were more than 1 m lower in some areas when the modified wind speeds were included, as opposed to wind speeds assuming open-ocean marine conditions. This implies that the effects of vegetation on wind speeds could significantly influence storm surge water levels.

14.6 DISCUSSION, DATA GAPS, AND CONCLUSIONS

Both empirical data and numerical models suggest that mangrove vegetation can reduce storm surge peak water levels where the mangrove canopy is present over sufficiently large areas, usually kilometers in width. Such large areas of mangroves are still present in many parts of the tropics that are affected by cyclones and storm surges, including Mexico, Central America, the Caribbean islands, the Continental Caribbean, Northern South America and Brazil, Florida, Bangladesh, India, Indonesia, and Australia. In these locations, the conservation and restoration of mangroves can contribute to a risk reduction strategy against storm surge inundation and damage.

Reported rates of water level reduction with distance during surge events are sparse. The data that are available give highly variable results. It is difficult to separate out the effects of storm characteristics, the local physical setting, and the characteristics of the mangrove forests themselves. These groupings are not independent of one another. Furthermore, nothing is known about the capacity of mangrove ecosystems to sustain/resist multiple events over time or where the threshold for the protective service may lie in areas prone to such multiple impacts. As a result, it appears that numerical models based on the underlying physics of wind forcing and water movement are best able to represent the impacts of storm surges (Resio and Westerink, 2008), precisely because the interrelationships between storm surge reduction, bathymetry, topography, distance from shore, and width of mangrove vegetation are so complex. Thus, for example, the CEST model used by Zhang et al. (2012) and Liu et al. (2013) demonstrates how such models can be used to better understand the effects of these factors on storm surge reduction rates, and thus to predict flooding extent for different storm types in the presence or absence of mangroves. However, a major challenge is to extend these studies to the full range of geomorphic settings (see Spencer and Möller (2013) for classifications), and mangrove vegetation types, occurring on tropical and subtropical coasts.

These studies also demonstrate an approach for quantifying storm surge reduction by mangroves that can help inform coastal managers, engineers, and insurers about the expected benefits of mangrove presence for risk reduction. However, a limitation of current modeling approaches is their inability to include spatial variation in mangrove characteristics, such as mangrove forest density or the presence of particular mangrove species. It is very likely that the ability of mangroves to reduce peak water levels depends on mangrove characteristics, with sparse, fragmented, or channelized areas reducing storm surge water levels less effectively than dense mangrove vegetation. Including plant and canopy architecture in models would probably improve the prediction of storm surge heights, and would therefore aid in planning the use of mangroves as a form of coastal defense.

Where extensive areas of mangroves currently exist, reducing the threats they face from sea level rise, coastal development, and other anthropogenic factors will help to maintain their coastal defense functions. In other areas, large-scale restoration or afforestation of mangroves can reduce risk from storm surges. Numerical storm surge models will generally be required to calculate the potential benefits of mangroves. Other considerations include the chances of successful mangrove planting, which depends both on the methods employed (Lewis, 2005; Lewis and Perillo, 2009; Twilley and Rivera-Monroy, 2005; Primavera et al., 2012) and on the social and legal frameworks present in a particular region. These frameworks may greatly influence the perception of the roles played by mangroves, and thus their future use and stability of tenure (Primavera and Esteban, 2008).

The most appropriate use of mangroves in coastal defense of high-value or high-vulnerability areas of human occupation is likely to be in combination with other risk reduction measures. Hybrid approaches may be best in many areas: for example, mangroves growing seaward of a sea wall or levee may reduce storm surge water levels and wave energy acting on the wall itself, reducing the likelihood of overtopping or breaching of the structure. This could significantly reduce the design specifications, and therefore the cost, of such engineered works. Such approaches require engineered structures to be built landward of existing mangroves, as planting mangroves in lower elevation settings is rarely successful (Lewis and Perillo, 2009). Mangroves have been used in this way in disaster risk preparedness projects in Vietnam (Figure 14.9), where they have provided significant economic benefits, both in terms of the reduced cost of dike repair and avoided losses to public infrastructure and private property (Jegillos et al., 2005; IFRC, 2011; Powell et al., 2011). Studies of the most extreme storm impacts in mangrove areas have all pointed to the critical importance of other key elements in risk reduction associated with human behavior—these have included long-term policies on coastal development to avoid high-risk settlement and infrastructure, the existence of early

FIGURE 14.9 Mangroves in front of a dike in Vietnam. *Photo: Mai Sỹ Tuấn.*

warning systems, and clear and well-understood evacuation procedures (Spalding et al., 2014; Das and Vincent, 2009; Williams et al., 2007).

Overall, the understanding of the role of mangroves in mitigating the impacts of large storm surges remains incomplete. Even as such understanding improves, it is likely that, from a perspective of coastal planning, it will be necessary to factor in levels of uncertainty associated with coastal protection performance from coastal ecosystems that may be higher than those derived from engineered coastal defenses. In both cases, however, it is important to remember that a level of residual risk will always be present. A further important consideration in coastal planning must be the cobenefits from mangroves that can be substantial, notably in terms of fisheries enhancement, timber, and other forest products. Such benefits may be of particular importance in postdisaster settings, through provision of food and building materials to coastal communities.

REFERENCES

Alongi, D.M., 2008. Mangrove forests: resilience, protection from tsunamis, and responses to global climate change. Estuarine, Coastal Shelf Sci. 76, 1–13.

Apotsos, A., Jaffe, B., Gelfenbaum, G., 2011. Wave characteristic and morphologic effects on the onshore hydrodynamic response of tsunamis. Coastal Eng. 58 (11), 1034–1048.

Badola, R., Hussain, S.A., 2005. Valuing ecosystem functions: an empirical study on the storm protection function of Bhitarkanika mangrove ecosystem, India. Environ. Conserv. 32 (1), 85–92.

Baird, A.H., Ballah, R.S., Kerr, A.M., Pelkey, N.W., Srinivas, V., 2009. Do mangroves provide an effective barrier to storm surges? Proc. Natl. Acad. Sci. USA. 106, E111.

Barbier, E.B., Hacker, S.D., Kennedy, C., Koch, E.W., Stier, A.C., Silliman, B.R., 2011. The value of estuarine and coastal ecosystem services. Ecol. Monogr. 81 (2), 169–193.

Cahoon, D.R., Hensel, P.R., Rybczyk, J., McKee, K.L., Proffitt, E.E., Perez, B.C., 2003. Mass tree mortality leads to mangrove peat collapse at Bay Islands, Honduras after Hurricane Mitch. J. Ecol. 91, 1093–1105.

Chatenoux, B., Peduzzi, P., 2007. Impacts from the 2004 Indian Ocean Tsunami: analysing the potential protecting role of environmental features. Nat. Hazards 40, 289–304.

Chen, R., Twilley, R.R., 1999. A simulation model of organic matter and nutrient accumulation in mangrove wetland soils. Biogeochemistry 44, 93–118.

Chen, Y., Liao, B., Li, M., Chen, B., Chen, Y., Zhong, C., Li, H., Lin, W., 2012. Wind-attenuation effect of *Sonneratia apetala* and Kandelia obovata plantations. Ying Yong Sheng Tai Xue Bao 23 (4), 959–964.

Chow, V.T., 1959. Open Channel Hydraulics. McGraw-Hill Book Company, New York, p. 680.

Chow, V.T., Maidment, D.R., Mays, L.W., 1988. Applied Hydrology. McGraw-Hill Book Company, New York, p. 572.

Church, J.A., et al., 2013. Sea level change. In: Climate Change 2013: the Physical Science Basis. Contribution of Working Group I to the Fifth Assessment Report of the Intergovernmental Panel on Climate Change. Cambridge University Press, Cambridge.

Cochard, R., Ranamukhaarachchi, S.L., Shivakoti, G.P., Shipin, O.V., Edwards, P.J., Seeland, K.T., 2008. The 2004 tsunami in Aceh and Southern Thailand: a review on coastal ecosystems, wave hazards and vulnerability. Perspect. Plant Ecol., Evol. Syst. 10, 3–40.

Conner, W.H., Day Jr., J.W., Baumann, R.H., Randall, J.M., 1989. Influence of hurricanes on coastal ecosystems along the northern Gulf. Wetlands Ecol. Manage. 1, 45–56.

Dahdouh-Guebas, F., Koedam, N., 2006. Coastal vegetation and the Asian Tsunami. Science 311, 37.

Danielsen, F., Sorensen, M.K., Olwig, M.F., Selvam, V., Parish, F., Burgess, N.D., Hiraishi, T., Karunagaran, V.M., Rasmussen, M.S., Hansen, L.B., Quarto, A., Suryadiputra, N., 2005. The Asian tsunami: a protective role for coastal vegetation. Science 310, 643.

Das, S., Vincent, J.R., 2009. Mangroves protected villages and reduced death toll during Indian super cyclone. Proc. Natl. Acad. Sci. 106, 7357–7360.

Dean, R.G., Bender, C.J., 2006. Static wave setup with emphasis on damping effects by vegetation and bottom friction. Coastal Eng. 53 (2–3), 149–156.

Done, T., 1993. On tropical cyclones, corals and coral reefs. Coral Reefs 12, 126.

Engle, V.D., 2011. Estimating the provision of ecosystem services by gulf of Mexico coastal wetlands. Wetlands 31 (1), 179–193.

FAO, 2007. The World's Mangroves 1980–2005, FAO Forestry Paper 153. Forest Resources Division. FAO (The Food and Agriculture Organisation of the United Nations), Rome, p. 77.

Feagin, R.A., Mukherjee, N., Shanker, K., Baird, A.W., Cinner, J., Kerr, A.M., Koedam, N., Srudha, A., Arthur, R., Jayatissa, L.P., Seen, D.L., Menon, M., Rodriguez, S., Shamsuddoha, M., Dahdouh-Guebas, F., 2010. Shelter from the storm? Use and misuse of coastal vegetation bioshields for managing natural disasters. Conserv. Lett. 3, 1–11.

Flather, R.A., 2001. Storm surges. In: Steele, J.H., Thorpe, S.A., Turekian, K.K. (Eds.), Encyclopaedia of Ocean Sciences, vol. 5. Academic Press, London and California, pp. 2882–2892.

Garrison, T., 1999. Oceanography: an Invitation to Marine Science, third ed. Wadsworth Publishing Company, Belmont, California. p. 471.

Gedan, K., Kirwan, M., Wolanski, E., Barbier, E., Silliman, B., 2011. The present and future role of coastal wetland vegetation in protecting shorelines: answering recent challenges to the paradigm. Clim. Change 106, 7–29.

Giri, C., Ochieng, E., Tieszen, L., Zhu, Z., Singh, A., Loveland, T., Masek, J., Duke, N., 2011. Status and distribution of mangrove forests of the world using earth observation satellite data. Global Ecol. Biogeogr. 20 (1), 154–159.

Granek, E.F., Ruttenberg, B.I., 2007. Protective capacity of mangroves during tropical storms: a case study from 'Wilma' and 'Gamma' in Belize. Mar. Ecol. Prog. Ser. 343, 101–105.

Gunawardena, M., Rowan, J.S., 2005. Economic valuation of a mangrove ecosystem threatened by shrimp aquaculture in Sri Lanka. Environ. Manage. 36, 535–550.

Hinkel, J., Lincke, D., Vafeidis, A.T., Perrette, M., Nicholls, R.J., Tol, R.S.J., Marzeiong, B., Fettweish, X., Ionescu, C., Levermann, A., 2014. Coastal Flood Damage and Adaptation Costs under 21st Century Sea-level Rise. Proc. Natl. Acad. Sci 111, 3292–3297. URL: www.pnas.org/cgi/doi/10.1073/pnas.1222469111.

Hutchison, J., Manica, A., Swetnam, R., Balmford, A., Spalding, M., 2014. Predicting global patterns in mangrove biomass. Conserv. Lett. 7, 233–240. http://dx.doi.org/10.1111/conl.12060.

IFRC (International Federation of Red Cross and Red Crescent Societies), 2011. Breaking the Waves. Impact Analysis of Coastal Afforestation for Disaster Risk Reduction in Vietnam. IFRC, Geneva, p. 51.

ITJSCE (Investigation Team of Japan Society of Civil Engineering), 2008. Investigation Report on the Storm Surge Disaster by Cyclone SIDR in 2007, Bangladesh (Transient translation). URL: http://www.jsce.or.jp/report/46/files/Bangladesh_Investigation.pdf.

Iverson, L.R., Prasad, A.M., 2007. Using landscape analysis to assess and model tsunami damage in Aceh province, Sumatra. Landscape Ecol. 22, 323–331.

Jegillos, S.R., Lunde, G., Kawate, H., Dzung, T.V., 2005. Vietnam Red Cross mangrove and disaster preparedness in the Red River Delta and northern coastal Vietnam (1994–2005). Final Evaluation Report for the Danish Red Cross. (Available from the Danish Red Cross).

Jevrejeva, S., Moore, J., Grinsted, A., 2012. Sea level projections to AD2500 with a new generation of climate change scenarios. Global Planet. Change 80–81, 14–20.

Jimenez, J.A., Lugo, A.E., Cintron, G., 1985. Mortality in mangrove forests. Biotropica 17, 177–185.

Kathiresan, K., Rajendran, N., 2005. Coastal mangrove forests mitigated tsunami. Estuarine, Coastal Shelf Sci. 65, 601–606.

Kerr, A.M., Baird, A.H., 2007. Natural barriers to natural disasters. Bioscience 57, 102–103.

Kerr, A.M., Baird, A.H., Campbell, S.J., 2006. Comments on "Coastal mangrove forests mitigated tsunami" by K. Kathiresan and N. Rajendran [Estuar. Coast. Shelf Sci. 65 (2005) 601–606]. Estuarine, Coastal Shelf Sci. 67, 539–541.

Khairoutdinov, M., Emanuel, K., 2013. Rotating radiative-convective equilibrium simulated by a cloud-resolving model. J. Adv. Model. Earth Syst. 5. http://dx.doi.org/10.1002/2013MS000253.

Komar, P.D., 1998. Beach Processes and Sedimentation. Prentice Hall, New Jersey, p. 544.

Krauss, K.W., Doyle, T.W., Doyle, T.J., Swarzenski, C.M., From, A.S., Day, R.H., Conner, W.H., 2009. Water level observations in mangrove swamps during two hurricanes in Florida. Wetlands 29 (1), 142–149.

Lacambra, C., Spencer, T., Moeller, I., 2008. Tropical coastal ecosystems as coastal defences. In: ProAct Network. The Role of Environmental Management and Eco-Engineering in Disaster Risk Reduction and Climate Change Adaptation. ProAct Network/United Nations International Strategy for Disaster Reduction. URL: http://proactnetwork.org/proactwebsite.

Laso Bayas, J.C., Marohn, C., Dercon, G., Dewi, S., Piepho, H.P., Joshi, L., van Noordwijk, M., Cadisch, G., 2011. Influence of coastal vegetation on the 2004 tsunami wave impact in west Aceh. Proc. Natl. Acad. Sci. 108, 18612–18617.

Lewis III, R.R., 2005. Ecological engineering for successful management and restoration of mangrove forests. Ecol. Eng. 24 (4), 403–418.

Lewis III, R.R., Perillo, G.M.E., 2009. Methods and criteria for successful mangrove forest restoration. In: Perillo, G.M.E., Wolanski, E., Cahoon, D.R., Brinson, M.M. (Eds.), Coastal Wetlands: an Integrated Ecosystem Approach. Elsevier, Oxford, pp. 787–800.

Liu, K., Fearn, M.L., 2000. Section 3.2: Holocene history of catastrophic hurricane landfalls along the Gulf of Mexico coast reconstructed from coastal lake and marsh sediments. In: Ning, Z.H., Abdollain, K.K. (Eds.), Current Stress and Potential Vulnerabilities; Implications of Global Change for the Gulf Coast Region of the United States. Gulf Coastal Regional Climate Change Council/Franklin Press, Baton Rouge, pp. 38–47.

Liu, H., Zhang, K., Li, Y., Xie, L., 2013. Numerical study of the sensitivity of mangroves in reducing storm surge and flooding to hurricane characteristics in southern Florida. Cont. Shelf Res. 64, 51–65.

Lugo, A.E., Sell, M., Snedaker, J.C., 1976. Mangrove ecosystem analysis. In: Patten, B.C. (Ed.), Systems Analysis and Simulation in Ecology. Academic Press, New York, pp. 113–145.

Lugo, A.E., Cintron, G., Goenaga, C., 1981. Mangrove ecosystems under stress. In: Barret, G.W., Rosenberg, R. (Eds.), Stress Effects on Natural Ecosystems. John Wiley & Sons Ltd, New York, pp. 129–153.

Mattocks, C., Forbes, C., 2008. A real-time, event-triggered storm surge forecasting system for the state of North Carolina. Ocean Model. 25 (3–4), 95–119.

Mazda, Y., Magi, M., Ikeda, Y., Kurokawa, T., Asano, T., 2006. Wave reduction in a mangrove forest dominated by *Sonneratia* sp. Wetlands Ecol. Manage. 14 (4), 365–378.

McCoy, E.D., Mushinsky, H.R., Johnson, D., Meshaka, W.E.J., 1996. Mangrove damage caused by Hurricane Andrew on the southwestern coast of Florida. Bull. Mar. Sci. 59 (1), 1–8.

McIvor, A.L., Möller, I., Spencer, T., Spalding M., 2012. Reduction of wind and swell waves by mangroves. Natural Coastal Protection Series: Report 1. Cambridge Coastal Research Unit Working Paper 40. Published by The Nature Conservancy and Wetlands International. p. 27. ISSN 2050-7941. URL: http://www.naturalcoastalprotection.org/documents/mangroves.

McIvor, A.L., Spencer, T., Möller, I., Spalding, M., 2013. The response of mangrove soil surface elevation to sea level rise. Natural Coastal Protection Series: Report 3. Cambridge Coastal Research Unit Working Paper 42. Published by The Nature Conservancy and Wetlands International. p. 59. ISSN 2050-7941. URL: http://coastalresilience.org/science/mangroves/surface-elevation-and-sea-level-rise.

Mendelsohn, R., Emanuel, K., Chonabayashi, S., Bakkensen, L., 2012. The impact of climate change on global tropical cyclone damage. Nat. Clim. Change 2, 205–209.

Ohira, W., Honda, K., Harada, K., 2012. Reduction of tsunami inundation by coastal forests in Yogyakarta, Indonesia: a numerical study. Nat. Hazards Earth Syst. Sci. 12, 85–95.

Paling, E.I., Kobryn, H.T., Humphreys, G., 2008. Assessing the extent of mangrove change caused by Cyclone Vance in the eastern Exmouth Gulf, northwestern Australia. Estuarine, Coastal Shelf Sci. 77, 603–613.

Perkins, R.D., Enos, P., 1968. Hurricane Betsy in the Florida-Bahama area – geologic effects and comparison with Hurricane Donna. J. Geol. 76, 710–717.

Powell, N., Osbeck, M., Tan, S.B., Toan, V.C., 2011. Mangrove Restoration and Rehabilitation for Climate Change Adaptation in Vietnam. World Resources Report Case Study. World Resources Report, Washington DC. URL: http://www.worldresourcesreport.org.

Primavera, J.H., Esteban, J.M.A., 2008. A review of mangrove rehabilitation in the Philippines – successes, failures and future prospects. Wetlands Ecol. Manage. 16 (5), 345–358.

Primavera, J.H., Savaris RV, J.P., Bajoyo, B.E., Coching, J.D., Curnick, D.J., Golbeque, R.L., Guzman, A.T., Henderin, J.Q., Joven, R.V., Loma, R.A., Koldewey, H.J., 2012. Manual on Community-Based Mangrove Rehabilitation. In: Mangrove Manual Series, vol. 1. Zoological Society of London, London pp. viii + 240.

Pugh, D.T., 1996. Tides, Surges and Mean Sea-Level (Reprinted with Corrections). John Wiley & Sons Ltd, Chichester, p. 486.

Putz, F.E., Chan, H.T., 1986. Tree growth, dynamics, and productivity in a mature mangrove forest in Malaysia. Forest Ecol. Manage. 17, 211–230.

Quartel, S., Kroon, A., Augustinus, P., Van Santen, P., Tri, N.H., 2007. Wave attenuation in coastal mangroves in the Red River Delta, Vietnam. J. Asian Earth Sci. 29 (4), 576–584.

Raupach, M.R., Thom, A.S., 1981. Turbulence in and above plant canopies. Annu. Rev. Fluid Mech. 13, 97–129.

Rego, J.L., Li, C., 2009. On the importance of the forward speed of hurricanes in storm surge forecasting: a numerical study. Geophys. Res. Lett. 36, L07609. http://dx.doi.org/10.1029/2008GL036953.

Resio, D., Westerink, J., 2008. Modelling of the Physics of Storm Surges. Physics Today. September, 33–38.

Roth, L.C., 1992. Hurricanes and mangrove regeneration: effects of Hurricane Joan, October 1988, on the vegetation of Isla del Venado, Bluefields, Nicaragua. Biotropica 24, 375–384.

Sathirathai, S., Barbier, E., 2001. Valuing mangrove conservation in Southern Thailand. Cont. Econ. Policy 19, 109–122.

Seto, K.C., 2011. Exploring the dynamics of migration to mega-delta cities in Asia and Africa: contemporary drivers and future scenarios. Global Environ. Change 21 (Suppl. 1), S94–S107.

Smith III, T.J., Robblee, M.B., Wanless, H.R., Doyle, T.W., 1994. Mangroves, hurricanes and lightning strikes. Bioscience 44 (4), 256–262.

Smith III, T.J., Anderson, G.H., Balentine, K., Tilling, G., Ward, G.A., Whelan, K.R.T., 2009. Cumulative impacts of hurricanes on Florida mangrove ecosystems: sediment deposition, storm surges and vegetation. Wetlands 29, 24–34.

Spalding, M., Kainuma, M., Collins, L., 2010. World Atlas of Mangroves. Earthscan, London and Washington DC pp. xvi + 319.

Spalding, M.D., McIvor, A.L., Beck, M.B., Koch, E.W., Möller, I., Reed, D.J., Rubinoff, P., Spencer, T., Tolhurst, T.J., Wamsley, T.V., van Wesenbeeck, B.K., Wolanski, E., Woodroffe, C.D., 2014. Coastal ecosystems: a critical element of risk reduction. Conserv. Lett 7, 293–301. http://dx.doi.org/10.1111/conl.12074.

Spencer, T., 2007. Coral reefs and the tsunami of 26 December 2004: Generating processes and ocean-wide patterns of impact. Atoll Res. Bull. 544, 1–36.

Spencer, T., Slaymaker, O., Embleton-Hamann, C., 2009. Landscape, landscape scale processes, and global environmental change: synthesis and new agendas for the twenty-first century. In: Slaymaker, O., Spencer, T., Embleton-Hamann, C. (Eds.), Geomorphology and Global Environmental Change. Cambridge University Press, Cambridge, pp. 403–423.

Spencer, T., Möller, I., 2013. Mangrove systems. In: Schroder, J.F. (Ed.), Treatise on Geomorphology, vol. 10. Academic Press, San Diego, pp. 360–391.

Stive, M.J.F., Wind, H.G., 1982. A study of radiation stress and set-up in the nearshore zone. Coastal Eng. 6, 1–25.

Storch, H., Woth, K., 2008. Storm surges: perspectives and options. Sustainability Sci. 3 (1), 33–43.

Strusinska-Correia, A., Husrin, S., Oumeraci, H., 2013. Tsunami damping by mangrove forest: a laboratory study using parameterized trees. Nat. Hazards Earth Syst. Sci. 13, 483–503.

Tanaka, K., 2008. Effectiveness and limitation of the coastal vegetation for storm surge disaster mitigation. In: Investigation Report on the Storm Surge Disaster by Cyclone Sidr in 2007, Bangladesh. URL: http://www.jsce.or.jp/report/46/files/Bangladesh_Investigation.pdf.

Tanaka, N., Sasaki, Y., Mowjood, M.I.M., Jinadasa, K.B.S.N., Homchuen, S., 2006. Coastal vegetation structures and their functions in tsunami protection: experience of the recent Indian Ocean tsunami. Landscape Ecol. Eng. 3, 33–45.

Tanaka, N., 2009. Vegetation bioshields for tsunami mitigation: review of effectiveness, limitations, construction, and sustainable management. Landscape Ecol. Eng. 5 (1), 71–79.

Twilley, R.R., Rivera-Monroy, V.H., 2005. Developing performance measures of mangrove wetlands using simulation models of hydrology, nutrient biogeochemistry and community dynamics. J. Coastal Res. 40, 79–93.

Veron, J.E.N., 1995. Coral Reefs in Space and Time: the Biogeography of the Scleractinia. Cornell University Press, Ithaca, NY, p. 323.

Wamsley, T.V., Cialone, M.A., Smith, J.M., Atkinson, J.H., Rosati, J.D., 2010. The potential of wetlands in reducing storm surge. Ocean Eng. 37 (1), 59–68.

Wells, S., Kapos, V., 2006. Coral reefs and mangroves: implications from the tsunami one year on. Oryx 40, 123–124.

Westerink, J.J., Luettich, R.A., Feyen, J.C., Atkinson, J.H., Dawson, C., Roberts, H.J., Powell, M.D., Dunion, J.P., Kubatko, E.J., Pourtaheri, H., 2008. A basin- to channel-scale unstructured grid hurricane storm surge model applied to southern Louisiana. Mon. Weather Rev. 136 (3), 833–864.

Williams, M.J., Coles, R., Primavera, J.H., 2007. A lesson from cyclone Larry: an untold story of the success of good coastal planning. Estuarine, Coastal Shelf Sci. 71, 364–367.

Woodruff, J.D., Irish, J.L., Camargo, S.J., 2013. Coastal flooding by tropical cyclones and sea-level rise. Nature 504, 44–52.

Xu, H.Z., Zhang, K.Q., Shen, J.A., Li, Y.P., 2010. Storm surge simulation along the US. East and Gulf Coasts using a multi-scale numerical model approach. Ocean Dyn. 60 (6), 1597–1619.

Zhang, K.Q., Liu, H., Li, Y., Hongzhou, X., Jian, S., Rhome, J., Smith III, T.J., 2012. The role of mangroves in attenuating storm surges. Estuarine, Coastal Shelf Sci. 102, 11–23.

Zhang, Y., Baptista, A., Myers, E., 2004. A cross-scale model for 3D baroclinic circulation in estuary-plume-shelf systems: I. formulation and skill assessment. Cont. Shelf Res. 24, 2187–2214.

FURTHER READING

US Army Corps of Engineers, 1963. Overland Surge Elevations Coastal Louisiana: Morgan City and Vicinity. US. Army Corps of Engineers, New Orleans District. File No. H-2–22758, Plate A-4.

Chapter 15

Coral Reef Systems and the Complexity of Hazards

Paul S. Kench and Susan D. Owen
School of Environment, The University of Auckland, Auckland, New Zealand

ABSTRACT

Coral reefs are unique coastal systems as they represent the balance between ecological and physical processes. Known for their high biological diversity, both the ecology and geomorphic structure of reef systems support a range of ecosystem services. This chapter explores the complexities of hazards in reef systems, underpinned by an understanding of the dynamic interplay between ecological and physical processes. Key drivers of impact in reef systems are examined, which include extreme events and slow-onset changes in the environmental boundary conditions of reefs. Natural hazards, incremental environmental change, and anthropogenic stresses can each drive significant impacts on reefs. Case studies indicate that the degree of impact is temporally and spatially variable dependent on the antecedent condition of reefs. Impacts include the catastrophic loss of living reef cover, erosion of adjacent coastlines, the formation of extensive rubble deposits on reefs, and slow deterioration in reef health, leading to structural collapse of reef systems. However, coral reef systems are resilient to natural and anthropogenic perturbations, and the recovery period of reefs to a range of impacts is highlighted. This chapter also discusses how the resilience of reefs can be compromised through the compounding effect of natural and anthropogenic stresses on reefs that can force major changes in reef health and structure over decadal timescales. A reef system model is used to highlight this complexity and allows for consideration of factors such as impact and recovery timescales in reef systems.

15.1 INTRODUCTION

Coral reefs are the most biologically diverse marine ecosystems in the world (Wilkinson, 1999). Less well recognized, coral reef systems are also complex three-dimensional geological structures that support the living veneer of biological diversity, which in turn contributes to the ongoing development of reef structure. Situated across tropical to temperate latitudes (largely bounded 28° north and south of the Equator), coral reefs provide a suite of

Coastal and Marine Hazards, Risks, and Disasters. http://dx.doi.org/10.1016/B978-0-12-396483-0.00015-7
Copyright © 2015 Elsevier Inc. All rights reserved.

ecosystem services to coastal communities that include biological and food resources, the physical substrate for island accumulation and human habitation, aggregates for construction, and protection from incident oceanic wave energy. Indeed, coral reefs provide the foundation for a number of midocean atoll nations and along continental coastlines millions of people live in close association with reef systems. However, at the global scale, coral reefs are considered to be in serious ecological decline as a consequence of anthropogenic impacts, natural stresses, and climate change (e.g., Hughes et al., 2003; Buddemeier et al., 2004). Broad-scale assessments (e.g., Wilkinson, 2004) have argued that 20 percent of the world's coral reefs have been destroyed and that 25 percent of reefs are under an imminent or long-term risk of collapse.

Reef systems are subject to a range of natural perturbations from short-term "pulse" events, such as cyclones and tsunamis, to longer-term pressures and shifts in environmental controls on reef state, such as sea-level change and changing ocean water chemistry. Added to these natural perturbations are a range of anthropogenic stressors that impact reef systems at event through to long timescales, and include dredging and construction activities, and exploitation of biological resources. Natural and anthropogenic stresses on reefs can be categorized into three types, based on the geographic relationship between the stressor and reef system. Some impacts are local and direct (e.g., cyclone impact or coral blasting), some are "proximal" (e.g., resort development, harbor construction), and others are distant but are then translated to the reef (e.g., sediment release from catchments). These differences influence the time lag and duration of stress on the reef system.

The effects of perturbations on reef systems, whether natural or anthropogenic, have the potential to alter the natural functioning of the biophysical system. These changes can lead to deteriorations in reef health and reef structure that will compromise the ecosystem services provided by coral reefs and promote exposure to hazards for reef-associated communities. The hazards faced by reef systems and communities are frequently a tangle of interconnected stressors, including single events and sustained pressures. The magnitude and persistence of these stressors influence the ability of reef systems to respond and recover.

This chapter presents an overview of the major hazards affecting coral reefs and associated human communities. An outline of the unique characteristics of reef systems, as a balance between ecological and physical processes and the ecosystem services afforded by reefs, is first presented as a basis to explore perturbations and impacts to reef systems. The chapter then examines the range of hazards affecting reefs with a focus on the influence between temporal (pulse versus slow onset) and spatial (local to distal) perturbations. A broad definition of hazards is adopted that encompasses any natural or anthropogenic process that can fundamentally alter the functioning of reef systems. By viewing hazard events in isolation, the connectivity of reef systems is quickly obscured. While individual hazard events can shock a reef system, this chapter explores how the interaction of multiple hazards and the cumulative effects of such events can impact reef system resilience (ability to recover). Understanding the complex human–ecosystem dynamics of coral reefs and the implications these relationships have for ecosystem resilience is of high importance. The chapter explores such interconnected and complex processes from the perspective of cumulative impacts, thresholds of collapse, and potential for recovery.

15.2 STRUCTURE AND FUNCTION OF CORAL REEFS

Coral reefs are unique coastal systems formed from the interaction between ecological processes, responsible for the growth of coral and other calcium carbonate producing organisms, and physical processes (waves, currents, and sea-level change) that modulate ecological processes and redistribute carbonate material within reef systems (Figure 15.1; Perry et al., 2012; Kench, 2013; Yap, 2013). Without the presence of carbonate-producing organisms, reef systems would not exist.

Coral reefs are three-dimensional structures, consisting of veneers of living coral and reef-associated organisms that overlie vast sequences of previously deposited calcium carbonate that can extend thousands of meters beneath midocean reef platforms. These structures evolve over geological (millennial) timescales and produce a number of characteristic landform types including atolls, barrier reefs, fringing reefs, and reef platforms (Kench, 2013). These reef structures vary in size from $<1\ km^2$, in the case of smaller patch reefs, to $>100\ km^2$ in extent, with some reef networks forming barrier complexes >2400 km in length, such as the Great Barrier Reef.

Coral reef surfaces also support a range of sedimentary landforms that are coherent accumulations of sediment deposited by wave and current processes on, or adjacent to, a coral reef structure (Kench, 2013). Of interest to the analysis of hazards is the formation of subaerial deposits, such as islands and coastal plains, which are geomorphically important at the human timescale as they form the foundation for coastal communities and provide the only habitable land in a number of midocean atoll nations, such as Tuvalu, Kiribati, and the Maldives. While the physical structure of reefs and sedimentary landforms vary (Kench, 2013), they serve similar habitat functions. However, their exposure to hazards can vary as a function of their structure and proximity to threats.

15.2.1 The Building Blocks of Reef Systems

To understand the effect of hazard events, it is necessary to highlight key ecological processes and relationships that underpin reef system health and

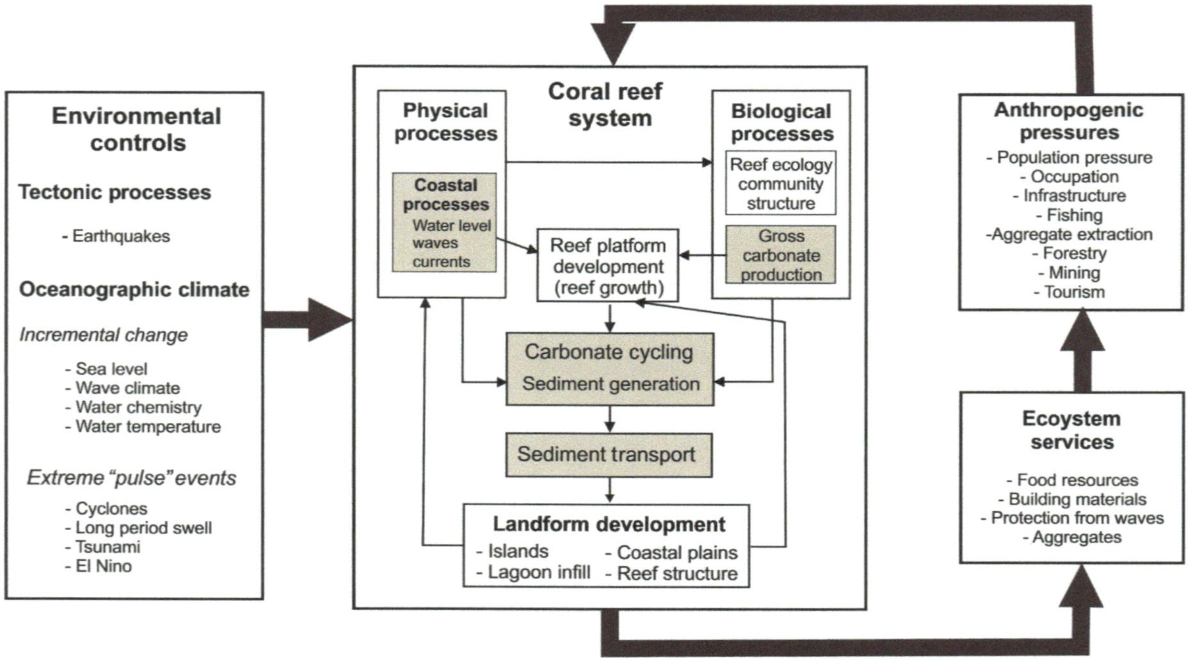

FIGURE 15.1 Conceptual diagram of the coral reef system and interaction between ecological and physical processes (central box), the processes driving change in the system, ecosystem services provided by reefs and anthropogenic impacts.

complexity (see Kench (2013) for an in-depth review). Central to the formation of reef structure and sedimentary landforms is the generation of calcium carbonate ($CaCO_3$) resulting from ecological processes. The major carbonate producers on reefs are typically divided into three groups. First, corals are considered the principal building blocks of coral reefs. These hermatypic corals are characterized by a symbiotic relationship between a coral animal and single-celled algae, zooxanthellae, which live within coral tissue. This relationship enables corals to secrete a rigid skeleton of calcium carbonate through a process known as calcification, a biologically mediated process that converts calcium and carbonate ions in supersaturated seawater to $CaCO_3$ (Kinzie and Buddemeier, 1996). Second, a range of encrusting organisms, such as calcareous algae, assists in the structural development of reefs and also bind loose sediment into the reef framework. These first two producers are known as primary producers as their growth can contribute directly to coral reef development. The third set of carbonate producers are benthic organisms that dwell on and within the reef. Such producers include molluscs, calcareous algae, foraminifera, bryozoans, and echinoderms. These organisms are known as secondary producers as they do not contribute directly to reef growth. However, once they die, their skeletal remains contribute to the detrital sediment reservoir.

In terms of the structural development of coral reefs, the most important consequence of reef metabolic processes is calcification. Rates of calcification on reefs are temporally and spatially variable and are dependent on a range of factors that include the density and growth rates of organisms across reefs. Typical rates of carbonate production range from $10\,kg\,m^{-2}\,year^{-1}$ on productive (coral rich) forereef zones to $<0.8\,kg\,m^{-2}\,year^{-1}$ in lagoons and rubble substrates (Kinsey, 1983).

While the geomorphic development of a reef and its associated sedimentary structures is dependent on the growth of carbonate-producing organisms, a suite of additional chemical, physical, and ecological processes also play a critical role in cycling carbonate sediment within reefs. These processes can aid the construction of reef landforms, others convert framework to detrital sediment, and some can destroy the reef framework (Figure 15.1). Constructive processes include sediment production by calcium carbonate-secreting organisms and precipitation of cements that bind and stabilize sediments (Scoffin, 1992). Destructive processes include bioerosion, the action of organisms in destroying reef framework through mechanical boring, etching and chemical dissolution (Perry and Hepburn, 2008), and physical processes, in which waves mechanically break the skeletal structure of carbonate material (Scoffin, 1993). Physical processes that erode, transport, and deposit carbonate sediment also control the distribution, structure, and morphology of reefs and sedimentary landforms such as islands and shorelines (Kench, 2013).

15.2.2 Environmental Limits on Coral and Reef Growth

Corals thrive in a relatively defined range of environmental limits. Understanding these limits is essential for contextualizing a range of hazard impacts on reef systems. The principle environmental controls on tropical $CaCO_3$ production are sea-surface temperature, light penetration, and the calcium carbonate saturation state of seawater (Table 15.1). Regional variations in these parameters govern the global distribution of coral reefs and the composition of carbonate sediment-producing biota. Coral reefs characteristically occur in shallow tropical and subtropical marine settings (between 28° N and 28° S) where sea-surface temperature rarely drops below 17–18 °C, or exceeds 33–34 °C, for prolonged periods.

Coral growth and reef development are also restricted by a number of other environmental thresholds (Kleypas et al., 1999; Perry and Larcombe, 2003). Corals depend on light for photosynthetic energy, but light levels reduce markedly with depth. Consequently, reef-building corals are limited to the "photic zone," the lower boundary of which is the depth of water at which surface light level is reduced to 1 percent. The depth of the photic zone also varies depending on turbidity levels and can range from >90 to <5 m in highly turbid environments. Salinity levels also pattern the distribution of hermatypic corals. While corals can endure a range of salinity levels (Table 15.1), reef

TABLE 15.1 **Environmental Parameters Governing the Distribution of Reef-Building (Hermatypic) Corals and Tropical Coral Reef Development. "Optimal" Values for Coral Growth Are Shown as Well as Recorded Upper and Lower Environmental Limits**

		Environmental Limits	
Environmental Parameter	**"Optimal" Levels**	**Lower**	**Upper**
Temperature (°C)[a]	21.0–29.5	16.0	34.4
Salinity (PSU)[b]	34.3–35.3	23.3	41.8
Nitrate (μmol l^a)[c]	<2.0	0.00	3.34
Phosphate (μmol l^a)[c]	<0.2	0.00	0.40
Aragonite saturation state (Ω-arag)[d]	~3.83	3.28	No data
Depth of light penetration (m)	~50	<10	~90

[a]*Weekly data.*
[b]*Monthly average data.*
[c]*Overall averages (1900–1999).*
[d]*Overall averages (1972–1978).*
Source: Kench et al. (2008).

development is constrained where salinities are low, such as river entrances, or zones of intense evaporation that elevate salinity. Nutrient levels of reefal waters are an additional regulator of coral growth. While corals thrive in low nutrient conditions, their growth can be inhibited where nutrient levels are elevated and in extreme cases may lead to replacement of coral communities with macroalgae. The constrained range of environmental parameters conducive to coral growth provides intrinsic thresholds against which deteriorations in coral health and hazards can be assessed.

15.2.3 The Ecomorphodynamic Framework

The concept of ecomorphodynamics provides a framework to conceptualize the dynamic balance between constructive and destructive processes that govern the structure and state of coral reefs and associated landforms (Figure 15.1; Kench, 2011a). The framework also provides a powerful lens to highlight how hazards can perturb the system leading to fundamental shifts in reef state. The framework highlights a number of key aspects of reef system interrelationship, which are critical to informing hazard analysis. First, the cycling of calcium carbonate is essential to supply the building blocks (corals, sands, and gravels) for reef and landform construction. Sediment transfer is a time-dependent process controlling morphological change in coastal morphodynamics (Cowell and Thom, 1994). As noted above, unique to coral reef systems, sediment is primarily produced by ecological processes and the rate of sediment production varies spatially between coral reef settings. Therefore, the "carbonate sediment factory" is a highly space- and time-dependent coupling mechanism. This time dependency emerges from the time lags in redistribution of sediment (as occurs in other coastal settings), and from the time lags related with organism growth, mortality, and conversion to sediment (Perry et al., 2008). Perturbations that influence the sediment factory may exacerbate hazards in the medium-term (decades).

Second, changes in boundary controls will force alterations in the health and physical state of the coral reef system. Such changes in boundary conditions can occur as a consequence of extrinsic factors (e.g., as sea-level rise, ocean temperature, and chemistry variations) or intrinsic factors such as anthropogenic impacts (e.g., coastal construction or resource exploitation). Third, the geomorphic sensitivity (magnitude, style, and timescales of geomorphic change) of reef systems varies between different geomorphic units. For example, the development of coral reef platforms is modulated by sea-level oscillations at millennial timescales (Montaggioni, 2005). In contrast, the dynamics of reef island shorelines occur at event to decadal timescales in response to changes in wave energy input (Maragos et al., 1973; Kench and Brander, 2006a).

Fourth, there are feedbacks in the system that can be temporally specific or cascade across timescales. For example, while sea-level oscillations govern the pattern of reef growth at millennial timescales, at shorter timescales, the reef structure modulates wave and current processes (Gourlay, 1988; Kench and Brander, 2006b). The characteristics of incident energy in turn control the structure of ecological communities (Chappell, 1980), reef morphology, sedimentation processes, and short-term geomorphic change of beach and island shorelines (Sheppard et al., 2005).

Fifth, feedbacks may be nonlinear and can be associated with significant time lags. For instance, changes in ecological condition of a reef may occur in response to short-term stresses, such as storms, human impact, or disease. However, depending on the magnitude and temporal scale of ecological change (severity, persistence, or ephemeral transition), alterations in the carbonate budget may, or may not, propagate through the system to yield detectable changes in the geomorphic system at decadal to centennial timeframes. An understanding of the dynamics of these feedbacks is necessary to exploring the linkages in changes to reef health, the propagation of hazard events, and their impacts and the longer-term resilience of reef systems.

15.2.4 Ecosystem Services

It has long been recognized that coral reefs provide a raft of ecosystem goods and services. Indeed, there has been considerable interest in quantifying the value of such services to neighboring communities to estimate the economic losses that would be sustained with catastrophic loss of reef function. Best and Bornbusch (2005) estimated that globally the goods and services provided by reefs exceed US$375 billion per annum. However, such a value is likely an underestimate in many reef regions, such as the central Pacific, where nonmonetary values, customary and subsistence life styles prevail (Laurans et al., 2013).

The goods and services afforded by reefs can be grouped into different categories. The most easily recognized is the income generated by extraction or use of the reef system. Tourism (including diving) constitutes the dominant source of income for many coral reef communities. For example, the Maldives receives >800,000 tourists per year, which generates approximately US$764 million annually, or 67 percent of Gross Domestic Product (GDP). At the local scale, tourism employs 58 percent of the nation's workforce. In another example, Sarkis et al. (2013) undertook a valuation of Bermuda's coral reefs, determining an average annual value of US$722.4 million per annum. Tourism comprised over half this value (56 percent), followed by coastal protection through the form of damage avoidance (37 percent). Laurans et al. (2013) observed lower estimates for tourism- and fisheries-related activities in a South Pacific context, reflecting lower population concentrations and infrastructure at risk. Fisheries and aggregate extraction for construction are two prevalent activities that also derive significant economic benefit from reef systems. In the Maldives, fishing accounts for 99 percent of all exports and is worth approximately US$170 million (16 percent of GDP). Perhaps equally important are the extractive practices of communities that harvest reef fish, shellfish, and aggregates for subsistence or community-scale trade, which are a critical source of food and income, but which is often not captured in national scale assessments of income (Laurens et al., 2013; Jaleel, 2013). Less well known, and more difficult to quantify, are the intrinsic values of coastal and marine biodiversity that emerge through the diversity of species and habitats.

Least recognized are the range of goods and services afforded by the three-dimensional structure of coral reefs (Moberg and Folke, 1999). First, reefs provide structural support for the living veneer of the reef system. Second, carbonate material is a source of aggregate for construction activities. Indeed, reef-derived sand and gravel for land reclamation and cement are the only viable source of aggregate in many low-lying atoll nations. Third, the physical structure of coral reefs also acts as a buffer to incident ocean swell, thereby affording protection for reefal habitat and adjacent shorelines (Sheppard et al., 2005; Kennedy et al., 2013).

Despite debates over actual "values," there is little doubt that the goods and services outlined above exist because of the existence of coral reef systems (Figure 15.1). Consequently, a shift in the health or physical state of an individual reef may impact negatively on the diversity and abundance of services provided. At one end of the spectrum, healthy reef systems are most likely to maintain goods and services, and such services are likely to be impacted in unhealthy or stressed reefs. Further, as the system is perturbed by natural or anthropogenic stresses the reef system will alter, which may be expressed as a change in the level of ecosystem services provided by reefs (Figure 15.1). However, the time lag for such changes to propagate through reef systems and be expressed in altered ecosystem services will vary depending on the magnitude and persistence of the stress event and whether multiple stressors are acting in concert.

15.3 IDENTIFYING HAZARDS AND KEY STRESSES ON CORAL REEF SYSTEMS

The major stresses on reef systems and the range of impacts such events generate are summarized in Table 15.2. When evaluating the impact of natural hazards and other stressors on reef systems, it is important to highlight a number of additional considerations. First, the concept of hazard often evokes natural climatic or tectonic events negatively impacting upon reef environments. However, studies suggest that more prevalent hazards on reef systems are anthropogenically or biologically induced. Second, the impacts of discrete perturbations on reef systems are temporally and spatially highly variable. Third, the magnitude of impact of stress events on reef systems can vary depending on whether the stress event is a short-lived "pulse" event or a slow-onset change in environmental conditions. Fourth, hazards may not affect

TABLE 15.2 Summary of Natural and Anthropogenic Stressors on Coral Reef Systems and the Range of Reef Impacts

	Stressor	Evidence of Stressor Impacts on Reef Structure and Function
Natural "hazard" events (pulse events)	Tectonic instability/ movement (volcanoes and earthquakes)	Changes to reef structure and stability, uplift or subsidence of reefs and islands (Baird et al., 2005; Aronson et al., 2012).
	Tsunami (earthquake/ landslide generated); storms; significant weather events (cyclones, typhoons, hurricanes); king tides.	Coral mortality, generation of rubble, loss of reef structure, erosion of shorelines, creation of new land, or coastal erosion (Maragos et al., 1973; Hagan et al., 2007; Kench et al., 2006; Kench, 2011b; Richmond et al., 2011), localized flooding and salination of crops and freshwater lens, terrestrial runoff, disruption of infrastructure (White et al., 2013).
Slow-onset natural impacts accentuated by anthropogenic activities	Increased CO_2 concentrations resulting in ocean acidification	Calcifying organisms reduced growth rates and increasingly fragile skeletons.
	Increasing sea-surface temperatures	Thermal stressors resulting in coral bleaching (Hughes et al., 2003). Predicted to increase in extent (van Hooidonk et al., 2013). Impact species specific and influenced by the level of anthropogenic disturbance (Polidoro and Carpenter, 2013).
	Sea-level rise	Increased water depth, enhanced wave energy across reefs, and impacting shorelines. Increased instability of islands. Coastal inundation, erosion, saltwater intrusion in aquifers (Mumby and Steneck, 2008).
	Biological disruption; invasive species, predator dominance, disease (also triggered by anthropogenic activity)	Increased diseases and changes in ecological state of reefs, reduced species diversity, and abundance, including phase shifts and collapse of corals.

TABLE 15.2 Summary of Natural and Anthropogenic Stressors on Coral Reef Systems and the Range of Reef Impacts—cont'd

	Stressor	Evidence of Stressor Impacts on Reef Structure and Function
Anthropogenic stressors	Increases in suspended sediment in receiving waters as a result of runoff from mining, deforestation, agriculture.	Shading and smothering of corals (Erftemeijer et al., 2012); increased macroalgal growth (Yap, 2013); reduced live coral (Fabricius, 2005); reduced species diversity (Jaleel, 2013).
	Diffuse and direct source inorganic and organic pollution resulting in nutrient enrichment/ eutrophication.	Eutrophication and heavy metal contaminant loading (Morrison et al., 2013; Edinger and Risk, 2000), increased bioerosion (related to pollution); increased macroalgal cover.
	Fishery exploitation (overfishing of resource and destructive fishing methods—blast fishing, poisoning).	Overfishing results in loss of predator and/or herbivores and changes to trophic chain, changed biodiversity function (Jaleel, 2013; Hughes, 1994; Ruppert et al., 2013).
	Coral and coral sand extraction (harvesting and mining for aggregate extraction); infrastructure development, reclamation, navigation channels, and port/ harbor expansion; military activities.	Coastal erosion due to loss of sedimentary buffer. Loss of reef structure and habitat (Kench et al., 2008). Modification of currents, wave energy, and processes across the reef platform. Influences water circulation and flushing of lagoon reef flat.

all components of the reef system equally. For example, a particular event may promote severe disruption to local communities, but may have more limited impact on the surrounding coral reef system. The transition of an event or activity in a reef system from a discrete, short-lived impact to an impact that can be deemed hazardous and disruptive to the wider reef system is highly context specific. Fifth, the dominant metric for reef impact has been a shift in reef health, as commonly measured by changes in the proportion of living coral. Such a metric has limited value in assessing other important services provided by reefs. Consequently, this chapter adopts a more encompassing definition of reef health that includes the structural integrity of reefs and its ability to maintain a positive carbonate budget state (e.g., Perry et al., 2008). Sixth, when identifying hazards to reef systems, it is necessary to also consider the relaxation period for reefs to recover or evaluate recovery is possible. The following sections provide an overview of the key types of natural and anthropogenic stressors that impact the functioning of reef systems and the ecosystem services they provide.

15.3.1 Environmental Stresses on Reef Systems

Changes in boundary environmental processes can pose severe stresses for coral reefs (Figure 15.1). Such changes can occur as extreme high magnitude and short-lived pulse events or occur more gradually through time and build the pressure on reef systems.

15.3.1.1 Extreme Events

There are multiple geological stresses on reef systems that can cause catastrophic change in reef systems. In tectonically active zones, vertical displacement of reefs can be caused by earthquakes (Baird et al., 2005). For example, the earthquake that generated the Sumatran tsunami in 2004 promoted differential uplift and subsidence along >1,000 km of reef at the plate boundary between the Andaman and Nicobar islands (Bilham, 2005, Figure 15.2(a)). One of the most dramatic effects on reef systems was the uplift of some fringing reefs by 1.6–2.0 m on the west coast of the Andaman Islands (Sieh, 2005; Searle, 2006). Uplift of this magnitude caused extensive mortality of corals due to subaerial exposure (Hagan et al., 2007) and shifted previously subtidal reef communities into the intertidal zone, with likely medium-term consequence for survival and adaption of these communities (Hagan et al., 2007). Development of cracks in the reef structure can also occur through uplift, which reduces the structural integrity of the reef system (Bahuguna et al., 2008). In contrast, other tracts of reef subsided by up to 2.5–3.0 m (Searle, 2006; Hagan et al., 2007), displacing reef communities to deeper reef zones. While this subsidence has created new accommodation space for coral growth on reef tops, there have also been changes in hydrodynamics and water temperatures that may affect species' survival and reef community structure in the medium term. Aronson et al. (2012) recorded the impact of the 2009 earthquake in the Caribbean on previously monitored sites on the Belize reef. They note that reef slope destruction accounted for the loss of habitat in 10 (of 21) sites, rendering any speculation about resilience to past environmental change pointless.

Tsunami, generated by tectonic processes, undersea slumps and bolide impacts can also cause major impacts on coral systems. On human timeframes, tsunamis are relatively rare, but their impacts can be significant. More than 1,040 tsunamis have been recorded during 1914–2014 (Scheffers and Kelletat,

FIGURE 15.2 Summary of the range of effects of tsunami on coral systems. (a) Uplifted coral reef, Andaman Sea *(source unknown)*. (b) Overturned table coral at a 3-m water depth, Ari atoll, Maldives. *(Source: Gischler and Kikinger (2006).)* (c) Tsunami deposited coral block, Bonnaire *(source, author)*. (d) Seaward front of inferred tsunami ridge, southern coast of Anguilla (dated at 1,500 years before present; *source: Scheffers et al. (2009)*). (e) Reef flat corals buried by sand following the Indian Ocean tsunami, Hulhudhoo Island, Baa atoll, Maldives *(source: author)*. (f) Coarse clast tsunami deposit, Caribbean. *Source: Scheffers et al. (2009).*

2003) although the majority of these events have been low in magnitude with negligible impacts. Catastrophic tsunami, having flow depths at the shoreline >10 m, comprise <2 percent of the centennial tsunami record. However, on geological timescales, tsunamis are regular occurrences, there having been >2,000 events during the past 4,000 years (NGDC, 2009). Of relevance to an assessment of reef vulnerability to tsunami is that coral reefs and their associated sedimentary landforms occur throughout the Caribbean Sea and tropical Indian and Pacific Oceans, surrounded by tsunamigenic seismic zones and have been the subject of multiple tsunami during the Holocene. The Sumatran tsunami in 2004 was the worst tsunami disaster in recorded history and provided the first megatsunami with which scientists could document impacts on coral reefs, and against which they could calibrate historical reconstructions of tsunami impact (Kench et al., 2007).

Tsunami impacts on reefs can be expressed in different ways including (1) mechanical damage to corals such as the breaking and fracturing of corals (Baird et al., 2005, Figure 15.2(b)); (2) physical overturning and transport of corals from deeper reef fronts to reef flats (Figure 15.2(c),(d),(f)). For example, Goto et al. (2007) identified >1,000 one-meter size coral blocks transported from the reef front to the reef flat surface at Pakanang Cape, Thailand, as a consequence of the 2004 Sumatran tsunami; (3) sedimentation through deposition of remobilized deeper marine sediments smothering corals and causing mortality (Baird et al., 2005; Kelletat et al., 2007, Figure 15.2(e)). A synthesis of posttsunami surveys indicates a number of preconditioning factors that make reefs susceptible to tsunami damage. These factors include relative exposure of reefs to wave impact; reef slope and nearshore bathymetry that control wave shoaling and breaking potential; the water depth across the reef surface; the ecological composition of reef communities; and the substrate type upon which corals are attached. Due to this combination of factors, damage to reef communities is highly variable and difficult to predict (Baird et al., 2005; Kench, 2011b; Worachananant et al., 2007).

While the ecological impacts of tsunami can be benign in some reef areas, adjacent coastal landforms can experience catastrophic wave forces that physically destabilize and alter landforms and cause major losses of infrastructure and life. Shoreline erosion (net loss of land) is a commonly cited effect of tsunami wave interaction with reef sedimentary landforms. For instance, along a 9.2-km stretch of coastline at Lhok Nga Bay, Banda Aceh, the mean rate of shoreline displacement was approximately 60 m and involved reworking of 276,000 m^3 of sediment from the 2004 Sumatran tsunami (Paris et al., 2009). However, in other reef sites, erosion was less prevalent. Like ecological impacts, the magnitude of reported erosion is spatially variable. For example, based on a comparison of pre-tsunami and post-tsunami surveys of reef islands, Kench et al. (2006) found that erosion of the vegetated core of islands in the Maldives ranged from 1 percent to 9 percent of island area. There are a number of explanations for this disparity in erosion response that include proximity (exposure) to tsunami source; elevation of the seaward margin; differences in the way tsunami interact with the shoreline; the volume of sediment present at shorelines to act as a buffer to tsunami impact; and the presence or absence of vegetation. On Maldivian reef islands, where plentiful volumes of sediment were stored in the beaches, this material was able to absorb impact of the tsunami and shoreline erosion was minimal (Kench et al., 2008).

A number of studies have also shown that an additional geomorphic outcome of tsunami is the creation and aggradation of coastal landforms through overwash sedimentation. Landward deposition of sand sheets has been observed on numerous reef-fringed shorelines and low-lying atoll islands, where sediment rich coasts have undergone net erosion. Boulder fields are perhaps the most commonly cited product of the landward transfer of nearshore material under tsunami flow. Paleo studies have identified boulders (reef framework and coral bommies) up to an 8-m diameter and weighing >260 tonnes (Scheffers et al., 2009, Figure 15.2(c)).

Cyclones/hurricanes (storms) are an additional extreme "pulse" event that can impart swathes of destruction to coral reefs with significant ecological and geological outcomes. Much attention has been focused on the role of extreme storms and cyclones in promoting change in coral reef ecosystems (Woodley et al., 1981; Done, 1992; Woodley, 1992) and geomorphic structure of reef systems (e.g., Stoddart, 1963; Maragos et al., 1973; Bayliss-Smith, 1988; Scoffin, 1993; Blanchon et al., 1997). While some of these impacts are similar to those of tsunami, there are significant differences in the nature of extreme weather impacts. During cyclone/hurricane events, the waves that are generated differ markedly to tsunami. In particular, waves have shorter periods (12–20 s) and larger heights, which can exceed 10 m. During an individual storm event, reefs are assaulted by thousands of these large waves, which shoal and break across the shallow forereef and reef edge, delivering extreme force and energy on the reef. This impact differs from tsunami events, which are characterized by a smaller number of waves (~1–20) in an individual episode. In addition, low atmospheric pressure conditions that accompany extreme events can superelevate water levels across reef systems, resulting in destructive wave energy propagating across the reef flat to adjacent coastlines. Such effects include wave overtopping leading to flooding, and remobilization of shoreline sediments (erosion and accretion). Forbes et al. (2013) make reference to the significant damage to Niue, a raised atoll, arising as a result of cyclones Ofa and Heta. They argue that despite the elevation of infrastructure the narrow reef width offered limited protection from high wave energy.

At the extreme end of the impact spectrum, storms can decimate the living coral veneer of reefs to wave base water depths (the water depth of physical effect of waves), effectively resetting the physical substrate ready for recolonization. In such instances, the time lag for recovery may be many decades and the resultant diversity and abundance of coral cover may shift to a new ecological state (White et al., 2013). There is growing evidence that, like tsunami, the effects of such extreme storms can be highly localized. For example, Perry et al. (2014) examined the ecological and geomorphic impacts of Tropical Cyclone Yasi, a 700-km-wide category 5 cyclone, on sections of the Great Barrier Reef that were directly impacted. They found impacts of TC Yasi to be site specific and spatially heterogeneous, strongly influenced by the orientation of the reef to the storm path. At the most affected sites, small-scale taxa-specific impacts were documented (breaking of branching *Acropora* corals), but other sites remained unaffected. However, geomorphic changes were minor and ecological impacts highly variable between sites, with no observed evidence of major reef structural change.

While storm damage is potentially disruptive to coral structures and can negatively impact on the reef systems (in the short to medium term), there is evidence of storms playing a significant role in the geological development of reefs. Geomorphically, storms can have both destructive and constructive effects on reef landforms. Storms can mechanically erode coral material, converting it to sand, gravel, and boulders, which can be deposited onto reef surfaces and islands (Scoffin, 1993, Figure 15.3). For example, Hurricane Bebe in 1972 deposited 1.4×10^6 m^3 of storm rubble (dead coral) on the windward reef flat of Funafuti atoll, Tuvalu (Maragos et al., 1973, Figure 15.3). Post-storm reworking of this rubble has increased the area of reef islands by >10 percent (Baines and McLean, 1976). In another example, Chivas et al. (1986) and Hayne and Chappell (2001) show that Lady Elliot Island in the Great Barrier Reef was formed through the sequential deposition of storm rubble

FIGURE 15.3 Examples of cyclone-driven changes on coral systems: (a) storm blocks and rubble tract, southwest reef flat, Funafuti atoll. (b) Cyclone deposited boulders on island surface, Funamanu, Funafuti atoll, Tuvalu. (c) and (d) Remnant rubble ridges deposited following cyclone Bebe (1972) on the eastern reef flat of Funafuti atoll, Tuvalu. *Source of all photographs, author.*

banks. Sequential storm deposition has also been highlighted as a key agent in the vertical development of coral reefs structures (Blanchon et al., 1997).

Extreme events can impart both erosional and constructional responses on reef islands and therefore are of primary concern for coastal communities. The likely geomorphic response is dependent upon the physical size of material comprising land and the frequency and intensity of storms (Bayliss-Smith, 1988). Where storm frequency is low, landforms are generally composed of sand-size sediments (e.g., sand cays), which are susceptible to erosion during storms. For example, Stoddart (1963) documented mass destruction of some reef islands in the Belize Barrier Reef as a consequence of Hurricane Hattie. The recovery time for such landforms to reform may be decades. In contrast, in storm dominated reef settings, islands are commonly composed of coarse rubble on their exposed reef margins while islands on leeward reefs are typically composed of sand. As noted above, in these settings, storms may provide episodic additions of material allowing landforms to expand.

15.3.1.2 Slow-Onset Environmental Changes

The environmental boundary controls on reef ecosystem function and state, such as ocean water chemistry and sea level (Figure 15.1), are expected to change over the coming centuries forcing transitions in reef ecological and physical state, such as the change from healthy to degraded reef condition and associated loss of reef structure (e.g., Figure 15.4). Rising sea levels will inundate reef surfaces and create additional accommodation space for vertical reef growth. Sea-level rise *per se* poses no specific threats to reef ecological systems. Increasing water depth may promote shifts in habitat zonation and stimulate recolonization of coral reef surface. Considerable debate exists concerning the capacity of reefs to be able to accrete vertically with sea-level rise, which is likely dependent on the health of a particular reef system. Geological evidence suggests that healthy reefs may be able to respond to current projections of sea-level change (Kench et al., 2008). However, impacted reef sites may have less capacity to respond leading to submergent reef surfaces. Increasing water depths across reef surfaces, as a consequence of sea-level rise, poses more significant consequences for reef-associated landforms such as reef islands. Reefs act as an effective buffer to incident wave energy, as their shallow depth forces waves to shoal and break releasing their energy at the reef edge. However, as water depth increases, greater wave energy can propagate across reefs and impact shorelines promoting flooding and coastal change and threatening coastal infrastructure and communities (Sheppard et al., 2005).

15.3.1.2.1 Ocean Temperatures and Coral Bleaching

Sea-surface temperature has a major role in modulating the distribution and growth of coral systems. There are some predictions that warming of the oceans might extend the region of reef growth into areas that are currently too cool to

FIGURE 15.4 Examples of transitions in reef health showing changes in biological diversity and structure. (a)–(c) Caribbean example of transition from Acropora dominated reef flats (a) to dead framework (b) and loss of reef structure (c). (d)–(f) Maldivian example of transition from diverse and healthy reef (d) to partially bleached corals (e) and denuded reef surface where branching corals have collapsed (f). *Source of photographs, author.*

sustain reef development. Within the reef seas, increased sea-surface temperatures might be expected to increase coral metabolism and increase photosynthetic rates of zooxanthellae, thus aiding calcification. For example, studies of the growth of the massive coral *Porites* indicated that growth rate increased with increasing water temperature (Lough and Barnes, 2000). However, such findings need to be tempered by (1) the raising of sea-surface temperatures to levels at which growth is terminated, temporarily or permanently, by coral bleaching and temperature-related outbreaks of coral diseases; and (2) ocean acidification.

Corals respond to thermal stress by whitening or "bleaching" (Figure 15.4(e)). Bleaching is the physical expression of harm to corals and reflects the damage or loss through expulsion of zooxanthellae from coral tissue. Loss of zooxanthellae inhibits carbon fixation, coral growth, and reproductive ability. Impacts due to increased ocean temperature change propagate at different temporal scales. In the medium-term (seasons to years) El Nino Southern Oscillation (ENSO) cycles can modulate water temperature. The impacts of such events vary depending on the magnitude and temporal scale of increased temperature. Temperature elevations of +1 to +2 °C above regional seasonal maxima, which might span several weeks, is often species- and/or reef location-specific and may be repaired after a few months. However, larger temperature variations of +3 to +4 °C, can in extreme cases produce "mass bleaching" of entire reef communities, with coral mortality >90 percent (e.g., Douglas, 2003; Hoegh-Guldberg et al., 2007). There have been a number of widespread and major bleaching events accompanying ENSO cycles in 1982–1983, 1987–1988, 1994–1995 and particularly in 1998 when 16 percent of the world's reef-building corals were estimated to have died (Walther et al., 2002).

Temperature-induced coral bleaching is expected to increase in response to an expected increase in ENSO events (Glynn, 1993). It is thought that these impacts reflect an early signal of global warming in the oceans, with ENSO causes of bleaching superimposed on a secular trend of increasing sea-surface temperatures of approximately 1–2 °C over the coming century (Williams and Bunkley-Williams, 1990; Hoegh-Guldberg, 1999).

The net impact of the increased frequency of thermal stress events through long and medium-term changes may cause reefs to experience annual bleaching events at some stage in the future. Such scenarios have driven "time to reef extinction" projections (e.g., Hoegh-Guldberg, 1999; Sheppard, 2003). However, evidence suggests, that like other impacts bleaching effects are spatially patchy (e.g., Figure 15.4(e)) and temporally variable, even within a single reef system (Berkelmans et al., 2004; McClanahan et al., 2007) and bleaching can vary dramatically between species (Marshall and Baird, 2000). Further offsetting projections of total reef collapse is the recognition that corals may be able to acclimate (an individual, physiological response) or adapt (a genetic response at the population level) to changed thermal regimes (e.g., Brown et al., 2002; Baker et al., 2004).

15.3.1.2.2 Ocean Acidification and Coral Reefs

Progressive declines in ocean pH (increasing acidity), resulting from enhanced oceanic uptake of carbon dioxide, are broadly expected to result in decreased calcium carbonate production in marine organisms (e.g., Orr et al., 2005; Royal Society, 2005; ISRS, 2008). Studies suggest that calcification rates in corals will decrease by 30 ± 18 percent within the next 30–50 years, offsetting a possible enhancement of calcification by increased sea-surface temperature. Similar changes are estimated for other calcifying organisms (e.g., coralline algae). Few studies have explored the consequences of reduced calcification on the ecology and structure of reef systems. Weaker carbonate skeletons will render corals more susceptible to storm damage, although this may stimulate a higher rate of sediment production. Extreme scenarios suggest changes in ocean acidity (and associated aragonite saturation state) may result in corals being unable to form skeletons, which would have serious implications for maintenance of reef structures (Kleypas and Langdon, 2006). Broader effects of ocean acidification on coral recruitment and demography are at largely unknown (ISRS, 2008). However, evidence suggests that lowered growth rates and differential responses on coral species may lead to a reduced ability to compete for space and transformation of reef community structure (Cooper et al., 2008; e.g., Figure 15.4). Further, reduced calcification coupled with an increased dissolution of carbonate will impact on the ability of reefs to structurally accrete and shift reefs to a state of net erosion. The geomorphic consequences of these changes are likely to be the progressive degradation of reef frameworks and reduced reef topographic complexity (Figure 15.4(c) and (f)).

15.3.2 Anthropogenic Stresses on Reef Systems

A myriad of review studies synthesize work undertaken on the impacts of anthropogenic activities on reef health (Fabricius, 2005; Hodgson, 1999; Jaleel, 2013; Kayanne et al., 2005; Ateweberhan et al., 2013). In particular, these studies document the impacts of events on reef health and speculate on the viability of reef recovery. There are few longitudinal studies that allow comparison with baseline environmental health.

15.3.2.1 Physical Construction and Extractions

For atoll nations and coastal communities located adjacent to reef systems, population growth and development pressures drive demands for building, transport, and infrastructure expansion, which is focused on reef systems. Construction-related activities in reef systems fall into three broad categories: (1) dredging and excavation for harbors and navigation channels and harbors; (2) relocation of reef-derived aggregates for land reclamation to support transport infrastructure (airports, causeways) particularly in smaller atoll nations; and (3) mining and aggregate extraction for construction of onshore facilities, breakwaters; and sea defenses (examples shown in Figure 15.5).

FIGURE 15.5 Examples of anthropogenic activities that impact coral reef systems. (a) Excavation for navigation channel, Maldives. (b) Aggregate extraction through removal of massive corals, Maldives. (c) Dredging and land reclamation occupying reef area, Maldives. (d) Example of structurally modified island, Maldives. (e)–(g) Coastal protection works occupying reef and altering nearshore processes. (h) Dense population pressures on surrounding reef Malé, Maldives.

Each of these forms of construction activity can not only have serious direct impacts on reefs but can also leave an enduring legacy promoting medium-term impacts. Reclamation represents a permanent loss of the buried reefs as well as additional injury to reefs at the excavation/dredging sites and reefs adjacent to the fill areas (Figure 15.5(c) and (h)). In a recent controversial example, reclamation for airport extension at Majuro atoll, Marshall Islands, has resulted in dredging and obliteration of a healthy lagoon reef (Majuro Coral Mining, 2014). Reclaimed areas and shoreline structures that have not been correctly built frequently fail. These failures can result in the release of sediments, chemicals, and other debris, which can be deposited on reefs, expanding the area of injury to healthy reefs. Construction of road causeways (linking islands) such as those in South Tarawa atoll (Kiribati), Majuro atoll (Marshall Islands), and Palmyra atoll have disrupted lagoon circulation, increased lagoon water temperatures, and caused shoreline erosion (Maragos and Williams, 2011). In the case of Palmyra, the long-term consequence of these structures has been to degrade ocean reefs, and its lagoon reefs remain dead 70 years after causeway construction (Maragos and Williams, 2011). Dredging and excavation of boat channels also carries a range of negative effects for reef systems including resuspension of fine sediments; changing lagoon, and reef circulation; lowering water levels exposing reef crests and permanently exposing lagoon reefs causing their death; and biodiversity loss (Erftemeijer et al., 2012). Depending on the scale and type of activity, impacts are recorded as varying from localized short-term impacts to longer-term ecosystem disruption.

In many midocean settings, sources of aggregate are constrained and limited to coral, reef rock, and sand (Forbes et al., 2013). Mining of coral reefs and beaches for building materials have been linked to localized erosion (Caras and Pasternak, 2009, Figure 15.5(a),(b),(f)) and habitat destruction (Jaleel, 2013; Caras and Pasternak, 2009). Caras and Pasternak (2009) observe the limited recovery of a reef postmining in Wakatobi marine park, Indonesia, even after two decades postmining activity, though Clifton et al. (2010) caution that wider consideration of stressors should also be accounted for. Coral harvesting is a more selective extraction of coral. Such practice has been documented in the Maldives where select removal of larger coral heads contributes to ecological change (Brown and Dunne, 1988), while also lowering reef surfaces and allowing higher energy waves to impact shorelines. In other areas, coral harvesting is driven by the aquarium trade, leading to localized coral depletion (Caras and Pasternak, 2009; Gomez, 1983).

15.3.2.2 Pollution and Changes in Water Quality

Anthropogenically driven pollution in reef systems is derived from multiple activities and can occur as point or diffuse inputs to reefs and be locally generated or originate from distal locations and transferred to reef systems.

While the deterioration of reef quality has been linked with sewage outfall, aquaculture, and agricultural runoff, determining a dominating influencing parameter is difficult (Edinger and Risk, 2000; Fabricius, 2005). Typical local point sources of pollution include sewage discharges that can locally alter water temperatures and the nutrient balance in the vicinity of the outfalls promoting ecological impacts. For example, sewage discharges in the lagoon of Kāne'ohe Bay, Hawaii, dating from the 1960s triggered the decline of corals and proliferation of green algae, which was stimulated by higher nutrient levels, that smothered corals (Maragos and Williams, 2011). Similar changes in nearshore nutrient conditions have been recorded from fertilizers derived from land-based agricultural practices.

Release of sediments to reef systems is generally through diffuse sources from livelihood activities. Clearing land for farming and agricultural production, deforestation, and mining activities have all been linked to observable increases in suspended sediment flows into local reef systems (Maina et al., 2013). For example, in Southeast Asia, land-based rain forest logging and poor agricultural practices increased sedimentation rates on reefs threatening the health of >20 percent or coral reefs in that region (Burke et al., 2002).

The impact of sedimentation on corals is exacerbated in environments characterized by more finely grained sediments and includes disruption of coral larvae, shading, smothering, and increased bacterial activity (Erftemeijer et al., 2012; Fabricius, 2005). Corals are light sensitive and while species have differing tolerance thresholds the impact of suspended sediments on light penetration can be marked (Erftemeijer et al., 2012). Prolonged light suppression has been observed to lead to reduced calcification and coral growth in some species (Fabricius, 2005). A decline in water clarity and quality has been associated with increased presence of macroalgal growth (Yap, 2013).

The effects of sedimentation on reef resilience is varied. The capacity of individual corals and reef systems to cope with increased sedimentation loads is dependent on the frequency and intensity of the episode and the ambient health of the reef (Erftemeijer et al., 2012). Estimates of critical thresholds of sedimentation rates for coral survival range from 10 to 300 mg cm^{-2} day^{-1}, and indicate a tolerance in some species for high short-term sedimentation rates. Despite historically high sedimentation rates on the Great Barrier Reef a diversity of coral has been observed (Erftemeijer et al., 2012); however, such tolerance is geographically variable and sustained sedimentation rates may result in a shift in species diversity and loss of sensitive species.

15.3.2.3 Exploitation of Biological Resources

Overfishing is perhaps the most acknowledged anthropogenic stress on reef systems and has a long history of impact on reef systems (Jackson et al., 2001).

Reef fisheries provide a key source of household protein and income for many communities. For example, Yap (2013) estimates coral reef-based fishing globally generates $5 billion annually. With increasing population pressures and changing methods of fishing, the sustainability of reef fisheries has been questioned. Combined with pressures of overfishing and the practice of the use of sodium cyanide in coral colonies to stun fish for the aquarium trade, a number of destructive methods such as blast fishing, where explosives are used to stun fish (Fox and Caldwell, 2006), have been documented as having localized ecosystem impacts. Blast fishing can result in a crater like reef landscape as substrate is destroyed (Yap, 2013). In longitudinal studies, Fox and Caldwell (2006) observed coral recovery in sites that have experienced a single blast but in sites with extensive blasting coral the unstable remains of coral rubble sheets limited the recovery of species.

A key concern with the impacts of overfishing is not exclusive to the methods of extraction. The loss of keystone species can result in potential disruption to trophic structures and alter benthic diversity. Studies variously link the loss of predators (Ruppert et al., 2013) and herbivores (Hughes, 1994) to observable shifts in reef systems.

15.4 THE COMPLEXITY OF MULTIPLE HAZARDS AND IMPLICATIONS FOR CORAL REEF RESILIENCE

This chapter has highlighted the varied and complex interrelations between internal and external drivers of coral reef health, reinforcing the need to recognize the complexity of coral system responses to environmental change (Figure 15.1). The previous section examined how reef systems might respond to natural and anthropogenic stressors that promote hazards that affect the health and geologic structure of reefs and impact coastal communities. It is apparent that impacts and the propagation of hazards in reef systems are neither geographically coherent nor homogenous across time (Mumby and Steneck, 2008). Much of this variability in the impact of hazards can be attributed to the antecedent conditions of a particular reef that determine its predisposition to impact and consequent hazards for coastal communities.

The impact of hazards on reef environments and the communities that depend on them is closely linked to the baseline health and structure of the reef. Reef health, as measured by the proportion of living coral, is a common metric for the condition of a reef (Roff and Mumby, 2012). However, the diversity and abundance of reefs vary markedly between reef regions (Bellwood et al., 2004). Regions with reduced diversity (e.g., Caribbean) may be dominated by a single species, which when lost can have a significant effect on reef health and may take considerable time periods to recover from stress events. However, zones of high biodiversity, such as the Indo-Pacific, are less susceptible to impacts on individual species and may rebound from stress events more rapidly.

The degree of anthropogenic modification or degradation may play a significant role in the magnitude of impact suffered by reefs. Reef systems that maintain their biophysical process interactions, and which might be considered healthy, are best placed to tolerate environmental stresses. Such reefs may absorb impacts or recover rapidly from a specific stress event. However, heavily degraded reefs in which ecological and physical process interactions are compromised (Figure 15.1) are less able to withstand environmental stressors leading to proportionately more severe impacts and longer recovery timescales (Figure 15.4).

15.4.1 Recovery Resilience and Hazard Propagation

Inherent in the conceptual model of coral reef systems (Figure 15.1) is a capacity to change or adapt to perturbations. A summary of impact and recovery timescales for different elements of reef systems (Table 15.2) suggests that recovery or resilience is dependent upon on the component of the reef that experiences stress. For example, a moderate temperature driven bleaching event will impact the living coral cover, but recovery may take as little as 12 months and have minor effects on local communities. In contrast, a cyclone event may promote severe infrastructural damage and recovery may take decades.

It is also important to note that not all exposure units in a reef system will be damaged equally as a consequence of any one hazard driver. As demonstrated by the impact of tsunami on midocean atoll islands, the structure and health of the reef systems experienced only minor impacts, whereas the sedimentary landforms had significantly greater exposure and impact. In the Maldives, the tsunami left an indelible imprint on islands through erosion of shorelines and flooding of island surfaces, which in some instance destroyed villages, freshwater lenses and killed vegetation (Kench et al., 2006). In contrast, the enduring effects of pollution may promote the degradation of living reef that might impact available food resources for coastal communities. However, the physical substrate of coastal communities may remain unchanged in the short to medium term.

15.4.2 Multiple Stressors and Cumulative Impacts on Reefs

It is increasingly apparent that examining the discrete effect of individual stresses on reefs, while instructive, belies a more complicated hazard spectrum that affects reef systems and compounds hazard impacts. For example, bleaching is known to degrade the proportion of living coral on reefs and may lead to geomorphic changes that affect ecosystem services. However, on some reefs, bleaching can be compounded by a range of anthropogenic stresses, the cumulative effects of which compromise the natural functioning of a reef. This notion has stimulated researchers to search for critical thresholds and "tipping points" at which entire reef systems may collapse. At the heart of such system, responses are conceptual notions of system recovery and resilience in response to the frequency and magnitude of stress episodes and to underlying shifts in environmental boundary conditions, whether forced through natural or anthropogenic processes (Figure 15.6). An individual reef evolves within a set of environmental conditions (Table 15.1) and can maintain its reef health within a defined range of biophysical tolerances (temperature, wave energy, etc.). Over time, a reef may be exposed to periodic extreme events, the severity of impact of which is dependent on the antecedent condition of the reef, together with the magnitude and persistence of the stress episodes. System recovery is also temporally dependent on the magnitude of impact and full recovery (years to decades) is conditional on the absence of a further stress event (i.e., recovery timescales fit within the occurrence frequency of an individual stress event). Where a system is perturbed by additional stressors, within the recurrence frequency of the initial stress event, system recovery is set back and may compound impacts ultimately leading to major changes in ecological condition, which will propagate through to the structural integrity of the reef and stability of adjacent landforms (Perry et al., 2008).

The coral reef "phase shift" is perhaps the best understood example of such multiple compound stressors that promote catastrophic changes in reef health and which over time can promote significant changes in reef structure that undermine the ecosystem services provided by reefs (Figure 15.4(a) and (b)). A well-documented example is the significant transition in the health of the north Jamaican reefs. The north Jamaican reefs were described as the most diverse in the Caribbean in the mid-twentieth century with high coral cover (~50–60 percent) dominated by the branching corals *Acropora palmata* and *Acropora cervicornis* (Goreau, 1959). A combination of factors have been implicated in the reduction of coral cover to <5 percent and dominance of fleshy macroalgae (Hughes, 1994). Key factors in this transition include overfishing, which reduced herbivory of algae; Hurricane Allen, which impacted Jamaica in 1980 had devastating impacts on the reefs reducing shallow water coral to rubble, with an immediate loss of three-dimensional reef structure (Woodley et al., 1981); the mass mortality of herbivorous urchins; the consequent dominance of fleshy algae; and further storms and bleaching episodes. In this example, removal of herbivores from the reef through overfishing and the mortality of urchins allowed fleshy algae to dominate and prevent coral recruitment and subsequent recovery.

In addition to the impact-recovery dynamics of reef systems to pulse events, it is increasingly recognized that such cycles are superimposed on longer-term changes in environmental boundary conditions, which place reef systems on trajectories beyond their normal ranges of environmental variability (Figure 15.6(c) and (d)). For example, the increasing number of bleaching episodes in the past decades is believed to be a symptom of underlying increases in global sea-surface temperatures. The capacity of

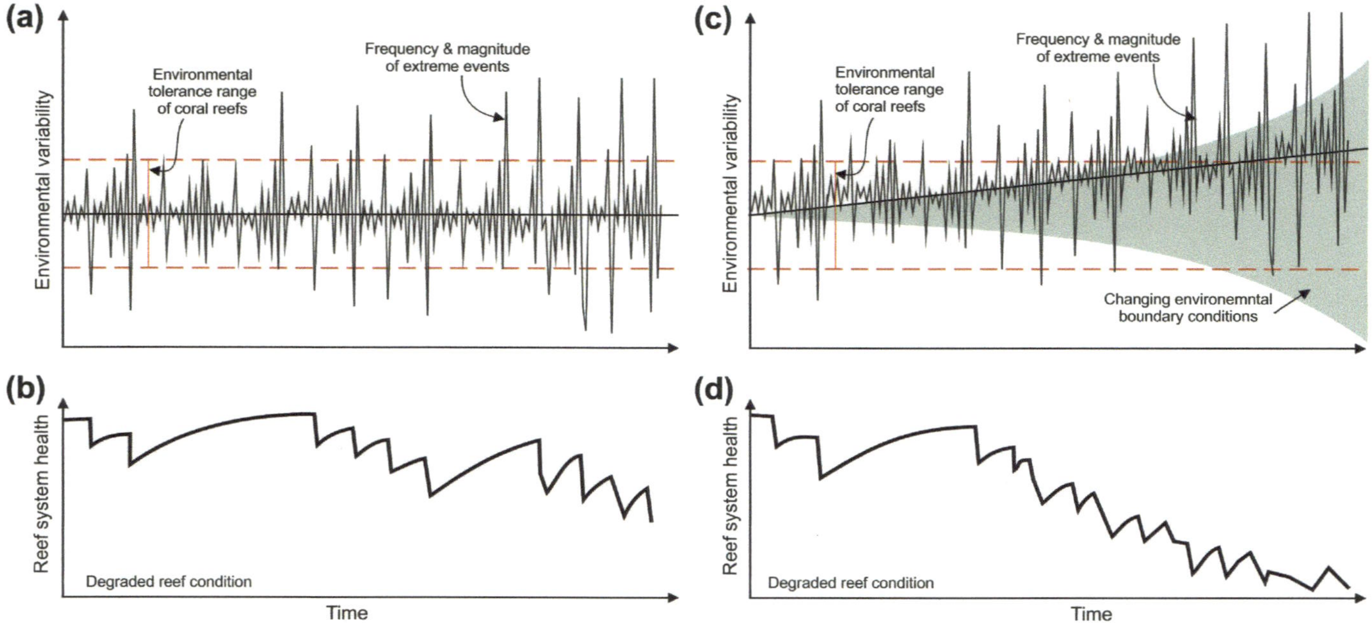

FIGURE 15.6 Conceptual diagram showing the variability in environmental stresses (a), (b) and reef system impact and recovery (b), (d). Variability changes can represent any natural or anthropogenic shock to the reef system. (a) Presents variability in pulse events that stress reefs and (b) the influence of changes in magnitude and frequency of these events on reef system health. (c) Shows variability in pulse events, their magnitude and frequency with changing environmental baselines, and their impact on reef condition (d). Note, area between dashed red lines reflects normal environmental tolerance range for coral reefs (i.e., temperature, salinity).

reef systems to acclimate to new tolerances ranges is not well resolved, with a number of studies projecting imminent collapse of reef ecological systems (Hoegh-Guldberg, 1999, Figure 15.4).

The geomorphic stability of low-lying reef islands provides an additional example of such compounding stress impacts. Low-lying reef islands are accumulations of sand and gravel deposited on reef surfaces through the focusing effect of wave and currents (Kench et al., 2014). The upper limit of island development is controlled by wave runup processes at the shoreline. In the case of sand cays, storms are known to promote short-term erosion with recovery cycles of months to decades depending on the sediment availability and process regime (Bayliss-Smith, 1988; Ford and Kench, 2014). However, an increased frequency of storms will constrain recovery and promote contraction of island shorelines. Added to these storm−recovery cycles are rising sea levels, which are expected to modify the process regime of island shorelines and sediment supply over the coming century. Increased water depths across reefs are expected to deliver higher wave energy promoting a phase of increased sediment transport. More extreme predictions suggest that water depth thresholds exist that will lead to disappearance of islands from reef surfaces (Dickinson, 1999; Woodroffe, 2008). While the precise impacts on sediment supply and processes are not well defined, it is expected that islands have begun a phase of increased instability and island migration on reef surfaces, which pose a major hazard to island dwelling communities (Webb and Kench, 2010).

15.5 CONCLUSIONS

Coral reefs are unique systems representing the balance between ecological and physical processes. Reefs are not only hotspots of biological diversity but they are also geological structures. Collectively, the ecology and geomorphic structure of reefs provide a host of ecosystem services including support for coastal communities. Understanding hazards in the reef system requires a complex understanding of the linkages between ecological and physical processes and how changes propagate through the reef system to impact reef biological and structural characteristics. This chapter has provided an overview of key event and incremental stressors that promote impacts on reef systems.

However, coral reef systems are resilient to a range of natural and anthropogenic perturbations. The use of a reef system model (Figure 15.1) elucidates this complexity and allows for consideration of factors that may be influential in future reef system resilience to hazards. Event-driven natural hazards impact local community ability to cope and respond but may have shorter term implications for reef health. However, placed in the context of reef degradation, these systems may be slower to recover from perturbations, placing greater stress on reliant communities.

Integrative studies are emerging that provide a greater understanding of the responses of reef ecology to these long-term and pulse events. This work is necessary to provide local context for reef recovery from hazards and to inform predictive models and improve understanding of the drivers of reef health. Under future scenarios of environmental change, it is evident that coral reef dynamics will be modified and ecosystems regimes may shift (Yamano et al., 2005). Such studies highlight a need to understand species resilience to changing environmental conditions (Hughes et al., 2003) and the dynamics of the terrestrial landforms upon which communities reside (Webb and Kench, 2010).

While effort is being directed at modeling future responses of coral and reef systems to climatic change, these predictions need to be evaluated in the context of the population and development pressures placed on the reef environment. Reef hazard impacts are frequently influenced by socioeconomic decisions, and for management of reef resilience there needs to be an understanding of the implications of shifts in land use and activity. Changes to agricultural practices or a shift from a fishing dominated economy can have unforeseen consequences for the reef environment (Bruggeman et al., 2012). While tourism is arguably a nonextractive activity, the impact of infrastructure development and population increases is well documented as having potential negative implications for the health of reefs. While risk and resilience to hazard exposure is a factor of the complexity of the biogeochemical interactions in a system, work highlights that anthropogenic impacts are a constraining factor in reef response and recovery.

REFERENCES

Aronson, R.B., Precht, W.F., Macintyre, I.G., Toth, L.T., February 2012. Catastrophe and the life span of coral reefs. Ecology 93 (2), 303–313.

Ateweberhan, M., Feary, D.A., Keshavmurthy, S., Chen, A., Schleyer, M.H., Sheppard, C.R.C., 2013. Climate change impacts on coral reefs: Synergies with local effects, possibilities for acclimation, and management implications. Marine Pollution Bulletin 74, 526–539.

Bahuguna, A., Nayak, S., Roy, D., 2008. Impact of the tsunami and earthquake of 26th December 2004 on the vital coastal ecosystems of the Andaman and Nicobar Islands assessed using RESOURCESAT AWiFS data. International Journal of Applied Earth Observation and Geoinformation 10, 229–237.

Baines, G.B.K., McLean, R.F., 1976. Sequential studies of hurricane bank evolution at Funafuti atoll. Mar. Geol. 21, M1–M8.

Baird, A.H., Campbell, S.J., Anggoro, A.W., Ardiwijaya, R.L., Fadi, N., Herdiana, Y., Kartawijaya, T., Mahyiddin, D., Mukminin, A., Pardede, S.T., Pratchett, M.S., Rudi, E., Siregar, A.M., 2005. Acehnese reefs in the wake of the Asian tsunami. Curr. Biol. 15, 1926–1930.

Baker, A.C., et al., 2004. Corals' adaptive response to climate change. Nature 430, 741.

Bayliss-Smith, T.P., 1988. The role of hurricanes in the development of reef islands, Ontong Java Atoll, Solomon Islands. Geogr. J. 154, 377–391.

Bellwood, D.R., Hughes, T.P., Folke, C., Nyström, M., 2004. Confronting the coral reef crisis. Nature 429, 827–833.

Berkelmans, R., et al., 2004. A comparison of the 1998 and 2002 coral bleaching events on the Great Barrier Reef: spatial correlation and predictions. Coral Reefs 23, 75–83.

Best, B., Bornbusch, A., 2005. Global Trade and Consumer Choices: Coral Reefs in Crisis. American Association for the Advancement of Science, Washington, D.C., p. 28.

Bilham, R., 2005. A flying start, then a slow slip. Science 308, 1126–1127.

Blanchon, P., Jones, B., Kalbfleisch, W., 1997. Anatomy of a fringing reef around Grand Cayman: storm rubble not coral framework. J. Sediment. Res. 67, 1–16.

Brown, B.E., Dunne, R.P., 1988. The impact of coral mining on coral reefs in the Maldives. Environ. Conserv. 15, 159–165.

Brown, B.E., et al., 2002. Experience shapes the susceptibility of a reef coral to bleaching. Coral Reefs 21, 119–126.

Bruggemann, J.H., Rodier, M., Guillaume, M.M.M., Andréfouët, S., Arfi, R., Cinner, J.E., Pichon, M., Ramahatratra, F., Rasoamanendrika, F., Zinke, J., McClanahan, T.R., 2012. Wicked social–ecological problems forcing unprecedented change on the latitudinal margins of coral reefs: the case of southwest Madagascar. Ecol. Soc. 17 (4), 47. http://dx.doi.org/10.5751/ES-05300-170447.

Buddemeier, R.W., Kleypas, J.A., Aronson, R.B., 2004. Coral Reefs and Global Climate Change: Potential Contributions of Climate Change to Stresses on Coral Reef Ecosystems. Pew Center on Global Climate Change, Arlington, p. 44.

Burke, L., Selig, E., Spalding, M., 2002. Reefs at Risk in Southeast Asia. World Resources Institute, Washington DC, p. 72.

Caras, T., Pasternak, Z., 2009. Long-term environmental impact of coral mining at the Wakatobi marine park, Indonesia. Ocean Coastal Manage. 52, 539–544.

Chappell, J., 1980. Coral morphology, diversity and reef growth. Nature 286, 249–252.

Chivas, A., et al., 1986. Radiocarbon evidence for the timing and rate of island development, beach-rock formation and phosphatization at Lady Elliot Island, Queensland, Australia. Mar. Geol. 69, 273–287.

Clifton, J., Cullen, L.C., Haapkylä, J., Unsworth, R.K.F., 2010. Ensuring appropriate and proportionate responses to environmental threats: a response to Caras and Pasternak. Ocean Coastal Manage. 53, 700–702.

Cooper, T.F., et al., 2008. Declining coral calcification in massive *Porites* in two nearshore regions of the northern Great Barrier Reef. Global Change Biol. 14, 529–538.

Cowell, P.J., Thom, B.G., 1994. Morphodynamics of coastal evolution. In: Carter, R.W.G., Woodroffe, C.D. (Eds.), Coastal Evolution: Late Quaternary Shoreline Morphodynamics. Cambridge University Press, Cambridge, pp. 33–86.

Dickinson, W.R., 1999. Holocene sea level record on Funafuti and potential impact of global warming on central pacific atolls. Quat. Res. 51, 124–132.

Done, T.J., 1992. Phase shifts in coral reef communities and their ecological significance. Hydrobiologia 247, 121–132.

Douglas, A.E., 2003. Coral bleaching—how and why? Mar. Pollut. Bull. 46, 385–392.

Edinger, E.N., Risk, M.J., 2000. Effects of land-based pollution on central Java coral reefs. J. Coastal Dev. 3 (2), 593–613.

Erftemeijer, P.L.A., Riegl, B., Hoeksema, B.W., Todd, P.A., 2012. Environmental impacts of dredging and other sediment disturbances on corals: a review. Mar. Pollut. Bull. 64, 1737–1765.

Fabricius, K.E., 2005. Effects of terrestrial runoff on the ecology of corals and coral reefs: review and synthesis. Mar. Pollut. Bull. 50, 125–146.

Fox, H.E., Caldwell, R.L., 2006. Recovery from blast fishing on coral reefs: a tale of two scales. Ecol. Appl. 16 (5), 1631–1635.

Forbes, D.L., James, T.S., Sutherland, M., Nichols, S.E., 2013. Physical basis of coastal adaptation on tropical small islands. Sustain. Sci. 8, 327–344.

Ford, M.R., Kench, P.S., 2014. Formation and recovery of typhoon impacted reef islands interpreted from remote imagery: Nadikdik Atoll, Marshall Islands. Geomorphology 214, 216–222.

Gischler, E., Kikinger, R., 2006. Effects of the tsunami of 26 December 2004 on Rasdhoo and northern Ari atolls, Maldives. Atoll Research Bulletin 544, 93–104.

Glynn, P.W., 1993. Coral reef bleaching: ecological perspectives. Coral Reefs 12, 1–17.

Gomez, E.D., 1983. Perspectives on coral reef research and management in the Pacific. Ocean Manage. 8, 281–295.

Goreau, T.F., 1959. The Ecology of Jamaican Coral Reefs I. Species Composition and Zonation. Ecology 40 (1), 67–90.

Goto, K., Chavanich, S.A., Imamura, F., Kunthasap, P., Matsui, T., Minoura, K., Sugawara, D., Yanagisawa, H., 2007. Distribution, origin and transport process of boulders deposited by the 2004 Indian Ocean tsunami at Pakarang Cape, Thailand. Sediment. Geol. 202, 821–837.

Gourlay, M.R., 1988. Coral cays: products of wave action and geological processes in a biogenic environment. In: Proceedings, Sixth International Coral Reef Symposium, Townsville, vol. 2, pp. 491–496.

Hagan, A.B., Foster, R., Perera, N., Gunawan, C.A., Silaban, I., Yaha, Y., Manuputty, Y., Hazam, I., Hodgson, G., 2007. Tsunami impacts in the Ache province and North Sumatra, Indonesia. Atoll Res. Bull. 544, 37–54.

Hayne, M., Chappell, J., 2001. Cyclone frequency during the last 5000 years at Curacao Island, north Queensland, Australia. Palaeogeogr., Palaeoclimatol., Palaeoecol. 168, 207–219.

Hodgson, G., 1999. A global assessment of human effects on coral reefs. Mar. Pollut. Bull. 38 (5), 345–355.

Hoegh-Guldberg, O., 1999. Climate change, coral bleaching and the future of the world's coral reefs. Aust. J. Mar. Freshwater Res. 50, 839–866.

Hoegh-Guldberg, O., et al., 2007. Vulnerability of reef-building corals on the Great Barrier reef to climate change. In: Johnson, J.E., Marshall, P. (Eds.), Climate Change and the Great Barrier Reef: a Vulnerability Assessment. GBRMPA/Australia Greenhouse Office, Australia, pp. 271–307.

Hughes, T.P., 1994. Catastrophes, phase shifts, and large scale degradation of a Caribbean coral reef. Science 265, 1547–1551.

Hughes, T.P., Baird, A.H., Bellwood, D.R., Card, M., Connolly, S.R., Folke, C., Grosberg, R., Hoegh-Guldberg, O., Jackson, J.B.C., Kleypas, J., Lough, J.M., Marshall, P., Nystrom, M., Palumbi, S.R., Pandolfi, S.R., Rosen, B., Roughgarden, J., August 15, 2003. Climate change, human impacts, and the resilience of coral reefs. Science 301, 929–933.

ISRS, 2008. Coral Reefs and Ocean Acidification, Briefing Paper 5. International Society for Reef Studies, p. 9. http://www.fit.edu/isrs/documents/ISRS_BP_ocean_acid_final28jan2008.pdf.

Jackson, J.B.C., Kirby, M.X., Berger, W., Bjorndal, K.A., Botsford, L.W., Bourque, B.J., Bradbury, R.H., Cooke, R., Erlandson, J., Estes, J.A., Hughes, T.P., Kidwell11, S., Lange1, C.B., Lenihan, H.S., Pandolfi1, J.M., Peterson, C.H., Steneck, R.S., Tegner1, M.J., Warner, R.R., 27 2001. Historical Overfishing and the Recent Collapse of Coastal Ecosystems. Science 293 (5530), 629–637.

Jaleel, A., 2013. The status of the coral reefs and the management approaches: the case of the Maldives. Ocean Coastal Manage. 82, 104–118.

Kayanne, H., Chikamori, M., Yamano, H., Yamaguchi, T., Yokoki, H., Shimazaki, H., 2005. Interdisciplinary approach for sustainable land management of atoll islands. Global Environ. Res. 9 (1), 1–7.

Kelletat, D., Scheffers, S.R., Scheffers, A., 2007. Field signatures of the SE-Asian mega-tsunami along the west coast of Thailand compared to Holocene paleo-tsunami from the Atlantic region. Pure Appl. Geophys. 164, 413–431.

Kench, P.S., 2011a. Eco-morphodynamics. In: Hopley, D. (Ed.), Encyclopedia of Modern Coral Reefs: Structure Form and Process. Springer, The Netherlands, pp. 359–363.

Kench, P.S., 2011b. Tsunami. In: Hopley, D. (Ed.), Encyclopedia of Modern Coral Reefs: Structure Form and Process. Springer, The Netherlands, pp. 1096–1110.

Kench, P.S., 2013. Coral systems. In: Shroder, John F., Sherman, D.J. (Eds.), Treatise on Geomorphology, Coastal Geomorphology, vol. 10. Academic Press, San Diego, pp. 328–359.

Kench, P.S., Brander, R.W., 2006a. Response of reef island shorelines to seasonal climate oscillations: south Maalhosmadulu atoll, Maldives. J. Geophys. Res. 111, F01001.

Kench, P.S., Brander, R.W., 2006b. Wave processes on coral reef flats: implications for reef geomorphology using Australian case studies. J. Coastal Res. 2, 209–223.

Kench, P.S., McLean, R.F., Brander, R.W., Nichol, S.L., Smithers, S.G., Ford, M.R., Parnell, K.E., Aslam, M., 2006. Geological effects of tsunami on mid-ocean atoll islands: the Maldives before and after the Sumatran tsunami. Geology 34, 177–180.

Kench, P.S., McLean, R.F., Nichol, S.L., Smithers, S.G., Brander, R.W., 2007. Impact of the Sumatran tsunami on the geomorphology and sediments of reef islands: south Maalhosmadulu atoll, Maldives. Atoll Res. Bull. 544, 105–134.

Kench, P.S., Perry, C.T., Spencer, T., 2008. Coral reefs. Chapter 7. In: Slaymaker, O., Spencer, T., Embleton-Hamann, C. (Eds.), Geomorphology and Global Environmental Change. Cambridge University Press, Cambridge, pp. 180–213.

Kench, P.S., Owen, S.D., Ford, M.R., 2014. Evidence for coral island formation during rising sea level in the Central Pacific ocean. Geophys. Res. Lett. 41, 820–827.

Kennedy, E.V., Perry, C.T., Halloran, P.R., Iglesias-Prieto, R., Schönberg, C.H.L., Wisshak, M., Form, A.U., Carricart-Ganivet, J.P., Fine, M., Eakin, C.M., Mumby, P.J., 2013. Avoiding coral reef functional collapse requires local and global action. Curr. Biol. 23, 912–918.

Kinsey, D.W., 1983. Standards of performance in primary production and carbon turnover. In: Barnes, D.J. (Ed.), Perspectives on Coral Reefs, pp. 209–220 (B. Clouston/AIMS, Manuka).

Kinzie III, R.A., Buddemeier, R.W., 1996. Reefs happen. Global Change Biol. 2, 479–494.

Kleypas, J., McManus, J., Menez, L., 1999. Environmental limits to coral reef development: where do we draw the line? Am. Zool. 39, 146–159.

Kleypas, J., Langdon, C., 2006. Coral reefs and changing seawater chemistry. In: Phinney, J.T., et al. (Eds.), Coral Reefs and Climate Change: Science and Management, AGU Monograph Series, Coastal and Estuarine Studies, vol. 61. American Geophysical Union, Washington D.C, pp. 73–110.

Laurans, Y., Pascal, N., Binet, T., Brander, L., Clua, E., David, G., Rojat, D., Seidl, A., 2013. Economic valuation of ecosystem services from coral reefs in the South Pacific: taking stock of recent experience. J. Environ. Manage. 116, 135–144.

Lough, J.M., Barnes, D.J., 2000. Environmental controls on growth of the massive coral Porites. J. Exp. Mar. Biol. Ecol. 245, 225–243.

McClanahan, T.R., et al., 2007. Predictability of coral bleaching from synoptic satellite and in situ temperature observations. Coral Reefs 26, 695–701.

Maina, J., de Moel, H., Zinke, J., Madin, J., McClanahan, T., Vermaat, E., 2013. Human deforestation outweighs future climate change impacts of sedimentation on coral reefs. Nat. Commun. 4 (1986), 1–7.

Majuro Coral Mining 2014. http://reefrelief.org/2014/03/update-majuro-coral-mining/.

Maragos, J.E., Baines, G.B.K., Beveridge, P.J., September 21, 1973. Tropical cyclone Bebe creates a new land formation on Funafuti atoll. Science, 1161–1163.

Maragos, J.E., Williams, G.J., 2011. Pacific coral reefs: an introduction. In: Encyclopedia of Modern Coral Reefs: Structure Form and Process. Springer, The Netherlands, pp. 753–776.

Marshall, P.A., Baird, A.H., 2000. Bleaching of corals on the Great Barrier Reef: differing susceptibilities among taxa. Coral Reefs 19, 155–163.

Moberg, F., Folke, C., 1999. Ecological goods and services of coral reef ecosystems. Ecol. Econ. 29, 215–233.

Montaggioni, L.F., 2005. History of Indo-Pacific coral reef systems since the last glaciation: development patterns and controlling factors. Earth Sci. Rev. 71, 1–75.

Morrison, R.J., Denton, G.R.W., BaleTamata, U., Grignon, J., 2013. Anthropogenic biogeochemical impacts on coral reefs in the Pacific Islands—an overview. Deep-Sea Res. II 96, 5–12.

Mumby, P.J., Steneck, R.S., 2008. Coral reef management and conservation in light of rapidly evolving ecological paradigms. Trends Ecol. Evol. 23 (10), 555–563.

NGDC, 2009. National Geophysical Data Center. Tsunami Hazard Database. www.ngdc.noaa.gov/hazard.

Orr, J.C., et al., 2005. Anthropogenic ocean acidification over the twenty-first century and its impact on calcifying organisms. Nature 437, 681–686.

Paris, R., Wassmer, P., Sartohadi, J., Lavigne, F., Barthomeuf, B., Desgages, E., Grancher, D., Baumert, P., Vautier, F., Brunstein, D., Gomez, C., 2009. Tsunamis as geomorphic crises: lessons from the December 26, 2004 tsunami in Lhok Nga, west Banda Aceh (Sumatra, Indonesia). Geomorphology 104, 59–72.

Perry, C.T., Larcombe, P., 2003. Marginal and non-reef building coral environments. Coral Reefs 22, 427–432.

Perry, C.T., Hepburn, L.J., 2008. Syn-depositional alteration of coral reef framework through bioerosion, encrustation and cementation: taphonomic signatures of reef accretion and reef depositional events. Earth Sci. Rev. 86, 106–144.

Perry, C.T., Spencer, T., Kench, P.S., 2008. Carbonate budgets and reef production states: a geomorphic perspective on the ecological phase shift concept. Coral Reefs 27, 853–866.

Perry, C.T., Edinger, E.N., Kench, P.S., Murphy, G.N., Steneck, R.S., Smithers, S.G., Mumby, P.J., 2012. Estimating rates of biologically driven coral reef framework production and erosion: a new census-based carbonate budget methodology and applications to the reefs of Bonaire. Coral Reefs 31 (3), 853–868. http://dx.doi.org/10.1007/s00338-012-0901-4.

Perry, C.T., Smithers, S.G., Kench, P.S., Pears, B., 2014. Impacts of Cyclone Yasi on nearshore, terrigenous sediment-dominated reefs of the central Great Barrier Reef, Australia, impacts of Cyclone Yasi on nearshore, terrigenous sediment-dominated reefs of the central Great Barrier Reef, Australia. Geomorphology 222, 92–105. http://dx.doi.org/10.1016/j.geomorph.2014.03.012.

Polidoro, B., Carpenter, K., 2013. Dynamics of reef recovery. Science 340, 34–35.

Richmond, B.M., Buckley, M., Etienne, S., Chague-Goff, C., Cark, K., Goff, J., Dominey-Howes, D., Strotz, L., 2011. Deposits, flow characteristics, and landscape change resulting from the September 2009 South Pacific tsunami in the Samoan islands. Earth Sci. Rev. 107, 38–51.
Roff, G., Mumby, P.J., 2012. Global disparity in the resilience of coral reefs. Trends Ecol. Evol. 27 (7), 404–413.
Royal Society, 2005. Ocean Acidification due to Increasing Atmospheric Carbon Dioxide. Policy document 12/05. The Royal Society, London, p. 223.
Ruppert, J.L.W., Travers, M., Smith, L.L., Fortin, M., Meekan, M.G., 2013. Caught in the middle: combined impacts of shark removal and coral loss on the fish communities of coral reefs. PLoS One 8 (9), 1–9.
Sarkis, S., van Beurkering, P.J.H., McKenzie, E., Brander, L., Hess, S., Bervoets, T., Looijenstijn-van der Putten, L., Roelfsema, M., 2013. Total economic value of Bermuda's coral reefs: a summary. In: Sheppard, C.R.C. (Ed.), Coral Reefs of the United Kingdom Overseas Territories, Coral Reefs of the World, vol. 4. Springer Science, Dordrecht, pp. 201–211.
Scheffers, A., Kelletat, D., 2003. Sedimentologic and geomorphologic tsunami imprints worldwide – a review. Earth Sci. Rev. 63, 83–92.
Scheffers, S.R., Haviser, J., Browne, T., Scheffers, A., 2009. Tsunamis, hurricanes, the demise of coral reefs and shifts in prehistoric human populations in the Caribbean. Quat. Int. 195, 69–87.
Scoffin, T., 1992. Taphonomy of coral reefs: a review. Coral Reefs 11, 57–77.
Scoffin, T., 1993. The geological effects of hurricanes on coral reefs and the interpretation of storm deposits. Coral Reefs 12, 203–221.
Searle, M., 2006. Co-seismic uplift of coral reefs along the western Andaman Islands during the December 26th 2004 earthquake. Coral Reefs 25, 2.
Sheppard, C., et al., 2005. Coral mortality increases wave energy reaching shores protected by reef flats: examples from Seychelles. Estuarine Coastal Shelf Sci. 64, 223–234.
Sheppard, C.R.C., 2003. Predicted recurrences of mass coral mortality in the Indian Ocean. Nature 425, 294–297.
Sieh, K., 2005. Aceh–Andaman earthquake: what happened and what's next? Nature 434, 573–574.
Stoddart, D.R., 1963. Effects of Hurricane Hattie on the British Honduras reefs and cays, October 30–31, 1961. Atoll Res. Bull. 95, 1–142.
van Hooidonk, R., Maynard, J.A., Planes, S., 2013. Temporary refugia for coral reefs in a warming world. Nature Climate Change 3, 508–511.
Walther, G., Post, E., Convey, P., Menzel, A., Parmesan, C., Beebee, T.J.C., Fromentin, J., Hoegh-Guldberg, O., Bairlein, F., 2002. Ecological responses to recent climate change. Nature 416, 389–395.
Webb, A.P., Kench, P.S., 2010. Vulnerability of atoll islands to sea level rise: multi-decadal analysis of island change in the central Pacific. Global Planet. Change 72, 234–246.
White, K.N., Ohara, T., Fujii, T., Kawamura, I., Mizuyama, M., et al., 2013. Typhoon damage on a shallow mesophotic reef in Okinawa, Japan. PeerJ 1, e151. http://dx.doi.org/10.7717/peerj.151.
Wilkinson, C.R., 1999. Global and local threats to coral reef functioning and existence: review and predictions. Mar. Freshwater Res. 50, 867–878.
Wilkinson, C.R. (Ed.), 2004. Status of Coral Reefs of the World: 2004. Australian Institute of Marine Science, Townsville, p. 557.

Williams Jr., E.H., Bunkley-Williams, L., 1990. The world-wide coral reef bleaching cycle and related sources of coral mortality. Atoll Res. Bull. 335, 1–71.

Woodley, J.D., et al., 1981. Hurricane Allen's impact on Jamaican coral reefs. Science 214, 749–755.

Woodley, J.D., 1992. The incidence of hurricanes on the north coast of Jamaica since 1870: are the classic reef descriptions atypical? Hydrobiologia 247, 133–138.

Woodroffe, C., 2008. Reef-island topography and the vulnerability of atolls to sea-level rise. Global Planet. Change 62 (1–2), 77–96.

Worachananant, S., Carter, R.W., Hockings, M., 2007. Impacts of the 2004 tsunami on Surin marine national park, Thailand. Coastal Manage. 35, 399–412.

Yamano, H., Kayanne, H., Chikamori, M., 2005. An overview of the nature and dynamics of reef islands. Global Environ. Res. 9 (1), 9–20.

Yap, H., 2013. Coral reef ecosystems. In: Orcutt, J. (Ed.), Earth System Monitoring: Selected Entries from the Encyclopaedia of Sustainability Science and Technology. Springer Science, New York, pp. 77–106.

Chapter 16

Threats to Marsh Resources and Mitigation

Virginia D. Hansen [1] and Kelly Chinners Reiss [2]
[1] *US Environmental Protection Agency, Gulf Breeze, FL, USA,* [2] *HT Odum Center for Wetlands, University of Florida, Gainesville, FL, USA*

ABSTRACT
Salt marshes inhabit low-energy, intertidal shorelines worldwide and are among the most abundant and productive coastal ecosystems. Salt-marsh ecosystems provide a wide array of benefits to coastal populations, including shoreline protection, fishery support, water quality improvement, wildlife habitat provision, and carbon sequestration. Historically, the major threat to salt marshes was filling for agricultural fields or urban construction, which continues as coastlines develop today. In recognition of salt-marsh value and loss, more recent wetland management and conservation policies in many countries have led to the protection and restoration of salt-marsh habitats; however, salt-marsh area and condition continue to decline globally. Currently, the major threats to salt-marsh resources include climate-change effects, pollution, land use change, and invasive species. In this chapter, we review our current state of knowledge regarding the risks to salt marshes from these threats, their effects on ecosystem services, and restoration and management measures designed to protect salt marshes.

16.1 INTRODUCTION

Salt marshes are most common and abundant in temperate coastal zones and also occur in arctic, boreal, and tropical latitudes (Adam, 2002). In this review, we focus on temperate salt marshes such as those along the coasts of North America, Europe, China, and Australia. The global areal coverage of salt marshes is not known but has been estimated to be 40–80 million hectares (Nellemann et al., 2009). The most extensive salt marshes in the world occur in the USA along the Atlantic and Gulf of Mexico coasts (Table 16.1). Dahl and Stedman (2013) estimated that >10 million hectares of salt marsh occur in these coastal areas. China is estimated to have more than 2 million hectares, the second highest salt-marsh area (Table 16.1; An et al., 2007a). Figure 16.1 shows a current estimate of the relative distribution of salt marshes globally.

Coastal and Marine Hazards, Risks, and Disasters. http://dx.doi.org/10.1016/B978-0-12-396483-0.00016-9
Copyright © 2015 Elsevier Inc. All rights reserved.

TABLE 16.1 Estimates of the Area of Salt Marshes Worldwide

Location	Area (ha)	Reference
Global	$40–80 \times 10^6$	Nellemann et al. (2009)
USA		
Atlantic coast	4.6×10^6	Dahl and Stedman (2013)
Gulf of Mexico coast	5.5×10^6	Dahl and Stedman (2013)
Pacific coast	0.5×10^6	Dahl and Stedman (2013)
Alaska	0.1×10^6	Hall et al. (1994)
Canada	44,000	Mendelssohn and McKee (2000)
China	2.1×10^6	An et al. (2007a)
Europe		
Atlantic and Baltic coasts	0.2×10^6	Bakker et al. (2002)
Africa		
South Africa	17,000	O'Callaghan (1990)

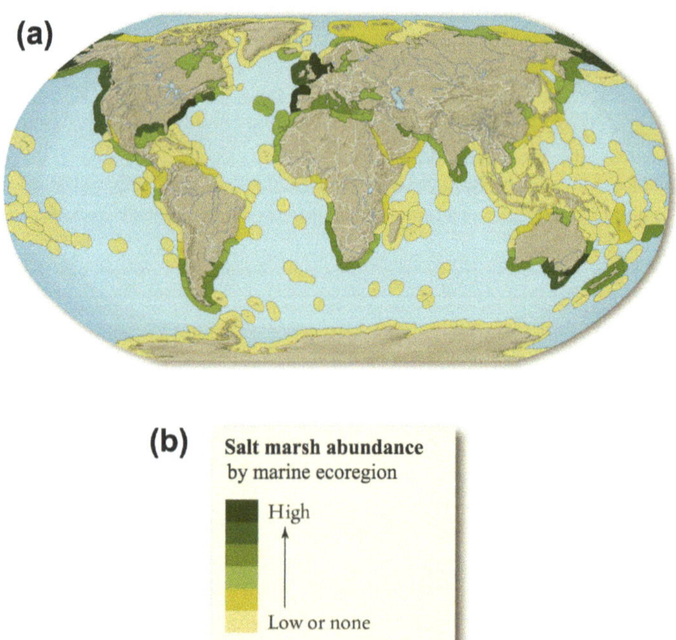

FIGURE 16.1 (a) Distribution of salt marshes worldwide. (b) Color scale represents relative abundance of salt marshes by marine ecoregion. *Source: Hoekstra et al. (2010) © The Nature Conservancy.*

Global estimates of salt-marsh loss are highly variable and uncertain. The *Global Biodiversity Outlook* 3 reports that 25 percent of historic salt-marsh area has been lost globally with an additional one to two percent of salt-marsh area lost annually (Secretariat of the Convention on Biological Diversity, 2010). Across North America, some 38 percent of coastal marshes have been lost since European settlement (Gedan and Silliman, 2009). Southeast Australia has lost 30 percent of its original salt-marsh area (Saintilan and Rogers, 2009). China's coastal wetlands have declined by >50 percent since 1950 (An et al., 2007a). Large losses of salt-marsh area have also occurred in Europe (Airoldi and Beck, 2007; Gedan et al., 2009).

Salt marshes are especially vulnerable to increasing human population density in the coastal zone. More than one-third of the world's population currently resides in coastal areas, which comprise only 4 percent of Earth's land area (UNEP, 2006). For example, population density in US coastal counties is more than six times greater than in US inland counties (NOAA, 2013). The increase in population in US coastal counties since 1970 corresponds to a decrease in salt-marsh area (Figure 16.2), which comes at a loss of ecosystem function and process. Salt marshes provide a wide range of ecosystem services (e.g., fishery support, storm surge protection, water quality improvement, hydrologic moderation, wildlife habitat provision and connectivity, recreational opportunities, carbon sequestration) that support human well-being in coastal communities. The pressures from coastal development

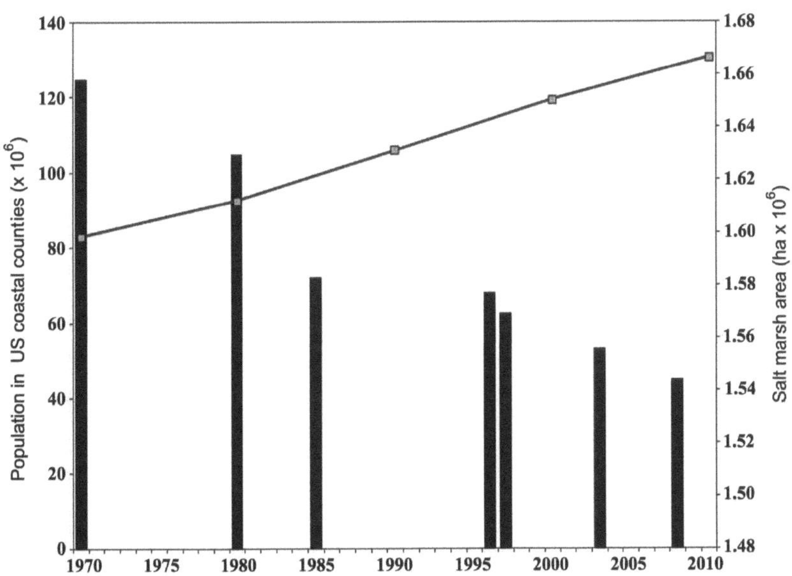

FIGURE 16.2 Comparison of estimates of the salt-marsh area in the conterminous USA (bars) to population estimates for US coastal counties (line). *Data sources: Dahl and Johnson (1991), Dahl (2000, 2006, 2011), NOAA (2013).*

and subsequent degradation and loss of salt marshes will reduce or remove the capacity of these wetlands to provide valuable ecosystem services.

The objectives of this review are to summarize the major threats to salt-marsh resources, discuss the current widely accepted causes of salt-marsh loss and degradation and effects on ecosystem services, and to highlight new and innovative approaches to mitigation and restoration of salt marshes. Historically, the major threat to salt marshes was land conversion for agriculture or development (Adam, 2002; Silliman et al., 2009; Valiela et al., 2009). As human populations settled along the coastlines, salt marshes were filled and converted to uplands for development, diked and ditched for navigation, agriculture, and mosquito control, and exploited for waste treatment. More recently, the primary threats to salt marshes include climate-change impacts (i.e., sea-level rise and increased storm intensity), pollution, and invasive species (Gedan et al., 2011). These coastal threats have complex and interactive impacts on salt-marsh vegetation and biogeochemical processes that can lead to marsh degradation and loss (Figure 16.3), which subsequently affect the provision of ecosystem services. In part because of these ecosystem services, salt-marsh conservation has moved beyond mere protection of existing marsh habitats to restoration and mitigation in many countries (Adam, 2002; Gedan et al., 2009).

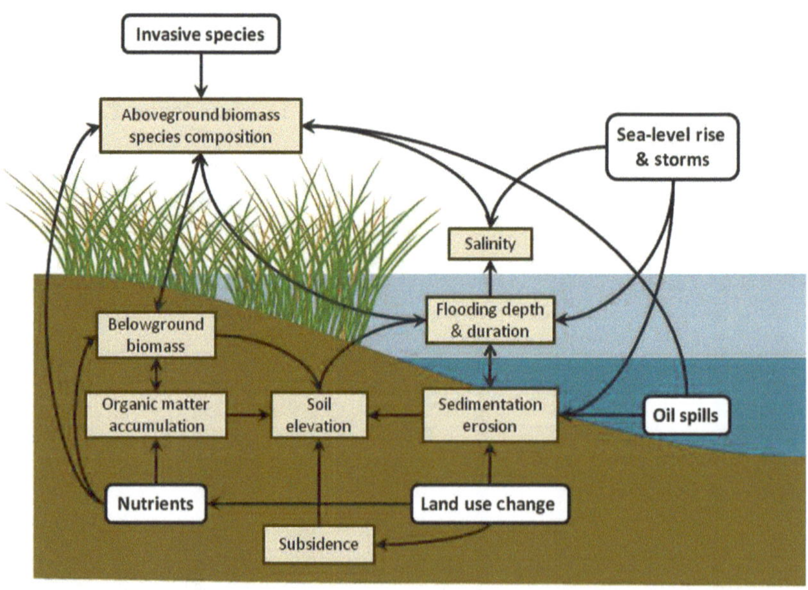

FIGURE 16.3 Conceptual model of threats to salt marshes and impacts on vegetation and biogeochemical processes. *Adapted from Figure 4.2 in Cahoon et al. (2009).* Spartina *image courtesy of Tracey Saxby, Integration and Application Network, University of Maryland Center for Environmental Science (ian.umces.edu/imagelibrary/).*

16.2 ECOSYSTEM SERVICES

Boyd and Banzhaf (2007) defined ecosystem services as the "components of nature, directly enjoyed, consumed, or used to yield human well-being." Salt marshes are among the most valuable coastal ecosystems, providing many benefits to human populations, including coastal shoreline protection, water quality improvement, fishery support, carbon sequestration, recreational opportunities, and provision of raw materials and food (Barbier et al., 2011). When salt marshes are degraded or lost, the socioeconomic impact to coastal communities can be understood in terms of the value of the ecosystem services that are lost as well. In strictly economic terms,[1] Woodward and Wui (2001) reported that on average tidal marshes provide \$16,500 ha^{-1} for four to five ecosystem services combined, but the values range widely from \$1 to 42,000 ha^{-1}.

Salt marshes protect coastal shorelines and communities from floods, storm surges, and erosion by stabilizing sediment, absorbing floodwaters, and attenuating wave energy (Gedan et al., 2011; Shepard et al., 2011; Arkema et al., 2013). In a metaanalysis, Shepard et al. (2011) found that salt-marsh size, vegetation density, and biomass production were positively correlated with wave attenuation and shoreline stabilization. The aboveground vegetation in salt marshes provides friction that reduces wave velocity and turbulence and promotes sedimentation, whereas the below-ground portion reduces erosion, promotes vertical accretion, and absorbs floodwaters (Barbier et al., 2011; Gedan et al., 2011). These relationships have been incorporated into storm surge models to quantify the reduction in storm surge height provided by salt marshes and to predict the risk from potential storm surge to coastal communities (Wamsley et al., 2010). The value of storm protection services provided by salt marshes is the highest in coastal areas with high wetland area, high storm probability, or high infrastructure and economic activity (Costanza et al., 2008). Along the US coastline, Arkema et al. (2013) found that coastal habitats provide the greatest storm protection value to the greatest number of people and economic property value in Florida, New York, and California. In an earlier study, Costanza et al. (2008) determined that the loss of 1 ha of salt marsh resulted in an average \$38,000 increase in storm damage costs along US Atlantic and Gulf of Mexico coasts, whereas the extensive loss of coastal wetlands in Louisiana translated to >\$923 million in lost storm protection services. King and Lester (1995) estimated that the complete loss of salt marshes in Essex, England, would cost >\$1.8 $\times 10^9$ to rebuild sea walls.

Many economically valuable commercial and recreational fisheries depend on salt marshes to provide suitable habitat for reproduction, nursery, shelter,

---

1. All monetary values in this paper were converted from original values to 2010 USD using the US Consumer Price Index (http://www.usinflationcalculator.com/) and same year currency converter.

and food. In the northern Gulf of Mexico, for example, commercial landings of shrimp have been linked to salt-marsh area (Turner, 1992), and many studies have shown that salt marshes contain higher densities of juvenile shrimp and fish than other estuarine habitats (see reviews by Deegan et al., 2000; Zimmerman et al., 2000). The northern Gulf of Mexico region has the greatest area of salt marsh in the USA and accounts for >75 percent of the total shrimp landings in the USA, which are worth >\$300 million year^{-1} (NOAA, 2011). Bell (1997) estimated the value of salt marsh for recreational fishing to be \$13,581 ac^{-1} and \$2059 ac^{-1} for the east and west coasts of Florida, USA, respectively. The link between fisheries and salt marshes is complicated, however, because most species also utilize other estuarine habitats and fishery landings are affected by other factors such as climatic variability and overfishing (Engle, 2011).

Salt marshes act as natural filters that can remove and retain nutrients and other pollutants from surface waters, thereby improving water quality in estuaries (Valiela and Cole, 2002; Sousa et al., 2008; Barbier et al., 2011). Water flowing through salt-marsh vegetation slows down, allowing sediments to settle within the marsh. Because salt marshes have high rates of denitrification and nitrogen burial, they can intercept up to 100 percent of land-derived nitrogen loads (Valiela and Cole, 2002; Valiela et al., 2009). The nutrient-removal capacity of salt marshes is not unlimited, however, and salt-marsh habitat will degrade as nutrient loads exceed thresholds for denitrification and burial (Deegan et al., 2012). Much of the nitrogen pollution filtering through wetlands originates in agricultural fields, and costs of reducing nutrient loss from the fields can inform valuation of salt-marsh ecosystem services such as denitrification.

Salt marshes may be more valuable than other wetlands as sinks for carbon due to high carbon sequestration rates and negligible greenhouse gas emissions (Chmura et al., 2003; Choi and Wang, 2004). Anaerobic biogeochemical processes in salt-marsh soils favor the long-term storage of carbon and inhibit the formation of methane (a potent greenhouse gas); however, the carbon sequestration potential of salt marshes is dependent on vertical accretion of sediments. The average global carbon accumulation rate for salt marshes is 218 g m^{-2} year^{-1} (calculated from estimates in Chmura et al., 2003). As climate-change policies consider carbon sinks to offset greenhouse gas emissions, the management of salt marshes to sequester carbon could be considered for carbon credits (Whiting and Chanton, 2001). The total value of carbon sequestered by Louisiana, USA, coastal wetlands, has been estimated from \$29.7 to \$44.5 million year^{-1}; continued annual loss of coastal wetlands in Louisiana would result in the loss of an estimated \$18–28 million worth of stored carbon (DeLaune and White, 2012). Hansen and Nestlerode (2014) estimated a potential loss of 700,000 metric tons of carbon from Gulf of Mexico estuarine emergent wetlands; applying DeLaune and White's (2012) value of \$10–15 per metric ton results in a value of \$7–10 million for this lost carbon.

Salt marshes also provide essential habitat for many wildlife species, which supports opportunities for recreation, tourism, education, research, and hunting (Barbier et al., 2011). Several North American bird species are endemic or restricted to salt-marsh habitat, including the endangered seaside sparrow (*Ammodramus maritimus mirabilis*) (Greenberg et al., 2006; Rush et al., 2009). Salt marshes are highly valued by recreational bird watchers who hope to catch a glimpse of these species. Even small patches of salt marsh, especially in urban settings, provide important habitat for wading birds (McKinney et al., 2009). Salt marshes have direct use value for waterfowl hunting that can be calculated as the cost of land sales and leases (e.g., estimates from sales and leases of marshes to hunters in England range from \$460 to \$1300 ac^{-1}) (King and Lester, 1995). Feagin et al. (2010) estimated the value of salt marsh in Galveston Bay, Texas, USA, for hunting and bird watching at \$4900 ha^{-1} $year^{-1}$. Bergstrom et al. (1990) estimated that the average value of fresh and salt marsh in coastal Louisiana for recreational hunting and fishing was \$450 ha^{-1} $year^{-1}$.

It is clear that salt marshes provide many ecosystem services that benefit human well-being; some of these services can be valued monetarily, while others simply have intrinsic value. Management and restoration policies need to consider the cumulative or total value of all of the services provided by salt marshes, rather than maximizing a single service (Gedan et al., 2009). Salt-marsh habitats often have a high total ecosystem service value but may have a low value for some individual services (Camacho-Valdez et al., 2013; deGroot et al., 2012). Estimates of total ecosystem service value of coastal wetlands (including salt marshes and mangroves) vary widely from \$65,000 ha^{-1} $year^{-1}$ for northern Mexico (Camacho-Valdez et al., 2013), \$15,000 ha^{-1} $year^{-1}$ globally (Costanza et al., 1997), and \$204,000 ha^{-1} $year^{-1}$ globally (deGroot et al., 2012).

16.3 LAND USE CHANGE

Salt-marsh systems are both directly and indirectly impacted by land use change, which results when natural lands are converted to agricultural fields, residential neighborhoods, commercial districts, or lands suitable for other human activities. Dredge and fill activities physically alter salt marshes and indirectly lead to sedimentation, increased nutrient concentrations, introduction of chemical pollutants, and exchange of genetic materials including nonnative and/or invasive species. Thus, land use changes have an important influence on ecosystem services provided by salt marshes in proximity to human development. Because salt marshes occupy the transition zone between the livable uplands and uninhabitable marine environments, salt marshes face impacts from the actions of coastal residents. McGranahan et al. (2007) report that nearly two-thirds of urban areas with populations greater than five million are located at least partly in the coastal zone. The United Nations Environmental

Programme (UNEP, 2007) further links urbanization and the environment when describing that nearly half of the global human population lives in towns and cities predominantly in coastal areas. In addition, because salt marshes occur in protected estuaries (e.g., partially enclosed basins where freshwater meets saltwater) in low-energy zones, salt marshes receive sediments and pollutants in the freshwater inputs from the watershed contributions of much of the remaining human population.

In modeling ecosystem processes as a function of multiple factors, Ellis and Ramankutty (2008) combined human population and land use along with the more commonly used ecological factors of biota, climate, terrain, and geology as part of the concept of anthromes (or anthropogenic biomes) where human actions are drivers in changing ecological processes. Although their anthromes focus on terrestrial systems, and no aquatic systems counterpart exists, this concept stresses the significance of land use change, for example, >75 percent of ice-free land on Earth shows evidence of alteration from anthropogenic land use (Ellis and Ramankutty, 2008). In one example, in a 40-year study of tidal salt marshes of the Bahía Blanca estuary in Argentina, Pratolongo et al. (2013) found that coastal areas were reshaped from human activities, with the loss of one-third of the *Sarcocornia perennis* salt marshes. Similarly, in the upper Newport River Estuary and Bogue Banks at Pine Knoll Shores in North Carolina, USA, Mattheus et al. (2010), suggested that in a relatively short period of time human activities have altered the development trajectory of fringing marsh implicating both agricultural and urban land use in increased suspended sediment and associated nutrient loading to the marsh.

Livestock impacts on salt marshes have been studied extensively around the world. An obvious direct impact is soil compaction from larger grazers (e.g., cattle, sheep), which reduces the rate of salt-marsh accretion (e.g., Elschot et al., 2013). In a nine-year study in Germany, Andresen et al. (1990) attributed eight changes in salt marshes to grazing, including reduced vegetation height, changes to community structure (e.g., plants, macro-invertebrates), decreased sedimentation rates, decreased plant species richness, and a shift in food web dynamics owing in part to shifts in litter production. All these changes threaten to reduce the ecosystem services that salt marshes provide. A study on salt-marsh soil properties and the microbial communities in the Ribble Estuary in northwest England identified strong significant differences in soil properties in grazed and ungrazed salt marsh (e.g., soil pH, nitrate concentration, and root biomass) (Ford et al., 2013). Other agricultural activities also threaten salt-marsh function and ecosystem services. For example, in the coastal marshes in the Great Lakes, USA, Morrice et al. (2008) found wetland water quality had a strong positive correlation to both proportion of cultivated land and intensity of agricultural chemical use.

To date, some tools have been developed to help mitigate salt-marsh loss, given sea-level rise forecasts and continued human development of coastal areas. The Coastal Squeeze Index, which was developed from data on wetlands in Maine, USA, and New Brunswick, Canada, uses surrounding topography and impervious surfaces to estimate the potential for a marsh to be prevented from migrating inland, that is, "coastal squeeze" (Torio and Chmura, 2013). This index can be used to rank potential restoration locations with the lowest threat of coastal squeeze to maximize economic return and/or ecosystem services (Torio and Chmura, 2013). Similarly, for salt marshes in Narragansett Bay, Rhode Island, USA, a loading index and an impact index, both based on correlations between land use and salt-marsh condition, have been suggested as rapid, remote-assessment tools to identify human disturbance and evaluate wetland condition (Brandt-Williams et al., 2013). Further, Clausen et al. (2013) proposed reestablishing well-managed wetlands to help counterbalance the expected threats of sea-level rise. Although none of these tools and strategies reflects a singular way forward in mitigating salt-marsh loss, in combination, they may offset some of the expected losses in ecosystem services associated with land use change.

16.4 CLIMATE CHANGE

Salt marshes are particularly vulnerable to the impacts of climate change, including sea-level rise and increased storm frequency and intensity, which change the delivery of freshwater, sediment, and nutrients (Day et al., 2008). Recent losses of salt marsh along the northern Gulf of Mexico, USA, have been attributed to major hurricanes (i.e., Katrina and Rita in 2005; Ike in 2008), whereas salt-marsh losses along the US Atlantic coast have been attributed to sea-level rise (DeLaune and White, 2012; Dahl and Stedman, 2013). Other impacts of climate change include increased temperature, elevated atmospheric carbon dioxide (CO_2), and changes in precipitation, which affect wetland hydrology, biogeochemical processes, and plant species composition and geographical distribution (Adam, 2002; Day et al., 2008; Erwin, 2009; Gedan et al., 2009).

16.4.1 Sea-Level Rise

Nicholls (2004) projected that 5–20 percent of coastal wetlands worldwide will be lost by 2080 under projected sea-level rise scenarios, and Clausen et al. (2013) predicted flooding of 15–44 percent of existent salt marshes because of sea-level rise by the end of the century. The most recent climate-change report by the Intergovernmental Panel on Climate Change (IPCC, 2013) projects global mean sea level will rise between 26 and 82 cm by 2100 but that relative sea-level rise and subsequent loss of coastal wetlands will vary regionally. Where salt marshes fringe large river deltas (e.g., Mississippi River, USA), high rates of subsidence lead to high relative sea-level rise, which has resulted in a global loss of salt marsh (Day et al., 2008).

Sea-level rise alone, though, is not likely to cause large-scale losses of salt marshes; where sea-level rise occurs in tandem with negative anthropogenic impacts, however, salt-marsh loss is predicted to be catastrophic (Scavia et al., 2002; Nicholls, 2004; Day et al., 2008). The regions that have suffered the most extensive salt-marsh losses in the twentieth century occur where human infrastructure engineering has impacted rates of subsidence and sediment delivery (Kirwan and Megonigal, 2013). In Louisiana, USA, for example, where vast areas of marsh have been lost due to a combination of climate- and human-induced impacts, 42–92 percent of the existing salt-marsh area is predicted to be lost if sea-level rise exceeds 75 cm by 2100 (Glick et al., 2013).

Salt marshes are uniquely adapted to gradual sea-level rise. Vertical accretion of sediment and organic matter leads to increased elevation and the landward migration of salt-marsh plants and has enabled marshes to remain in the intertidal zone as sea-level has risen (Michener et al., 1997; Adam, 2002; Morris et al., 2002). In the absence of additional human disturbance, if the rate of vertical accretion is equivalent to the rate of sea-level rise, salt marshes will continue to adapt and survive (Morris et al., 2002; Day et al., 2008; Erwin, 2009; Kirwan and Megonigal, 2013). Many other factors, however, affect the ability of salt marshes to keep pace with current projections of sea-level rise. Regional differences in the delivery of sediment from marine or upland sources, normal tidal ranges, vegetation, temperature, and anthropogenic alterations to the landscape will determine the regional response of salt marshes to sea-level rise (Michener et al., 1997; Morris et al., 2002; Kirwan and Megonigal, 2013; Weston, 2014). Salt marshes that are adapted to high tidal ranges and high sediment loads, for example, tend to be more resilient to sea-level rise than marshes with low tidal ranges and low sediment loads (Kirwan et al., 2010). The typical sea-level rise scenario for marshes with mesotidal (2–4 m) ranges (e.g., salt marshes along the coast of Georgia, USA) predicts that as the low salt marsh submerges, increased salinity will cause a decline in tidal freshwater marsh area, which then allows intermediate brackish marsh to migrate inland (Craft et al., 2009). The ability of coastal wetlands to migrate landward as sea level rises, however, will be compromised significantly by human modifications of the shoreline, including the construction of bulkheads, sea walls, and river-flow management structures (Michener et al., 1997; Scavia et al., 2002; Nicholls, 2004; Erwin, 2009; Kirwan and Megonigal, 2013). Reduction in sediment delivery to salt marshes because of shoreline hardening will prevent marshes from persisting with sea-level rise, which may be particularly relevant along urbanizing coastlines (Mattheus et al., 2010). Given the high density of human population found along the coast, this could be a significant concern moving forward.

16.4.2 Storms

Climate-change scenarios also predict an increase in the intensity and frequency of storms (IPCC, 2013). Although tropical storms and hurricanes can be immediately detrimental to salt marshes because of scouring and erosion from storm surge, major storms can also have long-term beneficial effects on salt marshes (Cahoon, 2006; Turner et al., 2006; Day et al., 2008; Gedan et al., 2009; DeLaune and White, 2012). Hurricanes and tropical storms commonly increase delivery of freshwater, sediment, and nutrients to salt marshes, which can increase marsh elevation and enhance productivity (Day et al., 2008). In the Mississippi River Delta (Louisiana, USA) where salt marshes no longer receive sediments from the river because of human engineering infrastructure, hurricanes have become the primary source of sediment delivery to salt marshes (Turner et al., 2006). Coastal wetlands have naturally evolved in response to specific patterns of hurricane frequency, intensity, and timing that may promote plant recruitment into nonvegetated areas, stimulate net primary productivity, and facilitate the development of ecotones in storm-affected areas (Michener et al., 1997). The effects of increased storm intensity and frequency on salt marshes will vary regionally, depending on wetland condition (Schuerch et al., 2013); marshes that are already degraded by human impacts will be most susceptible to loss of elevation because of erosion and compaction of soils resulting from storm surge (Cahoon, 2006; Gedan et al., 2009).

Controversy over the ability of salt marshes to ameliorate storm surge effects arose in the wake of the 2005 hurricanes (Katrina and Rita) in the northern Gulf of Mexico. Although these storms traversed a wide area of salt marshes in Louisiana, USA, and storm surge was somewhat reduced, these storms still caused catastrophic damages to coastal communities (Day et al., 2007; Resio and Westerink, 2008). Salt marshes protect coastal communities from storm surges because they promote sediment deposition that stabilizes shorelines, whereas marsh vegetation reduces velocity, height, and duration of storm-induced waves (Morgan et al., 2009; Gedan et al., 2011; Shepard et al., 2011; Spalding et al., 2014). The extent to which salt marshes are able to provide this service is affected by particular storm characteristics (e.g., tidal height, wave height and length, water depth, wind) as well as marsh characteristics (e.g., area and width, marsh species and density, local geomorphology) (Barbier et al., 2011, 2013; Engle, 2011; Shepard et al., 2011; Wamsley et al., 2010). Although widespread support remains for the paradigm that healthy salt marshes protect shorelines from erosion and coastal communities from surge-related damages, the effectiveness of salt marshes to perform this service decreases as salt-marsh condition deteriorates, or as wave height exceeds the height of the vegetation (Möller, 2006; Gedan et al., 2011). The combination of extreme storm surges from Hurricanes Katrina and Rita and the degraded condition of salt marshes in the Mississippi River Delta precluded these marshes from having much of a dampening effect.

The effects of sea-level rise and storms on other ecosystem services provided by salt marshes are complex. Salt marshes accumulate carbon through vertical accretion of organic matter, which is an adaptive response to sea-level rise; however, marshes that are deteriorating will likely lose carbon as they fail to keep pace with sea-level rise and are converted to open water (DeLaune and White, 2012). The extensive salt-marsh loss in Louisiana, USA, from major hurricanes like Katrina and Rita in 2005 resulted in an estimated loss of 15×10^6 mt of carbon to adjacent nearshore waters (DeLaune and White, 2012). The conversion of salt marsh to open water from sea-level rise and storm impacts will negatively affect the provision of habitat for bird species that are dependent on marsh vegetation for feeding, nesting, and nursery functions but may have positive effects on waterbirds (Hughes, 2004; Erwin et al., 2006; Rush et al., 2009). It is widely believed that commercial fishery species that are dependent upon salt-marsh habitat will be negatively affected by the loss of salt marsh because of sea-level rise. In the northern Gulf of Mexico, however, degradation and loss of coastal marsh have not led to a decline in the shrimp fishery (Zimmerman et al., 2000). Deteriorating marshes initially become fragmented, which increases marsh edge, thus providing enhanced nursery functions for juvenile shrimp in the short term (Zimmerman et al., 2000). In addition, shrimp and other marsh-dependent fishery species may be able to utilize other estuarine habitats (e.g., submerged aquatic vegetation) as marshes disappear (Zimmerman et al., 2000).

Many coastal management policies are aimed at mitigating salt-marsh loss specifically to aid in the defense of shorelines and coastal communities against sea-level rise and storm impacts (Gedan et al., 2009). Salt marshes are more vulnerable to sea-level rise where sediment delivery is limited, or hardened shorelines prevent their inland migration (Morris et al., 2002; Weston, 2014). Changes in the wetland area in the Mississippi River Delta, for example, have been directly linked to fluctuations in the sediment load resulting from human alterations to river flow and discharge (Day et al., 2005; Tweel and Turner, 2012). Restoration approaches to counter sea-level rise are designed to deliver sediment and nutrients back to these wetlands through the creation of freshwater diversions and hydrologic restoration (Day et al., 2005, 2007). Recognition of the value of salt marshes for shoreline protection following the 2005 hurricanes in the northern Gulf of Mexico, has led to public support for more aggressive approaches to restore these coastal wetlands (Day et al., 2007; Barbier et al., 2013).

Along the US Atlantic and Gulf of Mexico coasts, artificial "living shorelines" (e.g., oyster reefs) are being erected along the seaward margins of salt marshes; these structures are designed to become self-sustaining, provide the ecosystem service of shoreline protection, and promote the low-energy conditions necessary for salt-marsh restoration (Gedan et al., 2011). In some cases in Europe and the USA, restoration of salt marshes and intertidal habitats is viewed as a more sustainable approach to shoreline protection than constructing sea walls. "Managed realignment" is designed to restore previously reclaimed marshland by moving (or removing) constructed shoreline barriers to allow marshes to expand upland (Adam, 2002; Bakker et al., 2002; French, 2006; Esteves, 2013).

Although many countries have extensive wetland protection policies, support for replacing hardened shorelines with salt marshes is not universal (Esteves, 2013; Kirwan and Megonigal, 2013). Research has not conclusively demonstrated that salt-marsh restoration schemes such as managed realignment protect shorelines better or less expensively than "hard" engineered defenses (Spencer and Harvey, 2012; Esteves, 2013). Under current US coastal management policies, >60 percent of low-lying coastal area is expected to be developed, despite wetland protection measures (Titus et al., 2009); thus, the future fate of salt marshes may depend more on the actions taken to protect coastal communities from climate-change impacts than on the impacts of climate change itself (Kirwan and Megonigal, 2013). Salt-marsh restoration provides not only sustainable shoreline protection but also a wide array of additional ecosystem services and benefits; understanding how to maximize multiple ecosystem services through restoration will likely lend more support for these restoration approaches to protect shorelines (Gedan et al., 2009; Shepard et al., 2011; Spencer and Harvey, 2012; Spalding et al., 2014).

16.5 POLLUTION

16.5.1 Nutrients

Nutrient loads to coastal systems have increased worldwide (Seitzinger et al., 2002; Howarth, 2008). Salt marshes have long been valued for the ecosystem service of removing nutrients and other pollutants from land-derived sources, thereby reducing nutrient loads and eutrophication effects in estuaries (Valiela and Cole, 2002; Sousa et al., 2008). The nutrient-removal capacity of salt marshes is limited, however; as nitrogen loads increase above a threshold, salt marshes are unable to process and retain the excess nitrogen, and become sources of nitrogen to coastal waters (Valiela and Cole, 2002; Deegan et al., 2012). Because most salt-marsh plants are nitrogen limited, low nitrogen conditions favor plant species that are dominant below-ground competitors (Emery et al., 2001). When a salt marsh receives excess nitrogen, plant species no longer compete for nitrogen and those that are superior aboveground competitors become dominant (Bertness et al., 2002). Excess nutrients, therefore, result in increased above-ground biomass, decreased below-ground biomass, and reduced organic matter accumulation. These changes can lead to instability and eventual loss of marsh structure and elevation (Turner et al., 2009; Deegan et al., 2012).

The effects of excess nutrients on salt marshes are inextricably entwined with other threats to salt marshes (i.e., land use changes, sea-level rise, invasive species) and their effects on ecosystem services. Invasive species are more likely to out compete native species in salt marshes impacted by nutrient enrichment (Bertness et al., 2002; Tyler et al., 2007; Gedan et al., 2009). Nutrient enrichment may increase the vulnerability of salt marshes to sea-level rise and storm erosion (Deegan et al., 2012) and sea-level rise may reduce the nutrient retention capacity of salt marshes (Craft et al., 2009). Excess nutrients alter zonation and structure of salt-marsh plants and reduce the ability of marshes to accrete sediment and store carbon; these impacts may lead to marsh degradation and hamper the ability of marshes to accommodate sea-level rise (Morris and Bradley, 1999; Turner et al., 2009).

Much attention has been given to the problems caused by eutrophication in coastal waters worldwide (Boesch, 2002; Rabalais et al., 2009). Reduction in nutrient loads from fertilizer use, wastewater treatment, and atmospheric sources is the primary mechanism to reduce eutrophication in coastal waters. Because of their ability to uptake nutrients, salt marshes have been valued as natural sinks for nutrients and as aids to improving coastal water quality. With recent recognition of the limits on the capacity of salt marshes to remove nutrients and the deleterious effects of nutrient enrichment on salt marshes, however, reducing nutrient loads in watersheds before they reach the coastal fringe has become more important than ever (Turner et al., 2009; Deegan et al., 2012). Fertilizer reduction strategies and implementation of Best Management Practices (BMPs) may be effective means of protecting receiving coastal marshes.

16.5.2 Oil Spills

Major oil spills have widespread but variable impacts on salt-marsh habitats, depending on the type and amount of oil, time of year, marsh condition, and plant species sensitivity (Baker et al., 1994; Pezeshki et al., 2000; Mendelssohn et al., 2012). Physical effects of oil coating the leaves of marsh plants and soil surfaces include reduced photosynthesis and transpiration, impaired plant growth, and altered biogeochemical processes (Pezeshki et al., 2000; Lin and Mendelssohn, 2012; Mendelssohn et al., 2012). Chemical toxicity of oil to salt-marsh plants varies with the type of oil; lighter oils have higher toxicity, whereas heavier oils have lower toxicity but tend to cause more physical effects (Baker et al., 1994; Pezeshki et al., 2000). Salt-marsh plant species differ in their sensitivity to oil; in greenhouse studies with, exposures to south Louisiana crude oil, *Sagitarria lancifolia* growth was enhanced, whereas *Spartina patens* suffered more negative effects than *Spartina alterniflora* (Lin and Mendelssohn, 1996). After the "Deepwater Horizon" oil spill in the northern Gulf of Mexico, heavily oiled shorelines showed mortality of both *S. alterniflora* and *Juncus roemerianus*, whereas moderate oiling reduced the above-ground biomass of *J. roemerianus* but not *S. alterniflora* (Lin and Mendelssohn, 2012; Mendelssohn et al., 2012).

Recovery of salt marshes from major impacts of oil spills varies as well (Table 16.2). Forty years after the 1969 *Florida* oil spill in Buzzards Bay, Massachusetts, USA, *S. alterniflora* still showed reduced biomass and the marsh was subject to continued erosion (Culbertson et al., 2008). In Brittany, France, digital image analysis of salt marshes that were impacted by the 1978 *Amoco Cadiz* oil spill showed that marshes not subjected to clean-up operations

TABLE 16.2 Examples of Major Oil Spills that Have Impacted Salt Marshes

Location	Vegetation	Oil Type	Date	Recovery	References
Chile *Metula*	*Salicornia ambigua* *Suaeda argentinensis*	Arabian crude Bunker C	August 1974	>20 years	Wang et al. (2001)
Brittany, France *Amoco Cadiz*	*Salicornia* sp. *Suaeda* sp. *Halimione* sp. *Juncus maritimus*	Arabian light Iranian light crude Bunker C	March 1978	5–8 years	Gilfillan et al. (1995)
West Falmouth, MA *Florida*	*Spartina alterniflora* *Salicornia europaea* *Spartina patens*	No. 2 fuel	September 1969	>40 years	Peacock et al. (2005) Culbertson et al. (2008)
Buzzard's Bay, MA *Bouchard 65*	*S. alterniflora* *Salicornia virginica*	No. 2 fuel	October 1974	>25 years	Peacock et al. (2007)
Gulf of Mexico *Deepwater Horizon*	*S.alterniflora* *Juncus roemerianus* *Phragmites australis*	Macondo sweet crude	April 2010	Unknown	Silliman et al. (2012) Mendelssohn et al. (2012)

(i.e., sediment removal) had recovered within five to eight years (Gilfillan et al., 1995). The 2010 "Deepwater Horizon" oil spill impacted shorelines where salt marshes were already severely degraded; extensive plant mortality and sediment erosion occurred in heavily oiled marsh edges, but already some evidence exists of recovery, especially in moderately oiled areas further inland (Lin and Mendelssohn, 2012; Mendelssohn et al., 2012; Mishra et al., 2012; Silliman et al., 2012). It is a challenge to separate the impacts from the 2010 "Deepwater Horizon" oil spill from ongoing degradation and erosion in these salt marshes; however, long-term recovery and future impacts on ecosystem services have yet to be determined (Mendelssohn et al., 2012; Silliman et al., 2012).

Remediation of major oil spills in salt marshes generally involves clean-up activities, including mechanical removal, in situ burning, chemical applications, and bioremediation (see review by Pezeshki et al. (2000)). Determining the appropriate remediation strategy requires balancing the trade-offs between the potential damage to the marsh from clean-up activities and the benefits of ameliorating oil toxicity (Pezeshki et al., 2000). In some cases, the best course of action is to do nothing and let the marsh recover on its own (Mendelssohn et al., 2012), although this is not generally popular with the public. The activities associated with mechanical removal of oil, vegetation, or soil can have long-term adverse effects on salt-marsh vegetation and benthic fauna (Gilfillan et al., 1995; Mendelssohn et al., 2012). In contrast, in situ burning can effectively remove oil on the sediment surface and vegetation and marsh plants usually recover rapidly (Baustian et al., 2010). Chemical options include the use of dispersants or cleaners, which have been shown to reduce mortality in some marsh plant species; more information is needed, however, to determine whether the combined effects of oil and cleaners are worse than the effect of the oil itself (Pezeshki et al., 2000). Bioremediation is often augmented with additional nutrients to stimulate microbial degradation of oil, but Deegan et al. (2012) have shown the detrimental effects of excess nutrients on salt-marsh stability. Although this technique is commonly used on oiled beaches or rocky shores, the long-term effects of bioremediation on oiled salt marshes has yet to be determined (Pezeshki et al., 2000). During the "Deepwater Horizon" oil spill in 2010, approximately 800 km of salt-marsh shoreline was impacted by weathered oil, but only 71 km of marsh shoreline was subjected to cleanup activities, primarily mechanical removal (Michel et al., 2013). Although some recovery of marsh vegetation is already evident, the long-term impact on the sustainability of the affected marshes and the ecosystem services they provide is unknown (Mishra et al., 2012; Silliman et al., 2012; Mendelssohn et al., 2012).

16.6 INVASIVE SPECIES

The threat to salt-marsh sustainability from invasive plant species varies regionally. Invasive plant species can have both negative and positive effects on ecosystem services. In many regions, nonnative salt-marsh plant species that have been intentionally introduced to help buffer coastlines from the effects of sea-level rise and storms may have negative impacts on nutrient cycling and wildlife habitat. For example, the ability of nonnative common cordgrass, *Spartina anglica*, to spread rapidly, tolerate a wide range of environmental conditions, and accrete sediment, led to intentional planting for coastal defense throughout Europe (Daehler and Strong, 1996; Nehring and Hesse, 2008); however, *S. anglica* failed to meet expectations for coastal protection because it did not succeed in high-energy environments, preferring sheltered low-energy basins where additional shoreline protection was not needed (Nehring and Hesse, 2008). *Spartina anglica* and other invasive *Spartina* species have displaced native submerged aquatic vegetation, native marsh grasses, and open mudflats in Europe, the US Pacific coast, China, and Australia, and have had significant negative impacts on native bird and oyster populations (Daehler and Strong, 1996; Kriwoken and Hedge, 2000; Nehring and Hesse, 2008; Gan et al., 2010). Both *S. anglica* and smooth cordgrass, *S. alterniflora*, were introduced to China for agriculture and coastal engineering; however, *S. anglica* has declined, whereas *S. alterniflora* has expanded its coverage to occur along most of the China coast (An et al., 2007a,b).

In another example, common reed, *Phragmites australis*, has historically inhabited salt marshes in North America, although it was not dominant and was primarily limited to tidal freshwater upper marshes. More recently, a European M haplotype of *P. australis* has increased in abundance and expanded its coverage into higher salinity areas of salt marshes (Bertness et al., 2002; Saltonstall, 2002; Michinton and Bertness, 2003; Vasquez et al., 2005). The recent expansion of *P. australis* has been linked to shoreline modifications and increased nutrient loading (Chambers et al., 1999; Bertness et al., 2002; Michinton and Bertness, 2003; Silliman and Bertness, 2004). Increases in atmospheric carbon dioxide (CO_2) and temperature because of climate change may reduce the sensitivity of *P. australis* to higher salinities, allowing it to expand further into salt marshes (Eller et al., 2014). Although *P. australis* has higher accretion rates than native species, which can increase marsh elevations as sea-level rises, it is not a favorable habitat for many wildlife species (Zedler and Kercher, 2004). Other ecosystem services such as water quality improvement and shoreline protection do not appear to be reduced in wetlands dominated by the invasive haplotype of *P. australis* (Chambers et al., 1999).

Soil functions do appear to be threatened with *P. australis* invasion. For example, in a New Jersey, USA, salt marsh, Windham and Lathrop (1999) found that *P. australis* invasion changed soil properties in as little as 3–12 years, including decreased salinity, decreased water level, decreased maximum microtopographic relief, and increased redox potential. Ravit et al. (2003) also identified decreased sediment microbial diversity and changes in microbial biogeochemical functions with *P. australis* invasion. More generally, invasions can alter the soil biological community through changes in litter production, root exudates, release of new chemicals, changes in nutrient acquisition and cycling, and changes in root architecture and function, which may lead to indirect effects (e.g., changes in susceptibility to fire; Wolfe and Klironomos, 2005).

The management of invasive species is often complicated by multiple laws and policies, varying public opinion, and lack of research on cost-effective control methods (Kriwoken and Hedge, 2000). When invasive plant species are viewed as detrimental to native salt-marsh ecosystems, they are usually physically or chemically removed. Application of herbicides has been somewhat effective in controlling invasive salt-marsh plants but may have detrimental effects on adjacent coastal waters (Daehler and Strong, 1996; Kriwoken and Hedge, 2000; An et al., 2007b; Nehring and Hesse, 2008). Although reclamation of the coastal habitat in China has reduced the spread of invasive *S. alterniflora*, this practice has also led to the loss of native salt marsh. Currently, a combination of harvesting, flooding, diking, and aquaculture is used to control the spread of *S. alterniflora* while allowing native plant species to reestablish (An et al., 2007b). In addition to physical removal of seedlings, mowing, and herbicide application to remove *S. alterniflora* and *S. anglica* from US Pacific coast estuaries, biological control by the insect *Prokelesia marginata* shows some promise in controlling the spread of invasive *Spartina* (Daehler and Strong, 1996; Hedge et al., 2003). In contrast, *Phragmites* invasions have been particularly difficult to control although the timing of late summer application of herbicide followed by spring mowing has been effective in New England salt marshes (Warren et al., 2001). Because *Phragmites* succeeds in landscapes with high nutrient loads and modified shorelines, reducing nutrient loads and maintaining native vegetation buffers may limit its expansion in the long term (Michinton and Bertness, 2003; Silliman and Bertness, 2004).

Wolfe and Klironomos (2005) suggest restoration cannot be "aboveground-centric," focusing only on plant community structure, but technical and practical questions remain on how to measure microbial community structure and function. Ehrenfeld (2003) found that there are differences in soil nutrient cycling processes with plant species invasion, but the direction and strength of these differences are variable based on site-specific community characteristics or environmental factors. Such changes to physical, chemical, and biological soil parameters lead to the consideration of invasion as a driver to alternative states (after Suding et al., 2004). This can be true not only for plant invasions but also for animal invasions. In an example out of Hudson Bay, Canada, goose herbivory has been implicated in conversion of salt marsh to hypersaline mudflats. When goose herbivory is reduced, salt marsh is unlikely to return unassisted. To facilitate recovery, salt marshes are planted with creeping alkaligrass, *Puccinellia phryganodes*, plugs, fertilized, and treated with peat (Handa and Jefferies, 2000; as cited by Suding et al. (2004)). In another twist, in Cape Cod, Massachusetts, USA, overfishing led to a population explosion of native purple marsh crab, *Sesarma reticulatum*, which decimated large areas of marsh; however, after the invasion of green crabs, *Carcinus maenas*, a predator of *S. reticulatum*, the marshes have partially recovered (Bertness and Coverdale, 2013). Although many invasive species are considered threats to current ecosystem services, the potential for invasive species to increase ecosystem services remains. In a world faced with rising sea levels, a less predictable climate, and land use change, Schlaepfer et al. (2011) suggest that nonnative species may play a role in meeting conservation goals in the future.

16.7 MEASURING SALT-MARSH FUNCTION

The expression of ecosystem services varies widely between salt-marsh systems, but it is helpful to formulate quantitative means to measure impacts to salt marshes from natural and anthropogenic hazards and disasters and to assess progress in restoration or mitigation efforts. By far, hydrologic function is the most significant in tidally influenced systems. In a review of the management of tidally influenced marsh systems, Roman and Burdick (2012a) offer an insight into hydrologic restoration, including mechanisms and effects. Within the review, Roman and Burdick (2012b) present a simplified model of salt-marsh self-regulation with two drivers, that is, tidal restriction and sea-level rise. In practice, natural and anthropogenic hazards and disasters could modify these drivers (e.g., changes in sedimentation rates, establishment of new tidal connections), resulting in shifts in ecosystem services. Roman and Burdick (2012b) called for the restoration of natural processes to reestablish ecological resiliency and protect against changes in tidal restriction and sea-level rise.

At least two multifunction indices have been developed to assess salt-marsh function. A Restoration Performance Index (RPI), which combines structural and functional indicators of salt-marsh restoration projects, was designed by Moore et al. (2009) for New England, USA. The indicators roughly correlate to ecosystem services, and can inform restoration progress. Chmura et al. (2012) presented an example of the RPI for the Little River Marsh, New Hampshire, where tidal connectivity was restored. The RPI is calculated for four functions (i.e., hydrology, pore water, vegetation, and nekton) as a percent of the reference standard condition, and shifts in function can be tracked over time. Similarly, the hydrogeomorphic approach (HGM), developed a decade earlier, can be used to assess three hydrogeomorphic functions (i.e., tidal surge attenuation; tidal nutrient and organic carbon exchange; and sediment deposition) and five habitat functions (i.e., maintenance of characteristic plant community composition and structure; resident nekton utilization; nonresident nekton utilization; nekton prey pool; and wildlife habitat utilization) of salt marshes (Schafer and Yozzo, 1998). Both methods share a common framework, where current condition is assessed and compared to the expected or reference condition.

16.8 STRATEGIES MOVING FORWARD

It is clear that even with current policies that promote conservation and restoration of salt marshes, loss and degradation of salt-marsh ecosystems and their associated services will continue in the face of land use change, climate-change, pollution, and invasive species. New and innovative approaches are needed to mitigate threats to salt marshes from natural- and human-caused hazards and disasters in the future. One promising proposal is to incorporate natural energy sources, harnessing both natural and human engineering and ingenuity, through coastal ecological engineering. For example, Duarte et al. (2013) propose integrating vegetated coastal habitats (e.g., salt marsh, seagrass beds) with structural engineering approaches to mitigate impacts of climate change, describing ecoengineering as integrating human and ecological systems to the benefit of both. As Clausen et al. (2013) mentioned, reestablishing well-managed salt marsh may temporarily mitigate sea-level rise in the coming century, perhaps buying time for further research to determine achievable longer-term mitigation strategies. In a similar way, Wamsley et al. (2010) call for further research to improve models that integrate wetlands into coastal protection plans to better understand dynamics of flood attenuation and reduction of storm surge. In addition to shoreline protection, recognition of the value of multiple ecosystem services provided by salt marshes should be part of an integrated, ecosystem-based approach to manage and protect salt marshes in the future (Adam, 2002; Gedan et al., 2009). The coastal hazards that threaten salt marshes do not act in isolation nor do they impact only salt marshes; therefore, understanding the cumulative and synergistic impacts of these threats and the complex interdependence among all coastal habitats (i.e., estuaries, oyster reefs, and freshwater wetlands) is essential to sustainable coastal-resource management (Silliman et al., 2009).

No longer can marshes be viewed in scientific, conservation, social, and political circles as one of the most resilient and resistant ecological communities.

Silliman et al. (2009)

REFERENCES

Adam, P., 2002. Saltmarshes in a time of change. Environ. Conserv. 29, 39–61.

Airoldi, L., Beck, M.W., 2007. Loss, status and trends for coastal marine habitats of Europe. Oceanogr. Mar. Biol. 45, 345–405.

An, S., Li, H., Guan, B., Zhou, C., Wang, Z., Deng, Z., Zhi, Y., Liu, Y., Xu, C., Fang, S., Jiang, J., Li, H., 2007a. China's natural wetlands: past problems, current status, and future challenges. AMBIO 36, 335–342.

An, S.Q., Gu, B.H., Zhou, C.F., Wang, Z.S., Deng, Z.F., Zhi, Y.B., Li, H.L., Chen, L., Yu, D.H., Liu, Y.H., 2007b. *Spartina* invasion in China: implications for invasive species management and future research. Weed Res. 47, 183–191.

Andresen, H., Bakker, J.P., Brongers, M., Heydemann, B., Irmler, U., 1990. Long-term changes of salt marsh communities by cattle grazing. Vegetatio 89, 137–148.

Arkema, K.K., Guannel, G., Verutes, G., Wood, S.A., Guerry, A., Ruckelshaus, M., Kareiva, P., Lacayo, M., Silver, J.M., 2013. Coastal habitats shield people and property from sea-level rise and storms. Nat. Clim. Change Lett. 3, 913–918.

Baker, J.M., Adam, P., Gilfillan, E., 1994. Biological Impacts of Oil Pollution: Saltmarshes. International Petroleum Industry Environmental Conservation Association, London.

Bakker, J.P., Esselink, P., Dijkema, K.S., van Duin, W.E., de Jong, D.J., 2002. Restoration of salt marshes in the Netherlands. Hydrobiologia 478, 29–51.

Barbier, E.B., Hacker, S.D., Kennedy, C., Koch, E.W., Stier, A.C., Silliman, B.R., 2011. The value of estuarine and coastal ecosystem services. Ecol. Monogr. 81, 169–193.

Barbier, E.B., Georgiou, I.Y., Enchelmeyer, B., Reed, D.J., 2013. The value of wetlands in protecting southeast Louisiana from hurricane storm surges. PLoS One 8, e58715. http://dx.doi.org/10.1371/journal.pone.0058715.

Baustian, J., Mendelssohn, I., Lin, Q., Rapp, J., 2010. In situ burning restores the ecological function and structure of an oil-impacted coastal marsh. Environ. Manage. 46, 781–789.

Bell, F.W., 1997. The economic valuation of saltwater marsh supporting marine recreational fishing in the southeastern United States. Ecol. Econ. 21, 243–254.

Bergstrom, J.C., Stoll, J.R., Titre, J.P., Wright, V.L., 1990. Economic value of wetlands-based recreation. Ecol. Econ. 2, 129–147.

Bertness, M.D., Coverdale, T.C., 2013. An invasive species facilitates the recovery of salt marsh ecosystems on Cape Cod. Ecology 94, 1937–1943.

Bertness, M.D., Ewanchuk, P.J., Silliman, B.R., 2002. Anthropogenic modification of New England salt marsh landscapes. Proc. Natl. Acad. Sci. 99, 1395–1398.

Boesch, D.F., 2002. Challenges and opportunities for science in reducing nutrient over-enrichment of coastal ecosystems. Estuaries 25, 886–900.

Boyd, J., Banzhaf, S., 2007. What are ecosystem services? the need for standardized environmental accounting units. Ecol. Econ. 63, 616–626.

Brandt-Williams, S., Wigand, C., Campbell, D.E., 2013. Relationships between watershed energy flow and coastal New England salt marsh structure, function, and condition. Environ. Monit. Assess. 185, 1391–1412.

Cahoon, D.R., 2006. A review of major storm impacts on coastal wetland elevations. Estuaries Coasts 29, 889–898.

Cahoon, D.R., Reed, D.K., Kolker, A.S., Brinson, M.M., Stevenson, J.C., Riggs, S., Christian, R., Reyes, E., Voss, C., Kunz, D., 2009. Coastal wetland sustainability. In: Titus, J.G. (Ed.), Coastal Sensitivity to Sea-Level Rise: a Focus on the Mid-Atlantic Region. US Environmental Protection Agency, Washington, DC, pp. 57–72.

Camacho-Valdez, V., Ruiz-Luna, A., Ghermandi, A., Nunes, P.A.L.D., 2013. Valuation of ecosystem services provided by coastal wetlands in northwest Mexico. Ocean Coastal Manage. 78, 1–11.

Chambers, R.M., Meyerson, L.A., Saltonstall, K., 1999. Expansion of *Phragmites australis* into tidal wetlands of North America. Aquat. Bot. 64, 261–273.

Chmura, G.L., Anisfield, S.C., Cahoon, D.R., Lynch, J.C., 2003. Global carbon sequestration in tidal, saline wetland soils. Global Biogeochem. Cycles 17, 22.1–22.12.

Chmura, G.L., Burdick, D.M., Moore, G.E., 2012. Recovering salt marsh ecosystem services through tidal restoration. In: Roman, C.T., Burdick, D.M. (Eds.), Tidal Marsh Restoration: a Synthesis of Science and Management. Island Press, Washington, DC.

Choi, Y., Wang, Y., 2004. Dynamics of carbon sequestration in a coastal wetland using radiocarbon measurements. Global Biogeochem. Cycles 18, GB4016. http://dx.doi.org/10.1029/2004GB002261.

Clausen, K.K., Stjernholm, M., Clausen, P., 2013. Grazing management can counteract the impacts of climate change-induced sea level rise on salt marsh-dependent waterbirds. J. Appl. Ecol. 50, 528–537.

Costanza, R., d'Arge, R., deGroot, R., Farber, S., Grasso, M., Hannon, B., Limburg, K., Naeem, S., O'Neill, R.V., Paruelo, J., Raskin, R.G., Sutton, P., van den Belt, M., 1997. The value of the world's ecosystem services and natural capital. Nature 387, 253–260.

Costanza, R., Pérez-Maqueo, O., Martinez, M.L., Sutton, P., Anderson, S.J., Mulder, K., 2008. The value of coastal wetlands for hurricane protection. AMBIO 37, 241–248.

Craft, C., Cloug, J., Ehman, J., Joye, S., Park, R., Pennings, S., Guo, H., Machmuller, M., 2009. Forecasting the effects of accelerated sea-level rise on tidal marsh ecosystem services. Front. Ecol. Environ. 7, 73–78.

Culbertson, J.B., Valiela, I., Pickart, M., Peacock, E.E., Reddy, C.M., 2008. Long-term consequences of residual petroleum on salt marsh grass. J. Appl. Ecol. 45, 1284–1292.

Daehler, C.C., Strong, D.R., 1996. Status, prediction and prevention of introduced cordgrass, *Spartina* spp. invasions in Pacific estuaries, USA. Biol. Conserv. 78, 51–58.

Dahl, T.E., 2000. Status and Trends of Wetlands in the Conterminous United States, 1986 to 1997. U.S. Department of the Interior, Fish and Wildlife Service, Washington, DC.

Dahl, T.E., 2006. Status and Trends of Wetlands in the Conterminous United States, 1998 to 2004. U.S. Department of the Interior, Fish and Wildlife Service, Washington, DC.

Dahl, T.E., 2011. Status and Trends of Wetlands in the Conterminous United States, 2004 to 2009. U.S. Department of the Interior, Fish and Wildlife Service, Washington, DC.

Dahl, T.E., Johnson, C.E., 1991. Status and Trends of Wetlands in the Conterminous United States, Mid-1970's to Mid-1980's. U.S. Department of the Interior, Fish and Wildlife Service, Washington, DC.

Dahl, T.E., Stedman, S.M., 2013. Status and Trends of Wetlands in the Coastal Watersheds of the Conterminous United States 2004 to 2009. U.S. Department of the Interior, Fish and Wildlife Service and National Oceanic and Atmospheric Administration, National Marine Fisheries Service, Washington, DC.

Day, J.W., Barras, J., Clairain, E., Johnston, J., Justic, D., Kemp, G.P., Ko, J.-Y., Lane, R., Mitsch, W.J., Steyer, G., Templet, P., Yáñez-Arancibia, A., 2005. Implications of global climatic change and energy cost and availability for the restoration of the Mississippi delta. Ecol. Eng. 24, 253–265.

Day, J.W., Boesch, D.F., Clairain, E.J., Kemp, G.P., Laska, S.B., Mitsch, W.J., Orth, K., Mashriqui, H., Reed, D.J., Shabman, L., Simenstad, C.A., Streever, B.J., Twilley, R.R., Watson, C.C., Wells, J.T., Whigham, D.F., 2007. Restoration of the Mississippi delta: lessons from hurricanes Katrina and Rita. Science 315, 1679–1684.

Day, J.W., Christian, R.R., Boesch, D.M., Yáñez-Arancibia, A., Morris, J., Twilley, R.R., Naylor, L., Schaffner, L., Stevenson, C., 2008. Consequences of climate change on the ecogeomorphology of coastal wetlands. Estuaries Coasts 31, 477–491.

Deegan, L.A., Hughes, J.E., Rountree, R.A., 2000. Salt marsh ecosystem support of marine transient species. In: Weinstein, M.P., Kreeger, D.A. (Eds.), Concepts and Controversies in Tidal Marsh Ecology. Kluwer Academic Publishers, Dordrecht, The Netherlands, pp. 333–365.

Deegan, L.A., Johnson, D.S., Warren, R.S., Peterson, B.J., Fleeger, J.W., Fagherazzi, S., Wollheim, W.M., 2012. Coastal eutrophication as a driver of salt marsh loss. Nature 490, 388–392.

DeLaune, R.D., White, J.R., 2012. Will coastal wetlands continue to sequester carbon in response to an increase in global sea level?: a case study of the rapidly subsiding Mississippi River deltaic plain. Clim. Change 110, 297–314.

Duarte, C.M., Losada, I.J., Hendriks, I.E., Mazarrasa, I., Marbà, N., 2013. The role of coastal plant communities for climate change mitigation and adaptation. Nat. Clim. Change 3, 961–968.
Ehrenfeld, J.G., 2003. Effects of exotic plant invasions on soil nutrient cycling processes. Ecosystems 6, 503–523.
Eller, F., Lambertini, C., Nguyen, L.X., Brix, H., 2014. Increased invasive potential of nonnative *Phragmites australis*: elevated CO_2 and temperature alleviate salinity effects on photosynthesis and growth. Global Change Biol 20, 531–543.
Ellis, E.C., Ramankutty, N., 2008. Putting people in the map: anthropogenic biomes of the world. Front. Ecol. Environ. 6, 439–447.
Elschot, K., Bouma, T.J., Temmermanc, S., Bakker, J.P., 2013. Effects of long-term grazing on sediment deposition and salt-marsh accretion rates. Estuarine, Coastal Shelf Sci. 133, 109–115.
Emery, N.C., Ewanchuk, P.J., Bertness, M.D., 2001. Competition and salt-marsh plant zonation: stress tolerators may be dominant competitors. Ecology 82, 2471–2485.
Engle, V.D., 2011. Estimating the provision of ecosystem services by Gulf of Mexico coastal wetlands. Wetlands 31, 179–193.
Erwin, K.L., 2009. Wetlands and global climate change: the role of wetland restoration in a changing world. Wetlands Ecol. Manage. 17, 71–84.
Erwin, R.M., Sanders, G.M., Prosser, D.J., Cahoon, D.R., 2006. High tides and rising seas: potential effects on estuarine waterbirds. Stud. Avian Biol. 32, 214–228.
Esteves, L.S., 2013. Is managed realignment a sustainable long-term coastal management approach? J. Coastal Res. Special Issue No. 65, 933–938.
Feagin, R.A., Martinez, M.L., Mendoza-Gonzalez, G., Costanza, R., 2010. Salt marsh zonal migration and ecosystem service change in response to global sea level rise: a case study from an urban region. Ecol. Soc. 15, 14.
Ford, H., Rousk, J., Garbutt, A., Jones, L., Jones, D.L., 2013. Grazing effects on microbial community composition, growth and nutrient cycling in salt marsh and sand dune grasslands. Biol. Fertil. Soils 49, 89–98.
French, P.W., 2006. Managed realignment—the developing story of a comparatively new approach to soft engineering. Estuarine, Coastal Shelf Sci. 67, 409–423.
Gan, X., Choi, C., Wang, Y., Ma, Z., Chen, J., Li, B., 2010. Alteration of habitat structure and food resources by invasive smooth cordgrass affects habitat use by wintering saltmarsh birds at Chongming Dongtan, East China. Auk 127, 317–327.
Gedan, K.B., Silliman, B.R., 2009. Patterns of salt marsh loss within coastal regions of North America. In: Silliman, B.R., Grosholz, E.D., Bertness, M.D. (Eds.), Human Impacts on Salt Marshes: a Global Perspective. University of California Press, Berkeley, CA, pp. 253–265.
Gedan, K.B., Silliman, B.R., Bertness, M.D., 2009. Centuries of human-driven change in salt marsh ecosystems. Ann. Rev. Mar. Sci. 1, 117–141.
Gedan, K.B., Kirwan, M.L., Wolanski, E., Barbier, E.B., Silliman, B.R., 2011. The present and future role of coastal wetland vegetation in protecting shorelines: answering recent challenges to the paradigm. Clim. Change 106, 7–29.
Gilfillan, E.S., Maher, N.P., Krejsa, C.M., Lanphear, M.E., Ball, C.D., Meltzer, J.B., Page, D.S., 1995. Use of remote sensing to document changes in marsh vegetation following the Amoco Cadiz oil spill (Brittany, France, 1978). Mar. Pollut. Bull. 30, 780–787.
Glick, P., Clough, J., Polaczyk, A., Couvillion, B., Nunley, B., 2013. Potential effects of sea-level rise on coastal wetlands in southeastern Louisiana. J. Coastal Res. 63, 211–233.
Greenberg, R., Maldonado, J.E., Droege, S., McDonald, M.V., 2006. Tidal marshes: a global perspective on the evolution and conservation of their terrestrial vertebrates. Bioscience 56, 675–685.

deGroot, R., Brander, L., van der Ploeg, S., Costanza, R., Bernard, F., Braat, L., Christie, M., Crossman, N., Ghermandi, A., Hein, L., Hussain, S., Kumar, P., McVittie, A., Portela, R., Rodriguez, L.C., ten Brink, P., van Beukering, P., 2012. Global estimates of the value of ecosystems and their services in monetary units. Ecosys. Serv. 1, 50–61.

Hall, J., Frayer, W., Wilen, B., 1994. Status of Alaska Wetlands. U.S. Department of the Interior, Fish and Wildlife Service, Anchorage, AK.

Handa, I.T., Jefferies, R.L., 2000. Assisted revegetation trials in degraded salt-marshes. J. Appl. Ecol. 37, 944–958.

Hansen, V.D., Nestlerode, J.A., 2014. Carbon sequestration in wetland soils of the northern Gulf of Mexico coastal region. Wetlands Ecol. Manage. 22, 289–303.

Hedge, P., Kriwoken, L.K., Patten, K., 2003. A review of *Spartina* management in Washington State, US. J. Aquat. Plant Manage. 41, 82–90.

Hoekstra, J.M., Molnar, J.L., Jennings, M., Revenga, C., Spalding, M.D., Boucher, T.M., Robertson, J.C., Heibel, T.J., 2010. The Atlas of Global Conservation: Changes, Challenges, and Opportunities to Make a Difference. University of California Press, Berkeley, CA.

Howarth, R.W., 2008. Coastal nitrogen pollution: a review of sources and trends globally and regionally. Harmful Algae 8, 14–20.

Hughes, R.G., 2004. Climate change and loss of saltmarshes: consequences for birds. Ibis 146 (Suppl. 1), 21–28.

Intergovernmental Panel on Climate Change [IPCC], 2013. Climate Change 2013, the Physical Science Basis, Summary for Policymakers. Available from: http://www.ipcc.ch/report/ar5/wg1/#.UpUKRtK-q9E [November 15, 2013].

King, S.E., Lester, J.N., 1995. The value of salt marsh as a sea defence. Mar. Pollut. Bull. 30, 180–189.

Kirwan, M.L., Guntenspergen, G.R., D'Alpaos, A., Morris, J.T., Mudd, S.M., Temmerman, S., 2010. Limits on the adaptability of coastal marshes to rising sea level. Geophys. Res. Lett. 37, L23401. http://dx.doi.org/10.1029/2010GL045489.

Kirwan, M.L., Megonigal, P., 2013. Tidal wetland stability in the face of human impacts and sea-level rise. Nature 504, 53–60.

Kriwoken, L.K., Hedge, P., 2000. Exotic species and estuaries: managing *Spartina anglica* in Tasmania, Australia. Ocean Coastal Manage. 43, 573–584.

Lin, Q., Mendelssohn, I.A., 1996. A comparative investigation of the effects of south Louisiana crude oil on the vegetation of fresh, brackish and salt marshes. Mar. Pollut. Bull. 32, 202–209.

Lin, Q., Mendelssohn, I.A., 2012. Impacts and recovery of the deepwater horizon oil spill on vegetation structure and function of coastal salt marshes in the northern Gulf of Mexico. Environ. Sci. Technol. 46, 3737–3743.

Mattheus, C.R., Rodriguez, A.B., McKee, B.A., Currin, C.A., 2010. Impact of land-use change and hard structures on the evolution of fringing marsh shorelines. Estuarine, Coastal Shelf Sci. 88, 365–376.

McGranahan, G., Balk, D., Anderson, B., 2007. The rising tide: assessing the risks of climate change and human settlements in low elevation coastal zones. Environ. Urbanization 19, 17–37.

McKinney, R.A., Charpentier, M.A., Wigand, C., 2009. Assessing the wildlife habitat value of New England salt marshes: II. model testing and validation. Environ. Monit. Assess. 154, 361–371.

Mendelssohn, I.A., McKee, K., 2000. Saltmarshes and mangroves. In: Barbier, M., Billings, W. (Eds.), North American Terrestrial Vegetation. Cambridge University Press, Cambridge, pp. 501–536.

Mendelssohn, I.A., Andersen, G.L., Baltz, D.M., Caffey, R.H., Carman, K.R., Fleeger, J.W., Joye, S.B., Lin, Q., Maltby, E., Overton, E.B., Rozas, L.R., 2012. Oil impacts on coastal wetlands: implications for the Mississippi River Delta ecosystem after the deepwater horizon oil spill. Bioscience 62, 562–574.

Michel, M.J., Owens, E.H., Zengen, S., Graham, A., Nixon, Z., Allard, T., Holton, W., Reimer, P.D., Lamarche, A., White, M., Rutherford, N., Childs, C., Mauseth, G., Challenger, G., Taylor, E., 2013. Extent and degree of shoreline oiling: deepwater horizon oil spill, Gulf of Mexico, USA. PLoS One 8 (6), e65087. http://dx.doi.org/10.1371/journal.pone.0065087.

Michener, W.K., Blood, E.R., Bildstein, K.L., Brinson, M.M., Gardner, L.R., 1997. Climate change, hurricanes and tropical storms, and rising sea level in coastal wetlands. Ecol. Appl. 7, 770–801.

Michinton, T.E., Bertness, M.D., 2003. Disturbance-mediated competition and spread of *Phragmites australis* in a coastal marsh. Ecol. Appl. 13, 1400–1416.

Mishra, D.R., Cho, H.J., Ghosh, S., Fox, A., Downs, C., Merani, P.B.T., Kirui, P., Jackson, N., Mishra, S., 2012. Post-spill state of the marsh: remote estimation of the ecological impact of the Gulf of Mexico oil spill on Louisiana salt marshes. Remote Sens. Environ. 118, 176–185.

Möller, I., 2006. Quantifying saltmarsh vegetation and its effect on wave height dissipation: results from a UK East coast saltmarsh. Estuarine, Coastal Shelf Sci. 69, 337–351.

Moore, G.E., Burdick, D.M., Peter, C.R., Leonard-Duarte, A., Dionne, M., 2009. Regional Assessment of Tidal Marsh Restoration in New England Using the Restoration Performance Index. Final Report. NOAA Restoration Center, Gloucester, MA.

Morgan, P.A., Burdick, D.M., Short, F.T., 2009. The functions and values of fringing salt marshes in northern New England, USA. Estuaries Coasts 32, 483–495.

Morrice, J.A., Danz, N.P., Regal, R.R., Kelly, J.R., Niemi, G.J., Reavie, E.D., Hollenhorst, T., Axler, R.P., Trebitz, A.S., Cotter, A.M., Peterson, G.S., 2008. Human influences on water quality in Great Lakes coastal wetlands. Environ. Manage. 41, 347–357.

Morris, J.T., Bradley, P.M., 1999. Effects of nutrient loading on the carbon balance of coastal wetland sediments. Limnol. Oceanogr. 44, 699–702.

Morris, J.T., Sundareshwar, P.V., Nietch, C.T., Kjerfve, B., Cahoon, D.R., 2002. Responses of coastal wetlands to rising sea level. Ecology 83, 2869–2877.

National Oceanic and Atmospheric Administration [NOAA], 2013. National Coastal Population Report: Population Trends from 1970 to 2020. Available from: http://stateofthecoast.noaa.gov [November 26, 2013].

National Oceanic and Atmospheric Administration [NOAA], 2011. The Gulf of Mexico at a Glance: a Second Glance. Available from: http://stateofthecoast.noaa.gov/gulfreport.html [March 17, 2014].

Nehring, S., Hesse, K., 2008. Invasive alien plants in marine protected areas: the *Spartina anglica* affair in the European Wadden Sea. Biol. Invasions 10, 937–950.

Nellemann, C., Corcoran, E., Duarte, C., Valdés, L., De Young, C., Fonseca, L., Grimsditch, G. (Eds.), 2009. Blue Carbon. A Rapid Response Assessment (United Nations Environment Programme, GRID-Arendal).

Nicholls, R.J., 2004. Coastal flooding and wetland loss in the 21st century: changes under the SRES climate and socio-economic scenarios. Global Environ. Change 14, 69–86.

O'Callaghan, M., 1990. Saltmarshes—a highly specialized environment. Custos 18, 58–60.

Peacock, E.E., Nelson, R.K., Solow, A.R., Warren, J.D., Baker, J.L., Reddy, C.M., 2005. The West Falmouth oil spill: ~100 kg of oil found to persist decades later. Environ. Forensics 6, 273–281.

Peacock, E.E., Hampson, G.R., Nelson, R.K., Xu, L., Frysinger, G.S., Gaines, R.B., Farrington, J.W., Tripp, B.W., Reddy, C.M., 2007. The 1974 spill of the Bouchard 65 oil barge: petroleum hydrocarbons persist in Winsor Cove salt marsh sediments. Mar. Pollut. Bull. 54, 214–225.

Pezeshki, S.R., Hester, M.W., Lin, Q., Nyman, J.A., 2000. The effects of oil spill and clean-up on dominant US Gulf coast marsh macrophytes: a review. Environ. Pollut. 108, 129–139.

Pratolongo, P., Mazzon, C., Zapperi, G., Piovan, M.J., Brinson, M.M., 2013. Land cover changes in tidal salt marshes of the Bahía Blanca estuary (Argentina) during the past 40 years. Estuarine, Coastal Shelf Sci. 133, 23–31.

Rabalais, N.N., Turner, R.E., Diaz, R.J., Justic, D., 2009. Global change and eutrophication of coastal waters. ICES J. Mar. Sci. 66, 1528–1537.

Ravit, B., Ehrenfeld, J.G., Haggblom, M.M., 2003. A comparison of sediment microbial communities associated with *Phragmites australis* and *Spartina alterniflora* in two brackish wetlands of New Jersey. Estuaries 26, 465–474.

Resio, D.T., Westerink, J.J., 2008. Modeling the physics of storm surges. Phys. Today 61, 33–38.

Roman, C.T., Burdick, D.M. (Eds.), 2012a. Tidal Marsh Restoration: a Synthesis of Science and Management. Island Press, Washington, DC.

Roman, C.T., Burdick, D.M., 2012b. A synthesis of research and practice on restoring tides to salt marshes. In: Roman, C.T., Burdick, D.M. (Eds.), Tidal Marsh Restoration: a Synthesis of Science and Management. Island Press, Washington, DC.

Rush, S.A., Soehren, E.C., Woodrey, M.S., Graydon, C.L., Cooper, R.J., 2009. Occupancy of select marsh birds within northern Gulf of Mexico tidal marsh: current estimates and projected change. Wetlands 29, 798–808.

Saintilan, N., Rogers, K., 2009. Coastal saltmarsh vulnerability to climate change in SE Australia. In: 18th NSW Coastal Conference Papers: Proceedings of the NSW Coastal Conference, Australia, pp. 1–12.

Saltonstall, K., 2002. Cryptic invasion by a non-native genotype of the common reed, *Phragmites australis*, into North America. Proc. Natl. Acad. Sci. 99, 2445–2449.

Scavia, D., Field, J.C., Boesch, D.F., Buddemeier, R.W., Burkett, V., Cayan, D.R., Fogarty, M., Harwell, M.A., Howarth, R.W., Mason, C., Reed, D.J., Royer, T.C., Sallenger, A.H., Titus, J.G., 2002. Climate change impacts on U.S. coastal and marine ecosystems. Estuaries 25, 149–164.

Wetlands Research Program Technical Report WRP-DE-16. In: Shafer, D.J., Yozzo, D.J. (Eds.), National Guidebook for Application of Hydrogeomorphic Assessment to Tidal Fringe Wetlands. US Army Corps of Engineers, U.S. Army Engineer Waterways Experiment Station, Wetlands Ecology Branch, Vicksburg, MS.

Schlaepfer, M.A., Sax, D.F., Olden, J.D., 2011. The potential conservation value of non-native species. Conserv. Biol. 25, 428–437.

Schuerch, M., Vafeidis, A., Slawig, T., Temmerman, S., 2013. Modeling the influence of changing storm patterns on the ability of a salt marsh to keep pace with sea level rise. J. Geophys. Res. Earth Surf. 118, 84–96.

Secretariat of the Convention on Biological Diversity, 2010. Global Biodiversity Outlook 3 (Montreal).

Seitzinger, S.P., Kroeze, C., Bouwman, A.F., Caraco, N., Dentener, F., Styles, R.V., 2002. Global patterns of dissolved inorganic and particulate nitrogen inputs to coastal systems: recent conditions and future projections. Estuaries 25, 640–655.

Shepard, C.C., Crain, C.M., Beck, M.W., 2011. The protective role of coastal marshes: a systematic review and meta-analysis. PLoS One 6, e27374. http://dx.doi.org/10.1371/journal.pone.0027374.

Silliman, B.R., Bertness, M.D., 2004. Shoreline development drives invasion of *Phragmites australis* and the loss of plant diversity on New England salt marshes. Conserv. Biol. 18, 1424–1434.

Silliman, B.R., Grosholz, E.D., Bertness, M.D., 2009. Salt marshes under global siege. In: Silliman, B.R., Grosholz, E.D., Bertness, M.D. (Eds.), Human Impacts on Salt Marshes: a Global Perspective. University of California Press, Berkeley, CA, pp. 391–398.

Silliman, B.R., van de Koppel, J., McCoy, M.W., Diller, J., Kasozi, G.N., Earl, K., Adams, P.N., Zimmerman, A.R., 2012. Degradation and resilience in Louisiana salt marshes after the BP-deepwater horizon oil spill. Proc. Natl. Acad. Sci. 109, 11234–11239.

Sousa, A.I., Lillebø, A.I., Caçador, I., Pardal, M.A., 2008. Contribution of *Spartina maritima* to the reduction of eutrophication in estuarine systems. Environ. Pollut. 156, 628–635.

Spalding, M.D., Ruffo, S., Lacambra, C., Meliane, I., Hale, L.Z., Shepard, C.C., Beck, M.W., 2014. The role of ecosystems in coastal protection: adapting to climate change and coastal hazards. Ocean Coastal Manage. 90, 50–57.

Spencer, K.L., Harvey, G.L., 2012. Understanding system disturbance and ecosystem services in restored saltmarshes: integrating physical and biogeochemical processes. Estuarine, Coastal Shelf Sci. 106, 23–32.

Suding, K.N., Gross, K.L., Houseman, G.R., 2004. Alternative states and positive feedbacks in restoration ecology. Trends Ecol. Evol. 19, 46–53.

Titus, J.G., Hudgens, D.E., Trescott, D.L., Craghan, M., Nuckols, W.H., Hershner, C.H., Kassakian, J.M., Linn, C.J., Merritt, P.G., McCue, T.M., O'Connell, J.F., Tanski, J., Wang, J., 2009. State and local governments plan for development of most land vulnerable to rising sea level along the US Atlantic coast. Environ. Res. Lett. 4, 044008.

Torio, D.D., Chmura, G.L., 2013. Assessing coastal squeeze of tidal wetlands. J. Coastal Res. 29, 1049–1061.

Turner, R.E., 1992. Coastal wetlands and penaeid shrimp habitat. In: Stroud, R.H. (Ed.), Stemming the Tide of Coastal Fish Habitat Loss. National Coalition for Marine Conservation, Inc., Savannah, GA, pp. 97–104.

Turner, R.E., Baustian, J.J., Swenson, E.M., Spicer, J.S., 2006. Wetland sedimentation from hurricanes Katrina and Rita. Science 314, 449–452.

Turner, R.E., Howes, B.L., Teal, J.M., Milan, C.S., Swenson, E.M., Goehringer-Toner, D.D., 2009. Salt marshes and eutrophication: an unsustainable outcome. Limnol. Oceanogr. 54, 1634–1642.

Tweel, A.W., Turner, R.E., 2012. Watershed land use and river engineering drive wetland formation and loss in the Mississippi River birdfoot delta. Limnol. Oceanogr. 57, 18–28.

Tyler, A.C., Lambrinos, J.G., Grosholz, E.D., 2007. Nitrogen inputs promote the spread of an invasive marsh grass. Ecol. Appl. 17, 1886–1898.

United Nations Environmental Programme [UNEP], 2006. Marine and Coastal Ecosystems and Human Well-being: a Synthesis Report Based on the Findings of the Millennium Ecosystem Assessment. United Nations Environment Programme, Nairobi.

United Nations Environmental Programme [UNEP], 2007. UNEP Annual Report 2007. United Nations Environment Programme, Nairobi.

Valiela, I., Cole, M.L., 2002. Comparative evidence that salt marshes and mangroves may protect seagrass meadows from land-derived nitrogen loads. Ecosystems 5, 92–102.

Valiela, I., Kinney, E., Culbertson, J., Peacock, E., Smith, S., 2009. Global losses of mangroves and salt marshes. In: Duarte, C. (Ed.), Global Loss of Coastal Habitats: Rates, Causes and Consequences. Fundación BBVA, pp. 109–142.

Vasquez, E.A., Glenn, E.P., Brown, J.J., Guntenspergen, G.R., Nelson, S.G., 2005. Salt tolerance underlies the cryptic invasion of North American salt marshes by an introduced haplotype of the common reed *Phragmites australis* (Poaceae). Mar. Ecol. Prog. Ser. 298, 1–8.

Wamsley, T.V., Cialone, M.A., Smith, J.M., Atkinson, J.H., Rosati, J.D., 2010. The potential of wetlands in reducing storm surge. Ocean Eng. 37, 59–68.

Wang, Z., Fingas, M.F., Sigouin, L., Owens, H., 2001. Fate and persistence of long-term spilled Metula oil in the marine salt marsh environment: degradation of petroleum biomarkers. Int. Oil Spill Conf. Proc. 2001 (1), 115–124.

Warren, R.S., Fell, P.E., Grimsby, J.L., Buck, E.L., Rilling, G.C., Fertik, R.A., 2001. Rates, patterns, and impacts of *Phragmites australis* expansion and effects of experimental *Phragmites* control on vegetation, macroinvertebrates, and fish within tidelands of the lower Connecticut River. Estuaries 24, 90–107.

Weston, N.B., 2014. Declining sediments and rising seas: an unfortunate convergence for tidal wetlands. Estuaries Coasts 37, 1–23.

Whiting, G.J., Chanton, J.P., 2001. Greenhouse carbon balance of wetlands: methane emission versus carbon sequestration. Tellus 53B, 521–528.

Windham, L., Lathrop, R.G., 1999. Effects of *Phragmites australis* (common reed) invasion on aboveground biomass and soil properties in brackish tidal marsh of the Mullica River, New Jersey. Estuaries 22, 927–935.

Wolfe, B.E., Klironomos, J.N., 2005. Breaking new ground: soil communities and exotic plant invasion. Bioscience 55, 477–487.

Woodward, R.T., Wui, Y.S., 2001. The economic value of wetland services: a meta-analysis. Ecol. Econ. 37, 257–270.

Zedler, J.B., Kercher, S., 2004. Causes and consequences of invasive plants in wetlands: opportunities, opportunists, and outcomes. Crit. Rev. Plant Sci. 23, 431–452.

Zimmerman, R.J., Minello, T.J., Rozas, L.P., 2000. Salt marsh linkages to productivity of penaeid shrimps and blue crabs in the northern Gulf of Mexico. In: Weinstein, M.P., Kreeder, D.A. (Eds.), Concepts and Controversies in Tidal Marsh Ecology. Kluwer, Dordrecht, pp. 293–314.

Chapter 17

Living with Harmful Algal Blooms in a Changing World: Strategies for Modeling and Mitigating Their Effects in Coastal Marine Ecosystems

Clarissa R. Anderson [1], Stephanie K. Moore [2], Michelle C. Tomlinson [3], Joe Silke [4] and Caroline K. Cusack [4]

[1] *Institute of Marine Sciences, University of California, Santa Cruz, CA, USA,* [2] *Environmental and Fisheries Sciences, Northwest Fisheries Science Center, National Marine Fisheries Service, National Oceanic and Atmospheric Administration, Seattle, WA, USA,* [3] *NOAA, National Centers for Coastal Ocean Science, East-West Highway, Silver Spring, MD, USA,* [4] *Marine Institute, Oranmore, County Galway, Ireland*

ABSTRACT

Harmful algal blooms (HABs) are extreme biological events with the potential for extensive negative impacts to fisheries, coastal ecosystems, public health, and coastal economies. In this chapter, we link issues concerning the key drivers of HABs with the various approaches for minimizing their negative impacts, emphasizing the use of numerical modeling techniques to bridge the gap between observations and predictive understanding. We review (1) recent studies on the environmental pressures that promote HABs; (2) prominent strategies for preventing or controlling blooms; (3) modeling methods, specifically addressing harmful algal species dynamics, and their use as a predictive tool to facilitate mitigation; and then (4) highlight several coastal regions where the mitigation of HABs is generally approached from a regional Earth system and observation framework. Lastly, we summarize future directions for "living with" HABs in an era of limited financial resources for ocean observing.

17.1 INTRODUCTION

Decades of research on harmful algal blooms (HABs) in the world's coastal, estuarine, and freshwater environments have revealed immense complexity in

Coastal and Marine Hazards, Risks, and Disasters. http://dx.doi.org/10.1016/B978-0-12-396483-0.00017-0
Copyright © 2015 Elsevier Inc. All rights reserved.

the conditions that promote bloom development and the diversity of HAB species. Just as the physical features of the coastal zone cannot be represented by a single model across spatial and temporal scales, the biological variability within aquatic ecosystems requires a regional perspective, one that considers indigenous communities (from plankton to humans), habitat connectivity, and the influence of large-scale drivers of change (Cloern et al., 2010). Although levels of devastation experienced by coastal communities during HAB events might not approximate those of many natural disasters, the economic losses are often of great importance to local seafood industries (Imai et al., 2006; Jin et al., 2008; Dyson and Huppert, 2010) as are the risks to public health (Van Dolah et al., 2001). Ecosystem functioning and wildlife populations are also often negatively impacted by HABs, with legacy effects that compound over time (Sekula-Wood et al., 2009, 2011; Paerl et al., 2011; Montie et al., 2012). Understanding the ecological role of harmful algae and their seeming rise to prominence in phytoplankton communities requires that the role of natural variability be teased apart from human disturbance (Hallegraeff, 1993, 2010; Figure 17.1). The field of HAB science has made significant advances in this area, and this ecological knowledge is now informing methods for mitigating the harmful effects of HABs on natural resources and human populations, and in some instances, pushing forward technological advancements with broad application (Anderson et al., 2012b).

A major struggle in the study and management of HABs has been the sheer breadth of species, life histories, ecosystems, and impacts involved. The phytoplankton that are categorized as potentially harmful do not belong to a single, evolutionarily distinct group. Rather, they span the majority of algal taxonomic clades, including eukaryotic protists (armored and unarmored dinoflagellates, raphidophytes and diatoms, euglenophytes, cryptophytes, haptophytes, and chlorophytes) and microbial prokaryotes (the ubiquitous, sometimes nitrogen-fixing cyanobacteria that occur in both marine and freshwater systems). Interestingly, dinoflagellates account for the majority (75 percent) of HAB species (Smayda, 1997). The list of potential impacts from HABs include (1) the production of dangerous phycotoxins that enter food webs, the atmosphere (if aerosolized), fisheries, and the potential contamination of water supplies from freshwater reservoirs or desalination plants; (2) the depletion of dissolved oxygen and/or the smothering of benthic biota as algal biomass decays; and (3) physical damage to fish gill tissue. HABs fall under the umbrella term Ecosystem Disruptive Algal Blooms (EDABs; Sunda and Shertzer, 2012; Sunda et al., 2006), and all HABs or EDABs may impact local ecosystems and economies (e.g., fisheries, tourism, recreation). These impacts include noxious or nuisance blooms such as "brown tides" of pelagophytes *Aureoccocus anophagefferens* and *Aureoumbra lagunensis* (Gobler and Sunda, 2012) or the surfactant-producing *Akashiwo sanguinea* (Jessup et al., 2009). Given this diversity, no single set of conditions or approach to mitigation will apply to all harmful algae, nor is the often-used

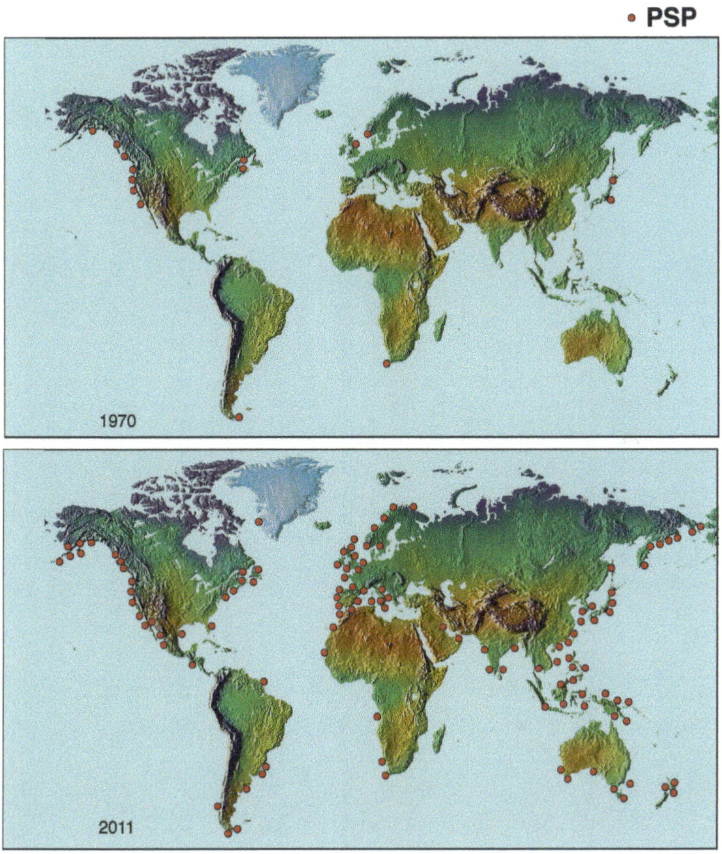

FIGURE 17.1 The expansion of global cases of Paralytic shellfish poisoning (PSP) from 1972 to 2011. PSP is associated with the marine dinoflagellates *Alexandrium* and *Pyrodinium,* several species of which produce saxitoxin, a dangerous neurotoxin that makes its way into the food web and can be lethal. *Map used with permission from the National Office for Harmful Algal Blooms at Woods Hole Oceanographic Institution.*

term "red tide" appropriate for phenomena with a broad range of pigment and spectral qualities generally undetectable to the human eye (Dierssen et al., 2006).

The suite of epidemiological syndromes associated with phycotoxin exposure is itself impressive (see Table 17.1 for symptoms and acronyms); more details on the symptoms associated with these syndromes and the geographic locations where illnesses have been reported can be found in reviews of phycotoxin poisonings (Fleming et al., 2002; Backer et al., 2005; Backer and Moore, 2012). New toxins and syndromes are continually discovered, such as the ecosystem-disruptive yessotoxin (De Wit et al., 2014) produced by the dinoflagellates *Gonyaulax spinifera* (Rhodes et al., 2006), *Protoceratium reticulatum* (Paz et al., 2004; Alvarez et al., 2011), and *Lingulodinium polyedrum* (Howard et al., 2008, Figure 17.2), a bioluminescent

TABLE 17.1 Human Syndromes Caused by Ingestion or Exposure to Marine HAB Toxins

Syndrome	Toxin(s)	Causative Organism	Symptoms
CFP	Ciguatoxins	*Gambierdiscus* spp.[b]	Nausea, vomiting, diarrhea, numbness of the mouth and extremities, rash, and reversal of temperature sensation. Neurological symptoms may persist for several months.
PSP	Saxitoxin and its derivatives	*Alexandrium* spp. *Pyrodinium* spp. *Gymnodinium* spp.	Numbness and tingling of the lips, mouth, face, and neck; nausea; and vomiting. Severe cases result in paralysis of the muscles of the chest and abdomen possibly leading to death.
ASP	Domoic Acid	*Pseudo-nitzschia* spp. *Nitzschia navis-varingica*	Nausea, vomiting, diarrhea, headache, dizziness, confusion, disorientation, short-term memory deficits, and motor weakness. Severe cases result in seizures, cardiac arrhythmia, respiratory distress, coma, and possibly death.
AZP	Azaspiracid and its derivatives	*Azadinium* spp.[a]	Nausea vomiting, severe diarrhea, and abdominal cramps
NSP	Brevetoxin	*Karenia* spp.	Nausea, temperature sensation reversals, muscle weakness, and vertigo
DSP	Okadaic acid and its derivatives	*Dinophysis* spp. *Prorocentrum* spp.	Nausea vomiting, severe diarrhea, and abdominal cramps

DSP[e]	Yessotoxin	*Gonyaulax spinifera* *Protoceratium reticulatum* *Lingulodinium polyedrum*	Nausea, vomiting, abdominal cramps, reduced appetite, cardiotoxic effects, respiratory distress
DSP[e]	Cooliatoxin[c]	*Coolia* spp.[b]	Nausea, vomiting, abdominal cramps, reduced appetite, cardiotoxic effects, respiratory distress
Palytoxicosis	Palytoxin and its derivatives[d,f]	*Ostreopsis* spp.[b]	Nausea; vomiting; diarrhea; abdominal cramps; lethargy; tingling of the lips, mouth, face, and neck; lowered heart rate; skeletal muscle breakdown; muscle spasms and pain; lack of sensation; respiratory distress
Lyngbyatoxicosis	Lyngbyatoxin-A and its derivatives	*Lyngbya majuscula*[d,g]	Weakness, headache, lightheadedness, salivation, gastrointestinal inflammation, potent tumor promoter

Note that aside from the diatom *Pseudo-nitzschia* and the cyanobacteria *Lyngbya majuscula* (now *Moorea* spp.), the causative organisms are all dinoflagellates. Freshwater groups such as the hepatotoxin-producing *Microcystis* spp. are not included here. ASP, amnesic shellfish poisoning; AZP, azaspiracid shellfish poisoning; CFP, ciguatera fish poisoning; DSP, Diarrhetic shellfish poisoning; DA, Domoic acid.

[a]*Azaspiracid was first thought to be associated with* Protoperidinium *(Yasumoto 2001; James et al. 2003) but was later shown to be produced by* Azadinium *spp. (Tillmann et al., 2009).*

[b]*Benthic epiphytes.*

[c]*A monosulfated analog of yessotoxin (Rhodes et al., 2000); complete structure uncharacterized (Van Dolah et al., 2013).*

[d]*Produces aerosolized toxins with known health consequences (Osborne et al., 2001; Ciminiello et al., 2010).*

[e]*Yessotoxins and cooliatoxins are grouped with DSP syndrome (Draisci et al., 2000) but may be more like PSP since yessotoxin exposure does not lead to diarrhea (Paz et al., 2008).*

[f]*One of the most toxic natural substances known.*

[g]*Lyngbya majuscula newly classified as* Moorea producens *(Engene et al., 2012).*

Adapted from Table 17.2 in Marques et al. (2010).

FIGURE 17.2 Sonoma County, California. In 2011, a mass mortality of red abalone, urchins, sea stars, chitons, and crabs (right) was the largest invertebrate die-off recorded for the region (De Wit et al., 2013). Yessotoxin was implicated as the causative agent (De Wit et al., 2014) and is produced by a number of common "red tide" dinoflagellates (inset) in coastal California (left). Red tide photo taken by Kai Schumann.

dinoflagellate common to the US West Coast. Widespread bird mortality caused by blooms of the dinoflagellate *A. sanguinea* is a new threat along the US West Coast (Jessup et al., 2009; Berdalet et al., 2013). Azaspiracid shellfish poisoning, caused by the dinoflagellate *Azadinium,* is another burgeoning disease with a possible worldwide distribution (Salas et al., 2011) after first being noticed in Northern European coastal communities (Krock et al., 2009) and now recently detected in Puget Sound, WA, USA (Trainer et al., 2013). Palytoxicosis is an emerging issue in the Mediterranean where palytoxin, the most toxic marine compound known, has caused extensive seafood poisoning after bioaccumulating in commonly consumed crustaceans and fish that have grazed upon the benthic dinoflagellate *Ostreopsis* (Amzil et al., 2012).

Discussion of HABs in the literature has traditionally focused on the disruptive or even "catastrophic" nature of "red tides" as toxic and/or high-biomass blooms (Margalef, 1978). However, the caveat is often made that such blooms are not new, unnatural phenomena (Cullen, 2008; Hallegraeff, 2010), and they have long been part of a region's local ecology, primary productivity, and important biogeochemical cycling. That said, there is increasing recognition that the effects of HABs on public health, marine and freshwater ecosystems, economies (Hoagland and Scatasta, 2006), and human social structures (Hatch et al., 2013) are worsening (Heisler et al., 2008; Anderson, 2009; Hallegraeff, 2010; Anderson et al., 2012b, Figure 17.1) and require new solutions from collaboration among scientists, the private sector, and governing bodies (Green et al., 2009). The potential causes for this trend have been thoroughly vetted elsewhere (e.g., Hallegraeff, 1993, 2010; Glibert et al., 2006; Anderson et al., 2002, 2008; Heisler et al., 2008; Paerl et al.,

2011). Eutrophication, climate change, ballast water dispersal, and improved monitoring are the most cited factors for the increased frequency of reported blooms.

At the interface between HABs and human communities is the socioeconomic outfall around which the majority of impacts are contextualized. The interaction between HABs and humans involves both positive and negative feedbacks to the blooms themselves and to the ability of society to mitigate adverse effects (Figure 17.3). Hoagland (2014) carefully illustrates this process for toxic blooms of *Karenia brevis* on Florida's Gulf coast and describes how "legacies" of indigenous and modern human behavior and the complex history of mitigation strategies inform past and future "policy responses" to HAB events. Ultimately, how these policies are implemented will depend on the cost-effectiveness of mitigation strategies that range from the reduction of exposure risk and illness to fisheries regulation (Heil and Steidinger, 2009; Heil, 2009; Hoagland, 2014). Significant overlap occurs with oil spill response strategies (Liu et al., 2011) that integrate local community impacts with particle tracking models, remote detection techniques, wildlife biology, and regional management mandates. Bringing these socioeconomic, governmental, and traditional science realms together is a challenging but crucial goal for next-generation coastal marine hazard mitigation.

In this chapter, we link issues concerning the key drivers of HABs with the various approaches for minimizing their negative impacts, emphasizing the use of numerical modeling techniques to bridge the gap between observations and predictive understanding. First, we review recent studies on the environmental pressures that promote HABs (Section 17.2); prominent strategies for preventing or controlling blooms (Section 17.3); and modeling methods, specifically addressing harmful algal species dynamics, and their use as a

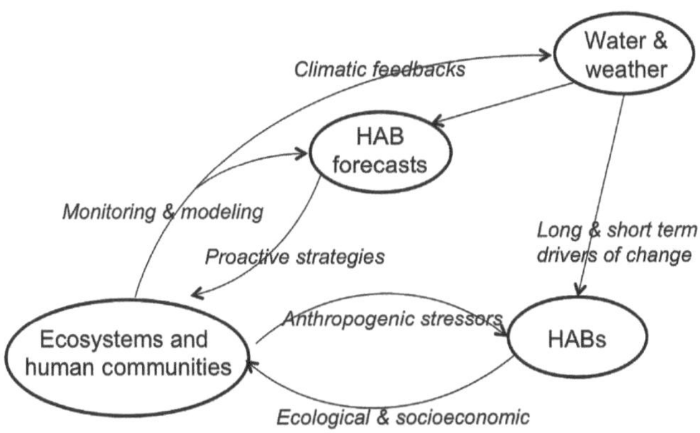

FIGURE 17.3 Schematic diagram illustrating the dynamic links that couple nature (e.g., water and weather conditions), HABs, and human communities. *Modified from Hoagland (2014).*

predictive tool to facilitate mitigation (Section 17.4). Next, several coastal regions are highlighted where the mitigation of HABs is generally approached from a regional Earth system and observation framework (Section 17.5). Such a framework ideally merges traditional monitoring methods, networked arrays, satellite observations, autonomous platforms, predictive models, and local to regional governance to mitigate impacts on human populations and ecosystems (Figure 17.3). In some instances, this approach may necessitate adaptive management for optimal resource use (Section 17.5.4). Lastly, we summarize future directions for "living with" HABs in an era of limited financial resources for ocean observing (Section 17.6).

17.2 ENVIRONMENTAL FORCING OF HABs

Research on the ecological processes that cause HABs and identification of the factors responsible for their worldwide increase has led to the development of predictive tools and mitigation strategies (GEOHAB, 2003, 2006). Highlights from recent studies are summarized in the following subsections to introduce the state of the science rather than duplicate the many exhaustive reviews (e.g., Heisler et al., 2008; Hallegraeff, 2010; Anderson et al., 2012b).

17.2.1 Eutrophication

The ecosystem response to eutrophication (i.e., biomass increases as a result of nutrient overenrichment) in coastal waters is complex and depends on the concentrations of macro- and micronutrients, the chemical form of those nutrients (organic vs inorganic), and the ratio of nutrient supply (Anderson et al., 2002, 2008; Heisler et al., 2008; Glibert and Burkholder, 2011; Kudela et al., 2010). These can all select for phytoplankton functional type (dinoflagellate, diatom, flagellate, cyanobacteria) as well as promote toxicity in toxigenic HAB species (Howard et al., 2007; Cochlan et al., 2008; Kudela et al., 2008). One compelling line of evidence from eutrophication studies is that land-based runoff and associated alteration of nutrient ratio supply (particularly Si:P and Si:N) away from the mean Redfield ratios selects for flagellates relative to diatoms (Smayda, 1997). This resource-mediated community composition shift is well-documented (reviewed in Anderson et al., 2002; Glibert and Burkholder, 2006) and now buttressed by increasing recognition that organic nutrients and reduced forms of nitrogen such as urea can modulate phytoplankton growth and toxicity (reviewed in Glibert et al., 2006; Kudela et al., 2010). This is important when we consider that industrial nitrogenous fertilizers are now predominantly composed of urea over nitrate (Glibert and Burkholder, 2006; Glibert et al., 2006). The role of groundwater in driving and regulating bloom development is also an important but understudied theme (Paerl, 1997). For example, Liefer et al. (2009, 2013) showed that dense blooms of toxigenic *Pseudo-nitzschia* species in the Northern Gulf of Mexico cluster near rivers known to transport high volumes of nitrate-rich discharge.

Davidson et al. (2012) challenged the rationale of some of the most canonical studies (e.g., "red tides" in Hong Kong; Hodgkiss and Ho, 1997) that link the process of nutrient enrichment with the effect of eutrophication and increasing HABs (Smayda, 2008). Although somewhat selective in its critique, the review provides a useful summary of the theoretical controls on nutrient uptake kinetics. It also reminds us of the caveats in applying nutrient limitation models to field scenarios where the role of organic nutrients (Howard et al., 2007), cell quotas/thresholds (Flynn, 2010), mixotrophy (Stoecker, 1998; Mitra and Flynn, 2010), "luxury" consumption of nutrients (Roelke et al., 1999), and interspecific competition for limiting resources are still poorly understood. Indeed, the interplay between cellular nutrient stoichiometry, exogenous nutrient pulses, and toxin production is nicely illustrated for *Alexandrium tamarense,* a paralytic shellfish poisoning (PSP)-causing organism that may have a high capacity for luxury phosphorous storage, thereby altering its response to ambient N:P ratios depending on its prior nutrient history (Van de Waal et al., 2013).

Despite this physiological complexity, nutrient loading from terrestrial environments into coastal and freshwater systems that are experiencing severe N and/or P limitation often appears directly related to the development of algal blooms (e.g., Glibert et al., 2001; Beman et al., 2005; Glibert, 2006; Paerl et al., 2011). The extent to which these blooms manifest as dense accumulations of biomass or as sources of harmful toxins depends on ecosystem responses and interactions. For instance, algal proliferation is heavily regulated by grazing pressure from zooplankton, with trophic cascades representing an often understudied component of bloom development and persistence (e.g., Gobler et al., 2002; Turner and Graneli, 2006; Smayda, 2008) relative to bottom-up effects or the pervasive influence of physical processes (Franks, 1992; Donaghay and Osborn, 1997; Ryan et al., 2008; Stumpf et al., 2008; Pitcher et al., 2010). Eutrophication may exert an indirect effect on zooplankton grazing efficiency such that at higher nutrient levels, grazing control of phytoplankton becomes saturated (Kemp et al., 2001). Mitra and Flynn (2006) further demonstrate that high nutrient conditions not only promote HAB species but also suppress grazing by enhancing the production of toxin grazing deterrents, a positive feedback that intensifies negative impacts of HABs (Sunda et al., 2006). Although we should be cautious about implicating the increase in HAB events specifically to eutrophication or to changes in nutrient ratios and specific nutrient compounds, it is clear that nutrient availability strongly modulates many aspects of HAB ecology. Ultimately, investigators will need to integrate nutrient dynamics at the land–sea interface, coastal and estuarine physics, and food web interactions to successfully model, predict, and forecast coastal HABs in a changing climate (Glibert et al., 2010).

17.2.2 Climate Change

The recently released Fifth Assessment Report by the Intergovernmental Panel on Climate Change (IPCC) verifies the role the ocean has played as a major heat sink, absorbing 90 percent of Earth's net energy increase over the past 40 years with an almost 4 °C increase in the upper 75 m of the water column (IPCC, 2013). Although internal variability remains a dominant governing force of regional climates, warming of the top 100 m of the ocean by as much as 2 °C is expected by the end of the twenty-first century (Stocker et al., 2013). Moore et al. (2008), Hallegraeff (2010), and Anderson et al. (2012b) examine the observed and expected consequences of warming sea surface temperatures, climate trends, and large-scale variability on phytoplankton. These consequences range from changing phenologies, "match–mismatch" in marine food webs, proliferation of HAB species into newly primed environments, potential adaptation to rapid adjustments in physicochemical conditions, and surprising range expansions. For the latter, debate still exists about whether observed expansions are driven by climate-mediated ocean circulation patterns or ship ballast water dispersal (Hallegraeff, 1993, 2010; Smayda, 2007). Warmer temperatures are projected to broaden the seasonal period over which phytoplankton can grow, i.e., phenology, thereby enhancing the risk of negative impacts and exposure to dangerous toxins (Moore et al., 2008, Figure 17.4). Natural decadal cycles of variability such as the El Niño Southern Oscillation (ENSO), North Atlantic Oscillation, Pacific Decadal Oscillation, North Pacific Gyre Oscillation, and the Madden–Julian Oscillation are also known regulators of phytoplankton primary production through their modulation of atmospheric patterns, water column mixing, stratification, circulation, and surface

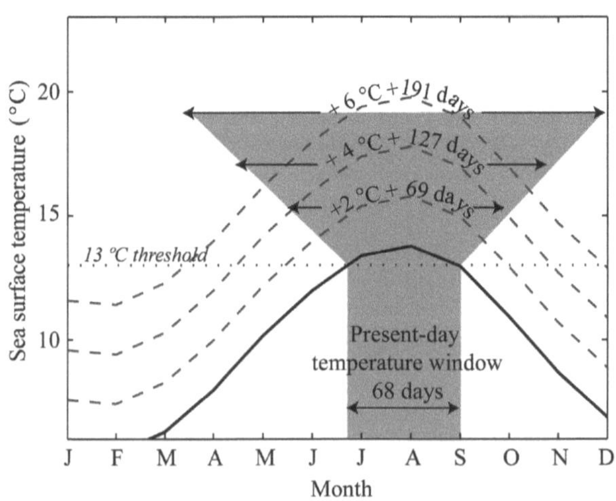

FIGURE 17.4 Puget Sound, Washington. The annual temperature window for accelerated growth of *Alexandrium catenella* for the present-day and in response to a 2, 4, and 6 °C increase in sea surface temperature. *Modified from Moore et al. (2008).*

nutrient delivery (Barton et al., 2003; Waliser et al., 2005; Di Lorenzo et al., 2008; Moore et al., 2008; Cloern et al., 2010). In the absence of long-term data, however, decadal and subdecadal oscillations in phytoplankton abundance and species composition (Jester et al., 2009) can camouflage secular trends.

17.2.3 Ocean Acidification

Anthropogenic CO_2 inputs to the atmosphere are overwhelming the buffering capacity of the ocean's carbonate system, leading to a corrosive environment for calcified organisms (e.g., Fabry et al., 2008; Feely et al., 2008). More counterintuitive is the effect that this change in aquatic pCO_2 will have on noncalcareous phytoplankton. Laboratory experiments demonstrate an increase in toxicity by the domoic acid (DA)-producing diatoms *Pseudo-nitzschia multiseries* and *Pseudo-nitzschia fraudulenta* and the saxitoxin-producing *Alexandrium catenella* after simulating projected pCO_2 levels in semicontinuous cultures (Sun et al., 2011; Tatters et al., 2012, 2013). This is caused by currently unexplained mechanisms tied to growth and toxin production. The effect will need to be verified and extended to other toxigenic HAB organisms, given the potentially complex, multifactorial response expected for natural ecosystems. As ocean acidification alters the saturation states of CO_2, HCO_3^-, and CO_3^{2-}, it will also interact with variability in temperature, salinity, and nutrient fields, leading to difficult-to-predict consequences for the phytoplankton (Moore et al., 2008), not to mention possible biophysical feedbacks that could amplify greenhouse gas emissions (Woods and Barkmann, 1993; Paerl et al., 2011). Cyanobacterial HABs that span a range of environments are expected to respond favorably to rising global temperatures, preferentially growing in warmer waters and outcompeting other phytoplankton for carbon because of their enhanced ability to acquire aqueous CO_2 over the more energetically expensive HCO_3^- and CO_3^{2-} (Paerl et al., 2011). While we are reminded that natural variations experienced by many coastal environments already expose phytoplankton to pH and pCO_2 concentrations well beyond long-term projections for the open ocean (Talmage and Gobler, 2009), pH levels in the Arctic, Southern Ocean, and coastal California are now on the verge of exceeding their "preindustrial variability envelopes" (Hauri et al., 2013). The synergistic effects of ocean acidification and eutrophication (Cai et al., 2011) on HABs (Figure 17.5) are severely stressing nearshore fin- and shell fisheries (Waldbusser et al., 2011).

17.3 BLOOM CONTROL AND PREVENTION

The desire to protect valuable fisheries and natural resources has motivated extensive research on methods for directly modifying blooms. Kim (2006) classifies these mitigation strategies for HABs into two categories,

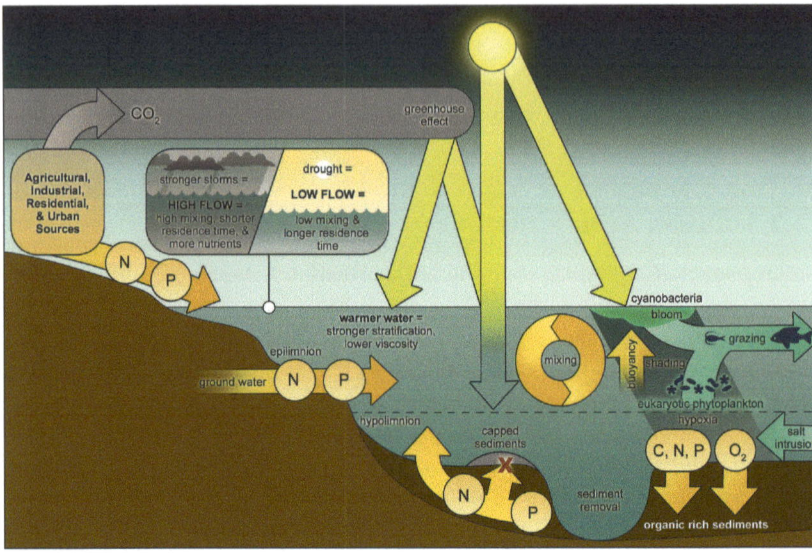

FIGURE 17.5 Conceptual diagram of cyanobacterial bloom development that can be generalized to a wide variety of algal blooms including HABs. The arrows indicate relationships between major biogeochemical processes found in both marine and freshwater environments; humans influence HAB development through modulation of nutrient sources at the land–sea interface and in the benthic zone where some mitigation strategies target the remineralization of limiting nutrients back into the water column. *Figure reproduced with permission from Paerl et al. (2011).*

precautionary impact preventions and *bloom controls*. Precautionary impact preventions refer to monitoring, predictive, and emergent actions. Bloom control involves both direct controls applied after an HAB has begun and indirect controls dealing with preventive strategies, including management of land-derived nutrient inputs. In this section, the distinction is made between (1) approaches to prevent and control a bloom and its impacts (Section 17.3.1) and (2) prediction, detection, and modeling capabilities (Section 17.3.2), which will form the backbone of future mitigation strategies within a regional Earth system framework (Section 17.3).

17.3.1 Biological and Chemical Control Methods

Biological and chemical controls refer to direct application or stimulation/suppression of factors that modify the biological (e.g., growth, grazing, mortality) or chemical (e.g., pH, inhibitors) composition or function of the ecosystem. These controls are often administered as emergency measures for suppressing blooms that threaten aquaculture facilities, or other spatially restricted regions, and their use can significantly accelerate the demise of a bloom or rid the water of toxins. These methods are most successful over small spatial scales within confined fish farms, reservoirs, desalination plants, or lakes and involve the manipulation of the environment and/or causative organism (Anderson et al., 2001; Kim, 2006). Biological agents such as grazers, parasites (Kim et al., 2008; Mazzillo et al., 2011), viruses (Nagasaki et al., 1999), and algicides (e.g., Jeong et al., 2003; Kim et al., 2009) are often host specific (Kodama et al., 2006) targeting a particular HAB species. Other moieties such as clays are used to promote flocculation and settling of algal particles to the sediment. Everything from microbial biosurfactants called sophorolipids (Sun et al., 2004; Lee et al., 2008) to algicidal bacteria (Imai et al., 1998; Doucette et al., 1999; Gumbo et al., 2008; Kang et al., 2008; Roth et al., 2008; Kim et al., 2009) and fungi (Jia et al., 2010) can be effective, at least in laboratory settings. The most extensively studied biocontrols target the PSP-producing *Alexandrium* spp. (Nakashima et al., 2006; Amaro et al., 2005; Bai et al., 2011; Su et al., 2007, 2011; Wang et al., 2010, 2012) or the fish-killing *Cocholidinium* spp. (Jeong et al., 2003; Kudela and Gobler, 2012), *Heterosigma akashiwo* (Nagasaki and Yamaguchi, 1997; Lovejoy et al., 1998; Imai et al., 1998; Jin and Dong, 2003; Kim, 2006), and *Chatonella* spp. (Imai et al., 2001). Zhou et al. (2008) achieved 80 percent inhibition of several species of *Alexandrium* in culture after applying garlic extract above 0.04 percent and attributed this effect to the active ingredient, diallyl trisulfide. This sort of "environment–friendly" approach to bloom control is appealing given the uncertainty and risk surrounding the use of toxic chemical agents that endanger a variety of aquatic flora and fauna. These compounds also minimize the issues associated with more environmentally damaging mitigation methods such as the use of copper sulfate ($CuSO_4$) on *K. brevis* blooms in the 1950s (Rounsefell and Evans, 1958 as cited in Kim, 2006). However, $CuSO_4$ and chlorination are still used routinely to rid drinking water reservoirs of nuisance algae and toxins (McKnight et al., 1983; Zamyadi et al., 2012).

Clay minerals such as kaolinite and loess compounds have been used effectively to control blooms in Asia, Europe, and the United States. Suspensions of the clay are sprayed onto the surface layer of a bloom (Figure 17.6), resulting in scavenging and flocculation of algal cells with over

FIGURE 17.6 Southern Sea of Korea. Clay dispersal used to mitigate blooms of *Cochlodinium polykrikoides*. *Photos by S. Moore.*

80 percent removal efficiency from surface waters in some cases (Sengco and Anderson, 2004). Phoslock® (lanthanum-modified bentonite) and chitosan have been applied to cyanobacterial blooms but prove too costly and impractical for routine management in the United States (Sellner et al., 2013), and in the case of Phoslock®, can lead to phosphorous limitation and increased ammonium regeneration (Sellner et al., 2013), further promoting cyanobacteria that respond to both P and N inputs (Paerl et al., 2011). Bloom removal is often successful only at very high clay and chitosan concentrations, with HAB species, pH (i.e., time of day), growth phase, and chitosan quality influencing results (Sellner et al., 2013). In lakes and ponds, barley straw and its extract can be cost-effective alternatives to controlling toxic cyanobacterial blooms and subsequent regrowth (Sellner et al., 2013; references in Brownlee et al., 2003), but it may have limited use in coastal marine environments where only a few dinoflagellate species appear susceptible (Terlizzi et al., 2002; Brownlee et al., 2003; Hagstrom et al., 2010). Peroxide additions are also effective (e.g., Matthijs et al., 2012) but limited due to cost and hazardous chemical permitting (particularly in the United States). In addition, little is known about the effects that this removal of toxic phytoplankton to the benthos has on the biota (Shumway et al., 2003) or on the potential for anoxic conditions in deeper waters (Imai et al., 2006). The list of physical disturbance methods now being tested is long, and most do not translate well to open coastal zones despite success in lakes and fjords; these include sediment capping (Pan et al., 2012), dredging (Lurling and Faassen, 2012), and even solar-powered circulation (Hudnell et al., 2010). A novel and potentially environmentally benign approach to control of blooms of cyst-forming HAB species (e.g., *Alexandrium*) in shallow, localized systems is being explored wherein manual mixing of bottom sediments can bury cysts uniformly throughout the disturbed layer, greatly reducing the number of cysts in the oxygenated surface layer, and thus the potential inoculum for future blooms (D. Anderson and D. Kulis, pers. comm.).

Viral and bacterial lysis appear to play a natural role in regulating phytoplankton communities and carbon flux (e.g., Fuhrman and Azam, 1980; Salomon and Imai, 2006). Capitalizing on this natural pathogenicity seems like a logical, cost-effective solution to HAB control. However, society has grown weary of runaway experiments with nature that introduce foreign, potentially invasive species or irreversibly alter natural assemblages in an ecosystem (Sanders et al., 2003; Secord, 2003). Given how poorly we understand phytoplankton community ecology, let alone viral and bacterial systematics and ecological interactions, Secord (2003) warns of the possibility for evolving host specificity in introduced viral and bacterial biocontrols that may not only prey switch but also could become less effective as their HAB hosts start to develop resistance. In a thought-provoking review on algicidal bacteria, Mayali and Azam (2004) considered the broader ecological context of microbial interactions in phytoplankton communities. Despite the many laboratory studies demonstrating the harmful predatory effects of heterotrophic bacteria on algal species, they argued that most field studies have failed to show conclusively the causal relationship between the decline of a bloom in natural ecosystems and the behavior of an introduced, algicidal bacterium. Moreover, translation from laboratory conditions to the field is inherently complex, given the flexibility of predator–prey dynamics mediated by the presence or absence of other algal species (Mayali and Azam, 2004) and the potential for toxicity effects due to HAB–microbe interactions (Moore et al., 2008).

17.3.2 Preventive Measures

The ultimate management strategy for preventing many algal blooms, particularly cyanobacterial blooms, is the reduction of nutrient inputs and the promotion of biodiversity. The rise of toxic *Nodularia spumigena* blooms in the Baltic Sea and their subsequent control after the Helsinki Convention in 1974 remains one of the strongest supporting narratives for curbing land-based nutrient pollutants (Elmgren and Larsson, 2001). The Baltic Sea is a complex network of contiguous basins bordering 12 nations. It has a long and ongoing history of hypoxia and fish kills associated with cyanobacteria blooms that are modulated by long-term climatic change and human land use practices (Zillen et al., 2008). Regions of both N and P limitation are separated in space and time with internal sources of phosphorous, an important regulator of offshore Baltic biogeochemistry (Vahtera et al., 2007). This not only necessitates but also complicates the dual reduction of N and P inputs to the system (Elmgren and Larsson, 2001; Conley et al., 2002; Vahtera et al., 2007). The largest source of nitrogen entering the Baltic is agriculture, but point-source discharge of sewage makes up a significant fraction. Sweden has adaptively managed sewage outflow by removing 80–90 percent of N and 95.5 percent of P to bring down overall phytoplankton biomass (Figure 17.7). They intermittently release more N into surrounding waters when there is a high risk of encouraging potentially toxic blooms of cyanobacteria species (called N_2 fixers or diazotrophs). Because diazotrophs can "fix" nitrate from elemental nitrogen in the atmosphere, they respond to low N:P ratios and thus will likely not bloom if additional N is supplied to the system (Elmgren and Larsson, 2001). This ecological strategy is stated in the joint initiatives management plan of the forward-thinking Helsinki Commission that advocates both N and P reductions and maintenance of biological diversity (HELCOM-BSAP, 2007) with the goal of returning the Baltic to a pristine state (Ronneby Declaration of 1990; Ehlers, 1994). The perennial problem noted by Elmgren and Larsson (2001) is the minimal involvement of local and regional stakeholders in decision making by most European Union (EU) directives and the lack of a clear end goal for determining restoration. One lesson learned is that dual reduction of N and P loads (Figure 17.7; see reviews by Conley et al., 2009; Paerl et al., 2011) as well as periodic control of N:P ratios appears appropriate for this region despite the theoretical limitations that

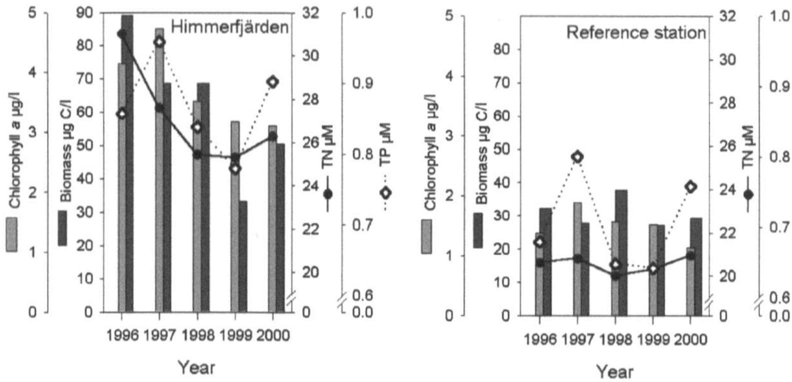

FIGURE 17.7 Baltic Sea. Reduction in the annual mean phytoplankton biomass in the upper mixed layer of Himmerfjärden, Sweden (left) after 1997 following N removal from the sewage treatment plant and subsequent declines in total nitrogen (TN) and dissolved inorganic nitrogen. As N:P ratios decreased, populations of N_2-fixing cyanobacteria rose in summer leading to an adaptive management strategy to control potentially toxic N_2 fixers. In contrast, phytoplankton biomass did not decrease at the open coastal reference station (right), nor did the annual mean TN and total phosphorus (TP). *Figure reproduced with permission from Elmgren and Larsson (2001).*

an exogenous nutrient ratio approach has been shown to impose on nutrient uptake by phytoplankton (Flynn, 2010).

Lake Erie, the shallowest, warmest, and most anthropogenically impacted of the Laurentian Great Lakes, poses another unique condition. Although it is a freshwater system, its large size and far-reaching impacts make it a good case study for marine HAB mitigation. In the mid-1960s to the 1970s, extensive cyanobacterial blooms, with associated hypoxic/anoxic conditions, were indicators of eutrophication within the shallow stratified portions of the western lake (Millie et al., 2009). Assemblages of other cyanobacteria species do occur, but the predominant bloom species in this region is *Microcystis aeruginosa,* a producer of the hepatotoxin microcystin. Phosphorus abatement strategies in the late 1970s successfully terminated blooms of cyanobacteria. However, following an invasion of foreign *Dreissena* mussels (zebra/quagga), cyanobacterial blooms began to reoccur in 1995 (Budd et al., 2001; Juhel et al., 2006). Zebra mussels were purportedly responsible for increased water clarity in the lake, but the consequence is that they selectively prey on eukaryotic phytoplankton, leaving cyanobacteria to thrive. Like clockwork, summer–fall blooms of *M. aeruginosa* have plagued the western basin on an annual basis ever since (Brittain et al., 2000; Vanderploeg et al., 2001), significantly impacting Ohio's beaches and water suppliers, with occasional effects in Michigan and Canada. In 2013, Carroll Township's water treatment facility in Ohio detected microcystin at concentrations more than threefold higher than the World Health Organization threshold of 1.0 part per billion in finished drinking water, forcing a shutdown of the municipal water supply (Henry, 2013). The chronic effects of human exposure to microcystin are poorly documented, and acute exposure is routinely implicated in deaths of domestic dogs and livestock shortly after exposure (Backer et al., 2013). As a result of exposure through recreational contact with water, contact dermatitis, nausea, and respiratory irritation (through inhalation of contaminated lake water) have been reported (Backer and McGillicuddy, 2006). The watershed surrounding the western basin is primarily represented by agricultural areas and drains into western Lake Erie by the Maumee River. The effluent of the Maumee River has elevated nutrients (particularly phosphorus), further exacerbating the cyanobacterial blooms (Stumpf et al., 2012). Unfortunately, as a consequence of climate change and resulting increases in water temperature, it is anticipated that toxic cyanobacterial events will increase in magnitude and frequency (Paerl and Huisman, 2008). Efforts to launch an operational forecasting system for cyanoHABs in Lake Erie are currently underway at the US National Oceanic and Atmospheric Administration (NOAA) as part of the Harmful Algal Bloom Operational Forecasting System (NOAA HAB-OFS; Wynne et al., 2013).

Biological diversity, although more difficult to assess, is an important determinant of water quality. Cardinale (2011) demonstrated that enhanced niche partitioning by benthic diatoms increased nitrogen uptake, providing a natural "buffer" against nutrient enrichment relative to less diverse communities associated with spatially homogenous environments. Promoting algal biodiversity and habitat preservation may then facilitate greater nutrient uptake capacity, particularly in protected environments where physical advection processes do not dominate phytoplankton turnover rates. Allelopathic interactions introduced when algae exude dissolved phycotoxins into the environment are an indicator of interspecific competition for limiting resources (Graneli and Hansen, 2006). It may also be that as species diversity increases, the ability of a given toxic species to dominate its competitors is suppressed by the wider array of competitive strategies present in the community. As marine ecosystem models become more sophisticated and include realistic phytoplankton biodiversity (Follows et al., 2007; Goebel et al., 2010), varying management strategies can be assessed in relation to community composition, competitive interactions, and nutrient dynamics.

17.4 MONITORING AND MODELING HABS

17.4.1 Ocean Observing

Once the far-reaching pressures of global climate change are superimposed on human impacts at regional scales (Figure 17.5), the projected response by phytoplankton communities becomes a seemingly intractable problem that can only be tackled through vigilant observation. This need for constant monitoring is a recurring mantra in the scientific and resource management communities. Baseline patterns cannot be distinguished from secular or decadal trends without consistent time series (McQuatters-Gollop et al., 2011). In particular, observations of species composition, phycotoxin loads throughout the food web, and ancillary measures of physical and chemical constituents are needed. These measurements are broadly defined into point, transect, and synoptic categories, all of which are necessary and require thoughtful integration for adequate HAB tracking and prediction (Stumpf et al., 2010; Jochens et al., 2010). It has been argued that at least 30 years of consistent monitoring data of HABs are required to discern climate-scale effects (Dale et al., 2006). One such record is provided by the 75-year time series of phytoplankton captured by the Continuous Plankton Recorder (CPR) in the North Sea. Using CPR data, a large-scale regime shift in open ocean phytoplankton was identified in the mid-1980s (McQuatters-Gollop et al., 2007). This "alternate resilient state" typified by anomalously high chlorophyll concentrations was found to be closely tied to climatic variability in the North Atlantic and decoupled from the significant reductions in nutrient loading implemented by the EU in the 1980s and the 1990s. Trophic cascades initiated by overfishing may also contribute to this rise in biomass (McQuatters-Gollop et al., 2007). Only via a fully integrated assessment of these pressures (using a combination of models and time series analysis) can the various factors be teased apart (Stumpf et al., 2010; Tett et al., 2013).

Efforts to codify public policy on reducing the impacts of HABs on human populations, wildlife, fisheries, aquatic ecosystems, aquaculture facilities, and drinking water supply (Bauer, 2006, Jewett et al., 2008) have been part of a growing movement by scientists and managers in the United States to "harness" monitoring and prediction capabilities through targeted research priorities aimed at holistic mitigation (HARRNESS, 2005). Federal investment in short- and long-term studies was mandated by the Harmful Algal Bloom and Hypoxia Research and Control Act (HABHRCA) of 1998, followed by the Harmful Algal Bloom and Hypoxia Amendments Act of 2004. While severe reduction in funding these programs has disrupted regional HAB monitoring in the United States, HABHRCA was recently reauthorized through 2018 indicating renewed interest in supporting HAB research. Many of the current and future efforts to apply technological advancements and Earth system frameworks in ocean observing to HAB ecology leverage the US Integrated Ocean Observing System (US IOOS) to bridge regional monitoring networks and sensor arrays with biological measurements (Green et al., 2009; Jochens et al., 2010; IOOS, 2013; Kudela et al., 2013).

Integrated observing systems to address HABs have been developed in several countries (See Section 17.5.3; Stumpf et al., 2010; Bernard et al., 2014). At the international level, the Global Ocean Observing System (GOOS) sponsored by the International Oceanographic Commission (United Nations Educational, Scientific and Cultural Organization) offers near real-time measurements of the state of the ocean (e.g., the successful Argo float program). It is part of a "permanent global collaborative system" with regional alliances comprising government and nongovernmental entities (GOOS, 2013). Fundamental gaps exist with respect to HABs in the initial design of most observing systems since there is greater emphasis on physics and meteorology than on biology (Frolov et al., 2012; Kudela et al., 2013; see also Section 17.6.1). The focus will be on leveraging those existing assets to create an end-to-end predictive system for HABs and other coastal hazards (Kudela et al., 2013), since the regional ocean observing networks now represent the best option for sustained HAB monitoring and forecasting in coastal waters.

17.4.2 Numerical Approaches to HAB Prediction

The number of approaches for monitoring, detecting, predicting, and forecasting the onset, fate, and demise of algal blooms is arguably comparable to the diversity of species being studied. Over the past two decades, there has been an increasing desire to apply our heuristic understanding of bloom ecology toward practical, numerical methods that will alert managers and communities of impending dangers (see McGillicuddy, 2010). An ideal alert system provides quantitative predictions of HAB likelihood, intensity, and movement or potential landfall along coastal margins. These approaches rely on a range of platforms from space-based, airborne, and in-water optical sensors, to traditional environmental sampling, to purely computational methods. Here, we focus our summary on the prediction of HABs using models or creative combinations of models, satellite observations, and in situ sampling (Table 17.2). We do not address the large body of work that directly associates aquatic optical properties with algal constituents nor the development of remote sensing indices for HAB detection (see recent review chapters in Pettersson and Pozdnyakov, 2013; the "HABWatch" volume, Babin et al., 2008; Stumpf and Tomlinson, 2005). Several regions are examined in detail in Section 17.5 to provide examples of how geographical variation in HAB species, monitoring programs, available satellite and modeling products, and resource management issues dictate the most effective mitigation strategy.

17.4.2.1 Empirical Models

Empirical or statistical methods range from fairly simple, steady-state regression techniques to more deterministic numerical solutions that draw from machine learning, such as artificial neural networks (ANN) and genetic programming (GP), or logic and rule-based reasoning, such as fuzzy logic. Some successful applications of linear regression to the prediction of HABs are found for toxigenic *Pseudo-nitzschia* populations (amnesic shellfish poisoning organism), beginning with a study that built several models of cellular DA concentration from cultures (with some field data) of *Pseudo-nitzschia pungens* using stepwise multiple regression (Blum et al., 2006). Anderson et al. (2009, 2010) and Lane et al. (2009) achieved similar success (~75 percent accuracy)

TABLE 17.2 Summary of Numerical Models Used to Predict Target HAB Species; in Some Cases, These are Forced with Output from (or Coupled to) 3D Circulation Models, and a Few are Involved in (or Moving Toward) Operational Use

Type of Model	Target Species	Region	Specific Approach	Forced with Other Regional Models, Observations	Source
Empirical/ statistical	Cyanobacteria	Japan, Finland, Australia	ANNs		Recknagel et al. (1997)
	Skeletonema spp.	Hong Kong	ANNs		Lee et al. (2003)
	Pseudo-nitzschia	Cardigan Bay, Canada	Multiple linear regression to predict toxins (DA)		Blum et al. (2006)
	Nodularia spumigena, Alexandrium minutum, Dinophysis spp., Karenia mikimotoi	Baltic sea, Gulf of Finland, Sweden, Ireland, United Kingdom, Netherlands	Fuzzy logic	HABES project	Laanemets et al. (2006), Blauw et al. (2006)
	Phaeocystis globosa	Dutch coast, Netherlands; United Kingdom	Decision tree; nonlinear regression; fuzzy cellular automata; fuzzy logic	Delft3D-WAQ (HABES project)	Chen and Mynett (2004), Chen and Mynett (2006), Blauw et al. (2006), Blauw et al. (2010)
	Dinophysis acuminata	Western Andalucia, Spain	ANNs		Velo-Suarez and Gutierrez-Estrada, (2007)

	Lyngbya majuscula	Deception Bay, Queensland, Australia	Bayesian model averaging		Hamilton et al. (2009)
	Pseudo-nitzschia spp.	Santa Barbara Channel, CA, USA; Monterey Bay	GLM (logistic), multiple linear regression to predict abundance and toxin concentration (DA)	ROMS-CoSiNE (CCS), HYCOM-CoSiNE (CCS), MODIS ocean color, HFR	Anderson et al. (2009, 2011), Lane et al. (2009), Anderson et al. (in review)
	Pseudo-nitzschia spp.	Chesapeake Bay	GLM (logistic)	ChesROMS-Fennel ecosystem model	Anderson et al. (2010)
	Karenia brevis	Gulf of Mexico	Supported vector machine learning		Gokaraju et al. (2011)
	Karlodinium veneficum, Microcystis aeurginosa, Prorocentrum minimum	Chesapeake Bay	ANNs, GP, GLM (logistic)	ChesROMS-Fennel ecosystem model	Brown et al. (2013)
Mechanistic	*Alexandrium fundyense*	Gulf of Maine	Deterministic cyst germination and growth model	HYCOM-ROMS	Stock et al. (2005), He et al. (2008), Mcgillicuddy et al. (2005, 2011)
	Karenia brevis	West Florida Shelf	Deterministic nutrient-limited growth model	HYCOM with MODIS FLH and LCS method	Olascoaga et al. (2008)

(Continued)

TABLE 17.2 Summary of Numerical Models Used to Predict Target HAB Species; in Some Cases, These are Forced with Output from (or Coupled to) 3D Circulation Models, and a Few are Involved in (or Moving Toward) Operational Use—cont'd

Type of Model	Target Species	Region	Specific Approach	Forced with Other Regional Models, Observations	Source
	Gambierdiscus spp.	Hawaii, Big Island	Deterministic nutrient-limited growth and export model		Parsons et al. (2010)
	Pseudo-nitzschia seriata; Psuedo-nitzschia spp.	Cultured from Scottish waters; Monterey Bay, CA	Deterministic nutrient-limited growth-mortality-toxin production model		Terseleer et al. (2013), Anderson et al. (2013)
Physical indices and Lagrangian particle tracking	*Psuedo-nitzschia spp.*	Galician Coast, Spain; Lisbon Bay, Portugal	Upwelling index; SST and UI; wind current patterns	AVHRR SST and SeaWiFS chlorophyll	Sacau-Cuadrado et al. (2003), Palma et al. (2010)
	Karenia brevis	Texas Shelf; Tampa Bay, FL, USA (Gulf of Mexico)	Passive tracer advection diffusion; trajectory/transport modeling from physics; LPT applied ex post facto to an identified *K. brevis* event	ROMS; HYCOM-ROMS-FVCOM with HFR; POM	Hetland and Campbell (2007), Weisberg et al. (2009), Havens et al. (2010)
	Dinophysis acuminate	Bantry Bay, Ireland; Bay of Biscay, Spain	Wind index; LPT (Ichthyop) to simulate *D. acuminate* bloom	MARS3D-Ichthyop	Raine et al. (2010), Velo-suarez et al. (2010)

	Pyrodinium bahamense (Phaeocystis globosa, Gymnodinium mikimotoi, Prorocentrum minimum)	South China Sea–Vietnam; Manila bay	Rudimentary growth model & passive tracer advection diffusion; LPT applied ex post facto to an identified HAB event	POM-SWAN; POM	Villanoy et al. (2006), Dippner et al. (2011)
Ecosystem/ biogeochemical	*Karenia brevis*	West Florida Shelf, Gulf of Mexico	Complex N-P-Z-D model with explicit *K. brevis* box and aerosolized brevetoxins	POM; FVCOMS, ROMS, HYCOM	Walsh et al., (2001, 2002)
	Potentially toxic cyanobacteria	Baltic Sea	Ensemble forecasting of C:Chl for cyanobacteria	Finnish Meteorological-Institute-coupled physical–biological model	Roiha et al. (2010)
	Pseudo-nitzschia spp.	Pacific Northwest	Particle tracking and wind index from a fully validated ecosystem model	ROMS (Eastern Pacific)	Giddings et al. (2013)

FLH, fluorescence line height; FVCOM, finite volume community ocean model; LCS, lagrangian coherent structures; MODIS, moderate-resolution imaging spectroradiometer; AVHRR, advanced very high resolution radiometer; C:Chl, carbon to chlorophyll ratio; HFR, high-frequency radar, SWAN, simulating waves nearshore; SST, sea surface temperature; UI, upwelling index; SeaWiFS, sea-viewing wide field-of-view sensor; WAQ, water quality. Empirical models relate the species distribution and abundance patterns of a particular algal taxonomic group to combinations of physical, chemical, biological, and optical environmental indices using varying levels of statistical complexity. Mechanistic models strive to numerically parameterize fundamental physiological and life history traits of the target organism to predict its abundance and/or toxicity. Physical methods range from statistical relationships between HAB species and physical indices to LPT methods that rely on sophisticated numerical solutions of the physical circulation to predict particle trajectories. LPT is a general method that is widely applied in ecological forecasting with some HAB examples cited here. The broad field of ecosystem or biogeochemical modeling has not historically focused on HAB prediction, but there are now several examples of direct incorporation of HAB species into model design or model analysis. For a comprehensive discussion of 3D physical–biological models applied to both HAB and non-HAB algal groups, see Petersson and Pozdynkaov (2013).

when applying a range of stepwise linear and logistic regression (as generalized linear models, GLMs) to time series of in situ physicochemical parameters to predict both DA and *Pseudo-nitzschia* blooms in the Chesapeake Bay (Anderson et al., 2010) and coastal California (Lane et al., 2009; Anderson et al., 2009; more in Section 17.6). An advantage of these simple models is their reproducibility and retuning by other investigators as data sets lengthen as well as the easily interpreted, ecological relationships between variables.

Somewhat more obscure are the numerical approaches that use artificial neural networks to model biological phenomena. ANN mimic complicated nonlinear neuronal connections, and thus are expected to capture the chaotic component of ecological patterns by deterministically modeling the inherent nonlinearity of the system. Time series data are generally divided into "learning" and validation sets for training the ANN to recognize patterns that connect the response and predictor variables, an approach also used for support vector machine learning techniques (Gokaraju et al., 2011; Ribeiro and Torgo, 2008). An early application of ANN was conducted by Recknagel et al. (1997) to predict algal blooms in four lake systems. Velo-Suarez and Gutierrez-Estrada (2007) were very successful ($r^2 = 94-96$ percent) in predicting *Dinophysis acuminata* blooms (diarrhetic shellfish poisoning (DSP) organism) over short timescales in Spanish coastal waters using ANN. Muttil and Lee (2005) applied GP evolutionary algorithms to chlorophyll data sets from Tolo Harbor, Hong Kong, a site with a long history of HAB events (Hodgkiss and Ho, 1997), and achieved good correspondence between observed and predicted chlorophyll (86 percent). Bayesian model averaging and similar techniques are becoming more popular in ecological studies due to their ability to stringently quantify uncertainty over all possible model forms and parameter estimates, as described by Hamilton et al. (2009) for *Lyngbya majuscula* blooms (now *Moorea producens)* in Australia. Fuzzy logic approaches include the HABs Expert System (HABES, http://habes.hrwallingford.co.uk) sponsored by the EU Fifth Framework Program. HABES predicted (using "Ecofuzz"; an open source model) a suite of HAB species at seven EU coastal sites (Blauw et al., 2006). The program illustrates the many regional considerations necessary when attempting to encompass regional diversity of HAB issues including *N. spumigena* in the Gulf of Finland (Laanemets et al., 2006) and nuisance blooms of *Phaeocystis globosa* along the Dutch coast (Blauw et al., 2010; Chen and Mynett, 2004, 2006).

17.4.2.2 Physical Models and Particle Tracking

A number of investigators have examined bloom formation and duration with hydrodynamic circulation models to constrain the physical processes controlling bloom dynamics. The numerically least intensive approaches use physical indices or relationships to predict conditions likely to promote HABs such as upwelling (Palma et al., 2010; Sacau-Cuadrado et al., 2003) or favorable winds (Raine et al., 2010). Using this empirical approach and recognizing that DSP-causing *Dinophysis* blooms (Table 17.1) on the southwestern Ireland coast occur during summer when offshore water is advected into the highly stratified nearshore, Raine et al. (2010) developed a model based on the wind index as a proxy for wind-driven exchange of water and HAB probability onto the shelf. This simple but elegant model has proven helpful for understanding the dynamics of DSP intoxications that have greatly impacted the shellfish in Bantry Bay.

Once a bloom has been positively identified through environmental sampling, satellite algorithms, or models, its trajectory can be mapped using particle transport (Lagrangian particle tracking (LPT)) coupled to either a two-dimensional or three-dimensional (3D) circulation model. Widely used in oil spill tracking and studies of fish larval transport, LPT is seeing growing popularity for HAB risk management. Because many blooms originate offshore and are advected into the nearshore environment via physical processes like mesoscale eddies, LPT can be a powerful tool for estimating the timing and spatial impact of landfall. Wynne et al. (2011) evaluated LPT applied to satellite data for cyanobacterial blooms in Lake Erie and confirmed that the model improved the accuracy of forecasted bloom locations. Another study tracked passive particle transport of a *K. brevis* bloom in Tampa Bay with LPT coupled to the Princeton Ocean Model (POM) to identify zones most likely to be affected, but was unable to adequately validate predictions with monitoring data (Havens et al., 2010; more on *K. brevis* particle tracking in Section 17.5). Velo-Suarez et al. (2010) determined the physical processes responsible for the demise of a *D. acuminata* bloom in the Bay of Biscay using an LPT model ("Ichthyop") coupled to a downscaled regional ocean model (MARS3D, Model for Application at Regional Scale), illustrating the importance of retention–dispersion patterns driven by the physics of the bay. Summer southwest monsoon patterns were shown to drive transport of HABs into sensitive aquaculture and coral reef zones along the Vietnamese coast of the South China Sea with a Lagrangian model coupled to the Hamburg Shelf Ocean Model (Dippner et al., 2011). Also focusing on the SW monsoon season, Villanoy et al. (2006) incorporated the physical–biological interaction into their LPT-POM coupled model by including a rudimentary individual-based growth model (IBM) for *Pyrodinium* cyst resuspension and transport in Manila Bay, similar to the treatment by McGillicuddy et al. (2003) to determine the offshore initiation of *Alexandrium fundyense* blooms from dormant cysts in the Gulf of Maine (both are PSP organisms). An advantage to IBMs is the ability to include diel vertical migration, a fundamental nutrient-acquisition strategy in dinoflagellates (Kamykowski et al., 1999; Peacock and Kudela, 2014) that may greatly affect passive tracer behavior if correctly applied to LPT models (Henrichs et al., 2013).

17.4.2.3 Coupled Physical–Biological Models

In the 17 years since Franks (1997) showcased the potential utility of coupled physical–biological models to HAB ecology, the fields of ecosystem modeling and data assimilation have advanced significantly. At the same time, a growing recognition has occurred that satellite observations for real-time HAB prediction are limited due to the poor temporal and spatial resolution of ocean color imagery; large uncertainty in chlorophyll-*a* (chl-a) estimates for coastal, optically complex waters; and the lack of taxonomic specificity that can be extracted from current sensors (e.g., Allen et al., 2008; Stumpf et al., 2009). Petersson and Pozdynkaov (2013) reviewed the current state of satellite methods and coupled physical-ecosystem models available for use in HAB studies. Many of these models predict bulk chlorophyll biomass rather than species-specific biomass pools (e.g., Allen et al., 2008; ERSEM model, European Regional Seas Ecosystem Model), highlighting the limitation that few models explicitly simulate HAB species dynamics. An additional take-home message is the tight symbiosis between observations and models, echoed by both Franks (1997) and Weisberg et al. (2009), who noted this joint utility in the design of ocean observing systems and the fine-tuning of predictive models. Despite model advances, some of the major hurdles outlined by Franks (1997) remain: (1) assimilation of biological and chemical observations to improve ecosystem model performance (a crucial role for satellite observations, see Gregg et al., 2009); and (2) uncertainty in initial conditions and multispecies interactions. Moreover, limitations still exist for most HAB species in understanding the mechanisms responsible for bloom initiation, termination, and toxicity—the factors most useful to managers. These limitations will persist so long as the large-scale observing systems continue to focus on variables with limited applicability to understanding species-specific dynamics (see Section 17.4.1).

The descriptive term often used for a wide variety of basic to complex deterministic formulations that examine the dynamical interaction of these *biogeochemical* compartments in zero-dimensional to 3D settings is nitrogen-phytoplankton-zooplankton-detritus (N-P-Z-D) model. Walsh et al. (2001) subdivided the phytoplankton state variable in an N-P-Z-D model into six functional/taxonomic groups, including an explicit HAB "box" for *Gymnodinium breve,* now classified as *K. brevis.* They then predicted (hindcasted) transport/landfall for the well-documented 1979 event on the West Florida Shelf by combining this one-dimensional model with a POM circulation model and light-mediated vertical migration behavior and proposed that the bloom was regulated by organic nutrients (Walsh et al., 2002). Olascoaga et al. (2008) applied a new technique, also for the West Florida Shelf, that isolates distinct regions in the flow termed Lagrangian coherent structures to back-calculate the origin of a satellite and field-detected *K. brevis* bloom. This latter approach is only possible where a bloom can be preverified and is in contrast to the forecasting efforts of Walsh et al. (2001) who also noted the "real-world" limitations of their complex model and a reliance on in situ biooptical sensors for model evaluation. Giddings et al. (2013) recently improved predictive skill of toxic DA events in the Pacific Northwest of the United States using a fully validated ecosystem model coupled to a Regional

Ocean Modeling System (ROMS) by tracking particle advection of simulated *Pseudo-nitzschia* particles (represented by the "large phytoplankton" size class in their model). They filtered out false positive values by sequentially applying a wind index and the presence of appropriate size classes of cells to classify favorable periods for onshore HAB transport.

17.4.2.4 Mechanistic HAB Models and Blended Dynamical Approaches

Mechanistic models that simulate HAB population dynamics or toxin production (as observed for a particular species or genus) are still rare (Parsons et al., 2010; Stock et al., 2005; Anderson et al., 2013; Terseleer et al., 2013), but they are arguably a key component to HAB forecasting from ecosystem models where, if not too complex, they can function as a tracer or a fully integrated state variable. In the Gulf of Maine, *A. fundyense* population dynamics is parameterized for cyst germination, growth, and mortality (McGillicuddy et al., 2005; Stock et al., 2005) and treated as a tracer within a regionally downscaled ROMS model (He et al., 2008) to make real-time, weekly *A. fundyense* forecasts during the bloom season as well as seasonal ensemble forecasts (McGillicuddy et al., 2011). Based on years of resting cyst abundance data for the region, models initiated with the previous year's cyst bed data were sufficient for estimating a climatological bloom horizon for the following year to alert stakeholders and resource managers of potential PSP outbreaks (Figure 17.8; Anderson et al., 2005; Li et al., 2009). McGillicuddy et al. (2011) candidly described the failure of this relationship to manifest a correct seasonal prediction for the western Gulf of Maine in 2010 after historically high cyst abundance in 2009 indicated otherwise (Figure 17.8). Nonlinearities in the dynamic system may be to blame, and the scenario is likened to that of the 1990s when poor ENSO model performance arose from shifts in the underlying ocean state far outside those used for model construction (McGillicuddy et al., 2011). This case study also demonstrates that ensemble forecasts from varying boundary conditions, while a potentially powerful management tool for creating model uncertainty envelopes and conducting sensitivity analyses (Roiha et al., 2010), are not immune to these nonlinearities. This emphasizes the need for advanced data assimilation techniques if ecosystem models are to be used operationally (McGillicuddy et al., 2011). Statistical models would likely also fail when the underlying ocean state shifts outside that used for model construction, highlighting the sensitivity of both these "simple" and more complex modeling approaches. Clearly, a need exists for close coordination between observation and modeling efforts.

Approaches that blend empirical and dynamic methods leverage the practicality of statistical HAB models with the sophistication of coupled hydrodynamic-ecosystem models (e.g., Anderson et al., 2010, 2011). The Chesapeake Bay Ecological Prediction System (CBEPS) is one such project, currently generating nonoperational nowcasts and 3-day forecasts for several HAB species (unpublished ANN and GLM empirical models for *Karlodinium*

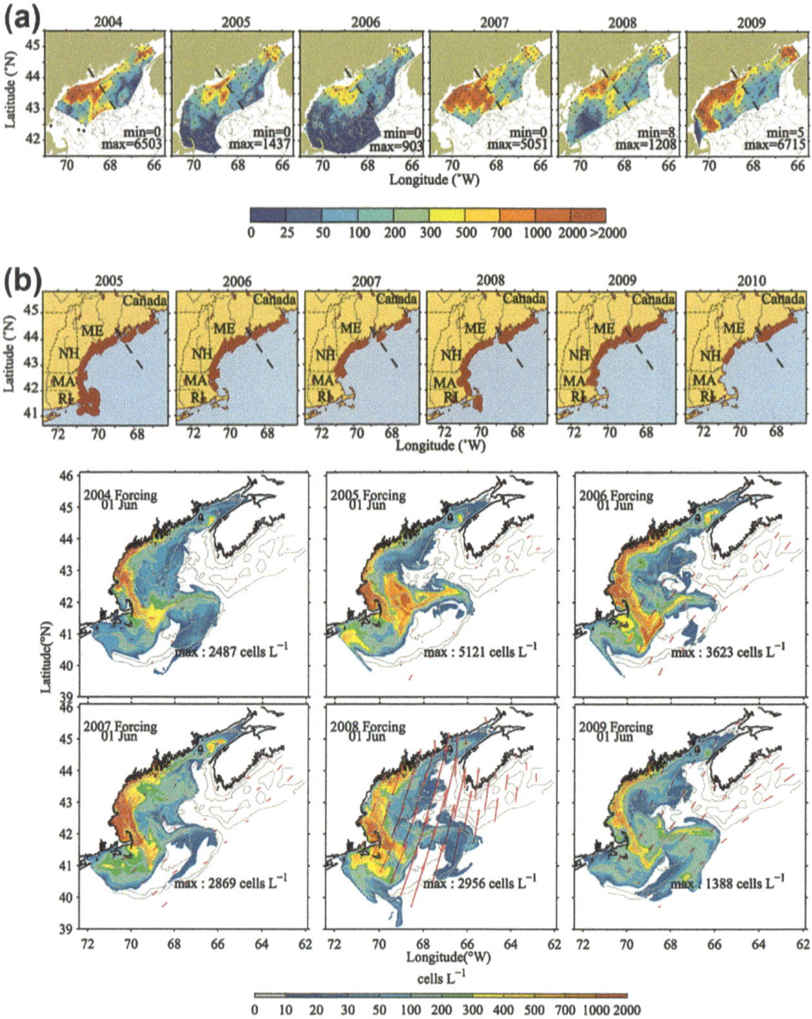

FIGURE 17.8 Gulf of Maine. Top panel: (a) Contour maps created from sediment samples of *Alexandrium fundyense* cysts collected from 2004 to 2009 (open circles). (b) Cyst abundance is paired with corresponding maps of paralytic shellfish poisoning (PSP) closures for the following year, i.e., 2005–2010. Bottom panel: Ensemble forecast of *A. fundyense* cell abundance generated from 2009 cyst data and a hydrodynamic (ROMS) model for 2004–2009 to constrain the variability in physical forcing each year while holding the biology (i.e., cyst distribution) constant. Predictive skill broke down in 2010 when water mass anomalies fell outside the "envelope of variability" used to train the model. *Figures reproduced with permission from McGillicuddy et al. (2011).*

veneficum, M. aeruginosa, and *Prorocentrum minimum),* stinging jellyfish called seanettles (Decker et al., 2007), pathogens such as *Vibrio cholerae* (de Magny et al., 2009), and dissolved oxygen content for the largest estuarine system in the United States (Brown et al., 2013). The CBEPS uses a downscaled ROMS

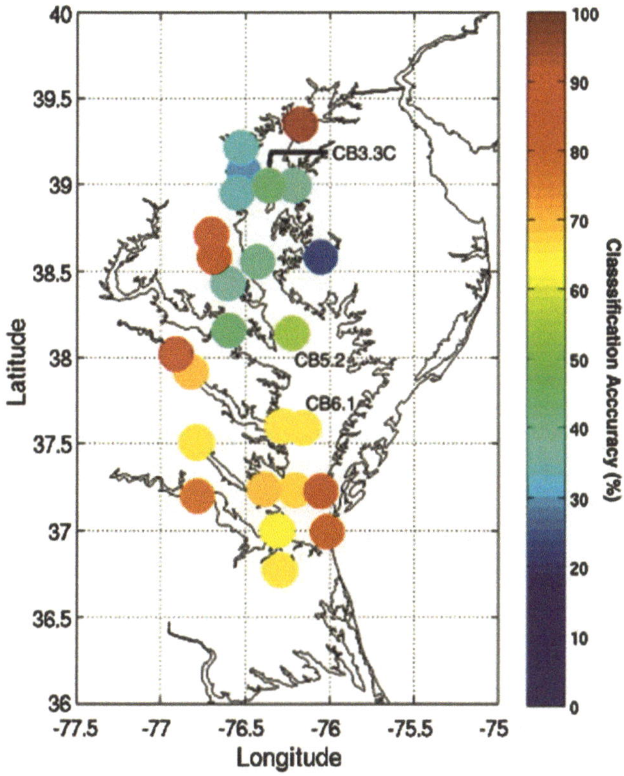

FIGURE 17.9 Chesapeake Bay. Skill assessment of *Karlodinium veneficum* simulations created from empirical HAB models and forced with a coupled ROMS and biogeochemical model for 2007–2009. Accuracy measures are based on comparisons with in situ data from 24 stations monitored by the Chesapeake Bay Program. *Figure reproduced with permission from Brown et al. (2013).*

("ChesROMS") coupled to an N-P-Z-D model (Fennel et al., 2006; Xu and Hood, 2006) that includes inorganic P and N, organic N, and dissolved oxygen. The ChesROMS configuration considers United States Geological Survey (USGS) river discharge and atmospheric deposition of nutrients, but does not currently run the real-time data assimilation routines developed for the region (Hoffman et al., 2012; Zhang et al., 2010). CBEPS is moving toward rigorous evaluation of model skill (Brown et al., 2013), a fundamental goal for all applied modeling systems (see Stow et al., 2009). Brown et al. (2013) report a mean accuracy of 59 percent for the *K. veneficum* nowcasts (Figure 17.9). Given that the only significant predictor variables are month, salinity, and temperature, all of which are well-validated for ChesROMS (Warner et al., 2005), this implies that the model may be too simple to sufficiently capture *K. veneficum* variability. Forthcoming endeavors to evaluate HAB models and assimilate biological data for this project and others, although extremely difficult, are imperative for strengthening the role of HAB forecasting in potential mitigation.

17.4.2.5 Valuation of Models for HAB Mitigation

The societal goal for all HAB modeling efforts should be the mitigation of negative impacts. The costs for developing an operational forecast system are ideally balanced by the socioeconomic gains and protection of living marine resources, or they should at least provide significant added value. These operational forecast systems also add value to the often significant investment in underlying observational infrastructure, costs incurred whether or not the data are used for HAB applications. One of the few (or perhaps only) such cost–benefit analyses that evaluates the relative investment of HAB model prediction looks at commercial finfish and shellfisheries in New England with respect to the *A. fundyense* forecast and tracking system discussed above for the Gulf of Maine (Jin and Hoagland, 2008). A unique advantage of a forecasting system for fishermen and shellfish growers is the fine-tuned spatial and temporal prediction of bloom or toxin presence and movement, which would enable targeted, proactive harvests and even geographic shifts in fishing effort (e.g., offshore to Georges Bank). In their study, Jin and Hoagland (2008) modeled the economic impacts of predictions in terms of (1) harvest loss if no prediction is made, (2) the value of HAB prediction over a range of possible skill levels, (3) the annual economic value to a public or private decision maker if action is or is not taken given a particular HAB prediction; and (4) the total value of a prediction. By examining a range of HAB frequency (from 2- to 30-year events) and model accuracy, the study elegantly estimated the variation in monetary value for responses to a given scenario. Not surprisingly, model accuracy is a leading factor driving prediction value, but so is the frequency of HABs, i.e., the value of a prediction increases when blooms are more common. For example, the model yields a 30-year maximum net value of $51.3 million when forecasts are completely accurate and PSP events occur every 2 years (Jin and Hoagland, 2008). Of course, one crucial aspect not captured in this study is the ecosystem service value of a functioning ecosystem and healthy wildlife populations, and as the authors note, spillover effects to other industries such as tourism or nontargeted fisheries.

17.5 REGIONAL EARTH SYSTEM FRAMEWORK

Whether predicting when a potentially dangerous bloom will strike or tracking its path, models should always be anchored to the regional chemistry, physics, and biology. Alert systems and mitigation strategies will be dictated by the history of human resource use in the region and will hinge on local to federal government mandates for protecting those resources. For these reasons, a "one-size-fits-all" approach for modeling HABs is not practical. This section highlights several regional narratives in the United States and EU that integrate observation networks and predictive models in idiosyncratic ways to meet society's challenge of mitigating the negative consequences of HABs.

17.5.1 Pacific Northwest of the United States

The Earth system framework for detecting and forecasting HABs in the PNW is composed of a diverse range of elements with varying degrees of complexity to overcome the challenges that are associated with the environment. For example, cloud and fog prevent bloom detection using satellites over much of the year. Furthermore, the toxic HAB species "bloom" in the PNW makes up only a small percentage of the total phytoplankton biomass. This renders the use of satellite-derived chlorophyll ineffective as a direct indicator of these events (e.g., Trainer et al., 2009). Direct observations of fish-killing HABs that form visible surface water accumulations are at some places made by small aircraft, but this is not effective for the toxic HABs that contaminate shellfish because they rarely discolor the water. Therefore, in situ observations are (necessarily) more commonly used for HAB detection, although progress is being made through the use of coupled satellite and modeling efforts (Giddings et al., 2013). In situ observations are obtained by manually collecting and analyzing samples using traditional methods, and also using advanced robotic sensors such as the Environmental Sample Processor (ESP). The coupled nature–human (CNH) system that contextualizes the impacts of HABs is also unique in the PNW. This is because of the cultural, spiritual, and economic significance of shellfish for over a dozen Native American tribes in the region. Shellfish feature so prominently in tribal customs that the native language of one coastal tribe includes a phrase that means "clam hungry." It stands to reason that the tribal people of the PNW may be disproportionately impacted by HABs and their toxins.

17.5.1.1 Puget Sound

Puget Sound is a large coastal estuary (2,330 km^2) in Washington State with long and branching basins and a complex coastline (Figure 17.10). A heuristic model of toxic blooms of *Alexandrium* was developed for Puget Sound using long-term records of PSP toxin concentrations in shellfish tissues. Blue mussels (*Mytilus edulis*) were used as a sentinel for toxic bloom activity because they readily acquire and accumulate toxins to high levels during a bloom, and they also rapidly depurate the toxins in the absence of toxic cells (Bricelj and Shumway, 1998). The model was based on toxin dynamics at "hot spot" sites where mussels most frequently attained the highest concentrations of toxin in their tissues. By examining daily time series of environmental conditions leading to the most toxic events at these hot spot sites, a specific combination of environmental conditions was identified that appeared to favor bloom development. These conditions are warm air and water temperatures,

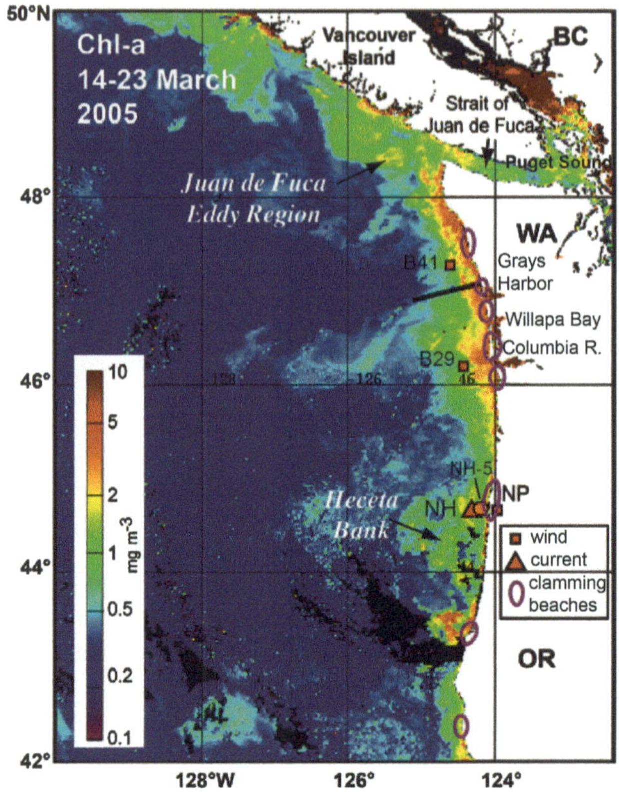

FIGURE 17.10 US Pacific Northwest. Coastal variability in algal blooms can be seen from this satellite-derived chlorophyll image (moderate-resolution imaging spectroradiometer-Aqua sensor) smoothed over a 10-day window. Major physical features such as the Juan de Fuca Eddy and Heceta Bank are sites of active research since they are associated with elevated primary production. Puget Sound is connected to coastal waters via the Strait of Juan de Fuca. Symbols denote observation platforms for monitoring coastal winds, surface currents, and clamming beaches to assess environmental forcing of HABs. *Modified and used with permission from Hickey et al. (2013); MODIS image courtesy of R. Kudela.*

weak winds, low streamflow, and small tidal variability. A temporal "window of opportunity" exists for toxic blooms of *Alexandrium* when these conditions occur in combination, and a wider window (i.e., more days) indicates increased bloom risk.

The window of opportunity for *Alexandrium* in Puget Sound can be used to evaluate past and future bloom scenarios (Moore et al., 2011). The caveat of this approach is the assumption that the environmental conditions that favor present-day blooms of *Alexandrium* have not changed from past conditions and will continue to favor blooms in the future. Moore et al. (2011) examined future *Alexandrium* bloom scenarios using the IPCC climate change projections for the PNW. Perturbations to the local environmental conditions that comprise the window of opportunity were calculated using climate projections from global climate models. The resulting forecast indicates that by the end of the twenty-first century, *Alexandrium* blooms may begin up to two months earlier in the year and persist for one month later compared to the present day. Changes to the duration of the bloom season (phenology) appear to be imminent and may be detectable within the next 30 years.

A framework that incorporates the heuristic model for *Alexandrium* with weather forecasts and in situ observations could inform managers and shellfish growers of increased HAB risk in Puget Sound (Figure 17.11). The advanced warning provided by this framework is of the order of a few days to a week, the timescale that has been identified by end users to be the most useful for putting mitigation measures in place to protect human health and reduce economic impacts. Mitigation measures for shellfish growers, and the cost savings associated with these measures, have been identified by Jin and Hoagland (2008) for the shellfish industry in the Gulf of Maine and include selectively harvesting different growing areas or increasing prebloom harvests to partially offset losses during bloom periods (Section 17.4.2.4). Public health managers can also better allocate limited resources to monitoring by targeting "hot spot" locations during time periods with increased risk for a bloom, or closing-growing areas during bloom periods more selectively than they would without a forecast.

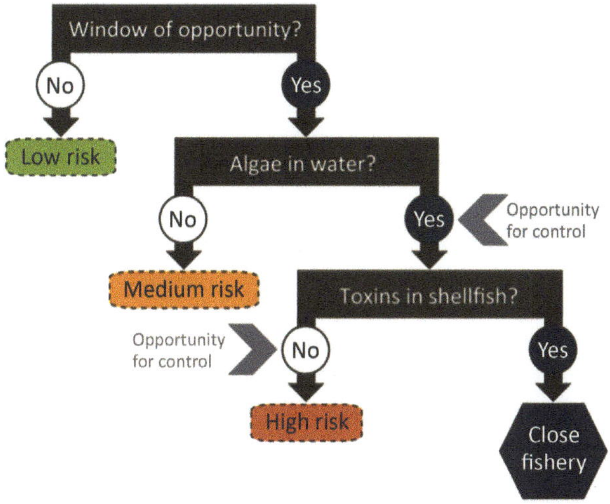

FIGURE 17.11 A risk-based approach to managing HABs in Puget Sound that provides advanced warning of outbreaks and identifies opportunities to mitigate impacts. The framework includes forecasts of the environmental conditions that favor bloom development (i.e., the window of opportunity) and timely observations of algae in the water to inform targeted and timely testing of shellfish for toxins.

A key component of the risk-based framework shown in Figure 17.11 is timely observations of harmful algae in the water. To overcome sampling challenges, in situ observations of harmful algae are provided by a citizen science program called SoundToxins (www.soundtoxins.org), which is a partnership of shellfish growers, fish farmers, tribes, environmental learning centers, and the general public. Participants are provided equipment and microscopes to collect and analyze water samples for HABs near their growing areas or near their homes, and the information is entered into a real-time database where all partners can assess the risk of imminent HABs. This program is an effective way to obtain observations of HABs from locations all around Puget Sound, but since participation is voluntary, the frequency of observations at most sites is limited to weekly. In contrast, autonomous in situ observations by an advanced biosensor, the Environmental Sample Processor (ESP), can provide high-frequency information on HAB abundance at key locations that complement the spatial coverage offered by SoundToxins observations. The ESP uses sensitive and specific molecular assays to quantitatively detect HABs and their toxins (Scholin et al., 2009). The results of the assays are captured in a photograph that is relayed via telemetry in near-real time. The entire process, from sample collection through results delivery, can occur in as little as 3 h. The ESP was first deployed in Puget Sound in 2012 and has been used to provide advanced warning of HABs at a commercial shellfish farm and a tribal shellfish hatchery as well as to answer research questions related to the ecology and oceanography of HAB species. The ESP is a new and powerful asset in the Earth system framework for detecting and forecasting HABs in the PNW.

17.5.1.2 Outer Washington–Oregon Coast

A mechanistic understanding of toxic *Pseudo-nitzschia* blooms on the outer Washington–Oregon coast underpins the approach used to forecast these HABs. *Pseudo-nitzschia* spp. are a common and natural component of the marine phytoplankton community in the PNW and are generally not toxic. However, when they are toxic they can have significant impacts; a single toxic bloom can cost as much as $22 million in lost revenue (Dyson and Huppert, 2010). A series of multiyear studies discovered two sources of toxic cells that are associated with retentive oceanographic features: the seasonal Juan de Fuca eddy located off the northern Washington coast and Heceta Bank located off the central Oregon coast (Hickey and Banas, 2003; Trainer et al., 2009). By examining measured and modeled (ROMS) ocean currents in relation to razor clam toxicity at coastal beaches, it has been shown that the Juan de Fuca eddy is the source of toxic cells of *Pseudo-nitzschia* in the summer and fall, whereas Heceta Bank is the source in the spring (Hickey et al., 2013). From the Juan de Fuca eddy, toxic cells can escape and be transported southward during periods of upwelling-favorable winds (MacFadyen et al., 2005) or transported onshore to coastal beaches during downwelling-favorable conditions (i.e., storm events; MacFadyen et al., 2005; Trainer et al., 2002). The transport of toxic cells from Heceta Bank to the Washington coast is complicated by the buoyant Columbia River plume, which during the summer and fall, acts as a barrier to onshore transport of toxic cells to Washington's beaches, but in the spring, acts primarily as a conduit (Hickey et al., 2013).

Data describing the environmental conditions that are known to bring toxic cells from the offshore source regions to the coastal beaches comprise the basis of the HAB forecasting framework for *Pseudo-nitzschia* on the outer Washington–Oregon coast. These data are examined and synthesized by local experts, and forecasts are based on the risk or likelihood of toxic cells being transported onshore. Risk factors are assessed using prior 10-day winds, observations, and numerical forecasts of regional weather, surface currents, and the Columbia River plume. A more detailed description of the complete set of models and environmental conditions that are used by the framework is given by Brown et al. (2012a). The overall risk of toxic cells contacting the coastal beaches is communicated via the PNW HAB Bulletin (http://www.pnwhabs.org/pnwhabbulletin/). The level of risk is conveyed using a "traffic light" approach, where a green symbol indicates a low risk of a toxic bloom occurring, and conversely, a red symbol indicates a high risk. The Bulletin is made available to managers and coincides with scheduled openings of the coastal razor clam beaches. As comanagers of the coastal clam resource, Washington State and Native American tribes use the Bulletin to inform their communications with stakeholders, such as sport razor clammers, coastal tourist businesses, and commercial crab fishers and processors, so that they may plan a productive season. The Bulletin also contains a narrative of the rationale behind the forecast which has helped to make the science accessible to managers.

A critical component of the PNW HAB Bulletin is information on *Pseudo-nitzschia* abundance and toxicity obtained from biweekly monitoring at six coastal beaches coordinated by the Olympic Region Harmful Algal Bloom (ORHAB) program (www.orhab.org; Trainer and Suddleson, 2005), a partnership of academic, federal, tribal, and state researchers. When concentrations exceed a size-dependent threshold (50,000 cells/l for large cells and 1 million cells/l for small cells), toxin testing of seawater and shellfish is triggered. The synergistic management approach provided by ORHAB and the PNW HAB Bulletin allows for effective and timely management of the coastal clam resource.

17.5.2 Gulf of Mexico

Over the past two decades, blooms of *K. brevis* have occurred almost annually along the Gulf Coast of Florida, intermittently along the Texas and east Florida coasts, and rarely along the coasts of Louisiana, Mississippi, and Alabama. As these blooms affect large sections of coastline, the most effective form of mitigation is to use a large synoptic framework to prevent exposure to the toxins. The Earth system framework for forecasting toxic *K. brevis* blooms comprises a combination of satellites, buoys, autonomous underwater vehicles (AUVs) and other remote technologies, along with shipboard observations to identify the initial presence of a bloom. Once *K. brevis* is identified and confirmed by water samples, blooms are monitored through the use of this system. Several models have been developed for understanding bloom dynamics and physical transport (see Table 17.1). These include an upwelling prediction model (Lanerolle et al., 2006), a biophysical model developed by Walsh et al. (2006), and a physical circulation model for transport (Weisberg et al., 2009). While complex ecosystem and high-resolution hydrodynamic models are extremely important for understanding bloom dynamics, they are not always conducive to real-time forecasting. Often, a simpler approach can provide improved guidance for mitigating bloom impacts, as the data streams and resolution are not always available in real time.

The two major human health concerns from *K. brevis* are neurotoxic shellfish poisoning (NSP) and respiratory irritation. Florida Department of Agriculture and Consumer Services (FDOACS) use the regulatory limit of 5,000 cells/l for shellfish closures to minimize the risk of NSP. In a comprehensive review on health effects associated with Florida red tide, Kirkpatrick et al. (2004) indicated that once this level is reached, beds remain closed until cell counts remain below this level for over 2 weeks and mouse bioassays indicate that levels in shellfish are below 20 MU/100 g (MU = mouse units). Current remote monitoring systems are not conducive to detection at that level and still rely on the state's system of field water sampling. As opposed to regulating NSP incidents, no regulatory limit exists for exposure to brevetoxin in seawater or air at this time, and a routine monitoring program for brevetoxin does not exist. In Florida, where the blooms occur annually, beaches are not closed even during active nearshore bloom events (Kirkpatrick et al., 2004). The abundance of *K. brevis* that typically leads to respiratory irritation in most people is >50,000 cells/l, densities that can be detected with remote sensing (Tomlinson et al., 2009; Tester et al., 1998). In an effort to predict the impact of *K. brevis* blooms, NOAA's Harmful Algal Bloom Operational Forecast System (NOAA HAB-OFS) was established and has provided HAB forecasts twice a week during the bloom season since 2004 (Stumpf et al., 2009, Figure 17.12). This system is the only operational HAB forecast system within the United States; however, several others have been developed and are being provided through crucial research programs (McGillicuddy et al., 2011). It is important to note that by being operational in the United States, the systems are "sustained, systematic, reliable and robust mission activities with institutional commitment to deliver appropriate, cost-effective products and services" (NOAA, 2008; Wilson, 2011). Although the investment in developing operational HAB forecast efforts is not trivial, significant costs can also occur in system maintenance. Operational systems can also be somewhat static, so planning improvements such as transition of new research tools into operations is crucial.

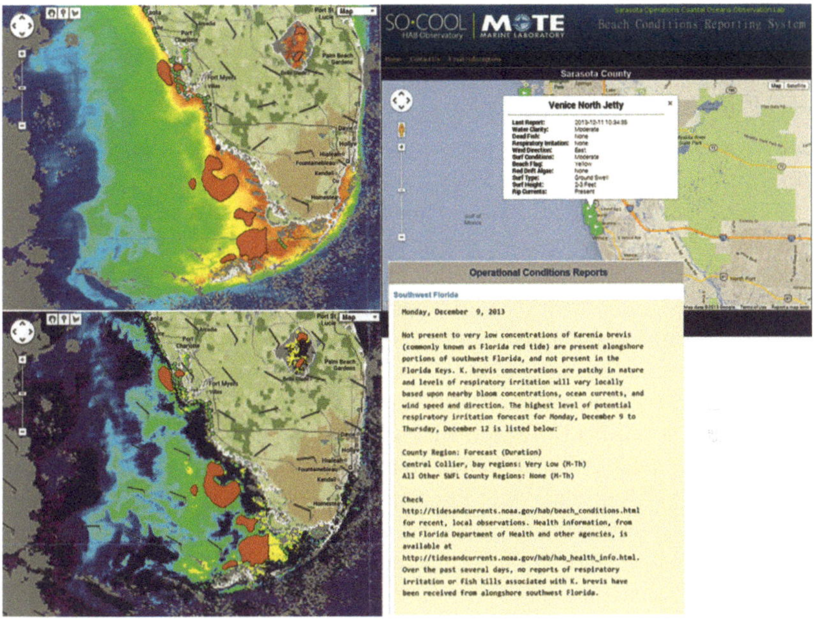

FIGURE 17.12 Western Gulf of Mexico. HAB Operational Forecast System for *Karenia brevis* blooms and brevetoxin events. Top left panel shows chl-*a* imagery from moderate-resolution imaging spectroradiometer-Aqua, with overlaid red polygons indicating likely *K. brevis* bloom due to anomalously high chlorophyll. Bottom right shows chlorophyll anomaly image used to create red polygon features as described in (Stumpf et al. 2013). The beach condition report shown in the top left, provided by Mote Marine Laboratory, provides information on respiratory irritation at individual beaches which, along with imagery and field samples, are used to produce the Operational Conditions Report in the bottom right panel.

The details regarding the development of the forecast have been addressed before (Tomlinson et al., 2004; Fisher et al., 2006; Stumpf et al., 2008, 2009; Wynne et al., 2005). The current forecast for Florida requires a combination of a satellite-derived chlorophyll anomaly product, wind speed and direction from NOAA's buoy system, and field validation and counts of *K. brevis* (provided by the Florida Fish and Wildlife Commission and shellfish bed sampling efforts by FDOACS). Since continental shelf circulation is mostly wind driven out to 50–60 m depth (Wiseman and Sturges, 1999), surface winds (rather than circulation models) are used to predict when the blooms will affect the coast, and how they will move along Florida's Gulf coast. In addition, a combination of cell concentration and wind speed and direction are used to predict the level of the respiratory effects at beaches (Stumpf et al., 2009). The forecasts are issued twice a week and the spatial resolution is at a half-"county" level. Ideally, forecasts would be issued on a beach-by-beach basis to account for the diurnal effect of sea breeze on aerosol contamination. This capability is limited by available technologies and resources (Fisher et al., 2006; Stumpf et al., 2009). Nevertheless, the current forecasts provide guidance for state sampling programs and give the public a general idea where blooms are anticipated to impact beaches. Using this information, individuals can reduce their exposure to the blooms by choosing to visit beaches that are not affected or traveling inland, remaining indoors, and/or changing vacation plans.

A unique aspect of this Earth system framework is that a human sentinel system has been established, where professional lifeguards are equipped with personal data assistants, to provide beach data regarding probable respiratory irritation intensity (i.e., audible coughing), dead fish, water color, wind direction and surf conditions at individual beaches (Kirkpatrick et al., 2008). These data are uploaded in near-real time and used in the analysis reported in the HAB forecasts. An instrument known as the BreveBuster (Kirkpatrick et al., 2000), which analyzes the optical signature of the water column, has been installed at Venice and Naples Piers, FL, USA, to facilitate monitoring. NOAA also purchased a Slocum glider equipped with a BreveBuster to locate the presence of the blooms offshore and subsurface. However, insufficient funding prevents the personnel-intensive AUV running, and these data have been unavailable for improving the forecasts. Even though weather forecasts are issued at an institutional level and significant resources are allocated for better accuracy of the forecasts, marine ecological forecasts are decades away from reaching this goal.

When looking at the cost of respiratory illnesses associated with *K. brevis,* Hoagland et al. (2009) found that just for Sarasota County, FL, USA the capitalized costs of emergency department visits were \$0.5–4 million, depending on the magnitude of the bloom. Since this estimate only takes into account the costs based on respiratory illness, it clearly underestimates the total costs for the region. In a study looking at the loss of revenue for restaurant and lodging sectors near the Ft. Walton Beach and Destin areas of Florida, red tide accounted for the most significant losses compared with tropical storms and precipitation events. For example, average restaurant revenue dropped by 2.9 percent per month during red tide events compared to 4.1–5.7 percent reduction due to hurricane or tropical storm events (Larkin and Adams, 2007). Direct losses to fishing and tourism sectors, including cleanup and subsequent costs, were \$18 million following a bloom event in Galveston County, Texas, in 2000 (Evans and Jones, 2001; Larkin and Adams, 2007). A single 1997–1998 event of *K. brevis* in North Carolina, USA, cost over \$30 million to the fisheries industries (Stumpf and Tomlinson, 2005). Additional costs associated with animal deaths, tourism, increased sampling efforts, to name a few, have not yet been accounted for. Mitigation strategies that communicate possible risk (through poison control center hotlines, signage, education, and outreach materials), facilitate natural resource management (shellfish bed closures, beach cleanup efforts), and guide nutrient control efforts all entail significant costs as well (Hoagland et al., 2009).

Through a CNH system, Hoagland (2014) used *K. brevis* blooms in Florida to describe human interactions with HABs. Although he acknowledged that the role of anthropogenic nutrients in stimulating *K. brevis* blooms continues to be a subject of scientific debate, policies associated with nutrient reduction were suggested to prevent blooms. Mechanisms for reducing nutrients through erosion control, fertilizer reductions, holding ponds for rainwater, and improvements to septic systems and tertiary municipal wastewater treatment plants may be effective. A second policy to reduce human exposure to phycotoxins involves monitoring and forecasts of blooms and increasing efforts to alert the public. The beach condition report that NOAA provides is just one aspect of such a policy that aims to reduce human exposure to brevetoxins in Florida. Other efforts are coordinated by Solutions to Avoid Red Tide and Mote Marine Laboratory to provide educational materials, effective beach signs, and public service announcements; the Aquatic Toxins Hotline established through collaboration with Florida Department of Health and Poison Control Center/Miami; and the "Beach Conditions Report" provided by Mote Marine Laboratory.

17.5.3 Northern European Continental Shelf

A EU project called Applied Simulations and Integrated Modelling for the Understanding of Toxic and Harmful Algal Blooms (HABs) (ASIMUTH) developed short-term forecasts of HAB events along Europe's Atlantic coast from 2010 to

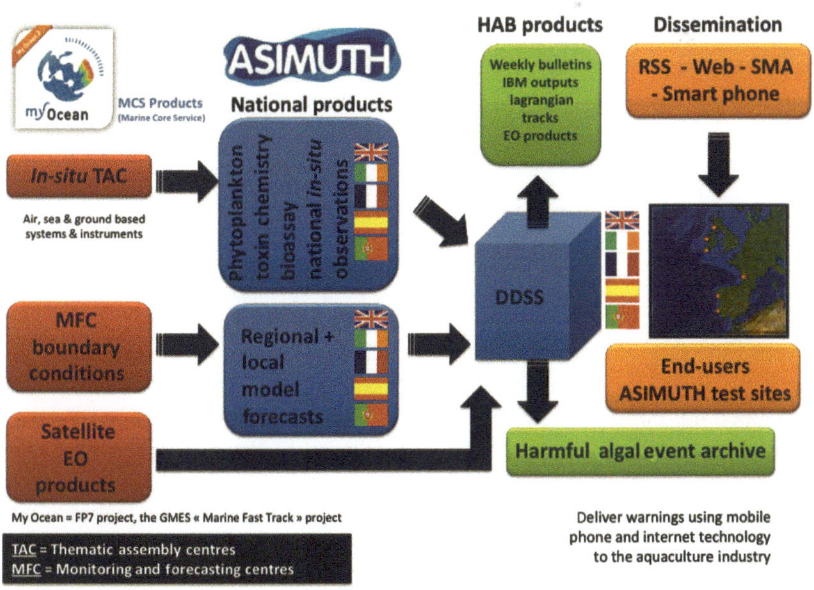

FIGURE 17.13 Northern Europe. Conceptual construct of the ASIMUTH forecasting system that compiles satellite Earth observation products with in situ monitoring and regional models to produce HAB forecasts with feedback to a network of end users in the aquaculture industry across five countries. EO, Earth Observations; RSS, Really Simple Syndication; DDSS, Distributed Decision Support System.

2013 (Figure 17.13), targeting a large range of HAB species and phycotoxins throughout the region (www.asimuth.eu). The project consortium was spread along the Atlantic coast of Europe—Portugal, Spain, France, Ireland, and Scotland—and included research agencies across the EU that are responsible for monitoring and regulating shellfish biotoxins and phytoplankton programs. Operational HAB model forecast systems were developed to track the regional distribution of selected phytoplankton species considered harmful and/or toxic to fin- and shellfish produced commercially by the aquaculture industry. Although local differences occur in existing model structures, physical oceanographic conditions, and species of interest, an important part of the process was the exchange of information (e.g., computer routines, methodologies, HAB alerts between subregions when blooms were likely to enter a neighboring domain) between partners in each EU member state. The operational system was run for a test period in 2013, and several partner countries continue to incorporate the system into operational activities past the lifetime of the ASIMUTH project.

The HAB groups common to all regions are *Dinophysis, Pseudo-nitzschia, Alexandrium,* and *Karenia mikimotoi.* The project sought to establish a new 3- to 4-day forecast service to the European Atlantic aquaculture industry making better use of historical data, current monitoring, and novel technologies. The forecast was delivered as a series of weekly HAB regional alert bulletins published online (http://www.asimuth.eu) with the overarching aim of providing the aquaculture industry and regulatory authorities with an assessment of areas at risk. To achieve this, a fusion of all relevant information from models (hydrodynamic, biogeochemical, biological), remote sensing (satellite and airborne), and in situ observations (monitoring stations) for the northeast Atlantic was assembled with local, experienced individuals in the field interpreting the information from the above data streams to produce a regular HAB bulletin for stakeholders.

17.5.3.1 Hydrodynamic Models

Each partner country developed hydrodynamic models to describe the physical processes in their selected geographical domain. In Ireland, France, and Spain, these models were based on ROMS; in Scotland, the Finite Volume Community Ocean Model (FVCOM) was used to account for Sea Lochs; and the Portuguese Coast Operational Model System/Modelo Hidrodinamico was developed in Portugal. Outputs from the French Mercator-Ocean (system name: PSY2V4), an operational model for the entire North Atlantic, were used for regional model boundary conditions and model initialization. This model provides daily predictions of ocean circulation, the mesoscale features, and ocean water properties (e.g., temperature, salinity, sea surface height). Regional models were run to produce hydrodynamic and ecological predictions, whereas Lagrangian (LPT) models are considered important in the interregional alert systems to track bloom progression between adjacent subregions. In the model domains, HABs were treated as particles and were tracked using the LPT approach, which has proven successful for tracking known HAB populations in the field (Section 17.4.2.2; Velo-Suarez et al., 2010; Mateus et al., 2012). In these cases, particle release and tracking were executed at monitoring stations and their progress modeled to aid assessment of risk to aquaculture operations. Upwelling events, coastal currents, and advection of water masses into aquaculture-producing bays were incorporated into the 3- to 4-day forecasts generated in each country.

17.5.3.2 Biogeochemical Models

Particle tracking is particularly useful for tracking HABs when little is known about aspects of HAB biological behavior. For example, sexual reproduction and resting life cycle stages are still unknown for many species. Where possible, efforts were made to produce models with biological behaviors to use in the forecast system. Ireland and Scotland developed *K. mikimotoi* functional models. Physical parameters including advection, diffusion, and shear were incorporated with biological processes of growth and mortality based on the algorithms developed by Gentien et al. (2007) to estimate progression of these blooms. The Irish model was subsequently modified to include additional behavioral procedures such as diel vertical migration, response to inorganic nutrients and oxygen, preference for ammonia as a source of nitrogen, an ability to kill other species of phytoplankton, and variation in swimming speeds depending on the temperature at depths where growth is most favorable. Mortality rates derived from turbulence, autotoxicity, and sinking were also included. Spain and Ireland use the biogeochemical model developed by Fennel et al. (2006) (also used in the Chesapeake Bay, CBEPS). In Portugal, a biomass-based pelagic biogeochemical model developed by Baretta-Bekker et al. (1995, 1997) is under development to account for all major phytoplankton groups, and in the context of this project, reflects the dynamics of the major HAB species/groups in Portuguese waters.

17.5.3.3 Satellite Remote Sensing

Satellites are an effective platform for obtaining a synoptic view of various water surface characteristics over large spatial areas, and to this end, the European Earth Observation Programme, operated by Copernicus, supports and develops downstream services such as those provided by the ASIMUTH project. Ocean color chl-*a* data are used to detect phytoplankton blooms from space from Moderate Resolution Imagine Spectroradiometer (MODIS) aboard National Aeronautics and Space Administration "Aqua" satellite. With chl-*a* as a proxy for phytoplankton biomass, the location of dense phytoplankton blooms can be detected. Since Northern Europe experiences considerable cloud cover, the North Atlantic merged chlorophyll product developed by French Research Institute for Exploration of the Sea (IFREMER) (Gohin et al., 2002; Saulquin et al., 2011) is used. This product interpolates daily cloudless

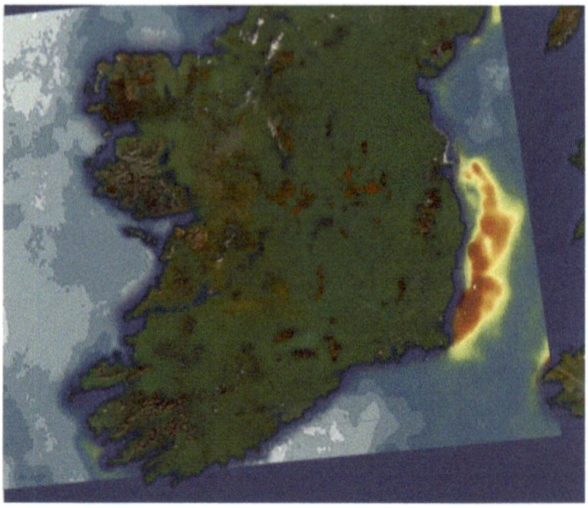

FIGURE 17.14 East Coast of Ireland. The spatial extent of a large *Phaeocystis* bloom in the Irish Sea is apparent in this satellite-derived chlorophyll-a image (moderate-resolution imaging spectroradiometer) from June 2013.

fields of chl-*a* to minimize gaps created by cloud interference, allowing ASIMUTH to identify high-biomass blooms with a potential to cause harm. Investigative work aboard research vessels and monitoring programs validated chl-*a* levels and identified the causative organism. Chl-*a* anomaly routines developed by the Gulf of Mexico *K. brevis* forecast system (Section 17.5.2; Stumpf et al., 2003; Tomlinson et al., 2004) were also used in a modified form to identify new blooms at the surface, whereas satellite-derived sea surface temperature helped define upwelling events and the location of oceanic frontal zones. An example of how Earth observation data were used in ASIMUTH is presented in Figure 17.14; high chl-*a* levels that extended from County Dublin on the east coast of Ireland as far south as County Wexford on the southeast were confirmed from shipboard observations to be *Phaeocystis* spp., and the extent confirmed by airborne observations. The bloom persisted for a month and was monitored using satellite tools. Although it was more of a nuisance (i.e., EDAB) than an HAB, coastal communities were informed of its presence and their concerns were addressed through local and national media. It is anticipated that higher spatial/temporal resolution and number of spectral bands that can detect phytoplankton composition (i.e., functional types) will be fulfilled with the launch of the European Space Agency's Sentinel-3 in 2014.

17.5.3.4 In situ Monitoring

Each European state has a legal obligation to monitor marine HAB species and associated phycotoxins in shellfish. Weekly marine phytoplankton and shellfish samples are collected along the Atlantic coastline, and laboratory data are stored by government institutions. Data were made available to the project by each national monitoring program and were assessed and incorporated into the national forecast bulletins by the local ASIMUTH expert. Additional monitoring is carried out for fish-killing phytoplankton by aquaculture operators

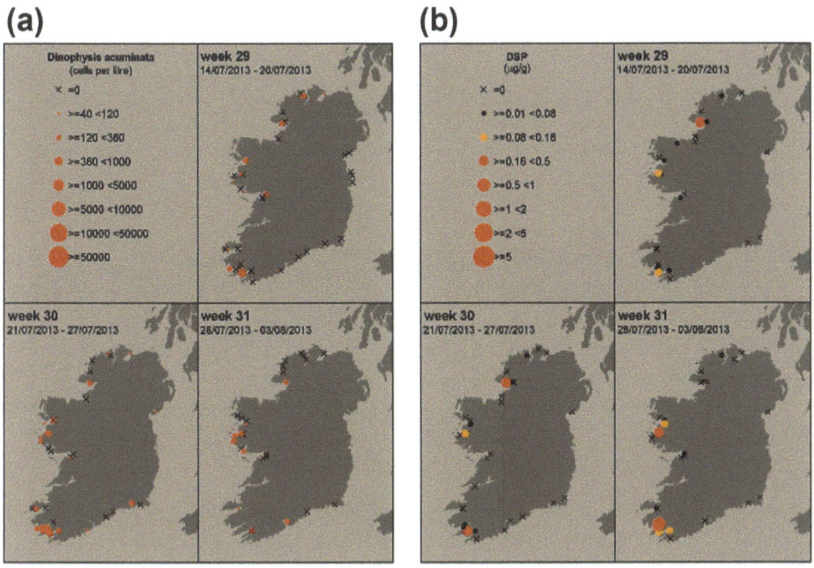

FIGURE 17.15 Irish Coast. Maps of *Dinophysis acuminata* abundance and its associated DSP-producing biotoxin, okadaic acid. (a) *Dinophysis acuminata* levels and (b) okadaic acid levels.

and private laboratories. ASIMUTH used these in situ marine data sets to complement satellite and modeled simulation products on the current state of the marine system, thereby producing time and space distribution maps for the public (Figure 17.15).

17.5.3.5 HAB Reports/Forecasts

A synopsis of each region was prepared using hydrodynamic and biogeochemical simulations, combined with satellite and in situ observations. These were assessed and weekly forecasts prepared, taking other information into account such as historical trends, extent of local events, bloom progression, and ad hoc information from social media streams. Bulletins were produced and made available on a frequent basis when an HAB risk was present (Figure 17.16), and custom-built weekly reports were produced for the target areas in Portugal, Spain, France, Ireland, and Scotland. Aquaculturists deem HAB forecasts a priority; therefore, the bulletin's primary intent was to give an assessment of the current HAB risk to the industry in order to assist in day-to-day management activities. Feedback indicated that: 80 percent of growers who filled out a satisfaction questionnaire were already using the information presented in the regional forecasts; 9 percent of respondents felt that the bulletin contained enough information to make it a useful tool; and 8 percent felt that the ideal forecast would be between 3 days and 1 week (as opposed to 3–4 days). The forecast was very well received, with 6 percent of respondents giving it a "good" or "very good" ranking and the remainder

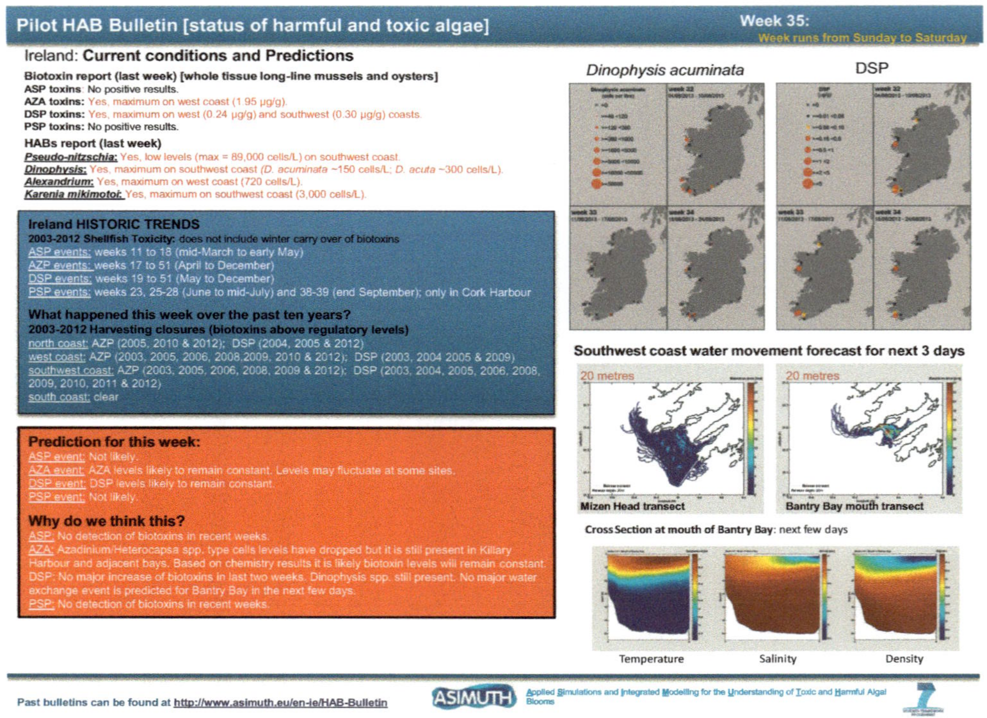

FIGURE 17.16 Excerpt of an ASIMUTH bulletin for the coast of Ireland. HAB and biotoxin summaries for the previous 3 weeks are provided along with summaries of modeled water movement for the region and vertical transects of physical properties measured in the HAB hot spot, Bantry Bay. Historic trends on harvest closures (blue box) and biotoxin predictions for the following 4 days (red box) provide context for end users to interpret forecast significance. ASP, amnesic shellfish poisoning; AZP, azaspiracid shellfish poisoning.

scoring it as "excellent." This was the first time that operational HAB forecasting was made available across the Atlantic countries of the EU, and industry end users confirmed that the ASIMUTH approach to forecasting and information dissemination is a tool that aquaculturists and industry find beneficial in planning and decision making.

17.5.4 California: A Testbed for Integrating Models, Optimal Network Design, and Adaptive Sampling

Research on HABs in coastal California has mostly focused on the DA-producing diatom *Pseudo-nitzschia* that is present throughout the California Current System (CCS; includes the PNW), although *Alexandrium, Dinophysis, Cochlodinium, Gonyaulax, Protoceratium, A. sanguinea, H. akashiwo,* and *Microcystis* (from coastal freshwater sources) also pose a threat. DA poisoning leads to particularly alarming phenomena in aquatic life when it enters marine food webs in the CCS (Scholin et al., 2000; Lefebvre et al., 2002; Bargu et al., 2012), resulting in seasonal strandings or beachings of sea lions, cetaceans, and birds on local beaches. This and the fact that *Pseudo-nitzschia* abundance and DA production can be high during spring and summer phytoplankton blooms has led to preferential treatment of *Pseudo-nitzschia* in HAB modeling studies (Anderson et al., 2009, 2006; Kudela et al., 2010; Schnetzer et al., 2007; Sekula-Wood et al., 2011). Recent research on the detrimental effects of chronic exposure to DA on the brains and reproductive fitness of sea lions has raised new concerns regarding human exposure to seasonally occurring low levels (i.e., detectable but below the regulatory limit for human consumption) of DA in shellfish (Montie et al., 2012). The California Department of Public Health (CDPH) Biotoxins Monitoring program periodically monitors ~50 shore stations for HAB species and phycotoxins (DA and saxitoxin) and also issues a proactive seasonal closure of recreational shellfish harvests. Commercial shellfish growers closely communicate with the CDPH, and some self-monitor their product. However, ever-tightening state and federal budgets have restricted how frequently the CDPH can sample, leading researchers to test alternative methods for alerting the public of HABs and for dually assessing impacts to the ecosystem at large.

Ocean color satellite imagery has been shown to provide good synoptic coverage of chl-*a* biomass in California, particularly for the southern portion of the state (Frolov et al., 2012), but no optical "smoking gun" has been found for inferring *Pseudo-nitzschia*- or DA-specific signatures from chl-*a* retrievals. Anderson et al. (2009, 2011) proposed a second-best solution by mining the statistical relationships between satellite reflectance measurements, *Pseudo-nitzschia* spp. abundance, and DA levels, while also accounting for the more physiologically relevant nutrient and physical properties (such as coastal upwelling) associated with toxic blooms (Section 17.4.2.1). From these and others studies (Anderson et al., 2006; Marchetti et al., 2004; Ryan et al., 2008;

Schnetzer et al., 2007; Trainer et al., 2000), several HAB biomass hotspots within retentive circulation features have been identified in the CCS. Research is now focused on blending satellite observations, hydrodynamic models, and statistical models for CA to spatially map and predict toxic blooms along the coast (Anderson et al., 2011). Physical circulation models already available for use are the ROMS (http://ourocean.jpl.nasa.gov/) and Navy Coastal Ocean Model (NCOM). This capability was evaluated using hindcasts of NCOM coupled to the ecosystem model Carbon, $Si(OH)_4$, Nitrogen Ecosystem model (CoSiNE) (Chai et al., 2002; Lui and Chai, in press), an approach that eliminates the need for satellite imagery and allows predictions to be based on empirical relationships between blooms and nutrient supply. Limitations in the realistic simulation of nutrients and physical fields hinder the immediate application of NCOM-CoSiNE, but a coupled physical–biological approach will be an asset for synoptic forecasts in the future (Figure 17.17(a)).

The modeling methods described here are only one aspect of an alternative approach to an HAB alert system that relies predominantly on shore-based monitoring. Two regional IOOS partnerships offer the ideal opportunity to merge oceanographic observation networks (moorings, HF radar, gliders/AUVs) and state-of-the-art models for managing coastal hazards such as HABs—these are the Central and Northern California Coastal Ocean Observing System (CeNCOOS) and the Southern California Coastal Ocean Observing System (SCCOOS) (Jochens et al., 2010). Kudela et al., (2013) recently identified these IOOS assets as the "most reasonable venue" for improving our HAB forecasting capability, contingent on future investment into IOOS resources. This assessment was partly based on the results of a study that evaluated the spatial and temporal variability of blooms in the CCS relative to existing monitoring from either satellite sensors, shore-based stations, or nearshore moorings (Frolov et al., 2012). Interestingly, the CDPH shore-based monitoring captures more of the regional variability in blooms than either long-term moorings (MBARI and LTER) or pier monitoring conducted by CeNCOOS and SCCOOS but with considerably more personnel effort to routinely cover those 50 stations. Frolov et al. (2012) propose an optimized network after calculating decorrelation scales and showing that blooms more than 4 km offshore are decoupled from those nearshore. The resulting network optimizes placement of offshore moorings (used for targeting the onset of HABs) and minimizes the number of shore-based stations (used for assessing threats to aquaculture) that are necessary to gain the most realistic representation of regional bloom dynamics (Figure 17.17(b)). Although new resources would be required to implement the proposed design, particularly if offshore moorings were to be equipped with expensive ESPs or a similar technology for identifying HAB species, the implication is that over time, this monitoring network would be more cost-effective and helpful for decision making than the current shotgun approach (Frolov et al., 2012).

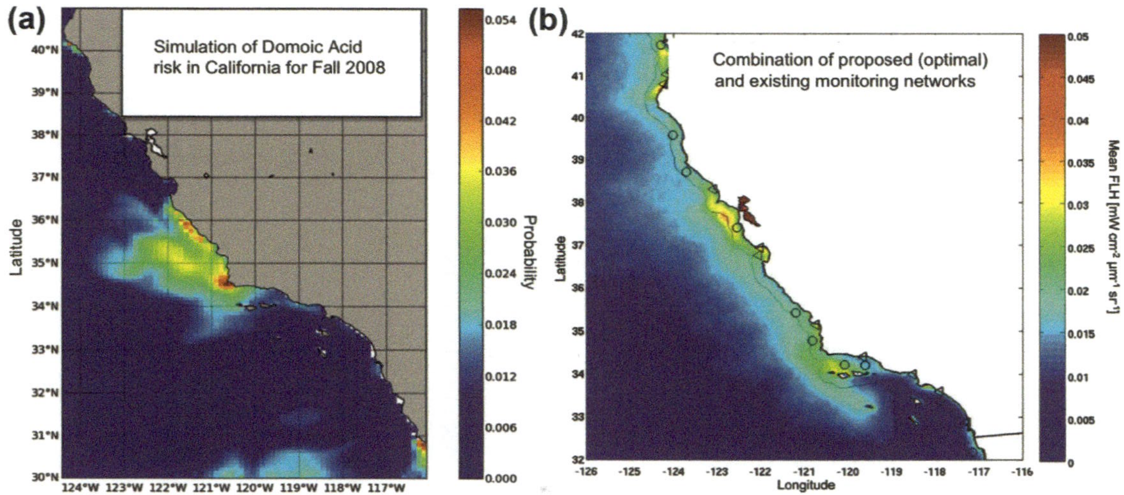

FIGURE 17.17 California. (a) Seasonal hindcast of the probability of a toxic DA event from *Pseudo-nitzschia* blooms created by forcing empirical toxin models with output from a hydrodyanmic model (NCOM) coupled to a complex biogeochemical model (CoSiNE; Anderson et al., 2012a). (b) A map of the existing network of monitoring stations (triangles) and those proposed based on an optimal sampling analysis (open circles) overlaid on the mean Fluorescence Line Height (FLH) (a proxy for phytoplankton blooms) derived from moderate-resolution imaging spectroradiometer- Aqua imagery. An optimal observation network design maximizes how much realistic bloom variability is captured while minimizing effort and cost. *Modified and used with permission from Frolov et al. (2013).*

The combination of an optimal observing system for HABs and operational forecast models with data assimilation (aided by on-going efforts of CeNCOOS and SCCOOS) will open the door for a fully operational early warning system in California. Recognizing the importance of this tight symbiosis between observations and model predictions, the Controlled, Agile, and Novel Observing Network (CANON) is merging these realms during regional process studies that bring together autonomous underwater vehicles, ESP sensors, drifters, numerical models, field experiments, and decision support systems (http://www.mbari.org/canon/). At the event scale, adaptive sampling techniques are being tested with the goal of tuning the models and better directing resources toward the focal point of a targeted feature such as an HAB. In order to smartly guide robotics for high-resolution coastal surveys, Das et al. (2010) culled the available CANON resources to simulate real-time bloom trajectories. They used HF radar surface current data and ROMS output for the Monterey Bay to project the movement of high-biomass features identified with ocean color imagery (Figure 17.18). This is just one example of how a fully integrated observation network could potentially be applied to adaptively sample and manage the impacts of an emergent HAB from its incipient stages offshore to its advection into nearshore environments through optimizing the efficiency of all available resources.

17.6 LIVING WITH HABs

We will likely never eradicate nuisance and harmful algae from aquatic ecosystems. As algal species expand into new territory and adapt to a changing climate, humans will also need to adapt and cope. Population levels along US coasts alone are projected to increase 10 percent by 2020 (NOAA, 2013). With the explosion in coastal development comes a higher likelihood for negative impacts of coastal marine hazards, including HABs. A state space method much like that used by Ramon Margalef in his phytoplankton "mandala" (Margalef, 1978) has recently been proposed to holistically manage our marine resources and promote ecosystem resilience at the regional level (Tett et al., 2013). This approach advocates the use of numerical models to understand decadal trends and to define the boundaries of a useful and functioning ecosystem for environmental managers. At the same time, it is important to realize that traditional boundaries, such as the land–sea interface, are "leaky," providing not just fluxes of nutrients to the coast through both fluvial and groundwater discharge but also transport of freshwater toxins, whereas marine toxins are "transported" inland via the food web and human consumption.

What lessons can we learn from the case studies presented in this chapter? First, all of the successful semi-operational and operational forecasting efforts rely on a close relationship between observations and models. Second, these programs rely on years (if not decades) of research leading to the development of a forecasting capability. Third, despite this research, there is a clear need for

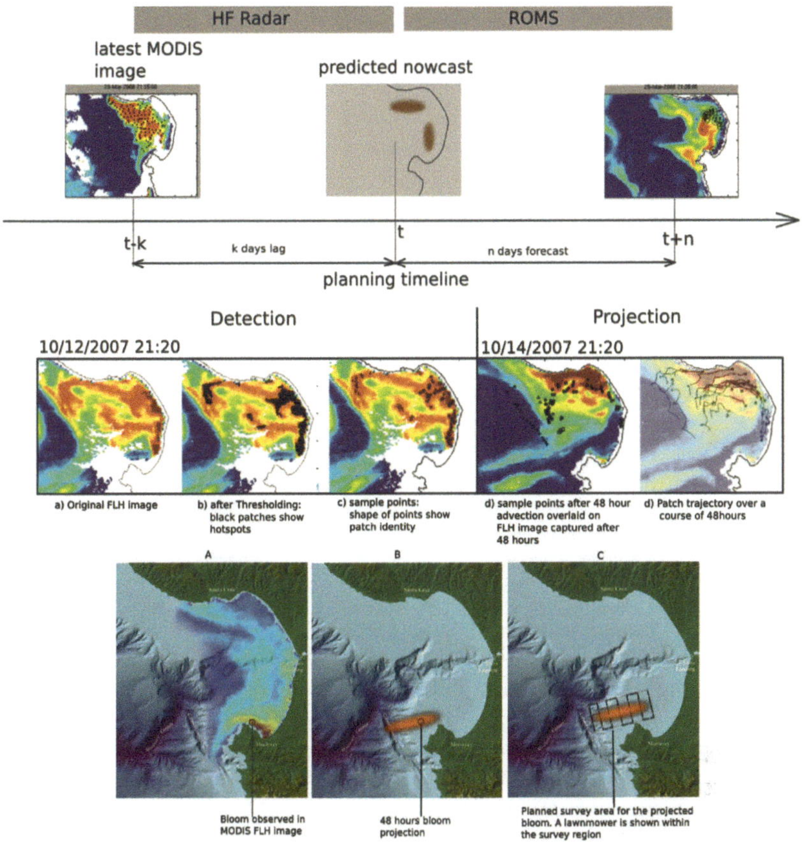

FIGURE 17.18 Monterey Bay, California. Top panel: Conceptual workflow for combining satellite ocean color imagery (MOIDIS-FLH), HF radar, and model (ROMS) platforms to identify a bloom, predict its current location, and project the advection of a high-biomass patch. Middle panel: This method is used to project an October 2007 bloom. Bottom panel: A hypothetical scenario for applying the methodology to adaptively sample a bloom using AUV robotics optimally positioned over a bloom footprint. *Modified and used with permission from Das et al. (2010).*

improved understanding of the underlying ecophysiological mechanisms leading to HABs; this will not be achieved through reliance on the bulk indices (e.g., chl-a, temperature, salinity) initially targeted by GOOS, IOOS, and other regional observing efforts. And finally, although there are many examples for individual regions, we clearly need a comprehensive assessment of the true cost and added value of these systems to adequately articulate the societal benefit of maintaining these efforts.

In summary, models, together with time series observations, are increasingly viewed as tightly coupled to management practices. Building sophisticated decision tools around an Earth system framework that properly accounts for model uncertainty and feedbacks to end user needs requires significant investment from funding agencies. However, many of the advancements are already in place if we consider economies of scale (Green et al., 2009; Weisberg et al., 2009). In the case of HAB mitigation, an integrated Earth observation-modeling framework would not only enhance our understanding of fundamental ecological relationships for constructing better models but also form the backbone of an early warning system, the goal of which is the protection of livelihoods and living resources.

ACKNOWLEDGMENTS

We thank Raphael Kudela, Kevin Sellner, Richard Stumpf, Vera Trainer, Corinne Gibble, and Greg Piniak, whose careful review and comments greatly improved this contribution. Cynthia Heil provided invaluable insight during manuscript development. California Sea Grant (NOAA award# NA100AR4170060) and the National Aeronautics and Space Administration (NASA award# NNX13AL28G) provided funding to C. Anderson. The funding source for the ASIMUTH program was the EC FP7 Programme (Space Theme, Grant Agreement No.: 261860).

REFERENCES

Allen, J., Smyth, T.J., Siddorn, J.R., Holt, M., 2008. How well can we forecast high biomass algal bloom events in a eutrophic coastal sea? Harmful Algae 8 (1), 70–76.

Alvarez, G., Uribe, E., Diaz, R., Braun, M., Marino, C., Blanco, J., 2011. Bloom of the Yessotoxin producing dinoflagellate *Protoceratium reticulatum*(Dinophyceae) in Northern Chile. J. Sea Res. 65 (4), 427–434.

Amaro, A.M., Fuentes, M.S., Ogalde, S.R., Venegas, J.A., Suarez Isla, B.A., 2005. Identification and characterization of potentially algal-lytic Marine bacteria strongly associated with the toxic dinoflagellate *Aalexandrium catenella*. J. Eukaryot. Microbiol. 52 (3), 191–200.

Amzil, Z., Sibat, M., Chomerat, N., Grossel, H., Marco-Miralles, F., Lemee, R., Nezan, E., Sechet, V., 2012. Ovatoxin-a and palytoxin accumulation in seafood in relation to *Ostreopsis* cf. *ovata* blooms on the French Mediterranean coast. Marine Drugs 10 (2), 477–496.

Anderson, Donald M., et al., 2005. *Alexandrium fundyense* cyst dynamics in the Gulf of Maine. Deep Sea Res. Part II: Topical Stud. Oceanogr 52 (19), 2522–2542.

Anderson, C.R., Brzezinski, M.A., Washburn, L., Kudela, R., 2006. Circulation and environmental conditions during a toxigenic *Pseudo-nitzschia australis* bloom in the Santa Barbara Channel, California. Mar. Ecol. Prog. 327, 119–133.

Anderson, C.R., Edwards, C.A., Goebel, N.L., Kudela, R.M., 2013. Forecasting the terrestrial influence on domoic acid production: a mechanistic approach. In: 7th Symposium on Harmful Algae in the US.

Anderson, C.R., Penta, B., Shulman, I., Chao, Y., Benitez-Nelson, C., Ellis, T., Siegel, D., Kudela, R., 2012a. What is the role of coupled physical-biological models for harmful algal bloom prediction in California? In: Ocean Sciences Meeting, Salt Lake City, UT.

Anderson, C.R., Kudela, R.M., Benitez-Nelson, C., Sekula-Wood, E., Burrell, C.T., Chao, Y., Langlois, G., Goodman, J., Siegel, D.A., 2011. Detecting toxic diatom blooms from ocean color and a regional ocean model. Geophys. Res. Lett. 38. http://dx.doi.org/10.1029/2010GL045858.

Anderson, C.R., Sapiano, M.R.P., Prasad, M.B.K., Long, W., Tango, P.J., Brown, C.W., Murtugudde, R., 2010. Predicting potentially toxigenic diatom blooms in the Chesapeake Bay. J. Mar. Sys. 83 (3–4), 127–140. http://dx.doi.org/10.1016/j.jmarsys2010.04.003.

Anderson, C.R., Seigel, D.A., Kudela, R., Brzezinski, M.A., 2009. Empirical models of toxigenic *Pseudo-nitzschia* blooms: potential use as a remote detection tool in the Santa Barbara Channel. Harmful Algae 8, 478–492. http://dx.doi.org/10.1016/j.hal.2008.10.005.

Anderson, D.M., 2009. Approaches to monitoring, control and management of harmful algal blooms (HABs). Ocean Coast. Manag. 52 (7), 342–347.

Anderson, D.M., Andersen, P., Bricelj, V.M., Cullen, J.J., Rensel, J.J., 2001. Monitoring and Management Strategies for Harmful Algal Blooms in Coastal Waters. UNESCO.

Anderson, D.M., Burkholder, J.M., Cochlan, W.P., Glibert, P.M., Gobler, C.J., Heil, C.A., Kudela, R.M., Parsons, M.L., Rensel, J.E.J., Townsend, D.W., Trainer, V.L., Vargo, G.A., 2008. Harmful algal blooms and eutrophication: examining linkages from selected coastal regions of the United States. Harmful Algae 8, 39–53.

Anderson, D.M., Cembella, A.D., Hallegraeff, G.M., 2012b. Progress in understanding harmful algal blooms: paradigm shifts and new technologies for research, monitoring, and management. Ann. Rev. Mar. Sci. 4, 143–176.

Anderson, D.M., Glibert, P.M., Burkholder, J.M., 2002. Harmful algal blooms and eutrophication: nutrient sources, composition, and consequences. Estuaries 25 (4B), 704–726.

Anderson, D.M., Stock, C.A., Keafer, B.A., Bronzino Nelson, A., Thompson, B., McGillicuddy, D.J., Keller, M., Matrai, P.A., Martin, J., 2005. *Alexandrium fundyense* cyst dynamics in the Gulf of Maine. Deep Sea Res. Part II, Topical Studies in Oceanography 52 (19), 2522–2542.

Babin, M., Roesler, C.S., Cullen, J.J., 2008. In: Real-time Coastal Observing Systems for Marine Ecosystem Dynamics and Harmful Algal Blooms. Monographs on Oceanographic Methodology Series. UNESCO, Paris.

Backer, L., McGillicuddy, D., 2006. Harmful algal blooms. Oceanography 19 (2), 94.

Backer, L.C., Moore, S.K., 2012. Harmful algal blooms: future threats in a warmer world. In: Nemr, A.E. (Ed.), Environmental Pollution and Its Relation to Climate Change. Nova Science Publishers, New York, pp. 485–512.

Backer, L.C., Rogers, H.S., Fleming, L.E., Kirkpatrick, B., Benson, J., 2005. Phycotoxins in marine seafood. In: Dabrowski, W., Sikorski, Z.E. (Eds.), Toxins in Food. CRC Press, Boca Raton.

Backer, L.C., Landsberg, J.H., Miller, M., Keel, K., Taylor, T.K., 2013. Canine cyanotoxin poisonings in the United States (1920s–2012): review of suspected and confirmed cases from three data sources. Toxins 5 (9), 1597–1628.

Bai, S.J., Huang, L.P., Su, J.Q., Tian, Y., Zheng, T.L., 2011. Algicidal effects of a novel marine actinomycete on the toxic dinoflagellate *Alexandrium tamarense*. Curr. Microbiol. 62 (6), 1774–1781.

Baretta-Bekker, J.G., Baretta, J.W., Ebenhoh, W., 1997. Microbial dynamics in the marine ecosystem model ERSEM II with decoupled carbon assimilation and nutrient uptake. J. Sea Res. 38 (3), 195–211.

Baretta-Bekker, J.G., Baretta, J.W., Koch Rasmussen, E., 1995. The microbial food web in the European regional seas ecosystem model. Neth. J. Sea Res. 33 (3), 363–379.

Bargu, S., Goldstein, T., Roberts, K., Chunyan, L., Gulland, F.M.D., 2012. *Pseudo-nitzschia* blooms, domoic acid, and related California sea lion strandings in Monterey Bay, California. Mar. Mamm. Sci. 28 (2), 237–253. http://dx.doi.org/10.1111/j.1748-7692.2011.00480.x.

Barton, A.D., Greene, C.H., Monger, B.C., Pershing, A.J., 2003. The continuous plankton recorder survey and the North Atlantic oscillation: interannual-to multidecadal-scale patterns of phytoplankton variability in the North Atlantic Ocean. Progr. Oceanogr. 58 (2), 337–358.

Bauer, M. (Ed.), 2006. Harmful Algal Research and Response: A Human Dimensions Strategy. National Office for Marine Biotoxins and Harmful Algal Blooms. Woods Hole Oceanographic Institution, Woods Hole, MA, 58 pp.

Beman, J.M., Arrigo, K.R., Matson, P.A., 2005. Agricultural runoff fuels large phytoplankton blooms in vulnerable areas of the ocean. Nature 434 (7030), 211–214.

Berdalet, E., McManus, M.A., Ross, O.N., Burchard, H., Chavez, F.P., Jaffe, J.S., Jenkinson, I.R., Kudela, R., Lips, I., Lips, U., 2014. Understanding harmful algae in stratified systems: Review of progress and future directions. Deep Sea Res., Part II, 101, 4–20.

Bernard, S., Kudela, R., Velo-Suarez, L., 2014. In: Djavidnia, S., Cheung, V., Ott, M., Seeyave, S. (Eds.), Developing Global Capabilities for the Observation and Prediction of Harmful Algal Blooms, Chapter 6 in Oceans and Society: Blue Planet. Cambridge Scholars Publishing, ISBN: 978-1-4438-5639-3.

Blauw, A.N., Anderson, P., Estrada, M., Johansen, M., Laanemets, J., Peperzak, L., Purdie, D., Raine, R., Vahtera, E., 2006. The use of fuzzy logic for data analysis and modelling of European harmful algal blooms: results of the HABES project. Afr. J. Mar. Sci. 28 (2), 365–369.

Blauw, A.N., Huisman, H., Peperzak, L., 2010. Nuisance foam events and *Phaeocystis globosa* blooms in Dutch Coastal waters analyzed with fuzzy logic. J. Mar. Sys. 83, 115–126.

Blum, I., Rao, D.S., YouLian, P., Swaminathan, S., Adams, N.G., Subba Rao, D.V., 2006. Development of statistical models for prediction of the neurotoxin domoic acid levels in the pennate diatom *Pseudo-nitzschia pungens* f. *multiseries* utilizing data from cultures and natural blooms. Algal Cultures, Analogues of Blooms and Applications 2, 891–916.

Bricelj, V.M., Shumway, S.E., 1998. Paralytic shellfish toxins in bivalve molluscs: occurrence, transfer kinetics, and biotransformation. Rev. Fish. Sci. 6 (4), 315–383.

Brittain, S.M., Wang, J., Babcock-Jackson, L., Carmichael, W.W., Rinehart, K.L., Culver, D.A., 2000. Isolation and characterization of microcystins, cyclic heptapeptide hepatotoxins from a Lake Erie Strain of *Microcystis aeruginosa*. J. Great Lake. Res. 26 (3), 241–249.

Brown, C.W., Hickey, B.M., Jacobs, J.M., Lanerolle, L.W., Moore, S.K., Schwab, D.J., Trainer, V.L., Trtanj, J., Turner, E., Wood, R.J., Wynne, T.T., 2012. Towards operational forecasts of algal blooms and pathogens. In: Morain, S.A., Budge, A.M. (Eds.), Environmental Tracking for Public Health Surveillance. CRC Press, pp. 345–368.

Brown, C.W., Hood, R.R., Long, W., Jacobs, J., Ramers, D.L., Wazniak, C., Wiggert, J.D., Wood, R., Xu, J., 2013. Ecological forecasting in Chesapeake Bay: using a mechanistic-empirical modelling approach. J. Mar. Sys, 125, 113–125.

Brownlee, E.F., Sellner, S.G., Sellner, K.G., 2003. Effects of barley straw (*Hordeum vulgare*) on freshwater and brackish phytoplankton and cyanobacteria. J. Appl. Phycol. 15 (6), 525–531.

Budd, J.W., Drummer, T.D., Nalepa, T.F., Fahnenstiel, G.L., 2001. Remote sensing of biotic effects: zebra mussels (*Dreissena polymorpha*) influence on water clarity in Saginaw Bay, Lake Huron.

Cai, W.-J., Hu, X., Huang, W.-J., Murrell, M.C., Lehrter, J.C., Lohrenz, S.E., Chou, W.-C., Zhai, W., Hollibaugh, J.T., Wang, Y., 2011. Acidification of subsurface coastal waters enhanced by eutrophication. Nat. Geosci. 4 (11), 766–770.

Cardinale, B.J., 2011. Biodiversity improves water quality through niche partitioning. Nature 472 (7341), 86–89.

Chai, F., Dugdale, R.C., Peng, T.H., Wilkerson, F.P., Barber, R.T., 2002. One-dimensional ecosystem model of the equatorial Pacific upwelling system. Part I: model development and silicon and nitrogen cycle. Deep Sea Res., II 49 (13–14), 2713–2745. http://dx.doi.org/10.1016/S0967-0645(02)00055-3.

Chen, Q., Mynett, A.E., 2004. Predicting *Phaeocystis globosa* bloom in Dutch coastal waters by decision trees and nonlinear piecewise regression. Ecol. Model. 176 (3), 277–290.

Chen, Q., Mynett, A.E., 2006. Modelling algal blooms in the Dutch coastal waters by integrated numerical and fuzzy cellular automata approaches. Ecol. Model. 199 (1), 73–81.

Ciminiello, P., Dell'Aversano, C., Fattorusso, E., Forino, M., 2010. Palytoxins: a still haunting Hawaiian curse. Phytochemistry Reviews 9 (4), 491–500.

Cloern, J.E., Hieb, K.A., Jacobson, T., Sanso, B., Di Lorenzo, E., Stacey, M.T., Largier, J.L., Meiring, W., Peterson, W.T., Powell, T.M., Winder, M., Jassby, A.D., 2010. Biological communities in San Francisco Bay track large-scale climate forcing over the North Pacific. Geophys. Res. Lett. 37 (21), L21602. http://dx.doi.org/10.1029/2010GL044774.

Cochlan, W.P., Herndon, J., Kudela, R.M., 2008. Inorganic and organic nitrogen uptake by the toxigenic diatom *Pseudo-nitzschia australis*(Bacillariophyceae). Harmful Algae 8 (1), 111–118.

Conley, D.J., Humborg, C., Rahm, L., Savchuk, O.P., Wulff, F., 2002. Hypoxia in the Baltic Sea and basin-scale changes in phosphorus biogeochemistry. Environ. Sci. Technol. 36 (24), 5315–5320.

Conley, D.J., Paerl, H.W., Howarth, R.W., Boesch, D.F., Seitzinger, S.P., Havens, K.E., Lancelot, C., Likens, G.E., 2009. Controlling eutrophication: nitrogen and phosphorus. Science 323 (5917), 1014–1015.

Cullen, J.J., 2008. Observation and prediction of harmful algal blooms. In: Babin, M., Roesler, C.S., Cullen, J.J. (Eds.), Real-time Coastal Observing Systems for Marine Ecosystem Dynamics and Harmful Algal Blooms: Theory, Instrumentation, and Modelling. UNESCO, Paris, pp. 1–41.

Dale, B., Edwards, M., Reid, P.C., 2006. Climate change and harmful algal blooms. In: Ecology of harmful algae. Springer, pp. 367–378.

Das, J., Rajany, K., Frolovy, S., Pyy, F., Ryany, J., Caron, D.A., Sukhatme, G.S., 2010. Towards Marine Bloom Trajectory Prediction for AUV Mission Planning. IEEE, pp. 4784–4790.

Davidson, K., Gowen, R.J., Tett, P., Bresnan, E., Harrison, P.J., McKinney, A., Milligan, S., Mills, D.K., Silke, J., Crooks, A.-M., 2012. Harmful algal blooms: how strong is the evidence that nutrient ratios and forms influence their occurrence? Estuarine Coastal Shelf Sci, 115, 399–413.

de Magny, G.C., Long, W., Brown, C.W., Hood, R.R., Huq, A., Murtugudde, R., Colwell, R.R., 2009. Predicting the distribution of Vibrio spp. in the Chesapeake Bay: a *Vibrio cholerae* case study. EcoHealth 6 (3), 378–389.

De Wit, P., Rogers-Bennett, L., Kudela, R.M., Palumbi, S.R., 2014. Forensic genomics – a novel tool for identifying the causes of mass mortality events, Nat. Commun. 5.

Decker, M.B., Brown, C.W., Hood, R.R., Purcell, J.E., Gross, T.F., Matanoski, J.C., Bannon, R.O., Setzler-Hamilton, E.M., 2007. Predicting the distribution of the scyphomedusa *Chrysaora quinquecirrha* in Chesapeake Bay. Mar. Ecol. Progr. 329, 99–113.

Di Lorenzo, E., Schneider, N., Cobb, K.M., Franks, P., Chhak, K., Miller, A.J., McWilliams, J.C., Bograd, S.J., Arango, H., Curchitser, E., 2008. North Pacific Gyre Oscillation links ocean climate and ecosystem change. Geophys. Res. Lett. 35 (8).

Dierssen, H.M., Kudela, R.M., Ryan, J.P., Zimmerman, R.C., 2006. Red and black tides: quantitative analysis of water-leaving radiance and perceived color for phytoplankton, colored dissolved organic matter, and suspended sediments. Limnol. Oceanogr. 51 (6), 2646–2659.

Dippner, J.W., Nguyen-Ngoc, L., Doan-Nhu, H., Subramaniam, A., 2011. A model for the prediction of harmful algae blooms in the Vietnamese upwelling area. Harmful Algae 10 (6), 606–611.

Donaghay, P.L., Osborn, T.R., 1997. Toward a theory of biological-physical control of harmful algal bloom dynamics and impacts. Limnol. Oceanogr. 42 (5, part 2), 1283–1296.

Doucette, G.J., McGovern, E.R., Babinchak, J.A., 1999. Algicidal bacteria active against *Gymnodinium breve* (Dinophyceae). I. Bacterial isolation and characterization of killing activity. J. Phycol. 35, 1447–1454.

Draisci, R., Lucentini, L., Mascioni, A., 2000. Pectenotoxins and yessotoxins: chemistry, toxicology, pharmacology, and analysis. In: Food Science and Technology. Marcel Dekker, New York, pp. 289–324.

Dyson, K., Huppert, D.D., 2010. Regional economic impacts of razor clam beach closures due to harmful algal blooms (HABs) on the Pacific coast of Washington. Harmful Algae 9 (3), 264–271.

Ehlers, P., 1994. The Baltic Sea area: convention on the protection of the marine environment of the Baltic Sea area (Helsinki Convention) of 1974 and the revised convention of 1992. Mar. Poll. Bull. 29 (6), 617–621.

Elmgren, R., Larsson, U., 2001. Nitrogen and the Baltic Sea: managing nitrogen in relation to phosphorus. Sci. World J. 1, 371–377.

Engene, N., Rottacker, E.C., Kastovsky, J., Byrum, T., Choi, H., Ellisman, M.H., Komarek, J., Gerwick, W.H., 2012. *Moorea producens* gen. nov., sp. nov. and *Moorea bouillonii* comb. nov., tropical marine cyanobacteria rich in bioactive secondary metabolites. Int. J. Syst. Evol. Microbiol. 62 (Pt 5), 1171–1178.

Evans, G., Jones, L., 2001. Economic Impact of the 2000 Red Tide on Galveston County, Texas: A Case Study. College Station: Department of Agricultural Economics, Texas A&M University.

Fabry, V.J., Seibel, B.A., Feely, R.A., Orr, J.C., 2008. Impacts of ocean acidification on marine fauna and ecosystem processes. ICES J. Mar. Sci.: J. Cons. 65 (3), 414–432.

Feely, R.A., Sabine, C.L., Hernandez-Ayon, J.M., Ianson, D., Hales, B., 2008. Evidence for upwelling of corrosive "acidified" water onto the continental shelf. Science 320 (5882), 1490–1492.

Fennel, K., Wilkin, J., Levin, J., Moisan, J., O'Reilly, J., Haidvogel, D., 2006. Nitrogen cycling in the Middle Atlantic Bight: results from a three-dimensional model and implications for the North Atlantic nitrogen budget. Global Geochem. Cycles 20 (GB3007), GB3007. http://dx.doi.org/10.1029/2005GB002456.

Fisher, K.M., Allen, A.L., Keller, H.M., Bronder, Z.E., Fanstermacher, L.E., Vincent, M.S., 2006. Annual Report of the Gulf of Mexico Harmful Algal Bloom Operational Forecast System (HAB-OFS): October 1, 2004 to September 30, 2005 (Operational Year #1), pp. 25. NOAA Technical Report NOS CO-OPS 047.

Fleming, L.E., Backer, L.C., Rowan, A., 2002. The epidemiology of human illnesses associated with harmful algal blooms. In: Adams, D.J., Baden, D., Bloomquist, J.R., Ehrich, M., Guilarte, T.R., Harvey, A. (Eds.), Neurotoxicology Handbook, Natural Toxins of Marine Origin, vol. 1. Humana Press, Totowa, NJ, pp. 363–381.

Flynn, K.J., 2010. Do external resource ratios matter? Implications for modelling eutrophication events and controlling harmful algal blooms. J. Mar. Syst. 83, 170–180.

Follows, M.J., Dutkiewicz, S., Grant, S., Chisholm, S.W., 2007. Emergent biogeography of microbial communities in a model ocean. Science 315, 1843–1846.

Franks, P.J.S., 1992. Sink or swim: accumulation of biomass at fronts. Mar. Ecol. Prog. Ser. 82 (1–12).

Franks, P.J.S., 1997. Models of harmful algal blooms. Limnol. Oceanogr. 42 (5, part 2), 1273–1282.

Frolov, S., Kudela, R.M., Bellingham, J.G., 2013. Monitoring of harmful algal blooms in the era of diminishing resources: a case study of the US west coast. Harm. Algae, 21, 1–12.

Fuhrman, J.A., Azam, F., 1980. Bacterioplankton secondary production estimates for coastal waters of British Columbia, Antarctica, and California. Appl. Environ. Microbiol. 39 (6), 109–120.

Gentien, P., Lunven, M., Lazure, P., Youenou, A., Crassous, M.-P., 2007. Motility and autotoxicity in *Karenia mikimotoi* (Dinophyceae). Phil. Trans. R. Soc. B Biol. Sci. 362 (1487), 1937–1946.

Gentien, P., Pitcher, G., Cembella, A.D., Glibert, P.M. (Eds.), 2003. GEOHAB: Global Ecology and Oceanography of Harmful Algal Blooms. SCOR and IOC, Baltimore and Paris.

Giddings, S.N., MacCready, P., Banas, N.S., Davis, K.A., Siedlecki, S.A., Hickey, B.M., Trainer, V.L., Kudela, R.M., Pelland, N., 2014. Hindcasts of harmful algal bloom transport on the Pacific Northwest coast. J. Geophys. Res.119(4), 2439–2461

Glibert, P.M., 2006. Global Ecology and Oceanography of Harmful Algal Blooms: Eutrophication and HABs. IOC and SCOR, Paris and Baltimore.

Glibert, P.M., Burkholder, J.M., 2006. The complex relationships between increases in fertilization of the earth, coastal eutrophication and proliferation of harmful algal blooms. In: Ecology of Harmful Algae. Springer, pp. 341–354.

Glibert, P.M., Burkholder, J.M., 2011. Harmful algal blooms and eutrophication: "strategies" for nutrient uptake and growth outside the Redfield comfort zone. Chin. J. Oceanol. Limnol. 29 (4), 724–738.

Glibert, P.M., Allen, J., Bouwman, A.F., Brown, C.W., Flynn, K.J., Lewitus, A.J., Madden, C.J., 2010. Modeling of HABs and eutrophication: status, advances, challenges. J. Mar. Syst. 83 (3), 262–275.

Glibert, P.M., Harrison, J., Heil, C., Seitzinger, S., 2006. Escalating worldwide use of urea-a global change contributing to coastal eutrophication. Biogeochemistry 77 (3), 441–463.

Glibert, P.M., Magnien, R., Lomas, M.W., Alexander, J., Fan, C., Haramoto, E., Trice, M., Kana, T.M., 2001. Harmful algal blooms in the Chesapeake and Coastal Bays of Maryland, USA: comparisons of 1997, 1998, and 1999 events. Estuaries 24, 875–883.

Gobler, C.J., Sunda, W.G., 2012. Ecosystem disruptive algal blooms of the brown tide species, *Aureococcus anophagefferens* and *Aureoumbra lagunensis*. Harmful Algae 14, 36–45.

Gobler, C.J., Renaghan, M.J., Buck, N.J., 2002. Impacts of nutrients and grazing mortality on the abundance of *Aureococcus anophagefferens* during a New York brown tide bloom. Limnol. Oceanogr. 47 (1), 129–141.

Goebel, N.L., Edwards, C.A., Zehr, J.P., Follows, M.J., 2010. An emergent community ecosystem model applied to the California Current System. J. Mar. Syst. 83, 221–241. http://dx.doi.org/10.1016/j.marsys.2010.05.002.

Gohin, F., Druon, J.N., Lampert, L., 2002. A five channel chlorophyll concentration algorithm applied to SeaWiFS data processed by SeaDAS in coastal waters. Int. J. Remote Sens. 23 (8), 1639–1661.

Gokaraju, B., Durbha, S.S., King, R.L., Younan, N.H., 2011. A machine learning based spatio-temporal data mining approach for detection of harmful algal blooms in the Gulf of Mexico. IEEE J. Sel. Top. Appl. Earth Observ. Remote Sens. 4 (3), 710–720.

GOOS: Global Ocean Observing System (GOOS), 2013. Regional Policy Report. IOC-UNESCO, Paris.

Graneli, E., Hansen, P.J., 2006. Allelopathy in harmful algae: a mechanism to compete for resources?. In: Ecology of Harmful Algae. Springer, pp. 189–201.

Green, D., Uccellini, L., Colton, M., Turner, E., Scheurer, D., Valette-Silver, N., Matlock, G., Brown, C., Wilson, D., 2009. Toward a Marine Ecological Forecasting System. IEEE.

Gregg, W.W., Friedrichs, M.A., Robinson, A.R., Rose, K.A., Schlitzer, R., Thompson, K.R., Doney, S.C., 2009. Skill assessment in ocean biological data assimilation. J. Mar. Syst. 76 (1), 16–33.

Gumbo, R.J., Ross, G., Cloete, E.T., 2008. Biological control of *Microcystis* dominated harmful algal blooms. Afr. J. Biotechnol. 7 (25).

Hagstrom, J.A., Sengco, M.R., Villareal, T.A., 2010. Potential methods for managing *Prymnesium parvum* blooms and toxicity, with emphasis on clay and barley straw: a review. JAWRA 46 (1), 187–198.

Hallegraeff, G.M., 1993. A review of harmful algal blooms and their apparent global increase. Phycologia 32 (2), 79–99.

Hallegraeff, G.M., 2010. Ocean climate change, phytoplankton community responses, and harmful algal blooms: a formidable predictive challenge. J. Phycol. 46 (2), 220–235.

Hamilton, G., McVinish, R., Mengersen, K., 2009. Bayesian model averaging for harmful algal bloom prediction. Ecol. Appl. 19 (7), 1805–1814.

HARRNESS: Harmful Algal Research and Response (HARRNESS), 2005. In: Ramsdell, J.S., Anderson, D.M., Glibert, P.M. (Eds.), National Environmental Science Strategy 2005–2015. Ecological Society of America, Washington, DC.

Hatch, M., Moore, S.K., Rhodes, L., Nilsson, W., Scholin, C., Hallam, S., Cox, A., Mueller, K., Storm, M., 2013. Real-time detection of harmful algae at a tribal marine aquaculture site. In: 7th Symposium on Harmful Algae in the U.S. Sarasota, FL.

Hauri, C., Gruber, N., Vogt, M., Doney, S.C., Feely, R.A., Lachkar, Z., Leinweber, A., McDonnell, A.M., Munnich, M., Plattner, G.-K., 2013. Spatiotemporal variability and long-term trends of ocean acidification in the California Current System. Biogeosciences 10 (1), 193–216.

Havens, H., Luther, M.E., Meyers, S.D., Heil, C.A., 2010. Lagrangian particle tracking of a toxic dinoflagellate bloom within the Tampa Bay estuary. Mar. Pollut. Bull. 60 (12), 2233–2241.

He, R., McGillicuddy, D.J., Keafer, B.A., Anderson, D.M., 2008. Historic 2005 toxic bloom of *Alexandrium fundyense* in the western Gulf of Maine: 2. Coupled biophysical numerical modeling. J. Geophys. Res.: Oceans (1978–2012) 113 (C7).

Heil, C.A., Steidinger, K.A., 2009. Monitoring, management, and mitigation of *Karenia* blooms in the eastern Gulf of Mexico. Harmful Algae 8 (4), 611–617.

Heil, D.C., 2009. *Karenia brevis* monitoring, management, and mitigation for Florida molluscan shellfish harvesting areas. Harmful Algae 8 (4), 608–610.

Heisler, J., Glibert, P.M., Burkholder, J.M., Anderson, D.M., Cochlan, W., Dennison, W.C., Dortch, Q., Gobler, C.J., Heil, C.A., Humphries, E., Lewitus, A., Magnien, R., Marshall, H.G., Sellner, K., Stockwell, D.A., Stoecker, D.K., Suddleson, M., 2008. Eutrophication and harmful algal blooms: a scientific consensus. Harmful Algae 8, 3–13.

Henrichs, D.W., Hetland, R.D., Campbell, L., 2013. Predicted origins of *Karenia brevis* bloom formation along the coast of Texas using an individual-based model. In: 7th Symposium on Harmful Algae in the U.S., Sarasota, FL.

HELCOM, B.S.A.P., 2007. The HELCOM Baltic Sea Action Plan. Helsinki Commission for the Protection of the Baltic Marine Environment Helsinki. Finland. Baltic Sea Environ. Proc. 111.

Henry, T., September 15, 2013. Carroll Township's Scare with Toxin a, 'Wake-up Call'. The Toledo Blade. Retrieved from. http://www.toledoblade.com/local/2013/09/15/Carroll-Township-s-scare-with-toxin-a-wake-up-call.html.

Hetland, R.D., Campbell, L., 2007. Convergent blooms of *Karenia brevis* along the Texas coast. Geophys. Res. Lett 34 (19).

Hickey, B.M., Banas, N.S., 2003. Oceanography of the US Pacific Northwest coastal ocean and estuaries with application to coastal ecology. Estuaries 26 (4), 1010–1031.

Hickey, B.M., Trainer, V.L., Michael Kosro, P., Adams, N.G., Connolly, T.P., Kachel, N.B., Geier, S.L., 2013. A springtime source of toxic *Pseudo-nitzschia* cells on razor clam beaches in the Pacific Northwest. Harm. Algae, 25, 1–14.

Hoagland, P., 2014. Coupled nature-human (CNH) systems: generic aspects of human interactions with blooms of Florida red tide (Karenia brevis) and implications for policy responses. In: Rossini, G.P. (Ed.), Toxins and Biologically Active Compounds from Microalgae, Biological Effects and Risk Management, vol. 2. CRC Press, Boca Raton, FL, pp. 502–537.

Hoagland, P., Scatasta, S., 2006. The economic effects of harmful algal blooms. In: Ecology of Harmful Algae. Springer, pp. 391–402.

Hoagland, P., Di Jin, L.Y.P., Kirkpatrick, B., Kirkpatrick, G., Fleming, L.E., Reich, A., Watkins, S.M., Ullmann, S.G., Backer, L.C., 2009. The costs of respiratory illnesses arising from Florida gulf coast *Karenia brevis* blooms. Environ. Health Perspect. 117 (8), 1239.

Hodgkiss, I.J., Ho, K.C., 1997. Are changes in N: P ratios in coastal waters the key to increased red tide blooms? Springer, pp. 141–147.

Hoffman, M.J., Miyoshi, T., Haine, T.W., Ide, K., Brown, C.W., Murtugudde, R., 2012. An advanced data assimilation system for the Chesapeake Bay: performance evaluation. J. Atmos. Oceanic Technol. 29 (10), 1542–1557.

Howard, M.D.A., Cochlan, W.P., Ladizinksy, N., Kudela, R.M., 2007. Nitrogenous preference of toxigenic *Pseudo-nitzschia australis*(Bacillariophyceae) from field and laboratory experiments. Harmful Algae 6, 206–217.

Howard, M.D., Silver, M., Kudela, R.M., 2008. Yessotoxin detected in mussel (*Mytilus californicus*) and phytoplankton samples from the US west coast. Harmful Algae 7 (5), 646–652.

Hudnell, H.K., Jones, C., Labisi, B., Lucero, V., Hill, D.R., Eilers, J., 2010. Freshwater harmful algal bloom (FHAB) suppression with solar powered circulation (SPC). Harmful Algae 9 (2), 208–217.

Imai, I., Kim, M.C., Nagasaki, K., Itakura, S., Ishida, Y., 1998. Relationships between dynamics of red tide-causing raphidophycean flagellates and algicidal micro-organisms in the coastal sea of Japan. Phycol. Res. 46 (2), 139–146.

Imai, I., Sunahara, T., Nishikawa, T., Hori, Y., Kondo, R., Hiroishi, S., 2001. Fluctuations of the red tide flagellates *Chattonella* spp. (Raphidophyceae) and the algicidal bacterium *Cytophaga* sp. in the Seto Inland Sea, Japan. Mar. Biol. 138 (5), 1043–1049.

Imai, I., Yamaguchi, M., Hori, Y., 2006. Eutrophication and occurrences of harmful algal blooms in the Seto Inland Sea, Japan. Plankton Benthos Res. 1 (2), 71–84.

IOOS, U.S., 2013. U.S. IOOS Summit Report: A New Decade for the Integrated Ocean Observing System. Interagency Ocean Observation Committee and U.S. Integrated Ocean Observing System, Herndon, Virginia.

IPCC, 2013. Contribution of Working Group I to the Fifth Assessment Report of the Intergovernmental Panel on Climate Change. In: Stocker, T.F., Qin, D., Plattner, G.-K., Tignor, M., Allen, S.K., Boschung, J., Nauels, A., Xia, Y., Bex, V., Midgley, P.M. (Eds.), Climate Change 2013: The Physical Science Basis. Cambridge University Press, Cambridge, UK and New York, NY, USA, 1535 pp.

James, K.J., Moroney, C., Roden, C., Satake, M., Yasumoto, T., Lehane, M., Furey, A., 2003. Ubiquitous "benign" alga emerges as the cause of shellfish contamination responsible for the human toxic syndrome, azaspiracid poisoning. Toxicon 41 (2), 145–151.

Jeong, S.-Y., Ishida, K., Ito, Y., Okada, S., Murakami, M., 2003. Bacillamide, a novel algicide from the marine bacterium, Bacillus sp. SY-1, against the harmful dinoflagellate, *Cochlodinium polykrikoides*. Tetrahedron Lett. 44 (43), 8005–8007.

Jessup, D.A., Miller, M.A., Ryan, J.P., Nevins, H.M., Kerkering, H.A., Mekebri, A., Crane, D.B., Johnson, T.A., Kudela, R.M., 2009. Mass stranding of marine birds caused by a surfactant-producing red tide. PLoS One 4 (2), e4550.

Jester, R., Lefebvre, K., Langlois, G., Vigilant, V., Baugh, K., Silver, M., 2009. A shift in the dominant toxin-producing algal species in central California alters phycotoxins in food webs. Harmful Algae 8 (2), 291–298. http://dx.doi.org/10.1016/j.hal.2008.07.001.

Jewett, E.B., Lopez, C.B., Dortch, Q., Etheridge, S.M., Backer, L.C., 2008. Harmful Algal Bloom Management and Response: Assessment and Plan. Interagency Working Group on Harmful Algal Blooms, Hypoxia, and Human Health of the Joint Subcommittee on Ocean Science and Technology. Washington, DC.

Jia, Y., Han, G., Wang, C., Guo, P., Jiang, W., Li, X., Tian, X., 2010. The efficacy and mechanisms of fungal suppression of freshwater harmful algal bloom species. J. Hazard. Mat. 183 (1), 176–181.

Jin, D., Hoagland, P., 2008. The value of harmful algal bloom predictions to the nearshore commercial shellfish fishery in the Gulf of Maine. Harm. Algae 7 (6), 772–781.

Jin, D., Thunberg, E., Hoagland, P., 2008. Economic impact of the 2005 red tide event on commercial shellfish fisheries in New England. Ocean Coast. Manag. 51 (5), 420–429.

Jin, Q., Dong, S., 2003. Comparative studies on the allelopathic effects of two different strains of *Ulva pertusa* on *Heterosigma akashiwo* and *Alexandrium tamarense*. J. Exp. Mar. Biol. Ecol. 293 (1), 41–55.

Jochens, A.E., Malone, T.C., Stumpf, R.P., Hickey, B.M., Carter, M., Morrison, R., Dyble, J., Jones, B., Trainer, V.L., 2010. Integrated ocean observing system in support of forecasting harmful algal blooms. Mar. Technol. Soc. J. 44 (6), 99–121.

Juhel, G., Davenport, J., O'halloran, J., Culloty, S.C., O'riordan, R.M., James, K.F., Furey, A., Allis, O., 2006. Impacts of microcystins on the feeding behaviour and energy balance of zebra mussels, *Dreissenapolymorpha*: a bioenergetics approach. Aquat. Toxicol. 79 (4), 391–400.

Kamykowski, D., Milligan, E.J., Reed, R.E., Liu, W., 1999. Geotaxis/phototaxis and biochemical patterns in *Heterocapsa (=Cachonina) illdefina* (Dinophyceae) during diel vertical migrations. J. Phycol. 35 (6), 1397–1403.

Kang, Y.-K., Cho, S.-Y., Kang, Y.-H., Katano, T., Jin, E.-S., Kong, D.-S., Han, M.-S., 2008. Isolation, identification and characterization of algicidal bacteria against *Stephanodiscus hantzschii* and *Peridinium bipes* for the control of freshwater winter algal blooms. J. Appl. Phycol. 20 (4), 375–386.

Kemp, W.M., Brooks, M.T., Hood, R.R., 2001. Nutrient enrichment, habitat variability and trophic transfer efficiency in simple models of pelagic ecosystems. Marine ecology Progress Series 223, 73–87. Oldendorf.

Kim, H.G., 2006. Mitigation and controls of HABs. In: Ecology of Harmful Algae. Springer, pp. 327–338.

Kim, S., Park, M.G., Kim, K.Y., Kim, C.H., Yih, W., Park, J.S., Coats, D.W., 2008. Genetic diversity of parasitic dinoflagellates in the genus Amoebophrya and its relationship to parasite biology and biogeography. J. Eukaryot. Microbiol. 55 (1), 1–8.

Kim, Y.S., Lee, D.-S., Jeong, S.-Y., Lee, W.J., Lee, M.-S., 2009. Isolation and characterization of a marine algicidal bacterium against the harmful raphidophyceae *Chattonella marina*. J. Microbiol. 47 (1), 9–18.

Kirkpatrick, B., Currier, R., Nierenberg, K., Reich, A., Backer, L.C., Stumpf, R., Fleming, L., Kirkpatrick, G., 2008. Florida red tide and human health: a pilot beach conditions reporting system to minimize human exposure. Sci. Tot. Environ. 402 (1), 1–8.

Kirkpatrick, B., Fleming, L.E., Squicciarini, D., Backer, L.C., Clark, R., Abraham, W., Benson, J., Cheng, Y.S., Johnson, D., Pierce, R., 2004. Literature review of Florida red tide: implications for human health effects. Harmful Algae 3 (2), 99–115.

Kirkpatrick, G.J., Millie, D.F., Moline, M.A., Schofield, O., 2000. Optical discrimination of a phytoplankton species in natural mixed populations. Limnol. Oceanogr. 45 (2), 467–471.

Kodama, M., Doucette, G.J., Green, D.H., 2006. Relationships between bacteria and harmful algae. In: Ecology of Harmful Algae. Springer, pp. 243–255.

Krock, B., Tillmann, U., John, U., Cembella, A.D., 2009. Characterization of azaspiracids in plankton size-fractions and isolation of an azaspiracid-producing dinoflagellate from the North Sea. Harmful Algae 8 (2), 254–263.

Kudela, R.M., Gobler, C.J., 2012. Harmful dinoflagellate blooms caused by *Cochlodinium* sp.: global expansion and ecological strategies facilitating bloom formation. Harmful Algae 14, 71–86.

Kudela, R.M., Frolov, S., Anderson, C.R., Bellingham, J.G., 2013. Leveraging Ocean Observatories to Monitor and Forecast Harmful Algal Blooms: A Case Study of the U.S. West Coast.

Kudela, R.M., Lane, J.Q., Cochlan, W.P., 2008. The potential role of anthropogenically derived nitrogen in the growth of harmful algae in California, USA. Harmful Algae 8, 103–110. http://dx.doi.org/10.1016/j.hal.2008.019.

Kudela, R.M., Seeyave, S., Cochlan, W., 2010. The role of nutrients in regulation and promotion of harmful algal blooms in upwelling systems. Prog. Oceanogr. 85, 122–135.

Laanemets, J., Lilover, M.-J., Raudsepp, U., Autio, R., Vahtera, E., Lips, I., Lips, U., 2006. A fuzzy logic model to describe the cyanobacteria *Nodularia spumigena* blooms in the Gulf of Finland, Baltic Sea. Hydrobiologia 554 (1), 31–45.

Lane, J.Q., Raimondi, P.T., Kudela, R.M., 2009. Development of a logistic regression model for the prediction of toxigenic *Pseudo-nitzschia* blooms in Monterey Bay, California. Mar. Ecol. Prog. Ser. 383, 37–51.

Lanerolle, L.W., Tomlinson, M.C., Gross, T.F., Aikman, F., Stumpf, R.P., Kirkpatrick, G.J., Pederson, B.A., 2006. Numerical investigation of the effects of upwelling on harmful algal blooms off the west Florida coast. Estuarine, Coastal Shelf Sci. 70 (4), 599–612.

Larkin, S.L., Adams, C.M., 2007. Harmful algal blooms and coastal business: economic consequences in Florida. Soc. Nat. Res. 20 (9), 849–859.

Lee, J.H., Huang, Y., Dickman, M., Jayawardena, A.W., 2003. Neural network modelling of coastal algal blooms. Ecol. Model 159 (2), 179–201.

Lee, Y.-J., Choi, J.-K., Kim, E.-K., Youn, S.-H., Yang, E.-J., 2008. Field experiments on mitigation of harmful algal blooms using a sophorolipid—yellow clay mixture and effects on marine plankton. Harmful Algae 7 (2), 154–162.

Lefebvre, K.A., Bargu, S., Kieckhefer, T., Sliver, M.W., 2002. From sanddabs to blue whales: the pervasiveness of domoic acid. Toxicon 40, 971–977.

Li, Y., Ruoying, H., McGillicuddy Jr., D.J., Anderson, D.M., Keafer, B.A., 2009. Investigation of the 2006 *Alexandrium fundyense* bloom in the Gulf of Maine: in-situ observations and numerical modeling. Cont. Shelf Res. 29, 2069–2082. http://dx.doi.org/10.1016/j.csr.2009.07.012.

Liefer, J.D., MacIntyre, H.L., Novoveska, L., Smith, W.L., Dorsey, C.P., 2009. Temporal and spatial variability in *Pseudo-nitzschia* spp. in Alabama coastal waters: a "hot spot" linked to submarine groundwater discharge? Harmful Algae 8 (5), 706–714.

Liefer, J.D., MacIntyre, H.L., Su, N., Burnett, W.C., 2013. Seasonal alternation between groundwater discharge and benthic coupling as nutrient sources in a shallow coastal lagoon. Estuaries Coasts, 1–16.

Liu, Y., Weisberg, R.H., Hu, C., Zheng, L., 2011. Tracking the deepwater horizon oil spill: a modeling perspective. Eos Trans. AGU 92 (6), 45–52.

Lovejoy, C., Bowman, J.P., Hallegraeff, G.M., 1998. Algicidal effects of a novel marine *Pseudoalteromonas* isolate (Class Proteobacteria, Gamma Subdivision) on harmful algal bloom species of the genera *Chattonella, Gymnodinium*, and *Heterosigma*. Appl. Environ. Microbiol. 64 (8), 2806–2813.

Liu, G., Chai, F., 2009. Seasonal and interannual variation of physical and biological processes during 1994–2001 in the Sea of Japan/East Sea: a three-dimensional physical–biogeochemical modeling study. J. Mar. Sys. 78 (2), 265–277.

Lurling, M., Faassen, E.J., 2012. Controlling toxic cyanobacteria: effects of dredging and phosphorus-binding clay on cyanobacteria and microcystins. Water Res. 46 (5), 1447–1459.

MacFadyen, A., Hickey, B.M., Foreman, M.G., 2005. Transport of surface waters from the Juan de Fuca eddy region to the Washington coast. Cont. Shelf Res. 25 (16), 2008–2021.

Marchetti, A., Trainer, V.L., Harrison, P.J., 2004. Environmental conditions and phytoplankton dynamics associated with *Pseudo-nitzschia* abundance and domoic acid in the Juan de Fuca eddy. Mar. Ecol. Prog. Ser. 281, 1–12.

Margalef, R., 1978. Life-forms of phytoplankton as survival alternatives in an unstable environment. Oceanol. Acta 1, 493–509.

Marques, A., Nunes, M.L., Moore, S.K., Strom, M.S., 2010. Climate change and seafood safety: human health implications. Food Res. Int. 43 (7), 1766–1779.

Mateus, M., Riflet, G., Chambel, P., Fernandes, L., Fernandes, R., Juliano, M., Campuzano, F., de Pablo, H., Neves, R., 2012. An operational model for the West Iberian coast: products and services. Ocean Sci. 8 (4), 713–732.

Matthijs, H.C., Visser, P.M., Reeze, B., Meeuse, J., Slot, P.C., Wijn, G., Talens, R., Huisman, J., 2012. Selective suppression of harmful cyanobacteria in an entire lake with hydrogen peroxide. Water Res. 46 (5), 1460–1472.

Mayali, X., Azam, F., 2004. Algicidal bacteria in the sea and their impact on algal Blooms. J. Eukaryot. Microbiol. 51 (2), 139–144.

Mazzillo, F.F., Ryan, J.P., Silver, M.W., 2011. Parasitism as a biological control agent of dinoflagellate blooms in the California Current System. Harmful Algae 10 (6), 763–773.

McGillicuddy Jr., D.J., 2010. Models of harmful algal blooms: conceptual, empirical, and numerical approaches. J. Mar. Syst. 83, 105–107. http://dx.doi.org/10.1016/j.jmarsys.2010.06.008 (Special Issue on Harmful Algal Bloom Modeling).

McGillicuddy Jr., D.J., Anderson, D.M., Lynch, D.R., Townsend, D.W., 2005. Mechanisms regulating large-scale seasonal fluctuations in *Alexandrium fundyense* populations in the Gulf of Maine: results from a physical-biological model. Deep Sea Res. II 52, 2698–2714.

McGillicuddy Jr., D.J., Townsend, D.W., He, R., Keafer, B.A., Kleindinst, J.L., Li, Y., Manning, J.P., Mountain, D.G., Thomas, M.A., Anderson, D.M., 2011. Suppression of the 2010 *Alexandrium fundyense* bloom by changes in physical, biological, and chemical properties of the Gulf of Maine. Limnol. Oceanogr. 56 (6), 2411.

McGillicuddy, D.J., Signell, R.P., Stock, C.A., Keafer, B.A., Keller, M.D., Hetland, R.D., Anderson, D.M., 2003. A mechanism for offshore initiation of harmful algal blooms in the coastal Gulf of Maine. J. Plankton Res. 25 (9), 1131–1138.
McKnight, D.M., Chisholm, S.W., Harleman, D.R., 1983. $CuSO_4$ treatment of nuisance algal blooms in drinking water reservoirs. Environ. Manag. 7 (4), 311–320.
McQuatters-Gollop, A., Raitsos, D.E., Edwards, M., Pradhan, Y., Mee, L.D., Lavender, S.J., Attrill, M.J., 2007. A long-term chlorophyll data set reveals regime shift in North Sea phytoplankton biomass unconnected to nutrient trends. Limnol. Oceanogr. 52 (2), 635.
McQuatters-Gollop, A., Reid, P.C., Edwards, M., Burkill, P.H., Castellani, C., Batten, S., Gieskes, W., Beare, D., Bidigare, R.R., Head, E., 2011. Is there a decline in marine phytoplankton? Nature 472 (7342), E6–E7.
Millie, D.F., Fahnenstiel, G.L., Bressie, J.D., Pigg, R.J., Rediske, R.R., Klarer, D.M., Tester, P.A., Litaker, R.W., 2009. Late-summer phytoplankton in western Lake Erie (Laurentian Great Lakes): bloom distributions, toxicity, and environmental influences. Aquat. Ecol. 43 (4), 915–934.
Mitra, A., Flynn, K.J., 2006. Promotion of harmful algal blooms by zooplankton predatory activity. Biol. Lett. 2 (2), 194–197.
Mitra, A., Flynn, K.J., 2010. Modelling mixotrophy in harmful algal blooms: more or less the sum of the parts? J. Mar. Syst. 83, 158–169.
Montie, E.W., Wheeler, E., Pussini, N., Battey, T.W.K., Van Bonn, W., Gulland, F.M.D., 2012. Magnetic resonance imaging reveals that brain atrophy is more severe in older California sea lions with domoic acid toxicosis. Harmful Algae 20, 19–29.
Moore, S.K., Mantua, N.J., Salathe Jr, E.P., 2011. Past trends and future scenarios for environmental conditions favoring the accumulation of paralytic shellfish toxins in Puget Sound shellfish. Harmful Algae 10 (5), 521–529.
Moore, S.K., Trainer, V.L., Mantua, N.J., Parker, M.S., Laws, E., Backer, L.C., Fleming, L.E., 2008. Impacts of climate variability and future climate change on harmful algal blooms and human health. Environ. Health 7 (Suppl. 2), S4. http://dx.doi.org/10.1186/1476-069X-7-S2-S4.
Muttil, N., Lee, J.H., 2005. Genetic programming for analysis and real-time prediction of coastal algal blooms. Ecol. Modell. 189 (3), 363–376.
Nagasaki, K., Yamaguchi, M., 1997. Isolation of a virus infectious to the harmful bloom causing microalga *Heterosigma akashiwo* (Raphidophyceae). Aquat. Microb. Ecol. 13 (2), 135–140.
Nagasaki, K., Tarutani, K., Yamaguchi, M., 1999. Growth characteristics of *Heterosigma akashiwo* virus and its possible use as a microbiological agent for red tide control. Appl. Environ. Microbiol. 65 (3), 898–902.
Nakashima, T., Miyazaki, Y., Matsuyama, Y., Muraoka, W., Yamaguchi, K., Oda, T., 2006. Producing mechanism of an algicidal compound against red tide phytoplankton in a marine bacterium y-proteobacterium. Appl. Microbiol. Biotechnol. 73 (3), 684–690.
NOAA, 2008. Policy on Transition of Research to Application. National Oceanic and Atmospheric Administration, Office of the Chief Administrative Officer, Silver Spring, MD.
NOAA, 2013. In: Crossett, K., Ache, B., Pacheco, P., Haber, K. (Eds.), National Coastal Population Report, State of the Coast Report Series. National Oceanic and Atmospheric Administration, Office of the Chief Administrative Officer.
Olascoaga, M.J., Beron-Vera, F.J., Brand, L.E., Kocak, H., 2008. Tracing the early development of harmful algal blooms on the West Florida Shelf with the aid of Lagrangian coherent structures. J. Geophys. Res. 113 (C12), C12014.
Osborne, N.J., Webb, P.M., Shaw, G.R., 2001. The toxins of *Lyngbya majuscula* and their human and ecological health effects. Environ. Int. 27 (5), 381–392.

Paerl, H.W., 1997. Coastal eutrophication and harmful algal blooms: importance of atmospheric deposition and groundwater as "new" nitrogen and other nutrient sources. Limnol. Oceanogr. 42 (5), 1154–1165.

Paerl, Hans W., Huisman, Jef, 2008. Blooms like it hot. Science-New York Then Washington -320.5872. 57.

Paerl, H.W., Hall, N.S., Calandrino, E.S., 2011. Controlling harmful cyanobacterial blooms in a world experiencing anthropogenic and climatic-induced change. Sci. Tot. Environ. 409 (10), 1739–1745.

Palma, S., Mourino, H., Silva, A., Barao, M.I., Moita, M.T., 2010. Can *Pseudo-nitzschia* blooms be modeled by coastal upwelling in Lisbon Bay? Harmful Algae 9 (3), 294–303.

Pan, G., Dai, L., Li, L., He, L., Li, H., Bi, L., Gulati, R.D., 2012. Reducing the recruitment of sedimented algae and nutrient release into the overlying water using modified soil/sand flocculation-capping in eutrophic lakes. Environ. Sci. Technol. 46 (9), 5077–5084.

Parsons, M.L., Settlemier, C.J., Bienfang, P.K., 2010. A simple model capable of simulating the population dynamics of *Gambierdiscus*, the benthic dinoflagellate responsible for ciguatera fish poisoning. Harmful Algae 10 (1), 71–80.

Paz, B., Daranas, A.H., Norte, M., Riobo, P., Franco, J.M., Fernandez, J.J., 2008. Yessotoxins, a group of marine polyether toxins: an overview. Mar. Drugs 6 (2), 73–102.

Paz, B., Riobo, P., Luisa Fernandez, M., Fraga, S., Franco, J.M., 2004. Production and release of yessotoxins by the dinoflagellates *Protoceratium reticulatum* and *Lingulodinium polyedrum* in culture. Toxicon 44 (3), 251–258.

Peacock, M.B., Kudela, R.M., 2014. Evidence for active vertical migration of two dinoflagellates experiencing iron, nitrogen, and phosphorous limitation. Limnol. Oceanogr. 59 (3), 660–673.

Pettersson, L.H., Pozdnyakov, D., 2013. Monitoring of Harmful Algae Blooms. Springer.

Pitcher, G.C., Figueiras, F.G., Hickey, B.M., Moita, M.T., 2010. The physical oceanography of upwelling systems and the development of harmful algal blooms. Prog. Oceanogr. 85, 5032.

Raine, R., McDermott, G., Silke, J., Lyons, K., Nolan, G., Cusack, C., 2010. A simple short range model for the prediction of harmful algal events in the bays of southwestern Ireland. J. Mar. Syst. 83 (3), 150–157.

Recknagel, F., French, M., Harkonen, P., Yabunaka, K.-I., 1997. Artificial neural network approach for modelling and prediction of algal blooms. Ecol. Model. 96 (1), 11–28.

Rhodes, L., Adamson, J., Suzuki, T., Briggs, L., Garthwaite, I., 2000. Toxic marine epiphytic dinoflagellates, *Ostreopsis siamensis* and *Coolia monotis* (Dinophyceae), in New Zealand. N. Z. J. Mar. Freshwater Res. 34 (2), 371–383.

Rhodes, L., McNabb, P., De Salas, M., Briggs, L., Beuzenberg, V., Gladstone, M., 2006. Yessotoxin production by *Gonyaulax spinifera*. Harmful Algae 5 (2), 148–155.

Ribeiro, R., Torgo, L., 2008. A comparative study on predicting algae blooms in Douro River, Portugal. Ecol. Model. 212 (1), 86–91.

Roelke, D.L., Eldridge, P.M., Cifuentes, L.A., 1999. A model of phytoplankton competition for limiting and nonlimiting nutrients: implications for development of estuarine and nearshore management schemes. Estuaries 22 (1), 92–104.

Roiha, P., Westerlund, A., Nummelin, A., Stipa, T., 2010. Ensemble forecasting of harmful algal blooms in the Baltic Sea. J. Mar. Syst. 83 (3), 210–220.

Roth, P.B., Twiner, M.J., Mikulski, C.M., Barnhorst, A.B., Doucette, G.J., 2008. Comparative analysis of two algicidal bacteria active against the red tide dinoflagellate *Karenia brevis*. Harmful Algae 7 (5), 682–691.

Rounsefell, G.A., Evans, J.E., 1958. Large-scale Experimental Test of Copper Sulfate as a Control for the Florida Red Tide. U.S. Fish and Wildlife Service. Special Scientific Report.

Ryan, J.P., Gower, J.F.R., King, S.A., Bissett, W.P., Fischer, A.M., Kudela, R.M., Kolber, Z., Mazzillo, F., Reinecker, E.V., Chavez, F.P., 2008. A coastal ocean extreme bloom incubator. Geophys. Res. Lett. 35. http://dx.doi.org/10.1029/2008GL034081.

Sacau-Cuadrado, M., Conde-Pardo, P., Otero-Tranchero, P., 2003. Forecast of red tides off the Galician coast. Acta Astronaut. 53 (4), 439–443.

Salas, R., Tillmann, U., John, U., Kilcoyne, J., Burson, A., Cantwell, C., Hess, P., Jauffrais, T., Silke, J., 2011. The role of *Azadinium spinosum* (Dinophyceae) in the production of azaspiracid shellfish poisoning in mussels. Harmful Algae 10 (6), 774–783.

Salomon, P.S., Imai, I., 2006. Pathogens of harmful microalgae. In: Ecology of Harmful Algae. Springer, pp. 271–282.

Sanders, N.J., Gotelli, N.J., Heller, N.E., Gordon, D.M., 2003. Community disassembly by an invasive species. Proc. Natl. Acad. Sci. 100 (5), 2474–2477.

Saulquin, B., Gohin, F., Garrello, R., 2011. Regional objective analysis for merging high-resolution MERIS, MODIS/Aqua, and SeaWiFS chlorophyll-a data from 1998 to 2008 on the European Atlantic shelf. IEEE Trans. Geosci. Remote Sens. 49 (1), 143–154.

Schnetzer, A., Miller, P.E., Schaffner, R.A., Stauffer, B.A., Jones, B.H., Weisberg, S.B., DiGiacomo, P.M., Berelson, W.M., Caron, D.A., 2007. Blooms of *Pseudo-nitzschia* and domoic acid in the San Pedro Channel and Los Angeles harbor areas of the Southern California Bight, 2003–2004. Harmful Algae 6 (3), 327–387.

Scholin, C.A., Gulland, F., Doucette, G.J., Benson, S., Busman, M., Chavez, F.P., Cordaro, J., DeLong, R., De Vogelaere, A., Harvey, J., Haulena, M., Lefebvre, K.A., Lipscomb, T., Loscutoff, S., Lowenstine, L.J., Marin III, R., Miller, P.E., McLellan, W.A., Moeller, P.D.R., Powell, C.L., Rowles, T., Silvagni, P., Silver, M.W., Spraker, T., Trainer, V.L., Van Dolah, F.M., 2000. Mortality of sea lions along the central California coast linked to a toxic diatom bloom. Nature 403, 80–83.

Scholin, C., Doucette, G., Jensen, S., Roman, B., Pargett, D., Marin III, R., Preston, C., Jones, W., Feldman, J., Everlove, C., 2009. Remote detection of marine microbes, small invertebrates, harmful algae, and biotoxins using the environmental sample processor (ESP). Oceanography 22.

Secord, D., 2003. Biological control of marine invasive species: cautionary tales and land-based lessons. Biol. Invasions 5 (1–2), 117–131.

Sekula-Wood, E., Benitez-Nelson, C., Morton, S.L., Anderson, C.R., Thunell, R., 2011. *Pseudo-nitzschia* and domoic acid fluxes in Santa Barbara Basin (CA) from 1993 to 2008. Harmful Algae 10, 567–575.

Sekula-Wood, E., Schnezer, A., Benitez-Nelson, C., Anderson, C.R., Berelson, W.M., Brzezinski, M.A., Burns, J.M., Caron, D.A., Cetinic, I., Ferry, J.L., Fitzpatrick, E., Jones, B.H., Miller, P.E., Morton, S.L., Schaffner, R.A., Siegel, D.A., Thunell, R., 2009. Rapid downward transport of the neurotoxin domoic acid in coastal waters. Nat. Geosci. 2 (4), 342–350. http://dx.doi.org/10.1038/NGEO472.

Sellner, K.G., Place, A.R., Paolisso, M., Gao, Y., Williams, E., VanDolah, F.M., 2013. Options in mitigating cyanobacterial blooms. In: 7th Symposium on Harmful Algae in the U.S., Sarasota, FL.

Sengco, M.R., Anderson, D.M., 2004. Controlling harmful algal blooms through clay flocculation. J. Eukaryot. Microbiol. 51 (2), 169–172.

Shumway, S.E., Frank, D.M., Ewart, L.M., 2003. Effect of yellow loess on clearance rate in seven species of benthic, filter-feeding invertebrates. Aquac. Res. 34 (15), 1391–1402.

Smayda, T.J., 1997. Harmful algal blooms: their ecophysiology and general relevance to phytoplankton blooms in the sea. Limnol. Oceanogr. 42 (5, part 2), 1137–1153.

Smayda, T.J., 2007. Reflections on the ballast water dispersal—harmful algal bloom paradigm. Harmful Algae 6 (4), 601–622.

Smayda, T.J., 2008. Complexity in the eutrophication-harmful algal bloom relationship, with comment on the importance of grazing. Harmful Algae 8 (1), 140–151.

Stock, C.A., McGillicuddy, D.J., Solow, A.R., Anderson, D.M., 2005. Evaluating hypotheses for the initiation and development of *Alexandrium fundyense* blooms in the western Gulf of Maine using a coupled physical-biological model. Deep Sea Res. Part II 52 (19), 2715–2744.

Stocker, T.F., Dahe, Q., Plattner, G.-K., 2013. Intergovernmental Panel on Climate Change. IPCC.

Stoecker, D.K., 1998. Conceptual models of mixotrophy in planktonic protists and some ecological and evolutionary implications. Eur. J. Protistol. 34 (3), 281–290.

Stow, C.A., Jolliff, J., McGillicuddy Jr., D.J., Doney, S.C., Allen, J., Friedrichs, M.A., Rose, K.A., Wallhead, P., 2009. Skill assessment for coupled biological/physical models of marine systems. J. Mar. Syst. 76 (1), 4–15.

Stumpf, R.P., Culver, M.E., Tester, P.A., Tomlinson, M., Kirkpatrick, G.J., Pederson, B.A., Truby, E., Ransibrahmanakul, V., Soracco, M., 2003. Monitoring *Karenia brevis* blooms in the Gulf of Mexico using satellite ocean color imagery and other data. Harm. Algae 2, 147–160.

Stumpf, R.P., Tomlinson, M.C., 2005. Remote sensing of harmful algal blooms. In: Remote Sensing of Coastal Aquatic Environments. Springer, pp. 277–296.

Stumpf, R.P., Litaker, R.W., Lanerolle, L., Tester, P.A., 2008. Hydrodynamic accumulation of *Karenia* off the west coast of Florida. Cont. Shelf Res. 28 (1), 189–213.

Stumpf, R.P., Tomlinson, M.C., Calkins, J.A., Kirkpatrick, B., Fisher, K., Nierenberg, K., Currier, R., Wynne, T.T., 2009. Skill assessment for an operational algal bloom forecast system. J. Mar. Syst. 76 (1), 151–161.

Stumpf, R.P., Fleming-Lehtinen, V., Graneli, E., 2010. Integration of data for nowcasting of harmful algal blooms.

Stumpf, R.P., Wynne, T.T., Baker, D.B., Fahnenstiel, G.L., 2012. Interannual variability of cyanobacterial blooms in Lake Erie. PLoS One 7 (8), e42444.

Su, J.Q., Yang, X.R., Zheng, T.L., Tian, Y., Jiao, N.Z., Cai, L.Z., Hong, H.S., 2007. Isolation and characterization of a marine algicidal bacterium against the toxic dinoflagellate *Alexandrium tamarense*. Harmful Algae 6 (6), 799–810.

Su, J., Yang, X., Zhou, Y., Zheng, T., 2011. Marine bacteria antagonistic to the harmful algal bloom species *Alexandrium tamarense*(Dinophyceae). Biol. Contr. 56 (2), 132–138.

Sun, J., Hutchins, D.A., Feng, Y., Seubert, E.L., Caron, D.A., Fu, F.-X., 2011. Effects of changing pCO_2 and phosphate availability on domoic acid production and physiology of the marine harmful bloom diatom *Pseudo-nitzschia multiseries*. Limnol. Oceanogr. 56 (3), 829–840.

Sun, X.-X., Choi, J.-K., Kim, E.-K., 2004. A preliminary study on the mechanism of harmful algal bloom mitigation by use of sophorolipid treatment. J. Exp. Mar. Biol. Ecol. 304 (1), 35–49.

Sunda, W.G., Shertzer, K.W., 2012. Modeling ecosystem disruptive algal blooms: Positive feedback mechanisms. Mar. Ecol. Progr. 447, 31–47.

Sunda, W.G., Graneli, E., Gobler, C.J., 2006. Positive feedback and the development and persistence of ecosystem disruptive algal blooms. J. Phycol. 42 (5), 963–974.

Talmage, S.C., Gobler, C.J., 2009. The effects of elevated carbon dioxide concentrations on the metamorphosis, size, and survival of larval hard clams *(Mercenaria mercenaria)*, bay scallops (*Argopecten irradians*), and Eastern oysters *(Crassostrea virginica)*. Limnol. Oceanogr. 54 (6), 2072.

Tatters, A.O., Fei-Xue, F., Hutchins, D.A., 2012. High CO_2 and silicate limitation synergistically increase the toxicity of *Pseudo-nitzschia fraudulenta*. PLoS One 7 (2). http://dx.doi.org/10.1371/journaLpone.0032116.

Tatters, A.O., Schnetzer, A., Fu, F., Lie, A.Y., Caron, D.A., Hutchins, D.A., 2013. Short-versus long-term responses to changing CO_2 in a coastal dinoflagellate bloom: implications for interspecific competitive interactions and community structure. Evolution 67 (7), 1879–1891.

Terlizzi, D.E., Ferrier, M.D., Armbrester, E.A., Anlauf, K.A., 2002. Inhibition of dinoflagellate growth by extracts of barley straw *(Hordeum vulgare)*. J. Appl. Phycol. 14 (4), 275–280.

Terseleer, N., Gypens, N., Lancelot, C., 2013. Factors controlling the production of domoic acid by *Pseudo-nitzschia* (Bacillariophyceae): a model study. Harm. Algae 24, 45–53.

Tester, P.A., Stumpf, R.P., Steidinger, K., 1998. Ocean color imagery: what is the minimum detection level for *Gymnodinium breve* blooms? In: Reguera, B., Blanco, J., Fernandez, M.L., Wyatt, T. (Eds.), Harmful Algae. Xunta de Galacia and Intergovernmental Oceanographic Commission of UNESCO.

Tett, P., Gowen, R., Painting, S., Elliott, M., Forster, R., Mills, D., Bresnan, E., Capuzzo, E., Fernandes, T., Foden, J., 2013. Framework for understanding marine ecosystem health. Mar. Ecol. Progr. 494, 1–27.

Tillmann, U., Elbrachter, M., Krock, B., John, U., Cembella, A., 2009. *Azadinium spinosum* gen. et sp. nov. (Dinophyceae) identified as a primary producer of azaspiracid toxins. Eur. J. Phycol. 44 (1), 63–79.

Tomlinson, M.C., Stumpf, R.P., Ransibrahmanakul, V., Truby, E.W., Kirkpatrick, G.J., Pederson, B.A., Vargo, G.A., Heil, C.A., 2004. Evaluation of the use of SeaWiFS imagery for detecting *Karenia brevis* harmful algal blooms in the eastern Gulf of Mexico. Rem. Sens. Environ. 91 (3–4), 293–303.

Tomlinson, M.C., Wynne, T.T., Stumpf, R.P., 2009. An evaluation of remote sensing techniques for enhanced detection of the toxic dinoflagellate, *Karenia brevis*. Rem. Sens. Environ. 113, 598–609.

Trainer, V.L., Suddleson, M., 2005. Monitoring approaches for early warning of DA events in Washington State. J. Oceanogr. 18 (2), 228–237.

Trainer, V.L., Adams, N.G., Bill, B.D., Stehr, C.M., Wekell, J.C., Moeller, P.D.R., Busman, M., Woodruff, D., 2000. Domoic acid production near California coastal upwelling zones, June 1998. Limnol. Oceanogr. 45 (8), 1818–1833.

Trainer, V.L., Hickey, B.M., Horner, R.A., 2002. Biological and physical dynamics of domoic acid production off the Washington coast. Limnol. Oceanogr. 47 (5), 1438–1446.

Trainer, V.L., Hickey, B.M., Lessard, E.J., Cochlan, W.P., Trick, C.G., Wells, M.L., MacFadyen, A., Moore, S.K., 2009. Variability of *Pseudo-nitzschia* and domoic acid in the Juan de Fuca eddy region and its adjacent shelves. Limnol. Oceanogr. 54 (1), 289.

Trainer, V.L., Moore, L., Bill, B.D., Adams, N.G., Harrington, N., Borchert, J., da Silva, D.A., Eberhart, B.-T.L., 2013. Diarrhetic shellfish toxins and other Lipophilic toxins of human health concern in Washington state. Mar. Drugs 11 (6), 1815–1835.

Turner, J.T., Graneli, E., 2006. "Top-Down" predation control on marine harmful algae. In: Ecology of Harmful Algae. Springer, pp. 355–366.

Vahtera, E., Conley, D.J., Gustafsson, B.G., Kuosa, H., Pitkanen, H., Savchuk, O.P., Tamminen, T., Viitasalo, M., Voss, M., Wasmund, N., 2007. Internal ecosystem feedbacks enhance nitrogen-fixing cyanobacteria blooms and complicate management in the Baltic Sea. AMBIO 36 (2), 186–194.

Van de Waal, D.B., Tillmann, U., Zhu, M., Koch, B.P., Rost, B., John, U., 2013. Nutrient pulse induces dynamic changes in cellular C: N: P, amino acids, and paralytic shellfish poisoning toxins in *Alexandrium tamarense*. Mar. Ecol. Progr. Ser. 493, 57–69.

Van Dolah, F.M., Zippay, M.L., Pezzolesi, L., Rein, K.S., Johnson, J.G., Morey, J.S., Wang, Z., Pistocchi, R., 2013. Subcellular localization of dinoflagellate polyketide synthases and fatty acid synthase activity. J. Phycol. 49 (6), 1118–1127.

Van Dolah, F.M., Roelke, D., Greene, R.M., 2001. Health and ecological impacts of harmful algal blooms: risk assessment needs. Hum. Ecolo. Risk Assess. 7 (5), 1329–1345.

Vanderploeg, H.A., Liebig, J.R., Carmichael, W.W., Agy, M.A., Johengen, T.H., Fahnenstiel, G.L., Nalepa, T.F., 2001. Zebra mussel (*Dreissena polymorpha*) selective filtration promoted toxic *Microcystis* blooms in Saginaw Bay (Lake Huron) and Lake Erie. Can. J. Fish. Aquat. Sci. 58 (6), 1208–1221.

Velo-Suarez, L., Gutierrez-Estrada, J.C., 2007. Artificial neural network approaches to one-step weekly prediction of *Dinophysis acuminata* blooms in Huelva (Western Andalucia, Spain). Harmful Algae 6 (3), 361–371.

Velo-Suarez, L., Reguera, B., Gonzalez-Gil, S., Lunven, M., Lazure, P., Nezan, E., Gentien, P., 2010. Application of a 3D Lagrangian model to explain the decline of a *Dinophysis acuminata* bloom in the Bay of Biscay. J. Mar. Sys. 83 (3), 242–252.

Villanoy, C.L., Azanza, R.V., Altemerano, A., Casil, A.L., 2006. Attempts to model the bloom dynamics of *Pyrodinium*, a tropical toxic dinoflagellate. Harmful Algae 5 (2), 156–183.

Waldbusser, G.G., Voigt, E.P., Bergschneider, H., Green, M.A., Newell, R.I., 2011. Biocalcification in the eastern oyster *(Crassostrea virginica)* in relation to long-term trends in Chesapeake Bay pH. Estuar. Coast. 34 (2), 221–231.

Waliser, D.E., Murtugudde, R., Strutton, P., Li, J.L., 2005. Subseasonal organization of ocean chlorophyll: prospects for prediction based on the Madden-Julian oscillation. Geophys. Res. Lett. 32 (23).

Walsh, J.J., Haddad, K.D., Dieterle, D.A., Weisberg, R.H., Li, Z., Yang, H., Muller-Karger, F.E., Heil, C.A., Bissett, W.P., 2002. A numerical analysis of landfall of the 1979 red tide of *Karenia brevis* along the west coast of Florida. Cont. Shelf Res. 22 (1), 15–38.

Walsh, J.J., Jolliff, J.K., Darrow, B.P., Lenes, J.M., Milroy, S.P., Remsen, A., Dieterle, D.A., Carder, K.L., Chen, F.R., Vargo, G.A., 2006. Red tides in the Gulf of Mexico: where, when, and why? J. Geophys. Res. Oceans (1978–2012) 111 (C11).

Walsh, J.J., Penta, B., Dieterle, D.A., Bissett, W.P., 2001. Predictive ecological modeling of harmful algal blooms. Hum. Ecol. Risk Assess. 7 (5), 1369–1383.

Wang, B., Yang, X., Lu, J., Zhou, Y., Su, J., Tian, Y., Zhang, J., Wang, G., Zheng, T., 2012. A marine bacterium producing protein with algicidal activity against *Alexandrium tamarense*. Harmful Algae 13, 83–88.

Wang, X., Li, Z., Su, J., Tian, Y., Ning, X., Hong, H., Zheng, T., 2010. Lysis of a red-tide causing alga, *Alexandrium tamarense*, caused by bacteria from its phycosphere. Biol. Contr. 52 (2), 123–130.

Warner, J.C., Geyer, W.R., Lerczak, J.A., 2005. Numerical modeling of an estuary: a comprehensive skill assessment. J. Geophys. Res. 110 (C5), C05001.

Weisberg, R.H., Barth, A., Alvera-Azcarate, A., Zheng, L., 2009. A coordinated coastal ocean observing and modeling system for the West Florida Continental Shelf. Harmful Algae 8 (4), 585–597.

Wilson, C., 2011. The rocky road from research to operations for satellite ocean-colour data in fishery management. ICES J. Mar. Sci.: J. Cons. 68 (4), 677–686.

Wiseman Jr., W.J., Sturges, W., 1999. Physical oceanography of the Gulf of Mexico: processes that regulate its biology. In: Kumpf, H., Steidinger, K., Sherman, K. (Eds.), The Gulf of Mexico Large Marine Ecosystem: Assessment, Sustainability and Management. Blackwell Science, New York, pp. 77–92.

Woods, J., Barkmann, W., 1993. The plankton multiplier—positive feedback in the greenhouse. J. Plankton Res. 15 (9), 1053–1074.

Wynne, T.T., Stumpf, R.P., Tomlinson, M.C., Ransibrahmanakul, V., Villareal, T.A., 2005. Detecting *Karenia brevis* blooms and algal resuspension in the western Gulf of Mexico with satellite ocean color imagery. Harmful Algae 4 (6), 992–1003.

Wynne, T.T., Stumpf, R.P., Tomlinson, M.C., Schwab, D.J., Watabayashi, G.Y., Christensen, J.D., 2011. Estimating cyanobacterial bloom transport by coupling remotely sensed imagery and a hydrodynamic model. Ecol. Appl. 21 (7), 2709–2721.

Wynne, Timothy T., et al., 2013. Evolution of a cyanobacterial bloom forecast system in western Lake Erie: Development and initial evaluation. J. Great Lakes Res. 39, 90–99.

Xu, J., Hood, R.R., 2006. Modeling biogeochemical cycles in Chesapeake Bay with a coupled physical-biological model. Estuarine Coastal Shelf Sci. 69 (1–2), 19–46.

Yasumoto, T., 2001. The chemistry and biological function of natural marine toxins. Chem. Rec. 1 (3), 228–242.

Zamyadi, A., MacLeod, S.L., Fan, Y., McQuaid, N., Dorner, S., Sauve, S., Prevost, M., 2012. Toxic cyanobacterial breakthrough and accumulation in a drinking water plant: a monitoring and treatment challenge. Water Res. 46 (5), 1511–1523.

Zhang, W.G., Wilkin, J.L., Arango, H.G., 2010. Towards an integrated observation and modeling system in the New York Bight using variational methods. Part I: 4DVAR data assimilation. Ocean Model. 35 (3), 119–133.

Zhou, L.H., Zheng, T.L., Chen, X.H., Wang, X., Chen, S.B., Tian, Y., Hong, H.S., 2008. The inhibitory effects of garlic *(Allium sativum)* and diallyl trisulfide on *Alexandrium tamarense* and other harmful algal species. J. Appl. Phycol. 20 (4), 349–358.

Zillen, L., Conley, D.J., Andren, T., Andren, E., Bjorck, S., 2008. Past occurrences of hypoxia in the Baltic Sea and the role of climate variability, environmental change and human impact. Earth Sci. Rev. 91 (1), 77–92.

FURTHER READING

Glibert, P.M. (Ed.), 2006. GEOHAB: Global Ecology and Oceanography of Harmful Algal Blooms. IOC and SCOR, Paris and Baltimore.

Index

Note: Page numbers followed by "f" and "t" indicate, figures and tables respectively.

I

J

K

L

P

Q

R

FSC
www.fsc.org
MIX
Papier | Fördert
gute Waldnutzung
FSC® C083411

Kolibri 360 GmbH
Ferdinand-Jühlke-Straße 7
99095 Erfurt, Deutschland
produktsicherheit@kolibri360.de